LAND USE, LAND-USE CHANGE, AND FORESTRY

The exchange of carbon between the atmosphere and biosphere is an important factor in controlling global warming and climate change. Vegetation exchanges carbon dioxide between the atmosphere and the terrestrial biosphere through photosynthesis and plant and soil respiration. This natural exchange has been occurring for hundreds of millions of years, but humans are changing this natural rate of exchange through land use, land-use change, and forestry activities. Consequently, it is important to examine how carbon flows between different pools and how carbon stocks change in response to afforestation, reforestation, and deforestation, and other land-use activities.

The Intergovernmental Panel on Climate Change (IPCC) Special Report on *Land Use, Land-Use Change, and Forestry* provides a comprehensive, state-of-the-art examination of the scientific and technical implications of carbon sequestration and the global carbon cycle. It also examines environmental and socioeconomic issues, conservation, sustainable resource management and development issues, as related to carbon sequestration.

The volume will be invaluable for government policymakers, business/industry analysts and officials, environmental groups, and researchers in global change, atmospheric chemistry, soil science, and economics.

Robert T. Watson is Chief Scientist and Director of Environmentally and Socially Sustainable Development at The World Bank and Chairman of the Intergovernmental Panel on Climate Change.

Ian R. Noble is Professor of Global Change Research in the Institute of Advanced Studies at the Australian National University and Chief Executive Officer of the Cooperative Research Centre for Greenhouse Accounting at the Research School of Biological Sciences, Australian National University.

Bert Bolin is a former Professor of Meteorology at the University of Stockholm and Director of the International Institute for Meteorology, and former Scientific Director at the European Space Research Organisation. Dr. Bolin served as Chairman of the Intergovernmental Panel on Climate Change from 1988–1997.

N.H. Ravindranath is a Principal Research Scientist at the Centre for ASTRA and Ecological Sciences, Indian Institute of Science.

David J. Verardo is an Environmental Scientist for the IPCC Working Group II Technical Support Unit, Washington DC, USA.

David J. Dokken is Project Administrator for the IPCC Working Group II Technical Support Unit, Washington DC, USA.

Land Use, Land-Use Change, and Forestry

Edited by

Robert T. Watson
The World Bank

Ian R. Noble
Australian National University

Bert Bolin
University of Stockholm

N.H. Ravindranath
Indian Institute of Science

David J. Verardo
IPCC Working Group II Technical Support Unit

David J. Dokken
University Corporation for Atmospheric Research

A Special Report of the IPCC

Published for the Intergovernmental Panel on Climate Change

PUBLISHED BY THE PRESS SYNDICATE OF THE UNIVERSITY OF CAMBRIDGE
The Pitt Building, Trumpington Street, Cambridge, United Kingdom

CAMBRIDGE UNIVERSITY PRESS
The Edinburgh Building, Cambridge CB2 2RU, UK, http://www.cup.cam.ac.uk
40 West 20th Street, New York, NY 10011-4211, USA, http://www.cup.org
10 Stamford Road, Oakleigh, Melbourne 3166, Australia
Ruiz de Alarcón 13, 28014 Madrid, Spain

First published 2000

Printed in the United States of America

Typeface Times 10/12 pt. *System* Quark XPress 4.1 [CS]

A catalog record for this book is available from the British Library

Library of Congress Cataloging-in-Publication Data available

ISBN 0 521 80083 8 hardback
ISBN 0 521 80495 7 paperback

Photo Credits
Left—Ground view of an agricultural field under conservation tillage (courtesy of The Monsanto Company).
Bottom—Aerial view of land-use change in the Department of Santa Cruz (courtesy of Hermes Justiniano and The Nature Conservancy).
Right—Aerial view of a forest stand (courtesy of the American Forest and Paper Association).

Contents

This report is dedicated to the memory of

Professor David Oakley Hall
14 November 1935 – 22 August 1999

David Oakley Hall, Professor of Life Sciences at King's College London, died in August 1999, aged 63, during the preparation of this report. David was a leading expert on the utilization of plants both as a clean and renewable source of energy and as a means of mitigating climate change. David was also an advocate for the many millions of people who rely on biomass energy—the rural poor of developing countries.

With his unique mix of energy and enthusiasm and his broad interests in science, technology, and policy, David made distinguished contributions not only to bio-energy but also to issues of global change, international energy and environmental policy, and plant physiology. His warmth, friendliness, approachability, and lack of pretension inspired affection and admiration from thousands of colleagues and students across the world.

David became aware of his illness around the time of the scoping meeting of this report in Rome, but he kept working until the end to promote his ideas. Simultaneously, he also contributed tirelessly to the IPCC *Third Assessment Report* and the Special Report on *Methodological and Technological Issues in Technology Transfer*. Some of David Hall's own words are here in Chapter 5. His ideas, his commitment, and his concerns have affected every contributor and influenced every chapter of this report.

In the end, there is no more fitting tribute to David Hall than the words inscribed on a memorial plaque placed in his honor in the Garden of Remembrance at Kearsney College in South Africa where David was a student from 1947 through 1952: "We will miss David the scientist, the teacher, the mentor, the environmentalist. We will also miss David the compassionate and caring human being who always had time to help his fellow man."

Foreword

The Intergovernmental Panel on Climate Change (IPCC) was jointly established by the World Meteorological Organization (WMO) and the United Nations Environment Programme (UNEP) in 1988 to assess available information on the science, the impacts, and the economics of, and the options for mitigating and/or adapting to, climate change. It also provides, on request, scientific/technical/socioeconomic advice to the Conference of the Parties (COP) to the United Nations Framework Convention on Climate Change (UNFCCC). Since then, the IPCC has produced a series of Assessment Reports, Special Reports, Technical Papers, methodologies, and other products that have become standard works of reference, widely used by policymakers, scientists, and other experts.

The Special Report on *Land Use, Land-Use Change, and Forestry* was prepared in response to a request from the United Nations Framework Convention on Climate Change (UNFCCC) Subsidiary Body for Scientific and Technological Advice (SBSTA). At its Eighth Session in Bonn from 2-12 June 1998, SBSTA requested a report examining the scientific and technical state of understanding for carbon sequestration strategies related to land use, land-use change, and forestry activities and relevant Articles of the Kyoto Protocol. The scope, structure, and outline of the Special Report was approved by the IPCC in plenary meetings during its Fourteenth Session in Vienna, Austria, from 1-3 October 1998.

This Special Report discusses the global carbon cycle and how different land use and forestry activities currently affect standing carbon stocks and emissions of greenhouse gases. It also looks forward and examines future carbon uptake and emissions that may result from employing varying definitional scenarios and carbon accounting strategies, linked to the Kyoto Protocol, within the forestry and land-use sectors.

As is usual in the IPCC, success in producing this document has depended on the enthusiasm and cooperation of volunteers dispersed worldwide who give freely of their professional and personal time. We would like to express our gratitude to all the Coordinating Lead Authors, Lead Authors, Contributing Authors, Review Editors, and Expert Reviewers. These individuals have expended considerable effort to produce this report and we are extremely grateful for their commitment to the IPCC process.

We would also like to express our sincere thanks to:

- Robert Watson—the Chairman of the IPCC and Chair of this Special Report
- Ian Noble, Bert Bolin, and N.H. Ravindranath—the Coordinators of this Special Report
- Neal Leary, Osvaldo Canziani, and Martin Manning (Working Group II); David Griggs, Fortunat Joos, and John Stone (Working Group I); and Bert Metz, Eduardo Calvo, and Peter Kuikman (Working Group III)—the Science Steering Committee for this Special Report
- David Verardo and the staff of the Working Group II Technical Support Unit
- N. Sundararaman, the Secretary of the IPCC, and the Secretariat staff.

G.O.P. Obasi

Secretary-General
World Meteorological
Organisation

K. Töpfer

Executive Director
United Nations Enviroment Programme
and Director-General
United Nations Office
in Nairobi

Preface

The Intergovernmental Panel on Climate Change (IPCC) Special Report on Land Use, Land-Use Change, and Forestry (SR-LULUCF) has been prepared in response to a request from the United Nations Framework Convention on Climate Change (UNFCCC) Subsidiary Body for Scientific and Technological Advice (SBSTA). At its eighth session in Bonn, Germany, 2–12 June 1998, the SBSTA requested a report examining the scientific and technical implications of carbon sequestration strategies related to land use, land-use change, and forestry activities. The scope, structure, and outline of this Special Report was approved by the IPCC in plenary meetings during its Fourteenth Session.

This Special Report examines several key questions relating to the exchange of carbon between the atmosphere and the terrestrial pool of aboveground biomass, below-ground biomass, and soils. Vegetation exchanges carbon dioxide between the atmosphere and the terrestrial biosphere through photosynthesis and plant and soil respiration. This natural exchange has been occurring for hundreds of millions of years. Humans are changing the natural rate of exchange of carbon between the atmosphere and the terrestrial biosphere through land use, land-use change, and forestry activities. Consequently, it is important to examine how carbon flows between different pools and how carbon stocks change in response to afforestation, reforestation, and deforestation (ARD) and other land-use activities.

The aim of the SR-LULUCF is to assist the Parties to the Kyoto Protocol by providing relevant scientific and technical information to describe how the global carbon cycle operates and what the broad-scale opportunities and implications of ARD and additional human-induced activities are, now and in the future. This Special Report also identifies questions that Parties to the Protocol may wish to consider regarding definitions and accounting rules.

This Special Report should be helpful in the implementation of relevant Articles in the Kyoto Protocol by providing information about measurement and monitoring techniques for assessing changes in carbon stocks in Annex I and non-Annex I countries, the applicability of the *Revised 1996 IPCC Guidelines for National Greenhouse Gas Inventories* for national and project-level accounting, the implications of Articles 3.3 and 3.4, and project activities relating to sustainable development.

This Special Report also estimates potential carbon yields from ARD and additional activities by evaluating changes in carbon stocks for different ecosystems, current land area converted per year (Mha yr^{-1}), and total land available for two different time periods: near term (between now and the end of the first commitment period) and longer term (1990–2040). Project experience is also provided for several projects, primarily in tropical countries.

Implementation of the Kyoto Protocol requires mutually acceptable definitions for a wide range of terms to ensure that effective sequestration strategies are planned and implemented. For instance, if key terms such as *forests, afforestation, reforestation,* and *deforestation* are not clearly defined or if carbon accounting principles are not clearly established, it becomes difficult to comprehend the implications of different land-use activities. Hence, the challenge is to derive a set of definitions that are simple and consistent with the aims of the UNFCCC and the Kyoto Protocol. To achieve this goal, definitions should be applicable to all Parties and be addressed using data that can be readily accessed. This process will enable Parties to estimate carbon stock changes that would need to be included in the calculation of assigned amounts.

In examining issues relating to land use, land-use change, and forestry, several critical scientific and technical questions present themselves. What are the implications of using different definitions or sets of definitions? Do the definitions need to be flexible enough to accommodate our present understanding of carbon dynamics while allowing for future innovations and advances? How do we distinguish among direct human-induced activities, indirect human-induced activities, and natural environmental variability that affects carbon uptake and release? How do we differentiate between pre- and post-1990 direct human activities? How do we measure changes in carbon stocks and flows in a transparent and verifiable manner over time? How permanent are carbon stocks? To what extent do we trade simplicity for accuracy in accounting?

In summary, the SR-LULUCF is written with a variety of questions in mind that examine the scientific and technical aspects of carbon sequestration in agricultural and forestry sectors as well as the implications of land use, land-use change, and forestry activities on environmental and socioeconomic issues, conservation, and sustainable resource management and development issues.

Robert T. Watson and David J. Verardo

SUMMARY FOR POLICYMAKERS

LAND USE, LAND-USE CHANGE, AND FORESTRY

A Special Report of the Intergovernmental Panel on Climate Change

This summary, approved in detail at IPCC Plenary XVI (Montreal, Canada • 1-8 May 2000), represents the formally agreed statement of the IPCC concerning current understanding of land use, land-use change, and forestry activities and their relationship to the Kyoto Protocol.

Based on a draft prepared by:

Robert Watson (USA), Ian Noble (Australia), Bert Bolin (Sweden), N.H. Ravindranath (India), David Verardo (USA), Ken Andrasko (USA), Michael Apps (Canada), Sandra Brown (USA), Graham Farquhar (Australia), Donald Goldberg (USA), Steven Hamburg (USA), Richard Houghton (USA), Paul Jarvis (UK), Timo Karjalainen (Finland), Haroon Kheshgi (USA), Thelma Krug (Brazil), Werner Kurz (Canada), Daniel Lashof (USA), Bo Lim (UNDP), Willy Makundi (Tanzania), Martin Manning (New Zealand), Gregg Marland (USA), Omar Masera (Mexico), Daniel Murdiyarso (Indonesia), Brian Murray (USA), Reidar Persson (Indonesia), Neil Sampson (USA), Jayant Sathaye (USA), Robert Scholes (South Africa), Bernhard Schlamadinger (Austria), Wim Sombroek (The Netherlands), Stephen Prisley (USA), John Stone (Canada), Raman Sukumar (India), and Riccardo Valentini (Italy)

CONTENTS

1. Introduction

1. Under Article 3.1 of the Kyoto Protocol, the Annex I Parties have agreed to limit and reduce their emissions of greenhouse gases between 2008 and 2012.

2. The Kyoto Protocol makes provision for Annex I Parties to take into account afforestation, reforestation, and deforestation and other agreed land use, land-use change, and forestry (LULUCF) activities in meeting their commitments under Article 3.

3. To implement the Kyoto Protocol, issues related to LULUCF will have to be considered. Relevant issues may include for example:
   - Definitions, including land-use change, forests, forestry activities, including afforestation, reforestation, and deforestation, carbon stocks, human-induced, and direct human-induced.
   - Methodological issues, such as:
     - ➢ Rules for accounting for carbon stock changes and for emissions and removals of greenhouse gases from LULUCF activities, including:
       - Which carbon pools to include.
       - How to implement "since 1990," "direct human-induced," and "human-induced."
       - How to address the risks and effects of events such as fires, pest outbreaks, and extreme meteorological events; baselines; permanence; interannual and decadal climate variability; and leakage.
       - Accuracy, precision, and uncertainties in tracking carbon stocks and greenhouse gases.
     - ➢ Approaches, such as geo-referencing and statistical sampling, associated with identifying lands with activities defined under Article 3.3, accepted under Article 3.4, or associated with project-based activities under the Kyoto Protocol, and measuring and estimating changes in carbon stocks and greenhouse gases.
     - ➢ Verification procedures.
   - Determination of how and which additional activities pursuant to Article 3.4 are included.
   - How to link the first and subsequent commitment periods.
   - Determination of how and which project-based activities are included.
   - What improvements, if any, are needed to the Revised 1996 IPCC Guidelines for National Greenhouse Gas Inventories and the Good Practice Guidance and Uncertainty Management in National Greenhouse Gas Inventories.
   - What are the implications of and what, if any, national and/or international sustainable development criteria could be associated with Articles 3.3 and 3.4 and project-based activities.

4. Therefore, to assist the Parties to the Protocol, this Summary for Policymakers (SPM) provides relevant scientific and technical information in three parts:
   - Part I describes how the global carbon cycle operates and provides a context for the sections on afforestation, reforestation, and deforestation (ARD) and additional human-induced activities.
   - Part II addresses important issues regarding definitions and accounting rules. It identifies a range of options and discusses implications and interrelationships among options.
   - Part III provides information that governments might find useful in considering these issues:
     - ➢ An assessment of the usefulness of models and of the usefulness and costs of ground-based and remotely sensed measurements and of monitoring techniques for assessing changes in carbon stocks.
     - ➢ The near-term (first commitment period) potential for carbon stock changes/accounting of activities in Annex I countries and globally.
     - ➢ Issues of special significance to project-based activities.
     - ➢ An evaluation of the applicability of the Revised 1996 IPCC Guidelines for National Greenhouse Gas Inventories for national and project-level accounting in light of the Kyoto Protocol.
     - ➢ Implications of Articles 3.3 and 3.4 and project activities on sustainable development (i.e., socioeconomic and environmental considerations).

Part I

2. Global Carbon Cycle Overview

5. The dynamics of terrestrial ecosystems depend on interactions between a number of biogeochemical cycles, particularly the carbon cycle, nutrient cycles, and the hydrological cycle, all of which may be modified by human actions. Terrestrial ecological systems, in which carbon is retained in live biomass, decomposing organic matter, and soil, play an important role in the global carbon cycle. Carbon is exchanged naturally between these systems and the atmosphere through photosynthesis, respiration, decomposition, and combustion. Human activities change carbon stocks in these pools and exchanges between them and the atmosphere through land use, land-use change, and forestry, among other activities. Substantial amounts of carbon have been released from forest clearing at high and middle latitudes over the last several centuries, and in the tropics during the latter part of the 20th century. [1.1.1.2][1]

6. There is carbon uptake into both vegetation and soils in terrestrial ecosystems. Current carbon stocks are much

[1] Numbers in brackets at the end of this and subsequent paragraphs indicate relevant sections of the Special Report containing details.

***Table 1**: Global carbon stocks in vegetation and soil carbon pools down to a depth of 1 m.*

Biome	Area (10^9 ha)	Global Carbon Stocks (Gt C)		
		Vegetation	*Soil*	*Total*
Tropical forests	1.76	212	216	428
Temperate forests	1.04	59	100	159
Boreal forests	1.37	88	471	559
Tropical savannas	2.25	66	264	330
Temperate grasslands	1.25	9	295	304
Deserts and semideserts	4.55	8	191	199
Tundra	0.95	6	121	127
Wetlands	0.35	15	225	240
Croplands	1.60	3	128	131
Total	15.12	466	2011	2477

Note: There is considerable uncertainty in the numbers given, because of ambiguity of definitions of biomes, but the table still provides an overview of the magnitude of carbon stocks in terrestrial systems.

larger in soils than in vegetation, particularly in non-forested ecosystems in middle and high latitudes (see Table 1). [1.3.1]

7. From 1850 to 1998, approximately 270 (± 30) Gt C has been emitted as carbon dioxide (CO_2) into the atmosphere from fossil fuel burning and cement production. About 136 (± 55) Gt C has been emitted as a result of land-use change, predominantly from forest ecosystems. This has led to an increase in the atmospheric content of carbon dioxide of 176 (± 10) Gt C. Atmospheric concentrations increased from about 285 to 366 ppm (i.e., by ~28%), and about 43% of the total emissions over this time have been retained in the atmosphere. The remainder, about 230 (± 60) Gt C, is estimated to have been taken up in approximately equal amounts in the oceans and the terrestrial ecosystems. Thus, on balance, the terrestrial ecosystems appear to have been a comparatively small net source of carbon dioxide during this period. [1.2.1]

8. The average annual global carbon budgets for 1980–1989 and 1989–1998 are shown in Table 2. This table shows that the rates and trends of carbon uptake in terrestrial ecosystems are quite uncertain. However, during these two decades, terrestrial ecosystems may have served as a small net sink for carbon dioxide. This terrestrial sink seems to have occurred in spite of net emissions into the atmosphere from land-use change, primarily in the tropics, having been 1.7 ± 0.8 Gt C yr^{-1} and 1.6 ± 0.8 Gt C yr^{-1} during these two decades, respectively. The net terrestrial carbon uptake, that approximately balances the emissions from land-use change in the tropics, results from land-use practices and natural regrowth in middle and high latitudes, the indirect effects of human activities (e.g., atmospheric CO_2 fertilization and nutrient deposition), and changing climate (both natural and anthropogenic). It is presently not possible to determine the relative importance of these different processes, which also vary from region to region. [1.2.1 and Figure 1-1]

9. Ecosystem models indicate that the additional terrestrial uptake of atmospheric carbon dioxide arising from the indirect effects of human activities (e.g., CO_2 fertilization and nutrient deposition) on a global scale is likely to be maintained for a number of decades in forest ecosystems, but may gradually diminish and forest ecosystems could even become a source. One reason for this is that the capacity of ecosystems for additional carbon uptake may be limited by nutrients and other biophysical factors. A second reason is that the rate of photosynthesis in some types of plants may no longer increase as carbon dioxide concentration continues to rise, whereas heterotrophic respiration is expected to rise with increasing temperatures. A third reason is that ecosystem degradation may result from climate change. These conclusions consider the effect of future CO_2 and climate change on the present sink only and do not take into account future deforestation or actions to enhance the terrestrial sinks for which no comparable analyses have been made. Because of current uncertainties in our understanding with respect to acclimation of the physiological processes and climatic constraints and feedbacks amongst the processes, projections beyond a few decades are highly uncertain. [1.3.3]

10. Newly planted or regenerating forests, in the absence of major disturbances, will continue to uptake carbon for 20 to 50 years or more after establishment, depending on species and site conditions, though quantitative projections beyond a few decades are uncertain. [1.3.2.2]

11. Emissions of methane (CH_4) and nitrous oxide (N_2O) are influenced by land use, land-use change, and forestry activities (e.g., restoration of wetlands, biomass burning, and fertilization of forests). Hence, to assess the greenhouse gas implications of LULUCF activities, changes in CH_4 and N_2O emissions and removals—the magnitude of which is highly uncertain—would have to be considered explicitly. There are currently no reliable global estimates

***Table 2**: Average annual budget of CO_2 for 1980 to 1989 and for 1989 to 1998, expressed in Gt C yr^{-1} (error limits correspond to an estimated 90% confidence interval).*

	1980 to 1989	1989 to 1998
1) Emissions from fossil fuel combustion and cement production	5.5 ± 0.5	6.3 ± 0.6[a]
2) Storage in the atmosphere	3.3 ± 0.2	3.3 ± 0.2
3) Ocean uptake	2.0 ± 0.8	2.3 ± 0.8
4) Net terrestrial uptake = (1) – [(2)+(3)]	0.2 ± 1.0	0.7 ± 1.0
5) Emissions from land-use change	1.7 ± 0.8	1.6 ± 0.8[b]
6) Residual terrestrial uptake = (4)+(5)	1.9 ± 1.3	2.3 ± 1.3

[a]Note that there is a 1-year overlap (1989) between the two decadal time periods.
[b]This number is the average annual emissions for 1989–1995, for which data are available.

of these emissions and removals for LULUCF activities. [1.2.2, 1.2.3, 3.3.2]

Part II

3. *Issues Associated with Definitions*

12. For purposes of this Special Report, in a given land area and time period, a full carbon accounting system would consist of a complete accounting for changes in carbon stocks across all carbon pools. Applying full carbon accounting to all land in each country would, in principle, yield the net carbon exchange between terrestrial ecosystems and the atmosphere. However, the Kyoto Protocol specifies, among other things, that attention focus onto those land areas subject to "direct human-induced" activities since 1990 (Article 3.3) or human-induced activities (Article 3.4). [2.3.2.5]

3.1. Forests, Afforestation, Reforestation, and Deforestation

13. There are many possible definitions of a "forest" and approaches to the meaning of the terms "afforestation," "reforestation," and "deforestation" (ARD). The choice of definitions will determine how much and which land in Annex I countries are included under the provisions of Article 3.3, lands associated with activities included under Article 3.3 (hereafter "lands under Article 3.3"). The amount of land included will have implications for the changes in carbon stocks accounted for under Article 3.3. [2.2.2, 2.2.3, 3.2, 3.5.2, 3.5.3]

14. Seven definitional scenarios were developed that combine definitions of forest and ARD and reflect a range of approaches that can be taken. The scenarios are not intended to be exhaustive. They can be split into two representative groups, which are discussed in the SPM: 1) scenarios in which only a forest/non-forest conversion (i.e., a land-use change) triggers accounting under Article 3.3 (e.g., IPCC Definitional Scenario), and 2) scenarios in which land-cover change or activities trigger accounting under Article 3.3 (e.g., FAO Definitional Scenario). [2.2.2, 2.2.3, 3.2, 3.5.2, 3.5.3, Table 3-4]

15. Countries have defined forests and other wooded lands, for a number of national and international purposes, in terms of (i) legal, administrative, or cultural requirements; (ii) land use, (iii) canopy cover, or (iv) carbon density (essentially biomass density). Such definitions were not designed with the Kyoto Protocol in mind and, thus, they may not necessarily suffice for the particular needs of Articles 3.3 and 3.4. [2.2.2, 3.2]

16. Forest definitions based on legal, administrative, or cultural considerations have limitations for carbon accounting as they may bear little relationship to the amount of carbon at a site. [2.2.2, 3.2]

17. Most definitions of forest are based in part on a single threshold of minimum canopy cover. However, such definitions may allow changes in carbon stocks to remain unaccounted under Article 3.3. For example, if a high threshold for canopy cover (e.g., 70% canopy cover) is used in the definition of a forest, then many areas of sparse forest and woodland could be cleared or could increase in cover without the losses or gains in carbon being counted under Article 3.3. If a low threshold is set (e.g., 10% canopy cover), then dense forest could be heavily degraded and significant amounts of carbon released, without the actions being designated as deforestation. Similarly, a forest, for example with 15% canopy cover, could be considerably enhanced without the actions qualifying as reforestation or afforestation under Article 3.3. Approaches to partly address these problems may include, *inter alia*, using

national, regional, or biome-specific thresholds (e.g., a low canopy cover for savannas and a high canopy cover for moist forests). [2.2.2, 3.2, 3.3.2]

18. Definitions of forests based on carbon-density thresholds have similar issues with respect to thresholds as canopy cover-based definitions. [2.2.2]

19. There are a number of approaches to definitions of afforestation, reforestation, and deforestation. One approach involves the concept of land-use change. Deforestation can be defined as the conversion of forest land to non-forest land. Reforestation and afforestation can be defined as the conversion of non-forested lands to forests with the only difference being the length of time during which the land was without forest. [2.2.3, 3.2]

20. An alternative definition of deforestation might be based on a decrease in the canopy cover or carbon density by a given amount or crossing one of a sequence of thresholds. Similarly, afforestation and reforestation could be defined in terms of an increase in canopy cover or carbon density. None of these definitions involves the concept of a land-use change. [2.2.2, 3.2]

21. Definitions of a forest based strictly on actual canopy cover without consideration of potential canopy cover could lead to harvesting and shifting agriculture being referred to as deforestation and to regeneration being referred to as reforestation, thus creating additional areas of lands under Article 3.3. If the definition of a forest was based on the potential canopy cover at maturity under planned land-use practices, harvesting/regeneration activities may not fall under Article 3.3. [2.2.2, 2.2.3, 3.2]

22. Some commonly used definitions of reforestation include the activity of regenerating trees immediately after disturbance or harvesting where no land-use change occurs. If, for example, the definition of deforestation or the accounting system do not include disturbance and harvesting, then emissions from a harvested stand will not be accounted for. In this particular example, uptake due to regeneration would be accounted for, resulting in potentially significant credits for which a corresponding net removal of carbon from the atmosphere would not occur. This issue could be considered when developing the accounting system. [2.2.3.2]

23. There are several consequences of using definitions that lead to the creation of lands under Article 3.3 by the harvest-regeneration cycle (i.e., where harvesting is included in the definition of deforestation, or regeneration is included in the definition of reforestation). For example, a forest estate managed on a sustainable-yield basis where an area of forest is cut in a regular cycle (e.g., 1/50th of the forest is harvested and regenerated each year on a 50-year rotation cycle) may be in approximate carbon balance. However, in this case, only those stands harvested or regenerated since 1990 would be considered lands under Article 3.3. The regrowth (carbon sink) on these lands will be less than the carbon emissions due to harvesting until all stands of the estate are lands under Article 3.3. Different definitional and accounting approaches would have different accounting consequences. For example:
    - If emissions from harvesting during a commitment period are counted (land-based approach I; see Table 3), then during the first and subsequent commitment periods a net debit could arise from a managed forest estate that is approximately in carbon balance.
    - If emissions from harvesting during a commitment period prior to regeneration are not counted (land-based approach II; see Table 3), then during the first and subsequent commitment periods a net credit would generally arise from a managed forest estate that is approximately in carbon balance. This may be offset to some extent by delayed emissions from soils and harvest residues.
    - If emissions from harvesting during a commitment period are not counted (activity-based approach; see Table 3), then during the first and subsequent commitment periods a net credit would arise from regeneration in a managed forest estate that is approximately in carbon balance. It would be practically very difficult to separate changes in soil carbon pools associated with harvesting and regeneration activities.

    In each of these approaches the accounted stock changes would generally be different from the actual net exchange of carbon between this example forest estate and the atmosphere during a commitment period. [3.2, 3.5.2]

24. Afforestation is usually defined as the establishment of forest on land that has been without forest for a period of time (e.g., 20-50 years or more) and was previously under a different land use. The precise period that distinguishes afforested from reforested land is not important in accounting for lands covered under Article 3.3 provided afforestation and reforestation are treated identically under the Protocol, as they are in the Revised 1996 IPCC Guidelines for National Greenhouse Gas Inventories.[2] [2.2.3, 3.3.2]

[2] The Glossary of the Revised 1996 IPCC Guidelines describes afforestation as "Planting of new forests on lands which, historically, have not contained forests. These newly created forests are included in the category Changes in Forest and Other Woody Biomass Stocks in the Land Use Change and Forestry module of the emissions inventory calculations" and reforestation as "Planting of forests on lands which have, historically, previously contained forests but which have been converted to some other use. Replanted forests are included in the category Changes in Forest and Other Woody Biomass Stocks in the Land Use Change and Forestry module of the emissions inventory calculations." Deforestation does not appear in the Glossary of the Revised 1996 IPCC Guidelines. The Revised 1996 IPCC Guidelines state, referring to land-use change, that "Conversion of forests is also referred to as 'deforestation' and it is frequently accompanied by burning." The Revised 1996 IPCC Guidelines were developed before the Kyoto Protocol was adopted and therefore provisions may not be sufficient to meet the needs of the Kyoto Protocol.

25. Article 3.3 encompasses ARD activities that have occurred since 1990 but recognizes only verifiable carbon stock changes in each commitment period. This has several implications. For example:
    - For lands deforested between 1990 and the beginning of the first commitment period only a fraction of carbon stock changes (such as those from delayed carbon emissions from soil and wood products if they are accounted) will occur during the commitment period and would be debited under Article 3.3. If these lands are subsequently reforested then there may be an increase in carbon stocks during the commitment period and a credit under Article 3.3. This would mean that the credit received would not match the actual carbon stock changes or the net exchanges of carbon with the atmosphere since 1990.
    - Another accounting issue could arise when land is reforested or afforested between 1990 and 2008 but stocks are reduced either by harvesting or natural disturbance during a commitment period. Even though the forest area and possibly carbon stocks may have increased since 1990, a debit could be recorded in a commitment period. This creates the possibility of a negative incentive for establishing forests well in advance of the first commitment period, because any stock increase prior to 2008 would not be credited but the later loss of this stock would be debited.

    Such outcomes could possibly be addressed through different combinations of definitional and accounting approaches. [3.3.2]

26. There are definitional and carbon accounting issues concerning drawing a clear boundary between natural phenomena and human-induced activities, when, for example, significant forest losses occur as a result of fires or disturbances such as pest outbreaks. In cases involving lands under Article 3.3 or 3.4 where fires or pest outbreaks occur in a forest, a question is whether accounting should, *inter alia*: (i) count neither the loss nor subsequent uptake of carbon (which reflects the actual net change in carbon stocks on those lands and exchange of carbon with the atmosphere in the long term, but creates problems in continuing to account for the area burnt/defoliated as lands under Article 3.3 or 3.4); (ii) count both the loss and subsequent uptake of carbon (which reflects the actual net change in carbon stocks on those lands and exchange of carbon with the atmosphere, but creates an initial carbon debit for the Party concerned); (iii) count only the loss of carbon (which would overestimate the actual losses of carbon stocks, not represent the exchanges of carbon with the atmosphere, and create future accounting problems), or (iv) count only the subsequent uptake (which would fail to reflect the actual changes in carbon stock and would not represent the exchanges of carbon with the atmosphere, and would provide carbon credits for the Party concerned). [2.2.3.3]

27. In cases involving lands that do not fall under Articles 3.3 or 3.4, where fires or pest outbreaks trigger land-use change, the consequences are similar to deforestation. If similar vegetation cover is allowed to regenerate, such disturbances may not lead to a long-term change in carbon stocks. [2.4.4, 2.2.3, 2.3.3]

3.2. Additional Activities[3]

28. When the inclusion of additional activities under Article 3.4 is considered, it is possible to interpret "activity" broadly (e.g., cropland management) or narrowly (e.g., change in tillage method, fertilization, or cover crops). Under either interpretation, it is, in principle, possible to choose either a land-based or an activity-based method of carbon accounting or a combination of both (see Section 4). These combined choices will affect the accuracy, feasibility, cost, transparency, and verifiability of monitoring and reporting of emissions and removals, including non-CO_2 greenhouse gases, and attributing them to specific activities. [2.3.2.2, 4.3.1, 4.3.2]

29. The term "broad activity" means an activity definition that is land- or area-based, where the net effect of all practices applied within the same area are included. A broad activity definition is likely to require land-based accounting (see paragraph 34). This definitional approach would capture the net emission or removal effects of practices that deplete carbon stocks as well as those that increase removals by sinks. Broad activity definitions, particularly in cases where land-use change is involved, may make it difficult to separate human-induced changes from naturally induced changes. [2.3.2, 4.3.2]

30. The narrow definition of "activity" is based on individual practices, such as reduced tillage or irrigation water management. The narrow definition may lend itself to activity-based accounting, but land-based accounting is also possible. Under activity-based accounting, discrete definitions and associated rates of emissions or removals are needed for each individual practice. Narrow definitions raise the potential for multiple activities to occur on a single land area, raising accounting issues (see paragraph 33). Narrow activity definitions may facilitate the separation of human-induced changes from natural influences (see paragraph 45). [4.2.1, 4.3.2, 4.3.4]

4. Carbon Acounting

31. A well-designed carbon accounting system would provide transparent, consistent, comparable, complete, accurate, verifiable, and efficient recording and reporting of changes in carbon stocks and/or changes in greenhouse gas emissions by sources and removals by sinks from applicable land use, land-use change, and forestry activities

[3] The technical issues addressed in paragraph 26 also apply to additional activities adopted under Article 3.4, but are not repeated here for conciseness.

and projects under relevant Articles of the Kyoto Protocol. Such data would be needed to assess compliance with the commitments under the Kyoto Protocol. Two possible accounting approaches towards meeting these requirements are outlined below, of which either one—or combination of the two—could be adopted (see Figure 1). [2.3.1]

32. A "land-based" approach to accounting would take as its starting point the change in carbon stock in applicable carbon pools on lands containing activities included under Article 3.3 or accepted under Article 3.4. This involves first defining the applicable activities, and in the next step identifying the land units on which these activities occur. Next, the change in carbon stocks on these land units during the relevant period is determined. In the land-based approach, it could be difficult to factor out the impact on stocks of indirect effects (see paragraph 44). Non-CO_2 greenhouse gas emission estimates would also need to be accounted for. Modifications could be made regarding, for example, baselines, leakage, timing issues, permanence, and uncertainties. Aggregate accounted CO_2 emissions and removals are the sum of carbon stock changes (net of any modifications) over all applicable land units over the specified time period. [2.3.2, 3.3.2]

33. An "activity-based" approach to accounting would start with the carbon stock change in applicable carbon pools and/or emissions/removal of greenhouse gases attributable to designated LULUCF activities. After defining the applicable activities, each applicable activity's impact on carbon stocks is determined per unit area and time unit. This impact is multiplied by the area on which each activity occurs and by the years it is applied or the years of the commitment period. Modifications could be made regarding, for example, baselines, leakage, timing issues, permanence, and uncertainties. Aggregate accounted emissions and removals are calculated by summing across applicable activities. Potentially a given area of land could be counted more than once if it is subject to multiple activities. If the effects of activities are not additive, this would result in inaccurate accounting. In this case, the carbon stock would be especially difficult to verify. Alternatively the Parties could decide that each land unit could contain no more than a single activity. In this case, the combined impact of multiple practices applied in the same area would be considered a single activity. [2.3.2, 3.3.2, 4.3.3]

34. The land-based approach to accounting could start either with the start of the activity or run for the entire commitment period, while the activity-based approach would start when the activity starts or at the beginning of the commitment period, whichever is later. Either accounting approach could end according to decisions that the Parties might adopt. In the activity-based approach, stock changes prior

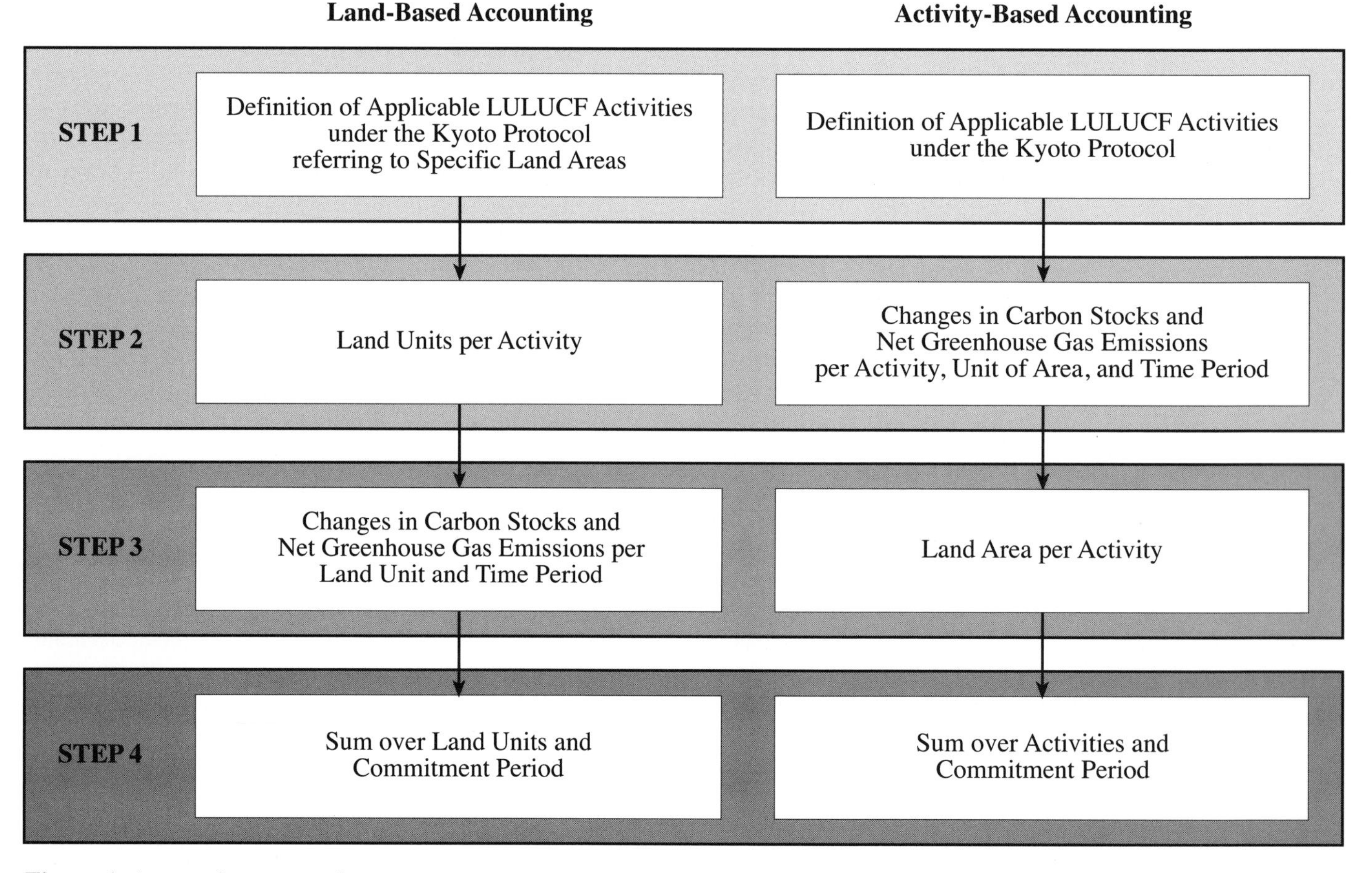

Figure 1: Accounting approaches.

to the start of the activity would not be accounted, even if they occur in a commitment period. [2.3.2]

35. Some activities must be persistently maintained to retain the stored carbon stocks, and this may influence the accounting methods required. Conservation tillage, for example, may increase carbon stocks on cropland if carried on continuously, but where it is practiced for a time, then interrupted by a year of intensive tillage brought on by, for example, a weather situation or crop change, much of the previous multi-year gain in soil carbon can be lost. Land-based estimates of the cropland estate should reflect the net effect of those gains and losses over the full area during the accounting period and give verifiable results, provided statistically representative sampling procedures are in place. If activity-based accounting occurs without sampling, it may report results inconsistent with actual stock changes during the accounting period. [2.3.2]

36. For technical reasons, only emissions and removals of CO_2 can be determined directly as changes in carbon stocks. Methane emissions and removals cannot in practice be directly measured as carbon stock changes, although CH_4 and N_2O can be determined by other means. Methane and nitrous oxide emissions from many land-use activities are included in Annex A of the Kyoto Protocol (e.g., rice cultivation, enteric fermentation, and agricultural soils) and in the Revised 1996 IPCC Reporting Guidelines for National Greenhouse Gas Inventories, and therefore they will be captured in national inventories. This is not the case, however, for emissions of these gases related to forestry activities and projects, which are not included in Annex A, although some of these forestry activities are discussed in the 1996 Revised IPCC Guidelines for National Greenhouse Gas Inventories. If the net emissions of CH_4 and N_2O are not considered, the full climate impact of forestry activities may not be reflected in the accounting system under the Kyoto Protocol. The treatment of CH_4 and N_2O emissions under Article 3.3 may deserve further consideration and clarification. For agreed activities, Article 3.4 leaves open how net greenhouse gas emissions will be accounted for in meeting the commitments under Article 3.1 of the Protocol. [2.3.2, 3.3.2]

37. Relevant carbon pools could include aboveground biomass, litter and woody debris, below-ground biomass, soil carbon, and harvested materials. The impact on these different carbon pools may vary significantly between activities and types of projects. While methods exist to measure all carbon pools, to date monitoring is not routinely performed on all pools and the costs vary significantly. A conservative approach that would allow for selective accounting of carbon pools to reduce monitoring costs could be to include all those pools anticipated to have reduced carbon stocks while omitting selected pools anticipated, with a sufficient level of certainty, to have unchanged or increased carbon stocks. Similar approaches could be used for fluxes of non-CO_2 greenhouse gases. Under this approach, verifiability would mean that only increases in carbon stocks and removal by sinks that can be monitored and estimated could potentially be credited. [2.3.7, 3.3.2, 4.2.1]

38. Accounting for LULUCF activities under Articles 3.3 and 3.4 includes different types of uncertainties, including measurement uncertainty, uncertainty in identifying lands under Article 3.3 or 3.4, and uncertainty in defining and quantifying baselines, if any. This uncertainty can be accounted for in several ways. One approach is to extend the application of good practice guidance in the choice of methods and handling of uncertainty in estimates which has been developed by the IPCC for other inventory categories. Another approach could be to adjust estimated stock changes in a conservative way—understating increases and overstating decreases in stocks. The latter option could allow tradeoffs between monitoring costs and the potential to receive increased carbon credits or reduced debits, but would not be consistent with established principles for estimation of emissions and removals in greenhouse gas inventories. [2.3.7]

39. Changes in carbon stocks in wood products could potentially be accounted as part of the activity that is the source of the wood products or as an independent wood products management activity. If management of wood products is treated as an additional activity under Article 3.4, then it may be necessary to exclude wood products from accounting under other Article 3.3 or 3.4 activities to avoid double-counting. Once wood products are in trade, they would be difficult in most instances to trace. The current IPCC default approach assumes that the wood product pool remains constant over time, and therefore does not account for it. However, if this pool is changing significantly over time, a potentially important pool may not be accounted for. [2.4.2, 3.3.2, 4.5.6, 6.3.3]

40. Enhancement of carbon stocks resulting from land use, land-use change, and forestry activities is potentially reversible through human activities, disturbances, or environmental change, including climate change. This potential reversibility is a characteristic feature of LULUCF activities in contrast to activities in other sectors. This potential reversibility and nonpermanence of stocks may require attention with respect to accounting, for example, by ensuring that any credit for enhanced carbon stocks is balanced by accounting for any subsequent reductions in those carbon stocks, regardless of the cause. [2.3.6, 3.3.2]

41. Contiguous commitment periods under the Kyoto Protocol would avoid incentives in subsequent periods to concentrate activities that reduce carbon stocks in time periods that were not covered. [2.3.2]

42. Policies by governments or other institutions (e.g., land tenure reform and tax incentives) may provide a framework and incentives for implementing LULUCF activities.

Changes in markets may also affect the economic conditions for land use, land-use change, and forestry activities. The ability to measure the impact of these conditions and incentives will depend, in part, upon the carbon inventory and monitoring system in each country. However, it may be very difficult for countries to assess the relative impact of policies by governments or other institutions compared to other human and natural factors that drive changes in carbon stocks. [2.3.5, 5.2.2]

43. Natural variability, such as El Niño cycles, and the indirect effects of human activity, such as CO_2 fertilization, nutrient deposition, and the effects of climate change, could significantly affect carbon stocks during a commitment period on lands under Article 3.3 or 3.4. The spatial distribution of the emissions and removals of greenhouse gases due to these factors is uncertain, as is the portion of them that may enter the accounting system. These emissions and removals could be potentially large compared to the commitments in the first commitment period. This could be a significant issue in the design of an accounting framework. [2.3.3]

44. The Kyoto Protocol specifies that accounting under Article 3.3 be restricted to "direct human-induced land-use change and forestry activities, limited to afforestation, reforestation, and deforestation" occurring since 1990. For activities that involve land-use changes (e.g., from grassland/pasture to forest) it may be very difficult, if not impossible, to distinguish with present scientific tools that portion of the observed stock change that is directly human-induced from that portion that is caused by indirect and natural factors. [2.3.4, 3.3.2]

45. For those activities where only narrowly defined management changes under Article 3.4 are involved (e.g., conservation tillage) and the land use remains the same, it may be feasible to partially factor out natural variability and indirect effects. One approach may be to subtract the stock changes on comparison plots where there have been no changes in management practice from changes measured on plots with modified management activities. In most cases experimental manipulation or paired plots can be used for this purpose, but they are likely to be expensive to apply over large areas. Ecosystem models can also be used but need further improvement to decrease uncertainties. Verifiability could be assisted by the application of a combination of models and measurements. [2.3.4, 4.3.4]

46. Baselines could be used in some cases to distinguish between the effects of LULUCF activities and other factors, such as natural variability and the indirect effects of human activities, as well as to factor out the effects of business-as-usual and activities undertaken prior to 1990 on carbon stock accounts and net greenhouse gas emissions. If the concept of a baseline was to be applied in national accounting for activities under Article 3.4, there are many options, which include: (i) the stock/flux change that would have resulted from "business-as-usual" activities; (ii) the stock/flux change that would have resulted from the continuation of 1990 activity levels; (iii) the stock/flux change that would result in the absence of active management; (iv) performance benchmarks or standard management practice; and (v) the rate of change of stocks/fluxes in 1990. The first three of these baseline options may involve the use of a counterfactual scenario. One difficulty with the use of counterfactual baselines is verification. [2.3.4, 4.6, 4.6.3.3]

47. Accounting under the terms land-use change and forestry in Article 3.7 will determine which emissions and removals of carbon will enter the 1990 base year or period for some countries. If the land-use change activities giving rise to these emissions and removals are not included under Article 3.3 or 3.4 during the commitment periods, then the inventories of countries subject to this clause in Article 3.7 would not be calculated on the same basis as their 1990 emissions base year or period. [3.3.2]

48. If different accounting rules are adopted for relevant Articles of the Kyoto Protocol, additional decision rules may be needed to determine which accounting rule applies to land that, over time, is subject to multiple types of activities. For example, one set of accounting rules could be given primacy in cases where more than one set could potentially apply and double-counting might result. [2.3.2, 3.3.2]

49. Leakage is changes in emissions and removals of greenhouse gases outside the accounting system that result from activities that cause changes within the boundary of the accounting system. There are four types of leakage: activity displacement, demand displacement, supply displacement, and investment crowding. If leakage occurs, then the accounting system will fail to give a complete assessment of the true aggregate changes induced by the activity. Although leakage is in many cases a negative effect, situations, such as the demonstration effect of new management approaches or technology adoption, may occur where the emissions reductions or removals of greenhouse gases extend beyond the accounting system boundaries (positive spillover effect). For some activities and project types, leakage may be addressed by increasing the spatial and temporal scale of the accounting system boundaries (i.e., by including areas where changes in removal and emissions of greenhouse gases may be induced). However, leakage may extend beyond any activity accounting boundaries (e.g., beyond national boundaries). Leakage is of particular concern in project-level accounting, but may also occur with activities under Articles 3.3 and 3.4. [2.3.5.2, 5.3.3]

Part III

5. Methods for Measuring and Monitoring

50. Lands under Articles 3.3 and 3.4 could be identified, monitored, and reported using geographical and statistical

information. Changes in carbon stocks and net greenhouse gas emissions over time can be estimated using some combination of direct measurements, activity data, and models based on accepted principles of statistical analysis, forest inventory, remote-sensing techniques, flux measurements, soil sampling, and ecological surveys. These methods vary in accuracy, precision, verifiability, cost, and scale of application. The cost of measuring changes in carbon stocks and net greenhouse gas emissions for a given area increases as both desired precision and landscape heterogeneity increase. [2.4, 3.4]

51. The spatial resolution of monitoring has important implications for accuracy and costs. If a small minimum resolvable land area is used, the task and cost of monitoring can become very demanding. If the spatial resolution is set at a coarse scale, the data demands can be modest, but significant areas subject to an activity may be lost in the averaging process. For example, if forests and deforestation are defined in terms of canopy cover and canopy cover is assessed over land areas of 100 ha, then deforestation of smaller areas within a unit may not take the canopy cover of the unit below the forest definition threshold. Thus, changes in carbon stocks may not be accounted and, likewise, afforestation or reforestation of small areas may not be accounted. Hence, there are clear tradeoffs between an accurate and precise assessment of changes in carbon stocks and cost. However, an appropriate design should result in a statistically reliable estimate. [2.2.2]

52. The technical capacity required by Annex I Parties to measure, monitor, and verify carbon stock changes and net greenhouse gas emissions under the Kyoto Protocol will be significantly affected by decisions of the Parties regarding definitions of key terms related to land use, land-use change, and forestry activities. It will also depend on decisions on, *inter alia,* additional activities that may be included under Article 3.4, and whether additional activities are defined broadly or narrowly. Depending upon decisions that may be made, establishing a monitoring, reporting, and verification system under Articles 3.3 and 3.4 is likely to involve a significant effort by Annex I Parties, given the technology, data, and resources required, and the short time available. [2.4.1, 3.4, 4.3.2, 4.3.5]

53. Annex I Parties generally have the basic technical capacity (soil and forest inventories, land-use surveys, and information based on remote-sensing and other methods) to measure carbon stocks and net greenhouse gas emissions in terrestrial ecosystems. However, few, if any, countries perform all of these measurements routinely, particularly soil inventories. Some Annex I Parties may use existing capacity with minimal modification to implement the various Articles in the Kyoto Protocol; however, some other Annex I Parties may need to significantly improve their existing measurement systems in order to develop operational systems. Non-Annex I Parties may require technical, institutional, and financial assistance and capacity building for measuring, monitoring, and verifying carbon stock changes as well as estimating net greenhouse gas emissions. [2.4.6, 3.4.3, 4.2]

54. Technical methods for measuring and estimating changes in forest carbon stocks in aboveground biomass over a 5-year commitment period may be deemed to be sensitive enough to serve the requirements of the Protocol. Sensitive methods for estimating below-ground carbon stocks also exist. However, changes in soil carbon stocks are in some instances small and difficult to assess accurately over a 5-year time period. This problem may be addressed by adoption of appropriate sampling techniques supported by modeling that take into account spatial variability. Methods that further improve estimates of soil and vegetation carbon stock will depend on future research and model development and are likely to be highly transferable between Parties. [2.4.2, 2.4.3, 4.2.2, 5.4.1]

6. *Estimates of Average Annual Carbon Stock Changes/Accounted for ARD Activities and Some Additional Activities*

6.1. *Afforestation, Reforestation, and Deforestation*

55. Different definitions and accounting approaches under Article 3.3 of the Kyoto Protocol produce different estimates of changes in carbon stocks. There are seven Definitional Scenarios described in Chapter 3 of the underlying report. Table 3 illustrates, with data and methods available at the time of the Special Report, the estimated carbon stock changes accounted from ARD activities under the IPCC and FAO Definitional Scenarios, assuming recent area conversion rates remain constant and excluding carbon in soils and wood products. Three different carbon accounting approaches have been applied to the FAO Definitional Scenario to illustrate the effect of different accounting approaches. [3.5.3, 3.5.4, Table 3-4, Table 3-17]

56. The IPCC Definitional Scenario yields estimates of average annual accounted carbon stock changes from afforestation and reforestation in Annex I Parties from 2008 to 2012 of 7 to 46 Mt C yr^{-1}. This would be offset by annual changes in carbon stocks from deforestation of about -90 Mt C yr^{-1}, producing a net stock change of -83 to -44 Mt C yr^{-1}. If hypothetically, for example, afforestation and reforestation rates were to be increased in Annex I Parties by 20%[4] for the years 2000 to 2012, estimated annual changes in carbon stocks would increase (from 7 to 46 Mt Cyr^{-1}) to 7 to 49 Mt C yr^{-1}. If hypothetically, for example, deforestation rates were to be decreased by 20%, estimated annual losses of carbon stocks due to deforestation would reduce (from -90 Mt C yr^{-1}) to -72 Mt C yr^{-1}. [3.5.4]

[4] The 20% is an arbitrary value chosen to show the sensitivity of the estimates to changes in practices.

Table 3: *Estimate of accounted average annual carbon stock change for ARD activities. The IPCC and FAO Definitional Scenarios and three accounting approaches under the FAO Definitional Scenario have been applied to illustrate with the available data the effect of different accounting approaches. Other Definitional Scenarios described in Chapter 3, Table 3-4, have not been included in this analysis. The figures and ranges of values in the table are illustrative, provide first-order estimates, and may not encompass the full range of uncertainties. Negative numbers indicate carbon emissions and positive numbers carbon removals. For details, see Table 3-17 in Chapter 3.*

			Area Change (Mha yr^{-1})		**Estimated Range of Accounted Average Annual Stock Change 2008–2012 (Mt C yr^{-1})** *Includes carbon in aboveground and below-ground biomass, excludes carbon in soils and in dead organic matter*			
Region	**Activity**	**AR Average Rate of Uptake (t C ha^{-1} yr^{-1}); D Average Stock (t C ha^{-1})**	Post-Harvest Regeneration	Conversion between Non-Forest and Forest	FAO Definitional Scenario, Land-Based I Accounting	FAO Definitional Scenario, Land-Based II Accounting	FAO Definitional Scenario, Activity-Based Accounting	IPCC Definitional Scenario
Boreal Region Total (=Annex I)	AR	0.4 to 1.2	3.1	0.1	-209 to -162	-56 to -8	5 to 48	0 to 2
	D	35		0.5	-18	-18	-18	-18
	Total ARD				-227 to -180	-74 to -26	-13 to 30	-18 to -16
Temperate Region Annex I	AR	1.5 to 4.5	5.4	0.5	-550 to -81	-134 to 303	81 to 519	7 to 44
	D	60		1.2	-72	-72	-72	-72
	Total ARD				-622 to -153	-206 to 231	9 to 447	-65 to -28
Annex I Total	AR		8.5	0.6	-759 to -243	-190 to 295	87 to 573	7 to 46
	D			1.7	-90	-90	-90	-90
	Total ARD				-849 to -333	-280 to 205	-3 to 483	-83 to -44
Temperate Region Total	AR	1.5 to 4.5	n/a	1.9	n/a	n/a	n/a	27 to 167
	D	60		2.1	-126	-126	-126	-126
	Total ARD				n/a	n/a	n/a	-99 to 41
Tropical Region Total	AR	4 to 8	n/a	2.6	n/a	n/a	n/a	170 to 415
	D	120		13.7	-1644	-1644	-1644	-1644
	Total ARD				n/a	n/a	n/a	-1474 to -1229
Global Total (summing regional totals)	AR		n/a	4.6	n/a	n/a	n/a	197 to 584
	D			16.3	-1788	-1788	-1788	-1788
	Total ARD				n/a	n/a	n/a	-1591 to -1204

Notes: n/a = no number is provided because the area of regeneration after harvest in the tropical region and part of the temperate region was not available. In addition, regeneration after selective cutting, as it is often used in the tropics, is difficult to capture with the FAO Definitional Scenario. It is assumed that recent area conversion rates ["recent" = for Annex I Parties AR late 1980s/early 1990s and for D 1980s (except for Canada and Russian Federation early 1990s); ARD in other regions 1980s] have applied since 1990, and will continue to do so until 2012. The IPCC Definitional Scenario includes transitions between forest and non-forest land uses under Article 3.3. For the purposes of this table, it is assumed that not only planting, but also other forms of stand establishment such as natural establishment, are considered AR activities. The FAO Definitional Scenario includes the harvest/regeneration cycle, because regeneration is defined as reforestation. Within the FAO Definitional Scenario, three accounting approaches are distinguished (see paragraph 23 and Section 3.3.2). Uptake rates are intended to span the range within which the average value for each region is expected to be. The lower bound of the estimated average annual stock change corresponds to the lower uptake rate in AR and the higher bound to the higher uptake rate. Trees have been assumed to grow according to a sigmoidal growth curve. Estimated area for conversion between non-forest and forest should be regarded as an upper limit for the temperate region total and the tropical region, because some countries may have reported plantations for 1990 but not for 1980, and because some of the plantations may not qualify as resulting from AR activities under the IPCC Definitional Scenario. Also, for tropical countries, the deforestation estimates are very uncertain and could be in error by as much as ±50%.

57. The three accounting approaches under the FAO Definitional Scenario yield different results. Estimated average annual carbon stock changes in Annex I Parties from afforestation and reforestation are -759 to -243 Mt C yr^{-1} under the FAO land-based I approach; -190 to 295 Mt C yr^{-1} under the FAO land-based II approach; and 87 to 573 Mt C yr^{-1} under the FAO activity-based approach. Estimated average annual carbon stock changes from deforestation are about -90 Mt C yr^{-1} in all three approaches, as in the IPCC Definitional Scenario. [3.5.4]

58. For comparison, the IPCC Definitional Scenario yields estimates of average annual accounted carbon stock changes from afforestation and reforestation globally from 2008 to 2012 of 197 to 584 Mt C yr^{-1}. This would be offset by annual changes in carbon stocks from deforestation of about -1788 Mt C yr^{-1}, producing a net stock change of -1591 to -1204 Mt C yr^{-1}. If, hypothetically, for example, afforestation and reforestation rates were to be increased globally by 20% for the years 2000 to 2012, estimated annual changes in carbon stocks would increase (from 197 to 584 Mt C yr^{-1}) to 208 to 629 Mt C yr^{-1}. [3.5.4]

59. In the IPCC Definitional Scenario and FAO Definitional Scenario with land-based I accounting approach, the accounted carbon stock changes are broadly consistent with the 2008–2012 actual changes in carbon stocks from land under Article 3.3. The IPCC and FAO Definitional Scenarios bring different amounts of land under Article 3.3, hence the estimated carbon stock changes in Table 3 differ.

60. In the FAO Definitional Scenario with land-based II and activity-based accounting approaches, the accounted carbon stock change is not consistent with the 2008–2012 actual changes in carbon stocks on land under Article 3.3, except in the case of short rotation cycles.

61. In neither of the two Definitional Scenarios is the accounted carbon stock change consistent with the 2008–2012 actual carbon stock changes, nor with the net exchanges with the atmosphere, at the national and global levels in part because the land under Article 3.3 is small in comparison with the national and global forest area. [3.3.2, 3.5.4]

6.2. *Additional Activities*

62. The magnitude of the stock changes from additional activities that might be included under Article 3.4 rests, *inter alia*, on any decisions that remain to be made in the process of implementing the Kyoto Protocol. A consideration of carbon stocks changes and net emissions of greenhouse gas emissions associated with additional activities on managed lands entails synthesizing available technical and scientific data, outlining the outcomes of one policy scenario, and assessing the aggregate impact of policies and other factors. The scientific literature to support such an analysis is currently quite limited. [4.3]

63. One such scenario is presented in Table 4, to illustrate in a general sense the potential scope for carbon stock increases through some broadly defined activities. It provides data and information on carbon stock changes for some candidate activities under Article 3.4 for the year 2010. This scenario relies on three components relating to the candidate activities: 1) an estimate of current relevant land areas (column 2); 2) an assumed percentage of those lands on which an activity would be applied in 2010 (column 3); and 3) a research-derived estimate of the annual rate of carbon stock increase per hectare (column 4). The uptake rate is multiplied by the applicable land area to approximately calculate the change in carbon stock in the year 2010 (column 5).

64. Table 4, rather than providing precise projections, reports calculated stock changes assuming an ambitious policy agenda that promotes the application of activities to a significantly greater share of the relevant land base than would have otherwise occurred. The assumed percentage of lands on which the activity is applied is derived from considered professional judgment based on existing literature of what a range of sustained and effective initiatives, which vary across countries, could achieve. The share of land on which the activity is actually applied in 2010 depends to a great extent on the accounting system under Article 3.4, the evolving economic and social aspects of the activity, and landowner response to incentives, among other factors. Thus, the total annual stock changes in Table 4 (column 5) are likely to be on the high side.

65. Table 4 estimates do not necessarily represent credits under Article 3.4 of the Kyoto Protocol, even if such levels of stock change are achieved, because the Protocol may include approaches that limit the applicability of these calculations.

66. Table 4 illustrates the estimated carbon stock changes from example additional activities within Annex I and globally, assuming roughly similar levels of policy support. For example, Table 4 suggests that although conversion of cropland to grassland can provide a relatively large carbon stock increase per hectare converted, forest management improvements, which can be applied over a larger land base, may provide relatively larger total annual increases. Very different estimates in changes of emissions and removals associated with options for additional land use, land-use change, and forestry activities would result from different definitions of additional activities that might be agreed under Article 3.4, different accounting approaches, and different decisions that might be taken on implementation rules for Article 3.4.

67. There is potential for carbon uptake into biomass, which may be stored over a time period of decades in wood products. Furthermore, biomass used for energy purposes, based on waste by-products of wood/crops or from trees/crops grown expressly for this purpose, has the potential to lead to a reduction in net greenhouse gas emissions by substituting for fossil fuels. [1.4.3, 1.4.4]

***Table 4**: Relative potential in 2010 for net change in carbon stocks through some improved management and changed land-use activities.*[a]

(1) Activity	(2) Total Area[b] (Mha)	(3) Assumed Percentage of Total Area of Column 2 under Activity in 2010 (%)	(4) Net Annual Rate of Change in Carbon Stocks per Hectare[b] ($t\ C\ ha^{-1}\ yr^{-1}$)	(5) Estimated Net Change in Carbon Stocks in 2010 ($Mt\ C\ yr^{-1}$)
A. Annex I Countries				
a) Improved Management within a Land Use[c]				
Forest Management	1900	10	0.5	100
Cropland Management	600	40	0.3	75
Grazing Land Management	1300	10	0.5	70
Agroforestry	83	30	0.5	12
Rice Paddies	4	80	0.1	<1
Urban Land Management	50	5	0.3	1
b) Land-Use Change				
Conversion of Cropland to Grassland	600	5	0.8	24
Agroforestry	<1	0	0	0
Wetland Restoration	230	5	0.4	4
Restoring Severely Degraded Land	12	5	0.25	1
B. Global Estimates				
a) Improved Management within a Land Use				
Forest Management	4050	10	0.4	170
Cropland Management	1300	30	0.3	125
Grazing Land Management	3400	10	0.7	240
Agroforestry	400	20	0.3	26
Rice Paddies	150	50	0.1	7
Urban Land Management	100	5	0.3	2
b) Land-Use Change				
Agroforestry	630	20	3.1	390
Conversion of Cropland to Grassland	1500	3	0.8	38
Wetland Restoration	230	5	0.4	4
Restoring Severely Degraded Land	280	5	0.3	3

[a] Totals were not included in the table for several reasons: i) The list of candidate activities is not exclusive or complete; ii) it is unlikely that all countries would apply all candidate activities; and iii) the analysis does not presume to reflect the final interpretations of Article 3.4. Some of these estimates reflect considerable uncertainty.

[b] A summary of reference sources is contained in Tables 4-1 and 4-4 of this Special Report. Calculated values were rounded to avoid the appearance of precision beyond the intent of the authors. The rates given are average rates that are assumed to remain constant to 2010.

[c] Assumed to be the best available suite of management practices for each land use and climatic zone.

68. Table 4 does not account for the possibly significant non-CO_2 greenhouse gas emissions and removals that could be influenced by the candidate activities. For example, the rates do not reflect net emissions of CH_4 or N_2O from agricultural practices or wetlands/permafrost management. The table also does not include the carbon stock impact of the use of biofuels and the changing wood product pools, and consideration of forest management does not include avoided deforestation, which is dealt with in Table 3.

7. *Project-Based Activities*

69. An LULUCF project can be defined as a planned set of activities aimed at reducing greenhouse gas emissions or enhancing carbon stocks that is confined to one or more geographic locations in the same country and specified time period and institutional frameworks such as to allow net greenhouse gas emissions or enhancing carbon stocks to be monitored and verified. Experience is being gained in Activities Implemented Jointly (AIJ) and other LULUCF

projects that are under initial stages of implementation in at least 19 countries.

70. Assessment of the experience of these projects is constrained by the small number, the limited range of project types, the uneven geographic distribution, the short period of field operations to date, and the absence of an internationally agreed set of guidelines and methods to establish baselines and quantify emissions and uptake. Generally, these projects do not report all greenhouse gas emissions or estimate leakage, and few have independent review.

71. However, through the experience of LULUCF projects aimed to mitigate climate change, it is possible in some cases to develop approaches to address some of the critical issues (seeTable 5).

72. There are 10 projects aimed at decreasing emissions through avoiding deforestation and improving forest management, and 11 projects aimed at increasing the uptake of carbon—mostly forest projects in tropical countries (see Table 5). [5.2.2]

73. Methods of financial analysis among these projects have not been comparable. Moreover the cost calculations do not cover, in most instances, *inter alia*, costs for infrastructure, monitoring, data collection and interpretation costs, opportunity costs of land and maintenance, or other recurring costs, which are often excluded or overlooked. Recognizing the different methods used, the undiscounted cost and investment estimates range from $US 0.1–28 per ton of carbon, simply dividing project cost by their total reported accumulated carbon uptake or estimated emissions avoided, assuming no leakage outside the project boundaries. [5.2.3]

74. Project-level financial analysis methods are widely used and fairly standardized in development assistance and private investment projects. But they have yet to be consistently applied to, and reported for, LULUCF projects aiming at mitigating climate change. Guidelines for developing methods of financial analysis may be needed in the future. [5.2.3]

75. LULUCF projects aiming to mitigate climate change may provide socioeconomic and environmental benefits primarily within project boundaries, although they may also pose risks of negative impacts. Experience from most of the pilot projects to date indicates that involvement of local stakeholders in the design and management of project activities is often critical. Other factors affecting the capacity of projects to increase carbon uptake and avoid greenhouse gas emissions and to have other benefits include consistency with national and/or international sustainable development goals, and institutional and technical

Table 5*: Carbon uptake/estimated emissions avoided from carbon stocks, assuming no leakage outside the project boundaries, by selected AIJ Pilot Phase and other LULUCF projects, in some level of implementation.*[a,b,c,d,e]

Project Type (number of projects)	Land Area (Mha)	Accumulated Carbon Uptake over Project Lifetime (Mt C)	Estimated Carbon Uptake per Spatial Unit during the Project Lifetime (t C ha^{-1})	Accumulated Estimated Emissions Avoided over the Project Lifetime (Mt C)	Estimated Emissions Avoided from Carbon Stocks per Spatial Unit during the Project Lifetime (t C ha^{-1})
		assuming no leakage outside the project boundaries			
Forest Protection (7)[f]	2.8			41 – 48	4 – 252
Improved Forest Management (3)	0.06			5.3	41 – 102
Reforestation and Afforestation (7)	0.1	10 – 10.4	26 – 328		
Agroforestry (2)	0.2	10.5 – 10.8	26 – 56		
Multi-Component and Community Forest (2)	0.35	9.7	0.2 – 129		

[a] Projects included are those for which we have sufficient data. Soil carbon management, bioenergy, and other projects are not included for this reason.

[b] "Some level of implementation"—Included projects have been partially funded and have begun activities on the ground that will generate increases in carbon stocks and reductions in greenhouse gas emissions.

[c] "Other LULUCF projects"—Refers to selected non-AIJ projects and projects within Annex I countries.

[d] Estimated changes in carbon stocks generally have been reported by project developers, do not use standardized methods, and may not be comparable; only some have been independently reviewed.

[e] Non-CO_2 greenhouse gas emissions have not been reported.

[f] Protecting an existing forest does not necessarily ensure a long-term contribution to the mitigation of the greenhouse effect because of the potential for leakage and reversibility through human activities, disturbances, or environmental change. Table 5 does not provide an assessment in relation to these issues. Sound project design and management, accounting, and monitoring would be required to address these issues.

capacity to develop and implement project guidelines and safeguards. [2.5.2, 5.6]

76. The accounting of changes in carbon stocks and net greenhouse gas emissions involve a determination that project activities lead to changes in carbon stocks and net greenhouse gas emissions that are additional to a without-project baseline. Currently there is no standard method for determining baselines and additionality. Approaches include determining project-specific baselines or generic benchmarks. Most AIJ projects have used a project-specific approach that has an advantage of using better knowledge of local conditions yielding more accurate prediction. A disadvantage is that project developers may choose scenarios that maximize their projected benefits. Baselines may be fixed throughout the duration of a project or periodically adjusted. Baseline adjustments would ensure more realistic estimates of changes in carbon uptake or greenhouse gas emissions but would create uncertainties for project developers. [5.3.2, Table 5-4]

77. Projects that reduce access to land, food, fiber, fuel, and timber resources without offering alternatives may result in carbon leakage as people find needed supplies elsewhere. A few pilot projects have been designed with the aim of reducing leakage by explicitly incorporating components that supply the resource needs of local communities (e.g., establishing fuelwood plantations to reduce pressures on other forests), and that provide socioeconomic benefits that create incentives to maintain the project. Due to leakage, the overall consideration of the climate change mitigation effects of a project may require assessments beyond the project boundary, as addressed in paragraph 49. [2.3, 5.3.3]

78. Project accounting and monitoring methods could be matched with project conditions to address leakage issues. If leakage is likely to be small, then the monitoring area can be set roughly equal to the project area. Conversely, where leakage is likely to be significant the monitoring area could be expanded beyond the project area, although this would be more difficult when the leakage occurs across national boundaries. Two possible approaches could then be used to estimate leakage. One would be to monitor key indicators of leakage, and the second would be to use standard risk coefficients developed for project type and region. In either case, leakage could be quantified and subsequently changes in carbon stock and greenhouse gas emissions attributed to the project could be reestimated. The effectiveness of these two approaches is untested. [5.3.3]

79. LULUCF projects raise a particular issue with respect to permanence (see paragraph 40). Different approaches have been proposed to address the duration of projects in relation to their ability to increase carbon stocks and decrease greenhouse gas emissions, *inter alia*: (i) They should be maintained in perpetuity because their "reversal" at any point in time could invalidate a project; and (ii) they should be maintained until they counteract the effect of an equivalent amount of greenhouse gases emitted to the atmosphere. [5.3.4]

80. Several approaches could be used to estimate the changes in carbon stocks and greenhouse gas emissions of LULUCF projects: (i) estimating carbon stocks and greenhouse gas emissions at a given point in time; (ii) estimating the average changes of carbon stocks or greenhouse gas emissions over time in a project area; or (iii) allowing for only a part of the total changes in carbon stocks or greenhouse gas emissions for each year that the project is maintained (e.g., tonne-year method). The year-to-year distribution of changes in carbon stocks and greenhouse gas emissions over the project duration varies according to the accounting method used. [5.4.2, Table 5-9]

81. LULUCF projects are subject to a variety of risks because of their exposure to natural and anthropogenic factors. Some of these risks particularly pertain to land-use activities (e.g., fires, extreme meteorological events, and pests for forests), while others are applicable to greenhouse gas mitigation projects in both LULUCF and energy sectors such as political and economic risks. Risk reduction could be addressed through a variety of approaches internal to the project, such as introduction of good practice management systems, diversification of project activities and funding sources, self-insurance reserves, involvement of local stakeholders, external auditing, and verification. External approaches for risk reduction include standard insurance services, regional carbon pools, and portfolio diversification. [5.3.5]

82. Techniques and tools exist to measure carbon stocks in project areas relatively precisely depending on the carbon pool. However, the same level of precision for the climate change mitigation effects of the project may not be achievable because of difficulties in establishing baselines and due to leakage. Currently, there are no guidelines as to the level of precision to which pools should be measured and monitored. Precision and cost of measuring and monitoring are related. Preliminary limited data on measured and monitored relevant aboveground and below-ground carbon pools to precision levels of about 10% of the mean at a cost of about US$1–5 per hectare and US$0.10–0.50 per ton of carbon have been reported. Qualified independent third-party verification could play an essential role in ensuring unbiased monitoring. [5.4.1, 5.4.4]

8. *Reporting Guidelines for the Relevant Articles of the Kyoto Protocol*

83. Under Article 5.2 of the Kyoto Protocol, the Revised 1996 Guidelines for National Greenhouse Gas Inventories provide the basis for the accounting and reporting of anthropogenic emissions by sources and removals by sinks of all greenhouse gases not controlled by the

Montreal Protocol. These Guidelines were developed to estimate and report national greenhouse gas inventories under the United Nations Framework Convention on Climate Change (UNFCCC), not for the particular needs of the Kyoto Protocol. However, the Guidelines do provide a framework for addressing the accounting and reporting needs of the Kyoto Protocol. Elaboration of the Land-Use Change and Forestry Sector of the Guidelines may be needed, reflecting possible decisions by the Parties for accounting and reporting LULUCF under the Kyoto Protocol, taking into account, *inter alia*:

- Any decisions made by Parties on ARD under Article 3.3 and on additional activities under Article 3.4. [6.3.1, 6.3.2]
- The need to ensure transparency, completeness, consistency, comparability, accuracy, and verifiability. [6.2.2, 6.2.3, 6.4.1]
- Consistent treatment of Land-Use Change and Forestry as other Sectors, with respect to uncertainty management and other aspects of good practice. [6.4.1]
- Any decisions adopted by Parties to address other accounting issues (e.g., permanence, the meaning of "human induced" and "direct human induced," wood products, and project based activities). [6.4.1]

9. Potential for Sustainable Development

84. Consideration would need to be given to synergies and tradeoffs related to LULUCF activities under the UNFCCC and its Kyoto Protocol in the context of sustainable development including a broad range of environmental, social, and economic impacts, such as: (i) biodiversity; (ii) the quantity and quality of forests, grazing lands, soils, fisheries, and water resources; (iii) the ability to provide food, fiber, fuel, and shelter; and (iv) employment, human health, poverty, and equity [2.5.1, 3.6]

85. For example, converting non-forest land to forest will typically increase the diversity of flora and fauna, except in situations where biologically diverse non-forest ecosystems, such as native grasslands, are replaced by forests consisting of single or a few species. Afforestation can also have highly varied impacts on groundwater supplies, river flows, and water quality. [3.6.1]

86. A system of criteria and indicators could be used to assess and compare sustainable development impacts across LULUCF alternatives. While there are no agreed upon set of criteria and indicators, several sets are being developed for closely related purposes, for example assessment of contributions to sustainable development by the UN Commission on Sustainable Development. [2.5.2]

87. For activities within countries or projects between countries, if sustainable development criteria vary significantly across countries or regions, there may be incentives to locate activities and projects in areas with less stringent environmental or socioeconomic criteria. [2.5.2]

88. Several sustainable development principles are incorporated in other multilateral environmental agreements, including the UN Convention on Biological Diversity, the UN Convention to Combat Desertification, and the Ramsar Convention on Wetlands. Consideration may be given to the development of synergies between LULUCF activities and projects that contribute to the mitigation or adaptation to climate change with the goals and the objectives of these and other relevant multilateral environmental agreements. [2.5.2]

89. Some of the more formal approaches to sustainable development assessment that could be applied at the project level are, for example, environmental and socioeconomic impact assessments. These methods have been applied across a wide range of countries and site-specific activities to date and could be modified to be applicable to LULUCF projects. [2.5.2.2]

90. Some critical factors affecting the sustainable development contributions of LULUCF activities and projects to mitigate and adapt to climate change include: institutional and technical capacity to develop and implement guidelines and procedures; extent and effectiveness of local community participation in development, implementation, and distribution of benefits; and transfer and adoption of technology. [5.5, 5.6]

Appendix I. Conversion Units

1 tonne (t)	1000 kilogram (kg)	10^6 gram (g)	1 Megagram (Mg)
1 Megatonne (Mt)	1,000,000 t	10^{12} g	1Teragram (Tg)
1 Gigatonne (Gt)	1,000,000,000 t	10^{15} g	1 Petagram (Pg)
1 hectare (ha)	10,000 square metre (m^2)		
1 square kilometee (km^2)	100 hectare (ha)		
1 tonne per hectare ($t\ ha^{-1}$)	100 gram per square metre ($g\ m^{-2}$)		
1 tonne carbon	3.67 tonne carbon dioxide ($t\ CO_2$)		
1 tonne carbon dioxide	0.273 tonne carbon (t C)		
1 tonne	0.984 imperial ton	1.10 US ton	2,204 pound
1 hectare (ha)	2.471 acre		
1 square kilometre (km^2)	0.386 square mile		
1 tonne per hectare ($t\ ha^{-1}$)	892 pound per acre		

Appendix II. Relevant Portions of Kyoto Protocol Articles Discussed in this Special Report

[Concepts in **bold** are discussed in the SPM]

*Article 2.1: Each Party included in Annex I in achieving its quantified emission limitation and reduction commitments under Article 3, in order to promote **sustainable development**, shall:*

(a) Implement and/or further elaborate policies and measures in accordance with its national circumstances, such as:
(ii) Protection and enhancement of sinks and reservoirs of greenhouse gases not controlled by the Montreal Protocol, taking into account its commitments under relevant international environmental agreements; promotion of sustainable forest management practices, afforestation and reforestation.
(iii) Promotion of sustainable forms of agriculture in light of climate change considerations.

(b) Cooperate with other such Parties to enhance the individual and combined effectiveness of their policies and measures adopted under this Article, pursuant to Article 4, paragraph 2(e)(i), of the Convention. To this end, these Parties shall take steps to share their experience and exchange information on such policies and measures, including developing ways of improving their comparability, transparency, and effectiveness. The Conference of the Parties serving as the meeting of the parties to this Protocol shall, at its first session or as soon as practicable thereafter, consider ways to facilitate such cooperation, taking into account all relevant information.

Article 3.1: "The Parties included in Annex I shall, individually or jointly, ensure that their aggregate anthropogenic carbon dioxide equivalent emissions of greenhouse gases listed in Annex A do not exceed their assigned amounts, calculated pursuant to their quantified emission limitation and reduction commitments inscribed in Annex B and in accordance with the provisions of this Article, with a view to reducing their overall emissions of such gases by at least 5% below 1990 levels in the commitment period 2008–2012."

*Article 3.3: The net changes in greenhouse gas emissions by sources and removals by sinks resulting from **direct human-induced land use change and forestry activities**, limited to **afforestation, reforestation and deforestation since 1990**, measured as **verifiable** changes in **carbon stocks** in each commitment period, shall be used to meet the commitments under this Article of each Party included in Annex I. The greenhouse gas emissions by sources and removals by sinks associated with those activities shall be reported in a **transparent** and **verifiable** manner and reviewed in accordance with Articles 7 and 8.*

*Article 3.4: Prior to the first session of the COP serving as the meeting of the Parties to this Protocol, each Party included in Annex I shall provide, for consideration by the SBSTA, **data to establish its level of carbon stocks in 1990** and to enable an estimate to be made of its **changes in carbon stocks in subsequent years**. The COP serving as the meeting of the Parties to this Protocol shall, at its first session or as soon as practicable thereafter, decide upon **modalities, rules and guidelines** as to **how, and which, additional human-induced activities** related to **changes in greenhouse gas emissions by sources and removals by sinks in the agricultural soils and the land-use change and forestry** categories shall be added to, or subtracted from, the assigned amounts for Parties included in Annex I, taking into account **uncertainties, transparency in reporting, verifiability**, the methodological work of the IPCC, the advice provided by the SBSTA in accordance with Article 5 and the decisions of the COP. Such a decision shall apply in the second and subsequent commitment periods. A Party may choose to apply such a decision on these additional human-induced activities for its first commitment period, provided that these activities have taken place **since 1990**.*

Article 3.7: In the first quantified emission limitation and reduction commitment period, from 2008 to 2012, the assigned amount for each Party included in Annex I shall be equal to the percentage inscribed for it in Annex B of its aggregate anthropogenic carbon dioxide equivalent emissions of the greenhouse gases listed in Annex A in 1990, or the base year or period determined in accordance with paragraph 5 above, multiplied by five. Those Parties included in Annex I for whom land use change and forestry constituted a net source of greenhouse gas emissions in 1990, shall include in their 1990 emissions base year or period, the aggregate anthropogenic carbon dioxide equivalent emissions minus removals in 1990 from land use change for the purposes of calculating their assigned amount.

Article 5.2:Methodologies for estimating anthropogenic emissions by sources and removals by sinks of all greenhouse gases not controlled by the Montreal Protocol shall be those accepted by the Intergovernmental Panel on Climate Change and agreed upon by the Conference of the Parties at its third session. Where such methodologies are not used, appropriate adjustments shall be applied according to methodologies agreed upon by the Conference of the Parties serving as the meeting of the Parties to this protocol at its first session. Based on the work of, inter alia, the Intergovernmental Panel on Climate Change and advice provided by the Subsidiary Body for Science and Technological Advice, the Conference of the Parties serving as the meeting of the parties to this Protocol shall regularly review and, as appropriate, revise such methodologies and adjustments, taking into account any relevant decisions by the Conference of the Parties. Any revision to methodologies or adjustments shall be used only for the purposes of ascertaining compliance with commitments under Article 3 in respect of any commitment period adopted subsequent to that revision.

Article 6.1: For the purpose of meeting its commitments under Article 3, any Party included in Annex I may transfer to, or acquire from, any other such Party emission reduction units resulting from ***projects*** *aimed at reducing anthropogenic emissions by sources or enhancing anthropogenic removals by sinks of greenhouse gases in any sector of the economy, provided that:*

Article 6.1(b): Any such project provides a reduction in emissions by sources, or an enhancement of removals by sinks, that is ***additional*** *to any that would otherwise occur.*

Article 12.2: The purpose of the clean development mechanism shall be to assist Parties not included in Annex I in achieving sustainable development and in contributing to the ultimate objective of the Convention, and to assist Parties included in Annex I in achieving compliance with their quantified emission limitation and reduction commitments under Article 3.

Article 12.3(a): Parties not included in Annex I will benefit from ***project activities*** *resulting in* ***certified emissions reductions****.*

Article 12.3(b): Parties included in Annex I may use the ***certified emissions reductions*** *accruing from such project activities to contribute to compliance with part of their quantified emission limitation and reduction commitments under Article 3, as determined by the Conference of the Parties serving as the meeting of the Parties to the Protocol.*

Article 12.5: Emissions reductions resulting from each project activity shall be certified by operational entities to be designated by the COP serving as the meeting of the Parties to this Protocol, on the basis of:

Article 12.5(b): Real, ***measurable****, and* ***long-term*** *benefits related to the mitigation of climate change.*

Article 12.5(c): Reductions in emissions that are ***additional*** *to any that would occur in the absence of the certified project.*

Appendix III. Glossary

[These definitions are provided solely for the purposes of this Special Report.]

Accuracy
The degree to which the mean of a sample approaches the true mean of the population; lack of bias.

Activity
A practice or ensemble of practices that take place on a delineated area over a given period of time.

Baseline
A reference scenario against which a change in greenhouse gas emissions or removals is measured.

Bias
Systematic over- or under-estimation of a quantity.

Biosphere
That component of the Earth system that contains life in its various forms, which includes its living organisms and derived organic matter (e.g., litter, detritus, soil).

Carbon Flux
Transfer of carbon from one carbon pool to another in units of measurement of mass per unit area and time (e.g., t C ha^{-1} yr^{-1}).

Carbon Pool
A reservoir. A system which has the capacity to accumulate or release carbon. Examples of carbon pools are forest biomass, wood products, soils, and atmosphere. The units are mass (e.g., t C).

Carbon Stock
The absolute quantity of carbon held within a pool at a specified time.

Flux
See "Carbon Flux."

Forest Estate
A forested landscape consisting of multiple stands of trees.

Forest Stand
A community of trees, including aboveground and below-ground biomass and soils, sufficiently uniform in species composition, age, arrangement, and condition to be managed as a unit.

Heterotrophic Respiration
The release of carbon dioxide from decomposition of organic matter.

Land Cover
The observed physical and biological cover of the Earth's land as vegetation or man-made features.

Land Use
The total of arrangements, activities, and inputs undertaken in a certain land cover type (a set of human actions). The social and economic purposes for which land is managed (e.g., grazing, timber extraction, conservation).

Permanence
The longevity of a carbon pool and the stability of its stocks, given the management and disturbance environment in which it occurs.

Pool
See "Carbon Pool."

Practice
An action or set of actions that affect the land, the stocks of pools associated with it or otherwise affect the exchange of greenhouse gases with the atmosphere.

Precision
The repeatability of a measurement (e.g., the standard error of the sample mean).

Regeneration
The renewal of a stand of trees through either natural means (seeded on-site or adjacent stands or deposited by wind, birds, or animals) or artificial means (by planting seedlings or direct seeding).

Reservoir
A pool.

Sequestration
The process of increasing the carbon content of a carbon pool other than the atmosphere.

Shifting Agriculture
A form of forest use common in tropic forests where an area of forest is cleared, or partially cleared, and used for cropping for a few years until the forest regenerates. Also known as "slash and burn agriculture," "moving agriculture," or "swidden agriculture."

Sink
Any process or mechanism which removes a greenhouse gas, an aerosol, or a precursor of a greenhouse gas from the atmosphere. A given pool (reservoir) can be a sink for atmospheric carbon if, during a given time interval, more carbon is flowing into it than is flowing out.

Source
Opposite of sink. A carbon pool (reservoir) can be a source of carbon to the atmosphere if less carbon is flowing into it than is flowing out of it.

Stand
See "Forest Stand."

Stock
See "Carbon Stock."

Soil Carbon Pool
Used here to refer to the relevant carbon in the soil. It includes various forms of soil organic carbon (humus) and inorganic soil carbon and charcoal. It excludes soil biomass (e.g., roots, bulbs, etc.) as well as the soil fauna (animals).

Uptake
The addition of carbon to a pool. A similar term is "sequestration."

Wood Products
Products derived from the harvested wood from a forest, including fuelwood and logs and the products derived from them such as sawn timber, plywood, wood pulp, paper, etc.

Land Use, Land-Use Change, and Forestry

A Special Report of the Intergovernmental Panel on Climate Change

This Special Report was accepted at IPCC Plenary XVI
(Montreal, Canada • 1-8 May 2000), but not approved in detail.

1

Global Perspective

BERT BOLIN (SWEDEN) AND RAMAN SUKUMAR (INDIA)

Lead Authors:
P. Ciais (France), W. Cramer (Germany), P. Jarvis (UK), H. Kheshgi (USA), C. Nobre (Brazil), S. Semenov (Russian Federation), W. Steffen (Australia)

Review Editors:
S. Linder (Sweden) and F. Joos (Switzerland)

CONTENTS

EXECUTIVE SUMMARY

The Global Carbon Cycle

- During the period 1850–1998, approximately 405 ± 60 Gt C has been emitted as carbon dioxide (CO_2) into the atmosphere as a result of fossil fuel burning and cement production (67 percent), and land use and land-use change (33 percent), predominantly from forested areas. As a result, the atmospheric CO_2 concentration has risen from 285 ± 5 ppmv to 366 ppmv (i.e., about a 28 percent increase). This increase in CO_2 concentration accounts for about 40 percent of these anthropogenic emissions, the remainder having been absorbed by the oceans and terrestrial ecosystems.
- CO_2 that is dissolved into the ocean will be transferred progressively to the deep ocean, and the carbon content of this reservoir is increased. The fate of CO_2 that is fixed on land depends on which ecosystem and which carbon pool is the repository (e.g., living biomass or soils). Carbon fixed into a pool with a turnover time of one year or less (leaves, fine roots) is returned to the atmosphere or transferred into pools with a longer turnover time of decades to centuries (stems, trunks, soil organic matter).
- The net global carbon flux between terrestrial ecosystems and the atmosphere is the result of a small imbalance between uptake by photosynthesis and release by various return processes. Plants, soil microbes, biochemical processes, animals, and disturbances contribute to the latter. Variations of climate and human activities have a major impact through land use and land-use changes, as well as indirectly through carbon dioxide fertilization, nutrient deposition, and air pollution.
- This global net carbon exchange has resulted in an uptake of CO_2 by the terrestrial biosphere amounting to 0.2 ± 1.0 Gt C yr^{-1} (90-percent confidence interval) over the 1980s (1980–89), and 0.7 ± 1.0 Gt C yr^{-1} during the most recent decade (1989–98) (see Table 1-2). It is unclear if the increase in the 1990s is a result of natural variability or, to some extent, also a trend induced by human activities.
- The direct effects of land use and land-use change are estimated to have led to a net emission of 1.7 ± 0.8 Gt C yr^{-1} during the 1980s, and 1.6 Gt C yr^{-1} during the 1990s. The difference between the net global terrestrial uptake and human-induced emissions as a result of land use and land-use change leaves a residual terrestrial uptake of 1.9 ± 1.3 Gt C yr^{-1} for the 1980s and 2.3 ± 1.3 Gt C yr^{-1} for the 1990s.
- The global net carbon flux varies from one year to another. These variations are on the order ±1 Gt C yr^{-1} and are correlated with variations in climate (e.g., El Niño/La Niña events) and major volcanic eruptions.

Present Knowledge about Global Terrestrial Ecosystems

- Gross Primary Productivity (GPP) is the uptake of carbon from the atmosphere by plants (global total approximately 120 Gt C yr^{-1}). Carbon losses as a result of plant respiration reduce this uptake to the Net Primary Productivity (NPP; global total approximately 60 Gt C yr^{-1}). Further losses occur because of decomposition of dead organic matter, resulting in Net Ecosystem Productivity (NEP; global total approximately 10 Gt C yr^{-1}). Additional losses are caused by disturbances, such as fire, wind-throw, drought, pests, and human activities. The resulting net imbalance of the terrestrial ecosystem can be interpreted as the Net Biome Productivity (NBP; presently approximately 0.7 ± 1.0 Gt C yr^{-1}, as a decadal average; see Figure 1-2).
- Forests contain a large part of the carbon stored on land, in the form of biomass (trunks, branches, foliage, roots etc.) and in the form of soil organic carbon (Table 1-1). On a time scale of years, most forests accumulate carbon through the growth of trees and an increase in soil carbon, until the next disturbance occurs. The net carbon uptake (NEP) may locally reach 7 t C ha^{-1} yr^{-1}, but losses may also be observed when soil carbon is decreasing or trees are overmature and mortality is occurring.
- In cropland ecosystems, carbon stocks are primarily in the form of below-ground plant organic matter and soil. Most of these ecosystems have large annual carbon uptake rates, but much of the gain is exported in the form of agricultural products and their associated waste materials; this gain is rapidly released to the atmosphere. Although carbon is recaptured during the succeeding cropping season, many agricultural soils are currently net sources of carbon. Shifting to low or no till cultivation is, however, increasingly being used to mitigate such trends.
- By far most of the carbon stocks in grassland and savannas, including rangelands and pasture, are found in the soils. These stocks are stable over long time spans, but losses can occur if grazing pressure exceeds carrying capacity or if the frequency of fires increases.
- Wetland stocks of carbon are found almost entirely in the soil as dead organic matter, which can be released by human activity, such as drainage. Afforestation may effectively compensate for such development. Soil carbon in subarctic wetlands may also be released as a result of reduction of permafrost resulting from climate warming.
- Globally, carbon stocks in the soil exceed carbon stocks in vegetation by a factor of about five (Table 1-1). This ratio ranges from about 1:1 in tropical forests to 5:1 in boreal

forests and much larger factors in grasslands and wetlands. Changes in soil carbon stocks are at least as important for carbon budgets as changes in vegetation carbon stocks.

Assessing Carbon Stocks and Their Change Over Time

- A sustainably managed forest comprising all stages in the stand life cycle operates as a functional system that maintains an overall carbon balance, retaining a part in the growing trees, transferring another part into the soils, and exporting carbon as forest products. Recently disturbed and regenerating areas lose carbon; young stands gain carbon rapidly, mature stands less so; and overmature stands may lose carbon. The Kyoto Protocol (Articles 3.3 and 3.4) focuses on only part of the stand life cycle for a few decades. During the early years of the life cycle, when trees are small, the area is likely to be a source of carbon; it becomes a sink when carbon assimilation exceeds soil respiration.
- Change in the carbon stocks of stands of forest trees over a 5-year period (NPP) can be assessed with good precision through standard inventory methods. Carbon stocks in soils have been determined by standard sampling techniques, but large numbers of samples are required to achieve adequate precision (see Figure 1-4).
- Changes in total carbon stocks—vegetation and soils—in forest stands have been assessed by direct determination of net sources and sinks (NEP) over periods of 1 or more years. This approach has been applied worldwide, predominantly to mature forest stands of a range of species and history. The annual NEP varies between approximately -1 and 2.5 t C ha^{-1} yr^{-1} for boreal forests, 2.5 to 7 t C ha^{-1} yr^{-1} for temperate forests, and 1 to 6 t C ha^{-1} yr^{-1} for humid tropical forests.
- At present, direct measurements of NEP alone are inadequate to provide estimates of NBP because of lack of data covering all stages in the life cycle from regeneration to harvest, as well as lack of data on the impacts of disturbances such as fire, wind-throw, drought, pollution, pests, and diseases. Thus, forest inventories and ecosystem process models, combined with assessments of harvest and disturbance-related carbon flows, must be used. These methodologies require further development and comparison with measurements of NEP.
- Measurements of NEP on young stands of trees and forest inventory yield tables indicate that in new forests planted since 1990 in relation to Article 3.3 of the Kyoto Protocol, sequestration of carbon is likely to continue for 20 to 200 years or more after establishment, depending on species and site conditions.
- Sequestration of carbon is stimulated by fertilization with atmospheric CO_2 and deposition of nutrients, particularly nitrogen and phosphorus. The quantitative contribution of these resources to carbon sequestration is difficult to determine and varies from region to region depending on the magnitude of nutrient inputs and inputs of associated, negatively acting, pollutants (e.g., acid precipitation or ozone).
- Ecosystem models indicate that the additional terrestrial sink arising from global climate change is likely to be maintained in the short term (over several decades) if management is appropriate and sustainable, but may gradually diminish in the medium term. One reason for this result is that the capacity of some ecosystems to sequester carbon may be approached; another is that photosynthesis will no longer increase as CO_2 concentration continues to rise, whereas respiration is expected to continue to increase with the rise in temperature. A third reason is that trees may begin to die as a result of climate change. The balance between forest photosynthesis and respiration is crucially dependent on the nutrient dynamics of the forest ecosystem, as well as other environmental variables. Because of current limitations on our understanding with respect to acclimation of the physiological processes, climatic constraints, and feedbacks amongst these processes, particularly those acting at biome scale, projections beyond a few decades are highly uncertain.

Impact of Human Activities

- Human activities modify carbon flows between the atmosphere, the land, and the oceans. Land use and land-use change are the main factors that affect terrestrial sources and sinks of carbon. Clearing of forests has resulted in a reduction of the global area of forests by almost 20 percent during the past 140 years. Management practices can restore, maintain, and enlarge vegetation and soil carbon stocks, however.
- Expansion of agriculture through conversion of forests and grassland during the past 140 years has led to a net release of about 121 Gt C, of which about 60 percent has been emitted in the tropics (primarily during the past half-century) and about 40 percent in middle and high latitudes (primarily before the middle of the 20th century). During the 1980s, more than 90 percent of the net release of carbon to the atmosphere was the result of land-use changes in the tropics.
- Reducing the rate of forest clearing can reduce carbon losses from terrestrial ecosystems. Establishing forests on previously cleared land provides an opportunity to sequester carbon in tree biomass and forest soils, but it will take decades to centuries to restore carbon stocks that have been lost as a result of land-use change in the past.
- Ecosystem conservation and management practices can restore, maintain, and enlarge soil carbon stocks. Forests and wetlands managed as nature reserves or recreation areas can sequester significant amounts of carbon.
- On cropland, soil carbon is lost through disturbances of soil by tillage. Management practices (e.g., irrigation and application of fertilizers) can enhance soil carbon stocks. Application of nitrogen fertilizer in croplands, however, probably is the largest human-induced source of the greenhouse gas nitrous oxide (N_2O) at present (see Section 1.2.3).

- Global emissions of the greenhouse gases methane (CH_4) and N_2O from land use-related activities [expressed as CO_2-equivalents (C-eq)] exceed net CO_2 emissions from land-use change. Anthropogenic emissions include 0.9 Gt C-eq yr^{-1} of N_2O from cultivated soils, 0.6 Gt C-eq yr^{-1} of CH_4 and 0.1 Gt C-eq yr^{-1} of N_2O from livestock, 0.3 Gt C-eq yr^{-1} of CH_4 from rice paddies, and 0.2 Gt C-eq yr^{-1} of CH_4 and 0.1 Gt C-eq yr^{-1} of N_2O from biomass burning. In addition, emissions from forest soils or wetlands, for example, which may be considered natural, are affected by land use-related activities that affect these ecosystems (see Table 1-3).
- Multiple uses of land, while enhancing carbon stocks and producing energy, may present opportunities to reduce greenhouse gas emissions with a minimum use of resources. Products made from biomass comprise carbon stocks and present an opportunity to substitute for materials that might otherwise lead to larger greenhouse gas emissions (see Section 1.4.3 and Chapter 4). Biomass energy can be produced continuously by planting and harvesting, thereby reducing the consumption of fossil fuels. The development of new technologies for efficient production of biomass energy is essential to keep costs low and to secure land in competition for its use for alternative services (see Section 1.4.4 and Chapter 4).

1.1. Introduction

The emergence of life on earth has led to the conversion of carbon dioxide in the atmosphere and carbon dissolved in the oceans into innumerable inorganic and organic compounds on land and in the sea. The development of different ecosystems over millions of years has established patterns of carbon flows through the global environmental system. Natural exchanges of carbon between the atmosphere, the oceans, and terrestrial ecosystems are now being modified by human activities, primarily as a result of fossil fuel burning and changing land use. This activity has led to a steady addition of carbon dioxide to the atmosphere and enhancement of the atmospheric concentration by more than 28 percent over the past 150 years.

We need to understand the global environmental system and in particular the circulation of carbon in nature, as well as how human activities have modified it, to assess how we may do so increasingly in the future. In addition to reducing emissions from fossil fuel use, we may also have an opportunity to reduce the rate of build-up of carbon dioxide in the atmosphere by taking advantage of the fact that carbon can accumulate in vegetation and soils in terrestrial systems—an opportunity that was brought into focus at the Third Conference of the Parties (COP) to the Framework Convention on Climate Change (FCCC) in Kyoto. This introductory chapter provides an overview of our present understanding of the fundamental natural processes at work, which is essential for an analysis of the opportunities, limitations, and implications of actions related to land use and land-use change.

Natural flows of carbon between the atmosphere, the oceans, and the terrestrial and freshwater systems vary from one part of the globe to another and in time (i.e., between seasons, from one year to the next, and over decades and centuries). It is often difficult to separate changes resulting from human interventions from these natural variations.

Some of the measures specified in the Kyoto Protocol are ambiguous because the terminology being used is not always adequately defined. Clarifications in this regard are required, and we need to analyze the implications of alternative interpretations as a basis for political agreements on how to proceed. The measures specified in the Protocol might also induce secondary changes that need to be evaluated. Knowledge about the functioning of global biogeochemical cycles, particularly the global carbon cycle, is essential in this context.

Information about human-induced disturbances of sources and sinks of other greenhouse gases is also provided. Methane emissions may be changed unintentionally when actions are taken to enhance carbon dioxide sinks, and sources and sinks of nitrous oxide will be modified if the cycle of nitrogen is disturbed. Changes in the nitrogen cycle will in turn influence terrestrial ecosystems and thereby the exchange of carbon dioxide between the terrestrial system and the atmosphere.

To judge the long-term consequences of the ways in which human activities disturb the circulation of carbon in nature and change its distribution between natural reservoirs, we need to analyze the carbon cycle in detail, particularly with respect to the terrestrial ecosystems. This analysis will also shed light on the implications of the measures specified in the Kyoto Protocol.

1.2. Biogeochemical Cycles of Greenhouse Gases

Terrestrial ecosystems are important components in the biogeochemical cycles that create many of the sources and sinks of carbon dioxide, methane, and nitrous oxide and thereby influence global responses to human-induced emissions of greenhouse gases (GHGs). The dynamics of terrestrial ecosystems depend on interactions between a variety of biogeochemical cycles, particularly the carbon cycle, the nutrient cycles, and the circulation of water—all of which may be modified indirectly by climate changes and by direct human actions (e.g., land-use/cover change).

1.2.1. The Global Carbon Cycle

1.2.1.1. Natural and Human-Induced Changes in the Past Carbon Cycle

Analyses of air bubbles in ice cores from Greenland and the Antarctic have given us a reasonably clear idea about variations in atmospheric CO_2 concentration since the end of the last glacial maximum. It was then about 200 ppmv; it rose gradually to about 250 ppmv 8,000 years ago and subsequently by 25 ppmv during the following 7,000 years. During the past millennium until the beginning of the industrial revolution, CO_2 varied between 275 and 285 ppmv. There seems to have been an increase of about 10 ppmv around 1300 AD, followed by a 10 ppmv decrease around 1600 AD (i.e., during the Little Ice Age) (Barnola *et al.*, 1995; Etheridge *et al.*, 1996; Indermühle *et al.*, 1999). All of these changes took place gradually, and the rate of change in the atmospheric reservoir probably seldom exceeded a few Gt C per decade (Ciais, 1999).

The CO_2 concentration has risen from the range noted above to a concentration of 366 ppmv in 1998 (Keeling and Whorf, 1999). The decadal rate of change over the past century has been persistent and more rapid than during any other period in the last millennium. This rate of change can be explained by the cumulative effects of emissions from fossil fuel combustion and land clearing and the response of the oceans and biosphere to this anthropogenic perturbation.

From 1850 to 1998, 270 ± 30 Gt C were emitted from fossil fuel burning and cement production (Marland *et al.*, 1999); 176 ± 10 Gt C accumulated in the atmosphere (Etheridge *et al.*, 1996; Keeling and Whorf, 1999). The cumulative ocean uptake during this time has been estimated (with the aid of ocean carbon cycle models) to be 120 ± 50 Gt C (Kheshgi *et al.*, 1999; Joos *et al.*, 1999). This estimate of ocean uptake is more uncertain than estimates of total emissions from fossil fuel burning and the accumulation in the atmosphere (Siegenthaler and Joos, 1992; Enting *et al.*, 1994). Nevertheless, balancing the carbon budget for this period yields a global net terrestrial source of about 26 ± 60 Gt C. In other words, it is likely that the terrestrial system has been a source during this period.

It is relevant to compare the magnitude of this global net terrestrial source with direct estimates of emissions during this time resulting from the expansion of cropland, deforestation, and other land-use changes (see Section 1.4.1). The area covered by cropland in temperate regions (particularly in North America and the former Soviet Union) reached a maximum by the middle of the 20th century (Ramankutty and Foley, 1998). The rate of increase of croplands in tropical regions (mainly Latin America), however, surpassed that of temperate regions around 1960 (Houghton, 1994, 1999).

During the period 1850–1998, net cumulative global CO_2 emissions from land-use change are estimated to have been 136 ± 55 Gt C (assuming that the relative uncertainty of land-use change emissions is the same as the estimate for the 1980s). Of these emissions, about 87 percent were from forest areas and about 13 percent from cultivation of mid-latitude grasslands (Houghton, 1999; Houghton *et al.*, 1999, 2000). A residual global terrestrial sink of 110 ± 80 Gt C is therefore required to reconcile the difference between the net terrestrial source estimated by balancing the carbon budget (26 ± 60 Gt C) and the larger terrestrial source estimated by accounting for the effects land-use change on carbon stocks (136 ± 55 Gt C). Thus, this residual terrestrial carbon sink—popularly referred to as the "missing carbon sink"—was comparable in size to the net ocean uptake over this period.

1.2.1.2. A More Detailed Analysis of the Carbon Budget and its Change during the Past 20 Years

Carbon in the form of inorganic and organic compounds, notably CO_2, is cycled between the atmosphere, oceans, and terrestrial biosphere (Figure 1-1). The largest natural exchanges occur between the atmosphere and terrestrial biota (GPP about 120 Gt C yr^{-1}, NPP about 60 Gt C yr^{-1}) and between the atmosphere and ocean surface waters (about 90 Gt C yr^{-1}). The atmosphere contains about 775 Gt C; the residence time for a CO_2 molecule in the atmosphere is therefore only about 2.5 years. The characteristic adjustment times between reservoirs in response

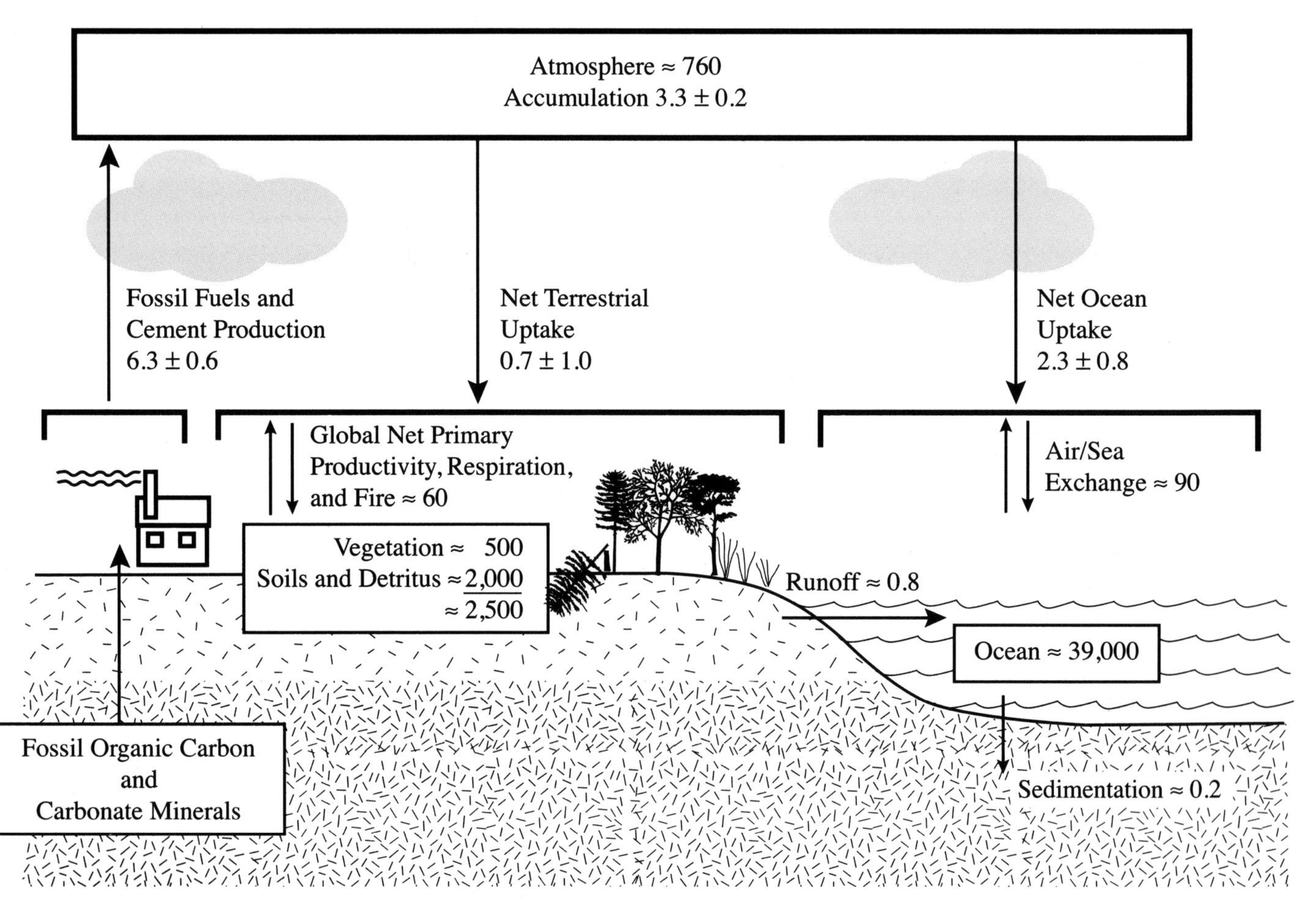

Figure 1-1: The global carbon cycle, showing the carbon stocks in reservoirs (in Gt C = 10^{15} g C) and carbon flows (in Gt C yr^{-1}) relevant to the anthropogenic perturbation as annual averages over the decade from 1989 to 1998 (Schimel *et al.*, 1996, Tables 2.1 and 2.2). Net ocean uptake of the anthropogenic perturbation equals the net air-sea input plus runoff minus sedimentation (discussed by Sarmiento and Sundquist, 1992).

***Table 1-1**: Global carbon stocks in vegetation and top 1 m of soils (based on WBGU, 1998).*

Biome	Area (10^6 km^2)	Carbon Stocks (Gt C) *Vegetation*	*Soils*	*Total*
Tropical forests	17.6	212	216	428
Temperate forests	10.4	59	100	159
Boreal forests	13.7	88	471	559
Tropical savannas	22.5	66	264	330
Temperate grasslands	12.5	9	295	304
Deserts and semideserts	45.5	8	191	199
Tundra	9.5	6	121	127
Wetlands	3.5	15	225	240
Croplands	16.0	3	128	131
Total	151.2	466	2011	2477

to perturbations to the system, however, are on the order of decades to centuries (Schimel *et al.*, 1996).

The oceans, vegetation, and soils are significant reservoirs of carbon; they actively exchange CO_2 with the atmosphere. Oceans contain about 50 times as much carbon as the atmosphere, predominantly in the form of dissolved inorganic carbon. Ocean uptake of carbon is limited, however, by the solubility of CO_2 in seawater (including the effects of carbonate chemistry) and the slow rate of mixing between surface and deep-ocean waters. Terrestrial vegetation and soils contain about three and a half times as much carbon as the atmosphere; the exchange is controlled by photosynthesis and respiration.

The amount of carbon stored globally in soils is much larger than that in vegetation (Table 1-1). Soil is a major carbon pool in all biomes, whereas carbon stocks in vegetation are predominantly in the forest biomes. Boreal forests have a larger proportion of carbon stored in soils than in trees, compared with temperate or tropical forests. There are wide local variations, however, in the amounts and proportions of carbon per unit ground area in vegetation and soil within each biome (see Section 1.3).

The average global carbon budget for the 1980s (1980 to 1989) (Schimel *et al.*, 1996) has been reassessed for the most recent decade from 1989 to 1998. There are some significant differences between the two decades (Table 1-2). Emissions from fossil fuel combustion and cement production have increased by about 0.8 Gt C yr^{-1} (based on estimates through 1996 by Marland *et al.*, 1999, and energy statistics for 1997 and 1998 by British Petroleum Company, 1999). There has been a slight decrease, however, in these emissions from Annex I countries in aggregate, with a marked decrease from Annex I countries with "economies in transition." The increase in these emissions from non–Annex I countries in aggregate has been about 0.9 Gt C yr^{-1}. The rate of increase in the atmospheric stock of carbon, on the other hand, has remained about the same (Keeling and Whorf, 1999). Although the net ocean uptake appears to have increased somewhat (Jain *et al.*, 1995; Harvey *et al.*, 1997), the difference between emissions resulting from the burning of fossil fuels and cement production, on the one hand, and atmospheric and oceanic uptake, on the other, has increased—with the result that the net terrestrial uptake of carbon for the period 1989–1998 was probably 0.7 ± 1.0 Gt C yr^{-1}.

Precise molecular oxygen (O_2) measurements in the atmosphere make it possible to quantify the net global terrestrial carbon flux and the oceanic uptake of carbon in an independent manner. Reconstruction of the mean atmospheric O_2 trend from air enclosed in bubbles in glacier ice (Battle *et al.*, 1996) and air archived in tanks yields a net terrestrial uptake of 0.6 ± 0.9 Gt C yr^{-1} (±1 standard deviation) for the 1980s. High-precision atmospheric observations (Keeling *et al.*, 1996b) yield a value of 0.9 ± 0.7 Gt C yr^{-1} for the period 1990–1997. Thus, there is satisfactory consistency between the estimates from the two approaches.

Factors that influence the net terrestrial uptake of carbon include the direct effects of land use and land-use change (e.g., deforestation and agricultural abandonment and regrowth) (see Section 1.4) and the response of terrestrial ecosystems to CO_2 fertilization, nutrient deposition, climatic variation, and disturbance (e.g., fires, wind-throws, and major droughts) (see Section 1.3). These natural phenomena may partially be indirect effects of other human activities: Many ecosystems are in some state of recovery from past disturbances. For the 1980s, the combination of estimates of the strength of these factors (Schimel *et al.*, 1995) yields a value for net terrestrial uptake that is consistent with, but more uncertain than, the residual calculated in line 4 of Table 1-2. For the 1980s, Houghton (1999) estimates the net CO_2 source from land-use change to be 2.0 ± 0.8 Gt C yr^{-1}, which was later revised to 1.7 ± 0.8 Gt C yr^{-1} considering newer regional data (Houghton *et al.*, 1999, 2000). Estimates for the most recent decade are 1.6 ± 0.8 Gt C yr^{-1} based on regional data up to 1995 (Houghton *et al.*, 1999, 2000). Yet from the revised carbon budget (Table 1-2) we can infer that the net global effect of all other factors has offset the source from land-use change, yielding a significant net terrestrial sink over the past 20 years. The residual terrestrial uptake for both decadal periods in Table 1-2 is comparable in size to the oceanic uptake.

***Table 1-2**: Average annual budget of CO_2 perturbations for 1980 to 1989 (consistent with values given in Schimel et al., 1996) and 1989 to 1998 (note the 1-year overlap in the two decadal periods). Flows and reservoir changes of carbon are expressed in Gt C yr^{-1}; error limits correspond to an estimated 90-percent confidence interval.*

	1980 to 1989	1989 to 1998
1) Emissions from fossil fuel combustion and cement production	5.5 ± 0.5	6.3 ± 0.6[a]
a) from Annex I countries[d]	3.9 ± 0.4[a]	3.8 ± 0.4[a]
i) from countries excluding those with economies in transition	2.6 ± 0.3	2.8 ± 0.3
ii) from countries with economies in transition[d]	1.3 ± 0.3[a]	1.0 ± 0.3[a]
b) from rest of world[d]	1.6 ± 0.3[a]	2.5 ± 0.4[a]
2) Storage in the atmosphere	3.3 ± 0.2	3.3 ± 0.2[b]
3) Ocean uptake	2.0 ± 0.8	2.3 ± 0.8[c]
4) Net terrestrial uptake = (1) – [(2)+(3)]	0.2 ± 1.0	0.7 ± 1.0
5) Emissions from land-use change	1.7 ± 0.8[e]	1.6 ± 0.8[f]
6) Residual terrestrial uptake = (4)+(5)	1.9 ± 1.3	2.3 ± 1.3

[a] Based on emission estimates through 1996 by Marland *et al.* (1999) and estimates derived from energy statistics for 1997 and 1998 (British Petroleum Company, 1999).
[b] Based on atmospheric CO_2 concentrations measured at Mauna Loa, Barrow, and South Pole (Keeling and Whorf, 1999).
[c] Based on ocean carbon cycle model (Jain *et al.*, 1995) used in the IPCC Second Assessment Report (IPCC, 1996; Harvey *et al.*, 1997) consistent with an uptake of 2.0 Gt C yr^{-1} in the 1980s.
[d] Annex 1 countries and countries with economies in transition (a subset of Annex 1 countries) defined in the FCCC. Emissions include emission estimates from geographic regions preceding this designation and include emissions from bunker fuels from each region.
[e] Based on land-use change emissions estimated by Houghton (1999) and modified by Houghton *et al.* (1999, 2000), which include the net emissions from wood harvesting and agricultural soils.
[f] Based on estimated annual average emissions for 1989–1995 (Houghton *et al.*, 1999, 2000).

1.2.1.3. Inter-Annual and Decadal Variability of Atmospheric CO_2 Concentrations

The uncertainty ranges in Table 1-2 result partly from our limited ability to determine accurately the gradual changes in the carbon balance resulting from human-induced emissions. In addition, however, variations in the atmospheric CO_2 growth rate that have been recorded since 1960 imply that global terrestrial and oceanic carbon sources and sinks may vary significantly in time (Conway *et al.*, 1994; Francey *et al.*, 1995; Keeling *et al.*, 1996a). Fossil fuel emissions, on the other hand, do not fluctuate much from one year to the next, whereas the exchange of atmospheric CO_2 with the oceans and the terrestrial biosphere responds to inter-annual climate variations. High atmospheric CO_2 growth rates have been recorded during three recent El Niño events—in 1983, 1987, and 1998—indicating a lower than normal uptake of atmospheric CO_2 by the terrestrial biosphere and the oceans (Gaudry *et al.*, 1987; Keeling *et al.*, 1989; Keeling and Whorf, 1999). Conversely, low atmospheric CO_2 growth rates were observed between 1991 and 1993, indicating enhanced uptake—particularly over the northern hemisphere (Ciais *et al.*, 1995a,b; Keeling *et al.*, 1996b).

Ocean carbon models and available data suggest that the oceans contribute less to observed year-to-year changes in atmospheric CO_2 concentration than does the terrestrial biosphere (Winguth *et al.*, 1994; Le Quéré *et al.*, 1998; Lee *et al.*, 1998; Feely *et al.*, 1999; Rayner *et al.*, 2000). The terrestrial biosphere therefore appears to drive most of the inter-annual variation in CO_2 flows. The way ecosystems respond to climate variability is not well understood, although the correlation and lag-correlation of inter-annual variability between CO_2 growth rates, climate, and the remotely sensed "greenness" normalized difference vegetation index (NDVI), which is related to photosynthesis, is illustrative (Braswell *et al.*, 1997; Myneni *et al.*, 1997).

When terrestrial biogeochemical models are forced with realistic year-to-year changes in temperature and precipitation, they can simulate changes in the global and regional biosphere and associated changes in CO_2 exchange with the atmosphere (Kindermann *et al.*, 1996; Tian *et al.*, 1998). These models can reproduce the magnitude and to some extent the phase of observed inter-annual variability of atmospheric CO_2 concentrations, though different processes have been implicated in attempts to explain the observed fluctuations (e.g., Heimann *et al.*, 1997). There are still differences in detail that have not been resolved.

Shifts in magnitude and phase of atmospheric CO_2 fluctuations on a decadal time scale suggest that seasonality of terrestrial biotic fluxes has been changing slowly at mid to high northern latitudes (Keeling *et al.*, 1996b; Randerson *et al.*, 1997). Remotely sensed data (Myneni *et al.*, 1997), as well as phenological observations (Menzel and Fabian, 1999), independently indicate a longer growing season in the boreal zone and in temperate Europe during recent decades.

1.2.1.4. Non–Land Use Influences on Sources and Sinks of CO_2

In addition to land use and land-use change, several other factors—many of which are anthropogenic in origin—affect large-scale sources and sinks of atmospheric CO_2. These factors contribute to the observed variability and upward trend in CO_2 concentration, thus complicate the determination of how much of the observed changes in carbon stocks in vegetation and soils during a commitment period should be attributed to the direct activities initiated in accordance with Articles 3.3 or 3.4 of the Kyoto Protocol. For example, changes in climate and climate variability influence sources and sinks of CO_2 from vegetation and soils and *vice versa,* and these may have natural or anthropogenic causes. Projections of future atmospheric CO_2 concentrations will, therefore, be quite uncertain, at least beyond the next few decades (Tans and Wallace, 1999).

Other factors that may have contributed to the rate of carbon sequestration into vegetation and soils include the increase in atmospheric CO_2 concentration during 1850–1998 from about 285 to 366 ppmv (see Section 1.3.2.3) and increasing atmospheric concentrations of NO_x (NO and NO_2) and NH_3 that enhance the atmospheric deposition of nitrogen—a limiting plant nutrient in many ecosystems (Schimel *et al.*, 1995) (see Section 1.3.2.4). On the other hand, emissions of NO_x and SO_2 also lead to atmospheric deposition of nitrogen and sulfur compounds that cause acidification of soils and waters, which may negatively affect plant growth and reduce carbon uptake. In addition, elevated concentrations of surface ozone that also reduce plant growth (Semenov *et al.*, 1998, 1999) result from NO_x emissions in the presence of volatile organic compounds (Houghton *et al.*, 1996). All of these factors affect the net removal of CO_2 from the atmosphere by terrestrial ecosystems, as given in Table 1-2, in addition to the direct effects of land use and land-use change.

1.2.2. Sources and Sinks of Methane

Methane (CH_4) is the most important greenhouse gas in the atmosphere after water vapor and CO_2. CH_4 concentrations have increased from about 700 ppbv in pre-industrial times to about 1700 ppbv today (Etheridge *et al.*, 1992; Prather *et al.*, 1995). About 550 Mt CH_4 yr^{-1} is emitted into the atmosphere from a variety of sources; chemical reaction with OH radicals and (to a smaller extent) uptake by soils remove approximately the same amount (Prather *et al.*, 1995). The small imbalance between global production and destruction of CH_4 resulted in an increase in the atmospheric concentration at a rate of 13 ppbv yr^{-1} during the early 1980s. This rate diminished, however, to 8 ppbv yr^{-1} in 1990 and dropped further to 4 ppbv yr^{-1} in 1996 (Steele *et al.*, 1992; Dlugokencky *et al.*, 1998). The lifetime of CH_4 with respect to the OH sink is about 9 years (Prinn, 1994), so the characteristic adjustment time of the atmospheric concentration to a perturbation in emissions is much shorter than for CO_2.

During pre-industrial times, wetlands (bogs at high northern latitudes and swamps in the tropics), termites, and wild animals (Chappelaz *et al.*, 1993) controlled atmospheric CH_4. Anthropogenic CH_4 sources are associated with rice cultivation, cattle breeding, biomass burning, waste treatment (landfills, sewage, and animal waste), and the use of fossil fuels, including natural gas and coal extraction as well as petroleum industry activities in general (Prather *et al.*, 1995) (Table 1-3). At present, anthropogenic sources represent about 70 percent of total CH_4

Table 1-3: *Global estimates (Prather et al., 1995) of recent sources of CH_4 and N_2O that are influenced by land-use activities.*

CH_4 Sources	**Mt CH_4 yr^{-1}**	**Gt C-eq yr^{-1} [a,b]**
Livestock (enteric fermentation and animal waste)	110 (85–130)	0.6 (0.5–0.7)
Rice paddies	60 (20–100)	0.3 (0.1–0.6)
Biomass burning	40 (20–80)	0.2 (0.1–0.5)
Natural wetlands	115 (55–150)	0.7 (0.3–0.9)
N_2O Sources	**Mt N yr^{-1}**	**Gt C-eq yr^{-1} [a,c]**
Cultivated soils	3.5 (1.8–5.3)	0.9 (0.5–1.4)
Biomass burning	0.5 (0.2–1)	0.1 (0.05–0.3)
Livestock (cattle and feed lots)	0.4 (0.2–0.5)	0.1 (0.05–0.13)
Natural tropical soils—wet forests	3 (2.2–3.7)	0.8 (0.6–1)
Natural tropical soils—dry savannas	1 (0.5–2)	0.3 (0.1–0.5)
Natural temperate soils—forests	1 (0.1–2)	0.3 (0.03–0.5)
Natural temperate soils—grasslands	1 (0.5–2)	0.3 (0.1–0.5)

[a] 12 Gt C-equivalent = 44 Gt CO_2-equivalent.
[b] Carbon-equivalent emissions based on CH_4 GWP of 21.
[c] Carbon-equivalent emissions based on N_2O GWP of 310.

emissions. Although the global CH_4 source is relatively well known, the magnitude of individual sources is still uncertain (Fung *et al.*, 1991; Prather *et al.*, 1995). Table 1-3 lists estimated land use-related emissions of CH_4 on both a CH_4- and a CO_2-equivalent basis.[1] The likely changes in these CH_4 sources and sinks associated with changes in land use and other modifications of terrestrial ecosystems are uncertain. There may also be indirect changes resulting from human activities in accordance with Articles 3.3 and 3.4 of the Kyoto Protocol.

1.2.3. Sources and Sinks of Nitrous Oxide

Nitrous oxide (N_2O), another major greenhouse gas, has no significant sinks on land and is destroyed by chemical reactions in the upper atmosphere. Land surfaces are the main source of atmospheric N_2O; thus, changes in land-use practices modify soil emissions and influence N_2O concentration in the atmosphere (Kroeze *et al.*, 1999). Uncertainty with respect to current magnitudes of sources and sinks of N_2O (Prasad, 1997) and its atmospheric lifetime limit an accurate budget: Existing data on fluxes of N_2O from soils and oceans are insufficient to quantify them in detail. Nevertheless, present-day global N_2O emissions have been estimated to be about 14 Mt N yr^{-1} (Prasad, 1997). Roughly half of the global N_2O emissions are anthropogenic (Davidson, 1991; Khalil and Rasmussen, 1992; Hutchinson, 1995; Prather *et al.*, 1995; Prasad, 1997).

Microbiological processes in soils are the primary sources of N_2O (Davidson, 1991, 1992; Shiller and Hastie, 1996). Table 1-3 lists estimated land use-related emissions of N_2O on both a nitrogen content and a CO_2-equivalent basis.[2] Emissions from soils are enhanced under warm and wet conditions (e.g., those present in the soils of moist tropical forests) and when nitrogen fertilizers are applied in agriculture (Conrad *et al.*, 1983; Winchester *et al.*, 1988; Khalil and Rasmussen, 1992). Thus, changes in agricultural soil management and tropical forestry may alter N_2O emissions from soils and influence its concentration in the atmosphere.

1.3. The Carbon Budget of Terrestrial Ecosystems

The carbon sequestration potential of terrestrial ecosystems depends on the type and condition of the ecosystem—that is, its species composition, structure, and (in the case of forests) age distribution. Also important are site conditions, including climate and soils, natural disturbances, and management. For the analysis of a carbon budget, the fundamental differences between GPP, NPP, NEP, and NBP must be recognized (see Figure 1-2). The justification of the quantitative global flux estimates as defined below is given in the succeeding sections of this chapter (see also Steffen *et al.*, 1998).

Gross Primary Production denotes the total amount of carbon fixed in the process of photosynthesis by plants in an ecosystem, such as a stand of trees. GPP is measured on photosynthetic tissues, principally leaves. Global total GPP is estimated to be about 120 Gt C yr^{-1}.

Net Primary Production denotes the net production of organic matter by plants in an ecosystem—that is, GPP reduced by losses resulting from the respiration of the plants (autotrophic respiration). Global NPP is estimated to be about half of the GPP—that is, about 60 Gt C yr^{-1}.

Net Ecosystem Production denotes the net accumulation of organic matter or carbon by an ecosystem; NEP is the difference between the rate of production of living organic matter (NPP) and the decomposition rate of dead organic matter (heterotrophic respiration, RH). Heterotrophic respiration includes losses by herbivory and the decomposition of organic debris by soil biota. Global NEP is estimated to about 10 Gt C yr^{-1}. NEP can be measured in two ways: One is to measure changes in carbon stocks in vegetation and soil; the other is to integrate the fluxes of CO_2 into and out of the vegetation (the net ecosystem exchange, NEE) with instrumentation placed above (Aubinet *et al.*, 2000). The precision of both of these methods is improving.

Net Biome Production denotes the net production of organic matter in a region containing a range of ecosystems (a biome) and includes, in addition to heterotrophic respiration, other processes leading to loss of living and dead organic matter (harvest, forest clearance, and fire, etc.) (Schulze and Heimann, 1998). NBP is appropriate for the net carbon balance of large areas (100–1000 km^2) and longer periods of time (several years and longer). In the past, NBP has been considered to be close to zero (Figure 1-2). Compared to the total fluxes between atmosphere and biosphere, global NBP is comparatively small; NBP for the decade 1989–1998 has been estimated to be 0.7 ± 1.0 Gt C yr^{-1} (Table 1-2)—about 1 percent of NPP and about 10 percent of NEP.

1.3.1. Carbon Stocks and Flows in Major Biomes

For the estimation of present and future carbon sequestration potential, it is necessary to consider broad vegetation types differentiated by climatic zones and water availability (i.e., tropical, temperate, and boreal regions). Table 1-1 lists the areas, current estimates of aboveground and below-ground carbon stocks, and NPP of the world's major regions or biomes. Within each biome, large additional variation exists resulting from local conditions and topography. In the tropics, for example, moist and dry forests have widely differing carbon stocks and NPP.

- Pristine forests (e.g., in the wet tropics or boreal region) were long believed to be mostly in a state of

[1]Emissions given as CO_2-equivalent are calculated, as specified in the Kyoto Protocol, using the Global Warming Potentials (GWPs) given in the IPCC Second Assessment Report (SAR) (Houghton *et al.*, 1996); the GWP for CH_4 was 21 (100-year time horizon).

[2]Emissions given as CO_2-equivalent are calculated, as specified in the Kyoto Protocol, using the GWP given in the IPCC SAR (Houghton *et al.*, 1996), which was 310 for N_2O (100-year time horizon).

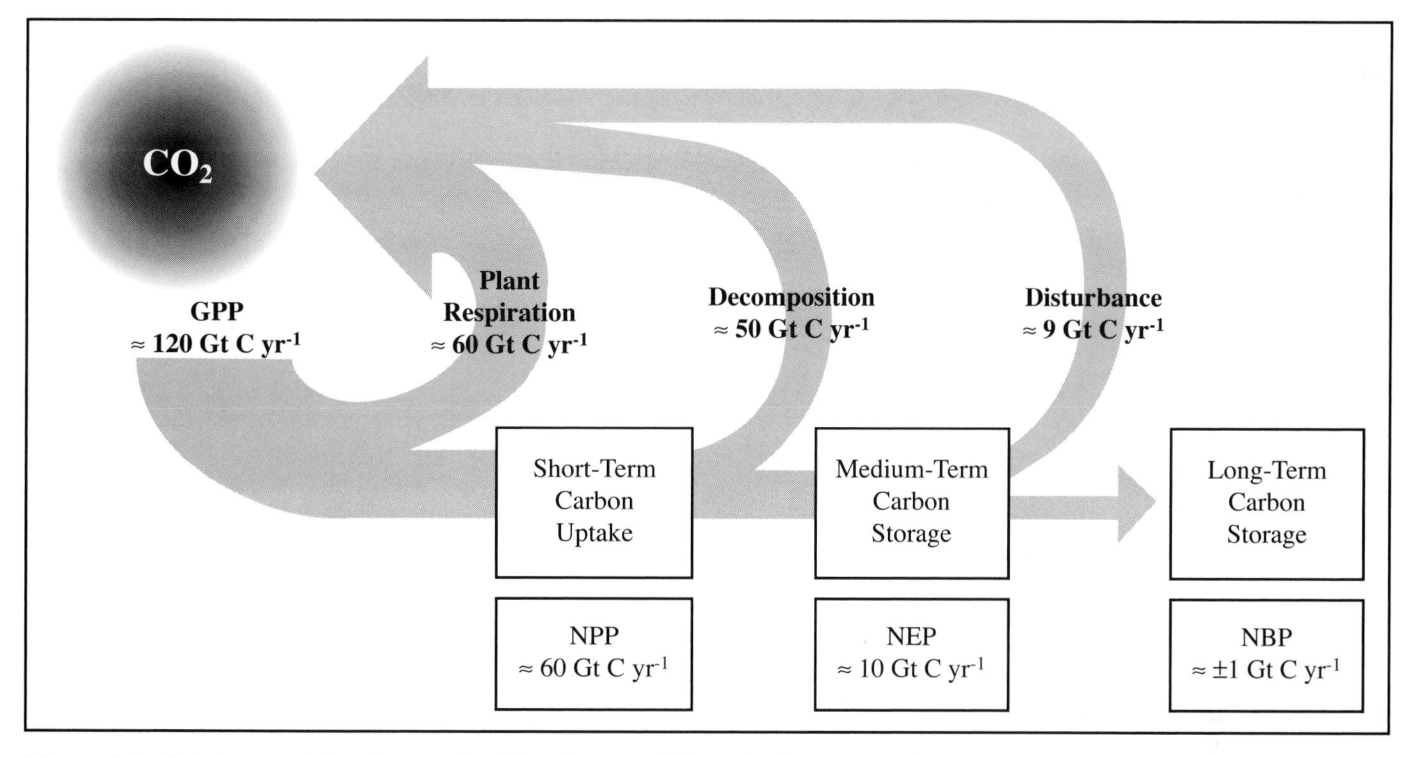

Figure 1-2: Global terrestrial carbon uptake. Plant (autotrophic) respiration releases CO_2 to the atmosphere, reducing GPP to NPP and resulting in short-term carbon uptake. Decomposition (heterotrophic respiration) of litter and soils in excess of that resulting from disturbance further releases CO_2 to the atmosphere, reducing NPP to NEP and resulting in medium-term carbon uptake. Disturbance from both natural and anthropogenic sources (e.g., harvest) leads to further release of CO_2 to the atmosphere by additional heterotrophic respiration and combustion—which, in turn, leads to long-term carbon storage (adapted from Steffen *et al.*, 1998).

equilibrium, such that over a period of several years their carbon balance would be neutral. This view has been challenged in more recent years by increasing evidence from sample plot studies that undisturbed areas of forests also sequester carbon (e.g., Lugo and Brown, 1993; Phillips *et al.*, 1998, for the tropics; Schulze *et al.*, 1999, for the Siberian boreal forest). These carbon quantities will eventually be returned to the atmosphere when patches of trees die for biological or climatic reasons, localized natural disturbance occurs, or compartments of the forest are cleared. Because of its importance to the Kyoto Protocol, carbon sequestration by managed and unmanaged forests is considered in some detail below (Section 1.3.2).

- Grassland ecosystems store most of their carbon in soils, where turnover is relatively slow (Table 1-1). In most grassland types, below-ground NPP is at least equal to or higher than aboveground production. Carbon accumulation by combined aboveground and below-ground NPP may be as much as 3.4 t C ha^{-1} yr^{-1} in tropical humid savannas and as little as 0.7 t C ha^{-1} yr^{-1} in tropical dry savannas and 0.5 t C ha^{-1} yr^{-1} in temperate steppe (Parton *et al.*, 1995).
- Wetlands are important reservoirs of carbon. Undrained peatlands in high latitudes have accumulated appreciable amounts of carbon from the atmosphere since the retreat of the ice and continue to be significant CO_2 sinks (0.2–0.5 t C ha^{-1} yr^{-1}), but they are also sources of methane (0.03–0.3 t CH_4 ha^{-1} yr^{-1}). By contrast, peatlands that are drained for agriculture or for afforestation release carbon as CO_2 because of accelerated decomposition of the aerobic peat (Cannell *et al.*, 1993), although they no longer release methane in significant amounts. The quantitative balance between these two processes is poorly understood (Cannell *et al.*, 1999), although it is important because the GWP of methane is 21 times that of CO_2 (see Section 1.2). Peatlands drained for agriculture continue to be a sustained carbon source as long as any peat remains in the soil. Peatlands drained for afforestation may continue to be a source of carbon in spite of forest biomass growth (Zoltai and Martikainen, 1985), but under certain climatic conditions they can also revert to carbon sinks (Minkkinen and Laine, 1998).
- In agricultural land, by far most of the carbon is stored below ground (see Table 1-1). Losses of carbon from terrestrial systems during the past 200 years, particularly until the middle of the 20th century, were mostly the result of the establishment of agriculture on grassland and land that was previously covered by forests. Regular plowing, planting, and harvesting led to enhanced oxidation of organic matter in the soils, which has been emitted into the atmosphere as carbon

dioxide. Today, agricultural lands are major sources of CO_2 in many countries as a result of past land-use changes (e.g., Cannell *et al.*, 1999). Soil organic carbon in cultivated soils is continuing to decline in many areas of the world. The use of fertilizers, high-yielding plant varieties, residue management, and reduced tillage for erosion control has contributed to the stabilization or increase in soil organic carbon (Cole *et al.*, 1993; Sombroek *et al.*, 1993; Blume *et al.*, 1998).

1.3.2. Components of the Carbon Balance of Ecosystems

To understand the fluxes of carbon between a specific ecosystem and the atmosphere over a specified period of time and to appreciate its sensitivity to current environmental conditions, disturbance, and climate change, we need to consider the principal component processes within the system that add up to give the overall NEP. The feedbacks among these processes, their speed of response, and their sensitivity to environmental change determine NBP—the future and long-term carbon sequestration potential of the ecosystem.

Figure 1-3 shows measured annual carbon fluxes in a typical boreal, temperate, and tropical forest. NEP is the difference between the gross input of carbon in photosynthesis, GPP, and the sum of the losses of carbon in autotrophic respiration (RA) and RH. The component fluxes accumulated over a year should add up to the annual NEP, which is measured independently as described below (see Section 1.3.2.3). A mass balance of the component fluxes above ground (photosynthesis; foliage, branch, and stem respiration; leaf, branch, and stem litter production; and aboveground NPP) enables an estimate to be made of the amount of carbon internally translocated below ground. A mass balance of the component fluxes below ground (the inputs of litter and translocate from above, root and heterotrophic respiration, fine root turnover, mycorrhizal and root system NPP) enables an approximate estimate to be made of net changes in the pool of soil carbon. There are appreciable errors, however, in measuring all component fluxes (belowground fluxes in particular), so close agreement between the two estimates of NEP is not to be expected.

The processes of photosynthesis and respiration are functions of several environmental and plant variables, including solar radiation, air and soil temperature and humidity, availability of water and nutrients, atmospheric ozone and other pollutants, leaf area, and foliar nutrition. Climate change therefore affects these processes in several ways. Photosynthesis is likely to be reduced by an increase in cloud cover but increased by enhanced global atmospheric CO_2 concentration and, on some sites, by atmospheric nutrient deposition. All respiratory processes are sensitive to temperature, as is the rate of population growth of respiring organs—particularly the fine roots and heterotrophic organisms in the soil. Thus, "soil respiration" is a function of soil temperature (e.g., Boone *et al.*, 1998; Rayment and Jarvis, 2000), which, if increased, leads in the short term to enhanced mineralization of soil organic matter and the release of nutrients—which, in turn, feed back to stimulate photosynthesis, increased leaf area, and tree growth. Evidence is accumulating, however, that in the longer term soil respiration acclimates to the rise in temperature and stabilizes at close to the original rate. In semi-arid and arid regions in particular, the availability of soil water—and thus changes in rainfall patterns—are also of utmost importance for the balance between carbon gains and losses.

1.3.2.1. Net Ecosystem Exchange on Different Time Scales

In assessing the carbon sequestration potential of terrestrial ecosystems now and in the future, we need to consider the time scale over which the carbon gain is measured or estimated. In the course of a day, carbon sequestration essentially follows solar radiation (provided there is no major constraint such as frozen or dry soil): Carbon is accumulated during the daylight hours and lost at night. Depending on the balance between these short-term gains and losses, for any vegetation there may be a net gain or loss of carbon over any day (24 hours) in the year. In boreal forest, for example, carbon may be gained in daytime by a stand of trees on about 75 percent of the days through a growing season but lost on 25 percent of the days, the latter being days of particularly low solar radiation input (e.g., days with clouds and rain, or smoke in the atmosphere), high temperatures, or drought (see, e.g., Jarvis *et al.*, 1997; Lindroth *et al.*, 1998).

Over a year, the length of the growing season has a major influence on carbon gain. In boreal coniferous forest, for example, one-third to one-half of the carbon gained in the summer months is lost during autumn and the long winter period when the ground is partly frozen. The length of the growing season in broad-leaved temperate forest is defined by bud burst in the spring and leaf senescence in the autumn, whereas the season of net photosynthetic gain in temperate coniferous forest is conditioned largely by the day length and daily total radiation input in the winter months, thus may be 10 months long in maritime climates. Vegetation in Mediterranean climates is generally sparse and strongly seasonal, with small NEP that is strongly constrained by water availability for a large part of the time. Only in moist tropical forests is the carbon gain nearly continuous throughout the year, reduced only by occasional short periods of low solar radiation (cloudiness), low temperature, or water deficit (Malhi *et al.*, 1998).

Annual dynamics are particularly important in forest systems because their carbon turnover times can be many decades, characterized by major changes in structure through a series of stages in the life cycle. The carbon sequestration potential of a young forest stand that is regenerating or regrowing after a disturbance such as fire or harvesting is critically dependent on the point of time within the life cycle. Initially the disturbed area is likely to be losing carbon to the atmosphere (the length of this period depends on species, site conditions, and degree of disturbance), but the trees that subsequently occupy the site fully will eventually replace the lost carbon (Krankina *et al.*,

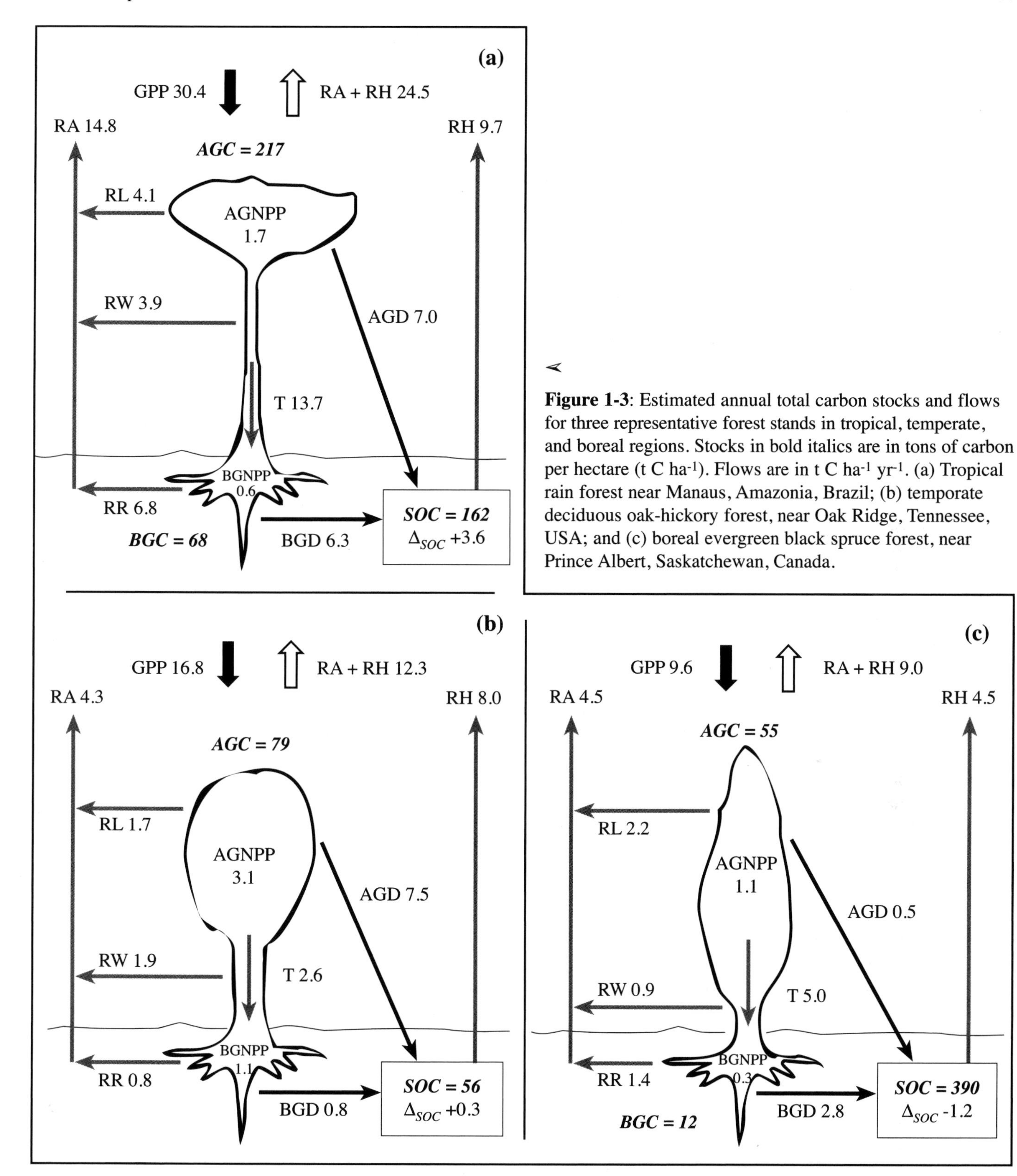

Figure 1-3: Estimated annual total carbon stocks and flows for three representative forest stands in tropical, temperate, and boreal regions. Stocks in bold italics are in tons of carbon per hectare (t C ha^{-1}). Flows are in t C ha^{-1} yr^{-1}. (a) Tropical rain forest near Manaus, Amazonia, Brazil; (b) temporate deciduous oak-hickory forest, near Oak Ridge, Tennessee, USA; and (c) boreal evergreen black spruce forest, near Prince Albert, Saskatchewan, Canada.

Notes: AGC = aboveground carbon stock; BGC = below-ground carbon stock; SOC = soil organic carbon stock; GPP = gross primary production; RA = autotrophic respiration; RH = heterotrophic respiration; RL = foliage respiration; RW = aboveground wood respiration; RR = root respiration; AGD = aboveground detritus (litter fall and mortality); BGD = below-ground detritus (fine root turnover, exudation, root mortality); T = translocation from above to below ground; AGNPP = aboveground net biomass carbon increment; BGNPP = below-ground net biomass carbon increment; and Δ_{SOC} = net increment in soil organic carbon.

1999). Carbon will then again accumulate during a phase of rapid growth that may last for centuries or at least decades, depending on the species of trees and site conditions (Buchmann and Schulze, 1999). Overmature forest stands take up carbon from the atmosphere at slower rates, but even as the growth increment of the trees approaches zero, carbon may continue to be funneled from the atmosphere to the soil via the trees in the form of aboveground and below-ground detritus. Nonetheless, forest stands can become net sources of carbon to the atmosphere—for example, if soil temperature abruptly increases (Peterjohn *et al.*, 1994) or the soil becomes more aerobic after drainage (Lindroth *et al.*, 1998), thus promoting oxidation of soil organic matter.

Over the long term, the sink capacity of any ecosystem is determined by the size of the pools (i.e., the aboveground and below-ground biomass) and their turnover times. Thus, additional carbon can be stored in an ecosystem only if more carbon is kept for the same period of time or the same amounts of carbon are kept over longer periods of time. A reduced rate of disturbance could therefore enhance carbon storage. Increased growth, on the other hand, will not add to the long-term sink if disturbances (or harvests) increase in frequency.

At any time during this life cycle, there may be appreciable interannual variation in NEP as a result of variability in the weather from year to year. Thus, a measurement of carbon uptake made for a single year at some arbitrary time in the life cycle, or even for a few years—such as the first 5-year "commitment period" of the Kyoto Protocol—may give a misleading picture of ongoing carbon sequestration as well as the long-term carbon sequestration potential of the vegetation, unless interannual variation is accounted for specifically. The results obtained from field measurement in recent years (as reported below) are still few, cover only selected areas of about 200 ha, and are not representative of all stages in the life cycle. Extrapolation to infer carbon sequestration over larger areas and longer time periods must be done with great care, and it is not as yet possible to obtain reliable average estimates for biomes as a whole from stand-scale measurements of NEP alone (see Section 1.3.2). Instead, forest inventory methods are generally being used to estimate standing stocks of carbon and the larger area and longer term carbon balances of these ecosystems and then extrapolated into the future using models.

1.3.2.2. Measured Carbon Fluxes in Different Ecosystems

At present, there are few carbon flux measurements in tropical forests over periods long enough to provide an annual estimate of NEP, although the number of such measurements are expected to increase appreciably over the next 5 to 10 years. Measurements in pristine, seasonal tropical rain forests in Amazonia indicate NEP of approximately 1.0 t C ha^{-1} yr^{-1} (Grace *et al.*, 1995a,b) and approximately 2.0 and 5.9 t C ha^{-1} yr^{-1} for dense, moist rain forest (Fan *et al.*, 1990; Malhi *et al.*, 1998, 1999). Sample plot studies also indicate net carbon sequestration rates within this range: 2.9 t C ha^{-1} yr^{-1} in selectively logged evergreen rain forest, 2.4 t C ha^{-1} yr^{-1} in heavily logged rain forest (Nabuurs and Mohren, 1993), and 0.7 to 1.5 t C ha^{-1} yr^{-1} in semi-evergreen tropical rainforest (Phillips *et al.*, 1998; Mahli *et al.*, 1999). Measurements of NEP for seasonally dry forest (savannas) indicate annual carbon sequestration rates of 0.12 and 0.75 t C ha^{-1} yr^{-1} for Sahelian and north Australian sites, respectively (Hanan *et al.*, 1998; Eamus *et al.*, n.d.).

Virtually all forests in the temperate region are managed to a greater or lesser extent, and there are only a few patches that might be regarded as pristine. Recent measurements of NEP by eddy covariance over one or more years in managed forests in Europe in the EUROFLUX network (Valentini *et al.*, 2000) and in North America (e.g., Greco and Baldocchi, 1996; Goulden *et al.*, 1996a,b; Baldocchi *et al.*, 1997) indicate rates of carbon sequestration in the range of 2.5 to 7 t C ha^{-1} yr^{-1} over approximately 20 sites differing in latitude, altitude, climate, species composition, and management intensity. The highest values are for young (20–50 years), fast-growing coniferous plantations and broad-leaved trees in old-field succession. Sample plots and yield tables, for example, lead to estimates of NPP of up to 4.5 t C ha^{-1} yr^{-1} for Sitka spruce in a maritime environment (Jarvis, 1981), 1.7 to 3.6 t C ha^{-1} yr^{-1} for European beech and mixed deciduous forests, and 2.5 to 3.4 t C ha^{-1} yr^{-1} for Norway spruce forests in Central Europe (e.g., Spiecker *et al.*, 1996) and 3.4 t C ha^{-1} yr^{-1} for Douglas fir in the northwest United States.

The extensive boreal forests across Siberia and Canada have been subjected to increasing exploitation over the past 150 years, but parts of the area standing can still be regarded as largely pristine. Most of the 12 or more boreal forest sites investigated are sequestering carbon (NEP) at annual rates of up to 2.5 t C ha^{-1} yr^{-1} (Black *et al.*, 1996; McCaughey *et al.*, 1997; Blanken *et al.*, 1998; Jarvis, 1998; Chen *et al.*, 1999). The values obtained depend particularly on latitude, soil type, and successional stage. However, NEP measurements over periods of up to 5 years in northern Canada in the BOREAS experiment (Sellers *et al.*, 1997), in Siberia (Schulze *et al.*, 1999), and in northern Europe (Valentini *et al.*, 2000) have demonstrated that a few old-growth coniferous stands may be carbon neutral (Goulden *et al.*, 1998) and in warm and cloudy years can be a carbon source (Lindroth *et al.*, 1998), losing carbon at a rate of up to 1.0 t C ha^{-1} yr^{-1}. Sample plots give average values for several species of aboveground and below-ground NPP of 4.7 t C ha^{-1} yr^{-1} for deciduous, broad-leaved species and 2.7 t C ha^{-1} yr^{-1} for coniferous species (Gower *et al.*, 1997).

Interannual variability in NEP may be considerable. The sequestration potential of tropical forests may vary by 10 percent from year to year depending on the length of the dry period and variation in solar radiation inputs and temperature, such as those caused by the eruption of Pinatubo (Grace *et al.*, 1995a,b) or those that occur during strong El Niño years (Tian *et al.*, 1998). The NEP of evergreen temperate forests (largely conifers) is similarly affected by variations in radiation input and temperature over the year and by the length of the photosynthetic season, particularly in oceanic climates (Valentini *et al*., 2000). The NEP of deciduous temperate forest (mostly broad-leaved trees) is also particularly sensitive to the length of the growing season as defined by the times of bud burst and leaf senescence, which may vary by 2 to 3 weeks from year to year (Goulden *et al*., 1996a,b). The NEP of boreal forests is even more sensitive to the onset of the spring thaw, which can vary from year to year by 2 to 3 weeks and may result in variation in annual carbon sequestration of 1.0 t C ha^{-1} yr^{-1} (Havranek and Tranquillini, 1995; Bergh *et al*., 1998; Goulden *et al*., 1998; Lindroth *et al*., 1998; Bergh and Linder, 1999; Chen *et al*., 1999).

1.3.2.3. Carbon Dioxide Fertilization

The impact of the slow progressive rise in the atmospheric concentration of CO_2 on the carbon sequestration capacity of stands of trees and other vegetation is difficult to measure. Some indications can be obtained, however, from the large number of recent experimental programs in which young trees have been exposed to double the current atmospheric CO_2 concentration over periods of up to 6 years. When rooted in the ground with unconfined root systems in open-top chambers, open-side chambers, closed chambers, and Free Air Carbon Dioxide Enrichment (FACE) rings, there is similar enhancement (up to 60 percent) of aboveground and below-ground growth rate (NPP) and carbon accumulation (Saxe *et al*., 1998; Norby *et al*., 1999). Trees grown in double the atmospheric CO_2 concentration translocate appreciably more carbon below ground than do trees grown in ambient CO_2 concentration. For example, 4-year-old birch trees grown in elevated CO_2 concentration translocated three times as much carbon below ground as trees grown in ambient CO_2 (Wang *et al*., 1998). Much of this carbon ends up as fine roots, microbes, and mycorrhizae that contribute detritus to the pool of soil organic matter (Rey and Jarvis, 1997).

Meta-analysis of the results from several experiments in Europe on deciduous and coniferous species gave an average increase of 54 percent (Medlyn *et al*., 2000), with no significant differences between the functional types or between stressed and unstressed groups of experiments. Similar results were found through meta-analysis of similar experiments in the United States (Curtis and Wang, 1998; Peterson *et al*., 1999). Moderate lack of nutrients and water reduces growth but has little effect on the relative impact of the increase in CO_2 concentration. Generally, the increase in CO_2 concentration speeds up development, so trees get bigger more quickly but otherwise remain very similar in most respects to trees of the same size growing in current ambient conditions. The rate of photosynthetic uptake of CO_2 is almost always higher in air with elevated CO_2 concentration than in ambient air. Meta-analysis of data from experiments largely on temperate trees has shown that the key enzymatic parameters that define the capacity for photosynthetic CO_2 as a result of growth in elevated CO_2 uptake are down-regulated on average by no more than 12 percent, in approximate proportion to the reduction in foliar nitrogen concentration; stomatal conductance is down-regulated by approximately 15 percent, with consequent increased efficiency of use of both nitrogen and water (Curtis, 1996; Curtis and Wang, 1998; Peterson *et al*., 1999; Medlyn *et al*., 2000).

Most experimental studies to date have been limited to exposure of young trees to elevated CO_2 concentrations for a few years of treatment. A key question that is yet to be resolved is the extent to which the responses of key parameters to increased CO_2 concentration will change when a stand of young trees reaches canopy closure. It is uncertain whether the primary result of the increased rate of growth is that maturity is approached more quickly or whether more carbon will also be finally stored in the trees and soil as a forest stand matures. Some experimental and observational studies suggest that the initial positive effects of elevated CO_2 on growth rate cited above are greatly reduced or disappear as a stand matures but that more carbon does remain sequestered (e.g., Hättenschwiler *et al*., 1997). These questions may be partially resolved through the long-term FACE experiments currently in progress on young stands (e.g., Ellsworth, 1999), but at present the long-term effect of CO_2 fertilization on carbon sequestration by forest stands remains an open question.

The phenological development and growth rate of agricultural crops and grassland are accelerated in elevated atmospheric CO_2 concentration. This leads to increases in crop productivity and increases in harvestable yield in some cases (e.g., Tubiello *et al*., 1999), though not in others (Wechsung *et al*., 1999). Tropical grasslands have very high NPP, especially below ground (House and Hall, 2000), but whether the amount of carbon stored in the soil is increasing is uncertain.

1.3.2.4. Nutrient Deposition and Mineralization

Production in the forests of the world, with the exception of lowland tropical forests (Vitousek and Sanford, 1986), is generally restricted by lack of nitrogen, particularly in northern temperate and boreal regions (Vitousek and Howarth, 1991), or lack of phosphorus, in the tropics (Lloyd *et al*., n.d.). Nitrogen fertilization is an efficient means for enhancing agricultural production and is very effective in enhancing the productivity of forest plantations in Mediterranean, temperate, and boreal climatic regions (e.g., McMurtrie and Landsberg, 1992; Linder *et al*., 1996). Addition of nitrogen and phosphorus promotes the activity of photosynthesis per leaf, and nitrogen in particular stimulates increases in the number and growth of leaves, increasing the area of leaves in vegetation canopies. Many experiments worldwide have demonstrated that the growth of temperate forests is very responsive to the application of fertilizers, particularly of nitrogen (Linder and Rook, 1984; Tamm, 1991). For example, recent long-term experiments have shown a four-fold increase in the growth of Norway spruce in response to annual, complete fertilizer applications (75 kg N ha^{-1} yr^{-1}) at 64°N over the past 12 years and a doubling of growth at 57°N over the past 10 years (Linder, 1995; Bergh *et al*., 1999). Maritime commercial spruce forest is also responsive to applications of nitrogen (Taylor and Tabbush, 1990; Wang *et al.*, 1991). Comparable information from tropical forests is not readily available. The additional growth of vegetation may also lead to an increase in the amount of organic matter in the soil—or at least minimize the decrease brought about by tillage, harvesting, and other agricultural and forestry management practices.

Inventory data from sample plots indicate that the growth of trees has been increasing across Europe (Spiecker *et al.*, 1996), and these trends have been observed elsewhere. Wet and dry deposition of nutrients from the atmosphere may be contributing to this enhancement of forest growth. In general, annual total (wet and dry) deposition of nitrogen (oxidized and reduced) to forests in rural areas is in the range of 5 to 40 kg ha^{-1} yr^{-1}; the smaller amounts are in more remote forests, particularly at high latitudes and in the tropics (e.g., Forti and Moreira-Nordemann, 1991; Eklund *et al.*, 1997; Freydier *et al.*, 1998). Larger amounts are deposited on forests close to cities and industrial centers from which there are substantial nitrogen emissions, as well as in the near vicinity of intensive agricultural pig and poultry enterprises; this deposition may lead to problems such as acidification and loss of biodiversity. It is likely that such nitrogen inputs are supporting additional growth of young forests of particular relevance to the Kyoto forests and hence carbon sequestration (see Section 1.4; Cannell *et al.*, 1999; Valentini *et al.*, 2000), although recently this effect has been disputed (Nadelhoffer *et al.*, 1999). A key uncertainty is to what extent and for how long the current annual rate of nitrogen deposition can sustain the growth rate and NEP of forests on Kyoto lands.

1.3.2.5. Spatial and Temporal Integration

Measurements of NEP are demanding of resources, therefore still are not numerous. They also are not always representative for the whole growing season, and they are never representative for the whole life cycle of longer lived plants. Few observational series extend over more than 5 years. Within decades, a forest site that was chosen for measuring NEP will likely enter a different phase or be subjected to natural disturbances such as fire, pests, and wind-throw or management operations such as thinning and felling. Although measurements made from satellites may suggest that NPP has been increasing over the past decades in some regions and measurements of NEP may seem to indicate that many forests are significant carbon sinks, when the effects of disturbances are included the sequestration of carbon into these forests may be significantly less or even, in some areas, close to zero. Importantly, the disturbances themselves may be influenced by climate, and a change in the disturbance regime may turn a positive NBP into a negative one—or vice versa. For example, measurements of NEP in Canadian boreal forests suggest that many of these forests are carbon sinks (see Section 1.3.2.3), but estimates of NBP from measurement of changes in disturbance regimes and consideration of forest stand dynamics indicate that NBP may have declined significantly over the past 3 decades and that these forests, over large areas, are close to being carbon neutral (Walker *et al.*, 1999). With the limited empirical data available, however, it is difficult to derive accurate local estimates of NBP for regions, biomes, countries, or continents from spatial and temporal integration of the constituent processes and disturbances (Schulze *et al.*, 1999; Houghton *et al.*, 2000). Our restricted ability to build estimates of NBP from its components at the present time defines a gap in our knowledge and the need to use other methodologies.

If the values in the preceding sections are considered representative for the major forest biomes—tropical, temperate, and boreal forests, respectively—and for the total area that they cover (approximately 40 x 10^6 km^2), total NEP for these systems would be about 10 Gt C yr^{-1}. However, global NBP, derived as the difference between the output resulting from fossil fuel burning, on the one hand, and the increase of atmospheric concentrations and net ocean uptake, on the other, is currently a little less than about 1 Gt C yr^{-1} (see Table 1-2). If the effects of land-use change are excluded from this estimate of NBP, the estimate of global NBP increases to about 2–3 Gt C yr^{-1}—about five times less than the total NEP (see Section 1.2.1.4). Thus, NEP values as reported so far are clearly not representative of the large-scale, long-term storage of carbon. This fact emphasizes the importance of viewing the activities defined in Articles 3.3 and 3.4 of the Kyoto Protocol in a large-scale and long-term perspective.

Recent attempts to determine the large-scale distribution of terrestrial sources and sinks (i.e., NBP) indirectly, on the basis of the observed spatial variation of atmospheric carbon dioxide concentrations, have interpreted low values as indications of net uptake and high values as the presence of net sources (Fan *et al.*, 1998; Bousquet *et al.*, 1999; Rayner *et al.*, 2000). In light of the relatively small magnitude of the regional NBP, this determination requires accurate knowledge about both the spatial distribution of carbon dioxide concentrations and its horizontal and vertical transport as a result of air motions. The inverse modeling analyses of Rayner *et al.* (2000) and Bousquet *et al.* (1999) agree reasonably closely in indicating a net terrestrial carbon sink in Siberia and comparable, smaller net sinks in North America and Europe, with small net sources in South America and Africa.

1.3.3. The Future of the Terrestrial Carbon Sink

Table 1-2 illustrates the considerable uncertainty about the relative importance of the present global oceanic and terrestrial sinks in acting as negative feedbacks on the rate of CO_2 increase. These sinks may also change in the future as atmospheric CO_2 continues to rise. The question therefore arises whether these feedbacks may reach some limit, so that no additional carbon can be stored on land. This possibility would imply that the airborne fraction of emissions would increase more rapidly in the future than at present.

How long the current carbon sink capacity of the terrestrial biosphere is likely to be maintained into the future is a matter of conjecture; several hypotheses proposed as the basis for quantitative explanation are discussed here. Consideration of the nature of the uncertainties for such projections is essential, however. The time scale of our immediate concerns with respect to the biospheric sink as a possible means of reducing the impact of GHG emissions is considered to be several decades, but longer time scales may become relevant if emissions of GHGs continue to rise. Projections of future sources and sinks of carbon in terrestrial ecosystems depend on multiple aspects of the future environment: climate (temperature, precipitation, humidity, and radiation), atmospheric CO_2 concentration, nutrient deposition, land use, and ecosystem management (Thompson *et al.*, 1996). Projections of most of these variables into the medium term are highly uncertain, particularly with respect to their future spatial patterns. The regional distribution of ecosystems is an important factor influencing the overall sink capacity. Therefore, improved regional scenarios for human land use are as important for better sink strength projections as they are for better regional climate scenarios.

We may assume that the total capacity for carbon storage in terrestrial ecosystems has an upper limit as a result of mechanical and physiological constraints on the amount of aboveground biomass and physiochemical constraints on the amount of carbon that can be held in soils. These fundamental limits are presently not known in quantitative terms, but it does not seem very likely that they will become of importance within a few decades. Over the past 10,000 years—since the last glaciation—carbon has accumulated in the boreal and north temperate forest ecosystems very largely as transfer from the atmosphere, via vegetation, to the soil organic carbon (SOC) pool. Experimental addition of nutrients has demonstrated the large capacity of these ecosystems for additional tree growth, production of detritus, and SOC. We may therefore conclude that this accumulation of carbon has been constrained since the retreat of the ice by the slow concurrent accumulation of nutrients, particularly nitrogen, in the ecosystem. The present small size of trees in the boreal region and their low nitrogen content suggest that the capacity of the system to store carbon will not by itself limit the transfer of carbon from the atmosphere to the tree and soil pools in the immediate future. Because an accelerated rate of climate change, as well as continuing atmospheric inputs, may increase the availability of nitrogen, it is possible that boreal ecosystems could increase their aboveground and below-ground carbon storage considerably.

By contrast, moist tropical forests in general have large standing stocks of carbon in the trees (Table 1-2). Therefore, there may be limited scope for additional storage of carbon in the trees, although that is less likely to be the case in sparse or secondary tropical forest (Phillips *et al.*, 1998). At present, moist tropical forests currently have high NPP and NEP (Section 1.3.2.2), so it is questionable whether this ongoing flow of carbon from the atmosphere into tropical forests can be maintained. Significant amounts of carbon leave tropical forests in floodwaters as particulate and dissolved organic carbon, some of which is subsequently stored in coastal sediments (Schlesinger, 1997). This and other removals of carbon from tropical forests may provide the opportunity for some increased uptake from the atmosphere and transfer to storage.

Research has demonstrated that in the short term, increased photosynthesis resulting from the rise in atmospheric CO_2 concentration diminishes at higher CO_2 concentrations, whereas RA and RH increase exponentially with increasing temperature (e.g., Boone *et al.*, 1998; Rayment and Jarvis, 2000). Thus, Scholes *et al.* (1998) have hypothesized that as atmospheric CO_2 concentration and temperature rise, the overall capacity to take up additional carbon from the atmosphere will progressively diminish so that at some point respiration will exceed photosynthesis and the carbon sink will become a source. Based on these two short-term physiological response functions, Scholes (1999) evaluated the global terrestrial carbon sink under various assumptions and concluded that the global sink would be likely to decline from its current level of approximately 2 Gt C yr^{-1} and become a source within a few decades. This approach assumes that acclimation to higher CO_2 concentrations and temperatures does not occur, that respiration is independent of photosynthesis, and that there are no feedbacks involving nutrition between the processes. A similar, parallel, hypothesis is that the present sink might be a temporary consequence of CO_2 fertilization that eventually will be overtaken by respiratory losses of carbon as temperature rises and respiration "catches up." It is supposed that while photosynthesis increases in response to the increased availability of CO_2, respiration will initially lag behind but will eventually catch up as the supply of substrate increases or the temperature lag resulting from the thermal inertia of the oceans declines.

To better understand the range of future terrestrial uptake of carbon, calculations with a suite of several more detailed terrestrial biosphere models have been based on the IPCC IS92a scenario of CO_2 increase and the associated climate change simulated by the HadCM2 General Circulation Model (GCM) (Cramer *et al.*, 2000). These models include integrated physiological effects of climate, CO_2, and nutrition on plant growth and plant population dynamics but include no consideration of land-use or management effects. Stimulation of carbon uptake as a result of enhanced uptake of soil nitrogen is included in several of the models. For the IPCC IS92a scenario, all of the models confirm that there is a terrestrial CO_2 sink of the same order of magnitude as above that explains the present-day overall balance; that this sink might increase in strength with further increases in CO_2 concentration; and that when the CO_2 concentration increases beyond about 500 ppmv CO_2 (i.e., around 2030–2050), the rate of increase of the sink approaches zero as the sink reaches a steady asymptote.

A similar result has been obtained using a well-validated local-scale ecosystem model (G'Day) that treats interactions and feedbacks between the effects of CO_2 and temperature on the pools and fluxes of carbon and nitrogen within vegetation and soil compartments of terrestrial ecosystems (Dewar *et al.*, 1999; McMurtrie *et al.*, n.d.). This model was parameterized for a boreal coniferous forest in northern Sweden; temperature and CO_2 concentration were increased according to the IPCC

IS92a scenario as above. The G'Day model indicates that, after initiation, NPP consistently exceeds RH and the carbon sink rises to an asymptotic maximum after about 60 years and remains there for another 50 years without significant decline (McMurtrie *et al.*, n.d.). The capacity at which the sink stabilizes depends on the availability of nitrogen and the fraction of nitrogen released in decomposition that is subsequently re-immobilized. Thus, with appropriate feedbacks and N:C relationships considered, this model also predicts that the carbon sink will be maintained well into the medium term.

On the other hand, sinks may disappear if climate change or deforestation would lead to widespread tree mortality (Friend *et al.*, 1997; Cao and Woodward, 1998; Walker *et al.*, 1999). Of the suite of models discussed above, one model (HYBRID) (Friend *et al.*, 1997) predicts a drastic decline in NPP that is primarily the result of drought-induced die-back of tropical forest in the Amazon basin. This finding highlights two points: First, changes in vegetation structure can have significant effects on terrestrial carbon sinks and indeed can be as important or more important than the physiological effects discussed above (see also Mooney *et al.*, 1999); second, projections of climate change impacts should be based on a range of predictions of future regional climates. The HadCM2 GCM produces somewhat drier climate change predictions for the Amazon than do some other GCMs, which might well lead to different results.

These model experiments and other analyses (e.g., Walker *et al.*, 1999) suggest that the current global terrestrial sink will be maintained over the short term of the next few decades unless serious mortality occurs but may diminish toward the end of the 21st century. With the present state of knowledge, projections beyond a few decades must be regarded as uncertain; this situation is improving rapidly, however.

This conclusion is based on consideration of the effects of future CO_2 and climate change on the present terrestrial sink only; it does not take into account future deforestation or possible actions taken to enhance the terrestrial sink through land use, land-use change, and afforestation or reforestation. Higher CO_2 concentration and warmer climate may be expected to modify the effectiveness of such activities to enhance the medium-term terrestrial sink. Improved understanding of these effects would enable these activities to be better adapted to future CO_2 and climate conditions.

1.4. The Influence of Land Use on Greenhouse Gas Sources and Sinks

Land use and land-use change directly affect the exchange of greenhouse gases between terrestrial ecosystems and the atmosphere. Changes such as the clearing of forests for use in agriculture or as settlements are associated with clear changes in land cover and carbon stocks. Much of the world's land area continues to be managed for food and wood production, human habitation, recreation, and ecosystem preservation without a change in land use. Management of these land uses affects sources and sinks of CO_2, CH_4, and N_2O. Furthermore, the resulting agricultural and wood products contain carbon. The carbon stocks held in these products are eventually released back to the atmosphere, after the products have served their use. Biomass carbon stocks are also used to produce energy that serves as a substitute for, and as complement to, fossil fuels.

1.4.1. Land-Use Change

Different factors and mechanisms drive land use and land cover transformation. In many cases, climate, technology, and economics appear to be determinants of land-use change at different spatial and temporal scales. At the same time, land conversion seem to be an adaptive feedback mechanism that farmers use to smooth the impact of climate variability, especially in extremely dry and humid periods (e.g., Viglizzo *et al.*, 1995). Land-use change is often associated with a change in land cover and an associated change in carbon stocks. For example, as Figure 1-4 shows, if a forest is cleared, the carbon stocks in aboveground biomass are either removed as products, released by combustion, or decay back to the atmosphere through microbial decomposition. Stocks of carbon in soil will also be affected, although this effect will depend on the subsequent treatment of the land. Following clearing, carbon stocks in aboveground biomass may again increase, depending on the type of land cover associated with the new land use. During the time required for the growth of the new land cover—which can be decades for trees—the aboveground carbon stocks will be smaller than their original value.

Houghton (1991) assessed seven types of land-use change for carbon stock changes: (1) conversion of natural ecosystems to permanent croplands, (2) conversion of natural ecosystems for shifting of cultivation, (3) conversion of natural ecosystems to pasture, (4) abandonment of croplands, (5) abandonment of pastures, (6) harvest of timber, and (7) establishment of tree plantations. We recognize that, depending on the temporal scope of the assessment, classes 6 and 7 may also be considered a land-use practice rather than land-use change.

When forests are cleared for conversion to agriculture or pasture (1,3), a very large proportion of the aboveground biomass may be burned, releasing most of its carbon rapidly into the atmosphere. Some of the wood may be used as wood products; these carbon stocks could thereby be preserved for a longer time. Forest clearing also accelerates the decay of dead wood and litter, as well as below-ground organic carbon (see Figure 1-4). Local climate and soil conditions will determine the rates of decay; in tropical moist regions, most of the remaining biomass decomposes in less than 10 years. Some carbon or charcoal accretes to the soil carbon pool. When wetlands are drained for conversion to agriculture or pasture, soils become exposed to oxygen. Carbon stocks, which are resistant to decay under the anaerobic conditions prevalent in wetland soils, can then be lost by aerobic respiration (Minkkinen and Laine, 1998).

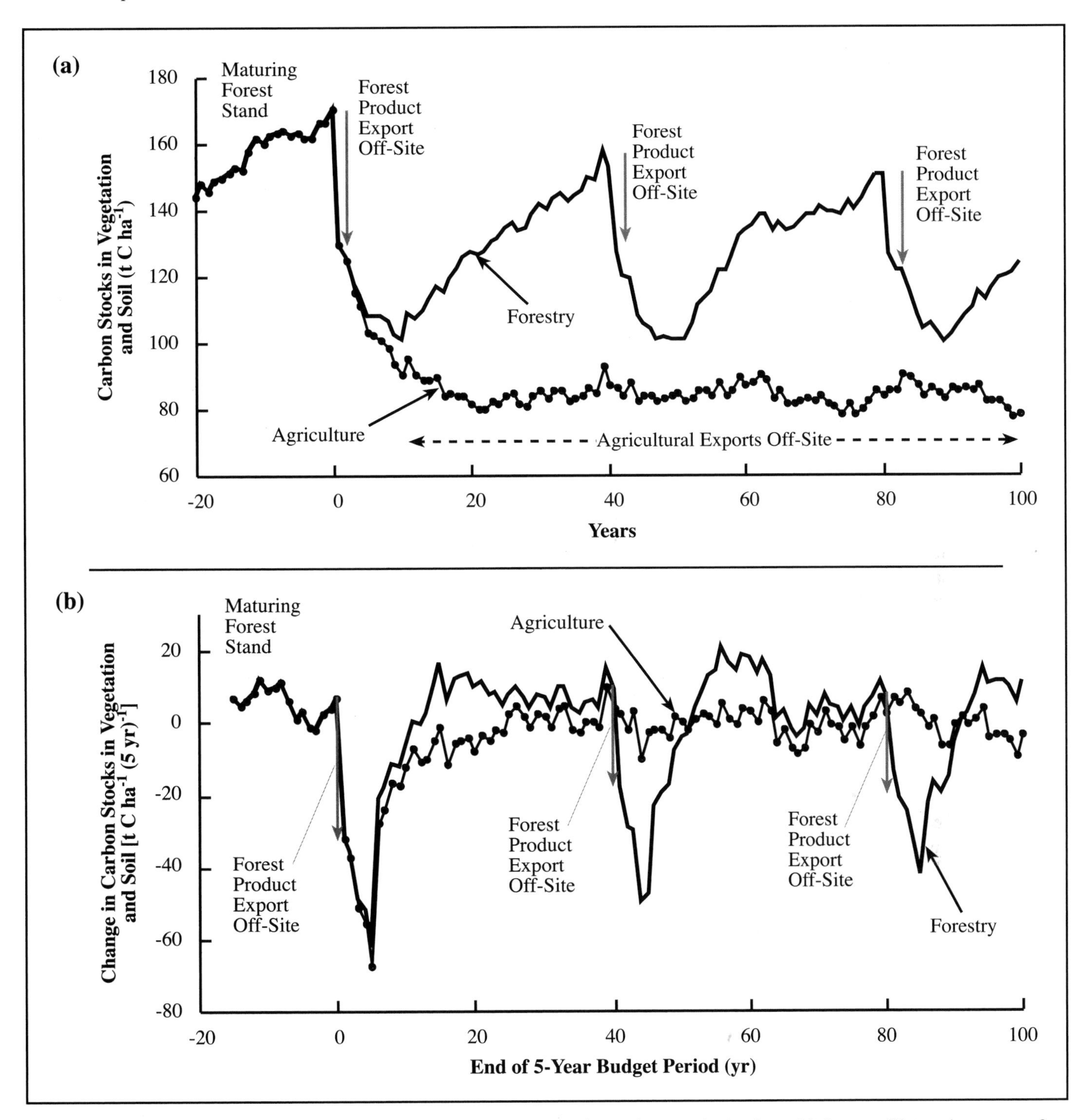

Figure 1-4: The hypothetical time-evolution of annual-average on-site carbon stocks is given (a) for two illustrative cases of land use, land-use change, and forestry. Both cases begin with a common maturing forest stand that is undisturbed for the first 20 years; natural variability, as well as an overall increase, in stocks occurs over this period. After 20 years, the forest stand is cleared and a fraction of the forest stocks is exported off-site as forest products. Following clearing, the land is used for forestry in one case and agriculture in the other. In both cases, carbon stocks continue to fall immediately following clearing due to loss of carbon from detritus and soils. In the cases subsequent to clearing, sustainable forestry is carried out with a 40-year stand cycle, and cultivation is carried out to yield crops with some further loss of soil carbon. For these cases, the change in carbon stocks over a running 5-year period (b) fluctuates due to both variability and human activities. Carbon stocks contained in forest and agricultural product exports off-site are not included in this illustration. Carbon stocks in products generally have a finite lifetime (see Section 1.4.3) and can be considered in a total carbon stock inventory (see Chapter 3).

Forest clearing for shifting cultivation (2) releases less carbon than permanent forest clearing because the fallow period allows some forest regrowth. On average, the carbon stocks depend on forest type and the length of fallow, which vary across regions. Some soil organic matter is also oxidized to release carbon during shifting cultivation—but less than during continuous cultivation (Detwiler, 1986). Under some conditions, shifting cultivation can increase carbon stocks in forests and soils, from one cut-regrowth cycle to another. Because shifting cultivation usually has lower average agricultural productivity than permanent cultivation, however, more land would be required to provide the same products. In addition, shorter rotation periods deplete soil carbon more rapidly.

Abandonment of cultivated land and pastures (4,5) may result in recovery of forest at a rate determined by local conditions (Brown and Lugo, 1982; Uhl *et al.*, 1988; Fearnside and Guimarães, 1996).

Selective logging (6) often releases carbon to the atmosphere through the indirect effect of damaging or destroying up to a third of the original forest biomass, which then decays as litter and waste in the forest (although there are techniques that may reduce these consequences). The harvested wood decays at rates dependent on their end use; for example, fuel wood decays in 1 year, paper in less than a few years, and construction material in decades. The logged forest may then act as a sink for carbon as it grows at a rate determined by the local soil and climate, and it will gradually compensate for the decay of the waste created during harvest. Clear-cutting of forest can also lead to the release of soil carbon, depending on what happens after harvesting. For example, harvesting followed by cultivation or intensive site preparation for planting trees may result in large decreases in soil carbon—up to 30 to 50 percent in the tropics over a period of up to several decades (Fearnside and Barbosa, 1998). Harvesting followed by reforestation, however, in most cases has a limited effect (±10 percent). This effect is particularly prevalent in the tropics, where recovery to original soil carbon contents after reforestation is quite rapid. There are also some cases in which soil carbon increases significantly, probably because of the additions of slash and its decomposition and incorporation into the mineral soil (Detwiler, 1986; Johnson, 1992).

If tree plantations are raised on land that has been specifically cleared (7), initially there would be net carbon emissions from the natural biomass and the soil. The plantations would then begin to fix carbon at rates dependent on site conditions and species grown. To estimate the time scale of carbon uptake in forest plantations, previous work has linked fixation rates to the growth rate over time (Nilsson and Schopfhauser, 1995). Nilsson and Schopfhauser summarize data suggesting the following rates of aboveground carbon accumulation in plantations: 10 t ha^{-1} yr^{-1} for coniferous plantations in Australia and New Zealand, 1.5 to 4.5 t ha^{-1} yr^{-1} in coniferous temperate plantations of Europe and the United States, 0.9 to 1.2 t ha^{-1} yr^{-1} in Canada and the former Soviet Union, and 6.4 to 10.0 t ha^{-1} yr^{-1} in tropical Asia, Africa, and Latin America. Even if soil carbon accumulation is considered, these numbers probably represent maximum rates achieved under intensive management that includes the use of fertilizers. However, tree plantations also go through a rotational pattern of harvest, and the long-term estimates of carbon uptake might therefore be much lower than suggested by the foregoing figures.

Changes in land use of the types listed above have led to an estimated net emission of CO_2 of 121 Gt C from 1850 to 1990 (Houghton, 1999; Houghton *et al.*, 1999, 2000), as well as an estimated 60 Gt C prior to 1850 (De Fries *et al.*, 1999). Prior to 1950, high- and mid-latitude Northern Hemisphere regions released substantial amounts of carbon from forest clearing and conversion to agricultural use, but this situation has since reversed as many forests presently seem to be in a stage of regeneration and regrowth (Kauppi *et al.*, 1992). The low-latitude tropical belts, on the other hand, have been experiencing high rates of deforestation in recent decades (Houghton, 1994). The wide variation in vegetation carbon density in the low latitudes, however, introduces considerable uncertainty in estimates of carbon stock changes resulting from land-use changes. An estimate of global net emissions of 1.6 ± 1.0 Gt C yr^{-1} from land-use changes from 1980–1989 (Houghton, 1994; Dixon *et al.*, 1994a) was judged to have been on the high side from newer data from the Brazilian Amazon (Schimel *et al.*, 1995). More recent analyses, however, have revised this estimate to even higher figures of 1.7 ± 0.8 Gt C yr^{-1} (Houghton *et al.*, 1999, 2000), 2.0 ± 0.8 Gt C yr^{-1} (Houghton, 1999), and 2.4 Gt C yr^{-1} (Fearnside, 2000). Most of the carbon emission in the 1980s was from tropical regions (tropical Asia alone accounted for 50 percent of this flux) where deforestation rates averaged about 15 Mha yr^{-1}. Of the major categories of land-use change, the clearing of forests for use as cropland accounted for the largest fraction of CO_2 emissions from net land-use change; emissions from conversion to pastures, harvest, and shifting cultivation were lower. These estimates, however, do not include sources and sinks of CO_2 caused by land-use management practices not associated with land-use change.

1.4.2. Land-Use Management

Management of forests, croplands, and rangelands affects sources and sinks of CO_2, CH_4, and N_2O. On land managed for forestry, harvesting of crops and timber changes land cover and carbon stocks in the short term while maintaining continued land use. Moreover, most agricultural management practices affect soil condition. A forest that is managed in a wholly sustainable manner will encompass stands, patches, or compartments comprising all stages from regeneration through harvest, including areas disturbed by natural events and management operations. Overall, a forest comprising all stages in the stand life cycle operates as a functional system that removes carbon from the atmosphere, utilizing carbon in the stand cycle and exporting carbon as forest products. Forests of such characteristics, if well managed, assure rural development through working opportunities at the beginning and establishment of forest industries in later stages of the development process. In addition, such forests provide other benefits, such as biodiversity, nature conservation, recreation, and amenities for local communities. For historical and economic reasons, however, many forests today depart from this ideal and are fragmented or have strongly skewed stand age distribution that influences their carbon sequestration capability.

Forest soils present opportunities to conserve or sequester carbon (Johnson, 1992; Lugo and Brown, 1993; Dixon *et al.*, 1994a). Several long-term experiments demonstrate that carbon can accrete in the soil at rates of 0.5 to 2.0 t ha^{-1} yr^{-1} (Dixon *et al.*, 1994b). Management practices to maintain, restore, and enlarge forest soil carbon pools include fertilizer use; concentration of agriculture and reduction of slash-and-burn practices; preservation of wetlands, peatlands, and old-growth forest; forestation of degraded and nondegraded sites, marginal agricultural lands, and lands subject to severe erosion; minimization of site disturbance during harvest operations to retain organic matter; retention of forest litter and debris after silvicultural activities; and any practice that reduces soil aeration, heating, and drying (Johnson, 1992).

Cropland soils can lose carbon as a consequence of soil disturbance (e.g., tillage). Tillage increases aeration and soil temperatures (Tisdall and Oades, 1982; Elliott, 1986), making soil aggregates more susceptible to breakdown and physically protected organic material more available for decomposition (Elliott, 1986; Beare *et al.*, 1994). In addition, erosion can significantly affect soil carbon stocks through the removal or deposition of soil particles and associated organic matter. Erosion and redistribution of soil may not result in a net loss of carbon at the landscape level because carbon may be redeposited on the landscape instead of being released to the atmosphere (van Noordwijk *et al.*, 1997; Lal *et al.*, 1998; Stallard, 1998). Although some the displaced organic matter may be redeposited and buried on the landscape, in general the productivity of the soil that is eroded—and its inherent ability to support carbon fixation and storage—is reduced. Losses through leaching of soluble organic carbon occur in many soils; although this leaching is seldom a dominant carbon flux in soils, it is a contributor to the transport of carbon from the terrestrial environment to the marine environment via runoff (Meybeck, 1982; Sarmiento and Sundquist, 1992; *cf.* runoff in Figure 1-1). Soil carbon content can be protected and even increased through alteration of tillage practices, crop rotations, residue management, reduction of soil erosion, improvement of irrigation and nutrient management, and other changes in forestland and cropland management (Kern and Johnson, 1993; Lee *et al.*, 1993; Cole *et al.*, 1996).

Livestock grazing on grasslands, converted cropland, savannas, and permanent pastures is the largest areal extent of land use (FAO, 1993). Grazing alters ground cover and can lead to soil compaction and erosion, as well as alteration of nutrient cycles and runoff. Soil carbon, in turn, is affected by these changes. Avoiding overgrazing can reduce these effects.

Croplands and pastures are the dominant anthropogenic source of CH_4 (Section 1.2.2) and N_2O (Section 1.2.3), although estimates of the CH_4 and N_2O budgets remain uncertain (Melillo *et al.*, 1996). Rice cultivation and livestock (enteric fermentation) have been estimated to be the two primary sources of CH_4. The primary sources of N_2O are denitrification and nitrification processes in soils. Emissions of N_2O are estimated to have increased significantly as a result of changes in fertilizer use and animal waste (Kroeze *et al.*, 1999). Alteration of rice cultivation practices, livestock feed, and fertilizer use are potential management practices that could reduce CH_4 and N_2O sources.

Ecosystem conservation may also influence carbon sinks. Many forests, savannas, and wetlands, if managed as nature reserves or/and recreation areas, can preserve significant stocks of carbon, although these stocks might be affected negatively by climate change. Some wetlands and old-growth forests exhibit particularly high carbon densities; other semi-natural ecosystems (e.g., savannas) may conserve carbon simply because of their large areal extent.

1.4.3. The Fate of Stored Carbon in Biomass Products

Carbon is sequestered in products made from biomass. These products can be traded among nations, lowering the carbon stock of the exporting country and adding to the carbon stock of the importing country. When carbon is sequestered into biomass for specific climate mitigation options—such as wood for fuel or industrial purposes—a life cycle analysis is needed to describe the fate of the stored carbon. Improvement of life cycle assessment methods and comprehensive application of these assessment methods to case studies are important ways to judge the industrial side of the forest carbon cycle. Such analysis would include the conversion efficiency from tree growth into wood products.

Several waste products of wood processing (sawdust, wood chips, bark or lignin from cellulose production, etc.) decay and emit carbon, although such products can be further processed to produce boards or pulp—thereby increasing the efficiency of conversion (Hall *et al.*, 1991)—or be used as energy sources. Wood products such as newsprint, fuel wood, paper, plywood, and sawn timber decay at rates that depend on the nature of their storage and use. Four life-span categories have been described for modeling carbon in forest products (Pussinen *et al.*, 1997). Half of the short life-span products (fuel wood, newsprint, some packing paper, paperboard, and printing and writing paper) manufactured in 1 year were assumed to have not yet decayed after 4 years. The respective half-life-spans were 13 years for medium-short life-span products (the rest of packing paper, paperboard, and printing and writing paper), 30 years for medium-long life-span products (part of sawn timber and plywood), and 65 years for long life-span products (rest of sawn timber and plywood) products. Waste by-products can, however, be used as biomass to produce energy (see Section 1.4.4).

1.4.4. Biomass Energy

Biomass energy can be used to avoid greenhouse gas emissions from fossil fuels by providing equivalent energy services:

electricity, transportation fuels, and heat. The avoided fossil fuel CO_2 emissions of a biomass energy system are equal to the fossil fuels substituted by biomass energy services minus the fossil fuels used in the biomass energy system. These quantities can be estimated with a full fuel-cycle analysis of the system. The net effect on fossil fuel CO_2 emissions is evident as a reduction in fossil fuel consumption.

For biomass energy to lead to an overall reduction in greenhouse gas emissions, land use and land-use change emissions of the biomass energy system must also be included (Marland and Schlamadinger, 1995). For example, if biomass is harvested and subsequently regrows without an overall loss of carbon stocks, there would be no net CO_2 emissions over a full harvest/growth cycle. In this way, land can be used continuously for the production of biomass energy to avoid fossil fuel CO_2 emissions. By contrast, using land to grow carbon stocks to be conserved thereafter can only be a temporary measure to limit fossil fuel use.

Biomass can originate as a co-product of forestry or from crops grown expressly for biomass energy. For example, logging and paper mill residues are being widely used for heat production in Sweden (Gustavsson *et al.*, 1995; Johansson and Lundqvist, 1999). In Brazil, sugar cane crops are used to provide ethanol for blending into motor vehicle fuels, and sugar cane residue (bagasse) is being used for electricity generation. The use of sugar cane biomass in Brazil led to the avoidance of fossil fuel CO_2 emissions of 9.2 Mt C yr^{-1} through the blending of ethanol with fossil fuels in 1997 and 1998 (Macedo, 1998)—approximately 11 percent of Brazil's fossil fuel CO_2 emissions (British Petroleum Company, 1999). During the next 10 years, the generation of an additional 3 GW of electrical power from sugar cane products is expected.

The use of land to hold stocks of carbon and to provide energy as a substitute for fossil fuels adds to the existing primary uses of land for agriculture, forestry, settlements, recreation, and conservation. The competition for land among these uses will partly determine the extent to which land can be used to reduce greenhouse gas concentrations. Analyses of scenarios of future development show that expanded use of biomass energy could lead to a significant reduction in atmospheric CO_2 concentrations (Edmonds *et al.*, 1996; Ishitani *et al.*, 1996; Leemans *et al.*, 1996). The development of technology for efficient production of biomass energy, as well as for competing land uses, will affect the amount of land available for alternative uses. Consideration of multiple uses of land to provide food and fiber while enhancing carbon stocks and producing energy may present further opportunities to reduce greenhouse gas concentrations with minimum use of resources.

References

Aubinet, M., A. Grelle, A. Ibrom, Ü. Rannik, J. Moncrieff, T. Foken, A.S. Kowalski, P.H. Martin, P. Berbigier, C. Bernhofer, R. Clement, J. Elbers, A. Granier, T. Grünwald, K. Morgenstern, K. Pilegaard, C. Rebmann, W. Snijders, R. Valentini, and T. Vesala, 2000: Estimates of the annual net carbon and water exchange of forests: The EUROFLUX Methodology. *Advances in Ecological Research,* **30,** 113–175.

Baldocchi, D.D., C.A. Vogel, and B. Hall, 1997: Seasonal variation of carbon dioxide exchange rates above and below a boreal jack pine forest. *Agricultural and Forest Meteorology,* **83,** 147–170.

Barnola, J.M., M. Anklin, J. Porcheron, D. Raynaud, J. Schwander, and B. Stauffer, 1995: CO_2 evolution during the last millennium as recorded by Antarctic and Greenland ice, *Tellus,* **47B,** 264–272.

Battle, M., M. Bender, T. Sowers, P.P. Tans, J.H. Butler, J.W. Elkins, J.T. Ellis, T. Conway, N. Zhang, P. Lang, and A.D. Clarke, 1996: Atmospheric gas concentrations over the past century measured in air from firn at the South Pole. *Nature,* **383,** 231–235.

Beare, M.H., P.F. Hendrix, and D.C. Coleman, 1994: Water-stable aggregates and organic matter fractions in conventional- and no-tillage soils. *Soil Science Society of America Journal,* **58,** 777–786.

Bergh, J. and S. Linder, 1999: Effects of soil warming during spring on photosynthetic recovery in boreal Norway spruce stands. *Global Change Biology,* **5,** 245–253.

Bergh, J., R.E. McMurtrie, and S. Linder, 1998: Climatic factors controlling the productivity of Norway spruce: a model-based analysis. *Forest Ecology and Management,* **110,** 125–139.

Bergh, J., S. Linder, T. Lundmark, and B. Elfving, 1999: The effect of water and nutrient availability on the productivity of Norway spruce in northern and southern Sweden. *Forest Ecology and Management,* **119,** 51–62.

Black, T.A., G. den Hartog, H. Neumann, P. Blanken, P. Yang, Z. Nesic, S. Chen, C. Russel, P. Voroney, and R. Stabeler, 1996: Annual cycles of CO_2 and water vapour fluxes above and within a boreal aspen stand. *Global Change Biology,* **2,** 219–230.

Blanken, P.D., T.A. Black, H.H. Neumann, G. den Hartog, P.C. Yang, Z. Nesic, R. Staebler, W. Chen, and M.D. Novak, 1998: Turbulent flux measurements above and below the overstory of a boreal aspen forest. *Boundary Layer Meteorology,* **89,** 109–140.

Blume, H.P., H. Eger, E. Fleischhauer, A. Hebel, C. Reij, and K.G. Steiner (eds.), 1998: Towards sustainable land use. *Advances in Geoecology,* **31,** 1625 pp.

Boone, R.D., K.J. Nadelhoffer, J.D. Canary, and J.P. Kaye, 1998: Roots exert a strong influence on the temperature sensitivity of soil respiration. *Nature,* **396,** 570–572.

Bousquet, P., P. Ciais, P. Peylin, M. Ramonet, and P. Monfray, 1999: Inverse modelling of annual atmospheric CO_2 sources and sinks 1. Method and control inversion. *Journal of Geophysical Research,* **104,** 26161–26178.

Braswell, B.H., D.S. Schimel, E. Linder, and B. Moore III, 1997: The response of global terrestrial ecosystems to interannual temperature variability. *Science,* **278,** 870–872.

British Petroleum Company, 1999: *BP Statistical Review of World Energy 1999.* British Petroleum Company, London, United Kingdom.

Brown, S. and A.E. Lugo, 1982: The storage and production of organic matter in tropical forests and their role in the global carbon cycle. *Biotropica,* **14,** 161–187.

Buchmann, N. and E.-D.Schulze, 1999: Net CO_2 and H_2O fluxes of terrestrial ecosystems. *Global Biogeochemical Cycles,* **13,** 751–760.

Cannell, M.G.R., R.C. Dewar, and D.G. Pyatt, 1993: Conifer plantations on drained peat in Britain: A net gain or loss of carbon? *Forestry,* **66,** 353–369.

Cannell, M.G.R., R. Milne, K.J. Hargreaves, T.A.W. Brown, M.M. Cruickshank, R.I. Bradley, T. Spencer, D. Hope, M.F. Billett, W.N. Adger, and S. Subak, 1999: National inventories of terrestrial carbon sources and sinks, the UK experience. *Climatic Change,* **42,** 505–530.

Cao, M. and F.I. Woodward, 1998: Dynamic responses of terrestrial ecosystem carbon cycling to global climate change. *Nature,* **393,** 249–252.

Chappelaz, J., I.Y. Fung, and A.M. Thomson, 1993: The atmospheric CH_4 increase since the last glacial maximum: (1) sources estimates, *Tellus,* **45B,** 228–241.

Chen, W.J., T.A. Black, P.C. Yang, A.G. Barr, H.H. Neumann, Z. Nesic, P.D. Blanken, M.D. Novak, J. Eley, R.J. Ketler, and R. Cuenca, 1999: Effects of climatic variability on the annual carbon sequestration by a boreal aspen forest. *Global Change Biology,* **5,** 41–53.

Ciais, P., 1999: Restless carbon pools. *Nature,* **398,** 111–112.

Ciais, P., P.P. Tans, J.W. White, M. Trolier, R. Francey, J. Berry, D. Randall, P. Sellers, J. Collatz, and D.S. Schimel, 1995a: Partitioning of the ocean and land uptake of CO_2 from delta ^{13}C measurements from the NOAA/CMDL global air sampling network. *Journal of Geophysical Research,* **100(D3),** 5051–5057.

Ciais, P., P.P. Tans, M. Trolier, J.W.C. White, and R.J. Francey. 1995b: A large northern hemisphere terrestrial CO_2 sink indicated by the $^{13}C/^{12}C$ ratio of atmospheric CO_2. *Science,* **269,** 1017–1188.

Cole, C.V. K. Paustian, E.T. Elliott, A.K. Metherell, D.S. Ojima, W.J. Parton. 1993: Analysis of agroecosystems carbon pools. *Water, Air and Soil Pollution,* **70,** 357–371.

Cole, C.V., C. Cerri, K. Minami, A. Mosier, N. Rosenberg, D. Sauerbeck, J. Dumanski, J. Duxbury, J. Freney, R. Gupta, O. Heinemeyer, T. Kolchugina, J. Lee, K. Paustian, D. Powlson, N. Sampson, H. Tiessen, M. Van Noordwijk, Q. Zhao, I.P. Abrol, T. Barnwell, C.A. Campbell, R.L. Desjardin, C. Feller, P. Garin, M.J. Glendining, E.G. Gregorich, D. Johnson, J. Kimble, R. Lal, C. Monreal, D.S. Ojima, M. Padgett, W. Post, W. Sombroek, C. Tarnocai, T. Vinson, S. Vogel, and G. Ward, 1996: Agricultural options for mitigation of greenhouse gas wmissions. In: *Climate Change 1995—Impacts, Adaptations and Mitigation of Climate Change: Scientific-Technical Analyses* [Watson, R.T., M.C. Zinyowera, and R.H. Moss (eds.)]. Cambridge University Press, Cambridge, United Kingdom and New York, NY, USA, 745–771.

Conrad, R., W. Seiler, and G. Bunse, 1983: Factor influencing the loss of fertilizers nitrogen, emission of nitrogen into the atmosphere as N_2O. *Journal of Geophysical Research,* **88,** 6709–6718.

Conway, T.J., P.P. Tans, L.S. Waterman, K.W. Thoning, D.R. Kitzis, K.A. Masarie, and N. Zhang, 1994: Evidence for interannual variability of the carbon cycle from the National Oceanic and Atmospheric Administration Climate Monitoring and Diagnostic Laboratory Global Air Sampling Network. *Journal of Geophysical Research,* ***99(D11)***, 22831–22855.

Cramer, W., A. Bondeau, F.I. Woodward, F.I., I.C. Prentice, R.A. Betts, V. Brovkin, P.M. Cox, V. Fisher, J.A. Foley, A.D. Friend, C. Kucharik, M.R. Lomas, N. Ramankutty, S. Sitch, B. Smith, A. White, and C. Young-Molling, 2000: Global response of terrestrial ecosystem structure and function to CO_2 and climate change: results from six dynamic global vegetation models. *Global Change Biology,* (in press).

Curtis, P.S. 1996: A meta-analysis of leaf gas exchange and nitrogen in trees grown under elevated carbon dioxide. *Plant, Cell and Environment,* **19,** 127–137.

Curtis, P.S. and Wang, X., 1998: A meta-analysis of elevated CO_2 effects on woody plant mass, form, and physiology. *Oecologia,* **113,** 299–313.

Davidson, E.A., 1991: Fluxes of nitrous oxide and nitric oxide from terrestrial ecosystems. In: *Microbial Production and Consumption of Greenhouse Gases: Methane, Nitrous Oxide and Halomethanes* [Rogers, J.E. and W.B. Witman (eds.)]. American Society of Microbiology, Washington, DC, USA, pp. 219–235.

Davidson, E.A., 1992: Sources of nitrous oxide following wetting of dry soils. *Soil Science Society of America Journal,* **56,** 95–102.

De Fries, R.S., C.B. Field, I. Fung, G.J. Collatz, and L. Bounoua, 1999: Combining satellite data and biogeochemical models to estimate global effects of human-induced land cover change on carbon emissions and primary productivity. *Global Biogeochemical Cycles,* **13,** 803–815.

Detwiler, R.P., 1986: Land use change and the global carbon cycle: The role of tropical soils. *Biogeochemistry,* **2,** 67–93.

Dewar, R.C., B.E. Medlyn, and R.E. McMurtrie, 1999: Acclimation of the respiration/photosynthesis ratio to temperature: insights from a model. *Global Change Biology,* **5,** 615–622.

Dixon, R.K., S. Brown, R.A. Houghton, A.M. Solomon, M.C. Trexler, and J. Wisniewski, 1994a: Carbon pools and flux of global forest ecosystems. *Science,* **263,** 185–190.

Dixon, R., J. Winjum, K. Andrasko, J. Lee, and P. Schroeder, 1994b: Integrated systems: assessment of promising agroforest and alternative land-use practices to enhance carbon conservation and sequestration. *Climatic Change,* **30,** 1–23.

Dlugokencky, E.J., K.A. Masarie, P.M. Lang, and P.P. Tans, 1998: Continuing decline in the growth rate of the atmospheric methane burden. *Nature,* **393,** 447–450.

Eamus, D., L.B. Hutley, and A.P. O'Grady, n.d.: *Daily and Seasonal Patterns of Carbon and Water Fluxes Above a North Australian Savanna,* unpublished manuscript.

Edmonds, J.A., M.A. Wise, R.D. Sands, R.A. Brown, and H.S. Kheshgi, 1996: *Agriculture, Land Use, and Commercial Biomass Energy: A Preliminary Integrated Analysis of the Potential Role of Biomass Energy for Reducing Future Greenhouse Related Emissions.* Research Report PNNL-111555, Pacific Northwest National Laboratory, Washington, DC, USA.

Eklund, T.J., W.H. McDowell, and C.M. Pringle, 1997: Seasonal variation of tropical precipitation chemistry: La Selva, Costa Rica. *Atmospheric Environment,* **23,** 3903–3910.

Elliott, E.T., 1986: Aggregate structure and carbon, nitrogen, and phosphorus in native and cultivated soils. *Soil Science Society of America Journal,* **50,** 627–633.

Ellsworth, D.S., 1999: CO_2 enrichment in a maturing pine forest: are CO_2 exchange and water status in the canopy affected? *Plant, Cell and Environment,* **22,** 461–472.

Enting, I.G., T.M.L. Wigley, and M. Heimann (eds.), 1994: *Future emissions and concentrations of carbon dioxide: Key ocean/atmosphere/land analyses.* Commonwealth Scientific and Industrial Research Organization (CSIRO) Division of Atmospheric Research, Technical Paper No. 31, Melbourne, Australia, 120 pp.

Etheridge, D.M., G.I. Pearman, and P.J. Fraser, 1992: Changes in tropospheric methane between 1841 and 1978 from a high accumulation rate Antarctic ice core. *Tellus,* **44,** 282–294.

Etheridge, D.M., L.P. Steele, R.L. Langenfelds, R.J. Francey, J.-M. Barnola, and V.I. Morgan, 1996: Natural and anthropogenic changes in atmospheric CO_2 over the last 1000 years from air in Antarctic ice and firn. *Journal of Geophysical Research,* **101,** 4115–4128.

Fan, S.-M., S.C. Wofsy, P.S. Bakwin,. and D.J. Jacob, 1990: Atmosphere-biosphere exchange of CO_2 and O_3 in the central Amazon forest. *Journal of Geophysical Research,* **95,** 16851–16864.

Fan, S., M. Gloor, J. Mahlman, S. Pacala, J. Sarmiento, T. Takahashi, and P. Tans, 1998: A large terrestrial carbon sink in North America implied by atmospheric and oceanic carbon dioxide data and models. *Science,* **282,** 442–446.

FAO, 1993: *1992 Production Yearbook.* Food and Agriculture Organization of the United Nations, Rome, Italy.

Fearnside, P.M., 2000: Global warming and tropical land-use change: greenhouse gas emissions from biomass burning, decomposition and soils in forest conversion, shifting cultivation and secondary vegetation. *Climatic Change,* (in press).

Fearnside, P.M. and R.I. Barbosa, 1998: Soil carbon changes from conversion of forest to pasture in Brazilian Amazonia. *Forest Ecology and Management,* **108,** 147–166.

Fearnside, P.M. and W.M. Guimarães, 1996: Carbon uptake by secondary forests in Brazilian Amazonia, *Forest Ecology and Management,* **80,** 35–46.

Feely, R.A., R. Wanninkhof, T. Takahashi, and P.P. Tans, 1999: Influence of El Niño on the equatorial Pacific contribution to atmospheric CO_2 accumulation, *Nature,* **398,** 597–601.

Forti, M.C. and L.M. Moeira-Nordemann, 1991: Rainwater and throughfall chemistry and in a "Terra Firme" rain forest: Central Amazonia. *Journal of Geophysical Research,* **96,** 7415–7421.

Francey, R.J., P.P. Tans, C.E. Allison, I.G. Enting, J.W.C. White, and M. Trolier, 1995: Changes in oceanic and carbon uptake since 1982. *Nature,* **373,** 326–330.

Freydier, F., R. Dupret, and J.P. Lacaux, 1998: Precipitation chemistry in intertropical Africa. *Atmospheric Environment,* **32,** 749–765.

Friend, A.D., A.K. Stevens, R.G. Knox, and M.G.R. Cannell, 1997: A process-based, terrestrial biosphere model of ecosystem dynamics (Hybrid v3.0). *Ecological Modelling,* **95,** 249–287.

Fung, I., J. John, J. Lerner, E. Matthews, M. Prather, L.P. Steele, and P.J. Fraser, 1991: Three dimensional model synthesis of the global methane cycle. *Journal of Geophysical Research,* **96,** 13033–13065.

Gaudry, A., P. Monfray, G. Polian, and G. Lambert, 1987: The 1982–83 El Niño: a 6 billion ton CO_2 release. *Tellus,* **39B,** 209–213.

Goulden, M.L., J.W. Munger, S-M. Fan, B. Daube, and S.C. Wofsy, 1996a: Exchange of carbon dioxide by a deciduous forest: response to interannual climate variability. *Science,* **271,** 1576–1578.

Goulden, M.L., J.W. Munger, S.M. Fan, B.C. Daube, and S.C. Wofsy, 1996b: Measurements of carbon sequestration by long-term eddy covariance: methods and a critical evaluation of accuracy. *Global Change Biology,* **2,** 169–182.

Goulden, M.L., S.C. Wofsy, J.W. Harden, S.E. Trumbore, P.M. Crill, S.T. Gower, T. Fries, B.C. Daube, S.M. Fan, D.J. Sutton, A. Bazzaz, and J.W. Munger, 1998: Sensitivity of boreal forest carbon balance to soil thaw. *Science,* **279,** 214–217.

Gower, S.T., J.G. Vogel, J.M. Norman, C.J. Kucharik, S.J. Steele, and T.K. Stow, 1997: Carbon distribution and aboveground net primary production in aspen, jack pine, and black spruce stands in Saskatchewan and Manitoba, Canada. *Journal of Geophysical Research,* **102,** 29029–029041.

Grace, J., J. Lloyd, J. McIntyre, A.C. Miranda, P. Meir, H. Miranda, J.B. Moncrieff, J. Massheder, I.R. Wright, and J. Gash, 1995a: Fluxes of carbon dioxide and water vapor over an undisturbed tropical rainforest in south-west Amazonia. *Global Change Biology,* **1,** 1–12.

Grace, J., J. Lloyd, J. McIntyre, A.C. Miranda, P. Meir, H. Miranda, C. Nobre, J.B. Moncrieff, J. Massheder, Y. Malhi, I.R. Wright, and J. Gash, 1995b: Carbon dioxide uptake by an undisturbed tropical rain forest in South-West Amazonia 1992–1993. *Science,* **270,** 778–780.

Greco, S., and D.D. Baldocchi, 1996: Seasonal variation of CO_2 and water vapor exchange rates over a temperate deciduous forest. *Global Change Biology,* **2,** 183–198.

Gustavsson, L., P. Börjesson, B. Johansson, and P. Svenningsson, 1995: Reducing CO_2 emissions by substituting biomass for fossil fuels. *Energy,* **20,** 1097–1113.

Hättenschwiler, S., F. Miglietta, A. Raschi, and C. Körner, 1997: Thirty years of in situ tree growth under elevated CO_2, a model for future forest responses. *Global Change Biology,* **3,** 463–471.

Hall, D.O., H. Mynick, and R. Williams, 1991: Cooling the greenhouse with bioenergy. *Nature,* **353,** 11–12.

Hanan, N.P., P. Kabat, A.J. Dolman, and J.A. Elbers, 1998: Photosynthesis and carbon balance of a Sahelian fallow savanna. *Global Change Biology,* **4,** 523–538.

Harvey, L.D.D., J. Gregory, M. Hoffert, A. Jain, M. Lal, R. Leemans, S.C.B. Raper, T.M.L. Wigley, and J.R. de Wolde, 1997: *An Introduction to Simple Climate Models Used in the IPCC Second Assessment Report.* IPCC Technical Paper, Intergovernmental Panel on Climate Change, Bracknell, United Kingdom, 250 pp.

Havranek, W.H. and W. Tranquillini, 1995: Physiological processes during winter dormancy and their ecological significance. In: *Ecophysiology of Coniferous Forests* [Smith, W.K. and T.M. Hinckley (eds.)]. Academic Press, London, United Kingdom, pp. 95–124.

Heimann, M., G. Esser, J. Kaduk, D. Kicklighter, G. Kohlmaier, D. McGuire, B. Moore III, C. Prentice, W. Sauf, A. Schloss, U. Wittenberg, and G. Würth, 1997: Interannual variability of CO_2 exchanges fluxes as simulated by four terrestrial biogeochemical models (AB0179). In: *Fifth International Carbon Dioxide Conference,* edited by Commonwealth Scientific and Industrial Research Organization (CSIRO), Cairns, Australia, pp. 129–130.

Houghton, J.T., L.G. Meira Filho, B.A. Callander, N. Harris, A. Kattenberg, and K. Maskell (eds.), 1996: *Climate Change 1995: The Science of Climate Change. Contribution of Working Group I to the Second Assessment Report of the Intergovernmental Panel on Climate Change.* Cambridge University Press, Cambridge, United Kingdom and New York, NY, USA, 572 pp.

Houghton, R.A., 1991: Tropical deforestation and atmospheric carbon dioxide. *Climate Change,* **19,** 99–118.

Houghton, R.A., 1994: The worldwide extent of land-use change. *BioScience,* **44,** 305–313.

Houghton, R.A., 1999: The annual net flux of carbon to the atmosphere from changes in land use 1850–1990. *Tellus,* **50B,** 298–313.

Houghton, R.A. J.L. Hackler, and K.T. Lawrence, 1999: The U.S. carbon budget: contributions from land-use change. *Science,* **285,** 574–578.

Houghton, R.A., D.L. Skole, C.A. Nobre, J.L. Hackler, K.T. Lawrence, and W.H. Chomentowski, 2000: Annual fluxes of carbon from deforestation and regrowth in the Brazilian Amazon. *Nature,* **403,** 301–304.

House, J.I. and D.O. Hall, 2000: Net primary production of savannas and tropical grasslands. In: *Terrestrial Global Productivity: Past, Present and Future* [Mooney, H., J. Roy, and B. Saugier (eds.)]. (in press).

Hutchinson, G.L., 1995: Biosphere-atmosphere exchange of gases. In: *Soils and Global Change* [Lal, R., J. Kimble, E. Levine, and B. Stewart (eds.)]. Lewis Publishers, Chelsea, United Kingdom, pp. 219–236.

Indermühle, A., T.F. Stocker, F. Joos, H. Fisher, H.J. Smith, M. Wahlen, B. Deck, D. Mastroaianni, J. Tschumi, T. Blunier, R. Meyer, and B. Stauffer, 1999: Holocene carbon cycle dynamics based on CO_2 trapped in ice at Taylor Dome, Antarctica. *Nature,* **398,** 121–126.

Ishitani, H., T.B. Johansson, and S. Al-Khouli: 1996: Energy supply. In: *Climate Change 1995: Contribution of Working Group III to the Second Assessment Report of the Intergovernmental Panel on Climate Change* [Watson, R.T., M.C. Zinyowera, and R.H. Moss (eds.)]. Cambridge University Press, Cambridge, United Kingdom and New York, NY, USA, pp. 587–647.

Jain, A.K., H.S. Kheshgi, M.I. Hoffert, and D.J. Wuebbles, 1995: Distribution of radiocarbon as a test of global carbon cycle models. *Global Biogeochemical Cycles,* **9,** 153–166.

Jarvis, P.G., 1981: Production efficiency of coniferous forest in the UK. In: *Physiological Processes Limiting Plant Productivity* [Johnson, C.B. (ed.)]. Butterworth Scientific Publications, London, United Kingdom, pp. 81–107.

Jarvis, P.G. (ed.), 1998: *European Forests and Global Change, the Likely Impacts of Rising CO_2 and Temperature.* Cambridge University Press, Cambridge, United Kingdom and New York, NY, USA, 380 pp.

Jarvis, P.G., J.M. Massheder, S.E. Hale, J.B. Moncrieff, M. Rayment, and S. Scott, 1997: Seasonal variation of carbon dioxide, water vapor and energy exchanges of a boreal black spruce forest. *Journal of Geophysical Research,* **102,** 28953–28966.

Johansson, J. and U. Lundqvist, 1999: Estimating Swedish biomass energy supply. *Biomass and Bioenergy,* **17,** 85–93.

Johnson, D.W., 1992: Effects of forest management on soil carbon storage. *Water, Air and Soil Pollution,* **64,** 83–120.

Joos, F., R. Meyer, M. Bruno, and M. Leuenberger, 1999: The variability in the carbon sinks as reconstructed for the last 1000 years. *Geophysical Research Letters,* **26,** 1437–1441.

Kauppi, P.E., K. Mielkäinen, and K. Kuusela, 1992: Biomass and carbon budget of European forests, 1971 to 1990, *Science,* **256,** 70–74.

Keeling, C.D. and T.P. Whorf, 1999: Atmospheric CO_2 records from sites in the SIO air sampling network. In: *Trends: A Compendium of Data on Global Change.* Carbon Dioxide Information Analysis Center, Oak Ridge National Laboratory, Oak Ridge, TN, USA.

Keeling, C.D., J.F.S. Chin, and T.P. Whorf, 1996a: Increased activity of northern vegetation inferred from atmospheric CO_2 measurements. *Nature,* **382,** 146–149.

Keeling, R.F., S.C. Piper, and M. Heimann, 1996b: Global and hemispheric CO_2 sinks deduced from changes in atmospheric O_2 concentrations. *Nature,* **381,** 218–221.

Keeling, C.D., R.B. Bacastow, A.F. Carter, S.C. Piper, T.P. Whorf, M. Heimann, W.G. Mook, and H.A. Roeloffzen. 1989: A three-dimensional model of atmospheric CO_2 transport based on observed winds: 1. analysis of observational data. In: *Geophysical Monographs,* **55,** American Geophysical Union, Washington, DC, USA, pp. 165–236.

Kern, J. and M. Johnson, 1993: Conservation tillage impacts on national soil and atmospheric carbon levels. *Soil Science Society of America Journal,* **57,** 200–210.

Khalil, M.A.K. and R.A. Rasmussen, 1992: The global sources and sinks of nitrous oxide. *Journal of Geophysical Research,* **97,** 14651–14658.

Kheshgi, H.S., A.K. Jain, and D.J. Wuebbles, 1999: Model-based estimation of the global carbon budget and its uncertainty from carbon dioxide and carbon isotope records. *Journal of Geophysical Research,* **104,** 31127–31144.

Kindermann, J., G. Würth, G.H. Kohlmaier, and F.-W. Badeck, 1996: Interannual variation of carbon exchange flux in terrestrial ecosystems, *Global Biogeochemical Cycles,* **10,** 737–755.

Krankina, O.N., M.E. Harmon, and A.V. Griazkin, 1999: Nutrient stores and dynamics of woody detritus in a boreal forest: modeling potential implications at the stand level. *Canadian Journal of Forest Research,* **29,** 20–32.

Kroeze, C., A. Mosier, and L. Bouwman, 1999: Closing the global N_2O budget: a retrospective analysis 1500–1994, *Global Biogeochemical Cycles,* **13,** 1–8.

Lal, R., J.M. Kimble, R.F. Follett, and C.V. Cole. 1998: *The Potential of U.S. Cropland to Sequester Carbon and Mitigate the Greenhouse Effect.* Ann Arbor Press, Chelsea, MI, USA, 128 pp.

Lee, J., D. Phillips, and R. Liu, 1993: The effect of trends in tillage practices on erosion and carbon content of soils in the US corn belt. *Water, Air, and Soil Pollution,* **70,** 389–401.

Lee, K., R. Wanninkhof, T. Takahashi, S.C. Doney, and D. Feely, 1998: Low interannual variability in recent oceanic uptake of atmospheric carbon dioxide. *Nature,* **396,** 155–159.

Leemans, R., A. van Amstel, C. Battjes, E. Kreileman, and S. Toet, 1996: The land cover and carbon cycle consequences of biomass as an energy source. *Global Environmental Change,* **6,** 335–357.

Le Quéré, C., J.C. Orr, and P. Monfray, 1998: Modeling the inter-annual variability of the air to sea flux of carbon dioxide in the years 1979–1993. *Proceedings of the International CO_2 Conference,* Cairns, Australia.

Linder, S., 1995: Foliar analysis for detecting and correcting nutrient imbalances in Norway spruce. *Ecological Bulletins (Copenhagen),* **44,** 178–190.

Linder, S., R.E. McMurtrie, and J.J. Landsberg, 1996: Global change impacts on managed forests. In: *Global Change and Terrestrial Ecosystems* [Walker, B. and W. Steffen (eds.)]. International Geosphere-Biosphere Programme (IGBP) Book Series No. 2, Cambridge University Press, Cambridge, United Kingdom and New York, NY, USA, pp. 275–290.

Linder, S. and D.A. Rook, 1984: Effects of mineral nutrition on carbon dioxide exchange and partitioning of carbon in trees. In: *Nutrition of Plantation Forests* [Bowen, G.D. and E.K.S. Nambiar (eds.)]. Academic Press, London, United Kingdom, pp. 212–236.

Lindroth, A., A. Grelle, and A.S. Morén, 1998: Long-term measurements of boreal forest carbon balance reveal large temperature sensitivity. *Global Change Biology,* **4,** 443–450.

Lloyd, J., M.I. Bird, E. Veenendaal, and B. Kruijt, n.d.: *Should Phosphorus Availability Be Constraining Moist Tropical Forest Responses to Increasing CO_2 Concentrations?* unpublished manuscript.

Lugo, A. and S. Brown, 1993: Management of tropical soils as sinks or sources of atmospheric carbon. *Plant and Soil,* **149,** 27–41.

Macedo, C., 1998: Greenhouse gas emissions and energy balances in bio-ethanol production and utilization in Brazil. *Biomass and Energy,* **14,** 77–81.

Malhi, Y., A.D. Nobre, J. Grace, B. Kruijt, M.G.P. Pereira, A. Culf, and S. Scott, 1998: Carbon dioxide transfer over a Central Amazonian rain forest. *Journal of Geophysical Research,* **D24,** 31593–31612.

Malhi, Y., D.D. Baldocchi, and P.G. Jarvis, 1999: The carbon balance of tropical, temperate and boreal forests. *Plant, Cell and Environment* **22,** 715–740.

Marland, G. and B. Schlamadinger, 1995: Biomass fuels and forest-management strategies: How do we calculate the greenhouse-gas emissions benefits? *Energy—The International Journal,* **20,** 1131–1140.

Marland, G., R.J. Andres, T.A. Boden, C. Johnston, and A. Brenkert, 1999: *Global, Regional, and National CO_2 Emission Estimates from Fossil Fuel Burning, Cement Production, and Gas Flaring: 1751–1996.* Report NDP-030, Carbon Dioxide Information Analysis Center, Oak Ridge National Laboratory, Oak Ridge, TN, USA.

McCaughey, J.H., P.M. Lafleur, D.W. Joiner, P.A. Bartlett, A.M. Costello, D.E. Jelinski, and M.G. Ryan, 1997: Magnitudes of seasonal patterns or energy, water and carbon exchanges at boreal young jack pine forest in the BOREAS northern study area. *Journal of Geophysical Research,* **102,** 28997–29009.

McMurtrie, R.E., B.E. Medlyn, and R.C. Dewar, n.d.: *Increased Understanding of Nutrient Immobilisation in Soil Organic Matter Is Critical for Predicting the Carbon Sink Strength of Forest Ecosystems Over the Next 100 Years,* unpublished manuscript.

McMurtrie, R.E. and J.J. Landsberg, 1992: Using a simulation model to evaluate the effects of water and nutrients on growth and carbon partitioning of *Pinus radiata*. *Forest Ecology and Management,* **52,** 243–260.

Medlyn, B.E., F-W. Badeck, D.G.G. de Pury, C.V.M. Barton, M. Broadmeadow, R. Ceulemans, P. de Angelis, M. Forstreuter, M.E. Jach, S. Kellomäki, E. Laitat, M. Marek, S. Philippot, A. Rey, J. Strassemeyer, K. Laitinen, R. Liozon, B. Portier, P. Roberntz, Y.P. Wang, and P.G. Jarvis, 2000: Effects of elevated CO_2 on photosynthesis in European forest species: A meta-analysis of model parameters. *Plant, Cell and Environment,* **22,** 1475–1495.

Melillo, J.M., I.C. Prentice, G.D. Farquhar, E.-D. Schulze, and O.E. Sala, 1996: Terrestrial biotic responses to environmental change and feedbacks to climate. In: *Climate Change 1995: The Science of Climate Change. Contribution of Working Group I to the Second Assessment Report of the Intergovernmental Panel on Climate Change* [Houghton, J.T., L.G. Meira Filho, B.A. Callander, N. Harris, A. Kattenberg, and K. Maskell (eds.)]. Cambridge University Press, Cambridge, United Kingdom, and New York, NY, USA, pp. 445–481.

Menzel, A., and P. Fabian, 1999: Growing season extended in Europe. *Nature,* **397,** 659.

Meybeck, M., 1982: Carbon, nitrogen and phosphorus transport by world rivers. *American Journal of Science,* **282,** 401–450.

Minkkinen, K. and J. Laine, 1998: Long-term effect of forest drainage on the peat carbon stores of pine mires in Finland. *Canadian Journal of Forest Research,* **28,** 1267–1275.

Mooney, H.A., J. Canadell, F.S. Chapin III, J.R. Ehleringer, C. Körner, R.E. McMurtrie, W.J. Parton, L.F. Pitelka, and E.-D. Schulze, 1999: Ecosystem physiology responses to global change. In: *Implications of Global Change for Natural and Managed Ecosystems. A Synthesis of GCTE and Related Research* [Walker, B.H., W.L. Steffen, J. Canadell, and J.S.I. Ingram (eds.)]. International Geosphere-Biosphere Programme (IGBP) Book Series No. 4, Cambridge University Press, Cambridge, United Kingdom and New York, NY, USA, pp. 141–189.

Myneni, R.B., C.D. Keeling, C.J. Tucker, G. Asrar, and R.R. Nemani, 1997: Increased plant growth in the northern high latitudes from 1981–1991. *Nature,* **386,** 698–702.

Nabuurs, G.J. and G.M.J. Mohren, 1993: *Carbon Fixation Through Forestation Activities.* IBN Research Report 93/4, Institute for Forestry and Nature Resources, Wageningen, The Netherlands, 205 pp.

Nadelhoffer, K.J., B.A. Emmett, P. Gundersen, O.J. Kjonaas, C.J. Koopmans, P. Schleppi, A. Tietema, and R.F. Wright, 1999: Nitrogen deposition makes a minor contribution to carbon sequestration in temperate forests. *Nature,* **398,** 145–148.

Nilsson, S. and W. Schopfhauser 1995: The carbon-sequestration potential of a global afforestation program. *Climatic Change,* **30,** 267–293.

Norby, R.J., S.D. Wullschleger, C.A. Gunderson, D.W. Johnson, and R. Ceulemans, 1999: Tree responses to rising CO_2 in field experiments: implications for the future forest. *Plant, Cell and Environment,* **22,** 683–714.

Parton, W.J., J.M.O. Scurlock, D.S. Ojima, D.S. Schimel, and D.O. Hall, 1995: Impact of climate change on grassland production and soil carbon worldwide. *Global Change Biology,* **1,** 13–22.

Peterjohn, W.T., J.M. Melillo, P.A. Steudler, K.M. Newkirk, F.P. Bowles, and J.D. Aber, 1994: Response of trace gas fluxes and N availability to experimentally elevated soil temperatures. *Ecological Applications,* **43,** 617–625.

Peterson, A.G., J.T. Ball, Y. Luo, C.B. Field, P.B. Reich, P.S. Curtis, K.L. Griffin, C.A. Gunderson, R.J. Norby, D.T. Tissue, M. Forstreuter, A. Rey, C.S. Vogel, and CMEAL Participants, 1999: The photosynthesis-leaf nitrogen relationship at ambient and elevated atmospheric carbon dioxide: a meta-analysis. *Global Change Biology,* **5,** 331–346.

Phillips, O.L., Y. Malhi, N. Higuchi, W.F. Laurance, R.M. Núñez, D.J.D. Váxquez, L.V. Laurance, S.G., Ferreira, M. Stern, S. Brown, and J. Grace, 1998: Changes in the carbon balance of tropical forests: evidence from long-term plots. *Science,* **282,** 439–442.

Prasad, S.S., 1997: Potential atmospheric sources and sinks of nitrous oxide 2. Possibilities from exited O_2 "embryonic" O_3, and optically pumped exited O_3. *Journal of Geophysical Research,* **102,** 21527–21537.

Prather, M., R. Derwent, D. Ehhalt, P. Fraser, E. Sanhueza, and X. Zhou, 1995: Other trace gases and atmospheric chemistry. In: *Climate Change 1994: Radiative Forcing of Climate Change and an Evaluation of the IPCC IS92 Emission Scenarios* [Houghton, J.T., L.G. Meira Filho, J. Bruce, H. Lee, B.A. Callander, E. Haites, N. Harris, and K. Maskell (eds.)]. Cambridge University Press, Cambridge, United Kingdom and New York, NY, USA, pp. 73–126.

Prinn, R (ed.), 1994: *Global Atmospheric-Biospheric Chemistry.* Plenum Press, New York, NY, USA.

Pussinen, A., T. Karjalainen, S. Kellomäki, and R. Mäkipää, 1997: Contribution of the forest sector in carbon sequestration in Finland. *Biomass and Bioenergy,* **13,** 377–387.

Ramankutty, N. and J.A. Foley, 1998: Characterizing patterns of global land use: an analysis of global croplands data. *Global Biogeochemical Cycles,* **12,** 667–685.

Randerson, J.T., M.V. Thompson, T.J. Conway, I.Y. Fung, and C.B. Field, 1997: The contribution of terrestrial sources and sinks to trends in the seasonal cycle of atmospheric carbon dioxide. *Global Biogeochemical Cycles,* **11,** 535–560.

Rayment, M.B. and P.G. Jarvis, 2000: Long-term measurement of photosynthesis, respiration and transpiration of black spruce. *Canadian Journal of Forest Research,* (in press).

Rayner, P.J., I.G. Enting, R.J. Francey, and R. Langenfelds, 2000: Reconstructing the recent carbon cycle from atmospheric CO_2, $d^{13}C$ and O_2/N_2 observations. *Tellus,* (in press).

Rey, A. and P.G. Jarvis, 1997: An overview of long term effects of elevated atmospheric CO_2 concentrations on the growth and physiology of birch (*Betula pendula* Roth.). *Botanical Journal of Scotland,* **49,** 325–340.

Sarmiento, J.L. and E.T. Sundquist, 1992: Revised budget of the oceanic uptake of anthropogenic carbon dioxide. *Nature,* **356,** 589–593.

Saxe, H., D.S. Ellsworth, and J. Heath. 1998: Tansley Review No. 98. Tree and forest functioning in an enriched CO_2 atmosphere. *New Phytologist,* **139,** 395–436.

Schimel, D., I. Enting, M. Heimann, T. Wigley, D. Raynaud, D. Alves, and U. Siegenthaler, 1995: CO_2 and the carbon cycle. In*: Climate Change 1994*: *Radiative Forcing of Climate Change and an Evaluation of the IPCC IS92 Emission Scenarios* [Houghton, J.T., L.G. Meira Filho, J. Bruce, H. Lee, B.A. Callander, E. Haites, N. Harris, and K. Maskell (eds.)]. Cambridge University Press, Cambridge, United Kingdom and New York, NY, USA, pp. 73–126.

Schimel, D., D. Alves, I. Enting, M. Heimann, F. Joos, D. Raynaud, T. Wigley, M. Prather, R. Derwent, D. Ehhalt, P. Fraser, E. Sanhueza, X. Zhou, P. Jonas, R. Charlson, H. Rodhe, S. Sadasivan, K.P. Shine, Y. Fouquart, V. Ramaswamy, S. Solomon, J. Srinivasan, D. Albritton, I. Isaksen, M. Lal, and D. Wuebbles, 1996: Radiative forcing of climate change. In: *Climate Change 1995. The Science of Climate Change. Contribution of Working Group I to the Second Assessment Report of the Intergovernmental Panel on Climate Change* [Houghton, J.T., L.G. Meira Filho, B.A. Callander, N. Harris, A. Kattenberg, and K. Maskell (eds.)]. Cambridge University Press, Cambridge, United Kingdom and New York, NY, USA, pp. 65–131.

Schlesinger, W.H., 1997: *Biogeochemistry and Analysis of Global Change* (2nd ed.). Academic Press, San Diego, CA, USA, 588 pp.

Scholes, R.J., 1999: Will the terrestrial carbon sink saturate soon? *Global Change Newsletter,* **37,** 2–3.

Scholes, R.J., E.-D. Schulze, L.F. Pitelka, and D.O. Hall, 1998: Biogeochemistry of terrestrial ecosystems. In: *Implications of Global Change for Natural and Managed Ecosystems: A Synthesis of GCTE and Related Research* [Walker, B.H., W.L. Steffen, J. Canadell, and J.S.I. Ingram (eds.)]. International Geosphere-Biosphere Programme (IGBP) Book Series No. 4, Cambridge University Press, Cambridge, United Kingdom and New York, NY, USA, pp. 271–303.

Schulze, E-D. and M. Heimann, 1998: Carbon and water exchange of terrestrial systems. In: *Asian Change in the Context of Global Change* [Galloway, J.N. and J. Melillo (eds.)]. Cambridge University Press, Cambridge, United Kingdom and New York, NY, USA, pp. 145–161.

Schulze, E-D., J. Lloyd, F.M. Kelliher, C. Wirth, C. Rebmann, B. Lühker, M. Mund, I. Milukova, W. Schulze, A. Ziegler, A. Varlagin, R. Sogachov, S. Valentini, S. Dore, O. Grigoriev, O. Kolle, and N.N. Vygodskaya, 1999: Productivity of forests in the Eurosiberian boreal region and their potential to act as a carbon sink—a synthesis of existing knowledge and original data. *Global Change Biology,* **5,** 703–722.

Sellers, P.J., F.G. Hall, R.D. Kelly, A. Black, D. Baldocchi, J. Berry, M. Ryan, K.J. Ranson, P.M. Crill, D.P. Lettenmaier, H. Margolis, J. Cihlar, J. Newcomer, D. Fitzjarrald, P.G. Jarvis, S.T. Gower, D. Halliwell, D. Williams, B. Goodison, D.E. Wickland, and F.E. Guertin, 1997: BOREAS in 1997: Experiment overview, scientific results, and future directions. *Journal of Geophysical Research,* **102,** 28731–28769.

Semenov, S.M., I.M. Kounina, and B.A. Koukhta, 1998: An ecological analysis of anthropogenic changes in ground-level concentrations of O_3, SO_2, and CO_2 in Europe. *Doklady Biological Sciences,* **361,** 344–347.

Semenov, S.M., I.M. Kounina, and B.A. Koukhta, 1999: Tropospheric ozone and plant growth in Europe. Meteorology and Hydrology Publishing Center, Moscow, Russia, 208 pp. (in Russian).

Shiller, C.L. and D.R. Hastie, 1996: Nitrous oxide and methane fluxes from perturbed and unperturbed boreal forest sites in northern Ontario. *Journal of Geophysical Research,* **101,** 22767–22774.

Siegenthaler, U. and F. Joos, 1992: Use of a simple model for studying oceanic tracer distributions and the global carbon cycle. *Tellus,* **44B,** 186–207.

Sombroek, W.G., F.O. Nachtergaele, and A.V. Hebel, 1993: Amounts, dynamics and sequestering of carbon in tropical and subtropical soils. *Ambio,* **22,** 817–826.

Spiecker, H., K. Mielikäinen, M. Köhl, and J.P. Skovsgaard (eds.), 1996: *Growth Trends in European Forests—Studies from 12 Countries.* Springer-Verlag, Heidelberg, Germany, 354 pp.

Stallard, R.F., 1998: Terrestrial sedimentation and the carbon cycle: Coupling weather and erosion to carbon burial. *Global Biogeochemical Cycles,* **12,** 231–257.

Steele, L.P., E.J. Dlugockenky, P.M. Lang, P.P. Tans, R.C. Martin, and K.A. Masarie, 1992: Slowing down of the global accumulation of atmospheric methane during the 1980's. *Nature,* **358,** 313–316.

Steffen, W., I. Noble, J. Canadell, M. Apps, E.-D. Schulze, P.G. Jarvis, D. Baldocchi, P. Ciais, W. Cramer, J. Ehleringer, G. Farquhar, C.B. Field, A. Ghazi, R. Gifford, M. Heimann, R. Houghton, P. Kabat, C. Körner, E. Lambin, S. Linder, H.A. Mooney, D. Murdiyarso, W.M. Post, C. Prentice, M.R. Raupach, D.S. Schimel, A. Shvidenko, and R. Valentini, 1998: The terrestrial carbon cycle: Implications for the Kyoto protocol. *Science,* **280,** 1393–1394.

Tamm, C.O., 1991: Nitrogen in terrestrial ecosystems. Questions of productivity, vegetational changes, and ecosystem stability. In: *Ecological Studies 81*. Springer-Verlag, Berlin, Germany, New York, NY, USA, and London, United Kingdom, 116 pp.

Tans, P.P. and D.W.R. Wallace, 1999: Carbon cycle research after Kyoto. *Tellus,* **51B,** 562–571.

Taylor, C.M.A. and P.M. Tabbush, 1990: Nitrogen deficiency in Sitka spruce plantations. *Forestry Commission Bulletin 89,* 20 pp.

Thompson, M.V., J.T. Randerson, C.M. Malmström, and C.B. Field, 1996: Change in net primary production and heterotrophic respiration: How much is necessary to sustain the terrestrial sink? *Global Biogeochemical Cycles,* **10,** 711–726.

Tian, H., J.M. Melillo, D.W. Kicklighter, A.D. McGuire, J.V.K. Helfrich III, B. Moore III, and C.J. Vörösmarty, 1998: Effect of interannual climate variability on carbon storage in Amazonian ecosystems. *Nature,* **396,** 664–667.

Tisdall, J.M. and J.M. Oades, 1982: Organic matter and water-stable aggregates in soils. *Journal of Soil Science,* **33,** 141.

Tubiello, F.N., C. Rosenzweig, B.A. Kimball, P.J. Pinter, G.W. Wall, D.J. Hunsaker, R.L. La Morte, and R.L. Garcia, 1999: Testing CERES-wheat with free-air carbon dioxide enrichment (FACE) experiment data: CO_2 and water interactions. *Agronomy Journal,* **91,** 247–255.

Uhl, C., R. Buschbacher, and E.A.S. Serrao, 1988: Abandoned pastures in Eastern Amazonia. I. Patterns of plant succession. *Journal of Ecology,* **76,** 663–681.

Valentini, R., G. Matteucci, A.J. Dolman, E-D. Schulze, C. Rebmann, E.J. Moors, A. Granier, P. Gross, N.O. Jensen, K. Pilegaard, A. Lindroth, A. Grelle, C. Bernhofer, T. Grünwald, M. Aubinet, R. Ceulemans, A.S. Kowalski, T. Vesala, Ü. Rannik, P. Berbigier, D. Loustau, J. Gudmundsson, H. Thorgeirsson, A. Ibrom, K. Morgenstern, R. Clement, J. Moncrieff, L. Montagnani, S. Minerbi, and P.G. Jarvis, 2000: The carbon sink strength of forests in Europe: novel results from the flux observation network. *Nature,* (in press).

van Noordwijk, M., C. Cerri, P.L. Woomer, K. Nugroho, and M. Bernoux, 1997: Soil carbon dynamics in the humid tropical forest zone. In: *The Management of Carbon in Tropical Soils Under Global Change Science, Practice and Policy* [Elliott, E.T., J. Kimble, and M.J. Swift (eds.)]. *Geoderma,* **79,** 187–225.

Viglizzo, E.F., Z.E. Roberto, M.C. Filippín, and A.J. Pordomingo. 1995: Climate variability and agroecological change in the Central Pampas of Argentina. *Agriculture, Ecosystems & Environment,* **55,** 7–16.

Vitousek, P.M. and R.W. Howarth, 1991: Nitrogen limitation on land and in the sea: How can it occur? *Biogeochemistry,* **13,** 87–115.

Vitousek, P.M. and R.L. Sanford, Jr. 1986: Nutrient cycling in moist tropical forest. *Annual Review of Ecology and Systematics,* **17,** 137–167.

Walker, B.H., W.L. Steffen, and J. Langridge, 1999: Interactive and integrated effects of global change on terrestrial ecosystems. In: *Implications of Global Change for Natural and Managed Ecosystems: A Synthesis of GCTE and Related Research* [Walker, B.H., W.L. Steffen, J. Canadell, and J.S.I. Ingram (eds.)]. International Geosphere-Biosphere Programme (IGBP) Book Series No. 4, Cambridge University Press, Cambridge, United Kingdom and New York, NY, USA, pp. 329–375.

Wang, Y.P., P.G. Jarvis, and C.M.A. Taylor, 1991: PAR absorption and its relation to above-ground dry matter production of Sitka spruce *Journal of Applied Ecology,* **28,** 547–560.

Wang, Y-P., A. Rey, and P.G. Jarvis, 1998: Carbon balance of young birch trees grown in ambient and elevated atmospheric CO_2 concentrations. *Global Change Biology,* **4,** 797–807.

Wechsung, G., F. Wechsung G.W. Wall F.J. Adamsen B.A. Kimball P.J. Pinter R.L. La Morte R.L. Garcia, and T. Kartschall, 1999: The effects of free-air CO_2 enrichment and soil water availability on spatial and seasonal patterns of wheat growth. *Global Change Biology,* **5,** 519–530.

Winchester, J.W., F. Song-Miao, and L. Shao-Meng, 1988: Methane and nitrogen gases from rice fields of China—possible effects of microbiology, benthic fauna, fertilizer and agricultural practice. *Water, Air and Soil Pollution,* **37,** 149–155.

Winguth, A.M.E., M. Heimann, K.D. Kurz, E. Maier-Reimer, U. Mikolajewicz, and J. Segschneider, 1994: El Niño-southern oscillation related fluctuations of the marine carbon cycle. *Global Biogeochemical Cycles,* **8,** 39–63.

Wissenschaftlicher Beirat der Bundesregierung Globale Umweltveränderungen (WBGU), 1998: Die Anrechnung biologischer Quellen und Senken im Kyoto-Protokoll: Fortschritt oder Rückschlag für den globalen Umweltschutz. Sondergutachten 1998, Bremerhaven, Germany, 76 pp. (available in English).

Zoltai, S.C. and P.J. Martikainen, 1985: Estimated extent of forest peatlands and their role in the global C cycle. In: *Forest Ecosystems. Forest Management and the Global C Cycle* [Apps, M.J. and D.T. Price (eds.)]. NATO ASI series, Springer-Verlag, Heidelberg, Germany, 548 pp.

2

Implications of Different Definitions and Generic Issues

IAN NOBLE (AUSTRALIA), MICHAEL APPS (CANADA), RICHARD HOUGHTON (USA), DANIEL LASHOF (USA), WILLY MAKUNDI (TANZANIA), DANIEL MURDIYARSO (INDONESIA), BRIAN MURRAY (USA), WIM SOMBROEK (NETHERLANDS), AND RICCARDO VALENTINI (ITALY)

Lead Authors:
M. Amano (Japan), P.M. Fearnside (Brazil), J. Frangi (Argentina), P. Frumhoff (USA), D. Goldberg (USA), N. Higuchi (Brazil), A. Janetos (USA), M. Kirschbaum (Australia), R. Lasco (Philippines), G.J. Nabuurs (The Netherlands), R. Persson (Sweden), W. Schlesinger (USA), A. Shvidenko (Russia), D. Skole (USA), P. Smith (UK)

Contributors:
M. Cannell (UK), C. Cerri (Brazil), D. Goetze (Canada), H. Janzen (Canada), J. Kimble (USA), R. Lal (USA), P. Moura-Costa (Brazil), M. O'Brien (Australia), P. Sanchez (Kenya), T. Singh (India), R. Scholes (South Africa)

Review Editors:
K. MacDicken (USA) and M. Manning (New Zealand)

CONTENTS

EXECUTIVE SUMMARY

In implementing the Kyoto Protocol, the Parties are likely to make a series of policy decisions on definitions, accounting procedures, and methods for accounting for carbon stocks and changes in stocks. This chapter addresses the core definitional, accounting, and methodological issues that may be relevant to the interpretation of several Articles in the Protocol. It introduces many issues that cut across other chapters in this report.

This chapter has four main sections. Section 2.2 deals with definitional issues; Section 2.3 with accounting issues; Section 2.4 with methods available for measuring and verifying carbon stocks, emissions, and sinks; and Section 2.5 with sustainability issues.

Definitions and Their Implications

In a given land area, a "full" carbon accounting system would consist of a complete accounting for changes in carbon stocks across all carbon pools in a given time period. In principle, applying full carbon accounting to all land in each country would yield the net carbon exchange between terrestrial ecosystems and the atmosphere. However, the Kyoto Protocol specifies, among other things, that attention be restricted to land areas subject to "direct human-induced" (Article 3.3) or human-induced (Article 3.4) activities.

The Parties to the United Nations Framework Convention on Climate Change (UNFCCC) may wish to make decisions about the definitions of terms used in Articles 3.3 and 3.4—in particular, "land-use change," "forestry activities," "human-induced" and "direct human-induced," "afforestation," "reforestation," "deforestation," and "carbon stocks"—because these definitions may affect carbon accounting under the Protocol. These definitions are critical to the accounting of sources and sinks under the Protocol: They determine the scope of human activities that are accounted under Article 3.3 and which may be eligible under other Articles (such as Article 3.4) of the Protocol.

Definitions that already exist to serve other purposes will not necessarily suffice for the particular needs of Articles 3.3 and 3.4; therefore, this Special Report provides a series of definitional scenarios to illustrate the implications of possible sets of definitions. The basis of the definitional scenarios is described in this chapter; the scenarios are presented and analyzed in Chapter 3.

The revised Guidelines for National Greenhouse Gas Inventories—hereafter referred to as the IPCC Guidelines (IPCC, 1997)—describe a methodology for a comprehensive approach to measuring "anthropogenic emissions by sources and removals by sinks of greenhouse gases" that is feasible to implement in most nations. These guidelines include suggested definitions for many important terms and can form a starting point for reporting compliance with the Kyoto Protocol. Article 5.2 of the Kyoto Protocol and a decision (Decision 2/CP.3) at the Third Conference of Parties (COP) refers to the IPCC Guidelines; Article 5.2 notes that where the IPCC Guidelines are not used "appropriate adjustments shall be applied according to methodologies agreed upon by the" first COP.

Most definitions of "forest" used at the national and regional levels are based on land-use status or a minimum threshold of canopy cover and/or tree height. Definitions based solely on either land use or canopy cover have limitations for carbon accounting. Land use, for example, may be defined in terms of administrative or cultural purposes and may bear little relation to the actual vegetation or amount of carbon on a site. Similarly, there are times (e.g., during regeneration after disturbance or harvesting) when the canopy cover of forests may be negligible.

Deforestation can be defined (e.g., in the IPCC Guidelines) as the conversion of forest to non-forest; reforestation and afforestation are the conversion of non-forested lands to forest. Issues arise with respect to the timing, sequencing, and causes of these activities. However, some operational forestry definitions of reforestation include the regeneration of trees after harvesting. When such definitions of reforestation are used but the accompanying definition of deforestation does not include disturbance or harvesting, asymmetric accounting occurs. Under such definitions of reforestation, many forestry management practices would create large areas of lands under Article 3.3.

Definitions of ARD that are based only on canopy cover may allow activities that lead to significant carbon fluxes to remain unaccounted. For instance, definitions that are based on a single threshold for canopy cover alone may allow changes in carbon stocks to remain unaccounted under Article 3.3. If a high threshold for canopy cover (e.g., 70-percent canopy cover) is used in the definition of a forest, many areas of sparse forest and woodland could be cleared without the losses in carbon being counted under Article 3.3. If a low threshold is set (e.g., 10-percent canopy cover), dense forest could be heavily degraded and significant amounts of carbon released and the actions would not be designated as deforestation. This problem could be addressed by using national, regional, or biome-specific thresholds—for example, a low canopy cover for savannas and a high canopy cover for moist forests—or by definitions based on degradation or aggradation criteria.

Definitions that are based strictly on a canopy cover threshold may lead to an interpretation of conventional harvest/ regeneration cycles as deforestation followed by reforestation. Such interpretations lead to significantly greater areas of forest being included within lands under Article 3.3. These interpretations also exacerbate other issues that relate to the measurement of carbon stocks (e.g., which pools?) over time (e.g., how long and when?) and the attribution of change to human activity. Additional qualifiers that are based on the timing of activities or growth potential (see Chapter 3) can be introduced to the definition of a forest to prevent this problem.

Afforestation is usually defined as the establishment of forest on land that has been unforested for a long period (variously described as "in historical times," "many decades," etc.). The precise period that distinguishes afforestation from reforestation is not critical to lands under Article 3.3, provided that afforestation and reforestation are treated identically in the Protocol, as they are in the IPCC Guidelines.

Carbon content in vegetation (or woody component) is an alternative basis for the threshold for the definition of a forest, but it has similar limitations as canopy-cover criteria. Such definitions relate more directly, however, to carbon stocks and fluxes.

There are definitional issues concerning the boundary between natural phenomena and direct human-induced activities—for example, when significant forest losses occur as a result of natural or human-induced fires or disturbances such as pest outbreaks. Distinguishing between natural and anthropogenic factors that influence the vulnerability of the land to disturbance will be difficult. In cases involving non–Kyoto lands, where fires or pest outbreaks trigger land-use change, the consequences are similar to deforestation. If vegetation cover is allowed to recover, with no change in land use, such disturbances would not lead to a long-term change in carbon stocks. In this context, a key issue for policymakers to decide is whether accounting should count neither the loss nor subsequent sequestration of carbon (symmetric accounting as in the IPCC Guidelines); count both the loss and subsequent sequestration of carbon (symmetric accounting that would increase monitoring requirements); count only the loss of carbon (asymmetric accounting); or count only the subsequent sequestration (asymmetric accounting).

In cases involving forested Kyoto lands, a key issue for policymakers to decide is whether accounting should count neither the loss nor subsequent sequestration of carbon (symmetric accounting)—which, however, essentially changes the area burned to non–Kyoto land and leaves several accounting issues to be resolved; count both the loss and the subsequent sequestration of carbon (symmetric accounting); count only the loss of carbon (asymmetric accounting), which creates similar problems to counting neither the loss nor subsequent sequestration of carbon; or count only the subsequent sequestration (asymmetric accounting), which may create an incentive to burn forests.

This chapter provides a logical schema for examining a set of alternative forest definitions that are based on combinations of land use and variable canopy-cover thresholds. Together with appropriate ARD definitions, these forest definitions form a set of definitional scenarios to aid in the implementation of the Protocol. Each of these scenarios leads to a logical sequence of decisions to be made by policymakers, and each decision has implications for the accounting of carbon, especially under Article 3.3. These definitional scenarios are analyzed in Chapter 3.

Article 3.4 refers to additional activities that do not explicitly fall under ARD. These activities potentially include all forms of land management, including activities in both the agricultural sector and the forestry sector. In the broadest sense, these activities could incorporate all lands under Article 3.3, as well as all lands that are (or will be) managed by humans for the production of food, fiber, forage, and other purposes.

When inclusion of additional activities is considered, it is possible to interpret "activity" broadly (e.g., cropland management) or narrowly (e.g., change in tillage method, fertilization, or cover crops). The broad definition may lend itself more readily to land-based carbon stock change methods of measuring the net impact of the activity. The narrow interpretation of "activity" may lend itself to activity-based measures in some instances, where the area of practice reported is multiplied by research-derived estimates of changes in emissions or removals in sinks associated with the activity. In other instances, current scientific knowledge of carbon flux rates associated with particular activities is very limited, and these combinations could produce very high uncertainty in estimates.

Narrow definitions may be useful if a small number of discrete activities are added, whereas a broad interpretation can provide a comprehensive framework for full carbon accounting as discussed in the IPCC Guidelines. Under either definitional interpretation, it is possible to choose a land-based or activity-based method of accounting for carbon impacts. Those combined choices will have major impacts on the accuracy, transparency, and verifiability of reported amounts.

Carbon Accounting

A well-designed carbon accounting system would provide consistent, transparent, and verifiable recording and reporting of changes in carbon stocks from applicable land use, land-use change, and forestry (LULUCF) activities and projects under Articles 3.3, 3.4, 3.7, and other relevant Articles to demonstrate compliance with the commitments defined by the Kyoto Protocol.

A "land-based" approach to accounting would take as its starting point the total carbon stock change in applicable carbon pools on land units subject to Kyoto accounting requirements. This approach involves first identifying land units on which applicable activities occur. Next, the total change in carbon stocks on these land units during the commitment period is determined. Adjustments can then be made to reflect decisions that the Parties may adopt regarding baselines, leakage, uncertainties, and timing. Aggregate emissions or removals are the sum of stock changes (net of adjustments) over all applicable land units.

An "activity-based" approach would start with the carbon stock change attributable to designated LULUCF activities. First, each applicable activity's impact on carbon stocks is determined per unit area. This impact is multiplied by the area on which each activity occurs. The carbon stock change would be less verifiable than it would be under a land-based approach. Adjustments may also be made to reflect policy decisions by the Parties. Aggregate emissions or removals are calculated by summing across applicable activities. A given area of land could be counted more than once if it is subject to multiple activities. This extra counting could result in inaccurate accounting if the effects of activities are not additive.

Alternatively, the Parties could decide that each land unit could contain no more than one activity. In this case, the combined impact of multiple practices applied in the same area would be considered a single activity. The activity-based accounting starts when the activity starts and ends when the activity ends. Stock changes prior to the start of the activity may or may not be accounted, even if they occur in a commitment period. Another option is to use a combination of land- and activity-based approaches, depending on the activity involved. For example, an activity that was reported using activity-based methods could be verified by land-based methods.

Only emissions and removals of carbon dioxide can be measured as changes in carbon stocks. Methane and nitrous oxide cannot be measured this way. Methane and nitrous oxide emissions from many land-use activities are included in Annex A of the Kyoto Protocol and the IPCC Guidelines; therefore, they will be captured in national inventories. However, emissions of these gases related to forestry activities, which are not included in Annex A, may not be captured. If these effects are not considered, the full impact of forestry activities may not be reflected in the Kyoto accounting system. Similarly, accurate assessment of the climate benefits of projects in non-Annex I countries would require accounting for changes in methane and nitrous oxide emissions.

Relevant carbon pools are pools that are affected by activities and projects. Relevant pools include aboveground biomass, litter and woody debris, below-ground biomass, soil carbon, and harvested materials. The degree of impact may vary significantly between activities and types of projects, however. Although methods exist to measure all carbon pools, monitoring is far from routine and costs vary significantly. To reduce monitoring costs, the Parties may decide that selective accounting of carbon pools, consistent with the objectives of the UNFCCC, could be used by including all pools that are expected to have reduced carbon stocks while neglecting selected pools that are expected, with a sufficient level of certainty, to have increased carbon stocks. The verifiability requirements in the Protocol mean that only carbon pools that can be measured and monitored could be claimed as a credit.

Changes in carbon stocks in the wood products pool could be accounted as part of the activity that is the source of the wood products or as an independent wood products management activity. If management of wood products is treated as an additional activity under Article 3.4, it may be necessary to exclude wood products from accounting in other Article 3.3 or 3.4 activities to avoid double-counting. Once products are in trade, their source will seldom be traceable. If changes in wood product pools are not accounted and it is assumed that the carbon is lost immediately, the system will not account for a potentially important pool.

Measurement uncertainty could be addressed by adjusting estimated stock changes. A precautionary approach would adjust the estimate in the direction that is likely to understate increases and overstate decreases in stocks. This approach would allow individual Parties to make tradeoffs between increased monitoring costs and the potential to increase carbon credits or reduce debits. If "good practice" methodologies for measuring the impact of LULUCF activities are developed and adopted, it could be decided that estimates made in conformity with this methodology would not be subject to an uncertainty adjustment.

Enhancement of carbon stocks resulting from LULUCF activities is potentially reversible through human activities, disturbances, or environmental change. This reversibility can be accounted for by ensuring that any credit for enhanced carbon stocks would have to be balanced by accounting for subsequent reductions in those carbon stocks, regardless of the cause.

Contiguous commitment periods would avoid incentives to concentrate activities that reduce carbon stocks in time periods that were not covered.

Policies by governments or other institutions (e.g., land tenure reform, tax incentives, timber and agricultural subsidies) provide conditions and incentives for the implementation of LULUCF activities. The ability to measure the impact of these conditions and incentives will depend, in part, on the carbon inventory and monitoring system in each country. It may be very difficult, however, for countries to assess the relative contribution of such policies or institutions compared to other human and natural factors that drive changes in carbon stocks.

Natural variability and the indirect effects of human activity—such as carbon dioxide fertilization, nutrient deposition, or climate-induced changes in disturbance regimes—could significantly affect carbon stocks during a commitment period. If credits and debits were available for natural variability and indirect effects, the accounting implications for the Kyoto Protocol could be significant.

The Kyoto Protocol specifies that accounting under Article 3.3 be restricted to certain "direct human-induced" activities occurring since 1990. Therefore, the Parties may wish to consider whether to count the full carbon stock change on lands under Article 3.3 or only the part of the stock change that can be attributed to direct human-induced activity. For activities that involve land-use changes (e.g., from grassland to forest), distinguishing with present scientific tools the portion of the observed stock change that is directly human induced from the portion that is caused by indirect and natural factors is very difficult, if not impossible.

For activities for which only management changes (e.g., conservation tillage) are involved and land use remains the same, it may be feasible to largely "factor out" natural variability and indirect effects. One approach may be to subtract the stock changes on control (comparison) plots where there have been no changes in management practice from changes measured on plots with modified management activities. In most cases, experimental manipulation or paired plots can be used for this purpose, although they are likely to be expensive to apply over large areas. Ecosystem models can also be used but may need further improvement to decrease uncertainties.

In national accounting, there are many options for baselines that could be applied under Article 3.4. These potential baselines include the stock change that would have resulted from "business-as-usual" activities; the stock change that would have resulted from the continuation of 1990 activity levels; the stock change that would result in the absence of active management; performance benchmarks or standard management practice; and the rate of change of stocks in 1990. One difficulty with the use of counterfactual baselines is verification.

Another option for implementing Article 3.4 is a project-based approach that uses a baseline to limit credits to carbon benefits that are "additional" to what would have otherwise occurred. Article 3.4 refers to additional activities that are not project-based activities; unlike in Articles 6 and 12, there may be no requirement that these activities are "additional" to what would have occurred against a business-as-usual baseline. The use of a project-based approach to Article 3.4, as opposed to a national approach, would increase the potential for leakage of project benefits.

If different accounting rules are adopted for Article 3.1, 3.3, or 3.4 activities, additional decision rules may be needed to determine which accounting rule applies to land that over time is subject to multiple types of activities. For example, one set of accounting rules could be given primacy in cases in which more than one set of rules could apply and double-counting might result.

The spatial resolution of samples taken in monitoring has important implications. If the size of sampled land areas is small, monitoring becomes very demanding and costly. Hence, there are clear tradeoffs between an accurate assessment of changes in carbon stocks and cost. If the sampling resolution is set at a coarse scale, data demands can be modest, but significant areas subject to an activity may be lost in the averaging process. For example, if canopy cover is assessed over land areas of 100 ha, deforestation of smaller areas (i.e., several hectares) within a unit will not take the canopy cover of the unit below the forest definition threshold, and changes in carbon stocks will not be counted. Likewise, afforestation or reforestation of smaller areas will not be counted.

Leakage is the failure to capture greenhouse gas (GHG) changes outside the accounting system that result from GHG changes within the system. For instance, forest protection measures may be used to sequester or reduce the emission of carbon from lands within the accounting system. Market forces may simply shift the extractive activity elsewhere, however. If emissions from the diverted activity fall outside the accounting system, system measures will fail to give an accurate assessment of the true aggregate changes. As the spatial and temporal scale of accounting increases, leakage errors generally become less significant (i.e., there is less activity that falls outside the system). Leakage is of particular concern in project-level accounting, but it can also occur with activities under Articles 3.3 and 3.4. Leakage (from Annex I to non-Annex I) could also be an issue in other sectors, but it is not assessed in this report.

Methods

Methods are available for measuring carbon stocks and fluxes in all components of terrestrial ecosystems. These methods vary in accuracy, precision, cost, and scale of application.

Carbon accounting requires knowledge of the net change in carbon stocks over the land surface for the period 2008–2012. These changes can be measured with high precision (without distinguishing cause or attribution). The number of samples required might make such measurements expensive, but the methods exist. Most uncertain is the task of attributing measured changes in terrestrial carbon storage to direct human-induced activities. Experimental manipulation and paired plots can be used to distinguish direct human-induced effects on carbon stock changes, although these methods will be expensive to apply over large areas. Ecosystem models can also be used but need further improvement to decrease uncertainties.

The technical capacity required by Annex 1 Parties to measure, monitor, and verify carbon stock changes under the Kyoto Protocol will be significantly affected by decisions of the Parties regarding the definitions of key terms. This capacity will also be affected by decisions on additional activities that may be included under Article 3.4, as well as whether additional activities are defined broadly or narrowly.

Methods for measuring changes in forest carbon stocks in aboveground biomass over a 5-year commitment period are sensitive enough to serve the requirements of the Protocol. Sensitive methods for measuring below-ground carbon stocks also exist, but changes in soil carbon stocks over the 5-year commitment period may be difficult to measure in some soils because they may be small compared to background levels, which may have high spatial variability. This problem can be addressed by taking a large number of samples within a single commitment period or by measuring and accounting over multiple commitment periods. Methods that further improve soil and vegetation carbon stock measurements will require research and development; these methods are likely to be highly transferable between Parties.

Direct flux measurements integrate changes over all carbon stocks, including below-ground carbon, for areas of about 100 ha. These measurements can assist in the determination of the carbon budgets of forests and other terrestrial vegetation at a project level, in the validation of forest growth models, and in verifying studies of carbon balances determined by stock-change methods.

Regional and continental estimates of net carbon exchanges based on CO_2 concentration measurements and modeling techniques (such as inverse analysis) are not yet suitable for use in implementing the Protocol.

The cost of conducting biospheric carbon inventories is only weakly dependent on the size of the area inventoried; it is more dependent on the range of ecological conditions within the area. This is because the spatial scale over which both soil and biomass carbon varies is quite small: a few tens to hundreds of meters. The sample size needed to achieve the desired precision is similar for a small country or a distinct region within a large country. Both require approximately the same analytical equipment and statistical treatment. Thus, for a given level of precision, the cost to a country will depend more on the range of different biogeophysical regions that exist within its borders than on the country's size.

Independent methods also exist for the verification of measured changes in carbon stocks. For example, a range of existing flux measurement techniques may have application in verifying direct inventory measurements. Furthermore, methods applicable to the Kyoto Protocol are likely to improve and become less costly with time.

Sustainable Development

Article 2 of the Kyoto Protocol calls on Annex I Parties to promote sustainable development in the course of reducing GHG emissions, including the implementation of policies and measures undertaken in LULUCF. Article 12 explicitly identifies the achievement of sustainable development, along with the reduction of GHGs, as a central purpose and requirement of project activities undertaken in non-Annex I countries through the Clean Development Mechanism (CDM).

To implement the Protocol, the Parties will likely face a series of decisions and tradeoffs on a broad range of environmental and socioeconomic impacts on biodiversity; the quantity and quality of forest, grazing land, soil, and water resources; the ability to provide food, fiber, fuel, and shelter; and employment, poverty, and equity.

Although generalizing the impacts of LULUCF activities and projects on sustainable development across activities, locations, and time is difficult, the Parties may wish to consider a variety of broad environmental and socioeconomic synergies and tradeoffs:

- Converting non-forest land to forest will typically increase the diversity of flora and fauna, except in situations where biologically diverse non-forest ecosystems, such as native grasslands, are replaced by forests consisting of single or a few species. These negative impacts can be mitigated, however, by measures that lengthen rotations, maintain understory vegetation, use native tree species, and minimize chemical inputs. Avoiding deforestation can provide potentially large co-benefits, including protection of biodiversity, water, and soil resources and maintenance of non-timber forest products and services.
- Converting land from agricultural use to forest or avoiding the conversion of forests to agriculture on the margin enhances the socioeconomic role of the forest sector and diminishes the role of the agricultural sector. Where agricultural land is scarce (i.e., through high population growth and/or expanding cultivation), the economic and social impacts could be large unless forest expansion/conservation is coupled with agricultural intensification and related measures.
- Changing the management of forests to store more carbon (or release less carbon) will alter the size and structure of forests. More intensive forms of management that increase tree biomass may make more timber available for eventual harvest, but such management may also impart some negative consequences on other ecosystem components. Alternatively, if current commercial management practices are modified to leave more biomass in the forest for longer periods of time (e.g., extended rotation lengths, reduced-impact logging), other ecosystem functions may benefit while timber production is curtailed.
- Managing agricultural soils to sequester more carbon is likely to reduce soil erosion and increase soil fertility. Thus, the opportunity costs of these activities may be relatively low if they enhance rather than diminish food security for affected populations.

In implementing Articles 2 and 12 of the Kyoto Protocol, the Parties may decide whether to establish international or national criteria to ensure that LULUCF activities and projects are consistent with sustainable development objectives. With regard to activities within countries or projects between countries, if criteria vary significantly across countries or regions, there may be incentives to locate activities and projects in areas with less stringent environmental or socioeconomic criteria.

A system of criteria and indicators could be used to assess and compare sustainable development impacts across LULUCF alternatives. No agreed-upon set of indicators currently exists, however. Parties could draw from a variety of systems under development for closely related purposes, such as criteria and indicators for sustainable forest management or the United Nations Commission on Sustainable Development, to develop an assessment tool for LULUCF measures under the Protocol.

Several sustainable development objectives are the focus of other multilateral environmental agreements, including the UN Convention on Biological Diversity, the UN Convention to Combat Desertification, and the Ramsar Convention on Wetlands. Parties may need to decide whether and how to ensure that LULUCF definitions, activities, and projects are implemented in a manner that is consistent with the goals and objectives of these Conventions and other relevant international agreements.

There are a variety of ways to assess the sustainable development impacts of LULUCF projects, depending on who has responsibility for performing the assessment, what standards are used, and how the standards are developed. Assessment standards can be based on local, regional, sectoral, national, or international criteria.

More formal approaches to sustainable development assessment that could be applied at the project level include environmental and socioeconomic impact assessments, ISO 14000, and forest certification. These methods have been applied across a wide range of activities and countries, and they could be modified so that they are applicable to LULUCF projects. Their use could raise project costs and slow implementation, however.

Key logistical factors affecting the sustainable development contributions of LULUCF activities and projects include institutional and technical capacity to develop and implement guidelines and procedures; the extent and effectiveness of local community participation in development and implementation; and technology transfer and adaptation.

2.1. Introduction

Chapter 2 of this Special Report describes the components of the global carbon cycles and the issues to be considered in accounting for carbon fluxes to and from the atmosphere.

The IPCC Guidelines (IPCC, 1997) describe a methodology for a comprehensive approach to measuring "anthropogenic emissions by sources and removals by sinks of greenhouse gases" that is feasible to implement in most nations. These guidelines include suggested definitions for many important terms and can form a starting point for reporting compliance with the Kyoto Protocol. Article 5.2 and a decision at COP3 (2/CP.3) states that the "[m]ethodologies for estimating anthropogenic emissions by sources and removals by sinks of all greenhouse gases not controlled by the Montreal Protocol shall be those" in the IPCC Guidelines. A variety of ill-defined terms and ambiguities may cause problems, however, in putting the Protocol into operation. There are also other sets of definitions in use nationally and internationally, such as those suggested by the UN Food and Agriculture Organization (FAO) for forestry applications and land use (FAO, 1999). Article 5.2 states that where the IPCC Guidelines are not used, "appropriate adjustments shall be applied according to methodologies agreed upon by the" first COP.

This chapter outlines the generic issues associated with specific definitions and methodologies that Parties might wish to consider in agreeing on methodologies in accordance with Articles 5.2 and 7.4. The broad picture in this chapter is elaborated in subsequent chapters.

A central issue in identifying lands that may fall under Articles 3.3 and 3.4 is the definition of ARD, which in turn is related to the definition of a forest. Section 2.2 deals with this issue, and Chapter 3 elaborates on it, analyzing the implication of a series of definitional scenarios. Some activities that take place in forested land but do not fall within Article 3.3 may be eligible for inclusion under Article 3.4. These options are further discussed in Chapter 4.

Afforestation, reforestation, and deforestation are usually defined as activities that change a piece of land between a forested and non-forested state. The IPCC Guidelines do not provide an explicit definition of a forest. They refer to the FAO (1993a) usage for the tropics as example categories, but they note that "[n]ational experts are free, indeed encouraged, to use more detailed characterizations of ecosystems in their countries." In this section, we first discuss definitions of a forest because the precise definition used can have a significant effect on actions that are classified as ARD and thus on measured emissions and sinks of GHGs.

We then discuss alternative definitions of ARD and their relationship with land use and land-use change. The precise definitions affect which lands are lands under Article 3.3 activities and which may be considered under Article 3.4. The quantitative implications of different definitional scenarios are described in detail in Chapter 3; options for additional activities under Article 3.4 are described in Chapter 4.

Alternative definitions abound for some of the fundamental concepts relevant to the land-use change and forestry (LUCF) sector. For example, a recent survey (Lund, 1999) listed more than 200 definitions of "forest," 50 definitions of "a tree," and another 50 definitions of "reforestation." This Special Report discusses alternative definitions and their merits, demerits, and implications in a way that is not policy prescriptive. Considering all of the alternative definitions and their interactions is impossible, however, because the number of combinations quickly increases beyond practical limits. In selecting definitions for discussion, we have considered their utility in estimating carbon stocks and changes in stock in a manner that is relevant to the Kyoto Protocol.

2.2. Definitions of Terms Used in the Convention and Protocol

2.2.1. Land Use, Land-Use Change, and Forestry

The phrase "land-use change and forestry" was used in discussions leading up to the Protocol as the name for a "sector/source category." It is also used as the title of a chapter in the IPCC Guidelines. The Guidelines refer to sources and sinks associated with:

> "...greenhouse gas emissions from human activities which
> 1) Change the way land is used (e.g., clearing of forests for agricultural use, including open burning of cleared biomass), or
> 2) Affect the amount of biomass in existing biomass stocks (e.g., forests, village trees, woody savannas, etc.)."

The Guidelines conclude that:

> "the most important land-use changes that result in CO_2 emissions and removals are
> - Changes in forest and other woody biomass stocks—The most important effects of human interactions with existing forests are considered in a single broad category, which includes commercial management, harvest of industrial roundwood (logs) and fuelwood, production and use of wood commodities, and establishment and operation of forest plantations, as well as planting of trees in urban, village, and other non-forest locations.
> - Forest and grassland conversion—The conversion of forests and grasslands to pasture, cropland, or other managed uses can significantly change carbon stored in vegetation and soil.
> - Abandonment of croplands, pastures, plantation forests, or other managed lands that regrow into their prior natural grassland or forest conditions.
> - Changes in soil carbon."

Some confusion arises from the use of the term "land-use change" in the IPCC Guidelines because many of the stock changes (especially in forest pools) are not associated with a change in land use. For example, "commercial management [or] harvest of industrial roundwood (logs) and fuelwood" is a land use but not a land-use change. Similarly, changes in cropping practice, such as minimum tillage that reduces soil disturbances and consequential carbon loss, are described in the IPCC Guidelines even though they are not associated with changes in land use.

The Protocol uses the phrase "land-use change and forestry" on three occasions: in Articles 3.3, 3.4, and 3.7. (Article 12 does not explicitly mention this phrase, but we make no presumption about decisions by the Parties in relation to this Article.)

Article 3.3 refers to "greenhouse gas emissions by sources and removals by sinks resulting from direct human-induced land-use change and forestry activities, limited to afforestation, reforestation, and deforestation since 1990." Decision 9/CP.4 at COP4 clarified that the domain of the clause is restricted to the specified forestry activities: "The adjustment to a Party's assigned amount shall be equal to verifiable changes in carbon stocks during the period 2008 to 2012 resulting from direct human-induced activities of afforestation, reforestation, and deforestation since 1 January 1990." There is no consensus on definitions of ARD activities in the technical literature; some of the more important issues associated with their definition are discussed below and elaborated in Chapter 3.

Article 3.4 refers to "additional human-induced activities related to changes in greenhouse gas emissions by sources and removals by sinks in the agricultural soils and the land-use change and forestry categories." This clause is less restrictive; it requires definitions of agricultural soils, land use (to define land-use change), and forestry. These terms are also discussed below and again in more detail in Chapter 4.

The phrase "land-use change and forestry" appears again in Article 3.7 in the context of whether this sector was a net source of GHG emissions in 1990. Later in the clause, the phrase "land-use change" is used without mention of "forestry."

The title of this Special Report incorporates the term "land use" as well as "land-use change" [COP4 decision 9/CP.4 and Subsidiary Body for Scientific and Technological Advice (SBSTA8)]. This chapter and subsequent chapters deal with situations in which significant changes in carbon stocks might occur, regardless of whether they are associated with changes in land use. Wherever possible, however, we attempt to identify whether an activity is associated with an actual land-use change.

2.2.1.1. *Land Use*

The terms "land cover" and "land use" are often confused. *Land cover* is "the observed physical and biological cover of the earth's land, as vegetation or man-made features." In contrast, *land use* is "the total of arrangements, activities, and inputs that people undertake in a certain land cover type" (FAO, 1997a; FAO/UNEP, 1999). National categories of land use differ, but many have been harmonized under the influence of FAO's periodical World Census of Agriculture (Table 2-1). Categories of land cover/use systems are used in Chapter 4 (Section 4.4) to illustrate the expected potential for carbon sequestration from a change in system management (e.g., intensification or extensification) or upon conversion from one category to another.

2.2.1.2. *Forestry*

A common dictionary and glossary definition of *forestry* is "the science, art, and practice of managing and using for human benefit the natural resources that occur on and in association with forest lands." Thus *forestry* is a broader term than *silviculture*, which refers more specifically to the planting and tending of growing trees. For the purposes of this Special Report, forestry is taken to include a wide range of activities in addition to those associated with silviculture, including the production of non-timber products, watershed management, wildlife protection, and eco-tourism and extending to activities such as pest control and fire management. The question remains, however: What are forest lands?

Although *forest land* has many national definitions, many of these definitions have the following in common: "All lands bearing vegetative associations dominated by trees of any size, exploited or not, capable of producing wood or other forest products, of exerting an influence on climate or on water regime, or providing shelter for livestock or wildlife. Includes lands which have been clear-cut or burned but which will be reforested in the foreseeable future. Excludes orchards, shelter belts, groups of trees along roads or city parks" (Encyclopaedia Britannica, 1970). Woodlands and treed wetlands that fit the minimum definitions of forest are usually included.

These definitions, however, have been developed for a variety of purposes that rarely include aspects of measuring carbon stocks and changes in stocks as required by the Protocol. Thus, Section 2.2.2 deals in more detail with the definitions of a forest and associated terms in the context of the Protocol.

2.2.2. *Definitions of Forests*

Forests are usually composed of many individual stands in different stages of development and with different characteristics. Thus, a forest can include a range of different forested ecosystems composed of different species and different ages and having different carbon stock densities (t C ha^{-1}). In this Special Report, the term *forest* refers to the whole forest (as a landscape), including its forest stands (its component units). The term *stand* is used where necessary for clarity.

Many definitions of forest are in use throughout the world, reflecting wide differences in biogeophysical conditions, social structures, and economies. Lund (1999) listed about 240 such definitions. Most countries have developed very specific definitions that are suitable for their own administrative purposes and reflect their forests' ecological conditions. In seeking a set of definitions for the Protocol, it must be recognized that many national agencies have invested significantly in databases that are founded on their own definitions. Many agencies will be unwilling or unable to reformulate that information if arbitrary definitions devised to implement the Protocol are too dissimilar from their own.

2.2.2.1. Types of Forest Definitions

There are three broad categories of forest definition: administrative, land use, and land cover.

***Table 2-1**: Land-use categories recognized in FAO's World Census of Agriculture (FAO, 1986, 1995a; FAO/UNEP, 1999).*

In Sequence of Increasing Intensity of Use	Chapter 4 Equivalents
(a) *Deserts* (barren land and waste land)	–
(b) *Non-Forest Wooded Lands* (scrubland; may include national parks and wilderness recreational areas)	–
(c) *Wetlands, Non-Forest* (marshes)	Wetlands
(d) *Land under Forest* (natural forests and most non-managed woodlands)	Forest Land
(e) *Land under Forestry/Silviculture*	Forest Land
(f) *Land under Shifting Cultivation* (temporarily abandoned land that is not part of a holding)	Agroforestry Land
(g) *Land under Agroforestry* (permanent use of land at holding level, but with mixed crop growing, animal herding, and tree utilization)	Agroforestry Land
(h) *Land with Temporary Fallow* (resting for a period of time, less than 5 years, before it is planted again with annual crops)	Cropland
(i) *Land under Permanent Meadows and Pastures* [used for herbaceous forage crops that are either managed/cultivated (pastures) or growing wild (grazing land); trees and shrubs may be present or grown purposely, but foraging is the most important use of the area; grazed woodlands]	Rangeland/Grasslands
(j) *Land under Temporary Meadows and Pastures* (cultivated temporarily, for less than 5 years, for herbaceous forage crops, mowing, or pasturing, in alternation with arable cropping)	Rangeland/Grasslands
(k) *Land under Permanent Crops* (perennials; cultivated with long-term crops that do not have to be replanted for several years after each harvest; harvested components are not timber but fruits, latex, and other products that do not significantly harm the growth of the planted trees or shrubs: orchards, vineyards, rubber and oil palm plantations, coffee, tea, sisal, etc.)	Agroforestry Land
(l) *Land under Temporary Crops* (annuals; cultivated with crops with a growing cycle of under 1 year, which must be newly sown or planted for further production after harvesting; not only small grain crops such as beets, wheat, and soy bean but also bi-annuals that are destroyed at harvesting, such as cassava, yams, and sugarcane; bananas are transitional to the permanent crops category)	Cropland
(m) *Land under Temporary Crops Requiring Wetland Conditions* [wet-foot crops such as irrigated rice and jute (dry-foot crops with intermittent irrigation included in other categories)]	Wetlands
(n) *Land under Protective Cover* (greenhouses and other urban or peri-urban intensive use, formal or informal; vegetable growing, home gardening, residential parks, golf courses, etc.)	Peri-Urban Land
(o) *Land under Residential/Industrial/Transportation Facilities*	Peri-Urban Land

Administrative. Forests have been defined in terms of legal or administrative requirements. Typical examples follow: "Any lands falling within the jurisdiction of the Department of XYZ" or "any lands so mapped in the ordinance survey of XYZ." These definitions bear no relationship to the vegetation characteristics and associated carbon on that land. For this reason, definitions of deforestation and reforestation that are based on administrative or legal definitions of forests may not provide as full a picture of changes in carbon stocks as some of the alternatives discussed below. Nevertheless, many national statistics are based on such definitions. This situation may lead to a confounding of information on forests with that for other vegetation types.

Land use. Another set of definitions defines forests in terms of land use. A typical example follows: "An area managed for the production of timber and other forest products or maintained as woody vegetation for such indirect benefits as protection of catchment areas or recreation." Some definitions incorporate an element of potential or even desirable land use, such as that in the Swedish Forest Act of 1994: "For the purposes of this Act, forest land is defined as: (i) land which is suitable for wood production, and not used to a significant extent for other purposes; and (ii) land where tree cover is desirable in order to protect against sand or soil erosion, or to prevent a lowering of the tree line. Land that is wholly or partially unused shall not be regarded as forest land if, due to special conditions, it is not desirable that this land be used for wood production." Many definitions explicitly include areas that are temporarily not covered by large trees (e.g., stands of seedlings that are regrowing after clear-felling or disturbance) or small, included, non-treed areas (e.g., roads and other infrastructure). Again, these definitions may fail to reflect land cover. Treed areas used for purposes other than the forest land-use definition, such as grazed woodlands, may be excluded—along with their substantial carbon pools.

Finding a universally applicable definition of forest based on land use will be difficult. Such definitions may be related simply to the intent of management; they may require at least a minimum amount of management, or they may require frequent and intensive management.

Land cover. The third category defines a forest in terms of vegetative land cover. An example follows: "An ecosystem characterized by more or less dense and extensive tree cover." Typically, the cover is assessed as percent crown cover. Distinctions may be made between open- and closed-canopy forests (FAO, 1999). Other variants include the use of basal area, wood volume, proportion of land with trees above a minimum height, or proportion of land with tree biomass exceeding a minimum threshold, with no distinction made between single-stem or multi-stem tree forms. Different elements may be combined in the definition of forest. For example, the definitional scenarios in Section 3.2 use three of the foregoing indicators (minimum canopy cover, minimum height, and minimum biomass).

A variety of issues arise in relation to the Protocol for each of the categories of forest definitions because the key activities of ARD generally involve transitions between forest and non-forest. Forest definitions that are based solely on land-cover attributes (e.g., canopy projected cover, CPC)[1] may exclude stands of young trees that are regenerating after disturbance (e.g., harvest, fire, insect outbreak, wind-throw). Some existing definitions of forest explicitly deal with this regeneration phase with clauses that recognize it to be a temporary condition. An FAO definition includes the following clause: "Young natural stands and all plantations established for forestry purposes which have yet to reach a crown density of 10 percent or tree height of 5 m are included under forest, as are areas normally forming part of the forest area which are temporarily unstocked as a result of human intervention or natural causes but which are expected to revert to forest." (FAO/UNEP, 1999).

Chapter 3 examines the implications of using different components of the foregoing categories of forest definitions in combination with contingent definitions of ARD, using a set of definitional scenarios (Section 3.2, Table 3-1).

Minimum CPC is a common quantitative element in land cover-based definitions of forest. This indicator distinguishes between dense forests (closed canopy), in which virtually the entire land surface is covered by tree canopies (approaching 100 percent CPC), and woodlands (open canopy)—in which the crowns of scattered trees or groups of trees may cover only a few percent of the land surface. In definitions collected by Lund (1999), the minimum CPC to be included as a forest varied from 10 to 70 percent.

Globally, about 50 percent of wooded land has a canopy cover of less than 20 percent. This figure varies nationally, however, from 10 to about 70 percent. The amount varies greatly from region to region and country to country (Figure 2-1).

Minimum forest stand dimensions are usually included within forestry definitions to keep the task of monitoring forested areas feasible. For the purposes of forestry operations, the limit is often set as low as 0.5 ha (sometimes 0.01 ha), with minimum width of only 10 m. Although such resolution may be required at the scale of forestry operations, it creates practical difficulties in monitoring extensive areas for changes (such as those associated with ARD activities). The cost of monitoring rises sharply with increasing resolution. Thus, in practice, monitoring and reporting agencies will be constrained by the cost of measurement programs and by available resources.

[1]Canopy (or crown) projected cover is the proportion of ground area covered by tree crowns. CPC is estimated by vertically projecting outlines of tree crowns onto the horizontal plane that represents the forest stand area. It is also called "crown cover," "canopy closure," "cover closure," or similar terms such as "polnota." Another term—foliage projected cover (FPC)—is used occasionally. FPC is the proportion of a forest stand covered by foliage, which is usually a little less than CPC. FPC should not be confused with leaf area index (LAI), which is the ratio of the one-sided projection of the area of leaves (or needles) to the ground area—essentially, the average number of leaves above a point in the forest stand.

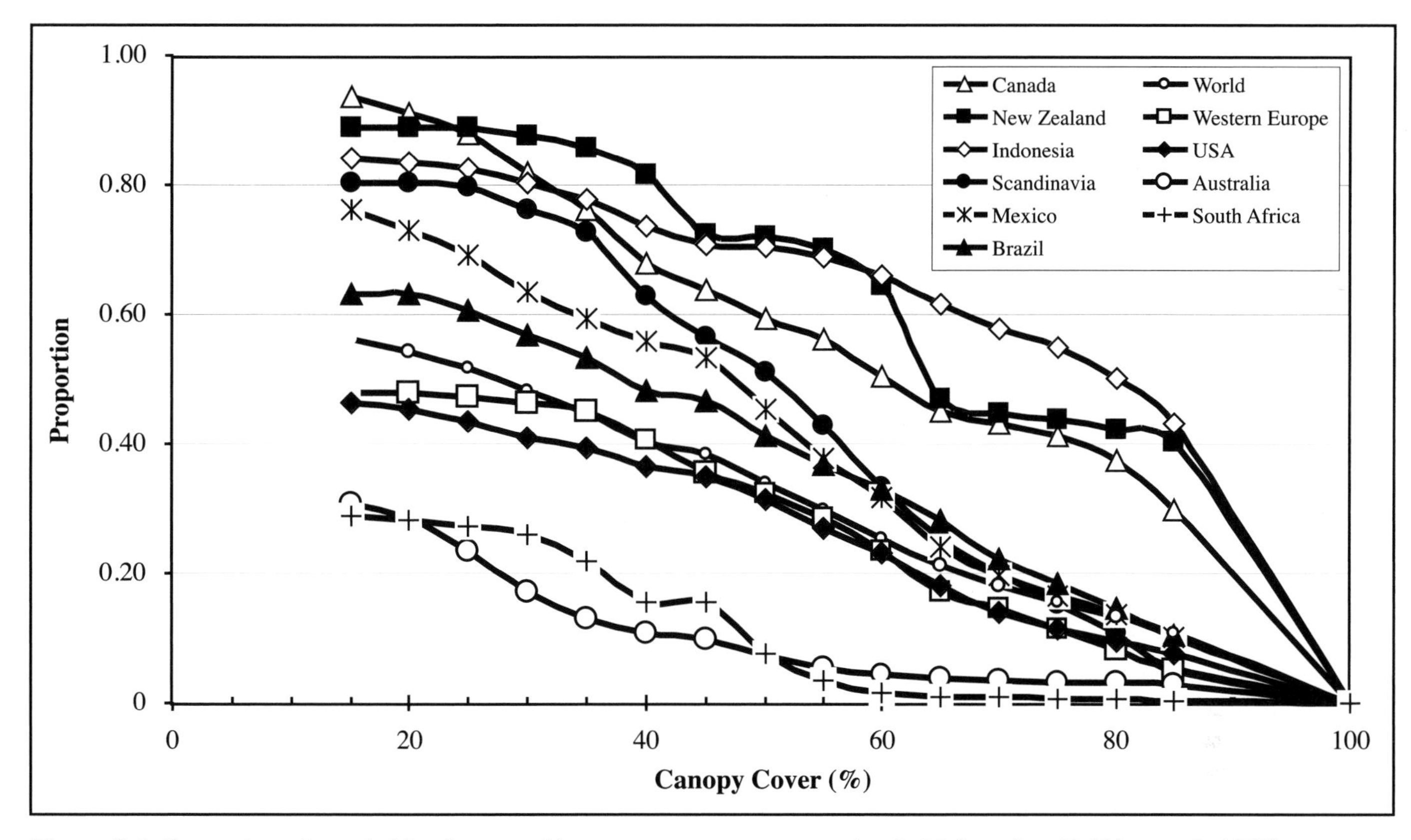

Figure 2-1: Proportion of wooded land captured by a percent canopy cover threshold (based on DeFries *et al.*, 1995).

If the upper limit of resolution is set too high, however, significant areas of treed land may be excluded from monitoring or reporting as forest. Similarly, areas of non-treed land might be reported as forest. Landscape units that consist of a mixture of forest and non-forest patches may be assigned a cover type that represents an average value for the area. If the spatial unit is very large, only very large scale activities will cause the average value to move below or above a definitional threshold of forest. Thus, smaller scale activities would be neither detected nor reported.

Ideally, the resolution should be compatible with the scale of human activities (e.g., clearing, planting, infrastructure). For example, consider an assessment over an area of 1 km^2 of dense forest using a forest definition incorporating a minimum CPC of 30 percent. Numerous human-induced openings on the scale of a few hectares could take place within this region without resulting in a decrease of the average CPC over the assessment area below the 30-percent definitional threshold; hence, these openings would not be recorded as deforestation. Similarly, forestation activities in a sparsely treed region may not be detected or recorded as either reforestation or afforestation.

The maximum spatial unit chosen will involve a tradeoff between practicality (cost) and the ability to identify areas where actual changes have taken place. The larger the spatial unit employed, the lower the proportion of ARD activities detected and recorded. Very large spatial units (coarse resolution) may result in the detection and recording of lower ARD activity rates. We note, however, that the detected change in cover resulting from activities may be captured under Article 3.4 (e.g., associated with aggradation or degradation, respectively)—a tradeoff that the Parties may wish to keep in mind. Chapter 3 analyzes in one of its definitional scenarios (see Table 3-1) an approach that captures forest degradation/aggradation activities within modified definitions of deforestation/reforestation.

The use of productivity, or potential timber-volume production, in the definition of a forest is relatively new in most countries. Such an approach could avoid the need for explicit rules relating to maximum height and canopy cover, which to a large extent are surrogates for timber volume and carbon content. From the perspective of the atmosphere, however, Net Biome Productivity (NBP) (not simply timber productivity) is a relevant measure (see Chapter 1). From the Kyoto perspective (which focuses on change caused by direct human-induced land-use change and forestry activities), the cascading changes in carbon in all of the affected components (living biomass, dead organic matter, and organic material transported off-site) would need to be included in the productivity estimate if it is to reflect an accurate appraisal of carbon gains and losses. We return to this point in Sections 2.2.5 and 2.3 and in Chapters 3 and 4.

2.2.3. *Afforestation, Reforestation, and Deforestation*

2.2.3.1. *Afforestation*

Afforestation and *reforestation* both refer to establishment of trees on non-treed land. Reforestation refers to establishment of forest on land that had recent tree cover, whereas afforestation refers to land that has been without forest for much longer. A variety of definitions differentiate between these two processes. Some definitions of afforestation are based on phrases such as "has not supported forest in historical time;" others refer to a specific period of years and some make reference to other processes, such as "under current climate conditions." The IPCC Guidelines define afforestation as the "planting of new forests on lands which, historically, have not contained forests."

Some definitions emphasize a change in land-cover or land-use designation—for example, "The establishment of a forest or stand in an area where the preceding vegetation or land use was not forest" (Helms, 1998)—although this definition could equally fit many definitions of reforestation (see below).

If afforestation and reforestation are defined in similar terms—distinguished only by the period of time that the land was without forest—then the actual time cut-off between them does not matter. Article 3.3 deals with afforestation and reforestation activities in exactly the same way. Other issues, however, affect the application of these terms. These issues relate to the sequence of human activities prior to, and the actual mode of establishment of, new trees. These issues are discussed in Section 2.2.3.2. The important point, however, is that if both actions are treated equivalently in terms of accounting, their precise distinction is not important for the implementation of the Protocol.

2.2.3.2. *Reforestation*

Reforestation refers to "the establishment of trees on land that has been cleared of forest within the relatively recent past." Operational forestry definitions of reforestation often include the establishment of trees on land that has just been harvested. With such a definition, areas subject to conventional forest harvesting and regeneration cycles would come within the domain of Article 3.3 of the Protocol at their first post-1990 harvest and regeneration cycle. This definition could have major implications for the size of the land base included as lands under Article 3.3 (see Chapter 3).

In some definitions, reforestation is the conversion of land use back to forest after a period of some other land use. The IPCC Guidelines, which were developed explicitly for carbon inventory, used such a definition of reforestation: "the planting of forests on lands which have, historically, previously contained forests but which have been converted to some other use."

Most of the definitions for reforestation collected by Lund (1999) arise from this forestry perspective and appear to include regeneration after harvesting. Some are very explicit: "Reforestation areas are temporarily unstocked areas caused by harvesting, wind breaks, natural disasters and so on. These areas have to be reforested artificially (usually within 3 years, under certain circumstances within maximal 8 years) or with methods of natural regeneration (usually within 8 years, under certain circumstances within maximal 11 years). In Austria, reforestation has always been recognized as a part of forest management and has never been linked to land-use change."

Such definitions of reforestation do not involve transition from a non-forest to forest state (unless, for example, a strict canopy-cover definition of forest is used). Instead, these definitions are activity-based; they do not reflect a change of land-use designation. This factor raises several options for combinations of definitions based on land use, land cover, and activities that are considered in the definitional scenarios of Chapter 3.

One must consider the different types of reestablishment methods. Decision 9/CP.4 includes the phrase "direct human-induced activities of afforestation, reforestation, and deforestation." Some definitions of reforestation restrict the activity to the planting of trees (e.g., the IPCC definition) or the "artificial establishment" of trees (see Chapter 3), which could be taken to exclude methods based on more natural regeneration. Most forest regeneration is carried out through "natural" reestablishment from seed remaining on the site or from retained seed-trees. Sometimes the seed-bed is prepared by mechanical or chemical treatment. Only a small percentage of forest area is reestablished by direct planting. If the definition is broadly encompassing with regard to regeneration methods, other questions still arise. For example, should land-use practices that lead to an increase in woody cover (e.g., invasion by woody weeds) be counted as reforestation? Chapter 3 deals with some aspects of this issue; Chapter 4 (Section 4.4.3) takes up other aspects.

2.2.3.3. *Deforestation*

Most of the definitions collected by Lund (1999) characterize deforestation as the long-term or permanent removal of forest cover and conversion to a non-forested land use. For example, deforestation is the "permanent removal of forest cover and withdrawal of land from forest use, whether deliberately or circumstantially." Similarly, the IPCC Guidelines emphasize the conversion of forests (to pasture, cropland, or other managed uses): "Conversion of forests is also referred to as 'deforestation,' and it is frequently accompanied by burning."

It is important to note that a common forestry definition of reforestation is not the antonym (mirror) to the common definition of deforestation. In forestry practice, reforestation commonly refers to any act involving reestablishment of trees regardless of whether it follows harvesting or a long period of deforestation, but harvesting is not equated with deforestation. For example: "Clear-cutting (even with stump removal), if shortly followed by reforestation, is not deforesting" (FAO, cited in Lund, 1999).

Nevertheless, many of the issues that arise with regard to certain definitions of reforestation also apply in defining deforestation. The period of changed land use required for reestablishment of trees to be treated as reforestation must be compatible with the period of changed land use that describes deforestation. Including a waiting period in the definition of deforestation (e.g., "the act of removing forest cover and not beginning to regenerate forest within X years") introduces accounting needs that can extend beyond a specific commitment period. For example, areas cleared of forest during the first commitment period would not be confirmed as deforested for some years after the clearing, which may extend beyond 2012.

Many definitions of deforestation include natural (non-anthropogenic) events such as landslides, volcanism, and so forth. Where these events are unambiguously not direct results of human activities, the Protocol and Decision 9/CP.4 are clear: If they do not meet the "direct human-induced LULUCF activities" requirement, they would not be included as deforestation under Article 3.3.

Not all natural events are so easily partitioned, however. In particular, several difficult definitional issues arise where significant forest losses occur as a result of fire (and, in some cases, landslides). Forest fires can be the result of natural events (notably lightning) or the direct (or indirect) result of human activities, including prescribed burning (and their escape), accidental fires, and arson. Even if the cause of the fire can be unambiguously attributed to prescribed burning, arson, or escape from deliberate fires (including camp fires and cigarettes), the act may or may not be deforestation in the sense defined above. In many—but not all—instances, forest cover loss is followed by natural (or aided) regeneration of new forest. To the extent that the regrowth is complete, the net release of carbon to the atmosphere may be completely recovered over time by the new forest. There is, however, an asymmetry in the rates (Kurz *et al*., 1995): The net release of carbon to the atmosphere is rapid, whereas the recapture takes decades or more. From a scientific perspective, however, asymmetrical accounting of changes in carbon stock over time arises if only one of the paired activities (forest loss and regrowth) is accounted as deforestation or reforestation, respectively.

Given that distinguishing between natural and anthropogenic factors that influence the vulnerability of the land to disturbance will be difficult, a key question for policymakers is whether the accounting should include only direct human activities (reforestation) or both the human activities and the event (fire) that makes the opportunity for reforestation possible.

In some cases, there are definitional and carbon accounting issues concerning drawing a clear boundary between natural phenomena and human-induced activities, when, for example, significant forest losses occur as a result of fires or disturbances such as pest outbreaks. In cases involving lands under Article 3.3 or 3.4 where fires or pest outbreaks occur in a forest, a question is whether accounting should, *inter alia*: (i) count neither the loss nor subsequent uptake of carbon (which reflects the actual net change in carbon stocks on those lands and exchange of carbon with the atmosphere in the long term, but creates problems in continuing to account for the area burnt/defoliated as lands under Article 3.3 or 3.4); (ii) count both the loss and subsequent uptake of carbon (which reflects the actual net change in carbon stocks on those lands and exchange of carbon with the atmosphere, but creates an initial carbon debit for the Party concerned); (iii) count only the loss of carbon (which would overestimate the actual losses of carbon stocks, not represent the exchanges of carbon with the atmosphere, and create future accounting problems), or (iv) count only the subsequent uptake (which would fail to reflect the actual changes in carbon stock and would not represent the exchanges of carbon with the atmosphere, and would provide carbon credits for the Party concerned).

In cases involving lands that do not fall under Articles 3.3 or 3.4, where fires or pest outbreaks trigger land-use change, the consequences are similar to deforestation. If similar vegetation cover is allowed to regenerate, such disturbances may not lead to a long-term change in carbon stocks.

Will forest protection practices (such as fire management systems that seek to prevent wildfires or ameliorate the damage they cause through suppression efforts) be rewarded (in a Protocol sense)? Will Parties be penalized (under the Protocol) for failing to manage wildfires? These issues remain to be resolved through negotiation by the Parties.

A further complication is the fact that fires in many ecosystems do not result in complete tree mortality, even though there are significant changes in the carbon balance of the affected areas. Individual trees and small patches of trees and understory can survive. In some forest types (e.g., eucalypt and some pine forests), even the most intense fires do not cause significant mortality of mature trees. Significant quantities of carbon may be released, but most trunks and large branches survive and regrow full canopies within a few years. Although measurement of these changes under the stock change method as recommended by Article 3.3 is possible, in practice the areas of forest meeting the deforestation requirement of Article 3.3 would be contentious and difficult to assess.

A less difficult situation arises when land-use change takes place subsequent to or as a result of fire. In many parts of the world, major forest fires (whether ignited by humans or naturally caused) are taken as the cue for land-use change to agricultural or pastoral uses. In these cases, the combination of fire and land-use change should be accounted as deforestation in a comprehensive and symmetric accounting system.

For several reasons, accounting procedures that always treat changes in carbon stocks that are attributable to fire as debits may be inconsistent with long-term objectives of increasing carbon stocks. First, fire has been used in some ecosystems as a forest management tool for enhancing forest growth over the long term, even though it might result in some carbon reductions over the short term (for example, during a commitment period). Second, fire is a natural part of many forest ecosystems, and management to eliminate fire can be futile or counterproductive. Natural or prescribed ground fires in such ecosystems may result in preservation of carbon stocks at the landscape scale by creating firebreaks—thus reducing the likelihood of widespread, intense fires.

In the long term and over large areas, there is a balance between carbon loss resulting from fires and carbon sequestered in regrowth, providing there is no change in the fire regime. This balance underlies the recommendation in the IPCC Guidelines to treat wildfires as carbon-neutral. Although this approach is pragmatic, there is some evidence that fire regimes are changing in parts of the world (e.g., Kurz and Apps, 1999). In some cases, these changes may be a result of deliberate human manipulation; in others, it is an indirect consequence of human actions—possibly including climate change. Moreover, because of asymmetry in the rates and timing of carbon loss from fire and its re-accumulation through regrowth, the complete balance is not achieved if fire regimes vary in response to changes in El Niño patterns or rapid climate change.

2.2.3.4. Forest Degradation

Although the term forest degradation is sometimes used in forestry, existing definitions are generally inadequate to capture actions that change carbon stocks because they lack specificity. These definitions commonly refer to reductions in the productive capacity of the forest. As an example, the FAO definition states "Changes within the forest class that negatively affect the stand or site and, in particular, lower the production capacity. Thus, degradation is not reflected in the estimates of deforestation" (FAO, 1995b).

If the forest definition adopted for the Protocol includes relatively sparse forests (woodlands), considerable amounts of carbon could be removed from a dense forest while the remnant land cover remains within the strict definition of a forest. These actions would not constitute deforestation and thus need not be reported under Article 3.3. We note however, that such actions could be captured as "additional human-induced activities related to...the LUCF categories" of Article 3.4, if the Parties so decide. Furthermore, provided the accounting rules for Articles 3.3 and 3.4 are identical (see Section 2.3.2.2), whether a clearing activity is classified as deforestation (Article 3.3) or as degradation (Article 3.4) would have no influence on the Protocol's outcome. Chapter 3 discusses some definitional scenarios that may capture such activities.

2.2.3.5. Forest Aggradation

Forest management can lead to increased carbon stock on a site—the opposite of degradation. This aggradation can occur, for example, through increases in the density or average size of trees in a stand. Where the change in total carbon stocks can be attributed to human actions, Parties may decide to include such changes under Article 3.4 (see Section 4.4.5). We note, however, that attributing cause from among directly human-induced change (e.g., as a result of improved silviculture), indirect influences (e.g., nitrogen or CO_2 fertilization), and natural causes (including natural successional processes) generally will be difficult. Section 2.4 discusses some methods that assist in distinguishing the increment that results from the activity from background increase.

2.2.4. A Schemata for Forest Definitions

In the context of the foregoing discussion, any set of definitions of the critical terms *forest* and ARD clearly introduces its own set of issues and questions. It is instructive to look systematically at the options on the basis of two primary defining criteria for forest: land use and a measure of land cover or biomass (Figure 2-2). A definition of forest could require that a patch of land:

- Meet both requirements (i.e., current land use is forestry *and* land cover or biomass exceeds a threshold, here labeled the "LU+LC" option)
- Require only that either one or the other of the criteria is met ("LU or LC")
- Require only one or the other of the requirements ("LU" or "LC").

Given one of these definitions for forest, deforestation can then be broadly defined as the conversion of a forest to a non-forested state, and reforestation (and afforestation) can be defined as the conversion of a non-forested area to a forest. The scenarios in Chapter 3 analyze the implications of these definitions. The scenarios also examine the implications of treating the harvest/regeneration cycle of operational forestry as deforestation followed by reforestation or as reforestation alone. Under such an interpretation (shown as D2 and R2 in Figure 2-2), deforestation and reforestation are regarded as activities that result in land change (possibly of short duration) but without reference to subsequent land-use change.

The possible states of a patch of land can be formalized as shown in Table 2-2. The rows indicate the land-use criteria, and the columns are land-cover criteria. Four cases are distinguished. The two diagonal elements (1 = forest, 4 = non-forest) are unambiguous classifications under all criteria.

The status of lands falling in the off-diagonal categories (2 and 3) depends on definition chosen. Neither category 2 nor 3 would be included under the "LU+LC" definition. Both categories would be included with the "LU or LC" definition. Category 3 would be captured by an "LC" definition, and 2 by an "LU" definition.

Areas that meet the land-cover but not the land-use criterion (category 3) include significant carbon pools. This circumstance is especially true for sparsely treed grazing lands (savanna and woodlands) that may have canopy covers of less than 20 percent. If the threshold for canopy cover is set too high to include these lands, many sparsely treed areas will fall outside the forest definition (appearing instead under category 4).

Areas that fulfill the land-use but not the land-cover criterion (category 2) include service and recreation areas within a broader forested estate. Most national definitions of forest include these areas, with adjustments being made for their extent in calculating forest production statistics. In some nations, the legally defined forest estate includes extensive areas that are not currently forested. The carbon content of such areas will usually be relatively low, and they are likely to have little significance to the Protocol. If these areas were to be accepted as forests under the Protocol, however, devising effective definitions for deforestation and reforestation would be difficult.

A significant issue in finding a simple, unambiguous definition of a forest is the treatment of young stands that are regenerating after disturbance or harvesting. Such stands may not have reached the canopy cover or carbon threshold but may be expected to do so if allowed to continue their growth.

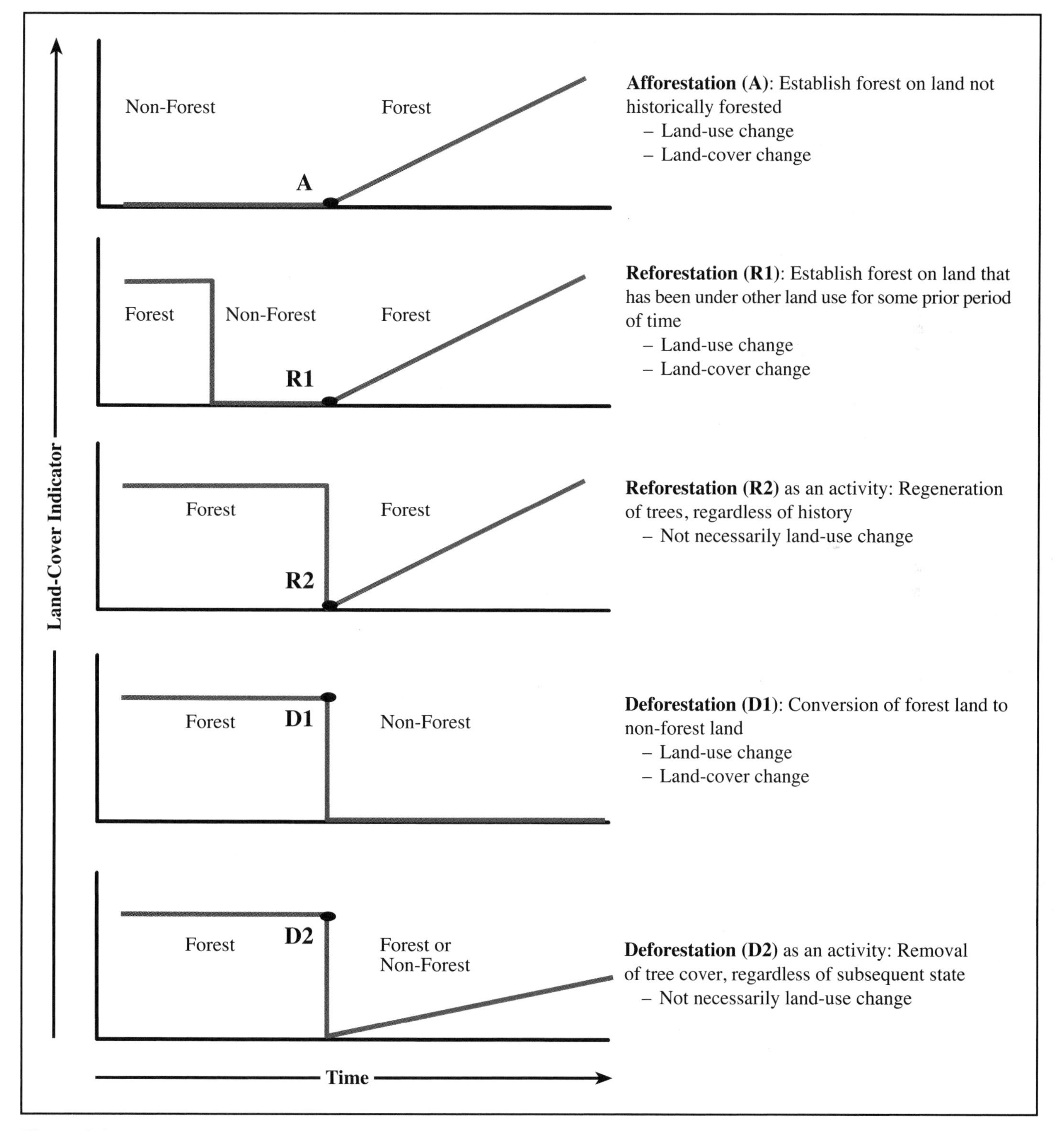

Figure 2-2: Alternative definitions of afforestation (A), reforestation (R), and deforestation (D) in terms of land cover and/or land use. Land-cover indicator (e.g., canopy cover) is plotted against time. In each panel, the dots represent the A, R, or D activity.

***Table 2-2**: Application of definitions and the state of a patch of land.*

Land Use	Canopy Cover or Biomass (C) *Above Threshold*	*Below Threshold*
Forestry	(1) Forest	(2) Young or regenerating stands; forest roads, service areas; legally defined under national laws, but not sufficiently wooded to meet the agreed threshold
Criteria Applying	*All*	*LU or LC, and LU*
Not Forestry	(3) Grazing lands; agroforestry; treed peri-urban areas	(4) Not forest, non-treed wetlands; croplands; rangelands, grasslands; non-treed peri-urban areas
Criteria Applying	*LU or LC, and LC*	*None*

In practice, distinguishing such stands from similarly affected lands that cannot—or will not be permitted—to regenerate will be difficult. Delaying recognition of such land as forest until crown cover develops could create accounting complications. These problems would be greatest in stands that are disturbed or harvested near the end of a commitment period: Such cases may require a retrospective analysis to be performed, with no guarantee that the necessary data were collected.

The complications are exacerbated by the wide variation in regeneration rates among different forest vegetation types. A regenerating boreal forest stand may require decades for its canopy cover to reach a definitional threshold. Again, specific amendments to definitions can be developed (see Chapter 3).

2.2.5. Land Cover, Land Use, and Agriculture-Related Definitions

This section examines existing national and international definitions of land and agriculture-related terms that are related to the Kyoto Protocol, especially Article 3.4.

2.2.5.1. Land

Land is internationally defined as "a delineable area of the earth's terrestrial surface, encompassing all attributes of the biosphere immediately above or below this surface, including those of the near-surface climate, the soils and the terrain forms, the surface hydrology (including shallow lakes, rivers, marshes, and swamps), the near-surface sedimentary layers and associated groundwater reserve, the plant and animal populations, the human settlement pattern and physical results of past and present human activity" (UN, 1994; CSD, 1996). Terrain forms that occur in a mosaic pattern are termed *landscapes* (similar terms are *land system units, landscape-ecological units,* or *unités de terroir),* which in turn constitute the building blocks of a watershed (catchment area) or a phytogeographic unit (biome).

2.2.5.2. Agricultural Lands

Although "agricultural land" is not mentioned in the Protocol, its definition is relevant to Article 3.4, which speaks of agricultural soils. In its narrowest sense, agricultural land is land that is arable and regularly tilled for the production of annual field crops, with or without irrigation. The word *agriculture* refers to a broad class of resource uses that includes all forms of land use for the production of biotic crops, whether animal or plants. In its broadest sense, agricultural land includes all land that provides direct benefits for mankind through the production of food, fiber, forage and fodder, biofuel, meat, hides, and skins, as well as timber. Only deserts; barren land; non-managed wetland, woodlands, and forests; and built-up areas are excluded. All categories included in the World Agricultural Census (FAO, 1995c) (summarized in Table 2-1) are included.

2.2.5.3. Carbon Pools

To fully account for carbon at a site, one must examine the forest, the crops, and the soils as a dynamic multi-component ecosystem, above- and below-ground, with changes in biomass and soil organic matter as key tracking mechanisms.

The most easily measurable pool is the total standing aboveground biomass of woody vegetation elements. The aboveground biomass comprises all woody stems, branches, and leaves of living trees, creepers, climbers, and epiphytes, as well as herbaceous undergrowth. In some inventories, dead fallen trees and other coarse woody debris, as well as the litter layer, are included in biomass estimates; in other inventories, these categories are considered as a separate dead organic matter pool. In practice, standing timber volumes per hectare are often taken as a proxy value, applying a locally tested conversion factor (see Section 2.4).

The below-ground biomass comprises living and dead roots, soil mesofauna, and the microbial community. There also is a large pool of organic carbon in various forms of soil humus (soil organic carbon, SOC). Other forms of soil carbon are charcoal from fires and consolidated carbon in the form of iron-humus pans or concretions. Many soils also contain a subpool of inorganic carbon in the form of hard or soft calcium carbonate (soil inorganic carbon, SIC).

Another major pool of carbon consists of forest products (timber, pulp products, non-timber forest products such as fruits and latex) and agricultural crops (food, fiber, forage, biofuels) taken off the site. Section 2.4 discusses their measurement and the monitoring of their routing and stability.

The components of the terrestrial carbon pools are illustrated in Figure 2-3.

2.2.5.4. *Forest Soils and Agricultural Soils*

The term "agricultural soils" is used explicitly in the Protocol (Article 3.4). The term "forest soils" is absent but may be considered as part of the forest ecosystem. There are significant differences between agricultural and forest soils that may affect inventory practices.

Fully developed forest soils are natural bodies with a vertical sequence of layers (FAO/ISRIC, 1990). At the top is an organic surface layer or "forest floor" (O horizon) with subdivisions of fresh, undecomposed plant debris (Oi horizon, formerly called L); semi-decomposed, fragmented organic matter (Oe horizon, formerly called F) and humus; and amorphous organic matter without mineral material (Oa horizon, formerly called H). Below this surface layer is a mineral surface horizon (A); a subsurface mineral horizon often leached (E); a subsurface mineral horizon with features of accumulation (B horizon); a mineral horizon penetrable by roots (C); and locally hard bedrock (R). The E, B, C, and R horizon may be lacking, or the B horizon may be modified by groundwater or stagnant water.

Agricultural soils associated with rangelands and grasslands often have similar vertical sequences. However, if they are being cultivated (arable land)—or have been in the past—they may lack the O horizon (unless peat soils are being used), and the A horizon may have been mixed with parts of the E and even the B horizon, resulting in a plow layer (A_p horizon). The B and/or C horizons may have been broken up by deep cultivation. The soils may have been so degraded by past human actions that they are no longer cultivatable. Such soils may still be classified as agricultural soils and used, for example, for grazing or non-cropping production.

The thick organic layers of wetlands, which may have peaty horizons of more than 30 cm up to several meters, are a special form of O horizon. These layers are important stores of organic carbon, which may be released as CO_2 and/or CH_4 if the land is drained and cultivated, artificially flooded, or subject to wildfires in dry years.

Both the topsoil and the subsoil are relevant in the context of carbon sequestration in agricultural soils. The topsoil is the layer with accumulation of more labile soil organic matter ("nutrient humus"). More stable humus ("structural humus") occurs in both topsoil and subsoil. The activity of soil biota such as rodents, earthworms, termites, or leaf-cutting ants leads to a dynamic interaction between these two layers and the substratum.

These three terms—*topsoil, subsoil,* and *substratum*—are likely to be used for carbon accounting in relation to Articles 3.3 and 3.4. Therefore, their definitions merit attention.

The definition of topsoil varies according to the focus, as well as national tradition. From a soil science perspective, topsoil is the surface plus subsurface mineral horizons (A, as well as E if present). Agronomically, the topsoil coincides with the plow layer. FAO/UNEP (1999) does not make a distinction between these two views: "The topsoil is the upper part of the soil, with the lower limit at 30 cm, or shallower if a root growth inhibiting layer is present above that depth." The subsoil comprises all densely rooted layers below 30 cm. The substratum can extend down to 10 m or more in well-drained tropical soils, then may still have living rootlets (Nepstad *et al.*, 1994). Agronomically, the substratum is the deeper layer not rooted by annual crops, below 100 cm.

There are many national soil classification systems, either soil science or land use oriented. The main internationally used systems are "Soil Taxonomy" (SSS, 1999) and the FAO/UNESCO terminology; the latter was recently updated as the World Reference Base for Soil Classification (WRB, 1998) and was recommended by the International Union of Soil Sciences and FAO. National soil scientists will be able to correlate their systems with the two international ones as far as required for carbon accounting.

Horizontally within a landscape, there may be large differences in soil organic matter and carbon storage at short distance, linked to differences in depth, texture, drainage condition, and slope position of the various components/facets of a landscape. Even if the landscape is homogeneously covered with high tropical forest, the total soil carbon stock can vary between 50 and 300 t ha^{-1} (Sombroek *et al.,* 1999).

2.2.5.5. *Soil Organic Matter and its Carbon*

Soil organic matter (SOM) is a generic term for all organic compounds in the soil that are not living roots or animals. SOM has been characterized in various ways: by origin, transformation stage, function, solubility, chemical constituents, elemental carbon:nitrogen (C/N) ratio, exchange capacity, functional activity level, or dynamics and stability (Parton *et al.*, 1987; Anderson and Ingram, 1993; Feller and Beare, 1997; Paustian *et al.*, 1997; Smith *et al.*, 1997a; Baldock and Nelson, 1999). In well-drained, non-acid soils—which occupy most agricultural lands—there is a balanced and dynamic composition of chemical compounds with a high degree of humification,

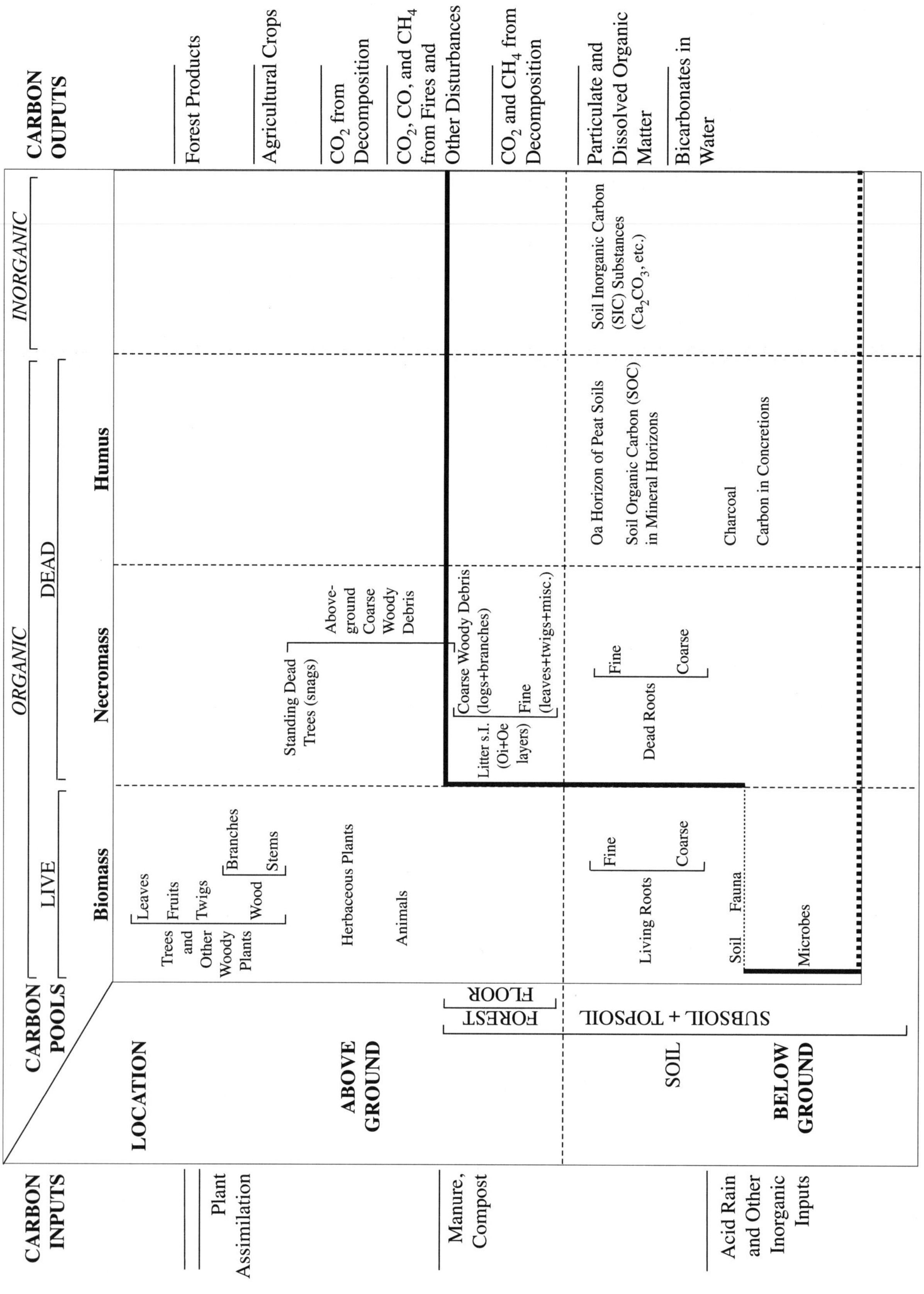

Figure 2-3: Components of the terrestrial carbon pool (compiled and amplified from Apps and Price, 1996).

resulting in medium to low C/N ratios (10–12). Characterization of SOM based on C/N ratio may be a convenient proxy for the composition and stability of SOM in topsoils. In deeper subsoils, where the amount of humus is normally small, the ratios are less reliable.

When grouping specifically on stability of the total pool of SOM, its dynamics and residence, or turnover time, one distinguishes inert or passive, stable or slow release, and labile functional subpools (Schlesinger, 1986; Smith *et al.*, 1997a; Batjes, 1999). Labile or active SOM largely consists of soil microorganisms and their immediate products, with a cycling or residence time of 1–10 years. Stable or slow-release SOM consists of neo-formations of polymeric substances, which can be extremely diverse in composition, depending on the litter source and other soil-forming conditions; it has a residence time of between 10 and 50 years (sometimes more). Inert or "recalcitrant" SOM, which is also diverse in composition, would not be destroyed for up to 500 years. The extent to which humic substances are stabilized against microbial mineralization largely depends on the type of bonding with clay minerals (Tate and Theng, 1980). In recent years, nuclear magnetic resonance (NMR) techniques have been used to establish the precise nature of these bondings (Kögel, 1997; Bayer *et al.*, 2000). The results of the application of this technique over the full range of the world's ecosystems and agricultural and anthropogenic soils are still awaited.

Parties may wish to agree on standardization of residence times, in view of the time horizons of IPCC and those mentioned in the Kyoto Protocol: 0–5 years, 5–50 years, 50–100 years, and >100 years (see Chapter 4 for details).

None of the current definitions and subdivisions of SOM provide fully quantified and universally accepted parameters to define SOM quality and quantity that are relevant to measure carbon sequestration in soils as a Kyoto Protocol activity. For accounting purposes, a simplification to SOC—which is easily measured through fine-earth samples and possibly C/N ratios—may be sufficient. Carbon isotope ratio measurements are very useful for research and modeling work on the dynamics of SOM, especially at the conversion of forest to grassland and *vice versa*. These techniques are not yet available for routine monitoring of the fate of soil carbon.

SIC (soft lime, calcium carbonate concretions or pans, primary carbonate of rock fragments or sediments) is much less mobile than SOC, except in irrigated agriculture (Schlesinger, 1999). Measurement of SIC, as well as of charcoal, will be required for monitoring and modeling of soil carbon changes, but the intensity of sampling can be much lower than with SOC.

2.2.5.6. *Degradation and Aggradation of Agricultural Lands*

Prevention of degradation and restoration of degraded lands and soils are potentially important activities under the Protocol. Human-induced land and soil degradation occur in various types and degrees, as indicated by the United Nations Environment Programme's Global Assessment of Soil Degradation (GLASOD), a provisional geographical database and international scheme (Oldeman *et al.*, 1991). This database—though based on country-level expert opinion only—has been used to conduct a global assessment of desertification (defined as land degradation—by human action or climate change—in arid, semi-arid, and dry subhumid areas) and is increasingly being applied and adapted for use at the national and regional levels (ASSOD, 1995, for Southeast Asia; Stolbovoi and Fischer, 1998, for the Russian Federation).

The GLASOD schema lists a variety of degradation processes, including water erosion (loss of topsoil, terrain deformation/mass movement); wind erosion (loss of topsoil, terrain deformation, overblowing); chemical deterioration (loss of nutrients and/or organic matter, salinization, acidification, pollution with hydrocarbons or heavy metals); and physical deterioration (soil compaction, sealing, and crusting; water logging; subsidence of organic soils). Avoidance or amelioration of these processes has potential as Kyoto activities. Bergsma *et al.* (1996) provide detailed definitions of these degradation types, as well as measures for their control and reclamation. The degree of degradation is given in four classes: "light," "moderate," "strong," and "extreme." Reclamation of SOC content is relatively easy for the first two classes (see Section 4.4).

Soil improvement (aggradation) has taken place in many parts of the world. The World Overview of Soil Conservation Approaches and Technologies (WOCAT) (Liniger *et al.*, 1988) gives spatial information on human-induced soil and land improvements for more sustainable agricultural land use, as well as the technologies applied. Restoration of SOM-depleted agricultural soils to near their original level appears to be a realistic opportunity, especially when terrain conditions permit mechanization of adapted cultivation practices (see Figure 2-4; Lal *et al.*, 1998; Paustian *et al.*, 1998; Batjes, 1999). Improvement of agricultural soils beyond the original level of soil organic matter, as a form of land and soil aggradation, may also be possible (Sombroek *et al.*, 1993; Sandor and Eash, 1995; Bridges and de Bakker, 1997; Blume *et al.*, 1998; Batjes, 1999; Gläser *et al.*, 1999; McCann *et al.*, 1999). The extra costs of such soil aggradation can become a feasible capital investment if collateral environmental and socioeconomic benefits are taken into account (e.g., control of desertification; increased security of food production; more sustainable agricultural settlement).

2.2.5.7. *Lateral Fluxes of Carbon*

In addition to vertical fluxes (to the atmosphere), there are a number of lateral fluxes of carbon and plant nutrients at the land surface or in the soil interior. A human-induced activity that leads to degradation in one location may result in aggradation elsewhere (e.g., nutrient transfers).

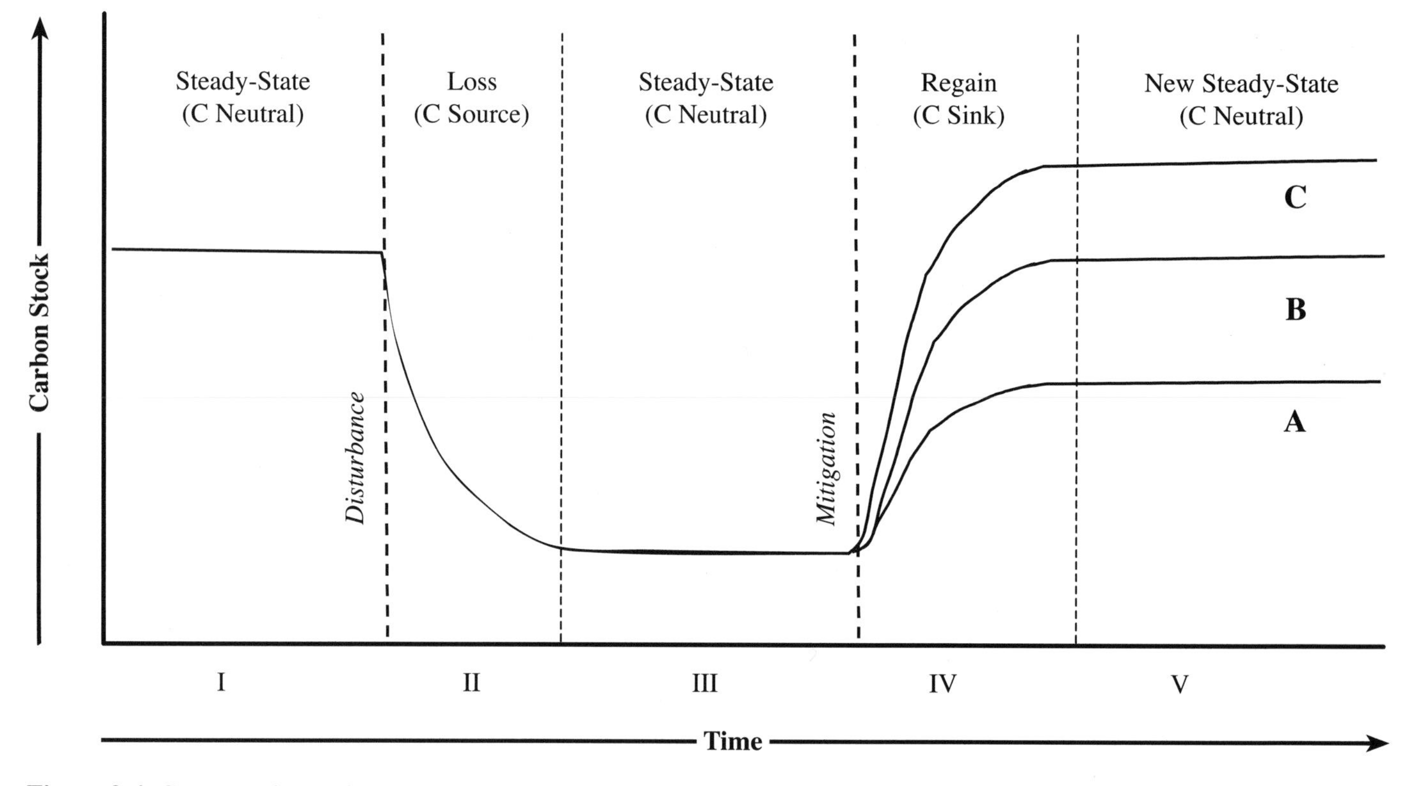

Figure 2-4: Conceptual model changes in ecosystem carbon stocks following disturbance or degradation, and mitigation through reclamation: (A) stabilization at lower stocks than original, (B) stabilization at original stocks, and (C) stabilization at higher stocks (after Johnson, 1995).

Examples at the biome/agro-ecosystem/agro-ecological zone level follow:

- Extraction and off-site distribution of products or organic materials derived from ecosystems (see Section 2.4.2.2).
- Large-scale biomass burning (natural vegetation or crop residues) and wind erosion can result in soot and dust being carried off-site by the wind. Transported materials include volatized soil organic matter, carbon-soot particulates and nitrogen compounds in subhumid areas, or calcium- and phosphorus-enriched particulates from desert areas (e.g., Nadelhoffer *et al.*, 1999).
- Human-induced water erosion in upper catchments of watersheds may result in downriver formation of new land with enriched nutrients and organic matter content.
- Within the soil profile itself, there can be slow but continuous percolation, and subsequent lateral transport through groundwater flows, of dissolved organic matter (DOM). An example is the catchment of the Rio Negro in the Amazon region. Its waters are acid and deep-brown colored, containing many fulvic acids ("rios-de-agua-preta;" Sioli, 1984). These DOMs may end up in sedimentary deposits when they come in contact with silt-loaded river waters ("rios-de agua-branca"), or at the riverine sedimentation front in seas and oceans (Richey *et al.,* 1980; Meybeck, 1982).

Examples of lateral carbon fluxes at the landscape level follow:

- Water erosion may result in accumulation of debris down-slope, as colluvia at the mouth of drainage ways, or as alluvia of the local river system. The particulate organic matter so displaced may become buried deep in the sedimentary layers of these deposits, effectively immobilizing the entrenched organic carbon matter, unless the new lands become incorporated in the local agricultural tillage system (see Section 1.4.1).
- Before the use of chemical fertilizers in Europe and Asia, there often was deliberate displacement of organic matter and nutrient-rich topsoil from forests and heathlands to the arable fields around villages, directly or through a cattle-stable or "night-soil" phase. This practice is still prevalent in many rural communities in sub-Saharan Africa (FAO, 1995a; Smaling, 1998) and elsewhere. In contrast, much of the organic effluent of industrialized countries is transported out of the terrestrial biosphere into aquatic systems.

If human activities that induce changes in soil organic matter are included as Protocol activities under Article 3.4, these lateral transfers (since 1990) will need to be taken into account, guided by a careful analysis of the local land-use system.

2.2.6. Ecologic-Economic Zoning and Globally Consistent Databases on Land Resources

The quantification of Articles 3.3 and 3.4 activities for accounting purposes will likely be undertaken by country institutions. There may be vast differences in carbon sequestration potential, however, within a Party's territory depending on prevailing socioeconomic conditions, legal-institutional structure, and bio-geophysical conditions. Identification and delineation of these variations will simplify national level reporting and reduce measurement sample requirements (Section 2.4). Such a classification system may also serve as a geographic base for land-based accounting (Section 2.3.2.2).

The socioeconomic conditions of rural areas are normally expressed in statistical data for each administrative unit (e.g., districts or provinces). Legal-administrative characteristics (e.g., land ownership type, national parks, indigenous reserves) are generally defined within historically defined geographical units. Bio-geophysical conditions within a country can be homogeneous or very diverse, depending on the country's size and variation in climatic zones.

The sequestration potential and the percentage of each biophysical zone within a country are important variables. One approach is to apply biome or eco-region concepts, with geographic domains (e.g., polar, humid temperate, dry, humid tropical), divisions (e.g., tundra, boreal forests, prairies, steppe, deserts, savannas, and rainforests), and subdivisions based on elevation. Bailey (1998) and Udvardy (1975) provide world maps that are based on these units; they are also defined in the "life-zones" scheme of Holdridge (1947), which is used in many Latin American countries for national zonations.

For Article 3.4 activities that involve agricultural soils, the agro-ecological zone (AEZ) concept developed by FAO (1996) may be used for within-country geographic subdivisions. This concept combines agroclimatic characteristics with elevation/topography and soil conditions. Agroclimatic criteria include major climates (boreal, arid, seasonally dry subtropics, humid tropics, etc.), annual growing periods, a dryness index, and the frost period. Initially applied only to developing countries, the AEZ program has been recently extended to all Annex I countries and made available in digital form, accompanied by a digital elevation model and a soil database that includes soil carbon (FAO/IIASA, 1999; Fischer *et al.*, 1999). The combination of the three data sets provides the spatially explicit AEZs. In a simplified form these zones can be used to construct tables of carbon sequestration potential per country (see Chapter 4).

A global database on forest cover and changes in forest cover since 1980 is available through FAO's decadal Forest Assessments, the latest being FRA 2000 (FAO, 1993a; FAO, 1997b; UN-ECE/FAO, 2000). FAO also maintains a global and country-level database on agricultural land use and commodity movements. The FAO/UNESCO world soils information (FAO, 1971–1981), although somewhat out of date, is based on correlation of national soil mapping information. Upgrades of the soil database per continent currently underway (SOTER program) (FAO, 1993b; Baumgardner, 1999) will provide spatial terrain data with associated soil profile measurements, climatic conditions, and vegetation or land-use information in relational digital form, based on country-wide geographic information at a medium intensity (scale 1:1 million). The Latin American-Caribbean component was completed in 1998 (FAO/UNEP/ISRIC, 1999); the other continents will be covered in the coming 3 to 6 years, with Africa as a priority region. Technical soil data are provided in terms of the World Reference Base for Soil Classification (WRB, 1998) and Soil Taxonomy (SSS, 1999), where appropriate. These databases can be used to compile country-wide maps on actual and potential soil carbon stock. For some regions, such as North America (Lacelle *et al.*, 1997), generalized soil carbon maps are already available.

In general, national and international geo-referenced databases on the 1990 state of forests and qualitative vegetative cover, soil and terrain conditions, actual land uses, and degree of soil degradation and improvement are approximate only.

2.3. Accounting and Reporting Issues

A carbon accounting system is needed to provide a consistent and transparent approach to recording and reporting of changes in carbon stocks from applicable activities for use in meeting commitments under Article 3 of the Kyoto Protocol. This section presents a general accounting framework and describes some of the generic issues that arise in bringing the LULUCF provisions of the Kyoto Protocol into operation. Key considerations include how to identify stock changes resulting from human-induced activities, the implications of different choices of system boundaries and the timing of stock changes, and how to account for incomplete data and uncertainty. Subsequent chapters in this Special Report present detailed analyses of the implications of these issues for specific provisions of the Protocol.

2.3.1. Objectives of an Accounting System

For the purposes of this Special Report, a carbon accounting system records, summarizes, and reports the quantity of carbon emissions by sources and removals by sinks through applicable LULUCF activities for a specific period of time. Through the accounting system, Parties will quantitatively demonstrate the extent to which LULUCF activities covered by the Kyoto Protocol affect their emission reduction commitments. Building on the principles established in the UNFCCC reporting guidelines for annual GHG inventories, an ideal accounting system possesses the following core objectives: transparency, consistency, comparability, completeness, and accuracy. Moreover, Article 3 of the Protocol requires that carbon stock changes be verifiable. Moreover, practical constraints on implementation indicate a need to consider the efficiency of an accounting system. These features are discussed in turn below.

2.3.1.1. *Transparency*

According to the UNFCCC inventory reporting guidelines, transparency implies that the assumptions and methods used are clearly explained so that users of the information can replicate and assess the information. For an accounting system, transparency means that reported information can be traced back to the underlying data through a logical set of procedures that summarize the data. For example, reported carbon fluxes may be estimated by a measurement method that accounts for all relevant pools; in turn, the measurement method is applied to data that represent carbon changes by pool.

2.3.1.2. *Consistency*

The ideal accounting system is consistent with the scientific principles of carbon processes and the institutional context in which the system is applied. Dimensions of scientific consistency include carbon coverage over space, pools, and time. Those aspects are addressed in the full carbon accounting discussion below. Institutional consistency refers to the system's correspondence to the objectives driving the need for the system in the first place. For the purposes of this Special Report, the institutional objective of the accounting system is to demonstrate compliance with the Kyoto Protocol.

2.3.1.3. *Comparability*

An accounting system should produce information that is comparable across Parties and over time. Because methods and data systems may differ across Parties and time, strict comparability may be difficult to maintain. Nonetheless, differences in methods and data should be made transparent so that the numbers are as consistent and comparable as possible. Comparability may require some form of standardization. Although such standards do not exist for a Protocol-specific accounting system, the IPCC Guidelines (IPCC, 1997) may provide an initial framework.

2.3.1.4. *Completeness*

Completeness refers to an accounting of all applicable sources and sinks. In the context of a Kyoto Protocol accounting system, the sources and sinks that are "applicable" would be defined by the provisions of the Protocol.

2.3.1.5. *Accuracy*

Accuracy refers to the general validity of the reported numbers from an accounting system. Accurate estimates are unbiased in that they do not systematically under- or overstate the true number. A related issue is precision. Precise estimates have small standard errors. Accuracy and precision can be independent. A system can be accurate (unbiased) but produce estimates of limited precision. On the other hand, extremely precise estimates can be biased if the system is not well designed.

2.3.1.6. *Verifiability*

The accounting system should enable a third party to verify the reported numbers. Verifiability requires that the accounting system is built on proper data collection, measurement, and reporting procedures. For instance, claims that relevant LULUCF activities sequestered a given quantity of carbon over a certain period of time should be based on substantiated data, models, and methods. The location and extent of the land on which the claimed carbon sequestration took place should also be clearly identified, preferably using a consistent geographic information system that would facilitate identification of possible double-counting and the overlay of information used for verification (e.g., remote-sensing data). In short, claims should be able to withstand reasonable scrutiny.

2.3.1.7. *Efficiency*

An accounting system is not free. Systems that are more accurate, precise, and verifiable may be more expensive to develop and operate. An efficient accounting system operates at the point where the marginal costs of increased accuracy, precision, and verifiability just equal the marginal benefits of achieving the improvements. When resources are not available to obtain an efficient system, the objective should be to obtain the most effective system given the available resources.

2.3.2. *Protocol-Specific Accounting Framework*

This section describes alternative structures for the accounting of GHG emissions by sources and removals by sinks from LULUCF activities. Policy questions that the Parties may wish to address in defining the accounting system are introduced, and illustrative accounting rules for LULUCF activities under the Protocol are presented and compared. Subsequent sections elaborate on options for addressing the key accounting questions.

Any accounting system for LULUCF activities under the Kyoto Protocol must address three fundamental questions:

- To what activities does Kyoto accounting apply?
- Will accounting be based on land units or activities?
- What carbon related to these activities will be counted?

2.3.2.1. Activities to Which the Accounting System Applies

The Kyoto Protocol identifies the following categories of possible activities:

- Afforestation, reforestation, and deforestation (ARD) occurring since 1 January 1990 (Article 3.3, Decision 9/CP.4)
- Additional agricultural soil and LULUCF activities (Article 3.4)
- Projects aimed at reducing GHG emissions from sources and enhancing GHG removals by sinks (Article 6 and potentially Article 12).

2.3.2.2. Land-Based versus Activity-Based Accounting

A carbon accounting system developed for the Kyoto Protocol must adhere to the basic scientific principles of carbon processes and the institutional terms and objectives of the UNFCCC. Two accounting approaches are discussed here that may meet these requirements. The Parties could decide to adopt either one of these approaches, or some combination of the two.

The first approach to accounting is land-based. Its starting point is the total carbon stock change in applicable carbon pools on land units subject to Kyoto activities. Implementing this rule involves first identifying land units on which applicable activities occur. Next, the total change in carbon stocks on these land units during the commitment period is determined. Adjustments can then be made to reflect decisions that the Parties may adopt regarding baselines, leakage, and timing issues, as discussed in the following sections. Aggregate emissions or removals are the sum of stock changes (net of adjustments) over all applicable land units.

The second approach is activity-based. Its starting point is the carbon stock change attributable to designated LULUCF activities. First, each applicable activity's impact on carbon stocks is determined per unit area. This impact is multiplied by the area on which each activity occurs. This equation may also include adjustments to reflect policy decisions by the Parties. Aggregate emissions or removals are calculated by summing across applicable activities. Potentially, a given area of land could be counted more than once if it is subject to multiple activities. This potential double-counting could result in inaccurate accounting if the effects of activities are not additive. Alternatively, the Parties could decide that each land unit could contain no more than one activity. In this case, the combined impact of multiple practices applied in the same area would be considered a single activity.

2.3.2.3. Illustrative Accounting Rules for the Protocol

Either of the generic rules described in Section 2.3.2.2 could be applied to the human-induced activities that are included in Kyoto accountings, or they may be included depending on decisions of the Parties. These approaches can be expressed by the equations in Table 2-3.

The accounting rules applied under Articles 3.3 and 3.4 need not be identical. For examples, the Parties may decide to apply a land-based accounting system for ARD activities and an activity-based accounting system to additional activities adopted under Article 3.4. Similarly, the Parties may decide on a different approach to baselines for ARD activities, additional activities, and projects. Baselines are discussed in Section 2.3.4.

If different accounting rules are adopted for ARD activities and Article 3.4 activities, decision rules may be needed to determine which accounting rule applies to land that, over time, is subject to both types of activity. For example, one set of accounting rules could be given primacy in cases where both sets could potentially apply. Similarly, if management of harvested material is treated as an additional activity under Article 3.4, it may be necessary to exclude changes in this pool from the accounting of other activities under Article 3.3 or 3.4 to avoid the potential for double-counting.

2.3.2.4 Activities and Projects

Chapter 5 defines a project as a planned scheme that integrates one or more types of activity, aimed at reducing GHG emissions or enhancing GHG sinks in the LULUCF and related sectors. A project is confined to a specific geographic location, time period, and institutional framework to allow GHG impacts to be adequately monitored and verified. Although Parties are free to select the projects they wish to report under Articles 6 and 12, Article 3.3 requires that the effects of ARD activities "shall" be reported everywhere they occur within the country. The approach to reporting any additional activities adopted under Article 3.4 is to be decided by the COP.

An important relationship exists between Article 6 projects and Article 3 activities. Article 3.11 requires Parties that host Article 6 projects to subtract from their assigned amounts all emissions reduction units (ERUs) transferred to other parties. If an Article 6 project in the LULUCF sector does not consist only of Article 3.3 or 3.4 activities, the host Party will not be able to make an adjustment to its assigned amount corresponding to the ERUs transferred by the project. Furthermore, because Article 6 projects must provide a reduction or removal that is "additional" to any that would otherwise occur, Article 3.3 or 3.4 activities used in such projects must be additional to business-as-usual.

Any ambiguities about what is an applicable activity under Articles 3.3 and 3.4 would raise an important compliance question. An international verification system would have a reasonable chance of detecting unreported decreases in stocks but might have considerably more difficulty identifying whether this decrease resulted from an applicable activity.

***Table 2-3**: Illustrative accounting equations.*

Land-Based Rule		Activity-Based Rule	
$Q = \sum_{i=1}^{M} \sum_{j=1}^{N} [S_{i,j}(TE) - S_{i,j}(TB)] - \sum_{k=1}^{R} A_k$		$Q = \sum_{i=1}^{M} \alpha_i L_i - \sum_{k=1}^{R} A_k$	
Symbol	**Definition**	**Symbol**	**Definition**
Q	Quantity of carbon emissions debited or removals credited	Q	Quantity of carbon emissions debited or removals credited
i	Indexes a landscape unit whose LUCF activity places it in Kyoto accounting system	i	Indexes activity
j	Indexes carbon pools (e.g., aboveground biomass, below-ground biomass, etc.)	k	Indexes adjustments
k	Indexes adjustments	M	Total number of activities in Kyoto accounting system
M	Total number of landscape units in Kyoto accounting system	L_i	Land area subject to activity *i*
N	Total number of carbon pools included in Kyoto accounting system	α_i	Carbon emissions or removals per unit land area from activity *i*
R	Total number of adjustments adopted for use in Kyoto accounting system	R	Total number of adjustments adopted for use in Kyoto accounting system
TE	Year at end of commitment period	A_k	Adjustment *k*, if adopted, to reflect, e.g., baselines, leakage, uncertainty, etc.
TB	Year at beginning of commitment period		
$S_{i,j}(t)$	Stock of carbon on landscape unit *i*, in carbon pool *j*, in year *t*		
A_k	Adjustment *k*, if adopted, to reflect, e.g., baselines, leakage, uncertainty, etc.		

Governments and private entities have a greater incentive to report activities that give rise to removals than those that give rise to emissions. Removals contribute to the fulfillment of obligations, whereas emissions make fulfillment more difficult. This factor may lead the Parties to resolve any ambiguities about what may be an applicable activity by overreporting activities that result in removals and underreporting activities that result in emissions. For the purposes of compliance, the Parties may want to adopt a rule that presumes that decreases in stocks result from applicable activities and increases in stocks do *not* result from such activities, unless the Party concerned demonstrates otherwise. Alternatively, for each activity, Parties might be required to specify in advance—as well as monitor and report on—the land base to which the activity could potentially apply. Parties would be required to report all increases and decreases occurring on these specified land bases.

2.3.2.5. Accounting Under the Kyoto Protocol Compared to Full Carbon Accounting

The term "full carbon accounting" can be used to imply complete accounting for changes in carbon stocks across all carbon pools, landscape units, and time periods. In this Special Report, "full carbon accounting" means complete accounting of stock changes in all carbon pools related to a given set of landscape units in a given time period. When complete coverage over the landscape and/or time is intended, that interpretation will be stated explicitly. Note that a carbon pool (e.g., forest products) can be related to a landscape unit without being physically located on the site. Policy options to account for these wood product pools are discussed in Section 6.2.2.

The land-based accounting rule in Table 2-3 can capture full carbon accounting over pools, landscape units, and time as a special case in which the summation covers all carbon pools and all landscape units, and contiguous commitment periods extend into the indefinite future. The Kyoto Protocol, on the other hand, mandates that accounting be restricted to certain "human-induced" activities. Full carbon accounting over pools and time could nonetheless be applied to landscape units that constitute Kyoto land by virtue of being subject to specified human-induced activities.

2.3.2.6. *Timing of Commitment Periods*

There is a 17-year gap between the end of the base year (1990) and the beginning of the first year of the first commitment period (2008). Although any relevant LULUCF activities that occur during that time span would enter into the Kyoto accounting system, the effect of those activities is confined to impacts on carbon stocks during the 2008–2012 commitment period. This situation raises the possibility that a Party could draw down carbon stocks prior to the commitment period (e.g., through forest clearing) and later obtain credit for carbon accumulated during the first and future commitment periods on forests established after 1990 (Schlamadinger and Marland, 1998). Conversely, for land forested after 1990 and harvested during a commitment period, the debits from the stock decrease could exceed the previously earned credits because stock increases prior to 2008 would not be counted. Options for addressing these situations are discussed in Chapter 3. Similar situations could arise if future commitment periods are not contiguous.

2.3.2.7. *Greenhouse Gases Other than CO_2*

The accounting rules described here focus only on emissions and removals of CO_2, which can be measured as changes in carbon stocks. Emissions of methane and nitrous oxide from many land-use activities are included in Annex A of the Kyoto Protocol and therefore will be captured in the national inventories of Annex I Parties. Emissions of these gases related to forestry activities may not be captured, however. As described in Chapter 1, forestry activities can affect emissions and removals of methane and nitrous oxide. If these effects are not considered, the full impact of forestry activities may not be reflected in the Kyoto accounting system. Similarly, to accurately assess the climate benefits of projects in non-Annex I Parties, changes in methane and nitrous oxide emissions/removals would have to be considered explicitly.

2.3.3. **Effect of Human-Induced LULUCF Activities versus Other Influences on Carbon Stocks**

This section describes the range of factors that can affect terrestrial carbon stocks. It focuses on the relationship between human-induced LULUCF activities and other factors. Section 2.3.4 discusses a range of options for addressing these factors in the accounting system.

2.3.3.1. *Natural versus Human-Induced*

Terrestrial ecosystems are dynamic. Carbon stocks change over the full range of time and space scales, for many different reasons. Although carbon stock changes can be measured directly with a variety of techniques (see Section 2.4), attributing a given change in carbon stocks to a particular cause can be much more challenging. Yet identifying the reason for a change in carbon stocks may be crucial to determining whether it should be included in the Kyoto accounting system, for two reasons. First, stock changes that are part of a natural cycle will not have a long-term effect on climate if increases during one period or at one location are offset by decreases at another time or place (recognizing that the 5-year commitment period defined by the Kyoto Protocol samples a period that is short compared to many natural cycles of disturbance and recovery). Second, policymakers may decide that stock changes from certain causes should not be credited or debited because doing so would be at odds with the underlying objectives of the Kyoto Protocol.

ARD activities under Article 3.3 and activities that might be added to the Kyoto Protocol under Article 3.4 must be "human-induced." The significance of this limitation can be understood by considering changes in carbon stocks at the global scale that are presumably not a result of human activity. Most pronounced is the annual cycle of increasing terrestrial carbon stocks (decreasing CO_2 concentrations) during the Northern Hemisphere spring and summer and decreasing carbon stocks during the autumn and winter (see Chapter 1). To the extent that this cycle is balanced on an annual basis between carbon uptake and release, there is no impact on average atmospheric GHG concentrations, and there would be no net change in carbon stocks observed during a 5-year commitment period. The Mauna Loa record, however, exhibits significant interannual variability around the long-term trend of increasing concentrations. These variations, which are associated in part with natural climate variations such as the El Niño Southern Oscillation (Keeling *et al.*, 1989), have resulted in residual carbon stock changes exceeding 4 Gt C during some 5-year intervals since 1980 (Keeling and Whorf, 1998)—which is similar in magnitude to the annual emissions from Annex I Parties. Thus, at a global level, significant carbon stock changes that are unrelated to human activities and unlikely to reflect long-term changes in carbon sequestration can be expected over a 5-year commitment period.

At the national, regional, or local scale, many natural factors could lead to significant carbon stock changes over a 5-year commitment period. These factors include natural cycles of disturbance and recovery (e.g., from fires, pests, and diseases). To the extent that such disturbances are random, their effect will tend to average out as larger spatial and temporal scales are considered. However, decisions about accounting for these stock changes, both during commitment periods and from one period to another, could result in an accounting imbalance—for example, if disturbances are not accounted for but recovery from disturbance is considered a management activity.

Although land-use change and forestry require human decisions and action, in some cases the decisions and actions may be largely driven by acts of nature. For example, a patch of forest may be destroyed by landslides or flood damage, leaving no option to regenerate forest on that site in the near future. Although a decision may be made to convert the area to cropping or grazing, this change arguably would not be human-induced because regeneration was not possible. In other cases, the argument for attributing the loss of forest and subsequent land-use change to human activity may be stronger. For example, a wildfire might destroy a tropical forest. Although the forest would regenerate naturally over the next few decades, local land managers or settlers may be opportunistic and convert the land to gardens or plantations.

In managed forests, natural occurrences such as wildfires and pest infestations may have contributing human factors. Indeed, management practices may specify that certain fires should be allowed to burn for ecological reasons. Management practices may also exacerbate the damage from fires, pests, or disease. Arguably, failure to prevent or suppress such events, when prevention or suppression is possible, amounts to human inducement. Thus, non-human-induced occurrences may be limited to places and events that are entirely removed from human activity. Management might be taken as a token of human inducement, and events in managed forests might be considered human-induced *per se*. The IPCC Guidelines provide for forests to be designated as managed or unmanaged, with stock changes on unmanaged forests excluded from the accounting framework. Alternatively, any event that is not directly attributable to an activity under Articles 3.3 or 3.4 might be considered to be outside the accounting system. This approach, however, could result in a significant divergence between carbon stock accounts and carbon on the ground, which could increase the difficulty of monitoring and verification.

Several commentators have suggested that only intentional activities should be regarded as human-induced. Intent could be very difficult to ascertain, however. It may not be possible to determine, for example, whether a fire was caused by a lightning strike, arson, or mere carelessness. Even if such a distinction were possible, there is no obvious policy rationale for doing so because Parties have a limited ability to control any of these events.

Even if such definitional issues can be resolved, in practice distinguishing natural stock changes from those that are directly or indirectly human-induced may be difficult. Suppose, for example, that a natural forest is placed under management to enhance its growth. Quantifying the change in stock that is directly attributable to management practices would be difficult; some growth would have occurred naturally, and some will be the result of CO_2 or nitrogen fertilization. Although the stock increase attributable to management may not be directly quantifiable, inferring it from the measured stock increase may be possible by using a comparison "control" plot or a model to estimate the stock increase that would have occurred without this management practice.

Policymakers should understand that the phrase "human-induced" has no scientific meaning, and guidance will be needed to ensure that Parties are consistent in their characterization of activities as natural or human-induced. Even with clear definitions, in practice distinguishing between changes resulting from human activity and changes caused by natural events may be difficult.

2.3.3.2. *Direct versus Indirect*

The Kyoto Protocol distinguishes between direct and indirect human-induced land-use change and forestry activities. The word "direct" precedes the phrase "human-induced" in Article 3.3 but not in Article 3.4. Temporal and spatial immediacy may indicate directness; the closer in time and space the activity is to the impact, the more direct it is. Intent and foreseeability also might be relevant in determining directness.

One of the most significant distinctions between direct activities and indirect influences relates to the effects of CO_2 fertilization and nitrogen deposition. CO_2 fertilization and nitrogen deposition are indirect because the removals are not geographically immediate—that is, they may occur thousands of miles from the site of the emissions. The fact that enhanced growth of biota is a completely unintended consequence of the polluting activity also argues in favor of treating it as an indirect activity. Moreover, CO_2 fertilization and nitrogen deposition cannot reasonably be described as land-use change or forestry activities.

At the global scale, carbon cycle studies suggest that terrestrial ecosystems that are not subject to tropical deforestation are sequestering an average of approximately 2.2 Gt C yr^{-1} through biomass regrowth resulting from natural regeneration and uptake of carbon dioxide, as well as nutrient fertilization and changing climate (see Chapter 1). The geographic distribution of this sink is uncertain, but if credit became available for the effects of carbon and/or nitrogen fertilization over a large fraction of the landscape, the implications would be profound. If Parties could obtain credit for 50 percent of the estimated sink from these factors, achieving the emission limitations of the first commitment period would not require any actions beyond business-as-usual projections (Lashof and Hare, 1999). Accounting inconsistencies also could arise if credit could be taken under the Kyoto Protocol for CO_2 and nitrogen fertilization.[2]

[2]In calculating the Global Warming Potentials (GWPs) of GHGs, it was assumed that CO_2 fertilization is not anthropogenic. The calculations use models that take into account the absorption of GHGs into the oceans and, in the case of the reference gas, CO_2, its uptake as a consequence of CO_2 fertilization. Thus, if CO_2 fertilization were subtracted from the estimate of total emissions (i.e., on the basis that it is anthropogenic and is an uptake), the effect of CO_2 fertilization would be double-counted. Although this effect could be corrected by recalculating all of the GWPs, this recalculation would have other policy repercussions. The main effect of such a recalculation would be to significantly reduce the GWPs of non-CO_2 GHGs.

On the other hand, completely excluding these effects from crediting may be impractical because current methods cannot precisely determine either the total or incremental amounts of uptake attributable to CO_2 and nitrogen fertilization. Indeed, estimates of carbon uptake from forest inventory data continue to suggest a much smaller current terrestrial carbon sink than that inferred from atmospheric measurements (Holland and Brown, 1999; Lashof and Hare, 1999). At a project level, however, an approach that compares a "with-project" scenario to a "without-project" scenario should control for much of the fertilization effect because both scenarios will be subject to fertilization. As Section 2.3.4 notes, a similar approach could be applied at the national level under Article 3.4, but use of a baseline may not be practical or important to accounting under Article 3.3.

2.3.4. Baselines

Once applicable activities and an accounting framework have been selected, it is necessary to determine which carbon related to these activities will be counted in determining Parties' compliance with their Kyoto commitments. This section introduces the concept of baselines—which can be used, if desired, to adjust carbon stock accounts to discriminate between the effects of LULUCF activities and other factors as well as to factor out the effects of business-as-usual and activity undertaken prior to 1990.

Baselines are the reference scenario against which a change in GHG emissions or removals is measured. There are many options for baselines, including the stock change that would have resulted from "business-as-usual" activities; the stock change that would have resulted from the continuation of current or 1990 activity levels; the stock change that would result in the absence of active management; performance benchmarks or standard management practices; or the rate of change of stocks in 1990.

Baselines could be set at a national, regional, or project level. National baselines presumably would be developed from analysis of regional trends and practices and could be based on a combination of measurements on control plots and models. Regional baselines could also be used as a constraint on the baselines for projects within the region. If national baselines were developed using a standardized methodology and agreed upon at the international level (e.g., by the COP), the design of incentives for individual landowners to enhance carbon sequestration could be purely a matter of domestic policy in each country.

Baselines need not necessarily be derived solely from observed trends and practices in a given country or region. Internationally agreed performance benchmarks have also been suggested as a means for establishing objective baselines (Lazarus *et al.*, 1999). For example, certain practices could be considered "standard management practice," and baselines might be set to reflect the level of carbon sequestration that would occur if these practices were universally applied. Credit would then be available only to the extent that there was an improvement compared to the results of applying these standard practices. The applicability of this approach to the LULUCF sector is unclear, however, because there is no set of internationally agreed "standard management practices," and such practices may be difficult to define for the wide variety of situations that will be encountered in the field.

Baselines can be reflected in the adjustment terms of the accounting equations presented in Table 2-3, if the Parties so decide. In the land-based rule, the baseline stock change may be subtracted from the gross change in carbon stocks during the commitment period on land included in the system. Similarly, in the activity-based rule a baseline area subject to the activity under business-as-usual might be subtracted from the total area where the activity takes place during a commitment period. In addition, the impact of the activity per unit area may implicitly be calculated relative to a baseline representing the absence of that activity.

Key policy questions that could be addressed through the use of appropriate baselines are as follows:

- Will the accounting system include adjustments for the effect of non-human-induced factors?
- Will the accounting system include adjustments for the effect of business-as-usual activity?
- Will the accounting system be adjusted for the effect of activities undertaken prior to 1990?

These questions may be addressed differently for ARD activities, Article 3.4 activities, and projects. Decision 9/CP.4 requires accounting for ARD activities as changes in stock during the commitment period without making reference to a baseline stock change (suggesting that the baseline adjustment term may be zero for Article 3.3 accounting). Articles 6 and 12, on the other hand, require that credit be based on benefits that are "additional" to what otherwise would have occurred, implying the use of a "business-as-usual" baseline. Project developers therefore have had to develop methods to define "business-as-usual" baselines. A more detailed discussion of these methods appears in Chapter 5. Article 3.4 requires the COP to "decide upon modalities, rules, and guidelines as to how" additional activities will be included. Thus, the Parties must decide whether to require a baseline and, if one is required, how it should be calculated. If the Parties determine that such a baseline requirement is not relevant to Article 3.4, the baseline adjustment term can be set to zero in the accounting equation.

The following three sections provide additional information to guide decisions regarding each of the policy questions relevant to the use of baselines. Distinctions are made between ARD activities, Article 3.4 activities, and projects where appropriate. More details related to each of these types of activity appear in subsequent chapters.

2.3.4.1. Human-Induced

Many factors will influence carbon stocks on any given area of lands under Article 3.3 or 3.4 (Section 2.3.3). In principle, accounting could focus on only stock changes resulting from human-induced activities, disregarding changes resulting from other causes, or on all stock changes during the commitment period.

It may be possible to use an activity-based approach to identify stock changes resulting only from human-induced activities by developing appropriate factors that represent the impact of these activities on carbon stocks.

Using a land-based approach for ARD activities factoring out the impact on stocks of indirect effects, such as CO_2 fertilization and acid rain, would be difficult, if not impossible, because in the absence of the ARD activity the land would be subject to fundamentally different conditions. Therefore, once an area has been designated as ARD land, the practical approach may be to count all stock changes that occur on that land, regardless of the cause.

Counting all stock changes during the commitment period ensures that debits and credits for emissions and removals correspond to actual stock changes. Stock changes during the commitment period will result from many factors. In practice, the fraction of the landscape designated as ARD land is likely to be small, and the nature of ARD implies that these activities are likely to be the dominant factor affecting carbon stocks on these lands. This approach is developed in Chapter 3.

Depending on the nature of additional activities adopted under Article 3.4, in some areas the influence of natural variability and indirect effects (such as CO_2 fertilization or climate change) could be the predominant factor affecting carbon stocks during a commitment period (see Section 3.3.3). Consider forest management as an example. If a carbon stock increase is measured in a forest stand that is managed with reduced-impact logging, this practice may not be responsible for all, or even most, of the stock increase. CO_2 or nitrogen fertilization could be playing a significant role, or the stock increase could be primarily caused by recovery following disturbance.

Although directly measuring the fraction of the observed stock increase on land attributable to the LULUCF activity may not be possible, the baseline stock increase that is not attributable to the activity could be estimated and subtracted. Appropriate baselines for this purpose might be estimated by using control plots that continue the current management regime or are not subject to active human management.

2.3.4.2. Business-as-Usual

A decision to broadly define additional activities under Article 3.4 has the potential to profoundly influence the environmental effectiveness of the Protocol. Such changes could result from allowing Parties to claim credit for a portion of ongoing carbon stock increases expected for Annex I Parties under "business-as-usual," as well from increasing the range of options available to Parties for adopting new measures to reduce emissions or increase removals. Annex I Parties have reported net GHG removals totaling 0.5 Gt C in 1990 from the land-use change and forestry sector, based on the reporting guidelines adopted under the UNFCCC. Although available projections made by Parties indicate that the rate of removal will decline over time, substantial net removals are still projected to occur during the commitment period under a business-as-usual scenario. Subtraction of a baseline during the first commitment period would reduce the potential for Parties to obtain credit for carbon stock increases expected under business-as-usual.

A business-as-usual baseline could also protect Parties from unexpected debits. Suppose, for example, that forest management is included as an additional activity under Article 3.4, and low-impact logging is the only practice applied to a particular patch of land. With a business-as-usual baseline, credits would be earned only for the carbon stock increase resulting from low-impact logging. Suppose that in a subsequent commitment period the forest is destroyed by wildfire. Without a business-as-usual baseline, debits could be equal to the entire stock of carbon destroyed by the fire. With a business-as-usual (or 1990 activity level) baseline, however, the Party is debited only for the increment of carbon attributable to low-impact logging because the rest of the forest would have been destroyed under business-as-usual. For further discussion of business-as-usual baselines, see Section 4.6 and Chapter 5.

2.3.4.3. Since 1990

Article 3.4 allows a Party to apply the decision made by the COP on additional activities to the first commitment period provided that the activities have taken place since 1990. This requirement is ambiguous because some activities may have started before and continued past 1990. Not all activities have a threshold, or point in time, when the activity may be said to begin.

In general, this Special Report treats ARD as events, the occurrence of which can be specified in time. Article 3.4 activities, whether they are defined narrowly or broadly, may be more in the nature of processes, which may make determining whether they occurred before or after 1990 more difficult. For example, "management" is a continuous process. Suppose that a forest has been managed using low-impact logging since 1980. The practice is ongoing, so the activity arguably has taken place "since 1990." Yet some of the observed stock increase undoubtedly is related to the practice of low-impact logging prior to 1990. If management or management practices are adopted as activities under Article 3.4, one option for applying a "since 1990" test (which is required for the first commitment period) would be to consider the effect of changes in management since 1990. This approach could be implemented by designating land on which there has been a change in management since 1990 as land under Article 3.4. A baseline for this land could be constructed by estimating the stock change that would have taken place under a continuation of the management regime in place as of 1990.

Similarly, the requirement that additional human-induced activities be "related to changes in" emissions and removals in the agricultural soils, land-use change, and forestry categories may suggest the need for a baseline against which to measure the change. Such a baseline could be the rate of change in carbon stocks on those lands in 1990.

2.3.5. System Boundaries

The boundaries that are set in defining LULUCF activities and projects can greatly influence the credit attributed to an activity and its true value in avoiding dangerous levels of GHGs in the atmosphere. Boundaries and other aspects of carbon accounting can affect decisions on allocating global warming mitigation funds between the energy and LULUCF sectors, as well as among different types of activities within the LULUCF sector.

2.3.5.1. Carbon Pools

One issue is defining which system components are included in the analyses. If system boundaries were defined to exclude some pools, there would be a risk of outcomes in which activities receive carbon credit when they actually result in net emissions. If one considers only the soil component, for example, converting forests or poorly managed pastures to well-managed pastures could increase carbon storage in Amazonian soils (e.g., Cerri *et al.,* 1996; Batjes and Sombroek, 1997). Forest conversions, however, lead to losses of carbon from biomass that much more than offset any potential gains in the soil. The same consideration applies to the question of whether conversion of natural savannas to improved pastures increases carbon stocks (Fisher *et al.,* 1994) or decreases them (Nepstad *et al.,* 1995).

Even within the soil sphere, the predominant pasture management system in Amazonia today results in substantial losses of soil carbon (Fearnside and Barbosa, 1998). In addition, if deep soil (below 1-m depth) is included in analyses and a long time horizon is considered, pasture can result in large emissions of soil carbon even if the stock in the surface soil has increased because trees have much deeper roots than pasture grasses: Some tree roots penetrate more than 8 m (Nepstad *et al.,* 1994). Roots supply carbon to the soil through exudates and root death (turnover); when deep-rooted trees are removed, the soil-carbon equilibrium in the deep soil shifts to a lower level over a period of decades (Trumbore *et al.,* 1995).

The question of whether subsidizing improved pasture management in Amazonia would result in carbon benefits is very important. Although some observers maintain that ranchers switching to improved pasture management will slow their rates of forest clearance (Faminow, 1998), evidence reviewed by Fearnside (1999a) indicates that increased capital supply to ranchers (from subsidies and from more profitable pastures) would have the opposite effect on deforestation. Although the impetus for the expansion of cattle ranching in Brazilian Amazonia currently comes largely from profit sources other than sale of beef (Fearnside, 2000a), saturation of the beef market will eventually reduce beef prices and contribute to limiting further deforestation for pasture. The possibility that this effect could form the basis of a cost-effective strategy for global warming mitigation has been questioned, however (Fearnside, 2000a).

2.3.5.2. Leakage

Leakage refers to the indirect impact that a targeted LULUCF activity in a certain place at a certain time has on carbon storage at another place or time. In spite of the linguistic implication of the term "leakage" that the flows involved are small and abnormal (as in water dripping from a leaky pipe), leakage may also include carbon flows that are large and predictable. The term "leakage" has generally been used in the context of project-based accounting to refer to impacts outside the project boundary (see Section 5.2.3), but leakage can also occur across other types of system boundaries. For example, action to reduce logging in Annex I Parties to reduce emissions reported under Articles 3.3 or 3.4 could result in leakage of benefits if a resulting reduction in timber supply led to increased deforestation in non-Annex I Parties. This effect would become much larger if tropical countries with high current rates of deforestation were to join Annex I. Similarly, the benefits of reduced logging could leak within Annex I Parties as a result of induced increases in activities that are not covered in the Kyoto accounting system, such as logging in areas that remain forested and fall outside the remit of Articles 3.3 and 3.4. The potential for leakage need not be a bar to undertaking LULUCF activities; unless these effects are either prevented or their magnitude quantified and deducted from the carbon benefits attributed to the activity, however, credit will be awarded in excess of the true benefits, and net GHG emissions may be higher than targeted levels. Although "leakage" has a negative connotation, in some cases positive effects can occur outside of a project area—as when a demonstration effect from a mitigation project leads to replication of the activities beyond the project boundaries (see Section 5.2.3). Leakage is not unique to LULUCF—the subject of this Special Report; it also can occur in energy-sector mitigation.

Leakage can be induced through several different mechanisms, such as activity displacement, demand displacement, and investment crowding. Activity displacement could occur, for example, if a silvicultural plantation or a forest reserve were created at a given location and the people who were formerly living at the site were displaced and continued to clear forest elsewhere. This kind of leakage can occur across international borders. An example is provided by the logging ban in Thailand instituted in 1989. Much of the logging activity formerly occurring in Thailand moved to neighboring Mynamar (Burma) and Kampuchea (Cambodia) (Leungaramsri and Malapetch, 1992). A similar effect is likely to result from the logging ban begun in 1998 in 18 provinces of the Peoples' Republic of China; the demand is likely to be satisfied by increased logging in other countries throughout Asia and beyond.

In addition to leakage at identifiable sites, a diffuse form of leakage occurs through global markets. Demand displacement occurs when a forest protection or management project reduces the supply of a marketed product, resulting in increased logging elsewhere to satisfy the demand for that product. Leakage can also result from supply displacement. For example, plantations that have been subsidized as global warming response options may have their carbon benefits negated when wood products derived from them simply replace products that would otherwise have come from elsewhere, or when output from subsidized plantations causes the price of plantation-produced wood to fall and unsubsidized plantations elsewhere consequently are cut and replaced with pasture or other low-biomass land uses (Fearnside, 1995). Leakage may also occur because of an investment crowding effect in which the targeted investment project (e.g., reforestation) crowds out the demand for other beneficial investments (e.g., replanting after harvesting) that are not targeted by projects.

Changes in national or international policies can lead to leakage—for example, when a government changes policy to lower the country's overall emissions but the emissions are displaced to other countries. This type of leakage is only a concern if the LULUCF activities are displaced to a country that does not have a full inventory of its emissions and a national cap (i.e., Annex I countries). For example, U.S. analysts assessed a scenario of 21-percent reduction of national forest harvests over the 2000–2040 period, which would lead to increases in wood imports from other countries. The net impact would depend on the relative efficiency of harvest of the imported wood compared to that of U.S. forest stands that the wood would replace (Andrasko, 1997). If the logging were shifted to Canada, the emissions might be captured in Canada's national inventory and consequent mitigation commitments, but if the logging were shifted to a country with no cap, carbon would be emitted but not accounted for.

Leakage is one effect of ARD activities that cannot be avoided through choices of definitions. Any large-scale establishment of new forests will create off-site effects, especially if the new forests generate a commercial wood supply. Additional wood supply on the world market will reduce the price of wood compared to prices without this additional supply. This reduction in wood prices will reduce the profitability of establishing or continuing forestry operations elsewhere (Adams *et al.*, 1993; IEA GHG R&D, 1999). Thus, reducing the establishment of other new forests would have a negative consequence, although where it occurs within Annex B countries it will be accounted for within those countries' adjustments to their assigned amounts. Any such effects in non-Annex B countries would not be accounted and are not readily prevented. If the lower wood price leads to reduced harvest from existing forests, on the other hand, it would be a beneficial side effect that further increases the atmospheric benefit. If Article 3.4 were to encourage increases in carbon stocks in existing forests, this effect could lead to lower harvest levels and counteract leakage effects from afforestation/reforestation.

As the scale of accounting increases, leakage errors should become less important. For instance, if accounting is based on observed changes in carbon stock levels at the national level, the data will implicitly capture leakage between sources within the nation. To the extent that national-level estimates omit certain LULUCF activities within the nation as well as indirect effects across nations, however, some leakage may still be a factor.

Program-level actions are generally much less prone to leakage than narrow projects that are tightly circumscribed in space, time, and subject matter. For example, broad policy initiatives are more likely to influence deforestation rates than are direct actions of limited scope. In tropical countries with large areas of remaining forest, reduction of deforestation has much greater potential climate benefits than other land-use change and forestry options (Fearnside, 1995). Reduction of deforestation also captures many more complementary benefits, such as maintaining biodiversity, watersheds, and water cycling. On the other hand, quantifying the direct effects of a program is much more difficult than quantifying the direct effects of more discrete activities. A probabilistic approach would be needed to compute the expected value of different options; under this approach, one would need to multiply the value associated with each outcome by the probability that the outcome will occur (Raiffa, 1968). For example, because of the great difference in potential benefits, investing in deforestation avoidance rather than relatively safe plantation silviculture options can be advantageous even in the face of a low probability of success for deforestation avoidance (Fearnside, 1999b). Optimal approaches are likely to include a mix of broad policy reforms and site-specific activities.

If accounting is to be adjusted for leakage, estimates would be needed of the magnitude of carbon benefits that are lost by each possible mechanism, and the carbon credit would have to be reduced accordingly. Because uncertainty in estimates of leakage magnitudes inevitably would be present, a further downward adjustment in carbon credit would be needed to assure a given certainty of achievement.

Global climate benefits may be reduced not only through leakage but also by other forms of project failure. The adjustments that would be needed to accurately represent net GHG benefits are similar regardless of the origin of the failure. Because the probability of success varies greatly among global warming response options, an adjustment of credit for these probabilities would be necessary to assure valid comparisons of the benefits for global climate associated with each option. Such adjustment is sometimes called "discounting," but we prefer to reserve this term for its traditional use as a time-preference weighting mechanism.

CDM projects that are undertaken to avoid tropical deforestation would face problems of minimizing, quantifying, and adjusting for leakage, as well as difficulties in establishing additionality and the possibility of reductions in credit resulting from uncertainties regarding the without-project baseline and the attribution of project effects. The large potential carbon and collateral benefits of avoiding tropical deforestation explain the high priority being given to achieving continued progress in addressing these matters. Many of these issues, including leakage, would cease to pose problems for crediting avoided deforestation, however, if tropical forest countries (other than Australia) were to join Annex I, thereby gaining access to emissions trading under Article 17, with the guarantee that deforestation reduction would be accounted for and potentially producing salable credits (Fearnside, 1999c).

2.3.6. *Timing Issues*

Timing issues involve the dynamics of emissions and removals, issues of duration (including ton-year accounting and risk), and time preference.

2.3.6.1. *Emissions versus Removals*

The inherent asymmetry in the timing of carbon emissions versus removals in biotic systems creates difficulties in initializing a carbon accounting system. Consider a forest that is managed for steady-state carbon stocks using a 100-year rotation. In any given year after the first century of management, 1 percent of the forest is logged while regrowth occurs in the other 99 percent of the forest area. In the area where logging occurs, most of the aboveground biomass carbon is immediately released or removed from the site. At that particular location, full recovery of the stock of biomass carbon will take 99 years. The carbon stock of the forest biomass as a whole remains constant, however, because regrowth is occurring throughout the rest of the forest. An accounting system that has been operating indefinitely will appropriately show zero net emissions or removals whether it treats the entire forest as a single unit or separately accounts for logging and regrowth. If, however, the accounting system only includes activities taking place after a given date, such as 1990, separate accounting of logging and regrowth would lead to reporting of net emissions because the emissions would be almost fully accounted for, but only a portion of the regrowth would be included. Conversely, if logging occurs periodically rather than on a continuous basis, net uptake of carbon could be reported under the accounting system if no logging happens to occur during the commitment period. The potential for accounting artifacts from these timing differences is a key issue that can be addressed through the definitions adopted for implementing Article 3.3 (see Chapter 3) and the approach to additional activities adopted under Article 3.4 (see Section 4.6).

2.3.6.2. *Duration of Sequestration*

Carbon sequestration in forest and other types of land cover is potentially reversible because carbon contained in terrestrial ecosystems is vulnerable to disturbances such as wildfires or pest outbreaks, as well as subsequent changes in management that would return some or all of the sequestered carbon to the atmosphere in addition to what would have been released if the sequestration activity had never taken place.[3] This situation contrasts with the case of avoided fossil fuel emissions because fossil fuels left in the ground in a given year will not be accidentally released in a subsequent year, even if the emission reduction activity itself is of a limited duration. For example, suppose that a homeowner replaces an incandescent bulb with a compact fluorescent bulb, avoiding one ton of emissions over the life of the compact fluorescent. This benefit is not reversed even if an incandescent bulb is installed at the end of the fluorescent's useful life.

It is important to understand that the logic behind considering either fossil fuel or forest as "permanent" is not based on the assumption that specific atoms of carbon will remain in the ground or in the forest forever. Instead, the effect of delaying for 1 year a given amount of fossil fuel burning or a given amount of deforestation will be to delay the release of carbon from the barrels of oil that would be burned or hectares of forest that would be deforested in subsequent years. To the extent that the emission displacement propagates forward until the end of the time horizon, the result is a "permanent" saving. Suppose that each ton of carbon is labeled ton_1 through ton_n, and a mitigation project avoids 1 t of emissions in year 1. As Table 2-4 illustrates, ton_1 is emitted in year 1 in the baseline scenario but is emitted in year 2 in the mitigation scenario. By the end of the time horizon (year n), n tons have been burned in the baseline scenario, but only n–1 tons have been burned in the mitigation scenario. The savings would not be permanent, on the other hand, if avoiding 1 t of emissions in year 1 results in additional emissions in any subsequent year before the end of the time horizon because of resource exhaustion or price feedbacks.

The potential reversibility of biotic carbon sequestration implies that substituting credit for LULUCF activities for reductions in fossil fuel emissions carries a risk of increasing long-term atmospheric CO_2 concentrations (Lashof and Hare, 1999). Fearnside (1999b) has argued, however, that this reasoning applies only to silvicultural plantations, and within the category of plantations it applies only to their role in carbon sequestration (as distinct from fossil fuel substitution). Lashof and Hare's argument is that, by allowing countries to emit more carbon from fossil stocks into the active carbon pool (biosphere + atmosphere), the increases in biotic carbon stocks that have been encouraged under the Kyoto Protocol as carbon offsets have a risk of subsequent release into the atmosphere through natural changes (which fossil carbon stocks do not have), and reduce the options

[3] Carbon sequestration must be distinguished from fossil fuel substitution, which can also be achieved by some LULUCF options (e.g., charcoal substitution for mineral coal). Fossil fuel substitution through forestry is just as permanent as avoided emissions through measures such as enhancing energy efficiency.

***Table 2-4**: Example of propagation of displaced emissions.*

	Year 1	Year 2	Year 3	....	Year n	Total Emissions (t C)
Baseline Scenario	ton_1	ton_2	ton_3	....	ton_n	n
Mitigation Scenario		ton_1	ton_2	....	ton_{n-1}	n-1

available for future responses in the forest sector because the capacity of these options to absorb carbon will have been saturated.

Fearnside (1999b) has argued that, in the case of avoiding tropical deforestation, the result is more like reducing fossil fuel carbon emissions than it is like carbon sequestration in plantations. Carbon stocks in areas of high-biomass old-growth forest, such as those in the moist tropics, are very unlikely to be allowed to regenerate to their present levels if these forests are cut down. Therefore, some of the carbon released by deforesting these areas is just as permanent an addition to what might be called the "most active carbon pool" (i.e., atmospheric carbon plus carbon in rapidly cycling stocks such as plantation biomass) as is release of fossil carbon. Activities by Annex I countries under the Kyoto Protocol to offset fossil fuel carbon emissions by helping tropical forest countries avoid deforestation keeps carbon out of this "most active pool" in the same way that avoiding fossil carbon emissions would and thus avoids carbon releases that would be just as irreversible as fossil fuel combustion (Fearnside, 1999b). These assumptions can break down if the area of remaining forest is small enough that it could be exhausted within the time horizon under consideration. If a country runs out of forest (or accessible or unprotected forest) within the time horizon, no net reduction in the atmospheric CO_2 concentration would accrue, although there may still be a benefit from the delayed emissions (see Section 2.3.6.3).

Concern regarding the duration of sequestration arises from the fact that forests are subject to degradation or destruction by forces such as climate change, extreme weather events under current climate regimes, insect outbreaks, diseases, and the entry of loggers or deforesters. These events can release all or part of the carbon contained in the affected forests. Table 2-5 shows the effect on carbon benefits in a hypothetical deforestation avoidance case. Here the hectares of forest cleared each year lose biomass from generalized degradation of the forest. Although a full hectare is gained by the cascading effect (as in the fossil fuel case in Table 2-4), the displaced hectare also degrades, proportionally reducing the benefit. In the example in Table 2-5, the total emission in the baseline scenario is 2.8 t C; in the mitigation scenario the total emission is 2.4 t C. If no degradation had taken place, the totals would have been 4 and 3 t C, respectively.

One approach to accounting for net impacts on the atmosphere is to treat removals and emissions as separate events. In this case, a full 1 t of credit would be awarded for each ton of carbon emissions avoided or removed in the year that it occurs; there would be an ongoing liability to account for any subsequent release of that carbon, however. The separate-events approach would require special attention in the context of the Kyoto Protocol: Any credit for carbon removed from the atmosphere by, for example, reforestation under Article 3.3 would have to be balanced by accounting for subsequent emission of that carbon regardless of the cause of the emission. In other words, once land enters the accounting system, full carbon accounting would have to be applied, even if it were to mean accounting for activities (e.g., forest degradation) that might not otherwise be included in the accounting system. A similar continuing liability would be required for project-based activities if unaccounted releases to the atmosphere are to be avoided. If full credit is awarded when removals occur, credits would have to be retired for any subsequent release of this carbon, even if it occurs after the end of the formal project period in a country that does not otherwise have quantified emission limits.

***Table 2-5**: Example of forest degradation effect on mitigation through avoided deforestation.*

	Year 1	Year 2	Year 3	Year 4	Total Emissions (t C)
Tons of Emissions per ha Deforested	1.0	0.8	0.5	0.4	
Baseline Scenario—Deforested Area					2.8
Mitigation Scenario—Deforested Area					1.8
Displaced Area					0.6

2.3.6.3. *Equivalence Time and Ton-Years*

An alternative approach is to compare activities that sequester (or release) carbon for different lengths of time by using an accounting convention or equivalency factor. The basic policy question that must be answered for any such system is how long carbon must be sequestered to be considered equivalent to "permanent" emission avoidance. (Article 3.3 of the Kyoto Protocol states that accounting should be based on verifiable changes in stocks in each commitment period—apparently precluding an equivalency factor approach). Several authors have analyzed the benefits of sequestration projects being accounted for on a ton-year basis rather than by requiring "permanent" sequestration. Ton-year accounting (Fearnside, 1995, 1997; Moura-Costa, 1996; Bird, 1997; Chomitz, 1998a; Dobes *et al.*, 1998; Tipper and de Jong, 1998; Moura-Costa and Wilson, 1999) would allow comparisons between avoided fossil fuel emissions and sequestration activities as well as among sequestration activities of different duration. Under a ton-year system, credit would be given for the number of tons of carbon held out of the atmosphere for a given number of years. A ton-year accounting system would provide a basis for temporary sequestration or delayed deforestation to be credited; the mitigation benefit from a given patch of land is greater the longer the carbon remains in place—which would be reflected in the credit earned.

As long as the policy time horizon is finite or a non-zero discount rate is applied to determine the present value of future emissions/ removals, even short-term sequestration will have some value. The explanation of this proposition is made clearer by considering the converse case: emission of 1 t CO_2 followed 20 years later by removal of 1 t CO_2. Although the net emission over the entire period is zero, there clearly has been an effect on the atmosphere. A ton-year equivalency factor can be used to determine the relative climate effect of different patterns of emissions and removals over time. For a given pattern, this factor will be a function of the time horizon and discount rate selected.

Alternative methodologies have been proposed to generate this equivalence time parameter. Tipper and de Jong (1998), for example, base their calculations on the difference between current atmospheric concentrations and the pre-industrial "equilibrium" concentration of CO_2 to derive a carbon storage period (Te) of 42–50 years following initial sequestration. Similar ranges have been proposed by Bird (1997; Te = 60 years) and Chomitz (1998b; Te = 50 years). Dobes *et al.* (1998) analyzed the effect of storage as a delay in emissions and calculated Te = 150 years.

A similar problem has been addressed through the use of GWPs to compare emissions of GHGs that have different residence times in the atmosphere (as well as different radiative forcing per molecule). Although this concept has limitations (IPCC, 1996; Smith and Wigley, 2000), it has been adopted for use in the Kyoto Protocol to account for the total emissions of covered GHGs on a CO_2-equivalent basis. Absolute Global Warming Potentials (AGWPs) are calculated by integrating the total radiative forcing of an emissions pulse over a 100-year time horizon with no discounting. Relative GWPs are the ratio of this integral for a given GHG to that of CO_2, which serves as the reference gas. This approach could be applied to compare carbon sequestration projects of different lengths, although there is no requirement in the Protocol to use the same conventions in this context. The reference is "permanent" (more than 100 years) removal (or emission) of 1 t CO_2. Based on the carbon cycle model used to calculate GWPs in the Second Assessment Report (SAR) (Joos *et al.*, 1996), this approach results in a reduction (or increase, in the case of an emission) in the cumulative CO_2-C loading of 46 ton-years relative to a reference scenario with no emission or uptake.[4]

A carbon sequestration project with a duration of 46 years, for example, can be analyzed as a removal of 1 t CO_2 in year zero followed by emission of 1 t CO_2 in year 46. As Figure 2-5 illustrates, the net reduction in atmospheric CO_2 burden from this project is the difference between the integrated effect of these two events (over a 100-year time horizon, for consistency with the GWP approach). The result is a reduction in atmospheric burden of 17 ton-years (the difference between the integrals of the two curves within the 100-year time horizon: 46 ton-years - 29 ton-years), or 37 percent of the effect of a "permanent" removal (or avoided emission). Note that the relative value of such a project would be higher with a shorter time horizon (100 percent for a time horizon of less than 46 years) and lower for a longer time horizon (6 percent for a time horizon of 500 years) (Fearnside *et al.*, 1999).

The foregoing calculations consider the difference between the mitigation and baseline scenarios by comparing the integrals of the atmospheric load of carbon over the time horizon. This approach focuses on carbon in the atmosphere for computing the benefits of LULUCF mitigation projects—as contrasted to some analyses that focus attention on carbon in trees (e.g., Moura-Costa and Wilson, 1999). The distinction between carbon in trees and carbon in the atmosphere is important because carbon in the atmosphere is subject to removal through natural processes that transfer it to sinks such as oceans and the biosphere, whereas carbon in trees is assumed to remain fixed. If carbon release is delayed because it is held in trees, the benefit is represented in this approach by the area of the tail of the second curve that is pushed beyond the end of the time horizon as a result of the delay in emissions (see Figure 2-6)—not by the larger area that would be described by a rectangle representing carbon stock in plantation biomass over the life of a mitigation project. In the approach proposed by Moura-Costa and Wilson (1999), an "equivalence time" is calculated as the point at which the area of the rectangle representing biomass carbon is

[4]The approximation of the output of the Bern model version used (but not published) in the SAR is given by:
$F[CO_2(t)] = 0.175602 + 0.137467 \exp(-t/421.093) + 0.185762 \exp(-t/70.5965) + 0.242302 \exp(-t/21.42165) + 0.258868 \exp(-t/3.41537)$.
where F is the fraction of CO_2 remaining in the atmosphere and t is the time after emission in years.

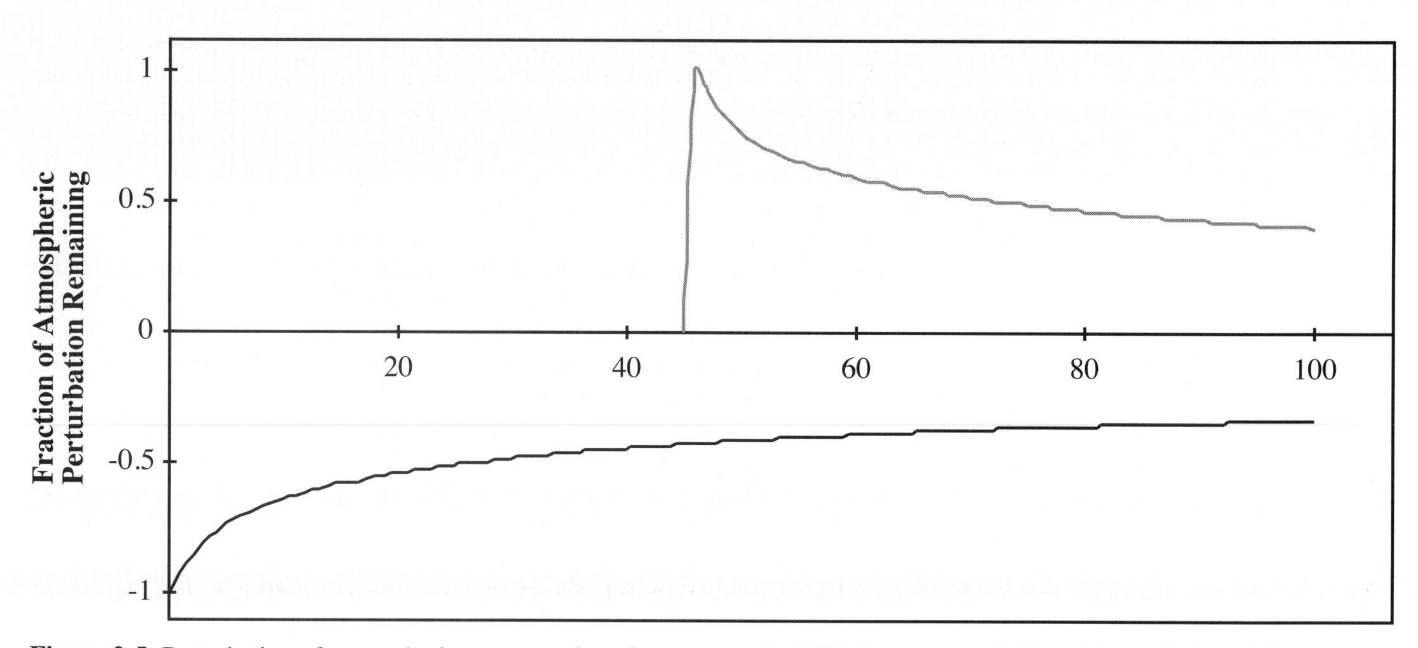

Figure 2-5: Perturbation of atmospheric concentrations from removal of CO_2 in year zero followed by emission of the same quantity of CO_2 in year 46. The initial change in concentration relaxes toward the unperturbed state, based on a reduced form of the carbon cycle model used to calculate Global Warming Potentials in the Second Assessment Report (Joos *et al.*, 1996). The net effect on the atmosphere over the 100-yr time horizon shown is the difference between the areas under the two curves, which is equal to the integral of the removal curve from year 54 to year 100.

equal to the area under the atmospheric carbon decay curve over the 100-year time horizon.

If the ton-year approach is adopted, incremental credit can be awarded for each year that carbon stocks remain sequestered. The cumulative award of credit would equal the credit from a "permanent" emission reduction of the same magnitude if the stocks remained intact for 100 years. If the stocks were released at any time prior to the 100-year time horizon, only the appropriate amount of partial credit would have been awarded (see Table 2-6).

Ton-year accounting requires some data that are not required by systems that are based solely on "snapshots" of carbon stocks at the beginning and end of each commitment period. However, for most LULUCF activities, such as plantation silviculture, the carbon stocks in intermediate years can be inferred from the snapshot data, supplemented with records of harvest and planting dates that should be readily available.

2.3.6.4. *Time Preferences and Discounting*

Various reasons have been suggested for applying discounting or some alternative form of time preference to determine the present cost/value to society of future carbon removals/emissions. Global warming initiates a change in the probability of occurrence of droughts, floods, and other unwanted events, rather than causing a one-time impact. If global warming impacts begin sooner rather than later, the number of lives that would be lost between the "sooner" and the "later" represents a net gain to be had from postponing global warming. Other arguments for discounting include expectations of changes in the wealth of the population suffering global warming impacts (wealthier people attributing less value to a given amount of monetary loss) (Azar and Sterner, 1996). The opposite relationship between wealth and value has been suggested for human life losses (Fankhauser and Tol, 1997), although it also has been strongly contested (Fearnside, 1998).

Discounting is the mechanism by which a value for time is normally translated into economic decisionmaking. Fearnside (1999b) has argued that postponing deforestation is a valid mitigation measure even if the forests in question are later cut,

Table 2-6*: Credit as a function of project duration.*[a]

Project Duration (yr)	Percentage of Full Credit
0	0.0
10	7.4
20	15.0
30	22.9
40	31.2
50	39.9
60	49.3
70	59.4
80	70.6
90	83.3
100	100.0

[a]Illustrates partial credit that would be received by projects that sequester carbon for various durations using the ton-year derived by analogy to 100-year Global Warming Potentials, as illustrated in Figure 2-4.

including cutting up to the theoretical maximum of clearing all forests in a country. The credit for such a delay depends on two key parameters: time horizon and discount rate (or another alternative time-preference scheme) (Fearnside, 2000b). From a carbon perspective, under some conditions postponing the clearing of a given number of hectares for a year is equivalent to avoided emissions by reduced combustion of fossil fuels.

Discounting can radically alter choices of energy sources and mitigation options (Price and Willis, 1993; Fearnside, 1995, 1997; Marland *et al.,* 1997). Discounting is needed for comparison of energy and forestry mitigation options and, within the forestry sector, to establish an equivalence between silvicultural plantations and avoided deforestation.

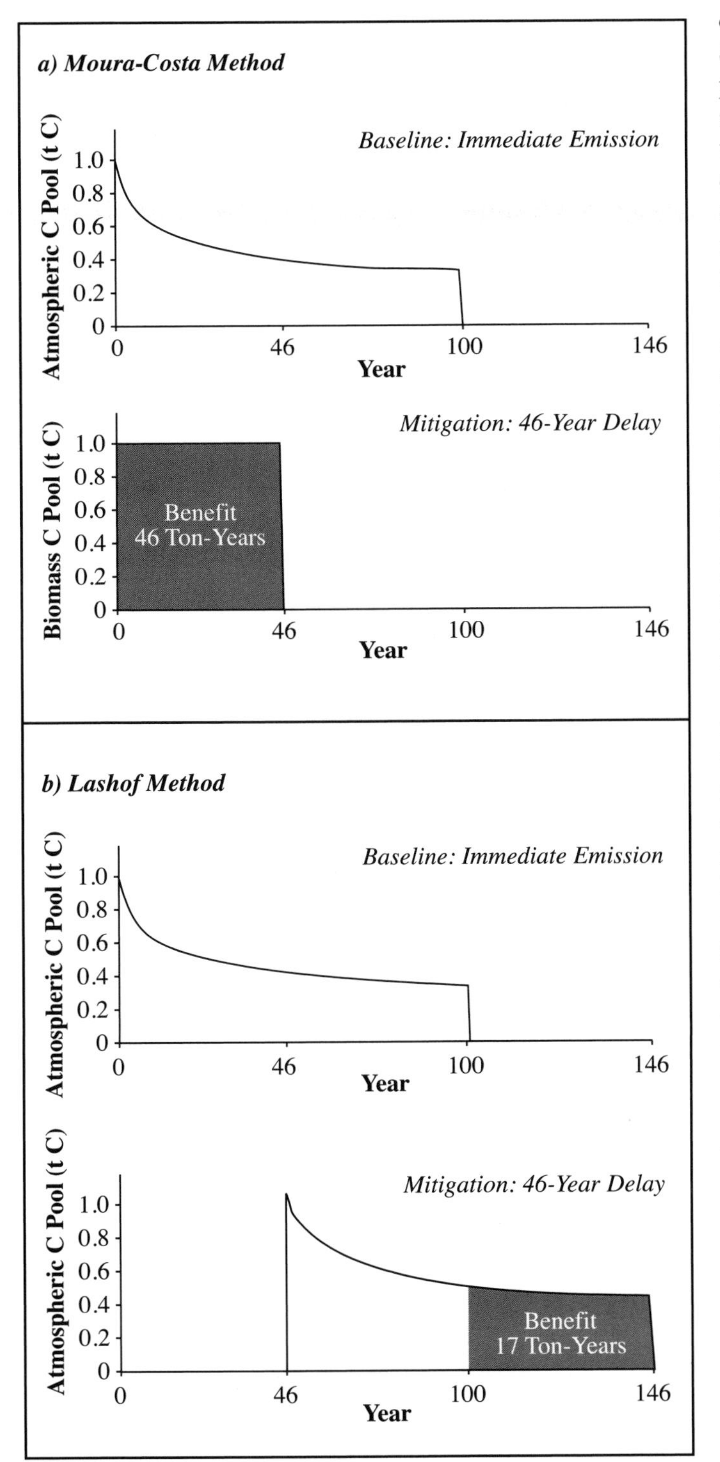

Figure 2-6: Alternative methods of crediting carbon sequestration.

The length of the time horizon has a strong effect on the importance of discounting. As time horizons become longer, the distortions become greater if no discounting is applied. In the case of forest sector options that can transfer carbon to very long-term pools, these pools can dominate the results if very long horizons are considered without discounting (Fearnside, 2000c). In the case of an infinite time horizon, equilibrium conditions will apply. Slow buildup of carbon in very slow turnover classes of wood products dominates the results at equilibrium but occurs at such remote times that it has little bearing on present decisions when discounting is applied. These problems also apply to calculations made under the assumption that the shadow price of carbon increases at the same rate as the discount rate for money, thereby allowing analysis without discounting carbon (Fearnside, 1995).

Agreement on a discount rate or other time-preference weighting arrangement for carbon facilitates comparison of forestry options with fossil fuel substitution (Fearnside, 1995, 1999b). Interpretation would be greatly simplified if the discount rate chosen were consistent with choices for global warming potentials (e.g., Lashof and Ahuja, 1990). GWPs adopted for use in the Kyoto Protocol are based on a time horizon of 100 years with no discounting (Section 2.3.6.3), which is a special case form of time preference. Discounting of carbon need not be the same as for money, although some observers advocate that the same rate should be applied (e.g., van Kooten *et al.,* 1997). The implications of discount rates as high as those for money are substantial for the relative impacts of different activities (Fearnside, 1997). The discount rates for other purposes—such as private investment decisions, public expenditures, and public regulation of renewable natural resources management—all have independent rationales. Because decisions are so sensitive to discount rate choices (e.g., the difference between a 3- and 6-percent annual discount rate is a factor of 20 over the course of a century), the consequences of allowing choices on global warming decisions to be determined by discount rates that are derived in other spheres could be severe.

2.3.7. Accounting for Uncertainty

Uncertainty in accounting for land-use change and forestry activities includes not only measurement uncertainty but uncertainty in defining and quantifying baselines (when they are used) and uncertainty related to the interpretation of the Protocol's requirements, including the definitions of key terms. Discussion of the magnitude of these uncertainties and approaches for minimizing them appear in Section 2.2 and in subsequent chapters related to specific Articles. This section briefly describes options for incorporating uncertainty into the accounting framework.

All GHG emissions and removals reported under the Kyoto Protocol will be subject to uncertainty to varying degrees. Dispersed emission sources, for example, face many of the same measurement challenges as LULUCF activities. Uncertainty enters the accounting system in a different way under the accounting framework adopted for LULUCF, however, than it does for sources that form part of the baseline for the Kyoto Protocol. In the latter case, systematic and random errors will be present in the 1990 emissions baseline and the emissions inventory during the commitment period. As long as consistent methods are used to estimate emissions in both periods, the potential to introduce bias into the accounting system will be minimized. On the other hand, LULUCF activities for which Annex I countries gain credit will only enter as a credit during the commitment period for most Annex I Parties. As a result, systematic errors are not offset through subtraction of the same error during the baseline period. Similarly, any change in measurement methods would affect only net emissions during the commitment period, without a compensating change in emission baselines (and thus assigned amounts under the Kyoto Protocol).

Accounting rules can be used to adjust for data limitations and uncertainty. For example, conservative estimates of carbon benefits can be applied by including all pools expected to have reduced carbon stocks and only a selection of pools expected to have increased carbon stocks; only the pools that are measured and monitored would be claimed as a carbon benefit (Sathaye *et al*., 1997). Similarly, measurement uncertainty could be accounted for by adjusting estimated fluxes, based on the uncertainty in this estimate, in the direction that is likely to understate removals and overstate emissions. This approach would provide an incentive to reduce uncertainties to the extent that it is cost-effective to do so, but it would not require expensive monitoring of carbon pools that do not significantly affect the overall carbon balance of a site.

2.4. Methods

2.4.1. Introduction

The amount of carbon held in the vegetation and soils of terrestrial ecosystems varies spatially and temporally as a result of natural processes and human activities. Sources and sinks of non-CO_2 GHGs (methane and nitrous oxide) are also affected by changes in land use. The issue addressed here is whether there are methods of measuring stocks, losses, and accumulations of carbon, as well as changes in the flux of CO_2 and non-CO_2 GHGs, in ways that are compatible with the requirements of the Kyoto Protocol—in particular, whether the precision and costs of such methods will enable changes in carbon stocks and changes in the flux of GHGs to be determined satisfactorily over the commitment period 2008–2012.

Measuring carbon sequestration or release during the commitment period as a result of human-induced activities depends on two factors: first, the type of activities to be included—namely, activities considered in Articles 3.3 (ARD), 3.4 (other LULUCF activities), and 3.7 of the Protocol—or full carbon accounting; and second, the scale of interest (project or national level). Table 2-7 presents an analysis of the various methods in terms of scale of applicability, suitability for project-level versus national-level carbon monitoring, costs, accuracy, and so forth.

Methods exist for measuring the amount of carbon in all components of terrestrial ecosystems, as well as for measuring changes in this amount. The methods vary in complexity, precision, accuracy, and cost. Different methods are appropriate for different pools and components of terrestrial carbon and for different temporal and spatial scales (see Table 2-7 and Figure 2-7). Methods used to measure carbon, or a change in carbon, are different from those used to attribute cause to an observed change in carbon (e.g., direct human activity versus natural causes). This distinction is important because the Protocol is concerned with human-induced, rather than total, changes in carbon. Even the most direct measurements on small plots do not distinguish mechanisms or yield attribution. Attribution must be inferred from controlled experiments or from ecosystem process models that are based on the mechanism thought to be held responsible for the change (e.g., land-use change versus CO_2 fertilization).

2.4.1.1. Uncertainty, Precision, Accuracy, and Costs

There is a widespread perception that accounting for carbon in the biosphere is inherently more difficult than accounting for carbon emitted by the burning of fossil fuels. This perception is only partly true. Determining the amount of carbon held in vegetation and soils does not require measurement of all of the carbon in all areas. Well-established statistical sampling techniques such as stratification and random sampling can be used to determine the carbon stocks of the biosphere as precisely and accurately as would be possible with complete measurement.

Neither accuracy (the absence of bias) nor precision (how well a measurement can be repeated) is compromised when appropriate statistical methods are used. Nevertheless, in statistical sampling there is a tradeoff between precision and effort (cost). A greater number of samples reduces error (i.e., increases precision) in the estimate but also requires more effort. In fact, each successive increment in precision requires a proportionately greater increase in effort. The decision about whether or how precisely to measure a carbon sink will depend on the effort required for that precision and the magnitude of the expected sink. Small changes in carbon or changes that require a large effort for measurement may not be worth the effort. Table 2-7 gives the costs of different methods in US$ per hectare for project-level measurements. Costs per ton of carbon are not given because the costs vary from country to country not only with the desired precision but with the types of land-use change and the magnitude of the change in carbon per unit area. Thus, costs per ton of carbon at the national level must be determined for

Table 2-7: *Characteristics of methods for determining changes in carbon storage.*

	Scale of Applicability	Time Span	Parameter Assessed	Suitable to Monitor ARD Activities	Applicable when Alternative Definitions for ARD are Chosen	Suitable to Monitor Soils and Additional Activities[1]	Suitable for Full Carbon Accounting	Sampling Density	Costs[2]	Accuracy[3]	Verifiability
Vegetation Inventory	0.01–10^9 ha	1–100 yr	Aboveground stemwood volume and increment, harvesting and mortality; derived from whole-tree biomass	Yes, but design of sampling must be adapted to cover Kyoto lands	Yes, provided forest definitions used in inventories are adapted	Mainly for specific additional measures that impact forest C stock, such as thinning, fertilization, etc.	No—usually excludes soils	Project basis: 400 plots on 5,000 ha; in national-scale inventories: 1 plot represents 1,000 ha	US$ 0.05–0.6 ha^{-1} in national scale inventories; US$11–18 at project levels (10,000 ha)	*Area*: s.e. = 0.4% *Growing Stock*: s.e. = 0.7% *Increment*: s.e. = 1.1% (Tomppo, 1996)	Relatively easy
Soil Inventory	0.1–10^3 ha	10–1,000 yr	SOC stock and changes over time	Yes	Yes	Yes	Assesses one compartment only	Depending on soil heterogeneity ~300 sample points per 10,000 ha; one sample for every 10-cm depth	US$ 3–20 per sample	2–3% error for analytical precision; total error much higher due to spatial heterogeneity and sampling error	Relatively easy
Eddy Flux	-20 ha	Day – 10 yr	Net Ecosystem Production	For verification only	For verification only	For verification only	No—excludes harvesting and decay of wood products	Required sampling density to obtain a large area representative flux must still be determined	US$ 100,000 per site initial costs; US$20,000 yr^{-1} running costs	10–20%	Relatively easy through forest inventory and soil analyses
Flask Measure-ments	-10^9 ha	Decades	Atmospheric CO_2 concentration	No	No	No	No—excludes wood products	~80 sites in Northern Hemisphere	Unknown	Sample analysis is very accurate	Verification of analysis is relatively easy

Table 2-7 (continued)

	Scale of Applicability	Time Span	Parameter Assessed	Suitable to Monitor ARD Activities	Applicable when Alternative Definitions for ARD are Chosen	Suitable to Monitor Soils and Additional Activities[1]	Suitable for Full Carbon Accounting	Sampling Density	Costs[2]	Accuracy[3]	Verifiability
Satellite Remote Sensing	0.05–10^9 ha	Day – decades	Area (sometimes derived estimates of biomass and NPP)	Yes	Yes, provided spatial resolution is adequate	Suitable for monitoring, e.g., fire management; in general, all area-related parameters of non-ARD	No, mainly to assess areas	Integral coverage through pixel size	US$ 0.0002 ha^{-1} for the picture and same amount for labor to process it; aircraft-derived pictures more expensive	Precise for area measurements (15%); for biomass, less precise	Relatively easy with ground truth data
Ecosystem Modeling	0.1–1 ha	Day – 100s of years	NPP, NEP per compartment	When validated with on-site data	Yes, provided model can be parameterized with new forest data	Yes, when management activities can be modeled	Yes, if all components of C cycle are included in model	Usually integral coverage	Cheap once model is developed	Uncertain; subject to many assumptions	Difficult in long term
Biome Models	Grid – 10^9 ha	Day – 100s of years	NPP, NEP per grid	No	No	Often soils are included in these models, management not	Yes, if all components of C cycle are included in model	Usually integral coverage	Cheap once model is developed	Uncertain; subject to many assumptions	Difficult

[1]Additional activities could be low tillage, drainage of peatlands, reduced-impact logging, thinning, wood products recycling.
[2]Costs cannot be scaled up to larger regions using these per hectare estimates. Average costs per hectare will vary more with heterogeneity than with absolute area.
[3]In the absence of a reference, estimates of accuracy are based on expert judgment.

each country individually. Similarly, the combination of methods most appropriate for national-level estimates of change will vary for each country; we do not attempt to suggest a single appropriate combination.

There are at least two significant generic problems with the estimation of change in terrestrial biospheric carbon stocks. The first problem relates to resolution (i.e., the smallest detectable change). Because the rate of change of most biospheric pools is slow, particularly in relation to the size of the pool, resolvable changes in stock are typically not easily obtained for the larger pools.

The second problem is practical. Most countries do not have the established infrastructure required for regular measurement of biospheric carbon (although all Annex I countries have a regular forest inventory in place). Where no infrastructure exists, measurement of carbon to the required degree of precision and accuracy is an expensive and logistically complex exercise. Most of the developed countries, as well as some less developed countries, have at least part of the required infrastructure already in place: certifiable analytical laboratories equipped to measure the carbon content of soils and biomass; a national forest and soil inventory system; accurate soil and vegetation maps on which to base the sample stratification; trained field, analytical, and statistical staff; and a physical infrastructure that allows access to remote sites. Even where this capacity exists, the incremental cost of performing a national-scale carbon inventory may be substantial. Australia, for example, is investing an additional $5 million annually in anticipation of upgrading its carbon accounting system for the Kyoto Protocol. The costs may be greater in countries in which the inventory

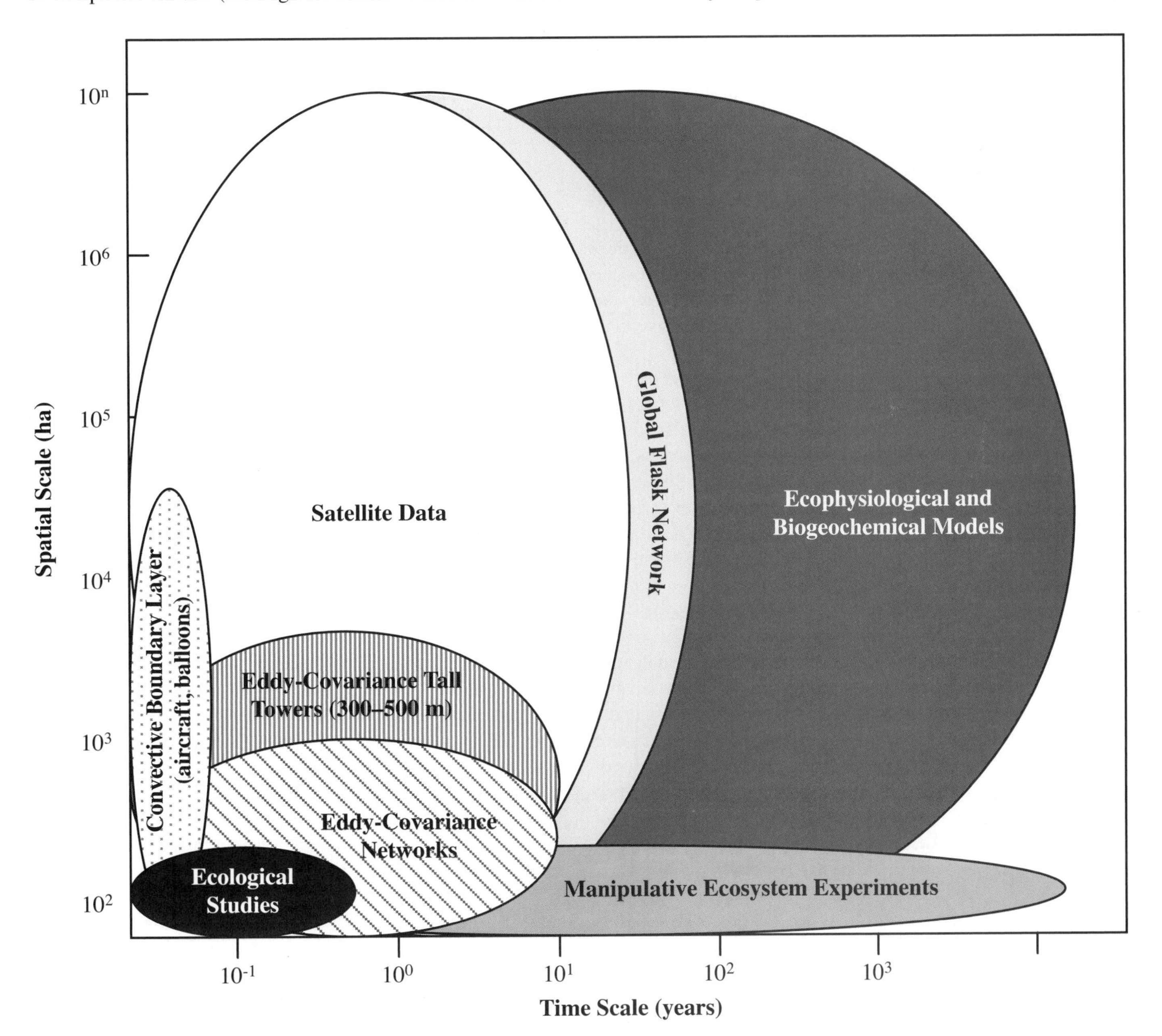

Figure 2-7: Range of temporal and spatial scales at which ecological processes occur, with related monitoring techniques. Forest inventories involve many of these methods and cover spatial scales from 1 to10^9 ha and temporal scales from 1 to 100 years.

infrastructure is less well developed. The use of models and stratified and multi-objective sample programs may reduce these costs, however.

The cost of conducting biospheric carbon inventories depends on the size of the area inventoried—but more on the range of ecological conditions within it because the spatial scale over which soil and biomass carbon varies is quite small: a few tens to hundreds of meters. The sample size needed to achieve the desired precision may be similar for a small country and a distinct region within a large country. Both need approximately the same analytical equipment and statistical treatment. Thus, the cost to a country will depend more on the range of different bio-geophysical regions that exist within its borders than on its actual size.

The total inventory costs for a single project will be substantially less than the inventory costs for an entire country, but the cumulative cost of inventorying several projects soon reaches the level of a national inventory. This is because the costs of forest inventory vary considerably depending on the desired accuracy, the accessibility of the forest, the availability of pre-information, the degree of automation employed, the availability of allometric relations, and so forth. At a typical project scale of tens of thousands of hectares, the costs are on the order of US$11–18 ha^{-1} (Nabuurs *et al.*, 1999). The typical cost of a national forest inventory, on the other hand, is on the order of US$0.05–0.6 ha^{-1}. Costs are declining with the increased use of automated data collection and analysis (see also Table 2-7).

The methods for determining changes in carbon storage are described briefly in the following sections. These methods are divided into methods for the measurement of stocks, the measurement of flux, and the measurement of area, as well as models. Table 2-7 summarizes the characteristics of different methods.

2.4.2. Measurement of Stocks

Two types of methods are used to measure losses or accumulations of carbon on land: methods that measure stocks of carbon and methods that measure fluxes. Measurement of stocks at the beginning of 2008 and at the end of 2012 (or at the date of commencement of the activity between 2008 and 2012) will yield the change in stocks that has occurred over the commitment period. Alternatively, measuring the flux of carbon into or out of an ecosystem over the 5-year period will also yield the net change.

2.4.2.1. Vegetation Inventory

The emphasis in this section is on forests. Nevertheless, there are appropriate and accurate techniques for measuring stocks of carbon in the vegetation of grasslands and other non-woody ecosystems.

2.4.2.1.1. Inventory techniques for stemwood volume

Techniques and methods for measuring terrestrial carbon pools that are based on commonly accepted principles of forest inventory, soil sampling, and ecological surveys are well established (Pinard and Putz, 1996; Kohl and Paivinen, 1997; MacDicken, 1997; Post *et al.*, 1999). All of these methods are suitable for project- and national-level surveys. Methods for measuring the fluxes of non-carbon GHGs are less well developed; at least for biomass burning, however, their magnitudes are often based on changes in carbon pools and corresponding emission ratios (Crutzen and Andreae, 1990).

Forest inventories are traditionally carried out to inform forest managers about the state of their forests in terms of area, species, age classes, growing stock (quantity of wood), and net annual increment (see Figure 2-8). These inventories traditionally have been carried out by ground measurements only; now, however, they usually include a combination of remotely sensed and ground data.

A continuous forest inventory typically consists of collecting pre-information, interpreting aerial photographs, designing the sample network, carrying out the field work, and processing data. The sample network consists of transects or grids of sample plots in the area to be inventoried (thus, they can be set up specifically for ARD projects). Along these transects, plots are selected in which at least the diameter of all trees and the height of some trees are measured. These data are processed and aggregated to the desired scale. Based on allometric relations and growth models, the total volumes and net annual increment can be assessed. When required, whole-tree biomass can be assessed using additional allometric relationships (or expansion factors). Thus, inventories can also detect changes in forest biomass that are caused by the adoption of an activity that falls under Article 3.4 of the Protocol—for example, changes in harvest intensity. A shortcoming of inventories, however, is that they do not cover very young stands accurately because measurements usually start when trees (seedlings) have reached a certain diameter at breast height. Below that limit, the stand is recorded as a clear-cut. This may be a problem in inventorying lands under Article 3.3.

The advantage of inventories is that they can be carried out for practically all scales of a project—from a single stand to a project covering millions of hectares. The sampling intensity of most national-scale inventories is such that individual plots represent large land areas. For example, 10 European countries have forest inventories in which a single field plot (~25 trees) represents around 200 to 1000 ha (Kohl and Paivinen, 1997).

There is an optimal size for plots in any given vegetation; this optimal size represents a compromise between statistical efficiency and sampling practicality. A common goal is to define the required precision of sampling, which then determines the number of plots required for a given plot size. The absolute uncertainty (e.g., 95-percent confidence) for biomass pools should be of similar magnitude to the absolute uncertainties of other components in the inventory. Inventories can be carried out so that they yield very accurate results; at the country level, for example, producing uncertainties (95-percent confidence) for forest land area of ±0.4 percent, growing stock ±0.7 percent, and total increment ±1.1 percent (Tomppo, 1996). Providing a relation between the number of plots on the one hand and costs and precision on the other is not simple, however, because that relation varies enormously between forests, countries, etc.

Continuous forest inventories have been carried out for all Annex I countries. There are considerable differences in precision and definitions between these countries, however. Therefore, variables reported by one country cannot always be compared to the same variables for another country (European Commission, 1997). The TBFRA 2000 project (UN-ECE/FAO, 2000) seeks to encourage data gathering based on a harmonized set of definitions.

Most developing countries possess only a forest area estimate assessed sometime between 1970 and 1990, although considerable progress has been made with remote-sensing techniques. Some time series on permanent plots of forest growth and biomass do exist, but most of the other data derive from scattered temporary plot measurements assessed under various research projects (Phillips *et al.*, 1998).

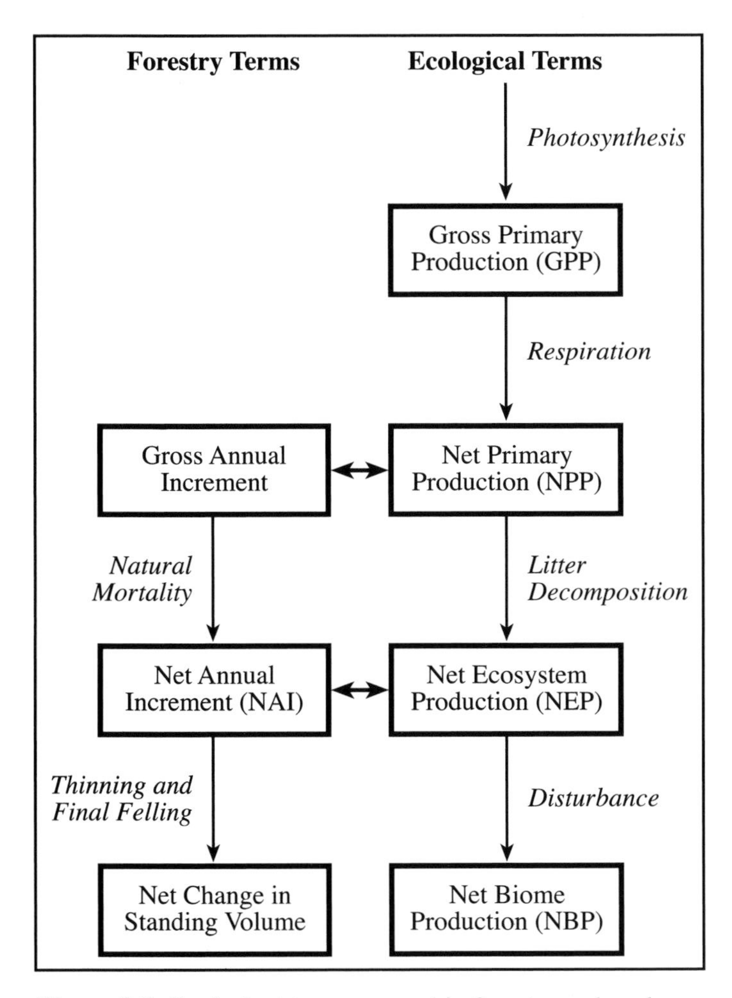

Figure 2-8: Equivalent terms as used in forestry and ecology. Note that forestry terms usually apply to the stemwood volume of forest ecosystems alone and ecological terms to the whole ecosystem.

Although national forest inventories would most likely detect lands under Article 3.3, the inventory sampling schemes will have to be adapted to cover forest dynamics with adequate precision and over the correct time period (2008–2012). Typical national inventories are carried out in a 10-year cycle and are often staggered, with different regions completed in different years. It may be possible to use models to adjust the estimates to the required reporting periods by using annual planting and felling data or to inventory more frequently. Sampling schemes may have to be adapted, however, where the area of lands under Article 3.3 constitutes only a small percentage of the total forest area. Detection of these lands may require increased frequency of remote-sensing data, and higher intensity of field data collection may be necessary to assess carbon stock changes accurately.

Most of the methodologies for assessing changes in forest area, whether based on terrestrial sampling or remote sensing, have limitations in detecting small changes in carbon and ARD activities on small patches of land. Terrestrial sampling allows for accurate measurement of a country's total forest area (error <1 percent); if only areas where ARD has occurred are to be considered, however, the error will be higher, especially if ARD areas are small in comparison to the total forest area. ARD activities on areas smaller than 1 ha might be difficult to assess, even with high-resolution satellites. The difficulty is not with the resolution of the satellite data but with determining the nature of the observed change: Is it related to ARD activities or is it the result of other factors, such as natural tree falls or the mortality of an individual tree?

2.4.2.1.2. Total tree biomass

For live tree biomass, diameters of a sample of trees are measured and converted to biomass and carbon estimates using allometric biomass regression equations. Such equations exist for many forest types; some are species-specific, whereas others, particularly in the tropics, are more generic in nature (e.g., Alves *et al.*, 1997; Brown, 1997; Schroeder *et al.*, 1997). Cutting and weighing a sufficient number of trees to represent the size and species distribution in a forest to generate local allometric regression equations with high precision, particularly in complex tropical forests, is extremely time-consuming and costly and may be beyond the means of most projects. The advantage of using generic equations, stratified by ecological zones (e.g., dry, moist, and wet; see Brown, 1997), is that they tend to be based on a larger number of trees (Brown, 1997) and span a larger range of diameters; these factors increase the precision of the equations. A disadvantage is that the generic equations may not accurately reflect the true biomass of trees in the project. Relatively inexpensive field measurements performed at the beginning of a project can be used to check the validity of the generic equations, however. It is very important that the database for regression equations contain large-diameter trees because such trees tend to account for more than 30 percent of the aboveground biomass in mature tropical forests (Brown and Lugo, 1992; Pinard and Putz, 1996). For plantation or agroforestry projects, developing or acquiring local biomass regression equations is less problematic because much work is done on plantation forestry (e.g., Lugo, 1997). Dead wood, both lying and standing, is an important carbon pool in forests that should be measured for an accurate representation of carbon stocks. Methods have been developed for this component and tested in many forest types. The non-tree component of the vegetation (understory, shrubs, mosses, lichens) may also be important in some forests and should be included in measurements.

The carbon content per unit mass of plant tissue varies little within a species and tissue type but can vary significantly between tissue types (e.g., fruits vs. wood) and function groups of plants (e.g., trees vs. grasses). Default values generally can be used, but they should be supported by a validated sample.

Roots are an important part of the carbon cycle because they transfer large amounts of carbon directly into the soil, where it may be stored for a long time. Most of the below-ground biomass of forests is contained in coarse roots—generally defined as >2 mm—but most of the annual growth is allocated to fine roots (Deans, 1981; Jackson *et al.*, 1997). Part of the carbon in roots is used to increase biomass, but carbon is also lost through exudation, respiration, and decomposition. Although some roots may extend to great depths (Canadell *et al.*, 1996), the overwhelming proportion of the total root biomass is generally found within 30 cm of the soil surface (Jackson *et al.*, 1996). Measuring the amounts of biomass in roots and their turnover is an extremely costly exercise. Therefore, regression equations are often used to extrapolate aboveground biomass to whole-tree biomass (Kurz *et al.*, 1996; Cairns *et al.*, 1997). The problem with this approach is that deforestation and harvests (as well as changing environmental factors) may change the relationship between aboveground and below-ground biomass. On the other hand, below-ground carbon might still be assessed from a known history of aboveground vegetation.

2.4.2.2. *Wood Products*

Wood products in use are part of a full forest sector carbon balance. Globally, about 3.4 billion m^3 of wood is harvested per year, excluding wood burned on-site (FAO, 1997b), and harvest rates are expected to increase at 0.5 percent per year (Solberg *et al.*, 1996). Of the total harvest, about 1.8 billion m^3 yr^{-1} is fuelwood, mainly from the tropics. Most of the harvest in the boreal and temperate zone is industrial roundwood. About one-half to two-thirds of the raw wood finds its way into wood products; the rest is used as energy or is otherwise lost. The annual production of wood materials therefore corresponds to a harvest flux of about 1.6 billion m^3, translating to a gross flux into wood products of about 0.3 Gt C yr^{-1}.

The long-term effectiveness of carbon storage in wood products depends on the uses of wood produced through project activities. In projects that stop logging (and perhaps those that stop deforestation, if some of the wood cut during deforestation entered the wood products market), the change in the wood products pool would be negative and would thus offset some of the carbon benefits from the project; this effect would have to be accounted for. In plantation projects, wood that goes into medium- to long-term products (e.g., sawn timber for housing, particle board, paper) represents additional carbon storage. Several (modeling) methods exist for accounting for the storage of long-lived wood products; these methods have been used to calculate the net changes in stocks in several countries (Pingoud *et al.*, 1996; Nabuurs and Sikkema, 1998; Skog and Nicholson, 1998; Winjum *et al.*, 1998; Apps *et al.*, 1999). These methods account for inputs of new products to the pool as well as decay, disposal in landfills, oxidation, and retirement of products from past use, accumulated as far back as individual country's records allow. An IPCC expert group for the land use and forestry sector of the guidelines for GHG inventories completed a report on a description and evaluation of the approaches available for estimating carbon emissions or removals from forest harvesting and wood products (Lim *et al.*, 1998).

If one of the additional activities (see Section 4.4.5) includes carbon in wood products, a method for monitoring changes in the carbon of that compartment will be needed. Monitoring of wood products is possible through national and international statistics on harvesting and trade. The net build-up of carbon in products might be derived from estimates of average lifetimes for different wood products. Keeping track of the wood products stemming from lands under Article 3.3 or 3.4 will be extremely difficult, however. The only practical approach may be to use estimates that are based on data from other forests.

2.4.2.3. *Soil and Litter*

2.4.2.3.1. *Woody debris*

Coarse and fine woody debris on the forest floor often accounts for about 20 percent as much carbon as living biomass. The amount of woody debris may increase markedly immediately following forest harvest, which typically leaves a large amount of organic debris on the soil surface (Covington, 1981). These residues will decompose within a few years following harvest, although there is little information on the carbon content of different types of litter. Sampling the volume of woody debris is as straightforward as it is for live trees, but assessing its mass is more problematic because decay affects wood density dramatically.

2.4.2.3.2. *Litter*

The litter layer—also known as the L and O horizons—is the layer of dead plant material that lies on top of the mineral soil. During forest regrowth, the litter layer may accumulate rapidly, so changes in its carbon content are an important component of a total carbon inventory in ecosystems (Richter and Markewitz, 1996). During a cycle of forest harvest followed immediately by regrowth, however, there is usually little overall change in carbon storage in the forest floor (Johnson, 1992).

Litter quantities change dramatically with the seasons. As a result, remeasurements must be made at the same time of the year as initial measurements. If litter were to be sampled together with the top layer of mineral soil, temporal variability in the standing crop of litter could confound detection of changes in soil carbon. Typically, dividing litter from mineral soil when sampling is not difficult.

The number of samples necessary to estimate the accumulation of surface litter will differ greatly between ecosystems. In some forests, as few as 10–15 randomly located samples may provide an accurate estimate of changes in the mass of the forest floor over large areas (e.g., Schiffman and Johnson, 1988); in shrub deserts, where the spatial variability of soils is enormous, it is often necessary to take a larger number of random samples in locations that are stratified by plant cover (e.g., Conant *et al.*, 1998). In any particular location, some preliminary sampling is probably required to determine the number of forest floor samples that is necessary to estimate the total mass to the desired level of accuracy.

2.4.2.3.3. Mineral soil horizons

The soil contains two major types of carbon: soil organic carbon and soil inorganic carbon (which occurs primarily as carbonates). Although SOC constitutes the majority of soil carbon in most soils globally (Batjes, 1996), soil carbon in arid and semi-arid regions may be dominated by SIC (Schlesinger, 1982). In fire-prone ecosystems, there may be significant amounts of charcoal present.

SOC held in the various layers, or horizons, of the mineral soil largely occurs as humus. Much of the humus is adsorbed to the soil mineral fraction, particularly to clay minerals (Richardson and Edmonds, 1987). A small amount of the total content of soil organic matter consists of dead roots, which become part of the soil organic matter upon their demise. In most cases, however, the amount of carbon associated with such roots is <5 percent of the total carbon in the soil profile (Ruark and Zarnoch, 1992).

There is a wide range of laboratory procedures available to measure SOC stocks, the most accurate of which is dry combustion (usually performed using a carbon analyzer). Changes in SOC stocks over a 5-year commitment period may be difficult to measure in some soils because although they are potentially large in absolute terms, they may be small compared with background levels (Batjes, 1999). Despite these difficulties, the rate of change in SOC stock during a commitment period can be measured (Lal *et al.*, 2000); because of high spatial variability, however, many sub-samples may be required to obtain a mean with an acceptable standard error (Izaurralde *et al.*, 1997; Garten and Wullschleger, 1999; Post *et al.*, 1999). Figure 2-9 (from Garten and Wullschleger, 1999), for example, shows the calculated minimum detectable difference in SOC as a function of variance and sample size for SOC changes after 5 years under a herbaceous bioenergy crop.

In some cases, the cost of demonstrating the change in stocks to the required level of accuracy and precision may exceed the benefits that accrue from the increase in stocks. As the size of the stock change decreases, there will be a point when the cost of demonstrating the change with sufficient confidence exceeds the value of the carbon offset provided. The cost of demonstrating a change in SOC stock could be decreased by developing and verifying locally calibrated models that can use more easily collected data (Section 2.4.5).

2.4.2.3.4. Sampling strategy

The number of soil pits or cores that must be sampled varies according to the topographic variation and spatial heterogeneity in the distribution of soils. At the large (regional/national) scale, a stratified random sample scheme is statistically efficient for soil data, provided that the basis for stratification accounts for a significant fraction of the variation in soil carbon contents. It is essential that the sample is accurately placed within the stratum, which requires the use of Geographic Information Systems (GIS) and Geographical Positioning Systems (GPS) or accurate field survey. For validation purposes, all sample locations should be geo-referenced as accurately as possible. The actual plot location should be marked to allow relocation of the plot reference point, for resampling in close proximity to prior sampling. High precision is requiring to avoid confounding

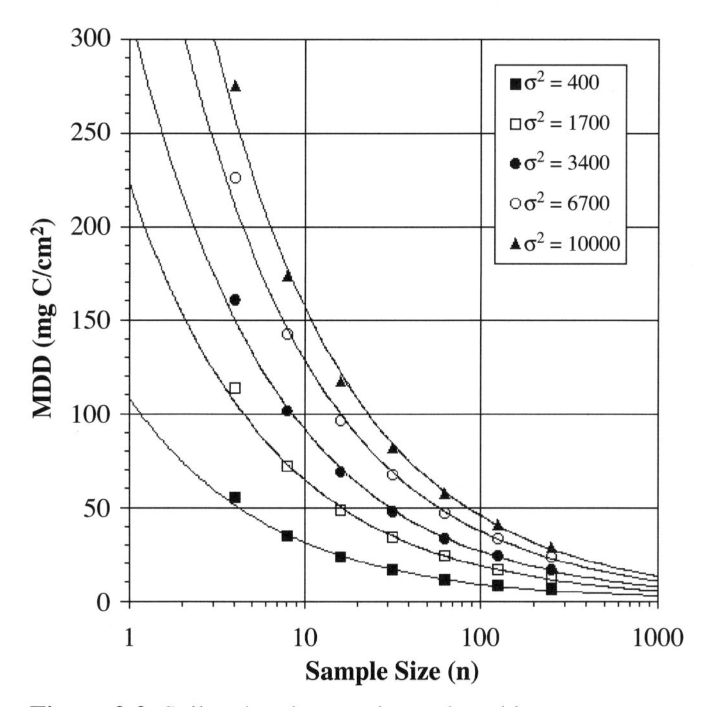

Figure 2-9: Soil carbon inventories under a bioenergy crop (Garten and Wullschleger, 1999).

by the high level of fine-scale variation that is common in soil carbon. Where an *a priori* randomly determined location is found to be inaccessible, it must be replaced with a sample having similar characteristics. The number of samples and the choice of location depend on the land-use and soil management systems.

At the project level, sampling strategies could be based on representative toposequences or catena for major land uses within the ecoregion of interest. The sampling strategy should be determined when the spatial distribution of the overlying plant community is known. Material should be collected at the end of each growing season (at the maximum stage of decomposition and with the maximum litter input), but at a minimum at the beginning and end of each commitment period. Uncertainty associated with differences in stock between two times can be substantially reduced by using paired samples; in other words, samples taken at time 2 should be from very close to the samples taken at time 1 and the difference calculated per pair, rather than on the accumulated averages (Lal *et al.*, 2000).

2.4.2.3.5. Sampling depth

Except in unusual circumstances and in peats, the amount of soil organic matter declines exponentially with depth (Nakane, 1976). Globally, only about one-third of the organic carbon in the surface 2 m is found at a depth of 1–2 m (Batjes, 1996). Losses or accumulations of soil carbon are greatest in the upper soil profile (0–15 cm), which should be sampled most intensively (Richter *et al.*, 1999). In the case of management changes on agricultural land (e.g., Smith *et al.*, 1997b, 1998), samples must be taken from lower in the profile because accumulations of carbon in the surface horizons may be balanced by losses of carbon at depth (Powlson and Jenkinson, 1981; Ismail *et al.*, 1994; McCarty *et al.*, 1998). Defining, *a priori,* a global soil depth to which carbon should be analyzed is not practical, however. The sampling depth should be below the depth that significant change in carbon is expected to occur.

Soil carbon changes in some situations may occur at substantial depths (e.g., 2–5 m in deep, tropical soils), whereas in cropped soils undergoing management change, almost all of the change may occur in the top 30 cm. The important issue is that the depth used in one inventory should match the depth used for that location or land-use type in the next inventory. The depth definition should be left open, subject to consistency between inventory times. Determining the effective sampling depth is not easy when the surface level is subsiding or aggrading (e.g., in highly organic bog soils, soils under arable agriculture, or soils that are subject to rapid erosion). Simultaneous measurements of bulk density are essential in all cases (Lal *et al.*, 2000); such measurements can be used to calculate an effective soil depth in mineral soils. Measuring soil depth as a function of mass, rather than distance from the surface, may be desirable because management activities may significantly change soil bulk density between measurements, substantially altering the amount of carbon found in different increments of distance from the soil surface. Failure to account for bulk density changes could lead to large artifacts in apparent carbon storage (Ellert and Bettany, 1995). In organic soils, an absolute reference level is required, which must be surveyed from a stable datum.

2.4.2.3.6. Analytical methods

Accurate laboratory assessments of soil carbon content require the use of carbon (combustion) analyzers. Analytical methods based on mass loss on ignition or wet digestion (see Nelson and Sommers, 1982; Tiessen and Moir, 1993) are prone to biases. Although these biases can be calibrated (Kalembasa and Jenkinson, 1973), they add another level of uncertainty to the estimate. Methods for sample preparation and analysis for SOC and SIC are well documented elsewhere (Lal *et al.*, 2000). The carbon stock per unit area of land can be calculated from the percentage of carbon and the soil bulk density. For transparency and verifiability purposes, soil samples can be geo-referenced and archived in a dried state in a dark, cool place for future carbon analysis, and original data can be retained. Some archived samples should be re-analyzed at the same time as repeat samples to confirm that there has been no drift in analytical precision. Methods also exist to measure different components of SOC (see Section 4.2).

Modern dry combustion procedures have an analytical precision of ±2–3 percent. The error associated with assessments at the regional and global scales also depends on errors introduced through spatial variability and sampling error and therefore will be much higher than 3 percent. Establishing quality protocols for all methods is important. The cost of analysis varies from region to region, by number of samples, and by efficiency. On a global scale, the per-sample cost of carbon analysis may range from US$3 (where labor costs are very low) to about US$20, with significant proportions of the total cost accounted for by sampling (high labor cost), sample preparation, bulk density measurement, inorganic carbon analysis (or pre-acidification of samples to remove carbonate), archival of samples, and quality control efforts. This cost range is broadly consistent with that reported for a Canadian study by Izaurralde *et al.* (1997).

Special consideration is necessary for peatlands and other wetland soils—collectively known as Histosols—in which the layer of undecomposed organic matter may be several meters in depth. These soils are vulnerable to large losses of organic matter upon changes in regional drainage or soil temperature. Particularly in areas that have been drained, accurate measurement of changes in carbon stock in peat soils requires reliable measurement of carbon loss through decay of organic material. Accurate measurement of changes in carbon in peat soils probably requires fixed reference points for measuring subsidence from decay. When subsidence is being measured, corrections should be made for depth changes resulting from changes in soil moisture content. Estimates of changes in the storage of carbon in peatlands must avoid biases that may result from changes in bulk density because of compaction of surface layers. Often, the age of peat at a given depth may be determined by ^{14}C or ^{210}Pb dating, allowing workers to assess additions or losses of organic carbon above a depth of known age.

2.4.3. Measurement of Flux

A range of methods exists for measuring and estimating the flux of GHGs from the land's surface over different spatial scales. Some of these methods are discussed in the following sections. The development of methods is an active area of research.

Flux measurements are the only practical way of measuring non-CO_2 GHGs. Unlike carbon, methane and nitrous oxide do not exist as stocks on land, and changes in inventories cannot be used to infer fluxes to the atmosphere. These fluxes must be determined through direct measurement of changes in the concentration of the gas in chambers that enclose a piece of an ecosystem, through emissions ratios that relate the non-CO_2 fluxes to CO_2 flux, or with models that calculate net exchange from knowledge of soil properties. Estimates of flux for these gases are generally more uncertain than fluxes of CO_2 because of their greater variability over time and space.

2.4.3.1. Local Scales (Less than 1 km^2)

Chambers may be used to enclose a small area, or an individual component, of an ecosystem (e.g., soils, stems, leaves). Changes in the concentration of CO_2 (or other gases) within the chamber or differences between the concentrations in incoming and outgoing air are used to calculate flux.

Several techniques exist for measuring the flux of CO_2 for an entire ecosystem (less than 1 km^2) without enclosures. The technique in most common use today is the eddy correlation/covariance technique (described below), which measures flux directly. Other techniques that are applicable at the same spatial scale infer a flux rather than measuring it directly. These techniques include the boundary layer gradient method (Griffith and Galle, 1999), tracer techniques (Leuning *et al.*, 1999), upwind-downwind measurements (Denmead *et al.*, 1998), and long-path infrared spectroscopy.

The *eddy covariance* method is a well-developed method for measuring the exchange of CO_2 between terrestrial ecosystems and the atmosphere—that is, Net Ecosystem Production (NEP) (e.g., Baldocchi *et al.*, 1988, 1996; Moncrieff *et al.*, 1996; Valentini *et al.*, 2000; Aubinet *et al.*, 2000). Measurements are continuous and semi-automatic (often with an hourly time step), and the net flux of CO_2 entering or leaving the ecosystem integrates an area typically on the order of 20 ha. The precision integrated annually has a confidence interval of ±30 g C m^{-2} yr^{-1} [0.3 t C ha^{-1} yr^{-1}]. A slight modification of the method, a *relaxed eddy correlation* technique, is applicable to non-CO_2 GHGs (Rinne *et al.*, 1999).

In some sites that are affected by complex topography, ecosystem nighttime respiration can be underestimated because of air advection (e.g., drainage flows), leading to an overestimation of the annual carbon balance (greater sink or smaller source of carbon). Methods have been developed to recognize when these cases occur, and corrections have been proposed (Aubinet *et al.*, 2000).

Automated eddy covariance measurements of CO_2 fluxes have now been made at more than 65 sites worldwide (FLUXNET16), with standard measurement protocols and storage systems. A typical cost for a complete eddy covariance system is on the order of US$50,000. The cost of site infrastructure is additional and will vary according to the remoteness of the site, the height of the vegetation (whether a tall tower must be built), and the existence of other facilities. The cost of a small mast over a pasture and an insulated container to house computers may be less than US$500; the typical cost of a 30-m walk-up scaffold tower for flux measurements is US$20,000.

The eddy covariance method determines overall net carbon exchanges at stand level. It is repeatable in time and is non-destructive. It can assist in the determination of carbon budgets of forests and other terrestrial vegetation on a project level, in the validation of forest growth models, and in verifying studies of carbon balances determined by stock change methods.

The rapidly expanding network of flux towers greatly assists in the understanding of land-atmosphere exchanges but does not yet provide reliable evidence of the magnitude and location of carbon sinks because a tower site does not operate over a large enough and sufficiently unbiased sample area to represent a land cover throughout its disturbance cycle. Furthermore, the method is limited to generally flat terrain with uniform vegetation. Nevertheless, the data confirm evidence for carbon sinks quantified by stock change measurements on a statistically representative sample.

2.4.3.2. Landscape or Regional Scales

Numerous techniques exist and are under development to infer mesoscale fluxes of CO_2 from spatial variations in atmospheric concentrations. At the smallest scales are methods that are based on measuring the build-up of CO_2 concentrations above an ecosystem during temperature inversions (usually nocturnal) (Woodwell and Dykeman, 1966; Gallagher *et al.*, 1994). At a larger scale (upward of 50 km^2), measurements of concentration are made from tall towers or balloons to approach or reach the top of the boundary layer (Bakwin *et al.*, 1995, 1997). The use of aircraft allows flux to be inferred at still larger scales (Choularton *et al.*, 1995; Hollinger *et al.*, 1995). Simultaneous measurement of CO_2 concentrations and other tracers (e.g., radon) and past trajectories of air for continuous measurement sites may also be used to infer fluxes of CO_2 and non-CO_2 gases.

An example of a technique that infers flux at this intermediate scale is the *convective boundary layer (CBL)* budgeting approach. The concentrations and fluxes of CO_2 in the atmosphere, higher than a few meters above the land surface, are the result of integrated individual contributions from the elements of the landscape at scales on the order of 50–100 km (Stull, 1988). The CBL budget approach exploits the natural integrating properties of the well-mixed atmospheric boundary layer, allowing average surface fluxes to be obtained over relatively large regions for time periods on the order of days rather than hours.

One of the main limits to widespread use of the CBL budgeting technique is the relatively high cost involved, because of the need to employ aircraft to sample above the top of the daytime CBL; these costs generally are on the order of US$1,000 per day. Another limitation is that the same air mass theoretically should be followed as it traverses the landscape. Practical experience has shown that, even with sophisticated weather forecasting tools, this tracking is difficult. Particularly when advection of air masses with different concentrations of the entity of interest occurs, failure to effectively track an air mass can lead to large errors in regional flux calculations using the CBL method. Additional limitations include that fact that the measurements are for relatively short, discontinuous periods and that the resulting flux includes an undefined area of land and all fluxes of CO_2 (fossil as well as biotic).

2.4.3.3. Continental Scales

Terrestrial sources and sinks of carbon (fluxes) can be estimated over very large areas by combining two types of information: spatial and temporal variations in atmospheric concentrations of CO_2 and models of atmospheric mixing or transport. At present, there are roughly 100 sites routinely operated by a few air sampling networks. Most measurement sites are located in the marine boundary layer, however—far from the direct influence of continental ecosystems. Better sampling of the continental air sheds is critical to reduce current uncertainties about terrestrial carbon sources and sinks. One also must make measurements through the air column to integrate the fluxes of CO_2 from ecosystems over a large spatial domain, typically between 100 and 1,000 km. CO_2 measurement by very high television towers (Bakwin *et al.,* 1995) and aircraft (Choularton *et al.,* 1995) appear to represent the most practical solutions to pursue that goal.

Atmospheric transport models are based on wind fields that are derived from numerical climate models or weather forecast models. Large-scale mathematical inversions of CO_2 data with these models provide an estimate of the net flux of carbon from all surface processes. These models cannot be applied to determine the carbon exchange for projects, and, in general, they cannot be used to attribute fluxes to specific activities.

Recent analyses show conflicting results in apportioning CO_2 fluxes over the continents. A debated issue is the partitioning of the Northern Hemisphere terrestrial CO_2 flux among North America, Europe, and Asia. For example, a study by Fan *et al.* (1998) found most of the northern mid-latitude sink in North America, whereas Bousquet *et al.* (1999) found most of it in Asia. The high estimate for North America is also inconsistent with analyses that are based on land-use change and forest inventories (Houghton *et al.,* 1999).

Current inversion analyses depend to some extent on *a priori* knowledge about surface fluxes and patterns of atmospheric transport. Thus, one must verify carefully that the solution is mostly constrained by atmospheric observations via the atmospheric transport, not by any of the model's artifacts. The current generation of inverse models of the global carbon cycle can generate intriguing hypotheses regarding the possible location of regional carbon fluxes, but these models are insufficiently constrained to be reliable for specific activities or inventory purposes. This situation is unlikely to change dramatically within the next decade, but in the longer term, inverse modeling in conjunction with large-scale flux measurements may pass from the research into the operational arena.

In summary, there are a range of flux measurement and flux estimation methodologies under active development that can be expected to improve the ability to determine average net surface fluxes over a range of spatial scales for CO_2 and non-CO_2 GHGs. In some cases, isotopic information can be used to partition fluxes between different source types (e.g., fossil fuel vs. biotic sources). Methods that are applicable at regional and larger scales are not expected to provide a basis for CO_2 flux estimation that is useful for the purposes of the Kyoto Protocol in the first commitment period. Nevertheless, information from these larger scale methods can provide constraints on total surface fluxes that may be useful for full carbon accounting and for validating process-based models of GHG fluxes.

2.4.3.4. Horizontal Fluxes of Carbon

Carbon may be lost from ecosystems in the form of dissolved or particulate organic compounds that are found in runoff waters and during the erosion of soils that often accompanies human land use. Some carbon may also be lost in the deep seepage of soil waters that enter groundwater. Typically, the loss of dissolved organic compounds is a small fraction of the pool of carbon in soils or the changes in soil carbon storage that result from human activities. Globally, about 1–10 g C m^{-2} yr^{-1} [0.01–0.10 t C ha^{-1} yr^{-1}] are lost in surface runoff from natural ecosystems (Schlesinger and Melack, 1981; Kortelainen *et al.,* 1997). Losses of soil carbon from agricultural soils are dominated by oxidation (Schlesinger, 1986). In the case of soil erosion, a large fraction of the organic matter that is lost may be deposited in floodplain and riparian habitats in downstream locations (Lal, 1995; Dean and Gorham, 1998; Stallard, 1998). The accumulation of soil carbon in these ecosystems can be assessed using sampling methods outlined above. It is important to recognize, however, that increments in these deposits do not necessarily represent net sequestration of atmospheric CO_2.

Horizontal fluxes of carbon are generally missed by atmospheric measurements of flux, which points to the fact that exchanges of carbon between the atmosphere and the land surface may not be equivalent to changes in the terrestrial storage of carbon.

2.4.4. Measurements Using Remote Sensing

Satellite remote sensing may be used to determine initial areas of different land-cover types, including forests; to determine the extent, rates of change, and locations of activities that result in forest clearing and regrowth; to determine the extent, rates of change, and locations of some of the additional activities that may be agreed on; and to determine other natural and human-induced changes in terrestrial ecosystems. The utility of data from satellites raises the question of whether such data will be used as a primary source of data by countries reporting sources and sinks of carbon or whether they will be used for verification.

2.4.4.1. *Determining Initial Conditions*

Satellite data (NOAA-AVHRR, synthetic aperture radar) have been used to create global maps of land cover and forest cover (Belward *et al.*, 1999; De Grandi *et al.*, 1999; Rosenqvist *et al.*, 2000). Existing maps generally have resolutions of 1 km, but future capabilities at 30-m resolution may be possible with data from Landsat-7 in conjunction with other satellites. National-level and global observation strategies have been proposed for monitoring changes in forest cover (Skole *et al.*, 1997).

Several active research efforts in laboratories in different countries continue to explore the potential applications of these data sets and seek to update them with new information as new optical measurements become available from SPOT-Vegetation and MODIS (e.g., Running *et al.*, 1994a,b; Townshend *et al.*, 1994; Belward *et al.*, 1999). One approach consists of direct parameterization of land-cover characteristics that are relevant to carbon studies, such as vegetation structure and percentage of woody vegetation (DeFries *et al.*, 1995, 2000). Another option is the development of a multi-resolution approach to global forest mapping. Such an approach would use coarse-resolution satellite data to determine a sampling frame within which higher resolution data would be nested. All of these methods are feasible within a scientific research agenda, but they would require further development—including calibration and evaluation with ground truth—for routine operational use.

2.4.4.2. *Determining Rates, Extent, and Locations of Forest Clearing and Regrowth*

Estimating sources and sinks of carbon from forest clearing and regrowth using remote-sensing requires repeated measurements of forest clearing over large areas, fine spatial and temporal scale analyses to document regrowth, and *in situ* data on carbon pool changes that are associated with changes in land cover.

Coarse-resolution optical data (>1 km) are useful in describing broad distributions of different types of land cover, including different forest types (Belward *et al.*, 1999). These coarse-resolution data generally are inadequate, however, for accurately quantifying changes in forest cover, such as clearing and regrowth. Justice and Townshend (1988) demonstrated a need for spatial resolution of optical data that is less than 250 m.

High-resolution data from the Landsat, SPOT, and similar series of Earth observation satellites have been employed to make regular regional measurements of forest clearing and regrowth (Skole and Tucker, 1993; INPE, 1999). Large amounts of these data exist in several national archives, dating back approximately 20 years. Thus, a continuous and consistent source of data is available from which a high-resolution, fine-scale (1:250,000 scale mapping) information system could be developed. Many countries routinely perform regular assessments of forest clearing and regrowth over large areas—in particular, Brazil, Thailand, and Indonesia (INPE, 1999). Accuracies that approach 15–25 percent of the area cleared have been demonstrated (Houghton *et al.*, 2000).

Because cleared areas may revegetate rapidly, they must be observed frequently—as close to annually as possible. Frequent measurements make it easier to co-register areas of clearing and regrowth and reduce the probability of missing clearing on patches that begin to regrow quickly. Frequent observations will help in attributing changes to specific activities.

Although traditional optical remote-sensing techniques can distinguish between cultivated areas, pastures, and secondary growth, they are limited in terms of mapping various stages of fallow and secondary forests (Sader *et al.*, 1989). On the other hand, synthetic aperture radars (SARs) operate independent of solar illumination, cloud cover, and smoke and can detect differences in forest structure and woody biomass associated with various stages of forest clearing and regrowth (Rignot *et al.*, 1997; Saatchi *et al.*, 1997). Use of both optical and SAR data can provide a better characterization of land cover. For example, a plot of newly cleared and partially burned tropical forest may contain a significant portion of dead woody debris (slash). Visible and near-infrared reflectance data will show that this area has been cleared but will provide little insight into the presence of the slash. The SAR data, however, would indicate a significant amount of biomass; as a result, this area could be confused with a secondary growth area with comparable radar cross-section. Using the optical and SAR data together would reveal that the site was deforested and not in secondary growth, yet still had a significant amount of residual woody biomass (Rignot *et al.*, 1997).

The difficulty with these high-resolution satellites is not that they fail to identify cleared areas or areas where trees are returning but that they may fail to distinguish such clearing and regrowth from other changes, such as harvests, natural disturbances (fire, insects, storms), or other changes that are unrelated to human activity.

Very high-resolution data (1-m panchromatic and 3-m multi-spectral) that are now available from the commercial IKONOS II satellite may be useful for determining the actual activities on the ground that have led to forest clearing. Although such data can detect very small clearings, the scientific community as yet has very little experience with these data.

To obtain annual estimates of forest clearing and regrowth, a stratified sampling scheme might be employed to determine deforestation rates between the complete inventory/census years, spaced 3–5 years apart. The stratification might be based on the last complete inventory/census, assuming that deforestation is spatially persistent over intervals of 3–5 years. Research with Landsat- or SPOT-scale spatial resolution in some areas of the tropics suggests that a sample of 30 percent or less of the total forest area would be sufficient, but further research is necessary to determine sampling densities for other regions.

The costs of using remote-sensing data vary greatly on a per-hectare basis. AVHRR, other coarse-resolution optical sensors, and coarse-resolution radar sensors typically have very low costs per hectare for access to data. The costs of acquiring Landsat data vary according to the year in which the data were originally acquired, as well as whether the Landsat system was under private or public management. The most recent data for Landsat-7 are also the least expensive (~US$600 per Level 1 scene) because the system is now operated as a public resource. SPOT data are somewhat more expensive than Landsat and have finer spatial resolution but do not have global coverage. Very high spatial resolution data are only publicly available from IKONOS II, at somewhat higher costs per hectare, but practical considerations regarding data volume are likely to inhibit their use for broad-area surveys.

Once the spatial extent and rates of forest clearing and regrowth have been quantified, accurate calculations of changes in carbon stocks require techniques for measuring pre- and post-disturbance biomass, rates of secondary growth formation and turnover, and rates of biomass accumulation in secondary growth. There have been several attempts to use SAR data to determine aboveground biomass directly, through the known sensitivity of the radar backscatter to total aboveground material, its structure, and its dielectric properties. With current techniques and wavelengths, however, there is little ability to discriminate biomass levels greater than 50–100 t ha^{-1}. The National Aeronautics and Space Administration's (NASA) planned Vegetation Canopy Lidar (VCL) mission—to be launched in 2001—is designed to provide data sets of the vertical distribution of vegetation canopies. These data may be a good proxy for aboveground biomass for many forested ecosystems, but full knowledge of the utility of the measurements will necessarily await the instrument's launch and subsequent experience in the scientific community.

2.4.4.3. Determining the Extent, Rates, and Location of Other Activities

Countries may determine that other activities will be allowed in addition to those that result in forest clearing and regrowth. Among the suite of activities currently under consideration in this Special Report, few appear to be suitable for the use of remotely sensed data. An exception might be wood harvest. Clear-cutting would most likely be observed, yet distinguishing between clear-cutting and clearing would be difficult. Satellite data have also been used to identify selective logging (Stone and Lefebvre, 1998; Souza and Barreto, 1999), although the techniques are more complex. In this regard, it may be important to recognize that satellite data are not routinely used in national forest inventories of most Annex I countries. Satellite data are not a substitute for *in situ* measurements; even changes in area are largely obtained though other methods. On the other hand, the use of satellite data in forest inventories is being explored (Kilpeläinen and Tokola, 1999).

Satellite data might also be important for activities other than ARD if it were decided that the locations of additional activities were to be required. In this case, high or very high spatial resolution remote-sensing data or the use of GPS information could be useful to document the locations of additional activities.

2.4.4.4. Determining the Extent and Degree of Natural Variability and Change

For full carbon accounting or for estimation of baselines, determining the degree of natural variability in fluxes of carbon between terrestrial ecosystems and the atmosphere may be important. Remotely sensed data can contribute to this estimation in three ways: by providing maps of land cover through time (see Section 2.4.4.1); by providing parameters for models that calculate net exchanges of gases between terrestrial ecosystems and the atmosphere; and by estimating the area affected by particularly large or frequent disturbances, especially fire and extensive pest outbreaks.

In the last 10 years, there has been tremendous progress in the scientific community's ability to simulate exchanges of carbon (and other material) between terrestrial ecosystems and the atmosphere (Schimel *et al.*, 1995). Many of these models use remotely sensed data to establish areas of land-cover types to which different attributes are assigned (e.g., production efficiency, surface roughness, albedo) or to determine parameters more directly (Sellers *et al.*, 1997). All of these models are research tools at this time; they would require extensive validation and further development to be used in an operational context.

Fires are one of the largest and most frequent perturbations of many ecosystems; they constitute a major source of GHGs and aerosols. Satellite data have been useful in measuring the occurrence of active fires and areas burned in the recent past (through fire scars observable from space). Daily fire monitoring can be provided with existing hot-spot detection algorithms working on thermal channel data from the National Oceanographic and Atmospheric Administration (NOAA) AVHRR and European Space Agency (ESA) ATSR sensors. Fine-resolution mapping of selected burned areas can be accomplished with data from Landsat, SPOT, and IRS satellites.

These satellites carry high-performance, well-calibrated sensors that include infrared, near-infrared, and shortwave infrared bands.

All of the satellite-based observational systems that are used to assess fire occurrence and burned area require *in situ* observations (or models) to assess fuel loads and thereby calculate GHG emissions. There currently is no direct way to measure GHG emissions from fire solely through the use of remotely sensed data. There also is no fully operational program for remote monitoring of fires, which would require observations throughout the day on a global basis, to characterize diurnal variability (Skole *et al.*, 1997). Nevertheless, several efforts are underway to develop operational protocols.

2.4.4.5. *Future Remote-Sensing Systems*

Many remote-sensing systems are expected to be implemented over the next several years, and plans exist for many more between now and the first commitment period. These systems include missions recently launched by the United States (including public and private systems), India, Brazil and China, Europe, Japan, and others. A detailed description of the plans for all of these systems is beyond the scope of this section.

Several points are important in the context of the first commitment period (and beyond), however. Currently, there is no published national or international commitment to ensure that remote-sensing measurements adequate to determine areas of forest clearing and regrowth will be collected during the first commitment period. Systems that currently provide such information—such as Landsat, SPOT, IRS, and CBERS—have current operational lifetimes that nominally end before the first commitment period begins. Ensuring that remote-sensing data adequate to this task will be available will require public and possibly private investment and a commitment by governments to ensure its availability. The situation with respect to remotely sensed data for ecosystem metabolism models and fire detection is less uncertain. Under current plans, these data will be included in joint U.S.-European meteorological missions that are projected to orbit during the first commitment period.

2.4.5. *Models*

Bookkeeping models have been used for years to calculate the effects of land-use change on terrestrial carbon storage (Moore *et al.*, 1981; Houghton *et al.*, 1983; Woodwell *et al.*, 1983). The same models may be used to calculate annual emissions and accumulations of carbon at any scale—from the project level to national, regional, and global levels. Two types of information are required: rates of land-use change (e.g., areas cleared annually for croplands, cubic meters of wood annually harvested) and per-hectare changes in carbon stocks following a change in land use (including changes in living vegetation, logging debris, harvested products, and soil carbon). The IPCC Guidelines are based on simplifications of these models. These models generally consider only fluxes of carbon resulting from land-use change, including forestry. They do not include the indirect effects of human-induced changes, such as climate, CO_2, or nitrogen deposition.

A similar set of bookkeeping models that include natural disturbances such as fire, insects, or wind-throw have been used to determine the carbon balance for large areas of forest (e.g., Kurz and Apps, 1995, 1999). These models require a third type of information—namely, areas annually disturbed by these factors, as well as by harvesting.

Process-based models do include the effects of environmental variables on the carbon dynamics of ecosystems; such models have some potential for estimating natural as well as human-induced changes in carbon stocks. These models also have the potential to extrapolate site-based measurements temporally and spatially; to attribute changes in carbon stock to direct human-induced and naturally occurring change; and to estimate emissions or consumption of trace gases such as NO_x, N_2O, and CH_4—although the current capacity to do so is not sufficiently accurate for emissions trading purposes. Before these models can be used to assess changes in carbon stocks, they must be thoroughly documented, evaluated against independent data sets (e.g., from long-term forestry or agronomic experiments), and, if necessary, calibrated for local conditions. Such models, when correctly parameterized, can be effectively used at the time scale of the commitment period, but their use over longer periods may be limited.

The range of available models enables simulation of carbon dynamics at various scales, from plot level (<1 ha) to global level. As the scale increases, the spatial resolution of the model decreases and the input data become more aggregated. Some models deal with single ecosystem components such as soil (e.g., RothC; Falloon *et al.*, 1998), whereas others simulate whole ecosystems (e.g., Century; Parton *et al.*, 1987). Other models operate at the biome level and are most often applied at the continental and global scale (Meentenmeyer *et al.*, 1985; Heimann *et al.*, 1998). To estimate and project changes in carbon stock for particular locations or specific combinations of environmental and management conditions, detailed biogeochemical process models can be employed. Such models have high data input requirements and may need particular calibration or parameters for individual locations. Model drivers include parameters such as soil clay content and topographic position, as well as climate data and land management information. Several reviews describe many of these models in detail (e.g., McVoy *et al.*, 1995; Smith *et al.*, 1997c; Heimann *et al.*, 1998).

The use of models to estimate changes in carbon stock may result in a loss of transparency because the modeling process can be very complex. If models are used, they must be documented and archived in the form in which they were used (including source code and input files), and the validation data must form part of the reporting requirement. With a view to using models for subsequent inventories, it is suggested that initial inventories gather data to evaluate model outputs in addition to measures that are required solely for inventory purposes. The incremental cost of doing so is small, and the data can be used for cross-validation.

2.4.6. *Preparations for and Operational Strategy during the First Commitment Period*

Although methods exist for measuring stocks of carbon and fluxes of CO_2 and non-CO_2 GHGs, implementation of these methods will require investments in human resources, technology, and infrastructure. First-time forest inventories, for example, will require calibration and validation of allometric relationships between merchantable wood volumes and total carbon in the ecosystem, including roots, small trees, and ground cover. Soil carbon models must be developed and tested against local or regional conditions. For remote-sensing data in particular, implementation of the Kyoto Protocol, even for the first commitment period, will require making observational systems operational, establishing international policies for sharing data, and building human resources and infrastructure.

Remote-sensing methods (Section 2.4.4) currently are largely confined to the scientific community. In only a few cases is remote-sensing information routinely being used by operational agencies to classify vegetation, to estimate rates and extent of forest clearing and regrowth, or to detect fires. Because methods do exist for many of the measurement tasks, however, a particular challenge that Parties may wish to consider is the development of an operational use of satellite data. Skole *et al.* (1997) provided an early view of what some of the requirements of such an operational system might be, from a technical and organizational perspective. Several critical challenges must be met, however, if such systems are to be routinely used in the context of the Kyoto Protocol.

Institutional and national commitments will be necessary to ensure that the basic means of making remote-sensing observations do, in fact, exist. The satellite systems and capabilities described in Section 2.4.4 currently are treated as research missions. The transition to operational status—balancing the need for continuous, comparable observations with the need to allow the technology to evolve—has not occurred for land observations to the same degree that it has occurred for basic meteorological measurements.

Provisions for archiving spatial data are also required. All of the methods and accounting systems described in this Special Report require maintenance of a historical record of what has occurred. If remote sensing is used to help in those determinations, access to historical data will certainly be necessary. In addition, the scientific community's ability to extract information from the data set should be expected to continue to evolve as algorithms themselves evolve. Maintenance of a data archive would ensure that nations are able to examine their records again as their understanding of their own situations continues to evolve.

Human resources and infrastructure required to exploit remotely sensed data will need to be established and made available to Parties. Although there has been substantial progress over the past decade in making new remotely sensed data and GIS tools for analysis available, as well as in the development of analytical methods and technologies, substantial hurdles still exist in terms of human capital. Education and training programs would be required to ensure that all Parties have sufficient access to expertise and methods for using these data sets.

Finally, the enormous capital expenses and consistent operational expenses are large enough that only a handful of nations may be expected to bear them. If other Parties also are to benefit from the observations, policies and procedures for sharing the data and information derived from them will have to be agreed on. Again, the question of how satellite data will be used for primary reporting and for verification must be addressed.

2.5. Sustainable Development Considerations

Article 2 of the Protocol calls on Annex I Parties to promote sustainable development in the course of reducing emissions of GHGs; this provision applies to the implementation and/or elaboration of policies and measures undertaken in LULUCF. Indeed, the adoption potential of climate mitigation measures is likely to be enhanced—and such measures therefore are more likely to meet their climate mitigation objectives—when they are designed to meet the sustainable development needs of those affected by the actions. Article 12 also explicitly identifies sustainable development, along with the reduction of GHGs, as a central purpose and requirement of project activities undertaken in non-Annex I countries through the CDM.[5]

This section provides information for policymakers about the sustainable development implications of LULUCF policies, measures, and definitions. Key issues related to sustainable development that the Parties may choose to consider include the following:

- What are the potential synergies and tradeoffs between climate mitigation and other sustainable development objectives?
- What are the environmental and socioeconomic implications of different types of activities under Article 3.3, potential additional LULUCF activities under Article 3.4, and project activities under Articles 6 and 12?
- What approaches are available for assessing the sustainable development impacts of LULUCF activities and projects?
  - Do these approaches differ if the Parties choose to develop criteria and indicators for assessing

[5]This Special Report makes no assumption about whether, or to what extent, LULUCF project activities will be approved for crediting under the CDM.

these impacts on a national rather than multi-national basis?
– Do these approaches differ if the Parties design policies and measures to be consistent with the objectives of other multilateral environmental agreements?

The first two issues are addressed at a broad level in Section 2.5.1. Specific impacts arising from Article 3.3 activities, potential Article 3.4 activities, and project activities are addressed in Chapters 3, 4, and 5, respectively. The third issue—approaches for assessing sustainable impacts of LULUCF activities and projects—is addressed in Section 2.5.2.

2.5.1. LULUCF Activities and Sustainable Development

The Brundtland Commission has suggested that sustainable development is "development that satisfies the needs of the present without compromising the needs of the future" (Brundtland, 1987). At its core, sustainable development seeks to increase the flow of goods and services generated by economic activity, while maintaining or increasing the stock and quality of natural and human capital engaged or affected by the activity (Maler, 1990; Munasinghe, 1993, 2000). Because capital has natural and human dimensions, sustainable development reaches beyond the traditional emphasis on growth in aggregate income (e.g., gross national product) to include environmental quality and social equity (OECD, 1998; Munasinghe, 2000). Thus, a core objective of sustainable development is to balance social, economic, and environmental activities and capital to improve current human welfare, while ensuring a sound foundation for future generations to maintain or improve their welfare (Solow, 1993).

Coincident with reducing atmospheric CO_2 concentrations, LULUCF mitigation activities are likely to generate associated impacts that could substantially affect sustainable development objectives. For the purposes of this Special Report, these associated impacts are categorized as primarily environmental or primarily socioeconomic in nature, although the distinctions between the two categories are rarely clear-cut. Environmental consequences of LULUCF activities include effects on biodiversity and land and water resources. Carbon mitigation objectives will generally be interwoven with traditional economic and social factors that affect land-use decisions, such as the demand for food, fiber, fuel, building materials, and habitable land. As a result, LULUCF activities may generate socioeconomic impacts through changes in producer and consumer welfare, employment, poverty, and equity. If the ultimate goal of sustainable development is to generate the optimal mix of all (climate-related and other) environmental and socioeconomic benefits, tradeoffs are likely. Areas and activities with the greatest CO_2 reduction benefits may not produce the optimal mix of non-CO_2 impacts for all stakeholders. For instance, areas with the greatest benefit for a given investment in carbon offsets may not always provide the greatest biodiversity benefits (Fearnside, 1995)—although there are several means by which the two objectives can be strongly linked, should the Parties be interested in doing so.

Because land-use decisions typically are driven by factors other than GHG mitigation, direct regulation or economic incentives often may be needed to incorporate carbon considerations into land-use decisions, particularly in the case of private lands. Land-use decisions typically are based on an assessment of costs and benefits that derives from consideration of a complex range of factors specific to each land unit; therefore, achieving global objectives relating to GHGs may require a mixture of regulatory and voluntary instruments. To the extent that economic incentives such as carbon credits compensate landowners for voluntarily changing their practices, those landowners can be at least as well off economically after the change. There are Parties other than the landowner, however, for whom the change in land use or management may affect their welfare either positively or negatively. For instance, modification of land-use or management practices may change employment opportunities or affect environmental quality either positively or negatively. Therefore, important indirect effects of LULUCF activities on the environment and socioeconomic welfare are likely to remain; the Parties may want to consider these effects in planning and implementing strategies for fostering LULUCF activities.

For the purposes of this discussion, LULUCF activities to mitigate climate change under the Protocol can be categorized as follows:[6]

- *Activities that increase or maintain the area of land in forests*. This category relates primarily to Article 3.3 activities of afforestation, reforestation, and (avoided) deforestation. From a sustainable development perspective, there may be important differences between activities that convert non-forest land to forest and those that protect existing forests from conversion. Reducing deforestation rates can avoid significant carbon emissions (especially in the tropics) and reduce associated environmental and social problems.
- *Activities that manage forests to store more carbon*. Changing the management of forests to store more carbon will alter the size and structure of forests. Although more intensive forms of management to increase tree biomass (e.g., more intensive efforts at regeneration, fertilization) may make more timber available for eventual harvest, they also may impart some negative consequences on other ecosystem functions. Alternatively, if current commercial

[6]The categorization of LULUCF activities in this chapter differs slightly from the categorization of LULUCF projects in Chapter 6. The primary distinction across activity categories here is the change in land use, whereas the primary distinction across project categories in Chapter 6 is the change in carbon.

management practices are modified to leave more biomass in the forest for longer periods of time (e.g., extended rotation lengths, selective logging), other ecosystem functions may benefit while timber production is curtailed.
- *Activities that manage non-forested lands to store more carbon.* As an alternative to afforestation or reforestation, carbon stocks in non-forested lands can be increased through changes in management practices—including agricultural soil management, grassland management, and agroforestry. These activities can have strong implications for sustainable development because of their interconnection with food production, rural poverty, and attendant consequences for the environment.
- *Activities that reduce dependence on fossil fuels through product substitution.* LULUCF activities can reduce dependence on fossil fuels primarily by providing a source of biomass that can be used as a renewable alternative to fossil fuels in generating energy and by supplying wood products that can substitute for other products requiring more energy to produce. Fossil fuel substitution will generally require investments in technology and infrastructure to enable the adoption of biofuels and less carbon-intensive products and processes.

Generalizing the impacts of LULUCF activities on sustainable development across activities, locations, and time, is difficult. Core environmental and socioeconomic aspects are discussed in the following sections. Chapters 3, 4, and 5 provide more detail on the impacts of specific activities and projects.

2.5.1.1. Potential Environmental Impacts of LULUCF Activities and Projects

This section discusses the potential associated impacts—positive and negative—of LULUCF activities and projects as they relate to several dimensions of environmental quality: biodiversity, soil quality, soil erosion, water quality and quantity, acidification, and climate feedbacks.

2.5.1.1.1. Biodiversity

Biodiversity refers to the variability among living organisms and the ecological complexes of which they are part. Biodiversity includes diversity within species (genetic diversity), diversity between species, and diversity of ecosystems (Glowka *et al.*, 1994). The goal of biodiversity protection is not met by simply maintaining the number of species but by ensuring that no species, genes, or ecosystems—especially those that are endemic, rare, or threatened—are lost. Some of these benefits are most prominent in tropical and subtropical forests: Tropical forests harbor an estimated 50–70 percent of all species. Anticipated climate change will greatly change the environments for many species, with the likely result that many species with restricted distributions—especially slow-growing, late-successional species or those with restricted seed dispersal—may be lost (Kirschbaum *et al.*, 1996).

The diversity of life forms and their living environments generally decreases along the spectrum of land use from primary forest to regenerated forest, plantation forest, and agricultural land (e.g., Marshall and Swaine, 1992; Whitmore and Sayer, 1992; Michon and de Foresta, 1995; Canaday, 1996). Conversely, biodiversity increases when degraded lands are forested, and it is usually assumed that biodiversity increases when agricultural land is converted to most types of forest.

Agricultural lands can also be managed to be species-rich, however; in the tropics, biodiversity may sometimes be greater in a mixed landscape of primary forest, regenerated forest, plantations, and agricultural lands than in primary forest alone (Lawton *et al.*, 1998). Afforestation can increase diversity only where it replaces land cover that is species-poor; in some places, afforestation can threaten valuable non-forest species and habitats.

Biodiversity benefits to society derive from the fundamental contribution of biodiversity to ecosystem function, the provision of diverse goods and services, and aesthetic and spiritual factors. These benefits are enjoyed locally and globally and are experienced by individuals directly and indirectly. Clearly, not all of these biodiversity benefits can be well quantified or monetized. Even if these values could be derived in principle (and there is much debate among economists, ecologists, philosophers, and others about whether they can be), the methods for estimating them are limited, especially for the more intangible services that biodiversity provides (Norton, 1988; Stirling, 1993). Although various indirect methods have been developed to estimate biodiversity's value in monetary terms, no single method is capable of producing a valid monetary equivalent to each of the aspects of biodiversity value (Price, 1997). Although the monetary value of biodiversity is recognized as being very high (Pearce and Moran, 1994; Costanza *et al.*, 1997; Pimentel *et al.*, 1997a), the willingness of the world at large to pay for it limits how much of this value can be translated into a monetary flow (Fearnside, 1999a). That willingness to pay generally has been increasing and may increase substantially in the future.

Existence value—the intrinsic value of species and natural ecosystems to human society, independent of "utilitarian" benefits—is an area in which some progress has been made toward quantification for decisionmaking purposes (Attfield, 1998). This class of value accrues mostly to populations who are either very close to the forest—such as indigenous communities—or those who are far removed from it, such as urban dwellers elsewhere.

Estimating biodiversity benefits is more tractable when those benefits are tied to the provision of commercial goods and services, such as non-timber forest products (Peters *et al.*, 1989; Vásquez and Gentry, 1989; Whitehead and Godoy, 1991;

Hecht, 1992; Richards, 1993; Grimes *et al.*, 1994; Pimentel *et al.*, 1997b). Local benefits also accrue from the stock of genetic material of plants and animals needed to produce a degree of adaptability to forest management and to agricultural systems that sacrifice biodiversity in nearby unprotected areas (Myers, 1989). The stock of useful chemical compounds and genetic materials for other than local use represents an investment in protecting future generations (including those in distant places) from the consequences of the absence of that material when it is needed one day. This factor is referred to as the "prospecting value" of genetic resources. A medicinal use, such as a cure for a disease, is worth more to society than the profits from selling the drug. This value also includes the value to consumers of the drug whose lives are saved or otherwise improved.

Several studies have shown substantial potential for identifying chemical models for pharmacological products based on tropical forest plants (Kaplan and Gottlieb, 1990; Elisabetsky and Shanley, 1994; Cordell, 1995). An estimate of the opportunity cost for the sale of medicinal products derived from rainforests in Mexico arrived at a figure of US$6.4 ha^{-1} yr^{-1}, with a range from US$1–90 ha^{-1} yr^{-1} (Adger *et al.*, 1995). Simpson *et al.* (1996), however, estimate somewhat lower values for willingness to pay for the medicinal benefits of biodiversity. Their estimates range from $0.20–20.63 ha^{-1} lump sum for incremental increases in land preserved in 18 biodiversity hot spots throughout the world.

Any major land-use change will change the *types* of habitats and species, irrespective of any change in diversity; the most serious negative impacts generally would result from deforestation of primary forest, especially in the tropics. Ongoing deforestation reduces the diversity of life forms by causing the extinction of organisms and the loss of genetic diversity in the remaining ones. Hence, with regard to biodiversity, the most beneficial outcomes generally can be achieved where deforestation can be halted or slowed.

Studies have systematically identified forest ecoregions throughout the world that are known to be both biologically important (containing an assemblage of species that is unusually rich, globally rare, or unique to that region) and under considerable threat of further loss or degradation (Dinerstein *et al.*, 1995; Olson and Dinerstein, 1998). Any LULUCF climate mitigation project that slows deforestation or degradation will help conserve biodiversity.

Protecting the most threatened ecosystems does not always provide the greatest carbon benefits, however. In Brazil, for example, the least well-protected and most threatened types of forest are along the southern boundary of Amazonia, where reserve establishment is relatively expensive and forests contain less biomass than in central Amazonia (Fearnside and Ferraz, 1995).

On the other hand, tradeoffs between carbon storage and maintenance of biodiversity can also occur in the creation of large areas of productive managed forest, especially monocultures of exotic species. High productivity demands high light interception, which suppresses ground flora and limits other life forms (Hill and Wallace, 1989). Rapid forest establishment means that there are no long periods of recovery that provide habitats following natural disturbance (Rochelle and Bunnell, 1979). Harvesting at the time, which maximizes timber yield, prevents the development of special habitats that occur in old-growth forests (Bull and Meslow, 1977), and plantations of single tree species are likely to have more limited and particular types of fauna and flora than natural forest stands (Kennedy and Southwood, 1984). There are, however, management options to address the tradeoffs between production and biodiversity—such as adopting longer rotation times; altering felling unit sizes; altering edge lengths; prolonging rotation lengths; creating a multi-aged patchwork of stands; minimizing chemical inputs; reducing or eliminating measures to clear understory vegetation; or using mixed species plantings, including native species (e.g., Allen *et al.*, 1995a,b; Da Silva *et al.*, 1995).

2.5.1.1.2. *Soil quality and organic carbon storage*

The relationship between land cover and SOC storage is briefly discussed below, with more detailed discussion in Section 4.2.2. Forest soils can store large amounts of carbon that could be released to the atmosphere through deforestation (Houghton *et al.*, 1983). For example, Brown and Lugo (1990) reported that in wet and moist life zones, sites cultivated after deforestation typically lose 60–70 percent of the initial carbon contained in mature forests. In the dry life zone, where initial soil carbon content is only about 50–60 percent of that in wet and moist life zone soils, soils may lose only 14 percent of their initial carbon under cultivation. Carbon may be lost by decomposition, with carbon released as CO_2 to the atmosphere, or by erosion—in which case carbon may be deposited somewhere else in the landscape, with only some lost as CO_2 (Lal, 1995).

There is ample evidence that when forests are converted to cultivated cropland, the organic layer is depleted, and soil carbon contents and cation exchange capacities can decrease (Detwiler, 1986; Mann, 1986; Schlesinger, 1986; Davidson and Ackerman, 1993). If cleared land is not cultivated or degraded, however, soil carbon amounts are mostly observed not to change over decades (e.g., Lugo and Brown, 1993; Christopher *et al.*, 1997)—although in some studies, soil carbon losses of up to 30 percent have also been reported (e.g., García-Oliva *et al.*, 1999). The outcome ultimately may depend more on the productivity of the vegetation (and hence nutrient supply and climate) than on vegetation type (Trumbore *et al.*, 1995). Thus, if there is no regular cultivation, the inter-conversion between forest and non-forest vegetation may not have a consistent effect on soil carbon amounts.

Soil organic matter (carbon) content is often related to soil fertility. Forests are distinct in that they develop an organic layer above the mineral soil. This layer generally improves physical (soil aeration, water retention, resistance to erodability, etc.) and biological properties (build-up of soil microorganisms, nutrients, etc.), which enhance the productive capacity of the soil.

The fertility of soils following slash-and-burn has been shown to decrease rapidly (Lal, 1987, 1996). Up to 80 percent of the total carbon and nitrogen content of the soil can be contained in macroaggregates, where it is partly protected from microbial action and thus more securely stored. Burning can reduce the amount of organic carbon associated with macroaggregates by 32 percent, and it can disrupt soil aggregate stabilization by changing the chemical nature of organic carbon (García-Oliva *et al.*, 1999).

When cultivated lands or soils that previously were low in organic matter are afforested or reforested, there can be substantial increases in the amount of soil organic matter (Ovington, 1959)—as occurs following shifting agriculture in the tropics. In addition, trees and their roots can play an important role in maintaining desirable soil physical properties (Young, 1989). Peatlands are exceptions: When drainage and disturbance to establish trees accelerates decomposition, the loss of carbon can eventually exceed the carbon store created by growing trees (Cannell *et al.*, 1993).

Grassland management may include practices that are beneficial for carbon storage and reduce the adverse environmental impacts of agricultural activity but would not be profitable without carbon compensation (e.g., allowing savanna thickening at the expense of livestock production). Chapter 4 provides more detail on the sustainable development impacts of changes in agricultural practices.

The introduction of forest operations such as harvesting and site preparation in areas that remain as forest have the potential to decrease the amount of soil carbon if litter inputs to the soil are reduced (Liski *et al.*, 1998; Thornley and Cannell, 2000). In most instances, however, detecting changes in soil organic matter observationally may be difficult (Johnson, 1992).

The Parties will have to decide which carbon pools to include as recognized carbon stocks (Section 2.3). If soil carbon pools were not included in the list of carbon stocks, there would be no incentive to maximize carbon storage in this important reservoir. Soil cultivation, for example, could be exercised to stimulate forest growth even though it could reduce soil carbon storage and thus lower total overall site carbon storage. Inclusion of all on-site carbon stocks for carbon accounting purposes would ensure that management options will have to consider effects on all of these pools.

Managing agricultural soils to store more carbon is likely to have ancillary benefits by reducing soil erosion; the use of cover crops, crop rotations, nutrient management, and organic amendments is likely to increase soil fertility and enhance food security for affected populations. Hence, practices such as improved crop productivity and conservation tillage may be warranted independent of their carbon sequestration benefits. As a result, the opportunity costs of these mitigation strategies may be fairly low. Another relevant activity is the removal of land from agricultural production. Such removal may be warranted for land that generates substantial carbon emissions in its current use or has high potential for carbon storage when it is left undisturbed. The opportunity costs for these set-asides depends on the land's relative productive potential in its current use.

2.5.1.1.3. *Soil erosion*

Deforestation in hilly areas often leads to decreased infiltration of water and consequently higher runoff and increased peak water discharges following rainfall events and reduced runoff during dry seasons. Most importantly, it results in increased soil erosion, gully and ravine formation, flooding risk, and siltation of reservoirs and irrigation schemes. For instance, deforestation in the Himalayas has been associated with a doubling in torrent width since 1990—and a downstream cost of more than US$1 billion (Government of India, 1983).

Forestation of denuded hilly land will reduce peak runoff; lessen the risk of flooding; conserve soils; and prevent severe siltation, gully formation, and landslides. In experimental work, up to 500-fold differences in erosion rates between forest and cultivated cropland have been recorded (Maass *et al.*, 1988). The presence of forest litter is critically important in facilitating infiltration rates and preventing soil/sediment movement by overland flow. Mono-specific plantations without an understory, however, may not provide such conditions. It is also important to conduct forest establishment activities sensitively to minimize negative impacts before a tree cover is established.

2.5.1.1.4. Water quality and water use

The watershed function of forests has particular public health importance because forests often are the primary determinant of water quality and quantity for household use in developing countries. In the Philippines, for instance, close to 20 million people live in upland watershed areas (Cruz and Zosa-Feranil, 1988). The importance of forests as watersheds may substantially increase in the next few decades because freshwater resources are projected to become increasingly scarce, particularly in developing countries (Mountain Agenda, 1997; Liniger and Weingartner, 1998).

Watershed areas are important sources of water for irrigation, hydro-electric power, and industrial use. Watershed protection is important in maintaining the quantity and quality of water supply, as well as flood avoidance. China's decision to halt logging in the Yangtze watershed in 1998 following severe flooding illustrates the magnitude of these effects. Trees in the Yangtze watershed were calculated to be worth at least three times as much for their water regulation functions as for their timber value (Smil and Yuchi, 1998, as cited by Abramovitz and Mattoon, 1999).

Deforestation and forest degradation associated with high-impact methods of logging and mining significantly erodes soils in many countries, with marked economic and environmental costs through consequent siltation of rivers, hydropower reservoirs, irrigation systems, and coastal ecosystems (Myers, 1997). In the Philippines, for instance, an estimated 8.3 Mha of a total land area of 30 Mha are severely eroded (EMB, 1990).

Forests generally are expected to use more water (the sum of transpiration and evaporation of water intercepted by tree canopies) than crops, grass, or natural short vegetation. This effect may be related to increased interception loss—especially if tree canopies are wet for a large proportion of the year (Calder, 1990)—or, in drier regions, to trees' greater root system, which allows water extraction and use during prolonged dry seasons.

Interception losses are greatest from forests that have large leaf areas throughout the year. Thus, such losses tend to be greater for evergreen forests than for deciduous forests (Hibbert, 1967; Schulze, 1982) and may be expected to be larger for fast-growing forests with high rates of carbon storage than for slow-growing forests. Consequently, afforestation with fast-growing conifers on non-forest land commonly decreases the flow of water from catchments and can cause water shortages during droughts (Hibbert, 1967; Swank and Douglass, 1974). Vincent (1995), for example, found that establishing high water-demanding species of pines to restore degraded Thai watersheds markedly reduced dry-season streamflows relative to the original deciduous forests. Although forests lower average flows, they may reduce peak flows and increase flows during dry seasons because forested lands tend to have better infiltration capacity and a high capacity to retain water (Jones and Grant, 1996). Forests also play an important role in improving water quality.

In many regions of the world where forests grow above shallow saline water tables, decreased water use following deforestation can cause water tables to rise, bringing salt to the surface (Morris and Thomson, 1983). In such situations, high water use by trees can be of benefit (Schofield, 1992).

In the dry tropics, forest plantations often use more water than short vegetation because trees can access water at greater depth and evaporate more intercepted water. Newly planted forests can use more water (by transpiration and interception) than the annual rainfall, by mining stored water (Greenwood *et al.*, 1985). Extensive afforestation in the dry tropics therefore can have a serious impact on supplies of groundwater and river flows. It is less clear, however, whether replacing natural forests with plantations, even with exotic species, increases water use in the tropics when there is no change in rooting depth or stomatal behavior of the tree species. In the dry zone of India, water use by *Eucalyptus* plantations is similar to that of indigenous dry deciduous forest: Both forest types essentially utilize all of the annual rainfall (Calder, 1992).

2.5.1.1.5. Acidification

When forests are planted on former non-forest lands, soils may become acidified through removal of base cations in harvested wood. In areas with high concentrations of polluted cloud water or reactive pollutant gases in the air (HNO_3, HN_3, and HCl), trees can also increase the transfer of those pollutants from the atmosphere to the ground (Fowler *et al.*, 1989). This transfer can result in acidification of soil and waters and increase aluminum in waters, especially in areas with base-poor soils. This transfer can have detrimental effects on fish, invertebrates, vegetation, and perhaps trees themselves (Ormerod *et al.*, 1989; Kreiser *et al.*, 1990). The risk of enhanced acidification is a well-known constraint on afforestation in parts of Europe and North America. It may be a consideration in any country with high levels of acidifying air pollutants or in areas with soils of low buffer capacity where cation extraction exceeds mineralization (Last and Watling, 1991).

2.5.1.1.6. Climate feedbacks

Water cycling is another major environmental service of forests. One of the expected impacts that would result from a significant expansion of the extent of deforestation in Amazonia and other parts of Brazil would be a reduction in rainfall, especially during the dry season (Lean *et al.*, 1996). Similar effects have been calculated for the effects of forests on rainfall in the Indian subcontinent (Harding, 1992), and tropical forest protection has been shown to generate drought mitigation and flood mitigation benefits in Indonesia (Pattanayak and Kramer, 2000) and Madagascar (Kramer *et al.*, 1997).

Anthropogenic wildland fires release significant quantities of GHGs and have considerable socioeconomic and ecological impacts. For example, the 1997–1998 wildland fires in southeast Asia resulted from extensive land clearing, exacerbated by unusually dry El Niño conditions. These fires adversely affected the health of an estimated 20 million people and produced extensive damage to the region's forests and biodiversity, at a total estimated cost of US$4.4 billion (EEPSEA/WWF, 1998, as cited in Levine *et al.*, 1999).

The relatively high water use by forests compared with non-forest lands transfers more water to the atmosphere, with potential effects on local and regional climate if forest areas are very extensive. Although the magnitude of these feedbacks is the subject of contention, extensive forestation may increase humidity, lower temperature, and increase rainfall in temperate and tropical regions (Harding, 1992; Blythe *et al.*, 1994).

Deforestation can lead to decreased local rainfall and increased temperature—most notably in Amazonia, where about 50 percent of rainfall originates from within the Amazon basin and the predicted decrease in rainfall and warming (after loss of forest) could make conditions unsuitable for subsequent regeneration of many rainforest species (Gash and Shuttleworth, 1991). Similarly, adverse impacts of reduced rainfall from deforestation have been documented in Asia (Chan, 1986; Meher-Homji, 1992).

2.5.1.2. Potential Socioeconomic Impacts of LULUCF Activities and Projects

This section addresses potential associated impacts on various socioeconomic dimensions of sustainable development, including provision of wood products and biofuels; agriculture; rural poverty alleviation; and aesthetic, spiritual, and recreational values.

2.5.1.2.1. Wood products and biofuels

The world's forests produced about 3.2 billion m^3 of harvested roundwood in 1994 and are projected to produce almost 3.7 billion m^3 per year by 2010 (FAO, 1997b). Of the 1994 total, about 1.7 billion m^3 (53 percent) was used as fuelwood and the remainder as industrial roundwood (e.g., sawlogs and pulpwood). Approximately half of the fuelwood was produced (and consumed) in Asia, one-quarter in Africa, and one-eighth in South America. World industrial roundwood production was concentrated in North America (40 percent), Europe (19 percent), Asia (18 percent), and the countries of the former Soviet Union (8 percent).

At the global level, forestry was estimated to account for 2 percent of the world's gross domestic product (GDP) and 3 percent of international trade. The estimated value of wood consumption in 1994 was more than US$400 billion, with industrial usage accounting for 75 percent. Several countries are using forest products as their main foreign exchange earners; any change in the status of forests in these economies is likely to affect their foreign trade and debt status (FAO, 1999). Globally, wood is expected to become increasingly scarce by 2050, assuming constant per capita wood use and a 2 percent annual volume growth increment (Solomon *et al*., 1996). The greatest shortfall will be in the tropics because of the projected increase in population that drives the conversion of forest to agricultural land (Zuidema *et al.,* 1994). There probably will be an associated shortage of fuelwood.

Aggregate temperate zone timber needs are likely to be met over the foreseeable future, given current projections of timber supply and demand (see, for instance, projections for the United States by Haynes *et al*., 1995). Trends in the boreal zone are less certain; they depend on the forest response to climate change. Climatic changes could lead to increased growth—if anticipated warmer conditions allow increase growth—or reduced timber production, if increasing incidences of pests, diseases, or fire were to increase tree mortality (Kirschbaum *et al*., 1996; Solomon *et al*., 1996).

Future changes in climate may affect forest productivity in ways that change the level and distribution of global timber supply. Perez-Garcia *et al*. (1997) modeled the effect of alternative climate scenarios on global timber markets and found, in general, that the expected rise in net primary productivity (NPP) would slightly expand timber supply and reduce timber prices. The net welfare effects varied across forest-product consumers and producers, as well as by region. Some Parties would benefit from enhanced timber stocks (e.g., consumers and timber importing countries), whereas some stand to lose (producers and timber exporting countries). Climatic changes projected by current climatological and ecological models may have relatively small effects on timber markets (Burton *et al*., 1997), but these effects can be quite sensitive to the climatological and ecological model used for selection and the extent to which species migration is captured (Sohngen and Mendelsohn, 1998; McNulty *et al*., 1999). These studies do not estimate the effects of climate change on non-timber and non-market outputs and therefore do not capture potential effects on biodiversity, watersheds, and other forest outputs. Moreover, possible climate change and associated impacts should be viewed within the context of many other activities and policies occurring in different sectors and possible mitigation responses by humans.

Stand-level analyses have shown that paying landowners to store carbon—although it still allows them to harvest timber—can elongate timber rotations significantly if carbon is priced sufficiently high (van Kooten *et al*., 1995; Hoen and Solberg, 1997), though the rotation age response has been found to be lower for commercial plantations than for natural forests in the United States (Murray, 1999). Longer rotations often co-produce environmental benefits, such as improved habitat for fauna, but collectively may lead to temporary constraints on aggregate timber supply until forests are fully regulated at longer rotation levels. Longer rotations and resultant production of larger logs could also result in more valuable timber. If forest carbon reserves are created, wherein no harvesting activity occurs, the corresponding supply constraint is permanent. Changes in timber availability directly affect forest industry employment but must also be viewed in the context of other factors that affect forest-based employment, such as capital-labor substitution and other forms of technical change.

Establishment of more forests also provides the opportunity to use that wood as a biofuel (Nakicenovic *et al*., 1996). In some circumstances, forest establishment could also decrease deforestation and degradation for fuelwood. Greater use of biofuel could also have the advantages of fossil fuel substitution by reducing the use of coal, kerosene, liquified petroleum gas (LPG), and so forth. Under steady-state conditions, the amount of carbon released through combustion of biofuels is offset by the amount of carbon taken up in biomass.

Establishment of plantations to produce charcoal is one prominent form of biofuel production. Plantations for charcoal may have substantial carbon advantages over traditional pulpwood plantations because of their ability to substitute for fossil fuels (Fearnside, 1995; Schlamadinger and Marland, 1996; Marland *et al*., 1997). Tree plantations can have mixed effects on the environment, depending on the characteristics of the land and the land cover being replaced by the plantation, as well as indirect socioeconomic effects through market processes. Although the use of biofuels in energy production is highly desirable from a carbon management point of view, it may not be without its own social and environmental side effects on air quality and waste disposal (Sutton, 1994; Fearnside, 1999d).

Savings in the emission of GHGs can also be achieved through material substitution. Typical building materials—such as steel, plastics and aluminum—have large energy requirements for mining, processing, smelting, and, with some materials, reduction of oxidized ore. These energy requirements lead to corresponding CO_2 emissions. Cement production also leads to additional direct CO_2 release during manufacturing. Wood leads to the lowest emissions because it requires only minor energy inputs in harvesting and sawing. Hence, any substitution of wood for other materials could reduce energy requirements and associated GHG emissions (Kirschbaum, 2000). Moreover, the production of metals and plastics generates higher volumes of air, water, and solid waste pollutants than wood products such as lumber—particularly so with toxic chemicals (USEPA, 1997).

2.5.1.2.2. *Agriculture, employment, and poverty*

Although expanding the area of forest generates many environmental and socioeconomic benefits for society, it generally comes at the cost of forgone opportunities for other activities, especially agriculture. Social impacts can include population displacement and loss by some section of society, but the net effect of land-use change on employment, income, and equity cannot generally be determined *a priori*: It must be evaluated on a case-by-case basis, with specific social conditions in each country taken into account in assessing overall impact. In principle, the decision to limit deforestation carries opportunity costs as well because it entails forgoing other potential land uses, such as agriculture. In many developed countries with agricultural surpluses—notably in Europe—or relatively abundant marginal agricultural land (e.g., the United States), there is considerable scope for increased afforestation on agricultural land with relatively little adverse effect on agricultural production, employment, or economic well-being.

Several studies in the United States have examined the scope of afforestation of agricultural land for carbon sequestration. Some of these studies (e.g., Moulton and Richards, 1990; Parks and Hardie, 1995; Stavins, 1999) have ignored the effects of afforestation on land markets and thereby on the agricultural sector. A study by Adams *et al.* (1993) found that large-scale afforestation programs in the United States could have potentially sizable impacts on the agricultural sector by diverting land from agriculture and increasing agricultural prices. Alig *et al.* (1997) developed an integrated model of the forest and agricultural sectors of the U.S. economy and used this model to assess the likely impact of a fairly crude afforestation program with limited long-term incentives. They found that much of the land converted from agriculture to forests could be offset simply by converting other land from forestry to agriculture to maintain equilibrium in the land market. Although this forest-to-agriculture conversion reduces the social cost of afforestation programs, it also reduces their carbon sequestration potential. Adams *et al.* (1999) demonstrated that the scope and timing of forest-based carbon sequestration activities have important effects on the social costs of the policy and the intersectoral impacts between forestry and agriculture. Their study suggested that carbon sequestration activities within existing land uses under Article 3.4 (e.g., changes in forest management) would induce less intersectoral spillover between the forest and agricultural sectors than would land-use change activities under Article 3.3.

In the tropics, afforestation could enhance or impoverish local agriculture and increase or decrease employment and wealth, depending on specific circumstances. On one hand, some of the biomass from new plantations could provide green manure, fodder, or fuelwood, which releases cow dung and crop residues to be incorporated into the soil. This biomass could increase overall agricultural production and reduce the pressure on natural forests (Ravindranath and Hall, 1995). Plantations can also provide employment, especially if they are organized through partnerships with local communities and if there are opportunities for the utilization of non-timber forest products (Indonesian First National Communication, 1999; TERI, 1999).

On the other hand, new plantations may take fertile land of high value from vulnerable groups, such as small farmers and landless rural people. Conversion of such land to forests could threaten the livelihoods of local groups, increase poverty, and perhaps accelerate deforestation elsewhere. The use of agroforestry may often provide a viable combination of carbon storage with the least effects on the food production capacity of the land. The inclusion of trees will typically increase the carbon density of sites relative to traditional agriculture, thereby providing climate change mitigation benefits. Agroforestry has also been demonstrated to produce many of the other environmental benefits of forests, however, as well as diversifying the income base of the individuals practicing it. Conversion of degraded croplands that are typical of smallholder farming in sub-Saharan Africa into agroforests has major positive impact on carbon storage, but it has also become key to food security and a way out of poverty for millions of African farmers (Sanchez *et al.*, 1997a,b; Sanchez, 2000).

Not only can changes in agricultural and agroforestry practices directly enhance carbon storage in soils, improvements in these practices on the forest margin can reduce deforestation pressure and corresponding emissions of carbon. The net effect of changes in agricultural practices on deforestation, however, may depend on the relative input intensity of the technologies that are applied. Capital and land-intensive technologies on the forest margin may further the rate of deforestation; labor-intensive technologies are less likely to do so.

Forests are valued for many goods and services, as reflected in recent forest policy and institutional changes in many countries. In contrast, carbon is a single commodity, and any form of carbon in trees and soils will suffice as long as it is securely stored. Promotion of carbon over other forest values could impoverish people who benefit from the diversity of products and non-carbon services. This problem will be greatest where livelihoods are at stake, especially where food security is threatened (Ogle, 1995), as well as where the interests of forest-dependent people are poorly represented (Tipper and de Jong, 1998). This factor is particularly important in countries with low per capita incomes that are not self-sufficient in terms of growing food and have a low capacity to finance food imports.

2.5.1.2.3. *Aesthetic, spiritual, and recreational values*

People throughout the world have a special relationship with their landscape; many indigenous communities sustain themselves physically and spiritually on forestlands. Depending on specific societal attitudes and the type of forest concerned, forests may be regarded as a desirable amenity, "unnatural," or undesirable. These perceptions must be considered when planning ARD activities.

Stewardship describes the responsibility of humans who have been entrusted with the management of all creation to take care of it. In India and several other countries, for example, sacred groves have long been identified for preservation—presumably to meet the need of noncommercial goods and services to human society. By and large, traditional sanctity has been effective to protect many such groves, notwithstanding pressures from population growth, urbanization, and industrialization. The sacred groves in their origin provide a code of conduct and associated restrictions or regulations through legends that prescribe the relationship of the community with their surroundings.

The aesthetic values of forests are somewhat more utilitarian when they are captured through recreational activity. Forests provide a wide range of recreational opportunities, including hiking, camping, wildlife viewing, and hunting. In the United States alone, these activities account for more than 4 billion participation days annually (Cordell *et al.*, 1997). Forest-based recreation and ecotourism have become significant sources of economic development in developed and developing countries worldwide.

2.5.2. *Options for Assessing and Strengthening the Sustainable Development Contributions of LULUCF Climate Mitigation Measures*

Potential LULUCF climate mitigation measures can have highly variable environmental and socioeconomic impacts (see Sections 2.5.1 and 5.5), depending on the measures and the means by which they are implemented. Article 2.1(a) of the Kyoto Protocol provides an impetus for Parties to assess these impacts to determine whether and how any LULUCF climate mitigation measures carried out within their borders may contribute to national sustainable development objectives. Because countries often have very different views about what constitutes sustainable development, such national objectives might be expected to vary widely. In addition, Article 2.1(b) provides that Parties might consider whether and how to establish some common approaches to promoting the sustainable development contributions of LULUCF measures—for example, to ensure that these impacts are evaluated consistently across Parties. The Parties might also develop common approaches to ensure that LULUCF activities and projects under the Protocol are consistent with the goals and objectives of other multilateral environmental agreements (Section 2.5.2.2). For projects between countries, the Parties may also decide to assess whether potential projects are consistent with the sustainable development objectives of investor and host countries.

Regardless of whether the Parties opt for a strictly national or multinational approach to sustainable development, a system of criteria and indicators would be useful for assessing and comparing sustainable development impacts across alternative LULUCF actions. An ideal set of sustainable development indicators would feature many of the same general characteristics as an ideal accounting system: transparency, consistency, comparability, completeness, and accuracy. No comprehensive set of indicators with these characteristics currently exists for the suite of LULUCF policies and measures considered in this Special Report. Several approaches have been developed for related purposes that the Parties might adapt to gauge the sustainable development contributions of LULUCF definitions, activities, and projects undertaken to mitigate GHG emissions. This section highlights approaches that have been developed explicitly for the purposes of assessing the environmental and/or socioeconomic impacts of policies or measures undertaken at international, national, or project levels. These approaches are not mutually exclusive; several could be adapted by the Parties to be used in concert. Criteria and indicators for sustainable forest management are most directly adaptable to specific LULUCF measures at the national and project levels (Section 2.5.2.4.1).

The following sections briefly describe each approach and discuss key factors that may affect its potential application to LULUCF activities and projects. In addition, potential tools—such as environmental screens and the application of environmental impact assessments—by which selected criteria and indicators could be applied to strengthening the sustainable development contributions of LULUCF activities and/or projects are identified and evaluated. As with criteria and indicators, Parties could adopt these tools on a national or multinational basis. More specific approaches to strengthen the national and local sustainable development contributions of LULUCF projects are described in Section 5.6.

2.5.2.1. *Compatibility with Internationally Recognized Principles and Indicators of Sustainable Development*

At the 1992 United Nations Conference on Environment and Development (UNCED), a broad framework for the objectives, activities, and means of achieving sustainable development was advanced in Agenda 21 (UNCED, 1992). Sections I and II of Agenda 21 ("Social and Economic Dimensions" and "Conservation and Management of Resources for Development") identified and described key program areas of sustainable development. From these program areas, the United Nations Commission on Sustainable Development (UNCSD) drafted a set of 134 specific social, economic, and environmental indicators (UN, 1996). The indicators were developed within a "Driving Force-State-Response" framework, each with a methodology for use at the national level on the understanding that countries would choose from among the indicators those that are relevant to their national priorities, goals, and targets. These indicators are being tested by countries from all geographic regions; a final set of national-level indicators is anticipated by 2001.

Table 2-8 provides a brief explanation of the framework, along with examples of relevant specific indicators within program areas. Chapters 10–15 of Agenda 21 encompass several program areas of particular relevance to LULUCF policies and measures:

- Integrated approach to the planning and management of land resources
- Combating deforestation

***Table 2-8**: Examples from UN Commission on Sustainable Development's working list of sustainable development indicators[1] relevant to LUCF policies and measures under the Kyoto Protocol.*

Program Area	Driving Force Indicators	State Indicators	Response Indicators
Combating Poverty	• Unemployment rate	• Head count index of poverty • Poverty gap index • Squared poverty gap index • Gini index of income inequality • Ratio of average female wage to male wage	(None listed)
Transfer of Environmentally Sound Technology, Cooperation, and Capacity-Building	• Capital goods imports • Foreign direct investments	• Share of environmentally sound capital goods imports	• Technical cooperation grants
Protection of Quality and Supply of Freshwater Resources	• Annual withdrawals of ground and surface water • Domestic consumption of water per capita	• Groundwater reserves	• Density of hydrological networks
Combating Desertification and Drought	• Population living below poverty line in dryland areas	• National monthly rainfall index • Satellite-derived vegetation index • Land affected by desertification	(None listed)
Combating Deforestation	• Harvesting intensity	• Forest area change	• Managed forest area ratio • Protected forest area as a percentage of total forest area
Promoting Sustainable Agriculture and Rural Development	• Use of agricultural pesticides and fertilizers • Irrigation percentage of arable land • Energy used in agriculture	• Arable land per capita • Land area affected by salinization or waterlogging	• Agricultural education
Conservation of Biological Diversity		• Threatened species as a percentage of total known native species	• Protected species as a percentage of total known native species

[1] *Driving Force Indicators* are key human activities, processes, and patterns that impact sustainable development; *State Indicators* provide information on the current status of sustainable development; and *Response Indicators* are intended to highlight policy options and elements that could be used to improve the state of sustainable development. The indicators listed here are not necessarily applicable to the evaluation of LUCF climate mitigation measures. For more information on this framework and specific indicators, see UN (1996).

- Managing fragile ecosystems: combating desertification and drought
- Sustainable mountain development
- Sustainable agriculture and rural development
- Conservation of biological diversity.

Other relevant program areas include transfer of environmentally sound technology; cooperation and capacity-building, and protection of the quality and supply of freshwater resources.

Assessing whether and to what extent the final set of national-level indicators could be effectively adapted to assess the sustainable development contributions of LULUCF measures under the Kyoto Protocol is difficult. Annex I and non-Annex I countries have reported significant challenges in implementing many of the indicators that have been drafted (UN, 1998). Several Annex I nations have reported that implementation of indicators was time-consuming and sometimes unclear with regard to their relevance to national development strategies. Developing countries have reported difficulty in implementing the indicators, citing problems such as inadequate data and insufficient national and local capacity. Some countries have reported effective progress in addressing these latter problems by cooperating ("twinning") in the implementation process (e.g., South Africa/ Finland, Tunisia/France and Brazil/Germany).

Other multilateral institutions are also developing broad sustainable development indicators. For example, the Organisation for Economic Cooperation and Development (OECD) has developed a core set of environmental indicators for use in the evaluation of the environmental performance of economic sectors either by function or institutions (OECD, 1993). The OECD framework uses a similar "pressure-state-response" model and attempts to identify indicators based on their policy relevance, analytical soundness, and measurability. The OECD has developed indicators for LULUCF-related issues such as forest resources, landscape, soil degradation, and biological diversity (Table 2-9).

The OECD is planning to develop and extend its environmental performance indicators to produce by 2001 a practical set of indicators that integrate key economic, environmental, and social elements of sustainable development. The OECD intends to develop these indicators in cooperation with non-OECD countries and other international organizations and expects the indicators to serve as a tool for policy analysis, monitoring, and evaluation of progress toward sustainable development at the local, national, and regional levels (OECD, 1998). The European Union (EU) also is developing a set of indicators for key human activities that affect the environment. A recent progress report lists 60 indicators for 10 policy areas—such as climate change, loss of biodiversity, and resource depletion—which are being evaluated for their quantitative usefulness and relative contribution to various economic sectors (Eurostat, 1999). Further analysis would be necessary should the Parties wish to assess whether and how these indicators might be operationalized for the purposes of the Protocol.

2.5.2.2. *Consistency with Goals and Objectives of Other Multilateral Environmental Agreements*

LULUCF policies and measures undertaken to reduce GHG emissions have significant potential to have positive or negative impacts on environmental and sustainable development objectives that are a central focus of other multilateral environmental agreements (MEAs). Key categories of potential impact and corresponding principal MEAs include biological diversity (UN Convention on Biological Diversity, CBD), desertification (UN Convention to Combat Desertification, CCD), and wetlands (Ramsar Convention on Wetlands). In addition, a broad range of issues relating to conservation, management, and sustainable development of forests were the focus of recently completed policy dialogues of the Intergovernmental Forum on Forests (IFF); IFF conclusions and associated proposals for actions will be taken up by the UNCSD (IISD, 2000).

***Table 2-9**: Examples of OECD environmental performance indicators[1] for sample issues relevant to LUCF policies and measures under the Kyoto Protocol.*

Issue	Driving Force Indicators	State Indicators	Response Indicators
Forest Resources	• Short-run sustained yield/actual harvests	• Share of disturbed/deteriorated forest in total forest area	• Percentage of harvest area successfully regenerated (including natural regeneration) or afforested
Biological Diversity and Landscape	• Habitat alteration and conversion of land from its natural state	• Threatened or extinct species as a share of known species	• Protected areas as a percentage of total area by ecosystem type
Soil Degradation (Erosion and Desertification)	• Erosion risk: potential and actual use of soil	• Degree of topsoil losses	• Rehabilitated areas

[1] See Table 2-8 for brief descriptions of performance indicators.

The Parties may need to determine whether and how to ensure that LULUCF definitions, activities, and projects under the Protocol are consistent with the goals and objectives of relevant MEAs. The foundation for such a decision has been laid by the recent SBSTA request that the UNFCCC Secretariat liase with the Secretariats of the CBD and the CCD; with the IFF; and with other international bodies of the United Nations, such as FAO. Relevant ecosystem protection components of the UNFCCC objectives are identified by Article 4, paragraphs 1(d), 1(e), and 8 and by Article 2.3 of the Kyoto Protocol.

There currently is no recognized set of indicators that could be used to assess the consistency of prospective LULUCF climate mitigation measures with the goals and objectives of other multilateral environmental agreements and processes. Despite this limitation, the Parties may need to decide whether and how to take steps to ensure that LULUCF definitions, activities, and projects are operationalized in a manner that is consistent with and synergistically supportive of them.

Brown (1998) proposes that there should be opportunities for LULUCF measures to support the objectives of multiple MEAs—for example, by identifying opportunities to slow deforestation in areas that are high in biodiversity and contain large carbon stores, by rehabilitating degraded rangelands, and by planting windbreaks to sequester carbon and reverse desertification. Indeed, many potential LULUCF measures could support carbon mitigation and help protect biodiversity, slow desertification, or support other environmental objectives (Sections 2.5.1 and 5.5). For potential LULUCF projects, the Parties may wish to consider whether the Global Environment Facility (GEF) is positioned to help non-Annex I countries develop and support projects that meet the objectives of multiple MEAs (GEF, 2000).

Some authors have pointed out that some potential LULUCF definitions, activities, and projects—such as those that support conversion of natural forests to plantations—might be considered inconsistent with the objectives of one or more MEAs (e.g., UNU, GEIC, and UNU/IAS, 1998). Similar arguments could be advanced for measures that expand plantations on native woodlands or grasslands or drain wetlands.

If the Parties wish to eliminate or otherwise restrict these activities from emissions reduction crediting, one option would be to consider the addition of limiting clauses on credits, such as that proposed in Chapter 3 to limit reforestation credits. Translating non-carbon environmental and socioeconomic concerns into quantitative limits on carbon credits may prove difficult, however. Alternatively, policymakers could develop a screening process to determine whether certain activities will receive carbon credit at all. Implementation of such a screening process for activities at the national level might present several challenges and complexities because of their broad geographic scope.

Several types of LULUCF climate mitigation projects also have the potential for negative impacts on native ecosystems, including forests, woodlands, grasslands, and wetlands (see Section 5.5). Brown *et al.* (1998) and Hardner *et al.* (2000) propose eliminating such projects through the application of environmental screens prior to project approval. Hardner *et al.* (2000) propose a filter for project approval that excludes from crediting any LULUCF project activities that convert native ecosystems to other land uses. The risk of negative environmental (and social) impacts of LULUCF projects might also be reduced through the application of sound environmental and social impact assessment methodologies (Section 2.5.2.5).

In the specific case of reforestation projects in non-Annex I countries, some observers have expressed concern that crediting of such projects could promote expansion of plantations that replace natural forests whose associated emissions would not be constrained by a national commitment (German Advisory Council on Global Change, 1998). Currently, the expansion of industrial plantations is a significant driver of natural forest loss in some regions (Potter and Lee, 1998). The Parties might choose to constrain such projects through the application of the Deforestation-Reforestation Rule proposed in Chapter 3 to projects in non-Annex I countries. Such projects could be further constrained through sound environmental and social screens prior to project approval (Brown *et al.*, 1998).

2.5.2.3. Consistency with Nationally Defined Sustainable Development and/or National Development Goals and Objectives

As an offshoot of Agenda 21, many countries have formulated their own national sustainable development goals and strategies. In response to the UN's call to form bodies similar to the UNCSD, for example, the Philippines established the Philippine Council for Sustainable Development (PCSD) in 1992 (PCSD, 1997a). One of the key outputs of the PCSD to date is the formulation of Philippine Agenda 21, which was formally adopted in 1996 as the national agenda for sustainable development in the 21st century (PCSD, 1997b).

In addition, national development goals and objectives are generally well defined for each country and are typically embodied in one or more documents. In the Philippines, national development goals are contained in its medium-term development plans. For instance, the plan for 1988–92 contained the following national development goals: alleviation of poverty, generation of more productive employment, promotion of equity and social justice, and attainment of sustainable economic development (NEDA, 1988). Currently being finalized is the medium-term development plan for 1999–2004.

The Parties may wish to ensure that LULUCF activities and projects are consistent with, and supportive of, national sustainability goals and the objectives of host countries. The broad set of national-level indicators being developed under the coordination of the UNCSD (Section 2.5.2.1), as well as those being developed for specific LULUCF sectors (Section 2.5.2.4), may be useful to Parties seeking to develop indicators with which to assess such consistency. A more detailed assessment of options for host countries to ensure that LULUCF projects are consistent with national and local goals and objectives is provided in Section 5.6.

2.5.2.4. *Consistency with Internationally Recognized Criteria and Indicators for Sustainable Forest Management and Sustainable Agriculture*

2.5.2.4.1. *Sustainable forest management*

Since the UNCED, several intergovernmental efforts have been initiated to develop criteria and indicators for sustainable forest management. These efforts include the Helsinki Process (covering 39 European countries), the Montreal Process (covering 12 non-European countries in the temperate and boreal zones), the Tarapoto Process (covering the eight countries in the Amazonian Cooperation Treaty), and the International Tropical Timber Organization (covering most forested countries in the tropics). In addition, several efforts to establish criteria and indicators at the national and subnational levels build on these international approaches and adapt them to national and local forest conditions (WCFSD, 1999).

These and similar (e.g., FAO, 1995a) criteria and indicators are generally moving beyond a narrowly defined focus on the productivity of timber and other commercial forest products to incorporate ecological and social dimensions of sustainability. For example, the broad forest values developed as criteria under the Montreal Process for the conservation and sustainable management of boreal and temperate forests follows:

- Conservation of biological diversity
- Maintenance of the productive capacity of forest ecosystems
- Maintenance of forest ecosystem health and vitality
- Conservation and maintenance of soil and water resources
- Maintenance of forest contribution to global carbon cycles
- Maintenance and enhancement of long-term multiple socioeconomic benefits to meet societal needs
- Effective legal, institutional, and economic framework for forest conservation and sustainable management.

The development of multiple national and international efforts to develop criteria and indicators has led some observers, including the World Commission on Forests and Sustainable Development (WCFSD), to propose that some elements of these approaches can be usefully harmonized (WCFSD, 1999). The WCFSD further suggests that these criteria and indicators should be based on a strategy for sustainable forest management that reflects several broadly applicable objectives, including the following:

- Indefinitely satisfying needs for timber, fiber, and non-timber forest products
- Ensuring conservation of soil and water
- Sustaining the resilience and renewal capacity of forests
- Supporting food security and livelihood needs of forest-dependent communities
- Conserving biological diversity
- Achieving the foregoing goals in a manner that is consistent with the incremental productive capacity of forests and requirements for ecological security
- Realizing a more equitable sharing of benefits from uses to which forests are put
- Increasing management, cultivation, harvesting, and utilization of minor forest products as potential pillars of sustainable forestry to sustain livelihoods from dwindling resources
- Securing tenural rights of forest-dependent populations as a means of promoting conservation.

Parties seeking to implement sustainable forest management in the context of LULUCF climate mitigation measures may be able to adapt the criteria and indicators developed under one or more of these international processes. It is important to recognize, however, that many of these general criteria address national-level policy and sustainability and are not intended to directly assess sustainability at the forest stand level. Indeed, some objectives will likely prove to be mutually contradictory, particularly when they are applied in small ecological units. For example, economically viable timber harvesting often may not be reconcilable with the conservation of mature forest-dependent biological diversity in the same forest tract (Section 2.5.1; Frumhoff, 1995; Bawa and Seidler, 1998).

For site-specific projects, Parties might find the criteria and indicators that the Center for International Forestry Research (CIFOR) has developed for the management of natural forests (CIFOR C&I Team, 1999; Prabhu *et al.*, 1999) particularly valuable. The CIFOR criteria and indicators were based on research in large-scale natural forests that are managed for commercial timber production in Indonesia, Cote d'Ivoire, Brazil, and Cameroon, with additional sites in Germany, Austria, and the United States. These criteria and indicators provide a useful framework for evaluating policy, environmental, social, and production aspects of sustainable forest management and are designed to be readily adaptable to local conditions. CIFOR is also planning to develop criteria and indicators for tropical plantations and community-managed forests.

One tool for encouraging voluntary application of sustainable forest management criteria and indicators to LULUCF projects that have timber or non-timber products involves forest product certification. Certification is a process that links market demands for sustainably produced forest products with producers who can meet those demands. Certification may reward the performance of companies that adopt sound forestry practices by enabling them to maintain or improve the marketability of wood or other forest products (FAO, 1997a; WCFSD, 1999). Currently, the Forest Stewardship Council (FSC) and its accredited certifiers offer one approach to product certification.

Parties might wish to consider whether and under what conditions encouragement of market certification of forest products could strengthen the capacity of managed forests to meet carbon mitigation and sustainable development goals.

The International Standards Organization (ISO), through its ISO 14000 series, also provides a framework for certifying forest management (and other environmental management) systems (ISO, 1996). Unlike FSC, the ISO does not identify performance standards and does not allow a label to be attached to forest products. Instead, ISO 14000 management standards are designed to allow the setting of specific environmental and sustainable development criteria for LULUCF projects. Projects could then be managed on an ongoing basis to attain those goals, and independent auditors could verify whether the management system was consistent with the standard.

If Parties wish to implement ISO 14000 management standards for forest or other projects under the Kyoto Protocol, they would need to define and periodically update the sustainable development guidelines for LULUCF project activities. Project participants could then employ the ISO 14000 standards as a means of assessing compliance with those guidelines, with independent auditing carried out by an existing pool of accredited private-sector agents. This Special Report does not analyze the potential cost and time implications of adopting ISO management standards for LULUCF projects.

2.5.2.4.2. *Sustainable agriculture*

The major objective of sustainable agriculture and rural development (SARD), as defined by Agenda 21 (UNCED, 1992), is to sustainably increase food production and enhance food security. Key elements of SARD (FAO, 1997b) include increasing agricultural production in ways that ensure access by all people to the food they need; helping people satisfy their social and cultural aspirations; and protecting and conserving the capacity of the natural resource base to continue to provide production, environmental, and cultural services. Potential LULUCF activities and projects to reduce carbon emissions or sequester carbon in agricultural systems (Section 4.4) could be designed to be consistent with these goals.

The FAO is helping countries evaluate the compatibility of policies with SARD objectives, advising on incentives, and developing indicators and guidelines for sustainable agricultural practices. Thus, the FAO may be a useful resource for Parties seeking to ensure that LULUCF climate mitigation measures in the agricultural sector are compatible with sustainable development objectives. The FAO recognizes several more specific issues associated with sustainable agriculture (FAO, 1997a; UNCSD, 1997), including the following:

- Adopting farmer-centered participatory approaches and carefully recording and assessing indigenous knowledge and technology
- Promoting use of environmentally friendly technologies to intensify production on high-potential land already converted to agriculture
- Promoting cycling and use of organic materials in low-input farming systems
- Rethinking priorities for conserving and using agro-biodiversity, including the use of locally adapted crop varieties and crop diversification.

In addition, the Committee on Sustainable Agriculture and the Environment in the Humid Tropics (NRC, 1993) proposed several specific land-use options to achieve sustainable agriculture in tropical regions, including the following:

- Intensive cropping systems under proper management that do not lead to resource degradation through, for example, nutrient loading from fertilizers or soil and water contamination from agro-chemicals
- Shifting cultivation systems, coupled with the use of local cropping systems, observation of sufficient fallow periods, diversification of cropping systems, maintaining continuous ground cover, and nutrient restoration through mulching
- Agro-pastoral systems combining crop and animal production, allowing for enhanced agro-ecosystem productivity and stability through integrated management of soil and water resources and crop and animal diversification
- Intensive animal husbandry (ranching), combined with sustainable pasture and rangeland management
- Agroforestry systems that involve various combinations of woody and herbaceous vegetation with agricultural crops, often practiced for multiple agronomic, environmental, and socioeconomic benefits.

2.5.2.5. *Consistency with International and National Environmental Impacts Standards and Guidelines*

For site-specific climate mitigation projects, including those in the LULUCF sector, one means of assessing and strengthening their contributions to sustainable development is through the application of environmental impact assessments (EIAs) prior to project approval. An EIA is a broad process that informs decisionmakers about a project's potential environmental and societal risks and impacts, as well as examining alternatives and identifying mitigative measures (Krawetz, 1991; Glasson *et al.*, 1994; Munn, 1994). Running in parallel with the process of identifying, designing, and implementing projects, effective application of an EIA to climate mitigation projects might help ensure that potential positive and negative environmental and societal impacts (Section 5.5) are effectively addressed in all phases of project development.

Currently, more than 100 countries have a national EIA system (Canter, 1996). One factor that affects their applicability to LULUCF climate mitigation projects is that EIA guidelines and their degree of rigor in application vary widely. For projects between countries, this factor could create an incentive for project investors to support carbon-offset projects in areas with the least rigorous standards.

One option for the Parties to address this concern would be to adopt internationally recognized EIA standards and guidelines for carbon-offset projects. For example, the World Bank has published its three volume *Environmental Assessment Sourcebook* (World Bank, 1991). The *Sourcebook* contains sectoral guidelines for natural forest, livestock, rangeland, and agricultural production management; plantation development/reforestation; and watershed development. For each of these sectors, the *Sourcebook* identifies several potential environmental and social impacts, as well as mitigating measures for negative impacts. Potential social impacts identified in the *Sourcebook* include impacts on the labor market and labor availability; a shift to more cash-based economy; alteration of daily living patterns and the political power structure; changes in access to resources, perhaps through changes in land tenure; changes in infrastructure and social services; and changes in population demography, such as increases in internal migration as a result of project activities. The Asian Development Bank (ADB) has a similar manual titled *Environmental Guidelines for Selected Agricultural and Natural Resources Development Projects* (ADB, 1990).

For projects between countries, this concern also could be addressed by ensuring that the national environmental standards and guidelines of donor and host countries be satisfied. The effectiveness of EIAs can be further strengthened by ensuring that they are carried out by independent, third-party experts; that they clearly integrate socioeconomic and environmental aspects of project assessment; and that recommendations be demonstrably incorporated into project activities.

A decision by the Parties to adopt requirements that climate mitigation projects undergo EIAs prior to project approval may affect project costs and the rate of project implementation. This Special Report does not analyze the potential cost and time implications of adopting national and/or international EIA standards and guidelines for LULUCF projects.

References

Abramovitz, J.N. and A.T. Mattoon, 1999: Reorienting the forest products economy. In: *State of the World 1999*. [Brown, L.R., C. Flavin, and H. French (eds.)]. W.W. Norton, New York, NY, USA, p. 259.

Adams, D.M., R.J. Alig, B.A. McCarl, J.M. Callaway, and S.M. Winnett, 1999: Minimum cost strategies for sequestering carbon in forests. *Land Economics,* **75 (3),** 360–374.

Adams, R.M., D.M. Adams, J.M. Callaway, C-C. Callaway, and B.A. McCarl, 1993: Sequestering carbon on agricultural land: social costs and impacts on timber markets. *Contemporary Policy Issues,* **11,** 76–87.

Adger, W.N., K. Brown, R. Cervigni, and D. Moran, 1995: Total economic value of forests in Mexico. *Ambio,* **24(5),** 286–296.

Alig, R., D. Adams, B. McCarl, J. Callaway, and S. Winnett, 1997: Assessing effects of mitigation strategies for global climate change with an intertemporal model of the U.S. forest and agriculture sectors. *Critical Reviews in Science and Technology,* **27(Special),** S97–S111.

Alves, D.S., J.V. Soares, S. Amaral, E.M.K. Mello, S.A.S. Almeida, O. Fernandes da Silva, and A.M. Silveria, 1997: Biomass of primary and secondary vegetation in Rondonia, western Brazilian Amazon. *Global Change Biology,* **3,** 451–462.

Anderson, J.M. and J.S.I. Ingram, 1993: *Tropical Soil Biology and Fertility: A Handbook of Methods*. CAB International, Wallingford, United Kingdom, 221 pp.

Andrasko, K., 1997: Forest management for greenhouse gas benefits: Resolving monitoring issues across project and national boundaries. *Mitigation and Adaptation Strategies for Global Change,* **2,** 117–132.

Apps, M.J., W.A. Kurz, S.J. Beukema, and J.S. Bhatti, 1999: Carbon budget of the Canadian forest product sector. *Environmental Science and Policy,* **2,** 25–41.

Apps, M.J. and D.T. Price, 1996: *Forest Ecosystems, Forest Management, and the Global Carbon Cycle*. NATO ASI Series, Subseries I: Global Environmental Change, Vol. 40. Springer-Verlag, Heidelberg, Germany, 452 pp.

Asian Development Bank (ADB), 1990: *Environmental Guidelines for Selected Agricultural and Natural Resource Development Projects*. Asian Development Bank, Manila, Philippines.

ASSOD, 1995: *The Assessment of the Status of Human Induced Soil Degradation in South and Southeast Asia*. International Soil Reference and Information Centre, Wageningen, The Netherlands.

Attfield, R., 1998: Existence value and intrinsic value. *Ecological Economics,* **24(2-3),** 163–168.

Aubinet, M., A. Grelle, A. Ibrom, U. Rannik, J. Moncrieff, T. Foken, A.S. Kowalski, P.H. Martin, P. Berbigier, Ch. Bernhoffer, R. Clement, J. Elbers, A. Granier, T. Grunwald, K. Morgenstern, K. Pilegaard, C. Rebmann, W. Snijders, R. Valentini, and T. Vesala, 2000: Estimates of the annual net carbon and water exchange of forests: the EUROFLUX methodology. *Advances in Ecological Research,* **30,** 113–175.

Azar, C. and T. Sterner, 1996: Discounting and distributional considerations in the context of global warming. *Ecological Economics,* **19(2),** 169–184.

Bailey, R.G., 1998: *Ecoregions, the Ecosystem Geography of the Oceans and Continents*. Springer-Verlag, New York, NY, USA, world maplet + 176 pp.

Bakwin, P.S., P.P. Tans, C. Zhao, W.I. Ussler, and E. Quesnell, 1995: Measurements of carbon dioxide on a very tall tower. *Tellus,* **47B,** 535–549.

Bakwin, P.S., D.F. Hurst, P.P. Tans, and J.W. Elkins, 1997: Anthropogenic sources of halocarbons, sulfur hexafluoride, carbon monoxide, and methane in the southeastern United States. *Journal of Geophysical Research,* **102,** 15915–15925.

Baldocchi, D.D., B.B. Hicks, and T.P. Meyers, 1988: Measuring biosphere-atmosphere exchanges of biologically related gases with micrometeorological methods. *Ecology,* **69,** 1331–1340.

Baldocchi, D., R. Valentini, S. Running, W. Oechel, and R. Dahlman, 1996: Strategies for measuring and modelling carbon dioxide and water vapour fluxes over terrestrial ecosystems. *Global Change Biology,* **2,** 159–168.

Baldock, J.A. and P.N. Nelson, 1999: Soil organic matter. In: *Handbook of Soil Science* [Summer, M.E. (ed.)]. CRC Press, Washington D.C., USA, pp. B25–B84.

Batjes, N.H., 1996: Total carbon and nitrogen in the soils of the world. *European Journal of Soil Science,* **47,** 151–163.

Batjes, N.H., 1999: *Management Options for Reducing CO_2 Concentrations in the Atmosphere by Increasing Carbon Sequestration in the Soil*. International Soil Reference and Information Centre, Wageningen, The Netherlands, 114 pp.

Batjes, N.H. and W.G. Sombroek, 1997: Possibilities for carbon sequestration in tropical and subtropical soils. *Global Change Biology,* **3(2),** 161–173.

Baumgardner, M., 1999: Soil data bases. In: *Handbook of Soil Science* [Summer, M.E. (ed.)]. CRC Press, Washington D.C., USA, pp. H1–H94.

Bawa, K.S. and R. Seidler, 1998: Natural forest management and conservation of biodiversity in tropical forests. *Conservation Biology,* **12(1),** 46–55.

Bayer, C., L. Martin-Neto, J. Mielniczuk, and C.A. Cerrata, 2000: Effect of no-till cropping systems on soil organic matter in a sandy clay loam Acrisol from Southern Brazil monitored by electron spin resonance and nuclear magnetic resonance. *Soil and Tillage Research,* **53,** 95–104.

Belward, A.S., J.E. Estes, and K.D. Kline, 1999: The IGBP-DIS global 1 km land cover data set DISCover: A project overview. *Photogrammetric Engineering and Remote Sensing,* **65,** 1013–1020.

Bergsma, E., P. Charman, F. Gibbons, H. Hurni, W.C. Moldenhauer, and S. Panchipong, 1996: *Terminology for Soil Erosion and Conservation (Concepts, Definitions and Multilingual List of Terms for Soil Erosion and Conservation in English, Spanish, French and German).* International Society of Soil Science/International Soil Reference and Information Centre, Wageningen, The Netherlands, 313 pp.

Bird, N., 1997: *Greenhouse Challenge Carbon Sinks Workbook: A Discussion Paper.* Greenhouse Challenge Office, Canberra, Australia.

Blume, H.P., H. Eger, E. Fleischhauer, A. Hebel, C. Reij, and K.G. Steiner, 1998: Towards sustainable land use. *Advances in Geoecology,* **31,** 741.

Blythe, E.M., A.J. Dolman, and J. Noilhan, 1994: The effect of forest on mesoscale rainfall: an example from HAPEX-MOBILHY. *Journal of Applied Meteorology,* **33,** 445–454.

Bousquet, P., P. Ciais, P. Peylin, M. Ramonet, and P. Monfray, 1999: Inverse modeling of annual atmospheric CO_2 sources and sinks. 1. Method and control inversion. *Journal of Geophysical Research,* **104,** 161–126, 178.

Bridges, E.M. and H. DeBakker, 1997: Soil as an artifact: human impacts on the soil resource. *The Land,* **1 (3),** 197–215.

Brown, P., 1998: *Climate, Biodiversity, and Forests.* World Resources Institute/International Union for Conservation of Natural Resources, Washington, DC, USA, 36 pp.

Brown, P., N. Kete, and R. Livernash, 1998: Forests and land use projects. In: *Issues and Options: The Clean Development Mechanism.* United Nations Development Programme, New York, NY, USA, pp. 163–173.

Brown, S. and A. Lugo, 1990: Effects of forest clearing and succession on the carbon and nitrogen content of soils in Puerto Rico and U.S. Virgin Islands. *Plant and Soil,* **124,** 53–64.

Brown, S. and A.E. Lugo, 1992: Aboveground biomass estimates for tropical moist forests of the Brazilian Amazon. *Interciencia,* **17,** 8–18.

Brown, S., 1997: *Estimating Biomass and Biomass Change of Tropical Forests: a Primer.* FAO Forestry Paper 134, Rome, Italy.

Brundtland, G.H., 1987: *Our Common Future: The UN World Commission on Environment and Development.* Oxford University Press, Oxford, United Kingdom.

Bull, E.L. and E.C. Meslow, 1977: Habitat requirement of the pileated woodpecker in northeastern Oregon. *Journal of Forestry,* **75,** 335–340.

Burton, D.M., B.A. McCarl, C.N.M. de Sousa, D.M. Adams, R. Alig, and S.M. Winnett, 1997: Economic dimensions of climate change impacts on southern forests. In: *The Productivity and Sustainability of Southern Forest Ecosystems in a Changing Environment.* Springer-Verlag, New York, NY, USA, pp. 777–794.

Cairns, M.A., S. Brown, E.H. Helmer, and G.A. Baumgardner, 1997: Root biomass allocation in the world's upland forests. *Oecologia,* **111,** 1–11.

Calder, I.R., 1990: *Evaporation in the Uplands.* John Wiley and Sons, Chichester, United Kingdom.

Calder, I.R., 1992: Water use of eucalyptus—a review. In: *Growth and Water Use of Forest Plantations.* [Calder, I.R., R.L. Hall, and P.G. Adlard (eds.)]. John Wiley and Sons, Chichester, United Kingdom, pp. 167–179.

Canaday, C., 1996: Loss of insectivorous birds along a gradient of human impact in Amazonia. *Biological Conservation,* **77,** 63–77.

Canadell, J.G., L.F. Pitelka, and J.S.I. Ingram, 1996: The effects of elevated CO_2 concentration on plant-soil carbon below-ground—a summary and synthesis. *Plant and Soil,* **187(2),** 391–400.

Cannell, M.G.R., R.C. Dewar, and D.G. Pyatt, 1993: Conifer plantations on drained peatlands in Britain: a net gain or loss of carbon? *Forestry,* **66,** 353–369.

Canter, L.W., 1996: *Environmental Impact Assessment.* McGraw Hill, New York, NY, USA, 660 pp.

Cerri, C.C., M. Bernoux, and B.J. Feigl, 1996: Deforestation and use of soil as pasture: Climatic impacts. In: *Interdisciplinary Research on the Conservation and Sustainable Use of the Amazonian Rain Forest and its Information Requirements* [Lieberei, R., C. Reisdorff, and A.D. Machado (eds.)]. Forschungszentrum Geesthacht GmbH (GKSS), Bremen, Germany, pp. 323.

Chan, N.W., 1986: Drought trends in northwestern peninsular Malaysia: is less rain falling? *Wallaceana,* **44,** 8–9.

Chomitz, K., 1998a: *Baselines for Greenhouse Gas Reductions: Problems, Precedents, Solutions.* Carbon Offsets Unit, World Bank., Washington, DC, USA.

Chomitz, K.M., 1998b: *The Performance and Duration Issue in Carbon Offsets Based on Sequestration.* World Bank Development Research Group, Washington, DC, USA.

Choularton, T.W., M.W. Gallagher, K.N. Bower, D. Fowler, M. Zahniser, and A. Kaye, 1995: Trace gas flux measurements at the landscape scale using boundary-layer budgets. *Philosophical Transactions of the Royal Society of London, Series A,* **351,** 357–369.

Christopher, N., J.M. Melillo, P.A. Steudler, C.C. Cerri, J.F.L. de Moreaes, M.C. Piccolo, and M. Brito, 1997: Soil carbon and nitrogen stocks following forest clearing for pasture in the southwestern Brazilian Amazon. *Ecological Applications,* **7,** 1216–1226.

CIFOR C and I Team, 1999: *The CIFOR Criteria and Indicators Generic Template.* Center for International Forest Research, Bogor, Indonesia, 53 pp.

Conant, R.T., J.M. Klopatek, R.C. Malin, and C.C. Klopatek, 1998: Carbon pools and fluxes along an environmental gradient in northern Arizona. *Biogeochemistry,* **43,** 43–61.

Cordell, G.A., 1995: Natural products as medicinal and biological agents: Potentiating the resources of the rain forest. In: *Chenistry of the Amazon: Biodiversity, Natural Products and Environmental Issues.* [Seidl, P.R., O.R. Gottlieb, and M.A.C. Kaplan (eds.)]. American Chemical Society, Washington DC, USA, pp. 315.

Cordell, H.J., J. Tealsley, G. Super, J.C. Bergstrom, and B. McDonald, 1997: *Outdoor Recreation in the United States: Results from the National Survey on Recreation and Environment.* Athens, GA, USA.

Costanza, R., R. Darge, R. Degroot, S. Farber, M. Grasso, B. Hannon, K. Limburg, S. Naeem, R.V. O'Neill, J. Paruelo, R.G. Raskin, P. Sutton, and M. Vandenbelt, 1997: The value of the world's ecosystem services and natural capital. *Nature,* **387(6630),** 253–260.

Covington, W.W., 1981: Changes in forest floor organic matter and nutrient content following clear cutting in northern hardwoods. *Ecology,* **62(1),** 41–48.

Crutzen, P.J. and M.O. Andreae, 1990: Biomass burning in the tropics: impact on atmospheric chemistry and biogeochemical cycles. *Science,* **250,** 1669–1678.

Cruz, M.C. and I. Zosa-Feranil, 1988: Policy Implications of Population Pressure in the Philippines. University of the Philippines, Los Baños.

Convention on Sustainable Development (CSD), 1996: *Progress Report on Chapter 10 of Agenda 21.* United Nations, New York, NY, USA.

Davidson, E.A. and I.L. Ackerman, 1993: Changes in soil carbon inventories following cultivation of previously untilled soils. *Biogeochemistry,* **20,** 161–193.

De Grandi, G.F., J.P. Malingreau, and M. Leysen, 1999: The ERS-1 Central Africa mosaic: A new perspective in radar remote sensing for the global monitoring of vegetation. *IEEE Transactions on Geoscience and Remote Sensing,* **37(3),** 1730–1746.

DeFries, R., M. Hansen, and J. Townshend, 1995: Global discrimination of land cover types from metrics derived from AVHRR pathfinder data. *Remote Sensing of Environment,* **54(3),** 209–222.

DeFries, R.S., M.C. Hansen, J.R.G. Townshend, A.C. Janetos, and T.R. Loveland, 2000: A new global 1-km data set of percentage tree cover derived from remote sensing. *Global Change Biology,* **6,** 247–254.

Dean, W.E. and E. Gorham, 1998: Magnitude and significance of carbon burial in lakes, reservoirs, and peatlands. *Geology,* **26,** 535–538.

Deans, J.D., 1981: Dynamics of coarse root production in a young plantation of Picea sitchensis. *Forestry,* **54,** 139–155.

Denmead, O.T., R. Leuning, D.W.T. Griffith, I.M. Jamie, M. Esler, H.A. Cleugh, and M.R. Raupach, 1998: *Estimating regional fluxes of CO_2, CH_4, and N_2O at OASIS Through Boundary-Layer Budgeting. Report to National Greenhouse Advisory Committee.* Commonwealth Scientific and Industrial Research Organization (CSIRO), Land and Water, F.C. Pye Laboratory, Canberra, Australia.

Detwiler, R.P., 1986: Land use change and global carbon cycle: The role of tropical soils. *Biogeochemistry,* **2,** 67–93.

Dinerstein, E., D.M. Olson, D.J. Graham, A.L. Webster, S.A. Primm, M.P. Bookbinder, and G. Ledec, 1995: *A conservation assessment of the terrestrial ecoregions of Latin America and the Carribean.* World Wildlife Fund/World Bank, Washington, DC, USA.

Dobes, L., I. Enting, and C. Mitchell, 1998: Accounting for carbon sinks: The problem of time. In: *Trading Greenhouse Emissions: Some Australian Perspectives* [Dobes, L. (ed.)]. Bureau of Transport Economics, Canberra, Australia.

EEPSEA/WWF, 1998: *The Indonesian Fires and Haze of 1997: The Economic Toll*. Economy and Environment Program for South East Asia and World Wildlife Fund, Singapore: Economy and Environment Program for Southeast Asia, International Development Research Centre (IDRC). http://www.idrc.org.sg/eepsea/specialrept/specreptIndofire.htm.

Elisabetsky, E. and P. Shanley, 1994: Ethnopharmacology in the Brazilian Amazon. *Pharmacology and Therapeutics,* **64(2),** 201–214.

Ellert, B.H. and J.R. Bettany, 1995: Calculation of organic matter and nutrients stored in soils under contrasting management regimes. *Canadian Journal of Soil Science,* **75,** 529–538.

EMB, 1990: *The Philippine Environment in the Eighties*. Environmental Management Bureau, Quezon City, Philippines, 302 pp.

Encyclopaedia Britannica, 1970: *Encyclopaedia Britannica 14th Edition*. Encyclopaedia Britannica, Chicago, IL, USA.

European Commission, 1997: *Study on European Forestry Information and Communication System*. Office for Official Publications of the European Communities, Luxembourg, 1,328 pp.

Eurostat, 1999: *Towards Environmental Pressure Indicators for the EU*. Eurostat, Luxembourg, 181 pp.

Falloon, P., P. Smith, J.U. Smith, J. Szabó, K. Coleman, and S. Marshall, 1998: Regional estimates of carbon sequestration potential: linking the Rothamsted carbon model to GIS databases. *Biology and Fertility of Soils,* **27,** 236–241.

Faminow, M.D., 1998: *Cattle, Deforestation and Development in the Amazon: An Economic, Agronomic and Environmental Perspective*. CAB International, Wallingford, United Kingdom, 253 pp.

Fan, S., M. Gloor, J. Mahlman, S. Pacala, J. Sarmiento, T. Takahashi, and P. Tans, 1998: A large terrestrial carbon sink in North America implied by atmospheric and oceanic CO_2 data and models. *Science,* **282,** 442–446.

Fankhauser, S. and R.S.J. Tol, 1997: The social costs of climate change: The IPCC second assesment report. *Mitigation and Adaptation Strategies for Global Change,* **1(4),** 385–403.

FAO, 1971-1981: *Soil Map of the World, 1:5,000,000 (Vol. II-X)*. UNESCO, Paris.

FAO, 1986: *Programme for the 1990 World Census of Agriculture*. FAO Statistical Development Series 2, Food and Agriculture Organization, Rome, Italy, 90 pp.

FAO, 1993a: *Forest Resources Assessment 1990: Tropical Countries*. Forestry Paper 112. Food and Agriculture Organization, Rome, Italy.

FAO, 1993b: *Global and National Soils and Terrain Digital Databases (SOTER)*. Food and Agriculture Organization, Rome, Italy.

FAO, 1995a: *Planning for Sustainable Use of Land Resources: Towards a New Approach*. Land and Water Bulletin 2. Food and Agriculture Organization, Rome, Italy, 60 pp.

FAO, 1995b: *Forest Resources Assessment 1990, Global Synthesis*. Food and Agriculture Organization of the United Nations, Rome, Italy, 46 pp. + appendices.

FAO, 1995c: *Programme for the World Census of Agriculture 2000*. FAO Statistical Development Series 5. Food and Agriculture Organization, Rome, Italy, 79 pp.

FAO, 1996: *Agro-ecological Zoning; Guidelines*. Food and Agriculture Organization, Rome, Italy.

FAO, 1997a: *State of the World's Forests*. Food and Agriculture Organization, Rome, Italy, 200 pp.

FAO, 1997b: *Agenda 21 Progress Report on Chapter 14. Prepared for the Special Session of the United Nations General Assembly from 23 to 27 June*. Available at: http://www.fao.org/WAICENT/FAOINFO/SUST-DEV/EPdirect/EPre0033.htm.

FAO, 1999: *State of the World's Forests*. Food and Agriculture Organization of the United Nations, Rome, Italy, 154 pp.

FAO/IIASA, 1999: World *Agro-Ecological Zoning*. Land and Water Digital Media Series. Food and Agriculture Organization, Rome, Italy, (in press).

FAO/ISRIC, 1990: *Guidelines for Soil Description*. Food and Agriculture Organization, Rome, Italy.

FAO/UNEP, 1999: *Terminology for Integrated Resources Planning and Management*. Food and Agriculture Organization/United Nations Environmental Programme, Rome, Italy and Nairobi, Kenya.

FAO/UNEP/ISRIC, 1999: *Soil and Terrain Databases for Latin America and the Carribean, Scale 1:5 m*. Land and Water Digital Media Series 5. Food and Agriculture Organization, Rome, Italy.

Fearnside, P.M., 1995: Global warming response options in Brazil's forest sector: comparison of project-level costs and benefits. *Biomass and Energy,* **8,** 309–322.

Fearnside, P.M., 1997: Greenhouse-Gas Emissions From Amazonian Hydroelectric Reservoirs—the Example of Brazil's Tucurui Dam as Compared to Fossil Fuel Alternatives. *Environmental Conservation,* **24(1),** 64–75.

Fearnside, P.M., 1998: The value of human life in global warming impacts. *Mitigation and Adaptation Strategies for Global Change,* **3(1),** 83–85.

Fearnside, P.M., 1999a: Biodiversity as an environmental service in Brazil's Amazonian forests: Risks, value and conservation. *Environmental Conservation,* **26(4),** 305–321.

Fearnside, P.M., 1999b: Uncertainty in land use chage and forestry sector mitigation options for global warming: plantation silvicluture versus avoided deforestation. *Biomass and Bioenergy*, (in review).

Fearnside, P.M., 1999c: The potential of Brazil's forest sector for mitigating global warming under the Kyoto Protocol's "Clean Development Mechanism." In: *Global Climate Change: Science, Policy and Mitigation/ Adaptation Strategies. Proceedings of the Second International Specialty Conference* [Kinsman, J.D., C.V. Mathai, M. Baer, E. Holt, and M. Trexler (eds.)]. Global Climate Change: Science, Policy and Mitigation/ Adaptation Strategies. Proceedings of the Second International Specialty Conference at Sewickley, PA, USA, 13–16 October 1998, AWMA: Air and Waste Management Association, pp. 634–646.

Fearnside, P.M., 1999d: Environmental and social impacts of charcoal in Brazil. In: *Os Carvoeiros: The Charcoal People of Brazil* [Prado, M. (ed.)]. Wild Images Ltd., Rio de Janeiro, Brazil, 182 pp.

Fearnside, P.M., 2000a: Can pasture intensification discourage deforestation in the Amazon and Pantanal regions of Brazil? In: *Patterns and Processes of Land Use and Forest Change in the Amazon* [Wood, C.H. (ed.)]. University of Florida, Gainesville, FL, USA, (in press).

Fearnside, P.M., 2000b: *Time Preference in Global Warming Calculations: A Proposal for a Unified Index*. unpublished manuscript.

Fearnside, P.M., 2000c: *Why a 100-Year Time Horizon Should Be Used for Global Warming Mitigation Calculations*. unpublished manuscript.

Fearnside, P.M. and R.I. Barbosa, 1998: Soil carbon changes from conversion of forest to pasture in Brazilian Amazonia. *Forest Ecology and Management,* **108(1-2),** 147–166.

Fearnside, P.M. and J. Ferraz, 1995: A conservation gap analysis of Brazil's Amazonian vegetation. *Conservation Biology,* **9,** 1134–1147.

Fearnside, P.M., D.A. Lashof, and P. Moura-Costa, 1999: Accounting for time in mitigating global warming. *Environmental Conservation*, (in press).

Feller, C. and M.H. Beare, 1997: Physical control of soil organic matter dynamics in the tropics. *Geoderma,* **79(1-4),** 69–116.

Fischer, G., H. Van Velthuizen, and F.O. Nachtergaele, 1999: *Global Agro-Ecological Zones Assessment: Methodology and Results*. Land and Water Bulletin 7. Food and Agriculture Organization, Rome, Italy.

Fisher, M.J., I.M. Rao, M.A. Ayarza, C.E. Lascano, J.I. Sanz, R.J. Thomas, and R.R. Vera, 1994: Carbon storage by introduced deep-rooted grasses in the South American savannas. *Nature,* **371(6494),** 236–238.

Fowler, D., J.N. Cape, and M.H. Unsworth, 1989: Deposition of atmospheric pollutants on forests. *Philosophical Transactions of the Royal Society, London,* **324B,** 247–265.

Frumhoff, P.C., 1995: Conserving wildlife in tropical forests managed for timber. *Bioscience,* **45(7),** 456–464.

Gallagher, M.W., T.W. Choularton, K.N. Bower, I.M. Stromberg, K.M. Beswick, D. Fowler, and K.J. Hargreaves, 1994: Measurements of methane fluxes on the landscape scale from a wetland area in North Scotland. *Atmospheric Environment,* **28,** 2421–2430.

García-Oliva, F., R.L. Sanford, and E. Kelly, 1999: Effects of slash-and-burn management on soil aggregate organic C and N in a tropical deciduous forest. *Geoderma,* **88(1-2),** 1–12.

Garten, C.T. and S.D. Wullschleger, 1999: Soil carbon inventories under a bioenergy crop (switchgrass): Measurement limitations. *Journal of Environmental Quality,* **28,** 1359–1365.

Gash, J.H.C. and W.J. Shuttleworth, 1991: Tropical deforestation: Albedo and the surface energy balance. *Climatic Change,* **19,** 123–133.

Global Environment Facility (GEF), 2000: Draft Operational Program #12: Integrated Ecosystem and Resource Management.

German Advisory Council on Global Change (WGBU), 1998: *The Accounting of Biological Sinks and Sources Under the Kyoto Protocol—a Step Forwards or Backwards for Global Environmental Protection?* WGBU, Bremerhaven, Germany, 75 pp.

Gläser, B., G. Guggenberger, L. Haumaier, and W. Zech, 1999: Burning residues as conditioner to improve sustainable fertility in highly weathered soils of the Brazilian Amazon. In: *Sustainable Management of Soil Organic Matter* [Rees, B. (ed.)]. CABI, Wallingford, United Kingdom, (in press).

Glasson, J., R. Therival, and A. Chadwick, 1994: *Introduction to Environmental Impact Assessment.* UCL Press Ltd., London, United Kingdom, 342 pp.

Glowka, L., F. Burhenne-Guilmin, H. Synge, J.A. McNeely, and L. Gundling, 1994: *A Guide to the Convention on Biological.* International Union for Conservation of Natural Resources, Gland, Switzerland, and Cambridge, United Kingdom, 172 pp.

Government of India, 1983: *Report on the Emergent Crises.* New Delhi, India, High level Commission on Floods, Ministry of Irrigation, New Delhi, India.

Greenwood, E.A.N., L.B. Klein, J.D. Beresford, and G.D. Watson, 1985: Differences in annual evaporation between grazed pasture and eucalyptus species in plantations on a saline farm catchment. *Journal of Hydrology,* **78,** 261–278.

Griffith, D.W.T. and B. Galle, 1999: Flux measurements of NH_3, N_2O and CO_2 using dual beam FTIR Spectroscopy and the flux gradient technique. *Atmospheric Environment,* (in press).

Grimes, A., S. Loomis, P. Jahnige, M. Burnham, K. Onthank, R. Alarcon, W.P. Cuenca, C.C. Martinez, D. Neill, M. Balick, B. Bennett, and R. Mendelsohn, 1994: Valuing the rain forest—the economic value of non-timber forest products in Ecuador. *Ambio,* **23(7),** 405–410.

Harding, R.J., 1992: The modification of climate by forests. In: *Growth and Wateruse of Forest Plantations* [Calder, I.R., R.L. Hall, and P.G. Adlard (eds.)]. John Wiley and Sons, Chichester, United Kingdom, pp. 332–346.

Hardner, J.J., P.C. Frumhoff, and D.C. Goetze, 2000: Prospects for mitigating carbon, conserving biodiversity, and promoting socio-economic development objectives through the Clean Development Mechanism. *Mitigation and Adaptation Strategies for Global Change,* **5(1),** (in press).

Haynes, R.W., D.M. Adams, and J.R. Mills, 1995: *The 1993 RPA Timber Assessment Update.* Technical Report RM-259, U.S. Department of Agriculture, Forest Service, Rocky Mountain Forest and Range Experiment Station, Fort Collins, CO, USA, p. 66.

Hecht, S.B., 1992: valuing land uses in Amazonia: Colonist agriculture, cattle and petty extraction in comparative perspective. In: *Conservation of Neotropical Forests: Working from Traditional Resource Use.* [Redford, K.H. and C. Padoch (eds.)]. Columbia University Press, New York, NY, USA, pp. 379–399.

Heimann, M., G. Esser, J. Kaduk, D. Kicklighter, G. Kohlmaier, D. McGuire, B. Moore III, C. Prentice, W. Sauf, A. Schloss, U. Wittenberg, and G. Wurth, 1998: Evaluation of terrestrial carbon cycle models through simulations of the seasonal cycle of atmospheric CO_2: First results of a model intercomaprison study. *GlobalBiogeochemical Cycles,* **12,** 1–24.

Helms, J.A., 1998: *The Dictionary of Forestry.* Society of American Foresters, Bethesda, MD, USA.

Hibbert, A.R., 1967: Forest treatment effects on water yield. In: *Forest Hydrology. Proceedings of an International Symposium on Forest Hydrology* [Sopper, W.E. and H.W. Lull (eds.)]. Forest hydrology. Proceedings of an international symposium on forest hydrology at Pergamon Press, pp. 527–543.

Hill, M.O. and H.L. Wallace, 1989: Vegetation and environment in afforested sand dunes at Newborough, Anglesey. *Forestry,* **62,** 249–267.

Hoen, H.F. and B. Solberg, 1997: Carbon dioxide taxing, timber rotations, and market implications. In: *Economics of Carbon Sequestration in Forestry* [Sedjo, R.A., R.M. Sampson, and J. Wisniewski (eds.)]. CRC Press, Boca Raton, FL, USA, pp. S151–S162.

Holdridge, L.R., 1947: Determination of world formations from simple climatic data. *Science,* **105,** 367–368.

Holland, E.A., and S. Brown, 1999: North American carbon sink. *Science,* **283,** 1815a.

Hollinger, D.Y., F.M. Kelliher, E.-D. Schulze, N.N. Vygodskaya, A. Varlagin, I.Milukova, J.N. Byers, A. Sogachov, J.E. Hunt, T.M. McSeveney, K.I. Kobak, G. Bauer, and A. Arneth, 1995: Initial assessment of multi-scale measurement of CO_2 and H_2O flux in the Siberian taiga. *Journal of Biogeography,* **22,** 425–431.

Houghton, R.A., J.L. Hackler, and K.T. Lawrence, 1999: The US carbon budget: Contributions from land-use change. *Science,* **285,** 574–578.

Houghton, R.A., J.E. Hobbie, J.M. Melillo, B. Moore, B.J. Peterson, G.R. Shave, and G.M. Woodwell, 1983: Changes in the carbon content of the terrestrial biota and soils between 1860 and 1990: a release of CO_2 to the atmosphere. *Ecol. Monogr.,* **53(3),** 235–262.

Houghton, R.A., D.L. Skole, C.A. Nobre, J.L. Hackler, K.T. Lawrence, and W.H. Chomentowski, 2000: Annual fluxes of carbon from deforestation and regrowth in the Brazilian Amazon. *Nature,* **403,** 301–304.

IEA GHG R&D, 1999: *Interraction Between Forestry Sequestration of CO_2 and Markets for Timber* [Sedjo R. and B. Sohngen (eds.)].

IISD, 2000: IFF-4 Final. *Earth Negotiations Bulletin,* **13(66),** 1–14.

Indonesian First National Communication, 1999: *Indonesia: Forest Sector.* Republic of Indonesia, The Centre for Environment Studies IPB for The State Ministry of Environment, Jakarta, Indonesia.

Instituto Nacional de Pesquisas Espaciais (INPE), 1999: Monitoramento da Floresta Amazonica Brasileira por Satelite/Monitoring of the Brazilian Amazon Forest by Satellite: 1997–1998. Available at: http://www.inpe.br.

Ismail, I., R.L. Bevins, and W.W. Frye, 1994: Long-term no-tillage effects on soil properties and continuous corn yields. *Soil Science Society of America Journal,* **58,** 193–198.

IPCC, 1996: Climate change impacts on forests. In: *Climate Change 1995: Impacts, Adaptations and Mitigation of Climate Change: Scientific-Technical Analyses. Contribution of Working Group II to the Second Assessment Report of the Intergovernmental Panel on Climate Change* [Watson, R.T., M.C. Zinyowera, and R.H. Moss (eds.)]. Cambridge University Press, Cambridge, United Kingdom, and New York, NY, USA, 879 pp.

IPCC, 1997: *Revised 1996 IPCC Guidelines for National Greenhouse Gas Inventories* [J.T. Houghton, L.G. Meira Filho, B. Lim, K. Tréanton, I. Mamaty, Y. Bonduki, D.J. Griggs, and B.A. Callander (eds.)]. Intergovernmental Panel on Climate Change, Meteorological Office, Bracknell, United Kingdom.

ISO, 1996: *ISO 14001: Environmental Managment Systems—Specification with Guidance for Use.* International Organization for Standardization Central Secretariat, Geneva, Switzerland, 14 pp.

Izaurralde, R.C. *et al.,* 1997: Scientific challenges in developing a plan to predict and verify carbon storage in Canadian prsoils. In: *Management of Carbon Sequestration in Soil* [Lal, R. (ed.)]. CRC Press, Boca Raton, FL, USA, pp. 433–446.

Jackson, R.B., J. Canadell, J.R. Ehleringer, H.A. Mooney, O.E. Sala, and E.D. Schulze, 1996: A global analysis of root distributions for terrestrial biomes. *Oecologia,* **108(3),** 389–411.

Jackson, R.B., H.A. Mooney, and E.D. Schulze, 1997: A global budget for fine root biomass, surface area, and nutrient contents. *Proceedings of the National Academy of Sciences of the United States of Science USA,* **94,** 7362–7366.

Johnson, D.W., 1992: Effects of forest management on soil carbon storage. *Water, Air and Soil Pollution,* **64,** 83–120.

Johnson, D.W., 1995: Role of carbon in the cycling of other nutrients in forested ecosystems. In: *Carbon Forms and Functions in Forest Soils* [Kelly J.M. and W.M. McFee (eds.)]. SSSA, Madison, WI, USA, pp. 299–328.

Jones, J.A. and G.E. Grant, 1996: Peak flow response to clear-cutting and roads in small and large basins, western Cascades, Oregon. *Water Resources Research,* **32,** 959–974.

Joos, F., M. Bruno, R. Fink, U. Siegenthaler, T. Stocker, C. Le Quere, and J.L. Sarmiento, 1996: An efficient and accurate representation of complex oceanic and biospheric models of anthropogenic carbon uptake. *Tellus,* **48 (3),** 623–637.

Justice, C.O. and J.R.G. Townshend, 1988: Selecting the spatial resolution of satellite sensors required for global monitoring of land transformations. *International Journal of Remote Sensing,* **9,** 187–236.

Kalembasa, S.J. and D.S. Jenkinson, 1973: A comparative study of titrimetric and gravimetric methods for the determination of organic carbon in soil. *J. Sci. Fd. Agric.,* **24,** 1085–1090.

Kaplan, M.A.C. and O.R. Gottlieb, 1990: Busca racional de principios ativos em plantas. *Interciencia,* **15(1),** 26–29.

Keeling, C.D., R.B. Bacastow, A.F. Carter, S.C. Piper, T.P. Whorf, M. Heimann, W.G. Mook, and H.A. Roeloffzen, 1989: A three dimensional model of atmospheric CO_2 transport based on observed winds: 1. Analysis of observational data. In: *Aspects of Climate Variability in the Pacific and Western Americas.* [Peterson, D.H. (ed.)]. American Geophysical Union, Washington, DC, USA, pp. 165–236.

Keeling, C.D. and T.P. Whorf, 1998: *Atmospheric CO_2 Concentrations—Mauna Loa Observatory, Hawaii, 1958–1997 (Revised August 1998).* Carbon Dioxide Information Research Analysis Center, Oak Ridge National Laboratory, Oak Ridge, TN, USA.

Kennedy, C.E.J. and T.R.E. Southwood, 1984: The number of species of insects associated with British trees: a re-analysis. *Animal Ecology,* **53,** 455–478.

Kilpeläinen, P. and T. Tokola, 1999: Gain to be achieved from stand delineation in LANDSAT TM image-based estimates of stand volume. *Forest Ecology and Management,* **124,** 105–111.

Kirschbaum, M.U.F., 2000: The role of forests in the global carbon cycle. In: *Criteria and Indicators for Sustainable Forest Managment.* [Freanc, A., D. Flinn, and J. Raison (eds.)]. CAB International, Wallingford, United Kingdom, (in press).

Kirschbaum, M.U.F., A. Fischlin, M.G.R. Cannell, R.V. Cruz, W. Galinski, and W.A. Cramer, 1996: Climate change impacts on forests. In: *Climate Change 1995: Impacts, Adaptations and Mitigation of Climate Change: Scientific-Technical Analyses. Contribution of Working Group II to the Second Assessment Report of the Intergovernmental Panel on Climate Change* [Watson, R.T., M.C. Zinyowera and R.H. Moss (eds.)]. Cambridge University Press, Cambridge, United Kingdom, and New York, NY, USA, pp. 94–129.

Kogel, I., 1997: ^{13}C and ^{15}N NMR Spectroscopy as a technique in soil organic matter studies. *Geoderma,* **80(3+4)(Special),** 243–270.

Kohl, M. and R. Paivinen, 1997: *Study on European Forestry Information and Communication System.* Office for Official Publications of the European Communities, Luxembourg, 1,328 pp.

Kortelainen, P., S. Saukkonen, and T. Mattsson, 1997: Leaching of nitrogen from forested catchments in Finland. *Global Biogeochemical Cycles,* **11,** 627–638.

Kramer, R., D. Richter, S. Pattanayak, and N. Sharma, 1997: Economic and ecological analysis of watershed protection in Eastern Madagascar. *Journal of Environmental Management,* **49,** 277–295.

Krawetz, N.M., 1991: *Social Impact Assessment: An Introductory Handbook.* Environmental Management Development in Indonesia Project, Jakarta, Indonesia, 220 pp.

Kreiser, A.M., P.G. Appleby, J. Natkanski, B. Rippey, and R.W. Batterbee, 1990: Afforestation and lake acidification. A comparison of four sites in Scotland. *Philosophical Transactions of the Royal Society, London B,* **327,** 377–383.

Kurz, W.A. and M.J. Apps, 1995: An analysis of future carbon budgets of Canadian boreal forests. *Water Air and Soil Pollution,* **82,** 321–331.

Kurz, W.A. and M.J. Apps, 1999: A 70-year retrospective analysis of carbon fluxes in the Canadian forest sector. *Ecological Applications,* **9(2),** 526–547.

Kurz, W.A., M.J. Apps, B.J. Stocks, and W.J.A. Volney, 1995: Global climatic change: Disturbance regimes and biospheric feedbacks of temperate and boreal forests. In: *Biospheric Feedbacks in the Global Climate System: Will the Warming Feed the Warming* [Woodwell, G.F. and F. Mackenzie (eds.)]. Oxford University Press, New York, NY, USA, pp. 119–133.

Kurz, W.A., S.J. Beukema, and M.J. Apps, 1996: Estimation of root biomass and dynamics for the carbon budget model of the Canadian forest sector. *Canadian Journal of Forest Research,* **26,** 1973–1979.

Lacelle, B., C. Tarnocai, S. Waltman, J. Kimble, F. Orozco-Chaves, and B. Jakobsen, 1997: *Soil Organic Carbon Map in North America.* USDA-NRCS/NSSC, Lincoln, NB, USA.

Lal, R., 1987: *Tropical Ecology and Physical Edaphology.* John Wiley and Sons, New York, NY, USA.

Lal, R., 1995: Global soil erosion by water and carbon dynamics. In: *Soils and Global Change. Advances in Soil Science.* [Lal, R., J. Kimble, E. Levine, and B.A. Stewart (eds.)]. Lewis Publishers, Chelsea, MI, USA, pp. 131–142.

Lal, R., 1996: Deforestation and land use effects on soil degradation and rehabilitation in Western Nigeria: II. Soil chemical properties. *Land Degradation and Development,* **7,** 87–98.

Lal, R., J. Kimble, R.F. Follett, and B.S. Stewart, 2000: *Methods of Assessment of Soil Carbon Pools and Fluxes.* (in press).

Lal, R., J.M. Kimble, R.F. Follett, and C.V. Cole, 1998: *The Potential of US Cropland to Sequester Carbon and Mitigate the Greenhouse Effect.* Ann Arbor Press, Chelsea, MI, USA, 128 pp.

Lashof, D.A. and D.R. Ahuja, 1990: Relative global warming potentials of greenhouse gas emissions. *Nature,* **344,** 529–531.

Lashof, D.A. and B. Hare, 1999: The role of biotic carbon stocks in stabilizing greenhouse gas concentrations at safe levels. *Environmental Science and Policy,* **2(2),** 101–109.

Last, F.T. and R. Watling, 1991: Acidic deposition. Its nature and impacts. *Proceedings of the Royal Society of Edinburgh, Section B,* 97.

Lawton, J.H., D.E. Bignell, B. Bolton, G.F. Bloemers, P. Eggleton, P.M. Hammond, M. Hodda, R.D. Holt, T.B. Larsen, N.A. Mawdsley, N.E. Stork, D.S. Srivastava, and A.D. Watt, 1998: Biodiversity inventories, indicator taxa and effects of habitat modification in tropical forest. *Nature,* **391,** 72–76.

Lazarus, M., S. Kartha, M. Ruth, S. Bernow, and C. Dunmire, 1999: *Evaluation of Benchmarking as an Approach for Establishing Clean Development Mechanism Baselines.* Tellus Institute, Boston, MA, USA, 135 pp.

Lean, J., C.B. Buntoon, C.A. Nobre, and P.R. Rowntree, 1996: The simulated impact of Amazonian deforestation on climate using measured ABRACOS vegetation characteristics. In: *Amazonian Deforestation and Climate* [Gash, J.H.C., C.A. Nobre, J.M. Roberts, and T.L. Victoria (eds.)]. John Wiley and Sons, Chichester, United Kingdom, pp. 549–576.

Leungaramsri, O. and P. Malapetch, 1992: Illegal logging. In: *The Future of People and Forests in Thailand after the Logging Ban* [Leungaramsri, O. and N. Rajesh (eds.)]. Bangkok, Thailand, pp. 29–55.

Leuning, R., S.K. Baker, I.M. Jamie, C.H. Hsu, L. Klein, O.T. Denmead, and D.W.T. Griffith, 1999: Methane emission from free-ranging sheep: a comparison of two measurement methods. *Atmospheric Environment,* **33,** 1357–1365.

Levine, J.S., T. Bobbe, N. Ray, R.G. Witt, and A. Singh, 1999: *Wildland Fires and the Environment: A Global Synthesis.* United Nations Environmental Programme, Division of Environmental Information, Assessment and Early Warning, Nairobi, Kenya, 46 pp.

Lim, B., S.B. Brown, and B. Schlamadinger, 1998: *Evaluating Approaches for Estimating Net Emissions of Carbon Dioxide from Forest Harvesting and Wood Products.* International Panel on Climate Change/Organization for Economic and Commercial Development/International Energy Agency Programme on National Greenhouse Gas Inventories, Dakar, Senegal, 20 pp. + appendices.

Liniger, H., D.B. Thomas, and H. Hurni, 1988: World overview of conservation approaches and technologies: preliminary results from eastern and southern Africa. In: *Towards Sustainable Land Use* [Blume, H.P., H. Eger, E. Fleishhauer, A. Hebel, C. Reij, and K.G. Steiner (eds.)]. pp. 1037–1046.

Liniger, H. and R. Weingartner, 1998: Mountains and freshwater supply. *Unasylva,* **195(49),** 39–46.

Liski, J., H. Ilvesniemi, A. Mäkelä, and M. Starr, 1998: Model analysis of the effects of soil age, fires and harvesting on the carbon storage of boreal forest soils. *European Journal of Soil Science,* **49(3),** 407–416.

Lugo, A.E., 1997: Rendimiento y Aspectos Silviculturales de Plantaciones Maderas en America Latina. Serie Forestal No. 9. Oficina Regional de la FAO para America Latina y el Caribe, Santiago, Chile. (in Spanish)

Lugo, A.E. and S. Brown, 1993: Management of tropical soils as sinks or sources of atmospheric carbon. *Plant and Soil,* **149,** 27–41.

Lund, H.G., 1999: *Definitions of Forest, Deforestation, Afforestation and Reforestation.* Forest Information Services, Manassas, VA, USA, Information Services.

Maass, J.M., C.F. Jordan, and J. Sarukhan, 1988: Soil erosion and nutrients losses in seasonal tropical agroecosystems under various management techniques. *Journal of Applied Ecology,* **25,** 595–607.

MacDicken, K.G., 1997: *A Guide to Monitoring Carbon Storage in Forestry and Agroforestry Projects.* Winrock International, Arlington, VA, USA, 87 pp.

Maler, K.G., 1990: Economic theory and environmental degradation: A survey of some problems. *Revista de Analisis Economico,* **5,** 7–17.

Mann, L.K., 1986: Changes in soil carbon storage after cultivation. *Science,* **142(5),** 279–288.

Marland, G., B. Schlamadinger, and P. Leiby, 1997: Forest/biomass based mitigation strategies—does the timing of carbon reductions matter. *Critical Reviews in Environmental Science and Technology,* **27(Special),** S 213–S 226.

Marshall, A.G. and M.D. Swaine, 1992: Tropical rain forest: disturbance and recovery. *Philosophical Transactions of the Royal Society, London B,* **335,** 323–457.

McCann, J.M., W.I. Woods, and D. Meier, 1999: Organic matter and anthrosols in Amazonia: interpreting the amerindian legacy. In: *Sustainable Management of Soil Organic Matter* [Rees, B. (ed.)]. CABI, Wallingford, United Kingdom.

McCarty, G.W., N.N. Lyssenko, and J.L. Starr, 1998: Short-term changes in soil carbon and nitrogen pools during tillage management transition. *Soil Science Society of America Journal,* **62,** 1564–1571.

McNulty, S.G., J.A. Moore, L. Iverson, A. Prasad, R. Abt, B. Smith, G. Sun, M. Gavazzi, J. Bartlett, B. Murray, R. Mickelr, and J.D. Aber, 1999: Development and application of linked regional scale growth, biogeography, and economic models for Southeastern United States pine forests. *World Resources Review,* (in press).

McVoy, C.W., K.C. Kersbaum, B. Diekkruger, and F. Sondgerath, 1995: Validation of agroecosystem models. *Ecological Modelling,* **81(Special),** 1–300.

Meentemeyer, V., J. Gardner, and E.O. Box, 1985: World patterns and amounts of detrital soil carbon. *Earth Surface Processes and Landforms,* **10(6),** 557–567.

Meher-Homji, V.M., 1992: Probable impact of deforestation on hydrological process. In: *Tropical Forests and Climate.* [Myers, N. (ed.)] pp. 163–174.

Meybeck, D.J.D., 1982: Carbon, nitrogen and phosphorous transport by world rivers. *American Journal of Science,* **282,** 410–450.

Michon, G. and H. de Foresta, 1995: The Indonesian agroforest model. Forest resource management and biodiversity conservation. In: *Conserving Biodiversity Outside Protected Areas: the Role of Traditional Agro-ecosystems* [Halliday, P. and D.A. Gilmour (eds.)]. International Union for Conservation of Natural Resources, Gland, Switzerland, pp. 90–106.

Moncrieff, J.B., J.M. Massheder, H.A.R. de Bruin, J.E. Elbers, T. Friborg, B. Heusinkveld, P. Kabat, S. Scott, H. Soegaard, and A. Verhoef, 1996: A system to measure surface fluxes of momentum, sensible heat, water vapour and carbon dioxide. *Journal of Hydrology,* **188/189,** 589–611.

Moore, B., R.D. Boone, J.E. Hobbie, R.A. Houghton, J.M. Melillo, B.J. Peterson, G.R. Shaver, C.J. Vörösmarty, and G.M. Woodwell, 1981: A simple model for analysis of the role of terrestrial ecosystems in the global carbon budget. In: *Carbon Cycle Modelling* [Bolin, B. (ed.)]. John Wiley and Sons, New York, NY, USA, pp. 365–385.

Morris, J.D. and L.A.J. Thomson, 1983: The role of trees in dryland salinity control. *Proceedings of Royal Society of Victoria,* **95,** 123–131.

Moulton, R.J. and K.R. Richards, 1990: *Costs of Sequestering Carbon Through Tree Planting and Forest Management in the United States.* U.S. Department of Agriculture, Forest Service, Washington, DC, USA, **WO-58:** 44.

Mountain Agenda, 1997: *Mountains of the World: Challenges of the 21st Century.* Mountain Agenda, Bern, Switzerland, 36 pp.

Moura-Costa, P.H., 1996: tropical forestry practices for carbon sequestration. In: *Dipterocarp Forest Ecosystems—Towards Sustainable Management* [Schulte, A. and D. Schone (eds.)]. World Scientific, Singapore, pp. 308–334.

Moura-Costa, P.H. and C. Wilson, 1999: An equivalence factor between CO_2 avoided emissions and sequestration—Description and applications in forestry. *Mitigation and Adaptation Strategies for Global Change,* (in press).

Munasinghe, M., 1993: *Environmental Economics and Sustainable Development.* World Bank, Washington, DC, USA.

Munasinghe, M., 2000: *Development, Sustainability and Equity in the Context of Climate Change.* IPCC Guidance Paper, Geneva, Switzerland.

Munn, R.E., 1994: *Keeping Ahead: The Inclusion of Long Term "Global" Futures in Cumulative Environmental Assessments.* Institute for Environmental Studies, University of Toronto, Ontario, Canada, 282 pp.

Murray, B.C., 1999: *Carbon Values, Reforestation, and Perverse Incentives Under Kyoto Protocol: An Empirical Analysis.* Center for Economic Research, Research Triangle Institute, Research Triangle Park, NC, USA, 41 pp. (in review).

Myers, N., 1989: Loss of biological diversity and its potential impact on agriculture and food production. In: *Food and Natural Resources* [Pimentel, D. and C.W. Hall (eds.)]. Academic Press, San Diego, CA, USA, pp. 49–68.

Myers, N., 1997: The world's forests and their ecosystem services. In: *Nature's Services: Societal Dependence on Natural Ecosystems* [Daily, G.C. (ed.)]. Island Press, Washington, DC, USA, pp. 215–235.

Nabuurs, G.J., A.V. Dolman, E. Verkaik, A. Whitmore, W. Daaman, O. Oenema, P. Kabat, and G.M.J. Mohren 1999: *Resolving Issues on Terrestrial Biospheric Carbon Sinks in the Kyoto Protocol.* Dutch National Research Programme on Global Air Pollution and Climate Change, Bilthoven, The Netherlands, 410 200 030: 100.

Nabuurs, G.J. and R. Sikkema, 1998: The role of harvested wood products in national carbon balances—an evaluation of alternatives for IPCC Guidelines. Institute for Forestry and Nature Research/Institute for Forest and Forest Products, **98/3,** 53.

Nadelhoffer, K.J., B.A. Emmett, P. Gundersen, O.J. Kjonaas, C.J. Koopmans, P. Schleppi, A. Tietema, and R.F. Wright, 1999: Nitrogen deposition makes a minor contribution to carbon sequestration in temperate forests. *Nature,* **398(6723),** 145–148.

Nakane, K., 1976: An empirical formulation of the vertical distribution of carbon concentration in forest soils. *Japanese Journal of Ecology,* **26,** 171–174.

Nakicenovic, N., A. Grubler, H. Ishitani, T. Johansson, G. Marland, J.R. Moreira, and H.-H. Rogner, 1996: Energy primer. In: *Climate Change 1995. Impacts, Adaptations and Mitigation of Climate Change: Scientific-Technical Analyses. Contribution of Working Group II to the Second Assessment Report of the Intergovernmental Panel on Climate Change* [Watson, R.T., M.C. Zinyowera, and R.H. Moss (eds.)]. Cambridge University Press, Cambridge, United Kingdom and New York, NY, USA, pp. 75–92.

National Research Council (NRC), 1993: *Sustainable Agriculture and the Environment in the Humid Tropics.* National Academic Press, Washington, DC, USA, 702 pp.

NEDA, 1988: *Medium Term Development Plan (1988–1992).* National Economic Development Authority, Manila, Philippines, 444 pp.

Nelson, D.W. and L.E. Sommers, 1982: Total carbon, organic carbon and organic matter. In: *Methods of soil analysis, Part 2.* [Page, A.L. (ed.)]. ASA/SSSA, Madison, WI, USA, pp. 539–579.

Nepstad, D.C., C.R. Decarvalho, E.A. Davidson, P.H. Jipp, P.A. Lefebvre, G.H. Negreiros, E.D. Dasilva, T.A. Stone, S.E. Trumbore, and S. Vieira, 1994: The role of deep roots in the hydrological and carbon cycles of Amazonian forests and pastures. *Nature,* **372(6507),** 666–669.

Nepstad, D.C., C. Klink, and S.E. Trumbore, 1995: Pasture soils as carbon sink. *Nature,* **376,** 472–473.

Norton, B., 1988: Commodity, amenity and morality: the limits of quantification in valuing biodiversity. In: *Biodiversity* [Wilson, E.O. (ed.)]. National Academy Press, Washington, DC, USA, pp. 200–205.

OECD, 1993: *OECD Core Set of Indicators for Environmental Performance Reviews.* Organization for Commercial and Economic Development, Paris, France, 39 pp.

OECD, 1998: *OECD Work on Sustainable Development.* Organization for Commercial and Economic Development, Paris, France, 53 pp.

Ogle, B., 1995: People's dependency on forests for food security—some lessons learnt from a programme of case studies. In: *Current Issues in Non-Timber Forest Products Research.* [Perez, M.R. and J.E.M. Arnold (eds.)]. Center for International Forestry Research, Bogor, Indonesia.

Oldeman, L.R., R.T.A. Hakkeling, and W.G. Sombroek, 1991: *World Map of the Status of Human-Induced Soil Degradation (2nd rev.).* International Soil Reference and Information Centre, Wageningen, The Netherlands, 30 pp.

Olson, D.M. and E. Dinerstein, 1998: *The Global 200: A Representation Approach to Conserving the Earth's Distinctive Ecoregions.* World Wildlife Fund, Washington DC, USA, 152 pp.

Ormerod, S.J., A.P. Donald, and S.J. Brown, 1989: The influence of plantation forestry on the pH and aluminium concentration of upland Welsh streams: a re-examination. *Environmental Pollution,* **62,** 47–62.

Ovington, J.D., 1959: The circulation of minerals in plantations of Pinus sylvestris L. *Annals of Botany,* **23,** 229–239.

Parks, P.J. and I.W. Hardie, 1995: Least cost forest carbon reserves: Cost effective subsidies to convert marginal agricultural land to forests. *Land Economics,* **71(1),** 122–136.

Parton, W.J., D.S. Schimel, C.V. Cole, and D.S. Ojima, 1987: Analysis of factors controlling soil organic matter levels in great plain grasslands. *Journal of the Soil Science Society of America,* **51,** 1173–1179.

Pattanayak, S. and R. Kramer, 2000: *Worth of Watersheds: A Producer Surplus Approach for Valuing Drought Control in Eastern Indonesia.* Environment and Development Economics, (in press).

Paustian, K., A.L. Agren, and E. Bosatta, 1997: Modelling litter effects on decomposition and soil organic matter dynamics. In: *Driven by Nature. Plant Litter Quality and Decomposition* [Cadisch G. and K.E. Giller (eds.)]. CABI, Wallingford, United Kingdom, pp. 315–334.

Paustian, K., E.T. Elliott, and M.R. Carter, 1998: Tillage and crop management impacts on soil C storage—Use of long-term experimental data. *Soil and Tillage Research,* **47(3-4),** R 7–R 12.

PCSD, 1997a: *The Philippine Council for Sustainable Development.* The Philippine Coucil for Sustainable Development, Manila, Philippines, 8 pp.

PCSD, 1997b: *Philippine Agenda 21.* The Philippine Coucil for Sustainable Development, Manila, Philippines, 188 pp.

Pearce, D. and D. Moran, 1994: *The Economic Value of Biodiversity.* Earthscan, London, United Kingdom, 172 pp.

Perez-Garcia, J., L.A. Joyce, C.S. Binkley, and A.D. McGuire, 1997: Economic impacts of climatic change on the global forest sector: An integrated ecological/ecnomic assessment. *Critical Reviews in Environmental Science and Technology,* **27,** S123–S138.

Peters, C.M., A.H. Gentry, and R.O. Mendelsohn, 1989: Valuation of an Amazonian rainforest. *Nature,* **339,** 655–656.

Phillips, O.L., Y. Malhi, N. Higuchi, W.F. Laurance, P.V. Nuriez, R.M. Vasquez, S.G. Laurance, L.V. Ferreira, M. Stern, S. Brown, and J. Grace, 1998: Changes in the carbon balance of tropical forests: evidence from long-term plots. *Science,* **282,** 439–442.

Pimentel, D., M. McNair, L. Duck, M. Pimentel, and J. Kamil, 1997a: The value of forests to world food security. *Human Ecology,* **25(1),** 91–120.

Pimentel, D., C. Wilson, C. McCullum, R. Huang, P. Dwen, J. Flack, Q. Tran, T. Saltman, and B. Cliff, 1997b: Economic and environmental benefits of biodiversity. *BioScience,* **47(11),** 747–757.

Pinard, M.A. and F.E. Putz, 1996: Retaining forest biomass by reducing logging damage. *Biotropica,* **28(3),** 278–295.

Pinard, M. and F. Putz, 1997: Monitoring carbon sequestration benefits associated with a reduced impact logging project in Malaysia. *Mitigation and Adaptation Strategies for Global Change,* **2,** 203–215.

Pingoud, K., I. Savolainen, and H. Seppala, 1996: Greenhouse impact of the Finnish forest sector including forest products and waste management. *Ambio,* **25,** 318–326.

Post, W.M., R.C. Izaurralde, L.K. Mann, and N. Bliss, 1999: Monitoring and verifying soil carbon sequestration. In: *Carbon Sequestration in Soils* [Rosenberg, N., R.C. Izaurralde, and E.L. Malone (eds.)]. Batelle Press, pp. 41–82.

Potter, L. and J. Lee, 1998: *Tree Planting in Indonesia: Trends, Impacts and Directions.* Center for International Forestry Research, Bogor, Indonesia, 76 pp.

Powlson, D.S. and D.S. Jenkinson, 1981: A comparison of the organic matter, biomass, adenosine triphosphate and mineralizable nitrogen contents of ploughed and direct-drilled soils. *Journal of Agricultural Science,* **97,** 713–721.

Prabhu, R., C.J.P. Colfer, and R.G. Dudley, 1999: *Guidelines for Developing, Testing and Selecting Criteria and Indicators for Sustainable Forest Management.* Center for International Forest Research, Bogor, Indonesia, 186 pp.

Price, C., 1997: Valuation of biodiversity: of what, by whom, and how? *Scottish Forestry,* **51(3),** 134-142.

Price, C. and R. Willis, 1993: Time, discounting and the valuation of forestry's carbon fluxes. *Commonwealth Forestry Review,* **72(4),** 265–271.

Raiffa, H., 1968: *Decision Analysis: Introductory Lectures on Choices Under Uncertainty.* Addison-Wesley, Reading, MA, USA, 312 pp.

Ravindranath, N.H. and D.O. Hall, 1995: *Biomass, Energy and Environment—a Developing Country Perspective from India.* Oxford University Press, New York, NY, USA, 376 pp.

Richards, M., 1993: The potential of non-timber forest products in sustainable forest management in Amazonia. *Commonwealth Forestry Review,* **72(1),** 21–27.

Richardson, J.L. and W.J. Edmonds, 1987: Linear regression estimations of Jenny's relative effectiveness of state factor equations. *Soil Science,* **144,** 203–208.

Richey, J.E., J.T. Brock, R.J. Naiman, R.C. Wissmar, and R.R. Stallart, 1980: Organic Carbon: oxidation and transport in the Amazon River. *Science,* **207,** 1348–1351.

Richter, D.D. and D. Markewitz, 1996: Carbon changes during the growth of lobolly pine on formerly cultivated soil: the Calhoun Experimental Forest, USA. In: *Evaluation of Soil Organic Matter Models* [Powlson, D.S., P. Smith, and J.U. Smith (eds.)]. Springer-Verlag, New York, NY, USA, pp. 397–407.

Richter, D.D., D. Markewitz, S.E. Trumbore, and C.G. Wells, 1999: Rapid accumulation and turnover of soil carbon in an aggrading forest. *Nature,* **400,** 56–58.

Rignot, E., W.A. Salas, and D.L. Skole, 1997: Mapping deforestation and secondary growth in rondonia, brazil, using imaging radar and thematic mapper data. *Remote Sensing of Environment,* **59(2),** 167–179.

Rinne, J., A. Delany, and A. Guenther, 1999: Development of a periodic eddy accumulator: A true eddy accumulation system for biogenic hydrocarbon fluxes. Abstract. *Transactions of the American Geophysical Union,* **80(46),** F129.

Rochelle, J.A. and F.L. Bunnell, 1979: Plantation management and vertebrate wildlife. In: *The Ecology of Even-aged Plantations* [Ford, E.D., D.C. Malcolm and J. Atterson (eds.)]. Institute of Terrestrial Ecology, Edinburgh, United Kingdom, pp. 373–411.

Rosenqvist, A., V. Taylor, B. Chapman , M. Shimada, A. Freeman, G. De Grandi, S. Saatchi, and Y. Rauste, 2000: The global rain forest mapping project—a review. *International Journal of Remote Sensing,* (in press).

Ruark, G.A. and S.J. Zarnoch, 1992: Soil carbon, nitrogen, and fine root biomass sampling in a pine stand. *Soil Science Society of America Journal,* **56,** 1945–1950.

Running, S.W., C.O. Justice, V.V. Salomonson, D. Hall, J. Barker, Y.J. Kaufman, A.R. Strahler, J.-P. Muller, V. Vanderbilt, Z.M. Wan, P. Teillet, and D. Carneggie, 1994a: Terrestrial remote sensing science and algorithms planned for the MODIS-EOS. *International Journal of Remote Sensing,* **15(17)**, 3587–3620.

Running, S.W., T.R. Loveland, L.L. Pierce, and E.R. Hunt Jr., 1994b: A remote sensing based vegetation classification for global land-cover analysis. *Remote Sensing of the Environment,* **51,** 39–48.

Saatchi, S.S., J.V. Soares, and D.S. Alves, 1997: Mapping deforestation and land use in amazon rainforest by using sir-c imagery. *Remote Sensing of Environment,* **59(2),** 191–202.

Sader, S.A., R.B. Waide, W.T. Lawrence, and A.T. Joyce, 1989: Tropical forest biomass and successional age class relationships to a vegetational index derived from Landsat TM data. *Remote Sensing of Environment,* **28,** 143–156.

Sanchez, P.A., 2000: Delivering on the promise of agroforestry: Environment, development and sustainability. Kluwer Academic Press, (in press).

Sanchez, P.A., K.D. Shepherd, M.J. Soule, F.M. Place, R.J. Buresh, A.-M.N. Izac, A.U. Mokwunye, F.R. Kwesiga, C.G. Ndiritu, and P.L. Woomer, 1997a: Soils fertility replenishment in Africa: an investment in natural resource capital. In: *Replenishing Soil Fertility in Africa.* [Buresh, R.J. (ed.)]. Soil Science Society of America, Madison, WI, USA, pp. 1–46.

Sanchez, P.A., R.J. Buresh, and R.R.B. Leakey, 1997b: Trees, soils, and food security. *Philosophical Transactions of the Royal Society of London, Series B,* **353,** 949–961.

Sandor, J.A. and N.S. Eash, 1995: Ancient agricultural soils of the Andes of Southern Peru. *Soil Science Society of America Journal,* **5,** 170–179.

Sathaye, J., W. Makundi, B. Goldberg, C. Jepma, and M. Pinard (eds.), 1997: International workshop on sustainable forestry management: Monitoring and verification of greenhouse gases: Summary statement. *Mitigation and Adaptation Strategies for Global Change,* **2(2-3),** 91–99.

Schiffman, P.M. and W.C. Johnson, 1988: Phytomass and detrital carbon storage during forest regrowth in the southeastern United States Piedmont. *Canadian Journal of Forest Research* **19,** 69-78.

Schimel, D., I. Enting, M. Heimann, T. Wigley, D. Raynaud, D. Alves, and U. Siegenthaler, 1995: CO_2 and the carbon cycle. In: *Climate Change 1994: Radiative Forcing of Climate Change and an Evaluation of the IPCC IS92 Emissions Scenarios* [Houghton, J.T., L.G. Meira Filho, B.A. Callander, N. Harris, A. Kattenberg, and K. Maskell (eds.)]. Cambridge University Press, Cambridge, United Kingdom, pp. 73–126.

Schlamadinger, B. and G. Marland, 1996: The role of forest and bioenergy strategies in the global carbon cycle. *Biomass and Bioenergy,* **10,** 275–300.

Schlamadinger, B. and G. Marland, 1998: The Kyoto Protocol: Provision and unresolved issues relevant to land use change and forestry. *Environmental Science and Policy,* **1,** 313–327.

Schlesinger, W.H., 1982: Carbon storage in the caliche of arid soils: a case study from Arizona. *Soil Science,* **133,** 247–255.

Schlesinger, W.H., 1986: Changes in soil carbon storage and associated properties with disturbance and recovery. In: *The Changing Carbon Cycle: a Global Analysis.* [Trabalka, J.R. and D.E. Reichle (eds.)]. Springer-Verlag, New York, pp. 175–193.

Schlesinger, W.H., 1999: Carbon and agriculture—carbon sequestration in soils. *Science,* **284(5423),** 2095.

Schlesinger, W.H. and J.M. Melack, 1981: Transport of organic carbon in the world's rivers. *Tellus,* **33,** 172–187.

Schofield, N.J., 1992: Tree planting for dryland salinity control in Australia. *Agroforestry Systems,* **20,** 1–23.

Schroeder, P., S. Brown, J. Mo, R. Birdsey, and C. Cieszewski, 1997: Biomass estimation for temperate broadleaf forests of the US using inventory data. *Forest Science,* **43,** 424–434.

Schulze, E.-D., 1982: Plant life forms and their carbon, water and nutrient relations. In: *Physiological Plant Ecology II. Water Relations and Carbon Assimilation* [O.L. Lange, C.B. Osmond, and H. Ziegler (eds.)]. Springer-Verlag, Berlin, Germany, pp. 615–676.

Sellers, P.J., F.G. Hall, R.D. Kelly, A. Black, D. Baldocchi, J. Berry, M. Ryan, K.J. Ranson, P.M. Crill, D.P. Lettenmaier, H. Margolis, J. Cihlar, J. Newcomer, D. Fitzjarrald, P.G. Jarvis, S.T. Gower, D. Halliwell, D. Williams, B. Goodison, D.E. Wickland, and F.E. Guertin, 1997: BOREAS in 1997: experiment overview, scientific results, and future directions. *Journal of Geophysical Research,* **102(D24),** 28,731–28,769.

Simpson, R.D., R.A. Sedjo, and J.W. Reid, 1996: Valuing biodiversity for use in pharmaceutical research. *Journal of Political Economy.* **104(1),** 163–185.

Sioli, H., 1984: *The Amazon, Limnology and Landscape Ecology of a Mighty Tropical River and Its basin: Wildlife Planning Glossary.* US Department of Agriculture Forest Service, Boston, MA, USA, 736 pp.

Skog, K.E. and G.A. Nicholson, 1998: Carbon cycling through wood products: The role of wood and paper products in carbon sequestration. *Forest Products Journal,* **48(7/8),** 75–83.

Skole, D. and C. Tucker, 1993: Tropical deforestation and habitat fragmentation in the Amazon: Satellite data from 1978 to 1988. *Science,* **260,** 1905–1910.

Skole, D.L., C.O. Justice, J.R.G. Townshend, and A.C. Janetos, 1997: A land cover change monitoring program: Strategy for an international effort. *Mitigation and Adaptation Strategies for Global Change,* **2,** 157–175.

Smaling, E.M.A., 1998: Nutrient balances as indicators of productivity and sustainability in Sub-Saharan African agriculture. *Agriculutral Ecosystems and Environment,* **71(1-3),** 346.

Smil, V. and M. Yuchi, 1998: *The Economic Costs of China's Environmental Degradation.* American Academy of Arts and Sciences, Cambridge, MA, USA.

Smith, P., D.S. Powlson, M.J. Glendining, and J.U. Smith, 1998: Preliminary estimates of the potential for carbon mitigation in european soils through no-till farming. *Global Change Biology,* **4(6),** 679–685.

Smith, P., J.U. Smith, D.S. Powlson, W.B. McGill, J.R.M. Arah, O.G. Chertov, K. Coleman, U. Franko, S. Frolking, D.S. Jenkinson, L.S. Jensen, R.H. Kelly, H. Kleingunnewiek, A.,S. Komarov, C. Li, J.A.E. Molina, T. Mueller, W.J. Parton, J.H.M. Thornley, and A.P. Whitmore, 1997a: A comparison of the performance of nine soil organic matter models using datasets from seven long-term experiments. *Geoderma,* **81(1-2),** 153–225.

Smith, P., D.S. Powlson, M.J. Glendining, and J.U. Smith, 1997b: Potential for carbon sequestration in European soils—preliminary estimates for five scenarios using results from long-term experiments. *Global Change Biology,* **3(1),** 67–79.

Smith, P., D.S. Powlson, J.U. Smith, and E.T. Elliott (eds.), 1997c: Evaluation and comparison of soil organic matter models using datasets from seven long-term experiments. *Geoderma,* **81,** 1–225.

Smith, S.J. and M.L. Wigley, 2000: Global warming potentials: 1. Climatic implications of emissions reductions. *Climatic Change,* **44(4),** 445–457.

Sohngen, B. and R. Mendelsohn, 1998: Valuing the impact of large-scale ecological change in a market—the effect of climate change on US timber. *American Economic Review,* **88(4),** 686–710.

Solberg, B., D. Brooks, H. Pajuoja, T.J. Peck, and P.A. Wardle, 1996: *Long Term Trends and Prospects in World Supply and Demand for Wood and Implications for Sustainable Forest Management. A Contribution to the CSD Ad Hoc Intergovernmental Panel on Forests.* European Forest Institute/Norwegian Forest Research Institute, p. 6:31.

Solomon, A.M., N.H. Ravindranath, R.B. Stewart, M. Weber, S. Nilsson, P.N. Duinker, P.M. Fearnside, P.J. Hall, R. Ismail, L.A. Joyce, S. Kojima, W.R. Makundi, D.F.W. Pollard, A. Shvidenko, W. Skinner, B.J. Stocks, R. Sukkumar, and D. Xu, 1996: Wood production under changing climate and land use. In: Climate Change 1995: Impacts, Adaptations and Mitigation of Climate Change: Scientific-Technical Analyses. Contribution of Working Group II to the Second Assessment Report of the Intergovernmental Panel on Climate Change. Cambridge University Press, Cambridge, United Kingdom, and New York, NY, pp.487–510.

Solow, R., 1993: *An Almost Practical Step Toward Sustainability: Resource Policy (Sept. 93).* Resources for the Future, Washington, DC, USA, pp. 162–172.

Sombroek, W.G., P.M. Fearnside, and M. Cravo, 1999: Geographic assessment of carbon stored in Amazonian terrestrial ecosystems and their soils in particular. In: *Global Climate Change and Tropical Ecosystems* [R. Lal, J.M. Kimble and B.A. Stewart (eds.)]. CRC Lewis, Boca Raton, FL, USA, pp. 375–389.

Sombroek, W.G., F.O. Nachtergaele, and A. Hebel, 1993: Amounts, dynamics and sequestering of carbon in tropical and subtropical soils. *Ambio,* **22(7),** 417–426.

Souza, C. and P. Barreto, 1999: An alternative approach for detecting and monitoring selectively logged forests in the Amazon. *International Journal of Remote Sensing,* (in press).

SSS, 1999: *Soil Taxonomy (2nd edition).* USDA/NRCS Soil Survey Staff, Agricultural Handbook 436, Washington DC, USA, 869 pp.

Stallard, R.F., 1998: Terrestrial sedimentation and the carbon cycle: Coupling weathering and erosion to carbon burial. *Global Biogeochemical Cycles,* **12,** 231–257.

Stavins, R., 1999: The cost of carbon sequestration: A revealed preference approach. *The American Economic Review,* **89(4),** 994–1009.

Stirling, A., 1993: Environmental valuation: How much is the emperor wearing? *The Ecologist,* **23(3),** 97–103.

Stolbovoi, W.G. and G. Fischer, 1998: A new georeferenced database of soil degradation in Russia. *Advances in Geoecology,* **31,** 143–152.

Stone, T.A. and P.A. Lefebvre, 1998: Using multi-temporal satellite data to evaluate selective logging in Para, Brazil. *International Journal of Remote Sensing,* **13,** 2517–2526.

Stull, R.B., 1988: *An Introduction to Boundary Layer Meteorology.* Kluwer Academic Press, Dordrecht, The Netherlands, pp. 680.

Sutton, A., 1994: *Slavery in Brazil—A Link in the Chain of Modernization.* Oxford University Press, Oxford, United Kingdom.

Swank, W.T. and J.E. Douglass, 1974: Streamflow greatly reduced by converting deciduous hardwood stands to pine. *Science,* **185,** 857–859.

Tate, K.R. and B.K.G. Theng, 1980: Organic matter and its interactions with inorganic soil constituents. In: *Soils With Variable Change* [Theng, B.K.G. (ed.)]. New Zealand Society of Soil Science, Lower Hutt, New Zealand.

Tata Energy Research Institute (TERI), 1999: *National Study on Joint Forest Management.* Ministry of Environment and Forests, Government of India, New Delhi, India.

Thornley, J.H.M. and M.G.R. Cannell, 2000: Managing forests for wood yield and carbon storage: a theoretical study. In: *Tree Physiology,* (in press).

Tiessen, H. and J.O. Moir, 1993: Total and organic carbon. In: *Soil sampling and methods of analysis*. [Carter, M.R. (ed.)]. Lewis Publishers, Boca Raton, FL. USA, pp. 187–199.

Tipper, R. and B.H. de Jong, 1998: Quantification and regulation of carbon offsets from forestry: Comparison of alternative methodologies, with special reference to Chiapas, Mexico. *Commonwealth Forestry Review,* **77,** 219–228.

Tomppo, E., 1996: Multi-source national forest inventory of Finland. In: *New Thrusts in Forest Inventory*. [Paivinen, R., J. Vanclay, and S. Miina (eds.)]. European Forest Institute, Joensuu, Finland, pp. 27–41.

Townshend, J.R.G., C.O. Justice, D. Skole, J.P. Malingreau, J. Cihlar, P. Teillet, F. Sadowski, and S. Ruttenberg, 1994: The 1 km resolution global data set—needs of the international geosphere biosphere programme. *International Journal of Remote Sensing,* **15(17),** 3417–3441.

Trumbore, S.E., E.A. Davidson, P.B. de Camargo, D.C. Nepstad, and L.A. Martinelli, 1995: Belowground cycling of carbon in forests and pastures of Eastern Amazonia. *Global Biogeochemical Cycles,* **9,** 515–528.

Udvardy, M.D.F., 1975: *A Classification of the Biogeographical Provinces of the World*. UNESCO's Man and the Biosphere Programme Project No. 8. IUCN Occasional Paper 18, Gland, Switzerland.

UN, 1994: *Convention on Desertification*. Information Program on Sustainable Development, United Nations, New York, NY, USA.

UN, 1996: *Indicators of Sustainable Development: Framework and Methodologies*. United Nations Publications, New York, NY, USA, 428 pp.

UN, 1998: *Report of the Fourth International Workshop on Indicators of Sustainable Development*. United Nations Division for Sustainable Development, Prague, Czech Republic. Available at: http://www.czp.cuni.cz/Zakl/CSD/Report.html.

UNCED, 1992: *Agenda 21: Earth Summit—United Nations Program of Action from Rio*. United Nations Publications, New York, NY, USA, 294 pp.

UNCSD, 1997: Overall progress achieved since the United Nations Conference on Environment and Development, Addendum 13: Promoting sustainable agriculture and rural development. United Nations Commission on Sustainable Development, Fifth Session, held 7–25 April, 1997. E/CN.17/1997/2/Add.13.

UN-ECE/FAO, 2000: *Temperate and Boreal Forest Resource Assessment 2000*. United Nations-Economic Commission for Europe/Food and Agriculture Organization, Geneva, Switzerland.

UNU, GEIC, and UNU/IAS, 1998: *Global Climate Governance: A Report on the Interlinkages Between the Kyoto Protocol and Other Multilateral Regimes*. United Nations University/Global Environment Information Centre/UNU Institute of Advanced Studies, Tokyo, Japan, 48 pp.

US Environmental Protection Agency (USEPA), 1997: *1995 Toxic Release Inventory: Public Data Release*. Office of Pollution Prevention and Toxics, Washington, DC, USA, 191 pp.

Valentini, R., G. Matteucci, A.J. Dolman, E.-D. Schulze, C. Rebman, E.J. Moors, A. Granier, P. Gross, N.O. Jensen, K. Pilegaard, A. Lindroth, A. Grelle, Ch. Bernhofer, T. Grünwald, M. Aubinet, R. Ceulemans, A.S. Kowalski, T. Vesala, U. Rannik, P. Berbigier, D. Loustau, J. Guomundsson, H. Thorgeirsson, A. Ibrom, K. Morgenstern, R. Clement, J. Moncrieff, L. Montagnani, S. Minerbi, and P.G. Jarvis, 2000: Respiration as the main determinant of European forests carbon balance. *Nature,* **404,** 861–865.

van Kooten, G.C., C.S. Binkley, and G. Delcourt, 1995: Effect of carbon taxes and subsidies on optimal forest rotation age and supply of carbon services. *American Journal of Agricultural Economics,* **77(2),** 365–374.

van Kooten, G.C., A. Grainger, E. Ley, G. Marland, and B. Solberg, 1997: Conceptual issues related to carbon sequestration: Uncertainty and time. *Critical Reviews in Environmental Science and Technology,* **27(Special),** S65–S82.

Vásquez, R. and A.H. Gentry, 1989: Use and misuse of forest-harvested fruits in the Iquitos area. *Conservation Biology,* **3(4),** 350–361.

Vincent, J.R., 1995: Timber trade, economics, and tropical forest management. In: *Ecology, Conservation and Management of Southeast Asia Rainforests* [Primack, B.R. and T.E. Lovejoy (eds.)]. Yale University Press, New Haven, CN, USA pp. 241–261.

Whitehead, B.W. and R. Godoy, 1991: The extraction of Rattan-like lianas in the new world tropics: A possible prototype for sustainable forest management. *Agroforestry Systems,* **16,** 247–255.

Whitmore, T.C. and J.A. Sayer, 1992: *Tropical Deforestation and Species Extinction*. Chapman and Hall, London, United Kingdom, 153 pp.

Winjum, J.K., S. Brown, and B. Schlamadinger, 1998: Forest harvests and wood products: Sources and sinks of atmospheric carbon dioxide. *Forest Science,* **44(2),** 272–284.

Woodwell, G.M. and W.R. Dykeman, 1966: Respiration of a forest measured by CO_2 accumulation during temperature inversions. *Science,* **154,** 1031–1034.

Woodwell, G.M., J.E. Hobbie, R.A. Houghton, J.M. Melillo, B. Moore, B.J. Peterson, and G.R. Shaver, 1983: Global deforestation: Contribution to atmospheric carbon dioxide. *Science,* **222,** 1081–1086.

World Bank, 1991: *Environmental Assessment Sourcebook*. World Bank Technical Paper no. 139-140:154, World Bank, Washington, DC, USA.

World Commission on Sustainable Forests and Sustainable Development (WCFSD), 1999: *Our Forests, Our Future*. Cambridge University Press, Cambridge, United Kingdom, 205 pp.

WRB, 1998: *World Reference Base for Soil Resources*. International Society of Soil Science/International Soil Reference and Information Centre/Food and Agriculture Organization, Rome, Italy.

Young, A., 1989: *Agroforestry for Soil Conservation*. CAB International, Oxford, United Kingdom, 276 pp.

Zuidema, G., G.J. van den Born, J. Alcamo, and G.J.J. Kreileman, 1994: Simulating changes in global land cover as affected by economic and climatic factors. *Water, Air, and Soil Pollution,* **76,** 163–198.

3

Afforestation, Reforestation, and Deforestation (ARD) Activities

BERNHARD SCHLAMADINGER (AUSTRIA)
AND TIMO KARJALAINEN (FINLAND)

Lead Authors:
R. Birdsey (USA), M. Cannell (UK), W. Galinski (Poland), A. Gintings (Indonesia), S. Hamburg (USA), B. Jallow (The Gambia), M. Kirschbaum (Australia), T. Krug (Brazil), W. Kurz (Canada), S. Prisley (USA), D. Schulze (Germany), K.D. Singh (India), T.P. Singh (India), A.M. Solomon (USA), L. Villers (Mexico), Y. Yamagata (Japan)

Contributors:
S. Beukema (Canada), G. Lund (USA), B. Murray (USA)

Review Editors:
E. Calvo (Peru) and J.D.G. Miguez (Brazil)

CONTENTS

EXECUTIVE SUMMARY

Chapter 3 focuses on Article 3.3 of the Kyoto Protocol. Article 3.3 identifies direct human-induced (DHI) land-use change and forestry activities for which Annex I Parties must account greenhouse gas (GHG) emissions by sources and removals by sinks in the first commitment period. These activities are afforestation, reforestation, and deforestation (ARD). The implementation of Article 3.3 requires definitions for several terms and decisions on carbon accounting rules. Chapter 3, which builds on the general concepts introduced in Chapter 2, identifies issues, describes various options to address these issues, and summarizes the implications of the options.

Definitions

The term "ARD land" is used in this report, for simplicity, to define areas on which ARD activities have occurred since 1990 and for which carbon stock changes are to be calculated. Key to the identification of ARD lands under Article 3.3 is the definition of a forest coupled with definitions of afforestation, reforestation, and deforestation. The individual definitions of these terms addressed in Sections 2.2.2 and 2.2.3 provide a basis for the discussion in this chapter.

This chapter introduces a series of definitional scenarios to illustrate the implications of several combinations of different definitions of forest and ARD. These scenarios were selected to illustrate the range of possible approaches that could be used to define the key terms necessary for implementing Article 3.3 and the implications of employing these definitions. Although many definitional scenarios could have been developed, seven have been chosen and are discussed in detail. Two of the definitional scenarios are based on existing Food and Agriculture Organization (FAO) definitions and definitions listed in the Glossary of the Intergovernmental Panel on Climate Change (IPCC) Guidelines. The other five scenarios were selected to represent a broad range of plausible combinations of definitions that could be applied under Article 3.3. All of these scenarios are listed in Table 3-1.

The definitions of *ARD* and *forest* used to implement Article 3.3 will affect the area of land covered by this Article. These definitions can also affect the area of land on which activities covered by Article 3.4 (see Chapter 4) could take place, assuming that no activity is to be counted under both Articles at the same time. For example, if the definition of reforestation does not include regeneration following clear-cut harvesting, ARD lands would be limited and more activities related to forest management could be considered under Article 3.4. The definitional scenarios illustrate how the harvest/regeneration cycle could be included or excluded from coverage under Article 3.3 through the use of different definitions of ARD and forest.

This chapter examines the implications of the definitional scenarios and identifies the key decisions. Eight such decisions, the options for each, and the implications of each option are outlined in Table 3-2. Although all combinations of options are possible, some combinations will create situations in which carbon stock changes reported under Article 3.3 will not reflect their actual contribution to the changes in the atmospheric concentration of carbon dioxide.

Carbon Accounting

Several carbon accounting issues influence the implications of the different definitions for key terms in Article 3.3 (see Table 3-3). These issues can be addressed either directly through selection of a definition or through the use of administrative or procedural rules that address the concerns Parties may have.

Discrepancies between Actual and Reported Stock Changes

Some definitions of ARD could result in Parties receiving "credits" (i.e., additions to their assigned amounts) for activities that result in an increase or no change in atmospheric CO_2 concentrations (type I discrepancy). An example is deforestation after 1990, followed by agricultural land use then reforestation to create credits in a commitment period. Although carbon emissions associated with the deforestation would not be reported, the sequestration associated with reforestation during a commitment period may be. Conversely, some definitions could result in Parties receiving debits (i.e., deductions from their assigned amounts) for emissions associated with activities that result in reductions or no change in atmospheric CO_2 concentrations (type II discrepancy). For example, a harvest during the commitment period on land afforested since 1990 could result in debits, despite the net carbon sequestered since 1990.

Several potential options would avoid these discrepancies:

- Define reforestation under Article 3.3 such that it specifies either a minimum period during which the land must have been non-forest prior to reforestation or a point in time prior to reforestation at which the land must have been non-forest (e.g., 1990).

- Limit the credits and debits available from reforestation/deforestation under Article 3.3. For example, credit for reforestation could be awarded only for carbon stock increases above the stocks on the land prior to deforestation (or in 1990). This approach avoids the type I discrepancy. Data on stocks in 1990 are required in Article 3.4; left unspecified, however, is whether this information should be reported at the country level or the activity level or whether the data should be site specific. Other data could be used as a proxy for geo-referenced stocks in 1990 or prior to deforestation—for example, crown-cover data from satellite imagery. To avoid the type II discrepancy, carbon debits on afforested or reforested lands could be limited to net credits (i.e., credits minus debits) received for carbon accumulating on the same land (A/R Debit Rule).

Land-Based versus Activity-Based Accounting

In an activity-based accounting system, the starting point is the carbon stock change attributable to ARD activities. This method

***Table 3-1**: Descriptions and implications of definitional scenarios.*

Definitional Scenario	Description	General Implications
FAO	Forest: Lands that have, or will have because of continued growth, more than 10% canopy cover. Deforestation is decline of canopy cover to below 10%, but excludes changes within the forest class; reforestation is artificial establishment of forest on lands that had them previously (including regeneration post-harvest); afforestation is artificial establishment of forest on lands that were not historically forest.	1) Deforestation between 1990 and 2008 followed by reforestation can create credits. 2) Degradation/aggradation unaccounted for if canopy cover threshold not crossed. 3) Harvest/regeneration cycle may create large areas of ARD lands. Many countries will report a debit for harvest/regeneration cycles, unless rotation periods are very short or if an activity-based carbon accounting approach is used. 4) If carbon accounting starts in 2008, conversion of high carbon-density forests, degradation, and aggradation in commitment period are covered. 5) Creates potential for inconsistency between Articles 3.3 and 3.7.
IPCC	Forest: As in FAO definition. Reforestation and afforestation are a land-use change from non-forest to forest through planting and differ only in that afforested lands never contained forest. Reforestation does not include regeneration post-harvest. Deforestation is conversion of forest to non-forest.	1 and 2 apply. 6) Harvest/regeneration cycle does not create ARD lands. 7) Only changes between forest and non-forest create ARD land.
Land Use	Forest: Defined administratively or based on specific land-use activities. Deforestation is conversion of forest to non-forest; reforestation and afforestation are the activities that lead to conversion of non-forest to forest.	1, 6, and 7 apply. 8) Carbon stock changes might not be considered as ARD activities if land-use classification remains unchanged.
Land Cover	Forest: As in FAO definition except that regrowing stands that are below the canopy cover threshold are not counted as forest. Deforestation is conversion of forest to non-forest; afforestation and reforestation are reestablishment of minimum canopy cover.	1, 2, and 5 apply 9) Kyoto land created only when canopy cover threshold is crossed; therefore, the time when ARD land is created differs from the FAO scenario. 10) Harvest/regeneration cycle may create large areas of ARD lands. Many countries will report a debit for harvest/regeneration cycles, unless rotation periods are very short (< 20 years). 11) Conversion of high carbon-density forests, degradation, and aggradation in commitment period are covered.

Table 3-1 (continued)

Definitional Scenario	Description	General Implications
Flexible	Forest: As in FAO definition, but countries have flexibility in choosing the threshold (e.g., based on carbon content of aboveground living woody biomass (t C ha^{-1}), tree height, and/or canopy cover). Afforestation, reforestation, and deforestation defined as in IPCC scenario, but natural regeneration is included in AR.	1, 2, 6, and 7 apply. 12) Gives countries flexibility in selecting a definition of forest. Allows use of existing data, reducing costs. Countries could choose threshold to maximize credits or minimize debits.
Degradation/ Aggradation	Forest: Defined in carbon density or canopy cover classes (e.g., 10 to <40%, 40 to <70% canopy cover). Deforestation is decrease in canopy cover or carbon density at maturity from one class to another. Reforestation is the reverse, and afforestation is establishment of forest on lands that were non-forest for a predefined period.	1, 6, and 11 apply. 13) Acknowledges differences in ecosystems by allowing creation of ARD land in cases that would be missed by use of a uniform threshold. 14) Implementation is complex; difficult to establish geographically specific land cover or carbon density at maturity for different times.
Biome	Forest: As in FAO definition, but threshold in the definition of forest is specified by biome through, for example, an international expert panel. Afforestation, reforestation, and deforestation defined as in IPCC scenario, but natural regeneration is included in AR.	1, 2, 6, 7, and 11 apply.

may require distinction of the activity effect from background or baseline carbon stock changes. It is conceivable to have more than one activity per unit land, but it will be difficult to separate the stock changes attributable to individual activities. The total ARD land area can increase or decrease because activities can have different durations.

In a land-based accounting system, the starting point is the total carbon stock change in applicable carbon pools on land units subject to an ARD activity. This method involves first identifying land units on which such activities occur, then determining the total change in carbon stocks on these land units. This approach assumes that all carbon stock changes will be considered (i.e., that there is no baseline). Once a unit of land enters the accounting system, the net effect of multiple activities (e.g., deforestation then reforestation) may be accounted, even though only one of these activities has triggered the land to become ARD land. ARD land area increases with each successive commitment period.

Another question under the land-based accounting system is whether carbon stock changes for ARD activities initiated after 1 January 2008 should be assessed beginning on 1 January 2008 or with the onset of the activity. If accounting begins with the activity, a stock change prior to that activity but in the commitment period will remain unaccounted. For example, if a harvest that is not defined as an activity under Article 3.3 precedes an event that is defined as reforestation under Article 3.3, there could be a discrepancy between actual and reported carbon stock changes during a commitment period. Similarly, if a forest is degraded from full stocking in 2008 to zero stocking in 2012, the carbon loss would be counted only from the time at which the threshold between forest and non-forest is crossed (e.g., in 2011). A significant part of the stock change might be missed. If accounting begins when the activity takes place and not at the beginning of the commitment period, data requirements will greatly increase.

This chapter analyzes three possible combinations of options:

- *Activity-based approach.* Carbon accounting begins with the start of the activity or on 1 January 2008, whichever is later. Any stock changes prior to the activity remain unaccounted because they do not result from the ARD activity. Any stock change that occurs after the start of the activity but does not result from the activity is not accounted. An example is decaying slash from a harvest that is followed by reforestation (FAO scenario).
- *Land-based approach I.* All carbon stock changes on ARD land are counted, beginning on 1 January 2008.
- *Land-based approach II.* All carbon stock changes on ARD land are counted, beginning with the start of the activity or with 1 January 2008, whichever is later. In this approach, for example, decaying slash would be counted as a negative carbon stock change.

Accounting of ARD Activities during Future Commitment Periods

Should there be conditions or events that would terminate the carbon stock reporting requirement of ARD activities? If not

***Table 3-2**: Issues, options, and implications related to definition of ARD activities.*

Issue	Options and Implications
What should be the basis for the definition of forest? (Section 3.2, Table 3-4, and Section 2.2.2.1)	*Vegetation characteristics* • Compatible with some countries' methods for identifying forests. • If based on actual vegetation characteristics, land status after clear-cut harvest is always non-forest. If potential vegetation characteristics are used, land status after clear-cut harvest is forest if vegetation is expected to return above threshold. • Assessment of forest/non-forest based on objective, measurable criteria.
	Land-use or administrative characteristics • Compatible with some countries' methods for identifying forests. • Land-use designation would not necessarily reflect actual carbon stocks on the land. • Change in land-use designation will create ARD land, which may or may not be associated with a change in carbon stocks.
Should the criterion that distinguishes forest from non-forest vary by biome or by country, or be the same for all Annex I countries? (Section 3.2 and Table 3-4)	*By biome* • Threshold values must be determined for major vegetation types and/or ecosystems. • Differences in vegetation types are taken into account. • More compatible with national vegetation surveys.
	By country • One threshold value chosen by each country, allowing use of existing data. • Some vegetation types and land uses that would be regarded as forest in some countries may be classified as non-forest in others.
	Same for all Annex I countries • Universal threshold would be applied for all ecosystem types and climatic zones. • Does not take into account differences between ecosystems and climatic zones.
Should a maximum spatial assessment unit be specified? (Sections 3.4.2 and 3.4.4)	*Yes* • If the spatial assessment unit is relatively large: – ARD activities affecting areas within the assessment unit may not be registered (e.g., deforestation of 89 ha within an assessment unit of 100 ha, using a crown cover threshold of 10%). This is "conservative" for AR, but underestimates emissions for D. – Monitoring costs would be low. • If the spatial assessment unit is relatively small: – Most ARD activities will be captured. – Monitoring costs could be high.
	No • The same issues will arise as under *Yes*, but countries' data will not be comparable because their decisions on the size of assessment units may vary.
Should the definition of "reforestation" include or exclude reestablishment of tree cover after clear-cut harvesting? (Section 3.2, Table 3-4, and Section 2.2.3.2)	*Include* • Area of ARD land will be large because areas will be added as they undergo a harvest/regeneration cycle. • Will result in unbalanced accounting at stand and landscape levels if harvesting is not considered deforestation; actual carbon stock changes will not match reported changes (see "Landscape Level" subsection in Executive Summary).
	Exclude • Relatively small areas of most Annex I countries will become ARD land. • There will be balanced accounting at stand and landscape levels.

Table 3-2 (continued)

Issue	Options and Implications
Should (re)establishment of forests through natural means be considered a form of afforestation or reforestation? (Section 3.2, Table 3-4, and Section 2.2.3.2)	*Yes* • Allows for consideration of natural regeneration, which accounts for 60% of all areas restocked in Annex I countries (82% when regeneration enhanced by planting is included; the rest is planting). • Will result in more ARD area compared to the *No* option. • May encourage natural regeneration methods, which lead to improved biodiversity.
	No • Restricts afforestation and reforestation to planting activities only. • Problems in distinguishing direct human-induced reforestation/afforestation are reduced because any planting or seeding activity is likely to be called direct human-induced. • Results in less ARD area compared to the *Yes* option.
Should Article 3.3 include degradation/aggradation of forest land? (Section 3.2, Table 3-4, and Sections 2.2.3.4 and 2.2.3.5)	*Yes* • Accounts for stock changes that otherwise would not be reported. • Encourages sustainable forest management. • Distinguishing silvicultural measures (e.g., thinning) from degradation/aggradation may be difficult, increasing the possibility that stock changes will be double-counted (under both Articles 3.3 and 3.4).
	No • Implementation is simplified. • Activities that significantly affect long-term forest cover without crossing a single crown-cover threshold, as well as associated stock changes, are ignored under Article 3.3. • These activities may still be covered by Article 3.4.
Should the qualifier "direct human-induced" refer only to ARD activities, or also to stock changes resulting from these ARD activities? (Sections 3.3.2.1, 2.3.3.1, and 2.3.3.2)	*Refer only to ARD activities* • All stock changes on ARD land are accounted. • Easier to implement.
	Refer to ARD activities and resultant stock changes • Only a part of the stock changes on ARD land are accounted. • Changes in stocks are very difficult to attribute to direct human-induced activities vs. indirect or natural effects (e.g., CO_2 fertilization, N deposition). • Large uncertainty associated with stock changes and great difficulty verifying them because estimates would heavily rely on the use of models.
How should "land-use change" in Article 3.7 be defined in relation to definition of ARD in Article 3.3? (Section 3.3.2.8)	*Define both ARD and "land-use change" as transitions between "forest" and "non-forest"* • Net-net approach (emissions and removals counted in commitment period and compared against those in 1990) for Article 3.3 activities for LULUCF source countries.
	Define ARD as transition between "forest" and "non-forest;" define "land-use change" as transition between, for example, the 15 land categories in Chapter 2 (see Table 2-1) • Net-net approach for Article 3.3 activities for LULUCF source countries. • Net-net approach for Article 3.4 activities possible for LULUCF source countries, if Article 3.4 activities are defined as "land-use changes."
	Define ARD broadly to include regeneration after harvest; define "land-use change" to include conversions either between "forest" and "non-forest" or between the 15 land categories • Net-net approach for ARD activities that are a forest/non-forest transition. • Gross-net approach (emissions and removals counted in commitment period but not compared against those in 1990) for all other ARD activities, such as regeneration after harvest.

***Table 3-3**: Issues, options, and implications related to carbon accounting rules.*

Issue	Options and Implications
When should stock change accounting for ARD activities initiated within a commitment period begin? (Sections 3.3.2.2, 3.3.2.4, 3.4, 3.5.1, and 3.5.2.4; Table 3-11; and Figure 3-6)	*Not before the initiation of an ARD activity* • Reported carbon stock changes will not always reflect actual stock changes over the commitment period (e.g., FAO scenario may credit reforestation after harvest without debit for prior loss of forest carbon; see "Landscape Level" subsection in Executive Summary). • Requires data on carbon stocks when activity begins. • In gradual deforestation, stock changes are counted only after non-forest threshold (e.g., 10% crown cover) is reached.
	With beginning of a commitment period • In all scenarios except the FAO scenario, accounting is balanced. The FAO scenario will yield a debit for many countries (see "Landscape Level" subsection in Executive Summary). • Requires determination of carbon stock at beginning of commitment period. • Not consistent with activity-based approach.
Does ARD land remain in that category indefinitely? (Section 3.3.2.7)	*Yes* • Ensures balanced carbon accounting in future commitment periods. • ARD land increases over time. • Not consistent with activity-based approach because activity and its impacts can have a defined end.
	No • Need to define criteria under which ARD land can leave that category. • Could reduce need to quantify future carbon stock changes on ARD land. • If no tight guidelines are given, could permit counting land under ARD during times of stock increases and remove it from that category during time of carbon losses.
Should dead organic matter and soil carbon pool changes be counted? (Section 3.3.2.5)	*Yes* • Stock changes resulting from ARD activities could be larger than otherwise would be reported. • Good match between actual and reported carbon stock changes. • Verification of stock changes difficult/complicated.
	No • Measurements of stock changes would be less costly and less uncertain. • Less accurate representation of actual carbon stock changes.
Should forest products derived from ARD land be included in the accounting of carbon stocks? (Section 3.3.2.5)	*Yes* • Requires use of wood products monitoring and accounting methodologies. • More complete accounting of all carbon pools. • Requires agreed carbon accounting approach for wood products.
	No • Assessment of carbon stock changes is easier. • Allows wood-product management to be an activity under Article 3.4 without double-counting wood products from ARD activities.
Should all future carbon stock changes on ARD land be counted? (Section 3.3.2.7)	*Yes* • Stock changes from natural disturbances (fire, wind, insects) on ARD land would be counted. • Assessment of stock changes will be easier to implement.
	No • Reduces reported carbon stock changes from non-ARD activities. • Could result in mismatch between real and reported stock changes.

Table 3-3 (continued)

Issue	Options and Implications
Should carbon credits available for R be limited if previous D occurred on the same land between 1990 and 2008? (Sections 3.3.2.2 and 3.3.2.3)	*Yes* • More balanced carbon accounting of stock changes between 1990 and 2012. • Additional data requirements (i.e., carbon stock in 1990 or prior to deforestation).
	No • Deforestation between 1990 and 2008, followed by reforestation, would result in credits.
Should AR between 1990 and 2008 limit carbon debits resulting from subsequent D, forest management, or natural disturbances on this land? (Sections 3.3.2.2 and 3.3.2.3)	*Yes* • More balanced carbon account. • Prevents potential penalty for AR activities.
	No • Parties might incur debit for activities that lead to long-term increases of carbon stocks.
How should activities under Article 3.4 be treated if they occur on land that has already undergone an ARD activity? (Sections 3.3.2.5, 3.3.2.7, and 3.3.2.10)	*Consider the land under Article 3.4 and no longer under Article 3.3* • Allows removal of stock changes associated with ARD activities from accounting under Article 3.3, which is not consistent with Article 3.3 ("shall"). • Might not be necessary in land-based approach because all stock changes on ARD land would be reported under Article 3.3.
	Consider the land under Article 3.3 • Avoids risk of double-counting. • Eliminates need to determine Article 3.4 activities on ARD land.
	Consider stock changes from ARD activity under Article 3.3 and those from Article 3.4 activity under Article 3.4 • Risk of double-counting. • Difficulty in separating stock changes according to cause.
How should ARD activities be treated if they occur on land that has already undergone an Article 3.4 activity? (Sections 3.3.2.5, 3.3.2.7, and 3.3.2.10)	*Consider the land under Article 3.3 and no longer under Article 3.4* • Allows accounting for ARD activities on land initially in Article 3.4.
	Consider the land under Article 3.4 • Not consistent with Article 3.3 ("shall"). • With activity-based approach, ARD activities are counted under neither Article 3.3 nor Article 3.4.
	Consider stock changes from ARD activity under Article 3.3 and those from Article 3.4 activity under Article 3.4 • Risk of double-counting. • Difficulty in separating stock changes according to cause.
Should emissions of non-CO_2 GHGs be considered under Articles 3.3 and 3.7? (Section 3.3.2.9)	*Yes* • Full impacts of ARD activities on atmospheric GHGs will be considered. • Increases monitoring costs. • Non-CO_2 GHGs cannot be measured as "changes in stocks."
	No • Changes in emissions of methane and nitrous oxide from AR activities (e.g., wetland drainage or use of fertilizers) or D activities will be missed.

all stock changes on ARD land have to be reported in future commitment periods, sequestered carbon could be released to the atmosphere through activities that are not covered under Article 3.3, without being reported. If the phrase "in each commitment period" in Article 3.3 suggests a reporting requirement of all stock changes on ARD land for all future commitment periods, the area of ARD land would increase over time. If the harvest/regeneration cycle is an ARD activity, eventually all managed forest area of a country can become ARD land.

Non-CO_2 Greenhouse Gases

Methane and nitrous oxide are among the six GHGs in Annex A of the Kyoto Protocol that can be affected by land use, land-use change, and forestry (LULUCF) activities (see Chapter 1). Although Annex A of the Kyoto Protocol lists the agricultural sector, including emissions from prescribed burning of savannas and field burning of agricultural residues, it does not include LULUCF. Only Articles 3.3 and 3.4 bring LULUCF into the accounting. Article 3.3 mentions GHG emissions from sources and removals by sinks, but they are to be measured as verifiable changes in stocks. This provision may suggest the exclusion of non-CO_2 GHG emissions because they cannot be measured or verified as changes in stocks. Article 3.7 mentions "CO_2-equivalents." Including non-CO_2 GHGs would greatly increase the data requirements and the complexity of the methods required for determining associated fluxes. Excluding them would neglect some of the GHG fluxes to and from the atmosphere associated with ARD activities.

Implications of Definitional Scenarios for Data Needs and Data Acquisition

The definitions chosen for forest and ARD, as well as the accounting approach adopted (activity-based versus land-based) will have an impact on the data required for implementation of Article 3.3. Activity-based accounting will require a system for accurate detection and reporting of activities, some of which can be reasonably met using remote-sensing data. A land-based approach lends itself better to the use of existing forest inventory systems. Chapter 2 discusses methods for collecting data on carbon stock changes, but decisions on Article 3.3 will determine which pools should be included in stock change reporting. Stock changes in some pools are substantially more measurable and verifiable (e.g., aboveground biomass); stock changes in other pools involve significantly higher costs or modeling (e.g., dead organic matter, wood products).

In most Annex I countries, methods (e.g., measurement techniques, statistical tools, models, and activity reporting systems) are available for identifying and measuring carbon stock changes associated with ARD lands that are based on accepted principles of forest inventory, statistics, soil sampling, and ecological surveys. These methods vary in accuracy, precision, cost, periodicity of survey, and scale of application.

Although most Annex I countries have the elements required for the routine measurement of carbon stocks, only a few have a comprehensive operational national system based on continuous forest inventories, remote sensing, or other methods. Some adaptations of current measurement systems might be necessary to implement Article 3.3. Such comprehensive systems are usually not available in non-Annex I countries.

Methods of measuring changes in carbon stocks in aboveground biomass over a 5-year commitment period are sensitive enough to serve the requirements of the Protocol. Sensitive methods for measuring below-ground carbon stocks also exist, but changes in soil carbon stocks over a 5-year commitment period may be difficult to measure in some soils because changes in soil carbon stocks on ARD land may be small compared to background levels and because of high levels of spatial variability. This challenge can be addressed by either taking a large number of samples within a single commitment period—possibly at significant cost and still limited accuracy—or by measuring and accounting over multiple commitment periods. Changes can also be assessed through appropriate comprehensive ecosystem models, but site-specific verification may be difficult.

Implications of Definitional Scenarios at Various Spatial Scales

Interactions between the definitional scenarios and rules by which carbon stock changes will be accounted and reported under Article 3.3 are assessed using numerical examples at the stand, landscape, regional, country, and global scales.

Stand Level

Afforestation and reforestation leads to slow changes in aboveground living biomass carbon pools, whereas deforestation results in a rapid change in the same pools. Activities that avoid deforestation can yield an immediate and significant reduction of atmospheric carbon emissions, whereas an afforestation or reforestation program can take several decades to make full use of the carbon sequestration potential on a piece of land.

At the stand level, the actual and reported carbon stock changes are the same in a commitment period for the IPCC, Land Use, Flexible, and Biome scenarios. In the FAO and Land Cover scenarios, harvest during a commitment period followed by regeneration results in large debits if the carbon stock change is calculated over the full commitment period. In the FAO scenario, starting the stock change accounting no earlier than the activity may result in a carbon credit or a carbon debit, depending on the decay rate of dead organic matter relative to the rate of increase in biomass and whether that decay is accounted. If there is a significant amount of slash left after harvest, a loss of carbon stocks is likely to occur for some years until the accumulation of carbon in regrowing vegetation turns the balance into a net sink. Over several commitment periods there will be large credits, even if the long-term (i.e., decades to centuries) average of the carbon stock in the stand does not change.

Landscape Level

To scale up from stand-level to landscape-level impacts of the definitional scenarios, a simulation model was used to assess nine cases of various human activities in a hypothetical landscape. The actual carbon stock changes, as simulated by the model, were compared to those reported for each definitional scenario.

The area of ARD land in the first commitment period is much less than the total area in the landscape. The ARD area is greater in definitional scenarios in which the harvest/regeneration cycle creates ARD land but is still less than the total area of forests, except where harvest cycles are shorter than 20 years.

Definitional scenarios that limit ARD land to areas with forest to non-forest change and in which the harvest/regeneration cycle does not create ARD land (IPCC, Land Use, Flexible, and Biome) report very similar amounts of carbon stock changes for the same amount of forest to non-forest change. These definitional scenarios are insensitive, however, to actual changes in the landscape-level carbon stock. Thus, although the same carbon stock change is reported for the ARD land area, the actual carbon stock change in the landscape could be positive, unchanged, or negative.

In two definitional scenarios (FAO and Land Cover), ARD land is created during the harvest/regeneration cycle. In the FAO scenario, the regeneration activity—not the harvest—creates ARD land. If land-based accounting approach I is used, regeneration following harvest that occurs during the commitment period establishes ARD land, and a carbon stock change over the commitment period is reported that includes the effects of harvesting. If land-based approach II or the activity-based accounting approach is used, the carbon stock change would start from the regeneration activity; because it would cover less than the 5-year commitment period, it would not account for the impacts of harvest on aboveground biomass.

These two definitional scenarios (i.e., FAO with land-based I accounting of stock changes, and Land Cover) report debits for ARD land in the early commitment periods even if the stocks in the entire landscape are not changing because regrowth on lands harvested prior to 1990 is not counted. In the FAO scenario with land-based II carbon accounting approach—and if dead organic matter and soil are included—there is still a carbon debit in the first commitment period, but this debit is 70 percent lower than with the land-based I accounting approach. If only aboveground biomass carbon is reported, small credits would be recorded as the regenerating seedlings accumulate carbon. Over longer time periods (decades to a century), large credits would be reported even if stocks on the landscape are unchanged because reforestation is credited whereas harvesting is not debited if the land-based II or activity-based accounting approach is applied to the FAO scenario.

If the intent is to report carbon stock changes that are accurate reflections of actual carbon stock changes in a landscape, only full carbon accounting will suffice. If the intent is the most accurate accounting of carbon stock changes in the ARD land area, a definitional scenario that includes only forest to non-forest transitions is best.

Differences between actual and reported carbon stock changes can be reduced further if forest management activities that alter landscape-level carbon stock are included using the mechanisms defined in Article 3.4.

Country, Regional, and Global ARD Potentials

If definitions of ARD include the harvest/regeneration cycle and stock changes are accounted over the full commitment period (land-based approach I), the reported carbon stock change in the commitment period will be negative for many Annex I countries. Debits will exceed credits, except in regions with forests that mature very quickly. If carbon accounting does not start before the start of the activity (land-based approach II or activity-based approach), reported carbon stock changes will be positive for most Annex I countries if only biomass is considered. Land-based approach II is likely to yield carbon debits initially if changes in dead organic matter and soil organic matter pool are also considered, as demonstrated in stand- and landscape-level examples.

The Temperate and Boreal Forest Resource Assessment 2000 (TBFRA 2000) (UN, 1999) provides estimates on area of extension of forest and natural colonization of forest and other wooded land that could be regarded as non-forest to forest change in all definitional scenarios. TBFRA 2000 provides areas of regeneration after harvest that could be regarded as reforestation in the FAO scenario. Estimates for annual loss to other uses, obtained from an earlier FAO report, could be regarded as deforestation in all definitional scenarios. Annex I countries have reported a regeneration area that was 14 times larger than the area of extension of the forest, and loss to other uses was approximately three times the extension of forest, with the consequence that deforestation dominates net changes in carbon stocks to be reported under Article 3.3. Some countries have not reported extension or regeneration of forest, nor losses to other uses. Because the terms and definitions in TBFRA 2000 do not fully correspond to the terms and definitions of ARD used here, there is additional uncertainty associated with these ARD estimates.

The maximum amount of carbon that can be sequestered by global afforestation and reforestation activities (IPCC Scenario) is 60–87 Gt C on 344 Mha between 1995 and 2050, with 70 percent in tropical forests, 25 percent in temperate forests, and 5 percent in boreal forests. The average annual carbon sequestration rate would be 1.1–1.6 Gt C yr^{-1}. The actual rate would be lower in the first commitment period (0.5 Gt C yr^{-1} above and below ground after 20 years of growth) because a large proportion of the forests established since 1990 would still be in a juvenile stage.

National studies indicate that change from non-forest to forest could provide credits on the order of 1–2 percent of 1990 CO_2 emissions in most Annex I countries and up to 50 or 100 percent for countries with low population density and substantial afforestation programs. Reforestation under the FAO scenario could contribute additional debits of up to 27 percent of 1990 CO_2 emissions for some countries with large reforested areas and relatively slow regrowth, if land-based accounting approach I is used. It could provide a credit of up to 25 percent when stock changes are calculated between reforestation and the end of the commitment period, if the effects of decaying slash are excluded (activity-based approach). Deforestation could contribute debits of up to 3 percent of 1990 CO_2 emissions.

Non-GHG Impacts of ARD Activities

Increasing tree cover or protecting it from being decreased can improve and protect soil quality, especially in vulnerable areas such as steep hillsides; it also can stabilize watershed flow patterns and thus reduce flooding and average water yield. Biodiversity and species composition is generally very different in undisturbed systems such as mature forests compared to highly disturbed ones such as pastoral landscapes, and the greatest benefits in biodiversity conservation can come from reduced deforestation and thus reduced loss of unique species.

Increasing the area of forest plantations can increase the supply of merchantable wood. This increased supply is likely to reduce the price of wood, with likely impacts on the economics of wood-supplying activities in other areas. Any increase in the area of land used to grow wood is likely to reduce the land available for agricultural activities, with likely impacts on agricultural markets.

3.1. Introduction

Article 3.3 of the Kyoto Protocol specifies that "The net changes in greenhouse gas emissions by sources and removals by sinks resulting from direct human-induced land-use change and forestry activities, limited to afforestation, reforestation and deforestation since 1990, measured as verifiable changes in carbon stocks in each commitment period, shall be used to meet the commitments under this Article of each Party included in Annex B. The greenhouse gas emissions by sources and removals by sinks associated with those activities shall be reported in a transparent and verifiable manner and reviewed in accordance with Articles 7 and 8."

Article 3.3 of the Kyoto Protocol describes land-use change and forestry activities that require GHG emissions by sources and removals by sinks to be accounted for by Annex I Parties in the first commitment period (see Figure 3-1). These activities are restricted to afforestation, reforestation, and deforestation. The term "ARD land" is used in this report, for simplicity, to define areas on which ARD activities have occurred since 1990 and for which carbon stock changes are calculated.

This chapter focuses on delineating issues related to the implementation of Article 3.3, options for implementation, and their implications. The United Nations Framework Convention on Climate Change (UNFCCC) and post-Kyoto clarifications are the starting point for this work. The fourth Conference of the Parties (COP) in November 1998 indicated that the adjustment to a Party's assigned amount of GHG emissions shall be equal to verifiable changes in carbon stocks during the period 2008 to 2012 resulting from DHI activities of ARD since 1 January 1990 (FCCC, 1998a; FCCC/CP/1998/L.5). The Subsidiary Body for Scientific and Technological Advice (SBSTA) noted at its 11th session the IPCC request to Parties to submit national data relevant to Article 3.3, preferably by the end of 1999. Some countries provided such data for use in this report.

The wording of Article 3.3 raises several questions and potential interpretations. How should "direct human-induced" be interpreted? Which definitions should be used for afforestation, reforestation, and deforestation activities? Which carbon pools should be considered? What carbon accounting rules should be employed?

We reviewed potential definitions of ARD and focused our analysis on considering the major differences among the potential approaches to implementing Article 3.3. We then developed a set of seven definitional scenarios (combinations of definitions of forest, afforestation, reforestation, and deforestation) that

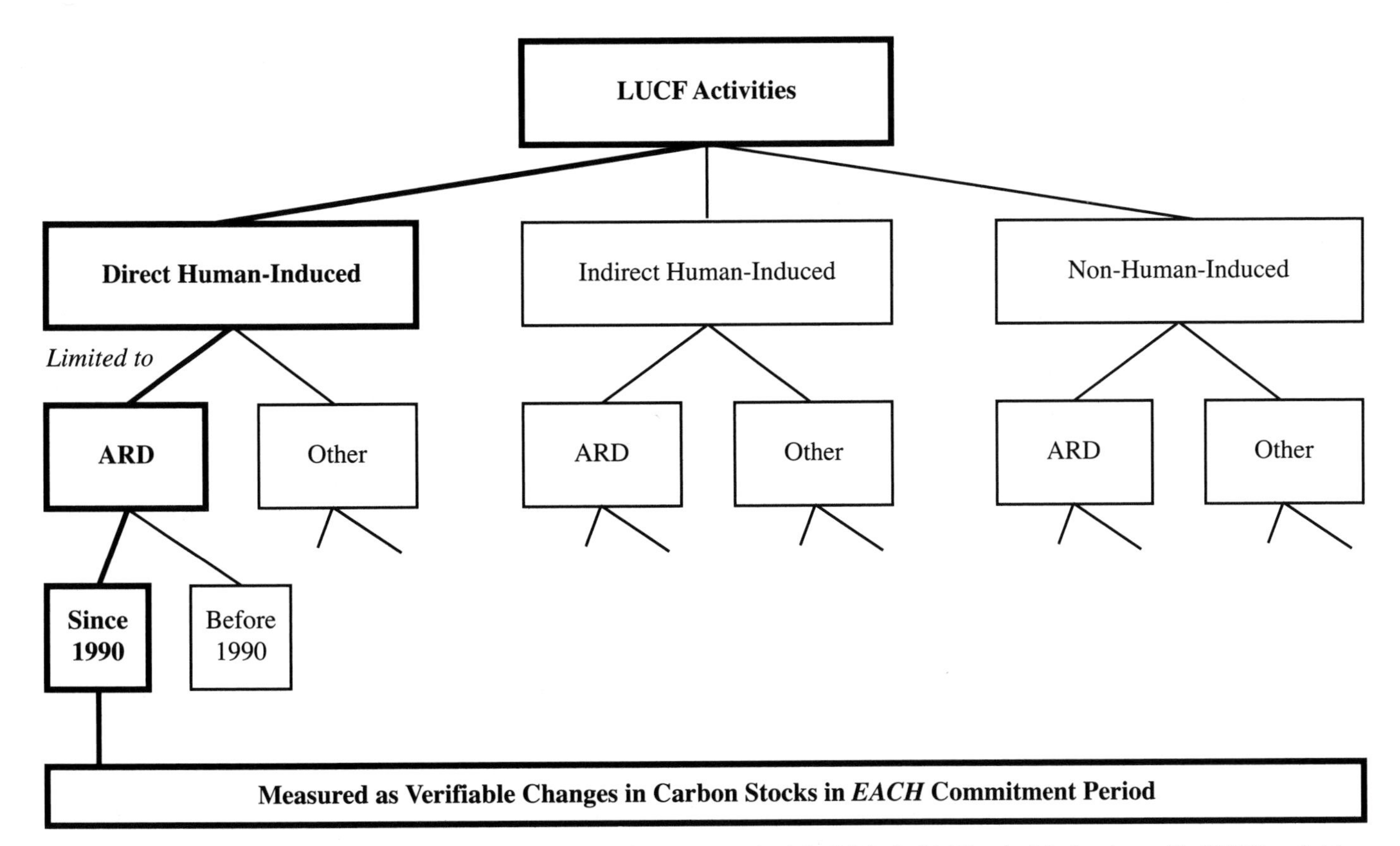

Figure 3-1: Sentence structure of Article 3.3, with parts that concern Article 3.3 in bold. The Article begins with *LUCF activities* (step 1), which must be directly induced by humans (step 2). Chapter 2 discusses whether and how to meaningfully distinguish *direct human-induced activities* from all activities in the LUCF sector. In step 3, Article 3.3 says that direct human-induced LUCF activities are *limited to ARD*. This language is seemingly straightforward except that ARD activities must be defined, which is key to implementing Article 3.3. Step 4 requires that only ARD activities *since 1990* can be counted, and step 5 demands that these activities be *measured as verifiable changes in carbon stocks in each commitment period*.

captured the breadth of the possible definitions. We then explored the implications of each scenario on GHG accounting; data needs; compatibility with available data; comparability of actual and reported stock changes at stand, landscape, and country levels; regional, country, and global potentials for ARD activities; and environmental and socioeconomic impacts. We have tried to support the decisionmaking process by highlighting the issues, options, and implications that are raised by different interpretations of Articles 3.3 and 3.7. This chapter builds on the general concepts and issues introduced in Chapter 2 (such as definitions for forest, afforestation, reforestation, deforestation, DHI activity, carbon pools, measurement methods, carbon accounting and reporting, associated impacts on environment, and associated socioeconomic impacts). There are interactions between Article 3.3 and other articles of the Kyoto Protocol; this chapter elaborates on some of these interactions—such as the possible counting of the same carbon under Articles 3.3 and 3.4 or interactions of Article 3.3 with Article 6.

3.2. Definitional Scenarios and Their General Implications

There are hundreds of variations of definitions for each term (e.g., forest, afforestation, reforestation, and deforestation—see Lund, 1999, 2000) that are relevant to implementing Article 3.3 of the Kyoto Protocol (see Sections 2.2.2 and 2.2.3 for discussions of the options). Examining the implications of each variant would not be fruitful; the result would be greater confusion, rather than the clarity we are seeking. Instead, the key to exploring the implications of these definitions is to classify them into broad groupings. Each of the groupings can then be explored in terms of the strengths and weaknesses of the broad approach. Section 2.2.2.1 discusses the broad categories of forest definitions (administrative, land use, and land cover), which are key to understanding the implications of different definitions of afforestation, reforestation, and deforestation. Sections 2.2.2 and 2.2.3 detail the range of plausible options for defining afforestation, reforestation, and deforestation. To facilitate discussion of the implications of different options for defining afforestation, reforestation, and deforestation, we have identified seven definitional scenarios that cover the range of definitions of forest, afforestation, reforestation, and deforestation. These scenarios are designed to provide a wide range of plausible combinations of definitions through which the implications of different options can be explored. We have explicitly avoided customizing definitions on the basis of biome type, specific management regimes, or socioeconomic conditions; instead, we explore definitions that are sufficiently robust to address diverse conditions over long periods of time.

This chapter explores how changes in land use and land cover influence carbon emissions and sequestration. The land-use status of an area of land characterizes how it is utilized to meet a human need; it represents a socioeconomic perspective on the status of the land. In contrast, land-cover characterizations of an area of land focus on what vegetation is, or is not, present.

Land use-based definitions of afforestation, reforestation, and deforestation are difficult to measure and verify because the intended use of the land may not be the same as the actual use. For example, if a nation designates a piece of land as a location where forest management should take place, is that designation sufficient to define the land as forest, even in the absence of trees? Land use can be determined administratively and culturally, further complicating the employment of land use-based definitions. Land cover, on the other hand, is easier to measure and verify, but transitions from one cover class to another can easily be misinterpreted. For example, following a clear felling, the lack of a forest canopy cover could be assumed to represent deforestation, when in fact tree regeneration is taking place. The use of existing land cover as the sole basis for defining afforestation, reforestation, and deforestation would cause harvest/regeneration cycles to be accounted under Article 3.3. This situation can be avoided by instead using the potential land cover of the existing vegetation at maturity.

A variant of the land-cover approach to define a forest is based on the carbon status of the land. Such an approach acknowledges that a carbon-based definition may avoid the geographical complexities and varying intents of the existing array of definitions of a forest. Given the variability in existing inventory data, any new approach to define a forest will need to consider regional/national differences in available data. Carbon-based definitions can rely on a variety of data (e.g., timber volume, basal area, stocking density and age) from which afforestation, reforestation, and deforestation events can be determined.

As Section 2.3.4.1 notes, it is not clear whether the term "direct human-induced" refers only to the ARD activities or also to the stock changes resulting from ARD activities; this issue is further discussed in Section 3.3.2.1. Here we discuss the difficulty in defining direct human-induced ARD activities. Implementation of ARD would be easier and unnecessary ambiguities could be avoided if any change in vegetation resulting from a decision by a land manager to change land-use practices were regarded as DHI. Another way of dealing with "direct human-induced" is to assume that ARD activities on certain lands (e.g., managed lands) are always DHI and that activities on the remaining lands (unmanaged lands) are not DHI. This rule has at least two logical exceptions. First, if afforestation or reforestation takes place on unmanaged lands, the burden of proof for "direct human-induced" is with the party who wants to claim credits. The land would subsequently become managed land. Second, if forest regeneration does not follow a natural disturbance on managed lands (e.g., hurricanes or landslides), one must decide whether this deforestation qualifies as DHI. With a subsequent agricultural or residential use, for example, the deforestation could be regarded as DHI.

3.2.1. Definitional Scenarios

From the definitions presented in Chapter 2, we assembled seven illustrative scenarios that reflect a range of possible approaches to implement Article 3.3 (see Table 3-4). This

***Table 3-4**: Definitional scenarios related to Article 3.3 and defining forest, afforestation, reforestation, and deforestation.*

	FAO Scenario	IPCC Scenario	Land Use Scenario	Land Cover Scenario	Flexible Scenario	Degradation/ Aggradation Scenario	Biome Scenario
Introduction	• Definitions of forest and ARD from Forest Resource Assessment Programme [FRA 2000 (UN-ECE/FAO, 1992)]. • Definitions do not distinguish direct human-induced changes, other than reforestation and afforestation that require "artificial establishment of forest..."—an action that may be interpreted as a direct human activity. • Planting of trees always qualifies as afforestation or reforestation, yet the site need not be deforested prior to being reforested. • Silvicultural planting following a harvest would qualify as reforestation, but the harvest would usually not count as deforestation (i.e., symmetry between deforestation and reforestation is not maintained).	• Based on reporting guidelines established by the Intergovernmental Panel on Climate Change (IPCC, 1997). • Definitions for ARD (specific definitions below) are land use-based, with specific activity requirements (e.g., planting versus regeneration through silvicultural activities). • No ambiguity in definition of deforestation, reforestation, or afforestation in that each involves direct human-induced activity. • Symmetry between deforestation and reforestation.	• Uses land use as basis for defining forest. • Determination of land-use status of a piece of land would need to be based on administrative procedures. • Based on a very simple structure and approach that would leave many decisions to national administrations. • Does not explicitly require symmetry between harvesting and replanting associated with forest management.	• Would lead to most intensive forest harvesting being considered as deforestation, and most forest regeneration being considered as reforestation (creates large area of ARD lands). • Most forest management activities would create ARD lands. • Definitions of reforestation and afforestation use the phrase "Activities that lead to..." so that credit can be given for increase in carbon stocks associated with reforestation and afforestation prior to reaching the Y% cover threshold required to define an area of land as forest.	• Allows countries to decide on the definition of a forest that they want to use, allowing maximum use of existing forest inventory data. • Selection of carbon-based definition is intended to give each country maximum flexibility in establishing its own criteria because it can convert cover or inventory data to a carbon per unit area basis. Countries may select threshold for crown cover, carbon, tree height, etc. Also, countries may select how long a land must have been non-forest to qualify as reforestation.	• Degradation and aggradation of a forest results in the creation of ARD lands. • Uses FAO definition of forest along with a forest cover class transition definition of deforestation and reforestation, which will result in large areas being considered ARD lands. • Selection of cover classes is based on thresholds in widest usage. • Inclusion of regenerating forest within definition of a forest means forest management will not result in creation of ARD lands unless there is long-term degradation or aggradation.	• Adjusts definition of a forest based on the biome involved (e.g., a closed canopy temperate forest will be treated differently than a woodland savanna). • Allows deforestation to be captured more realistically across diverse forest ecosystems and ensures consistency among countries. • Could allow for adjustments for data availability differences among biomes, as well as ease of data collection, both remotely and through inventory. • Specific biome-based thresholds could also be set to ensure that deforestation and reforestation activities are effectively captured and that creation of ARD lands based on cover change types is closely correlated with land-use activities.

Table 3-4 (continued)

	FAO Scenario	IPCC Scenario	Land Use Scenario	Land Cover Scenario	Flexible Scenario	Degradation/ Aggradation Scenario	Biome Scenario
Forest	• Land with tree crown cover (or equivalent stocking level) of more than 10% and area of more than 0.5 ha. Trees should be able to reach a minimum height of 5 m at maturity *in situ*. May consist of closed forest formations where trees of various storeys and undergrowth cover a high proportion of ground or open forest formations with a continuous vegetation cover in which tree crown cover exceeds 10%.[a]	• Within the Guidelines there exists no specific definition of forest; definition is left to the discretion of reporting countries. Because evaluating a definitional scenario without specific language defining each term is impossible, we have assumed the FAO definition of forest for purposes of our analysis, but acknowledge that the IPCC approach leaves countries free to choose their own definition of forest.	• An area that is being managed for forest values and/or forest products, or is designated as a forest by an appropriate governmental entity. • All ARD activities are assumed to be direct human-induced unless demonstrated not to be the case.	• An area of land that has >Y% (Y between 10 and 70%) cover of woody vegetation of (e.g., between 0 and 5 m) height.	• An area of land that has a minimum of Z (e.g., Z between 10 and 50) t C ha^{-1} aboveground living woody biomass (carbon threshold to be selected by each country) of woody vegetation, or would contain such at maturity of existing vegetation with continuation of current land use.	• Same as in FAO scenario.	• An area of land that has A% cover of woody vegetation with >B m in height, or would contain at maturity of existing vegetation with continuation of current land use. For each biome, a minimum crown cover and tree height would have to be established to determine what would qualify as a forest. Determination of biome types would have to be carried out systematically through the use of common criteria.
Deforestation	• Refers to change of land cover with depletion of tree crown cover to less than 10%. Changes within the forest class (e.g., from closed to open forest) that	• Conversion of forest to non-forest. • Note: No definition provided in the Glossary. However, the Reference Manual states on page 5.6 that "forest and grassland	• Conversion of forest to non-forest.	• Conversion of forest to non-forest.	• Conversion of forest to non-forest	• Decrease in potential crown cover from one cover class to another (e.g., 0 to <10, 10 to <40, 40 to <70, >70% crown cover).	• Conversion of forest to non-forest.

[a] Young natural stands and all plantations established for forestry purposes that have yet to reach a crown density of 10% or tree height of 5 m are included under forest, as are areas normally forming part of the forest area that are temporarily unstocked as a result of human intervention or natural causes but are expected to revert to forest. Definition includes forest nurseries and seed orchards that constitute an integral part of the forest; forest roads, cleared tracts, firebreaks, and other small open areas; forest in national parks, nature reserves, and other protected areas such as those of specific scientific, historical, cultural, or spiritual interest; windbreaks and shelter belts of trees with an area of more than 0.5 ha and width of more than 20 m; and plantations that are used primarily for forestry purposes, including rubberwood plantations and cork oak stands. Definition excludes land that is used predominantly for agricultural practices.

Table 3-4 (continued)

	FAO Scenario	IPCC Scenario	Land Use Scenario	Land Cover Scenario	Flexible Scenario	Degradation/ Aggradation Scenario	Biome Scenario
Deforestation (continued)	negatively affect the stand or site—and, in particular, lower the production capacity—are termed forest degradation.	conversion"—that is, conversion of forests and grasslands to pasture, cropland, or other managed uses—is referred to as "deforestation" (see footnote 7 in Guidelines).				Potential crown cover refers to state of existing vegetation at maturity under continuation of current land use.	
Afforestation	• Artificial establishment of forest on lands that previously did not carry forest within living memory. • Note that "natural extension" (e.g., abandoning agricultural land without direct planting), which is not included under "afforestation," also contributes to an increase in forest area (FAO, 1992). We have assumed such processes to be part of "afforestation;" their inclusion under Article 3.3 would be determined by the "direct human-induced" clause.	• Glossary: Planting of new forests on lands that historically have not contained forests. These newly created forests are included in the category "Changes in Forest and Other Woody Biomass Stocks" in the "Land-Use Change and Forestry" module of emissions inventory calculations.	• Activities that lead to conversion of non-forest to forest on lands that over the past X (e.g., 30, 50) years did not contain forest (30 years assumed throughout this report).	• Conversion of non-forest to forest on lands that over the past X (e.g., 30, 50) years did not contain forest (30 years assumed throughout this report).	• Conversion of non-forest to forest on lands that in 1990 [or over the past X (e.g., 10, 30, 50) years] did not contain forests.	• Conversion of non-forest to forest on lands that did not contain forest in 1990 [or over the past X (e.g. 30, 50) years].	• Conversion of non-forest to forest on lands that in 1990 [or over the past X (e.g., 10, 30, 50) years] did not contain forests.

Table 3-4 (continued)

	FAO Scenario	IPCC Scenario	Land Use Scenario	Land Cover Scenario	Flexible Scenario	Degradation/ Aggradation Scenario	Biome Scenario
Reforestation	• Artificial establishment of forest on lands that carried forest before. • "Artificial establishment" does not necessarily include natural regeneration. However, we assume that activity to be included because the separation could later be made through the "direct human-induced" clause.	• Glossary: Planting of forests on lands that have previously contained forests but that have been converted to some other use. Replanted forests are included in the category "Changes in Forest and Other Woody Biomass Stocks" in the "Land-Use Change and Forestry" module of the emissions inventory calculations. • Footnote 10 on page 5.14 of the Reference Manual implies a different definition of reforestation, but this definition has not been used here.	• Activities that lead to conversion of non-forest to forest on lands that contained forest at some time during the past X (e.g., 30, 50) years (30 years assumed throughout this report).	• Conversion of non-forest to forest on lands that contained forest at some time during the past X (e.g., 30, 50) years (30 years assumed throughout this report).	• Conversion of non-forest to forest on lands that in 1990 [or over the past X (e.g., 10, 30, 50) years] did contain forests.	• Increase in potential crown cover from one cover class to a higher cover class (e.g., 0 to <10, 10 to <40, 40 to <70, >70% crown cover) on lands that over the past X (e.g., 30, 50) years contained forest (30 years assumed throughout this report). Potential crown cover refers to the state of existing vegetation at maturity under continuation of current land use.	• Conversion of non-forest to forest on lands that in 1990 [or over the past X (e.g., 10, 30, 50) years] did contain forests.

approach was selected to highlight the complexity of combining definition elements and the possible implications of these combinations. These seven scenarios are not intended to be exhaustive of all possible scenarios. We selected two scenarios that use definitions found in FAO publications and the IPCC Guidelines. These two scenarios are intended to capture the existing frameworks in which Article 3.3 could be viewed.

A key difference between the FAO and most of the other scenarios is that the FAO definition of reforestation includes artificial regeneration of tree cover after a clearing. The State of the World's Forests 1999 report (FAO, 1999) defines reforestation as "establishment of a tree crop on forest land." The Forest Resource Assessment Terms and Definitions (FAO, 1998) define reforestation as "artificial establishment of forest on lands which carried forest before" but gives this definition under the headings "forest cover changes" and "new plantations." TBFRA 2000 (UN, 1999, p. 157) explains regeneration of forest land as "reforestation of land that has recently been forested." Such regeneration includes natural regeneration, natural regeneration enhanced by planting, coppice sprouting, and planting or seeding. We interpret the FAO definition of reforestation to include regeneration after clearing.

The IPCC scenario is extracted from the Glossary of the IPCC Reporting Guidelines. These guidelines do not include a definition of a forest, however, which limits their direct applicability. For the purposes of the IPCC scenario, we have assumed the FAO definition of forests. Neither FAO nor IPCC designed their definitions of forest, afforestation, reforestation, and deforestation with the Kyoto Protocol in mind; thus, the definitions are not necessarily optimal for meeting the requirements of Article 3.3.

To ensure that a full range of possible definitional approaches was explored, we developed five additional scenarios on the basis of the definitional framework for the terms forest, afforestation, reforestation, and deforestation, as outlined in Chapter 2. The Land Use scenario employs a land-use definition of a forest and bases the definitions of afforestation and reforestation on activities undertaken on a piece of land. Three of the scenarios—Land Cover, Flexible, and Biome—use different definitions of a forest but similar definitions of afforestation, reforestation, and deforestation. The differences in carbon amounts associated with differences in the definition of what is and is not a forest are very large, reflecting fundamental differences about how Article 3.3 could be implemented and the amount of credits or debits it would create. The Land Cover scenario uses a fixed threshold for what is a forest; the Flexible scenario entails a country-based definition that utilizes carbon or crown-cover criteria; and the Biome scenario uses a biome-based fixed criterion. The remaining definitional scenario, Degradation/Aggradation, attempts to capture in the definitions of deforestation and reforestation the incremental nature of many deforestation and reforestation events.

The differences among these definitional scenarios are summarized in Table 3-5. Note that the definitional scenarios are not intended to be discrete but illustrative. It is possible to combine definitional elements from one scenario with those from another. In combining definitional elements, it is important to recognize that the implications are not always transparent.

The definitional scenarios imply activities and conditions that invoke the creation of ARD lands under Article 3.3, as well as which activities are left for Article 3.4 if double-counting is to be avoided. For example, if ARD and forests are defined broadly to include the harvest/regeneration cycle (e.g., FAO or Land Cover scenarios), few if any forestry activities will be left for inclusion under Article 3.4. The Degradation/Aggradation scenario might also somewhat reduce the choice of activities under Article 3.4, though not as much. With the IPCC, Flexible, Biome, and Land Use scenarios, most forestry activities would fall outside Article 3.3 and thus would be candidates for Article 3.4.

In analyzing the implications of the definitional scenarios, the assumption has been made that all activities are DHI unless they have been demonstrated to be otherwise. The remainder of this chapter discusses how the definitional scenarios interact with a wide range of activities and conditions.

***Table 3-5**: Main attributes of the seven definitional scenarios utilized to illustrate choices associated with implementing Article 3.3. Six criteria are examined, as is how each scenario does or does not address these issues.*

Definitional Scenario	Is Afforestation a Land-Use Change?	Is Reforestation a Non-Forest/ Forest Change?	Is Deforestation a Forest/ Non-Forest Change?	Does Clear-Cutting Create ARD Land?	Does Regeneration after Clear-Cutting Create ARD Lands?	Can Articles 3.3 and 3.7 be Compatible?[a]
FAO	In most cases	Not if regenerating following harvest	Yes	No	Yes	No[b]
IPCC	Yes	Yes	Yes	No	No	Yes
Land Use	Yes	Yes	Yes	No	No	Yes
Land Cover	In most cases	Yes	Yes	Yes	Yes	No[c]
Flexible	Yes	Yes	Yes	No	No	Yes
Degradation/ Aggradation	In most cases	Not always if change from one forest class to another	Not always if change from one forest class to another	No	No	No[d]
Biome	Yes	Yes	Yes	No	No	Yes

[a] See also Section 3.3.2.8.

[b] Reforestation is not always a land-use change, and comparability between Articles 3.3 and 3.7 is more difficult to achieve. For example, harvest/regeneration in the commitment period would create debits under land-based approaches, but would not be counted in the 1990 baseline because it is not a land-use change. Hence, more emissions would be included in the commitment period than in the baseline.

[c] For example, deforestation includes harvest activities in which the 40% crown-cover threshold is crossed. Emissions from such harvests are counted in the commitment period but not in the 1990 baseline, because harvest does not constitute a land-use change.

[d] For example, degradation counts fully in the commitment period. However, it does not enter the 1990 baseline if degradation is not a land-use change.

3.3. Processes, Time Scales, and Carbon Accounting Rules

This section describes issues regarding processes, time scales, and carbon accounting rules that are specific to ARD activities. Where carbon accounting is different between definitional scenarios, we identify these differences.

3.3.1. Processes and Time Scales for ARD Activities

As described in Chapter 1, carbon accumulates in living biomass when forest stands are established. The accumulation rate depends on climatic conditions (especially temperature and precipitation), site conditions (slope, exposure, soil texture, fertility, etc.), tree species/genetic stock, and human activities (e.g., tree cutting, burning, thinning, litter removal). Dead organic matter (i.e., litter, coarse woody debris) and soil carbon may start to accumulate once living biomass produces litter as substrate for soil organic matter formation. The rate of soil carbon accumulation varies as a function of the type and amount of living biomass present and the site's land-use history. If the site was previously tilled, afforestation and reforestation are likely to lead to a significant accumulation of soil carbon.

A distinction between afforestation and reforestation under Article 3.3 is relevant only for assumptions concerning changes in soil and dead organic matter stocks. In general, the land-use history (tillage, type of cropping system, etc.) and site conditions are more relevant than the afforestation/reforestation labels for the determination of assumptions and methods to calculate stock changes on land converted from non-forest to forest.

Reforestation does not always lead to immediate increases in ecosystem carbon stocks. When the definitional scenario creates ARD land where a stand is cleared and a new forest crop is established (e.g., in the FAO definitional scenario), the loss of carbon from decaying slash, stumps, dead root systems, and soil organic matter may exceed the removal of carbon by young trees for several years. The choice of definitions and accounting rules for the implementation of Article 3.3 will determine whether such negative changes in carbon stocks will be reported (see numerical examples in Section 3.5.2).

Afforestation/Reforestation and Deforestation Differ in Their Dynamic Effects

The magnitude and rate of change in carbon stocks differ greatly between afforestation/reforestation activities and deforestation activities, as well as among carbon pools within a single stand. Afforestation and reforestation generally cause small annual changes in carbon stocks over long periods of time, whereas deforestation can cause large changes in carbon stocks over a short period of time. Deforestation causes an immediate reduction in aboveground biomass carbon stocks, followed by several years of decreases in other carbon stocks. The differing pace of response among different activities "since 1990" creates outcomes that are not intuitively obvious. Even if the total forest area and carbon stocks in a country remain constant ("normal forest management")—for example, if afforestation or reforestation in one area are matched by deforestation in another area—a decline in carbon stocks may be reported under Article 3.3. This imbalance occurs because most emissions from deforestation take place rapidly (except where a high percentage of the wood enters long-lived products), whereas carbon removal from afforestation and reforestation activities is a slow process. Afforestation and reforestation take decades to centuries to come to completion; under Article 3.3, however, only the portion of the uptake that originates from stands created "since 1990" is considered. In subsequent commitment periods, if the pattern of normal forest management is maintained, the Article 3.3 imbalance will diminish: A larger and larger number of parcels reforested or afforested since 1990 will make their cumulative carbon gain contributions in comparison to the constant number of parcels being deforested and having their carbon loss. This Article 3.3 imbalance is more pronounced for regions with low growth rates and long rotation periods, such as the boreal zone. In summary, ARD activities can result in a reported carbon stock change that may not reflect the actual changes in carbon stocks (see also Section 3.5.1). Avoiding this effect does not seem possible with the language of Articles 3.3 and 3.7.

An afforestation and reforestation program with a fixed amount of land afforested or reforested each year will yield large positive stock changes after an initial period of slow carbon accumulation. In the first few years, afforestation and reforestation tree growth is slow, and not all land in the program will have been afforested or reforested. Over time, more and more land will become part of the program, and benefits will continue to increase for decades (Nilsson and Schopfhauser, 1995). Because the carbon gain is low at the beginning of afforestation and reforestation programs, in most parts of the world such programs will yield only small carbon gains during the first commitment period.

Avoidance of deforestation under Article 3.3, however, will immediately produce a large amount of avoided emissions. The magnitude of carbon benefits from avoidance of deforestation is high initially, whereas the optimal level in afforestation and reforestation that is initiated at the same time is reached decades later.

3.3.2. Carbon Accounting of ARD Activities

Once it has been determined that an ARD activity has taken place, it should be known *where* the activity took place and *when* it began. Chapter 2 elaborates two carbon accounting options: activity-based accounting, whereby only the stock changes resulting from the activity are counted; and land-based accounting, whereby all carbon stock changes on a piece of ARD land are accounted. Another question for ARD activities initiated after 1 January 2008 is whether carbon stock changes should be assessed beginning on 1 January 2008 or beginning with the onset of the activity.

Three combinations of options are possible:

- *Activity-based approach*. Carbon accounting begins with the start of the activity or on 1 January 2008, whichever is later. Any stock changes prior to the activity remain unaccounted because they do not result from the ARD activity. Any stock change that occurs after the start of the activity but does not result from the activity is not accounted. An example is the decaying slash from a harvest that is followed by reforestation (FAO scenario).
- *Land-based approach I*. All carbon stock changes on ARD land are counted, beginning on 1 January 2008.
- *Land-based approach II*. All carbon stock changes on ARD land are counted, beginning with the start of the activity or with 1 January 2008, whichever is later. For example, the decaying slash would be counted as a negative carbon stock change.

These three options will be discussed further in the stand, landscape, and global level examples (Sections 3.5.1, 3.5.2, and 3.5.4).

3.3.2.1. Do We Need to Distinguish between Direct Human-Induced and Natural Factors that Impact Stock Changes on ARD Lands?

Does "direct human-induced" refer only to the ARD activities, or does it refer to stock changes that result from ARD activities as well? If it refers to stock changes, all indirect human-induced activities and natural impacts on stock changes would have to be excluded. For example, if a forest is growing at a rate of 3 t C ha^{-1} yr^{-1}, 1 t C might be related to the effects of atmospheric nitrogen and CO_2 fertilization. The DHI part would be 2 t C; only that amount would be credited. A converse case can be made for a forest growing at 2 t C ha^{-1} yr^{-1} that would grow at 3 t C in the absence of an adjacent pollution source. In other words, if "direct human-induced" refers to stock changes, the stock changes reported under Article 3.3 might be greater or smaller than the stock changes observed on the land. The exclusion of effects other than those that are DHI is technically difficult, if not impossible (see Section 2.3.3). Achieving this exclusion at a moderate uncertainty level will be costly and could create unbalanced accounting. For example, credits could be given for a growing stand, but subsequent emissions following a natural disturbance or local air pollution would not be counted. If the stand regrows after the disturbance, more credits could be gained even though there is no net GHG improvement for the atmosphere.

If "direct human-induced" refers only to the ARD activity, the technical problems involved in separating out parts of the stock changes disappear. In addition, only the afforestation and reforestation activity allows benefits from nitrogen or CO_2 fertilization, for example. With this approach, stock changes are measured irrespective of their attribution as DHI (e.g., fertilization or thinning), indirect human-induced (e.g., CO_2 fertilization), or entirely natural (e.g., wind-throw and avalanches) (see also Section 2.3.3).

3.3.2.2. Discrepancies between Actual and Reported Stock Changes

Although afforestation and reforestation ordinarily result in net carbon sequestration and deforestation results in net carbon emissions, these outcomes may not always be true in reporting under Article 3.3, depending on the definitions and carbon accounting approaches adopted. Certain definitions of ARD could result in Parties receiving "credits" (i.e., additions to their assigned amounts) for activities that cumulatively result in an increase or no change in atmospheric CO_2 (type I discrepancy). Conversely, certain definitions could result in Parties receiving debits (i.e., deductions from their assigned amounts) for activities that result in reductions or no change in atmospheric CO_2 (type II discrepancy). This section analyzes some of these possibilities. The discussion below is organized around the terms "forest," "reforestation," "degradation," and "deforestation":

a) *Forest*. If forests are defined on the basis of a threshold for canopy cover or biomass density, a certain number (e.g., 10 percent canopy cover of trees) will be the cut-off point between forest and non-forest land. Two type I discrepancies suggest themselves:
   1) Forests with canopy cover that is greater than the threshold value could be degraded to the defined canopy cover limit, releasing a large amount of carbon to the atmosphere or transferring it to non-forest carbon pools, without carbon debit. Subsequent deforestation of the 10 percent canopy cover forests could result in an accounting of carbon loss that matches the carbon in the 10 percent canopy forest rather than in the forest that existed previously. Only the Degradation/Aggradation and Biome definitional scenarios could avoid this problem—and only if the maximum crown cover or carbon content of vegetation at maturity is reduced. Otherwise, such degradation effects could be covered only by Article 3.4.
   2) Thinning or clear-cut harvesting operations on afforestation/reforestation lands could occur just before the first commitment period and between subsequent commitment periods to maximize carbon increments over the commitment period. This effect could be minimized through specification of contiguous commitment periods. Forest managers, however, are unlikely to base their management decisions mainly on considerations of carbon credits rather than silvicultural needs and timber market demands.

b) *Reforestation*

1) *FAO scenario.* If a stand is cleared and reforested between 1990 and 2008, carbon credits could be claimed even if the carbon stocks in the landscape are not increasing (type I discrepancy) (see Section 3.5.2). For a stand that is cleared and reforested between 2008 and 2012, we have to separate the three accounting options described in Section 3.2.2:
   - *Activity-based approach.* Carbon stock losses at the time of harvest are not accounted, nor are subsequent emissions from decaying dead organic matter. Only the accumulation of carbon in the new stand and in new dead organic matter is accounted because it results from the reforestation activity. Carbon credits could be claimed even if the carbon stocks in the landscape are not increasing (type I discrepancy) (see Section 3.5.2 for results on the landscape level).
   - *Land-based approach I.* The stock change over the full commitment period is measured, including stock losses during harvest, as well as delayed emissions from dead organic matter. The consequence is a large debit (except where rotation periods are below 20 years) (see Section 3.5.2). This is a type II discrepancy.
   - *Land-based approach II.* The carbon stock change between the beginning of the activity and the end of the commitment period is measured, including the decaying slash resulting from the harvest. In the short term, debits may be reported because of the effects of decaying slash (type II discrepancy). In the long term, credits will accrue even if stocks on the landscape are not increasing (type I discrepancy) (see Section 3.5.2 for results on the landscape level).
2) *All definitional scenarios.* Assume a case where land is deforested and used for agriculture before 2008, then reforested. Credits could be claimed even though carbon stocks in the commitment period are likely to be lower than in 1990 (type I discrepancy).
3) In the case of afforestation or reforestation between 1990 and 2008, carbon stocks can decrease in a commitment period—for example, because of harvesting or thinning. Debits would be assigned even though the afforestation or reforestation activity reduces atmospheric CO_2 in the long run and carbon stocks in the commitment period likely exceed those in 1990 (type II discrepancy).
4) *FAO scenario.* There can be an unbalanced treatment of different harvesting methods because partial cutting techniques do not result in reforestation under this scenario. This imbalance could result in an incentive to favor clear-cutting over other harvesting methods, possibly working against other considerations related to sustainable forest management.

c) *Degradation*

In all scenarios except Land Cover and FAO (in combination with land-based approach I), forestry activities could focus on high carbon-density forests because they can be harvested without carbon penalty as long as another forest replaces them. No carbon has to be accounted when forests of high carbon density are logged (and their large amounts of carbon released) and replaced by plantings with low carbon stocks (type I discrepancy).

d) *Deforestation*

FAO and IPCC scenarios. The forest definition contains a list of exceptions for lands that are forests without reaching the crown-cover threshold at maturity. For example, the determination of whether a new road results in deforestation would be based on the use of the road. If it is used to access forest stands for logging, it would not be considered deforestation, whereas a road that is used for transportation (such as a new highway) would be. For the purpose of determining long-term impacts on the atmosphere, this distinction is arbitrary.

3.3.2.3. Options for Reducing Discrepancies between Actual and Reported Carbon Stock Changes

Additional rules could be used to avert some of the type I and type II discrepancies described in Section 3.3.2.2, where actual and reported stock changes do not match. Two possible ways of doing this are as follows:

- By limiting "reforestation lands" that fall under Article 3.3 through the use of additional time rules. For example, reforestation could exclude establishment of tree cover on land that was non-forest for less than a specified period of time (e.g., 10 years) after deforestation or land that was forest at a specific point in time (e.g., 1990). If a densely stocked forest with 200 t C ha^{-1} is deforested in 1991, for example, and reforested 5 years later, no carbon credits could be claimed in the commitment period. One problem, however, is shown by the following example: Assume deforestation of a stand with little carbon (e.g., 20 t C ha^{-1} in 1991) and reforestation 5 years later with vigorously growing trees, with stocks increasing to 50 t C ha^{-1} in 2012. Even though this activity reduces atmospheric CO_2, it would be excluded by the use of a time rule. This approach addresses only the type I discrepancy.
- By not limiting "reforestation lands" but limiting the amount of credits and debits for these lands. There are at least two ways to do this:
  - Credit for reforestation in areas deforested after 1990 or 1997, for example, could be excluded.

With a stand of 200 t C ha^{-1} deforested after 1997, no credits would be possible for subsequent reforestation. In the foregoing example of an initial stand at 20 t C and the carbon stocks at 50 t C in the commitment period, no credits would be possible. This approach addresses only the type I discrepancy.

– Credits for reforestation in areas deforested after 1990 could be limited; debits for afforestation and reforestation lands could be limited. This approach addresses both type I and type II discrepancies.
  - *Avoiding the type I discrepancy.* For land deforested between 1990 and 2007, carbon credits for reforestation could be awarded only for increased carbon stocks above the level of carbon stocks present on the land prior to deforestation (or in 1990) (D-R Rule). This approach would eliminate an incentive to deforest land then reforest it to claim carbon credits during the commitment period, thereby avoiding that type I discrepancy in b2 on the facing page (note that ARD land is assumed to be able to switch from the "deforestation" to the "reforestation" category). This approach entails a moderate increase in data requirements because information about the carbon stock prior to deforestation or in 1990 is needed. The first sentence of Article 3.4 asks for data on carbon stocks in 1990, but it is not clear whether these data have to be geo-referenced; thus, it is not clear whether such data are of use here.
  - *Avoiding the type II discrepancy.* Carbon debits in afforestation and reforestation could be limited to the amount of net credits (i.e., credits minus debits) received for carbon accumulating on the same land (A/R Debit Rule). This approach would avoid the type II discrepancy in b3 on the facing page, and it does not create additional data needs. The purpose is to cap debits for reforestation activities (that sequestered carbon between 1990 and 2008) followed by harvesting in the first few commitment periods. It ensures that activities that increase carbon stocks in the long term are not counted as debits under Article 3.3. Particularly in the first few commitment periods, there are carbon stocks on afforestation/reforestation lands that would not be credited because they accrue before 2008. Any harvest loss in a commitment period, however, would count to the extent that carbon is lost (see Section 3.3.2.2). In the first commitment period, the A/R Debit Rule would disallow debits because no credits have been assigned previously and because the stocks are most likely higher than they were before the initiation of the afforestation/reforestation activity. As subsequent commitment periods occur, an increasingly greater share of carbon sequestering on afforestation/reforestation lands will have been credited, so the allowable debit for harvest also increases. In any event, the net credit (credit minus debit) over time for a unit of land would not be negative.
  - *The FAO scenario with land-based approach I* (stock changes over the full commitment period are accounted). With a harvest in 2009 and replanting in 2010, carbon losses from the harvest would be accounted through the 2008–2012 stock change. If the A/R Debit Rule were applied, however, there would be no debit in the first commitment period, even though stocks have decreased. In the FAO scenario with the land-based approach, the debits would be set to zero.

  The amount of land included in the "reforestation" category would not be affected by these two rules. Their main effect is on carbon stock changes, not all of which would necessarily count even if an area is "reforested since 1990."

The carbon discrepancies in items a1 and c1 could be avoided through the use of the Degradation/Aggradation definitional scenario or possibly the Biome scenario. The discrepancy in item a2 could be minimized through contiguous commitment periods. The problem outlined in b4 could be avoided by defining "reforestation" to include planting or regrowth of individual trees in an existing stand. Issue d1 could be addressed through the way a forest is defined.

3.3.2.4. *Necessary Steps for Determining ARD Lands and Stock Changes*

The steps required to determine the area of ARD land and the carbon stock changes on these lands differ between the definitional scenarios. Several questions need to be answered for each area of land sampled. Box 3-1 summarizes the steps to be taken. The sequence applies for the land-based and activity-based approaches, except where indicated otherwise.

3.3.2.5. *Carbon Pools that Could be Considered and How They are Impacted by ARD Activities*

Section 2.2.5.3 discusses various carbon pools that could be included in carbon accounting for the purposes of the Kyoto Protocol. This section discusses carbon pools from the specific perspective of ARD activities. These carbon pools are affected as follows (see Table 3-6):

- *Living biomass.* Aboveground (stems, branches, leaves, flowers, and fruits) and below-ground (roots)

Box 3-1. Steps for Determining ARD Lands and Stock Changes
[Text in square brackets indicates that one of the choices must be selected.]

For the Land Use Scenario:

1) Has the land-use designation for the area been changed?

For Other Scenarios:

2) Has the vegetation attribute [forest cover or carbon content—at maturity except for the Land Cover scenario] of the area passed a threshold [various possible definitions; or changed by more than a specified amount] since the last evaluation?
3) Is the activity that caused the crossing of the vegetation-attribute threshold an activity that results in the creation of ARD land (i.e., DHI)?
4) If the vegetation does not currently meet the criterion for forest, is it expected to grow past that threshold (all scenarios except Land Cover)?

For All Scenarios:

5) For ARD land where the ARD activity started before 2008: What was/is the carbon stock in the beginning of the commitment period?
   For ARD land where the ARD activity started after 1 January 2008: What was/is the carbon stock at the beginning of the activity (activity-based approach and land-based approach II)? What was/is the carbon stock at the beginning of the commitment period (land-based approach I)?
6) For ARD land: What is the carbon stock in the end of the commitment period?
7) Calculate carbon stock change as 2012 stocks from item 6 above minus [2008 stocks or pre-activity stocks] from item 5 above.

If the option of a D-R Rule and A/R Debit Rule is adopted, then:

8) For ARD land that was deforested between 1990 and 2007, limit credits from item 7 above to the extent by which the carbon stock in 2012 exceeds those [prior to deforestation or in 1990].
9) Limit any debit for afforestation and reforestation activities to credits received previously.

living biomass will sequester carbon at a rate and to a total magnitude that depends on species, climate, and site quality. Even if forests are allowed to continue to grow, carbon stock in living biomass will eventually reach a maximum. It takes at least decades and more commonly centuries for forests to reach their maximum carbon storage potential. That carbon storage potential also depends on the level of forest management and natural disturbances. The living biomass carbon in areas where new forest is created will increase. If below-ground living biomass is not included in the accounting of deforestation, emissions to the atmosphere would be underreported. In afforestation or reforestation, sequestration rates would be underreported.

- *Litter and debris*. Following creation of new forest, carbon in litter and woody debris (dead organic matter) increases over time. In deforestation, large amounts of debris are usually generated at the time of tree-felling. Amounts remaining on the ground may be low if logging debris is burned or removed from the site but can be substantial if debris is simply retained on site or if trees are not felled but simply killed by chemical injection. Amounts of litter and woody debris are usually higher in natural forest from which little or no timber has previously been harvested than in plantations (Ruhiyat, 1995).
- *Soil*. On previously cultivated land where new forest is created, soil organic matter is expected to increase for decades to centuries (O'Connell and Sankaran, 1997). Soil organic matter will decrease rapidly following deforestation if land is subsequently cultivated. Changes in soil organic matter are likely to be small if the soil is not cultivated.
- *Wood products/landfills*. Some ecosystem carbon removed during harvest or deforestation may be stored in forest products and landfills rather than immediately returning to the atmosphere.

If the harvest/regeneration cycle does not create ARD land, the proportion of the world wood supply derived from ARD land will be very small. In the FAO scenario, on the other hand, wood product stocks from ARD land could be quite significant, depending on whether the harvest that precedes reforestation is reported as a stock change. If wood products are included in carbon stock assessments, the initial loss of carbon following harvest on the stand level would be lower. The rate of increase in carbon stocks on ARD land also would not diminish as fast. The long-term effect would not be very different, however, from excluding wood products because only a portion of harvested wood usually is stored in long-lived timber products and because share also decays with time (Marland and Schlamadinger, 1999).

Whether any wood derived from ARD land results in a net increase in forest product and landfill carbon pools depends on future global harvest levels and the fate of the harvested material. In any case, it will be very difficult to track the fate of timber harvested from ARD land and distinguish it from all other harvested material. If (improved) wood product management is a separate activity under Article 3.4, wood products created under ARD lands would have to be excluded from Article 3.3 to avoid

Table 3-6*: Changes in components of terrestrial carbon stocks under different land-use changes. Note that, initially, carbon in litter and woody debris may increase following deforestation. Aboveground biomass and litter can also decrease (e.g., because of harvest or natural disturbances).*

	Biomass		**Litter/Woody Debris**		**Soil**	**Wood Products**
	Aboveground	Below-ground	Short-term	Long-term	**Organic Matter**	**and Landfills**
Cultivated land ➔ forest	↑	↑	–	↑	↑	↑[b]
Non-cultivated land ➔ forest	↑	↑	–	↑	?	↑[b]
Forest ➔ cultivated land	↓↓	↓	↓	↓	↓	–
Forest ➔ grazing land	↓↓	↓	↑[a]	↓	?	–

[a] It is assumed here that upon conversion of forest to grazing land, woody debris is not, or is only partly, removed. Dead roots, in particular, would not normally be removed. If woody debris is removed or burned, only dead roots would add to the short-term increase of woody litter.

[b] If forest is subsequently harvested and used for wood production.

double-counting. If all wood products from a particular ARD land were to be assigned to that land and counted under Article 3.3, this accounting approach would be the "production approach" (IPCC, 1999). As a consequence, these wood products would have to be excluded from the account of an importer of these products to avoid double-counting. If the importer were to count the carbon stock change from imported wood products, this would be the "stock change approach" (IPCC, 1999). The "atmospheric flow approach" in IPCC (1999) is not compatible with Article 3.3 because it does not provide a change in carbon stocks but gross flows of carbon to or from the atmosphere.

3.3.2.6. *What Are the Consequences of Simplified Reporting?*

A low-tier approach to determining a rough order of magnitude of ARD credits and debits could be based on default values and could be implemented without reference to specific land areas. Such a low-tier approach has been applied in country studies cited in Section 3.5.3. One option for implementation follows:

1) Determine the total increase of forest area (A_{new}) from afforestation and reforestation between 1990 and 2012 per country, biome, or region.
2) Determine the total loss of forest area (A_{loss}) from deforestation between 2008 and 2012 per country, biome, or region.
3) Determine the carbon stock at maturity (natural forests) or the average carbon stock over a rotation (managed forests), C_{equ}.
4) Determine the time to reach the carbon stock at maturity (natural forests) or the carbon stock averaged over a rotation (managed forests) for that country, biome, or region (T_{equ}).
5) Calculate the equilibrium change in carbon stocks from the area change, as area change times equilibrium stock change ($A_{new/loss}$ x C_{equ}).
6) Approximate credits from afforestation and reforestation, as A_{new} x C_{equ} x ($5/T_{equ}$). Only a share of the total equilibrium stock change would be attributed to the 5-year commitment period because afforestation and reforestation are gradual processes with an assumed duration of T_{equ}.
7) Approximate debits from deforestation, calculated as A_{loss} x C_{equ}. All or most of the equilibrium stock change would be attributed to the 5-year commitment period because deforestation, unlike afforestation and reforestation, is a process that results in stock changes over only a few years. One would miss the delayed emissions from pre-2008 deforestation, but this effect would cancel out post-2012 delayed emissions from deforestation in the commitment period.

This approach would reduce data needs and costs but would introduce lower accuracy. Use of conservative default values would be possible. Exact carbon accounting of ARD activities for future commitment periods is difficult, however, because ARD lands are not spatially tracked. This approach could be regarded as an activity-based approach with ARD lands not geographically referenced. Although the method of deriving ARD stock changes might be verifiable, the stock changes themselves would not.

3.3.2.7. *How Can Carbon Accounting of ARD Land during Future Commitment Periods Be Carried Out?*

Article 3.3 of the Kyoto Protocol is specific in that the allowable offsets from ARD activities must be "measured as verifiable changes in carbon stocks in each commitment period." Once a piece of land falls under the definitions of Article 3.3 during one commitment period, can it subsequently be removed from the accounting? It is logical to assume that land could change from the deforestation category to reforestation or vice versa. If a piece of land can be removed from the ARD lands category at later times (e.g., when the activity ends in an activity-based approach), large carbon stocks that were credited previously could be released without debits (Schlamadinger and Marland, 1998). The words "in each commitment period" in Article 3.3 suggest a reporting requirement for future commitment periods. Over time, the area of ARD land would increase because no land can be removed from this classification.

The inclusion of afforestation and reforestation since 1990 provides countries with an additional tool for achieving their assigned carbon emission targets during the first commitment period because most forests established since 1990 would still be growing rapidly in 2012. If the Kyoto Protocol were applied to future, contiguous commitment periods and if more and more post-1990 forests were harvested during these periods, credits would be followed by debits on the stand level (see also Section 3.3.2.3 on ways to limit such debits). On the national level, however, credits would only decrease over time as the forest estate approaches a biomass equilibrium. Carbon trends in these future commitment periods will partly depend on the definitions of stocks (see Section 2.3.6). On-site carbon stocks in post-1990 forests are likely to increase for decades (provided the total area of eligible forests does not decrease over time and forests are logged on a sustainable-yield basis). However, the rate of increase—hence the amount of allowable offsets—is likely to diminish over time.

3.3.2.8. Article 3.7: Adjusting the 1990 Baseline

The second sentence of Article 3.7 of the Kyoto Protocol states:

> *Those Parties included in Annex B for whom* ***land-use change and forestry*** *constituted a net source of greenhouse gas emissions in 1990 shall include in their 1990 emissions base year or period the aggregate anthropogenic carbon dioxide equivalent emissions by sources minus removals by sinks in 1990 from* ***land-use change*** *for the purposes of calculating their assigned amount.*

Article 3.3 specifies that carbon emissions and removals from ARD must be counted during the commitment period but are not included in the general baseline. Article 3.3 creates the following two "anomalies":

- Afforestation and reforestation allow credits to be gained during the commitment period even though the same activity in 1990 did not lower the 1990 baseline.
- Deforestation creates debits during the commitment period even though the same activity in 1990 did not cause an increase in the 1990 baseline.

Article 3.7 rectifies these anomalies for some countries by including emissions and removals from "land-use change" in the 1990 base year. The first part of the second sentence in Article 3.7 ("Those parties...emissions in 1990") determines to *which* countries the rest of the sentence ("shall include... assigned amount") applies. Land-use change and forestry (which is assumed here to be identical to LULUCF) was a net source in 1990 for all of these countries.

The second part of the second sentence of Article 3.7 ("shall include...assigned amount") states that CO_2-equivalent emissions by sources and removals by sinks in 1990 that result from "land-use change" shall be included in the 1990 base year. Parties may wish to clarify the nature of the linkage between Articles 3.3 and 3.7, particularly the scope and meaning of the term "land-use change" in Article 3.7. Options include applying "land-use change" narrowly to conversions between forest and non-forest (depending on whatever definition of "forest" the COP may adopt) or applying "land-use change" more broadly to conversions between, for example, the 15 categories listed in Table 2-1.

If ARD activities are defined as "land-use changes" (the IPCC, Flexible, Biome, and Land Use definitional scenarios), the first option (applying "land-use change" narrowly) counts carbon emissions and removals in 1990 from pre-1990 ARD activities in the 1990 base year, thereby rectifying the two anomalies. Equivalence between ARD in Article 3.3 and "land-use change" in Article 3.7 would be achieved under this option (see the right column in Table 3-5). Hence, for some countries a net-net approach for ARD is applied according to Article 3.7: All ARD activities would be considered in the commitment period and in the 1990 emissions base year. If ARD activities are not considered to be limited to "land-use changes" (the FAO, Land Cover, and Aggradation/Degradation definitional scenarios), equivalence between ARD in Article 3.3 and "land-use change" in Article 3.7 is not possible.

The option of applying "land-use change" broadly would include ARD and land-use conversions other than ARD under the term "land-use change." For countries with a net source from land-use change and forestry in 1990, this approach opens the possibility of using a net-net approach in the first commitment period for activities in Article 3.4, provided these activities are defined as "land-use changes."

3.3.2.9. Are Non-CO_2 Greenhouse Gas Emissions to Be Counted?

When forest land is cleared and biomass is burned, GHGs in addition to CO_2 can be emitted. Depending on fire intensity and other factors, such as moisture content, a variable fraction of carbon and nitrogen in biomass and the soil will be released as CH_4, CO, N_2O, and NO_x. Similarly, application of nitrogen fertilizers in afforestation or reforestation can result in significant emissions of N_2O. Draining peatlands, on the other hand, can lead to reduced CH_4 emissions—at the cost, however, of large carbon losses (Cannell *et al.*, 1993). Some studies suggest that water-level drawdown in peatlands decreases the greenhouse impact (Laine *et al.*, 1996).

Methane and nitrous oxide are among the six GHGs in Annex A of the Kyoto Protocol. Although Annex A lists the agricultural sector, including emissions from prescribed burning of savannas and field burning of agricultural residues, it does not include land-use change and forestry. Only Articles 3.3 and 3.4 are vehicles to bring land-use change and forestry activities into the accounting. Article 3.3 mentions GHG emissions from sources and removals by sinks, but these changes are to be measured as verifiable changes in stocks. This provision suggests the exclusion of non-CO_2 GHG emissions because they cannot be verified as changes in stocks. Conversely, given that Article 3.7 specifically mentions "CO_2-equivalents," non-CO_2 GHGs arguably could be included under Article 3.3 as well, if an equivalent treatment of emissions and removals between the 1990 baseline and the commitment period is to be achieved. Parties may wish to clarify the intent in the Protocol regarding non-CO_2 GHGs, with these gases universally ruled in or out for ARD activities. Including non-CO_2 GHGs would greatly increase data requirements and the complexity of methods required for determining associated fluxes. Excluding them would neglect some of the GHG fluxes to and from the atmosphere associated with ARD activities, such as wetland drainage, use of fertilizers, or prescribed burning.

3.3.2.10 Relation of ARD Activities to Activities under Article 3.4 and Projects under Article 6

Article 3.3 states that stock changes from ARD activities shall be used to meet commitments. If the COP accepts further activities under Article 3.4, how should these activities be treated if they occur on the same land for which stock changes are already counted under Article 3.3? There are three options:

- The land is no longer considered under Article 3.3 but included exclusively under Article 3.4: This option might violate Article 3.3, which stipulates that all stock changes resulting from ARD activities "shall" be reported. There is a possibility, however, that reporting stock changes from ARD activities under Article 3.4 would be regarded as sufficient to satisfy the "shall" requirement of Article 3.3.
- The land is considered under Article 3.3, and stock changes from the Article 3.4 activity are counted under Article 3.3 as well: This option would avoid the risk of possible double-counting. It eliminates the need to determine whether an Article 3.4 activity has occurred on ARD land.
- Stock changes from an ARD activity are counted under Article 3.3, and stock changes on the same piece of land from an Article 3.4 activity are counted under Article 3.4: This option poses a danger of double-counting, and separation of stock changes according to cause will be difficult.

Similarly, there might be lands that are already under the realm of Article 3.4 on which a subsequent ARD activity takes place. There are three options in this case as well:

- The land is no longer considered under Article 3.4 but counted under Article 3.3.
- The land continues to be considered under Article 3.4, so stock changes from the ARD activity are accounted under Article 3.4: This option would exclude some ARD activities from being counted under Article 3.3, however—which would be inconsistent with the wording of Article 3.3.
- Stock changes from an Article 3.4 activity are counted under Article 3.4, and stock changes on the same piece of land from an ARD activity are counted under Article 3.3.

Another issue involves how ARD activities relate to projects undertaken under Article 6. This issue is discussed separately for land-based and activity-based approaches:

- *Land-based approach.* All stock changes on a unit of land are accounted once the land undergoes an ARD activity. If the same activity also falls under Article 6, then—according to Articles 3.10 and 3.11—only the "additional" carbon credits [Article 6.1(b)] can be transferred to another country. In other words, assigned amounts are deducted from the host country's budget and added to the investor country's budget. For example, if the stock change from an ARD activity is 2 t C ha^{-1} yr^{-1}, but only 1.5 t C of that amount are "additional" (the rest coming from a longer term increase in soil carbon), the 2 t C are added to the host country's assigned amount via Article 3.3, and 1.5 t C are deducted via Article 3.11. The net gain for the host country, after the transaction, would be 0.5 t C.
- *Activity-based approach.* The stock changes that result from the ARD activity are accounted under Article 3.3. This approach seems compatible with the additionality requirement under Article 6. In the foregoing example, the credit under Article 3.3 would be 1.5 t C, all of which would be transferred via Article 3.11.

3.4. Data Needs and Methods for Implementing Article 3.3

Implementation of Article 3.3 may involve identification of ARD activities during specified time periods, as well as stock changes resulting from those activities. This combination translates into data requirements that are summarized in Table 3-7. The definitions chosen for forest, afforestation, reforestation, and deforestation and the interpretation of the phrase "direct human-induced" will have significant effects on the data required for implementation of Article 3.3. These impacts include the possible need to develop a procedure for recalculating forest carbon stocks from national forest inventory data or adjust national forest inventory systems to fit the needs of the Kyoto Protocol.

To assess carbon stock changes associated with activities defined under Article 3.3, two approaches could be developed: an activity-based approach or a land-based approach. The

***Table 3-7**: Data sources for Article 3.3 reporting requirements.*

Data Requirement	Remote Sensing	Forest Inventory	Activit Reporting
1990 forest extent	✔	✔[a]	
2012 forest extent	✔	✔[b]	
A/R activities undertaken			✔[c]
D activities undertaken	✔		✔[d]
1990 carbon stock		✔[b]	
2008 carbon stock		✔[b]	
2012 carbon stock		✔[b]	
C stock before deforestation		✔[b]	
C stock at beginning of activity that starts after 2008		✔[b]	✔
For forest extent			
Land use		✔	✔
Percent cover	✔	✔	
Regeneration potential		✔	✔[e]
Carbon density		✔	
Cover class	✔	✔	

[a] Where forest is present in inventories subsequent to 1990, it should be possible to determine if that forest originated since 1990. In some cases this will represent AR activities. Where forest is not present in inventories after 1990, it may be impossible to determine from sample plots whether forest was present in 1990.
[b] If inventory dates do not match specific years of interest, models can be used to project inventory conditions to the beginning and end year of commitment periods.
[c] It is assumed that activities beneficial to a country would be reliably reported in an activity reporting mechanism.
[d] Identification of deforestation through activity reporting relies heavily on enforcement of permitting or reporting regulations. Verifiability could be difficult to ensure.
[e] Some sustainable forestry verification methods require reporting of adequate regeneration.

activity-based approach requires identification of ARD activities (e.g., through reporting, legal permit systems, or remote sensing). The land-based approach requires identification of land areas where ARD has occurred (e.g., through remote sensing or forest inventory systems). Once direct human-induced ARD activities have been identified, carbon stock changes have to be estimated.

The timing for identification of ARD activities and estimation of carbon stock changes is different. This timing will depend on the type of carbon accounting system (land- or activity-based), as well as the definitions and rules accorded. These factors may also determine which stock estimates will be necessary. For instance, the Flexible scenario may require a stock estimate in 1990, and the D-R Rule (Section 3.3.2.4) may require stock estimates in 1990 or at other intermediate years between 1990 and 2012. For some countries, this information may be difficult to obtain. Likewise, changes in carbon stock may occur for many years after ARD activities (e.g., soil carbon loss under cultivation), so stock change assessments may be required for areas that have been in a non-forest condition for some time.

Depending on the definitions selected and the monitoring methodology used, accurate estimates of carbon stock changes may require knowledge of the specific year, or 5-year period, during which land becomes ARD land. In some monitoring methodologies, estimation of carbon changes during the reporting period may be based on models of temporal differences in carbon pools following disturbance, as described in the IPCC Guidelines (IPCC, 1997). If this methodology is used, accurate determination of the year of the disturbance is of prime importance.

Article 3.3 of the Kyoto Protocol limits activities to ARD resulting from "direct human-induced activities." The DHI clause establishes the need to determine—in every case in which a change in land use and/or land cover has occurred—whether the change resulted from a DHI activity. The difficulty in determining this differs between scenarios.

The most restrictive scenarios (with respect to determining DHI) are those that create ARD land only through change of land cover at maturity to or from forests but do not include reforestation after clearing (e.g., IPCC or Land Use scenarios). In these scenarios, the direct human involvement can be established where forests are replaced with a new form of land use, as well as where afforestation has created new forests. It is more difficult to determine DHI in cases such as *inter alia* exclusion of cattle grazing or fire suppression resulting in forest establishment, or where a natural disturbance (e.g., a severe forest fire followed by regeneration failure) caused the change in forest cover. Clear guidelines will be required to establish whether direct human involvement has occurred in such cases.

In less restrictive scenarios, harvest followed by reforestation creates ARD land (e.g., FAO or Land Cover scenarios). In these cases, two events need to be assessed: the cause of the reduction in forest cover (e.g., wildfire, wind-throw, harvest)—which needs to be assessed in the Land Cover scenario—and the cause of stand establishment (which needs to be assessed in both scenarios). Under FAO and Land Cover scenarios, for example, a wildfire caused by lightning in 2009 followed by reforestation through planting may create ARD land, for which the carbon stock change must be evaluated. In the FAO scenario with a land-based accounting approach, the decision to plant the area devoid of trees through natural disturbance will result in a carbon debit because the carbon stock in 2008 was greater than that in 2012 (land-based approach I) (see Section 3.3.2) or the carbon stock at the beginning of the reforestation was greater than that in 2012 because of decay of slash (land-based approach II) (see Section 3.3.2). This debit would penalize a decision to plant even though this reforestation activity might contribute to carbon sequestration.

3.4.1. *Identification of ARD Activities*

Regarding the determination of the area of ARD land, there are fundamentally two sources of information available: reports of activities (in countries where such reports exist) and monitoring of forest/non-forest change.

The activity reporting approach assumes that a well-defined set of human activities will determine ARD land and that information about these activities will be reliably reported and compiled. In some countries, for example, conversion of forest requires some form of legal permission. Because afforestation and reforestation activities require investments, it should be possible to create incentives for reporting of activities that would benefit the national carbon balance. An activity reporting approach is best if the ARD activities are well defined and are limited to changes to and from forest (e.g., the Land Use scenario) rather than changes within forest classes. An activity reporting approach cannot meet the data requirements for scenarios that include a transition from one forest class to another (e.g., the Degradation/Aggradation scenario).

Several techniques are available for monitoring forest/non-forest change. The simplest method is to identify the total area considered forest in 1990 and at the end of the commitment period. This information might be collected through independent samples at two points in time. A net increase in forest area represents an excess of afforestation and reforestation over deforestation; a net decline represents the reverse. This technique does not follow the fate of individual forest stands. Thus, it allows deforestation to the extent that it is compensated by afforestation and reforestation. A zero net change in forest area can be achieved by deforesting and afforesting the same amount of area. Because the two activities would be expected to contain very different carbon stocks per hectare, however, the zero net change in area would not result in a zero net change in carbon stocks. Therefore, an alternative technique that avoids this problem would be to identify specifically which areas were afforested, reforested, or deforested during the time period.

ARD activities can be monitored in a variety of ways. The two predominant methods are field sampling and remote sensing. These techniques are described in more detail in Section 2.4; both involve sampling the land (with either field plots or aerial/orbital images) at multiple points in time and subsequently identifying locations (or sample plots) where land-use or land-cover changes indicate that ARD activities have occurred. ARD activities could be identified by an assessment of forest extent in 1990 and in 2012; those areas that change from forest to non-forest or from non-forest to forest during this interval could be deemed to have undergone an ARD event. An assessment conducted only at the endpoints of this interval, however, may miss many ARD events that could result in changes in carbon stocks. For example, a 1990 forest that is deforested in 1991, only to be reforested in 2010, is in forest condition at both interval endpoints, yet it may be counted as ARD land.

3.4.2. *Monitoring Land-Cover Change*

Scenarios that create ARD land on the basis of a wide range of activities, including harvest/regeneration cycles and natural disturbances followed by regeneration (as in Land Cover or FAO scenarios), will result in a much larger area of ARD land. The data requirements for area determination under such scenarios may be met through approaches that are based on monitoring land-cover change, such as remote sensing. Forest definitions with a high canopy cover threshold likely would lead to more accurate identification of ARD activities with remote sensing than definitions with a low canopy cover threshold because sparse forest conditions are more frequently confused with some types of agriculture or scrub/shrub ecosystems in assessments based on the use of remotely sensed imagery. For the same reason, if regenerated areas with minimal crown cover are considered forest (e.g., FAO scenario), remote-sensing techniques would be unable to reliably detect them. If the forest definition is dependent on a biomass assessment (which is possible in the Flexible scenario), forest areas could not be

***Table 3-8**: Potential of current remote sensing systems to detect ARD activities.*

Definitional Scenario	Ability to Detect ARD	Notes
FAO	Low	a
IPCC	Low	a
Land Use	Low	
Land Cover	Moderate	a
Flexible	Low	b
Degradation/Aggradation	Low	c
Biome	Moderate	a,d

[a]Considerable misclassification may exist between sparse or young forest and agriculture or other vegetation types, leading to errors when forest cover thresholds are low.
[b]Biomass estimates will be of low accuracy.
[c]Transitions from one cover class at maturity to another within forest are difficult if not impossible to assess accurately.
[d]Highly variable, depending on defined threshold.

accurately estimated with current remote sensing systems. In this case, field inventory would almost certainly be required. Table 3-8 summarizes the potential of current remote-sensing systems to detect ARD activities.

When ARD activities are defined as a change between forest and non-forest, most such activities will be readily discernible using forest/non-forest classifications at least in 1990 and 2012. Some definitions of ARD, however, could mean that land that is in forest status in both 1990 and 2012 might still be included in Article 3.3 (Section 3.4.1), so simple forest classification at the two endpoints would be insufficient to determine ARD land. Thus, information beyond a forest/non-forest classification at different points in time is required. This information could be gathered through combinations of remote sensing and ground-based sampling. The information-gathering process includes two main tasks:

- Identification of land use/land cover in 1990 to serve as the baseline
- Monitoring of ARD activities between 1990 and 2012.

In many countries, reliable data on forest area, age, and species distribution result from detailed forest inventories. These inventories are usually conducted on periods of 10 to 20 years, however. Thus, assessment of changes between 1990 and 2012 will require application of approximation procedures that yield estimates based on assumptions or auxiliary information about the continuation or alteration of a trend.

3.4.3. *Changes in Carbon Stock (per Unit of Area)*

To assess carbon stock changes on lands in which ARD activities have taken place, it is necessary to either spatially delineate the land over which ARD activities have occurred and assess the stock changes therein or use a system of remeasured sample plots including the same areas.

The language of decision 9/CP.4 (FCCC, 1998b) implies that carbon stocks need to be measured only in 2008 and 2012 for Article 3.3 adjustments in the first commitment period. In Article 3.4, a stock assessment in 1990 is specifically called for. Incorporation of a D-R Rule (see Section 3.3.2.3), however, would require an additional carbon stock assessment at a specific year prior to 2008 to estimate the carbon stock at the specific time a parcel of land becomes part of the ARD land. Without such a rule, incentives exist for deforestation followed by reforestation after 1990; with the rule, a carbon stock assessment is required for a point in time prior to the deforestation (e.g., 1990).

Absent this rule, an assessment for Article 3.3 (under the land-based I approach) would entail accounting for ARD land at 2008 and 2012. If the measurements do not coincide with the endpoints of the commitment period, however, interpolation may be required.

Section 2.4.2 describes the data requirements for assessment of the carbon stock of a forest. The choice of scenario will not change the data necessary for carbon stock assessments. A carbon stock-based definition of forest implies, however, that a carbon stock assessment is required to determine whether an ARD activity has taken place; other scenarios involve carbon stock assessments only after the determination that land has become ARD land.

Existing forest inventory systems may not be adequate to assess stock changes resulting from ARD activities. For example, the U.S. National Forest Inventory uses a two-phase sampling system that determines the forest/non-forest condition of approximately 3 million plots (using aerial photographs) in the first phase and establishes 300,000 permanent field plots in a second phase. Although information collected from forested field plots should be quite adequate to estimate carbon stocks, there are insufficient data from non-forested areas to accurately assess stocks. An example would be a plot that was forested at one field visit, during which adequate carbon stock information was obtained. Subsequent conversion to a non-forest condition might mean the plot is no longer visited in the field, although it might still contain a substantial portion of the pre-conversion carbon stock. This plot could meet the criteria for Article 3.3, yet insufficient information would be available to assess the stock change. Similarly, application of a definitional scenario based solely on a low percentage of forest cover might result in inclusion of urban lands as forest. For example, an area of non-forest land may be developed for urban use and sufficient trees planted to exceed the minimum cover threshold. Because this land would be classified as forest under such a scenario, its carbon stock would be eligible for Kyoto Protocol purposes.

Most current forest inventories were developed to efficiently estimate merchantable wood production, not biomass or carbon content. Forest inventories generally provide estimates of main stem volume in a green condition. The use of conversion/expansion factors enables estimation of biomass and carbon from stemwood volume. This technique may mask significant gaps in data for some forest conditions, which may be important depending on the scenario. For example, wood volume in small trees and younger stands is rarely included in forest inventories, causing difficulty in reliably estimating forest biomass. This situation applies to any afforestation/reforestation activities involving young stands. In particular, scenarios in which regeneration after harvest constitutes reforestation (the FAO and Land Cover scenarios) will involve large afforestation/reforestation areas with young stands. For these cases, special regeneration surveys may be required. The lack of information on the dynamics of young stands (including soil carbon pools) is particularly critical in the first commitment period, when the old stands in the ARD lands will be at most 23 years old.

If the forest products pool is included in the carbon stocks to be reported under Article 3.3, a substantial body of additional data will be required, including estimates of carbon in wood products and the fate of this carbon. This type of data is unavailable in many countries, and collection of such data may be costly in relation to the stock change reported.

3.4.4. Uncertainty of Afforestation, Reforestation, and Deforestation Activities

The first determination needed for Article 3.3—identification of ARD land—requires classification of land as forest or non-forest for at least two points in time or an activity reporting mechanism to identify lands that have been subject to ARD activities. Regardless of the method chosen to identify where ARD activities have occurred, some uncertainty is expected to occur. The uncertainty involved in activity reporting is primarily caused by reporting errors or biases (underreporting or overreporting). In the land-use/land-cover classification case, for example, a remote-sensing identification of forest/non-forest areas may include some inaccuracy in classification, which is also a function of the type of orbital system used. Mantovani and Setzer (1997) compared the number of deforestation polygons using the Advanced Very High Resolution Radiometer (AVHRR) sensor (1.1-km resolution) with the number from the yearly deforestation project conducted by Instituto Nacional de Pesquisas Espaciais (INPE) using the Thematic Mapper (TM) sensor (30-m resolution). The comparative results indicated that only 49–57 percent of the total cases were correctly identified with the low-resolution sensor. This difference in ability to detect deforestation appears to be related to the size of the sites considered. As observed in the TM images, deforestation polygons less than 3.1 km^2 in size were usually not detected in the AVHRR imagery.

In addition, depending on the season in which the imagery is taken, forested areas can be confused with other vegetated areas in classifications. Areas in which forest cover may be sparse or very young may meet certain definitions of forest yet may appear in the image to be in a non-forest condition. Here, using definitions of forest that are based on low canopy cover thresholds or that provide for classification of early growth stages likely will result in more misclassifications (of forest as non-forest) than definitions that require high percentage cover. Ground-based inventories are more accurate in this respect than inventories that are based on remotely sensed data.

Different definitions for forest, afforestation, reforestation, and deforestation will lead to different levels of difficulty of data collection and different levels of uncertainty. Definitions that are based on objective and readily measurable variables will tend to yield lower uncertainties. For example, a forest definition that is based completely on the proportion of land covered by woody species (e.g., Land Cover scenario) will lead to more accurate estimates than a definition that involves purpose or history of land use or functional mechanisms used for establishing forests (e.g., Land Use scenario).

The uncertainty of stock change estimates will also vary with the magnitude of the stock in terms of spatial extent and density. Detecting large relative changes in a small stock is easier than detecting small relative changes in a large stock. This fact implies that countries with higher carbon density in forests will have a greater uncertainty for a given magnitude of stock change than countries with lower forest carbon density. Furthermore, changing the threshold of forest cover in a definition may influence the uncertainty in detecting ARD activities.

Because ARD activities are change processes, definitions and methods have to be consistent on successive occasions. In the framework of the Kyoto Protocol, it is convenient to group the scenarios presented in Section 3.2 in two broad classes:

- Scenarios for which country statistics and map systems with appropriate models exist
- Scenarios that require a new set of data (inventories, maps, results from models) following a common, consistent approach.

If a new data collection system has to be implemented, obtaining the necessary retrospective information will be difficult regardless of the scenario applied because, at a minimum, data are required on the status of lands in 1990. Use of archived remotely sensed data, inventory plots that may have been established for another purpose, and/or retrospective models may be the only available alternatives. Evaluating the uncertainty or verifying such estimates would be impossible.

Uncertainties in determining stock changes from ARD activities will be sensitive to various components of the data collection process. A crucial component is the sampling scheme used for detection of ARD activities, which will determine the minimum detectable change in forest condition. Difficulties arise if the sampling resolution is inconsistent with the assessment unit sizes in the definition adopted for forest. If a ground-based forest inventory is used to sample ARD activities, the sampling intensity—expressed as hectares of land represented by each plot—will give an indication of the ability of the system to detect ARD activities on small land areas. For assessments that are based on remotely sensed data, the spatial resolution of the sensor used will be crucial in determining the minimum detectable change.

Consistency in sampling schemes between assessments is also important: Erroneous conclusions can be reached if sampling intensity (or sensor resolution) changes substantially between assessments. For example, suppose that 1-km resolution data were used for a first assessment, and a 1-km^2 area was identified as forest. At a second assessment, 100-m resolution data might detect non-forest areas within the original 1-km^2 area, leading to an interpretation of deforestation. That 100-m opening may have existed within the 1-km^2 area during the first assessment, however, in which case the determination of deforestation would be in error.

Timing of measurements will also affect the ability to detect stock changes from ARD activities. Forest inventories can be time-consuming, multi-year processes, particularly in large areas. In the United States and many European countries, for instance, the national forest inventory cycle is about 10 years. With this type of measurement timing, it is impractical to expect accurate, verifiable estimates of stock changes during a

5-year commitment period and for a small but comprehensive subset of the inventory area. Some nations are beginning to use annual inventory systems, but the transition from periodic national inventories to annual inventories is costly and difficult. In some cases, new sampling protocols (timing and intensity of samples) may need to be adopted if national inventory data are to be useful for the carbon accounting required under Article 3.3.

Although the focus here has been on implications for the first commitment period, there are considerations for subsequent commitment periods. Errors in classification of forest/non-forest can have effects in subsequent commitment periods, either in the determination of areas requiring stock estimates for Article 3.3 or in assessments of stock changes between the first commitment period and later commitment periods. For example, consider a parcel that was forested in 1990 and was subsequently misclassified as non-forest after 1990. This misclassification could lead to an erroneous classification of deforestation and carbon debits for the assessment of the first commitment period. The error depends on the definitions applied and the method used to determine carbon stock changes. If the method involves some form of carbon stock measurement, the misclassification of ARD land will have no impact on the reported carbon stock change because there will be none in the misclassified area. A subsequent (correct) classification of the same area as forested would produce a second faulty determination—this time of reforestation and creditable carbon stock increase. Similarly, errors in estimates of stock changes can have persistent effects after the first commitment period. No matter what sampling scheme is employed, these types of errors will occur. The key issue is the identification of systematic problems that will generate biased estimates of ARD activities and random errors that offset one another, having little effect on ARD activities determinations.

3.4.5. Data for Verification

Verifiability of stock changes from ARD activities as required in Article 3.3 will require the ability to verify that the activities have taken place, as well as the ability to verify stock changes from those activities. The primary difficulty involves verification of the historical status of a resource that may no longer exist.

If an existing monitoring system uses accepted quality assurance/quality control procedures and reports statistics about the quality of on-site estimates, independent validation of each estimate may be simplified. Certifying that the methodology used includes quality assurance and quality control within some specified standards may be sufficient.

An advantage of remotely sensed data is that archives of images collected over time are frequently available. This archiving enables third parties, without on-the-ground inspections, to verify land-cover status at prior points in time (to the extent that land-cover types are accurately identifiable on the imagery). Even with extensive archives of imagery, however, weather conditions (cloud cover), sensor operating status, and other factors may preclude observation of large portions of the earth at various times. Hence, there is no guarantee that imagery will be available on or near desired points in time. In addition, although remote sensing may have a role in verification, use of imagery still requires ground verification in areas where there is doubt, as well as verification or causes of perceived ARD.

Even in situations in which historical imagery of acceptable quality is available, verification will require assessment of carbon stocks, which will rely primarily on sample data taken at appropriate times and the ability to spatially reference the sample locations. When such data are not available, stock changes cannot be accurately verified. For example, a non-reported deforestation activity might be detected through the use of archived remotely sensed imagery. Yet the only recourse to estimate the carbon stock loss (in the absence of site-specific field data prior to deforestation) would be to use averages for carbon stock by forest type and/or size class. Conversely, an afforested area could be sampled to obtain forest carbon stock estimates, but previous stocks (such as soil carbon) would have to be estimated without the use of field data. Thus, stock-based forest definitions (e.g., Flexible scenario) will result in unique difficulties in verification.

With good analysis of error or uncertainty and reliable quality control procedures, verification could be limited to confirmation that methods were applied correctly. Most forest inventories include quality control and sometimes report statistics on data quality. This procedure could be extended to carbon inventory.

Establishing institutional procedures may be necessary to verify that reported estimates were made using a transparent methodology that includes quality assurance and control procedures. The methodology could be monitored (or certified) by an independent authority.

Remote sensing would be useful as one among other data sources in establishing the initial land use/land cover, to ensure that results are verifiable. It also helps in detecting, delineating, and measuring area changes and can provide objective information on whether land-use and forestry activities are human-induced.

The Kyoto Protocol anticipates that Parties will have in place national and, where appropriate, regional, forest inventory systems for annual estimation and reporting of human-induced emissions by sources and removals by sinks from ARD activities in a transparent and verifiable manner. We briefly reviewed the applicability of remote sensing and forest inventory techniques for establishing the data and for monitoring and verifying changes in ARD land and carbon stocks in a statistically reliable manner.

A National Forest Inventory system that is based on continuous forest inventory concepts is an important component for obtaining and verifying information on changes in carbon stocks. In addition, transparent and verifiable reporting of land-use changes calls for use of interdependent remote sensing techniques. Further gains in precision and reductions of costs might result from integrating successive surveys for forest area and volume in a common statistical framework. Models and special studies are needed to estimate carbon above and below ground. Finally, institutional capacity-building is vital to ensure a high degree of consistency between successive survey and interpretation procedures, as well as imagery products (sensors, season, quality, etc.).

3.5. ARD Examples on Stand, Landscape, Country, Regional, and Global Levels

This section uses case studies to demonstrate the implications of applying the seven definitional scenarios (Section 3.2) at the scales of forest stands, landscapes, countries, regions, and the globe. Most of the country studies done to date have used only two of the definitional scenarios: the IPCC and FAO scenarios.

3.5.1. Stand Level

Figure 3-2 shows a hypothetical example of the stand-level carbon dynamics of afforestation and reforestation initiated on non-forest land. The diagram shows the accumulation of carbon in trees, dead organic matter, and soils until year 30, when the stand is harvested. At that time, carbon stocks are increasing in the dead organic matter and soil pool, but the stock of carbon in trees is reset to zero and starts accumulating again as the stand regrows. The carbon stocks in dead organic matter plus soils normally increase when new forests are established on non-forest land. The example also shows that

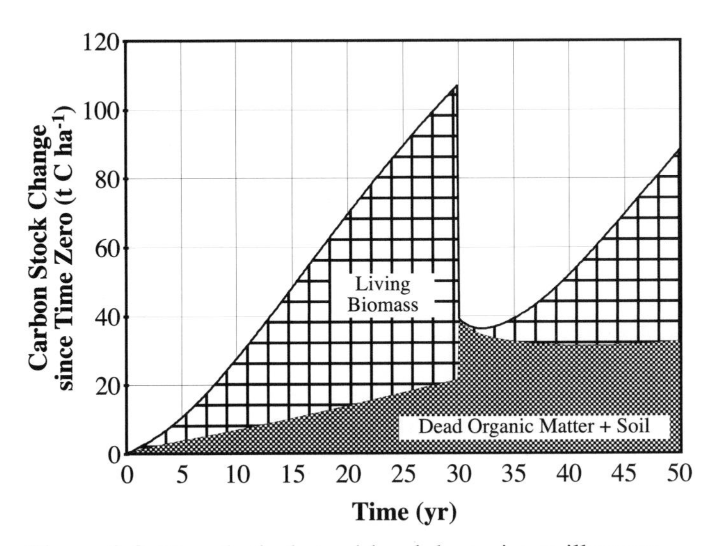

Figure 3-2: Hypothetical stand-level dynamics to illustrate carbon stock changes during reforestation or afforestation. The heavy line represents overall carbon stock change relative to initial stock at time zero. This example assumes planting of trees in an area in which dead organic matter and soil pools are low because there was no forest for some time. The GORCAM model used here is described in Marland and Schlamadinger (1999).

the harvest in year 30 produces a step increase in the dead organic matter and soil pool followed by a period of net emissions of carbon because the release from decomposing slash is greater than the carbon uptake in young trees. If the stand shown has been established in 1980 (time zero in the graph) and is harvested and replanted in 2010, this activity would qualify as reforestation in the FAO scenario.

In land-based approach I (see Section 3.3.2), one would report a carbon loss according to the top black line. The loss would be equal to -64 t C ha^{-1} yr^{-1} (carbon in 2010 [year 32] minus carbon in 2008 [year 28] = 36 - 100). In land-based approach II, the stock change would be counted beginning at the time of regeneration (i.e., year 30) and would amount to -3 t C ha^{-1} yr^{-1} (36 - 39). Under both accounting options, one would report a net loss of carbon, albeit at different rates. Only the exclusion of the decaying harvest slash from the accounting (activity-based approach) would yield a net gain in carbon (the tree growth between years 30 and 32). In the second and subsequent commitment periods, all three accounting approaches would lead to carbon credits. The time at which the carbon stock change on the land switches from negative to positive (year 32 in this example) depends on the decay rate of dead organic matter left on the site after harvest and the growth rate of the new forest.

Simplified versions of this stand-level graph demonstrate which land-cover transitions or events, combined with different definitional scenarios, will create ARD land. Figure 3-3 shows 10 cases of activities (or combinations of activities) that could qualify as ARD. For ease of calculation and demonstration purposes, stands are assumed to grow at a constant rate of 1 t C ha^{-1} yr^{-1}, and soil, woody debris, and wood product pools are excluded from the analysis. For each example, the stock changes reported under Article 3.3 are compared to real stock changes on the land.

In Table 3-9, the seven definitional scenarios are applied to the 10 cases in Figure 3-3. The first two rows illustrate whether the D-R Rule and the A/R Debit Rule come into effect. Definitional scenarios are grouped into two broad categories:

- The IPCC, Land Use, Flexible, Biome, and Aggradation/Degradation scenarios, which usually create small amounts of ARD land because ARD activities require a change in land classification (IPCC, Land Use, Flexible, and Biome scenarios) or forest management regime (Aggradation/Degradation scenario).
- The FAO and Land Cover scenarios, which include harvest-regeneration cycles.

In the FAO scenario, we distinguish three accounting rules (Section 3.3.2): land-based I and II, and activity-based. In the cases in Figure 3-3 and Table 3-9, there is no difference between land-based II and activity-based approaches because soil and dead organic matter pools are excluded. In both approaches, carbon accounting begins with the activity or in 2008, whichever is later.

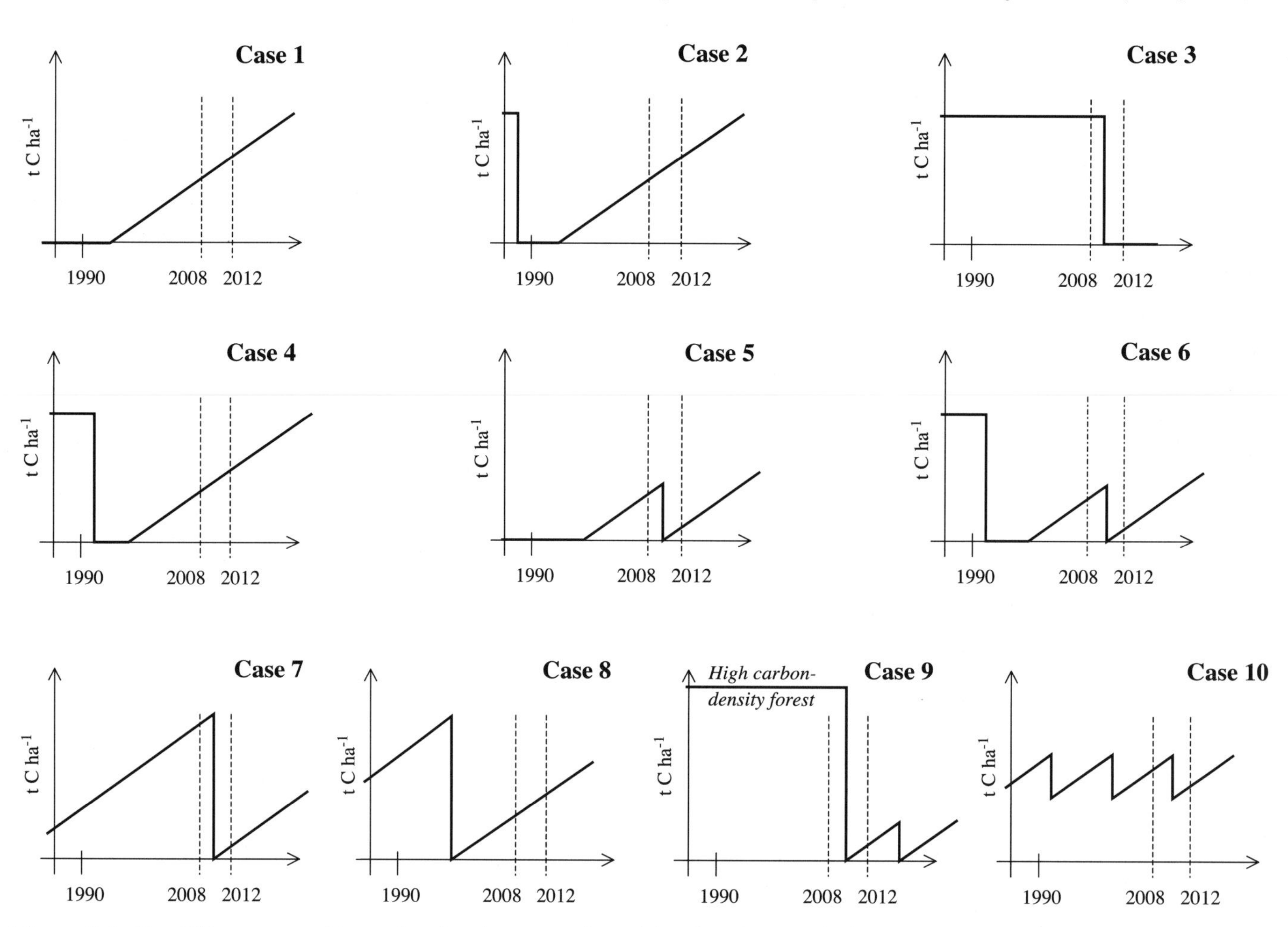

Figure 3-3: Ten different cases illustrating development of stand-level carbon stocks between 1990 and 2012. See Section 3.5.1 and Table 3-9 for descriptions and implications of the graphs.

For stands that are harvested and regenerated in the commitment period (Table 3-9, cases 7 and 9), significant debits accrue in combination with the Land Cover and FAO scenarios (FAO only with land-based approach I or II). This debiting is reflected in Table 3-9 in the FAO (land-based approach I) and Land Cover scenarios. It is not reflected in the FAO scenario with land-based approach II, however, because soil and dead organic matter are excluded from this analysis. In the FAO scenario with activity-based approach, carbon credits can be achieved as seen in Table 3-9, cases 7 and 9. Credits occur even though the stocks have decreased over the commitment period and between 1990 and 2012.

Table 3-9 shows no difference among the IPCC, Land Use, Flexible, and Biome definitional scenarios. Credits and debits in these scenarios are the same as the real 2008–2012 carbon stock changes on the land, except in cases 7 to 10; in the latter four cases, no ARD land is created in the IPCC, Land Use, Flexible, and Biome scenarios because a harvest/regeneration cycle takes place. Note that in cases 4 and 5, although there is indeed a match of real and reported carbon stock changes, the D-R Rule and A/R Debit Rule would alter the reported carbon stock changes. In case 4, this alteration removes the incentive for deforestation between 1990 and 2007, followed by reforestation to generate credits. In case 5, debits are limited to the amount previously credited (which is zero prior to the first commitment period). Thus, use of the D-R Rule and the A/R Debit Rule can prevent credits where deforestation previously lost carbon or debits where there is a long-term increase in carbon stocks through afforestation and reforestation. We note, however, that the D-R Rule will also remove the incentive for reforestation following a deforestation that occurred since 1990 or will occur in the future for reasons unrelated to the creation of carbon credits. Case 9 is of interest because it is a special situation: conversion of high carbon-density forest to plantation forest. Only the FAO scenario with stock change 2008–2012 (i.e., land-based approach I) and the Land Cover and Aggradation/Degradation scenarios manage to reflect at least some of the carbon loss that occurs at the stand level.

3.5.2. *Scaling Up from Stand to Landscape*

This section addresses the implications of the definitional scenarios introduced in Section 3.2 for reported changes in carbon stocks in the ARD land at the landscape level. We use a model that simulates the carbon dynamics in a hypothetical landscape. We simulate nine "cases" of human activities. For each case, we compare the "actual" change in carbon stocks of the landscape with that reported using the definitional scenarios.

The hypothetical landscape has an area of 150,000 ha in three land types: high- and low-productivity forest and agricultural land, each containing 50,000 ha. We assume that this landscape has been managed for some time. One percent of the forest area (i.e., 1,000 ha) has been harvested and planted annually for the past 100 years, so the forest age-class distribution of the landscape is stable. Each of 100 age-classes contains 1 percent of the forest area. We also assume that the carbon pools in agricultural land are constant.

To demonstrate the differences between the definitional scenarios, we simulate nine cases of various human activities in the landscape (Table 3-10). In all cases, we assume that the area harvested and planted prior to 1990 was as in case A. Starting in 1990, the areas affected by human activities are as described in Table 3-10 and are held constant throughout the simulation.

Case A represents a managed forest in which the rate of harvest is equal to the rate of forest growth. Thus, the standing wood volume in the landscape is in steady state. Case B is similar to A, except that 40 percent of the area harvested is allowed to reforest through natural regeneration that we assume (for the sake of illustration) does not involve DHI activity. In Case C, the rate of harvest is greater than the rate of forest growth, thus reducing wood volume. Case D is the opposite case of C: The harvest rate is less than the growth rate, allowing wood volume to increase. In both cases, the change in the harvest rate will result in a change in the age-class distribution of the forest. Cases E and F include human activities that result in degrading and aggrading forests, respectively, as a result of a change in the potential carbon stock at maturity. Activities in Case E convert high-productivity forest to low-productivity forest; in case F, low-productivity forest is converted to high-productivity forest. For these examples, the high- and low-productivity forest are in the same land-use category. In cases G and H, 100 ha yr^{-1} of each productive and degraded forest are converted to (G) or from (H) agricultural land in addition to the annual harvest of 500 ha yr^{-1} of each forest type. These changes are associated with a change in land use. Case I combines G and H: Every year 100 ha of forest land each of high- and low-productivity forest is converted to agricultural land, and 200 ha of agricultural land is converted to forest. Thus, the total forest area is constant, but it shifts in space. In cases G and I, we assume that deforestation is a random process that affects stands of all age-classes in the landscape; we represent this effect in the model by deforesting stands of the average biomass. Timber

***Table 3-9**: Characteristics of cases 1 to 10 in Figure 3-3 for each of the seven definitional scenarios. The definitional scenario FAO is split into two variants: 1) land-based I (accounting of stock changes between 2008 and 2012; see Section 3.3.2), and 2) land-based II and activity-based (accounting of stock changes between start of activity and 2012; see Section 3.3.2). For all other scenarios, this distinction makes no difference for the cases discussed here; thus, the accounting approaches are not shown separately. Changes in soil and dead organic matter carbon as well as carbon in wood products are generally not considered in this table.*

	Case 1	Case 2	Case 3	Case 4	Case 5	Case 6	Case 7	Case 8	Case 9[b]	Case 10
	Actual Change in Aboveground Carbon Stocks (t C ha^{-1})									
D-R Rule in effect	No	No	No	Yes	No	No	No	n/a	No	No
A/R Debit Rule in effect	No	No	No	No	Yes	No	No	No	No	No
Real ΔC 1990–2012	18	18	-60	-42	2.5	-57.5	-38	-38	-197	±0
Real ΔC 2008–2012	5	5	-60	5	-10	-10	-55	5	-197	-5
Definitional Scenario	*Reported C Stock Changes 2008–2012 (t C ha^{-1}) under Different Scenarios*									
FAO (land-based I)	5	5	-60	5(0)	-10(0)	-10	-55	5	-197	0
FAO (land-based II and activity-based)	5	5	-60	5(0)	-10(0)	-10	2.5	5	2.5	0
IPCC	5	5	-60	5(0)	-10(0)	-10	0	0	0	0
Land Use	5	5	-60	5(0)	-10(0)	-10	0	0	0	0
Land Cover	5[a]	5[a]	-60	5[a](0)	-10[a](0)	-10[a]	-55	5(0)	-197	0
Flexible	5	5	-60	5(0)	-10(0)	-10	0	0	0	0
Degradation/Aggradation	5	5	-60	5(0)	-10(0)	-10	0	0	-197	0
Biome	5	5	-60	5(0)	-10(0)	-10	0	0	0	0

Notes: Values in parentheses indicate that Deforestation-Reforestation Rule (D-R Rule)and Afforestation/Reforestation Debit Rule (A/R Debit Rule) considered; n/a = not applicable.

[a] In Land Cover scenario, a forest is only created—thus, a piece of land considered forested—if the land cover has reached the threshold. The longer the time between establishment of the stand and a commitment period, the more likely that ARD land would have been created.

[b] This scenario involves conversion of high carbon-density forest to plantation forest (could be referred to as degradation), with a reduction of biomass at maturity from 200 to 100 t C ha^{-1}.

Table 3-10: *Nine cases of different combinations of human activities operating in a hypothetical landscape that includes 1,500 land parcels of 100 ha each—with 500 parcels in productive forest, 500 in degraded forest, and 500 in agriculture. See Sections 3.5.2.4 and 3.5.2.5 for details of simulation for cases A to I.*

	Productive Forest		**Degraded Forest**		**Agricultural Land**		
Case	Harvest (ha yr^{-1})	Regenerate (ha yr^{-1})	Harvest (ha yr^{-1})	Regenerate (ha yr^{-1})	Add (ha yr^{-1})	Remove (ha yr^{-1})	**Comment**
A	500	500	500	500	0	0	Steady-state forest
B	500	300 P 200 N	500	300 P 200 N	0	0	Steady state [P = planted, N = natural regeneration]
C	600	600	600	600	0	0	Increased harvest/regeneration
D	400	400	400	400	0	0	Decreased harvest/regeneration
E	500	300	500	700	0	0	Degrading forest
F	500	700	500	300	0	0	Aggrading forest
G	500 defor 100	500	500 defor 100	500	200	0	Harvest/regeneration and land-use change to agriculture
H	500	500 affor 100	500	500 affor 100	0	200	Harvest/regeneration and land-use change from agriculture
I	500 defor 100	500 affor 100	500 defor 100	500 affor 100	200	200	Harvest/regeneration and land-use change to agriculture and afforestation

harvesting in the model affects the oldest stands with the highest biomass.

Table 3-11 summarizes, for the seven definitional scenarios, the activities that create ARD land. For this purpose, the definitional scenarios are divided into three broad groups: scenarios that consider forest change only, scenarios in which the harvest/regeneration cycle creates ARD land, and scenarios in which forest degradation or aggradation create ARD land.

3.5.2.1. Scenarios that Consider Forest Change Only

In the IPCC, Land Use, Flexible, and Biome definitional scenarios, the harvest/regeneration cycle does not create ARD land. ARD land is created only by changes between forest and non-forest categories.

We assume that the definitions of forest in the IPCC, Land Use, Flexible, and Biome scenarios are such that mature, unharvested stands in both the high- and low-productivity forest are considered forest. For the Land Use scenario, we assume that the observed change in forest cover is associated with an administrative reclassification of land use. Table 3-11 summarizes the human activities that create ARD land for each of the three groups of definitional scenarios.

3.5.2.2. Scenarios in Which the Harvest/Regeneration Cycle Creates ARD Land

In the FAO and Land Cover definitional scenarios, ARD land is created through harvest/regeneration cycles, albeit for different reasons. In the FAO scenario, harvest does not create ARD land; subsequent planting of forests in these areas creates ARD land. Under the FAO scenario, naturally regenerated areas do not create ARD land because, for illustrative purposes, we assume that natural regeneration was not considered DHI. In both scenarios, harvested areas that undergo land-use change are considered deforested and create ARD land.

In the Land Cover scenario, harvest causes the stand to cross the threshold of forest cover that is used to define forest. Because any area below the threshold is considered non-forest, harvest is considered deforestation and creates ARD land. Thereafter, it does not matter whether a stand is established through natural regeneration or through planting because the harvested area is already classified as ARD land.

The definition of reforestation in the Land Cover scenario specifies that a planted area becomes ARD land when the stand passes the threshold of forest cover that defines forest (40 percent cover in our example). This definition of reforestation will cause a delay between the planting activity and the creation of ARD land. It ensures that no carbon credit or debit

***Table 3-11**: Summary of human activities that create ARD land under conditions of groups of definitional scenarios. Cells marked Y indicate where ARD land is created.*

Activity	**Scenarios in which Forest Change Creates ARD Land** (IPCC, Land Use, Flexible, Biome)	**Scenarios in which Harvest/Regeneration Cycle Creates ARD Land** FAO	Land Cover	**Scenario** Degradation/ Aggradation	**Comment/Reason**
Harvest			Y		Cover passes forest threshold
Natural regeneration					Assumed to be not direct human-induced (for sake of illustration)
Replanting		Y			Any active reforestation
Replanting and grow past forest threshold			Y		Area may already be ARD land because of prior harvest
Change potential carbon at maturity				Y	Degrading or aggrading forest
Land-use change: deforestation	Y	Y	Y	Y	Any deforestation related to land-use change
Afforestation: plant	Y	Y		Y	Establishment of forest
Afforestation: grow past forest threshold			Y		Cover passes forest threshold

is obtained until the planted area passes the threshold defining a forest.

3.5.2.3. Scenarios in Which Forest Degradation or Aggradation Create ARD Land

Only in the Degradation/Aggradation definitional scenario is ARD land created when human activities aggrade or degrade forests (i.e., they change the potential carbon stock at maturity). In this scenario, human activities that result in land-use change also create ARD land. Forest harvesting followed by planting does not create ARD land if it does not alter the potential carbon stock at maturity.

3.5.2.4. Land- and Activity-Based Carbon Accounting Rules

Section 3.3.2 introduces three possible approaches to carbon accounting:

1) A land-based approach in which an activity creates ARD land and the carbon stock change on that land is reported for the entire commitment period (land-based approach I).
2) A land-based approach in which an activity creates ARD land and, if the activity occurred during the commitment period, the carbon stock change on that land is reported from the start of the activity to the end of the commitment period (land-based approach II).
3) An activity-based approach in which each activity is assigned a carbon stock change, and the reported carbon stock change of a landscape is calculated as the area of each activity times the stock change assigned to each activity. Only stock changes resulting from an ARD activity are accounted. There can be carbon stock changes on ARD land that do not result from an ARD activity and therefore are not accounted. This approach introduces a series of complications—such as multiple activities on the same land area, verification of ARD land area, verification of carbon stock changes, and so forth. This approach could be implemented without reference to a specific land area, but such implementation would aggravate such complications.

For the landscape-level analyses, we apply the carbon accounting rules of the three foregoing options to the FAO definitional scenario in which reforestation (and not harvest) creates ARD land. The difference between the three accounting rules becomes obvious when reforestation follows harvest during the commitment period. Under option 1, the carbon stock change from the start to the end of the commitment period is reported, thereby including the stock change effect of harvesting even though the subsequent reforestation creates ARD land. Under

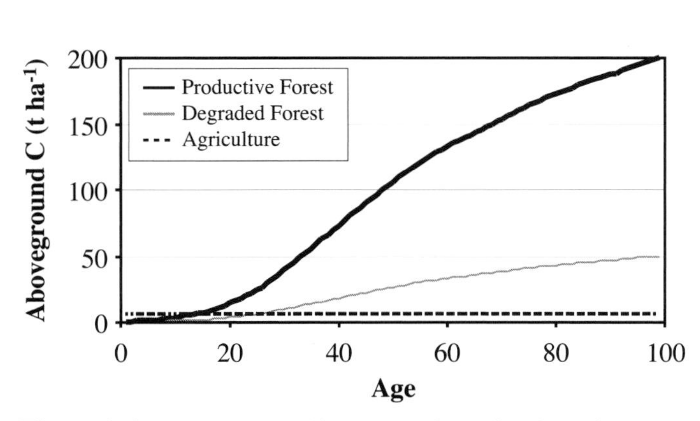

Figure 3-4: Aboveground biomass carbon simulated in the model as a function of age for productive and degraded forest and for agricultural land.

option 2, the carbon stock change from reforestation to the end of the commitment period is reported, including the loss of post-harvest slash. This reporting period includes less than the five-year commitment period. Under option 3, the carbon stock change is also reported from the reforestation to the end of the commitment period, but the carbon release from harvest slash and dead roots is not accounted because this loss does not result from the reforestation activity. This carbon accounting approach is approximated here by estimates of aboveground biomass changes obtained from option 2 and by ignoring all changes in soil and dead organic matter pools.

To conduct the quantitative analyses with the hypothetical landscape, we developed an analytical framework that employs a very simple model of ecosystem carbon dynamics and applies the definitions and accounting rules of the seven scenarios. The ecosystem model is driven by three curves that describe the aboveground biomass dynamics as a function of age for the high- and low-productivity forests and the agricultural land (Figure 3-4). These curves are applied to a landscape that contains 1,500 parcels of 100 ha each. Note that in the model, the agricultural land contains 5 t C ha^{-1} in aboveground biomass. A set of companion curves defines canopy cover as a function of stand age for the two forest types.

In the model, below-ground (i.e., living root) biomass is calculated as a proportion of aboveground biomass. Each year a proportion of the aboveground and below-ground biomass is transferred to a single dead organic matter and soil carbon pool. Carbon is released from this pool through decomposition. Harvest transfers carbon from biomass to the dead organic matter pool and to a pool of harvested material (which is not included in this model). The model accounts for annual growth and decomposition and for changes in cover type and land use associated with human activities.

3.5.2.5. *Landscape-Level Results*

The numerical values selected for all aspects of carbon dynamics in the hypothetical landscape are comparable to those of a temperate ecosystem. The actual values are not important here, nor are the absolute numerical values of the results. The purpose of these examples is to demonstrate qualitatively how, for each definitional scenario, the reported carbon stock changes in the ARD land portion of the landscape compare to the actual carbon stock changes in the simulated landscape. Although the simulations extend over several commitment periods, for clarity

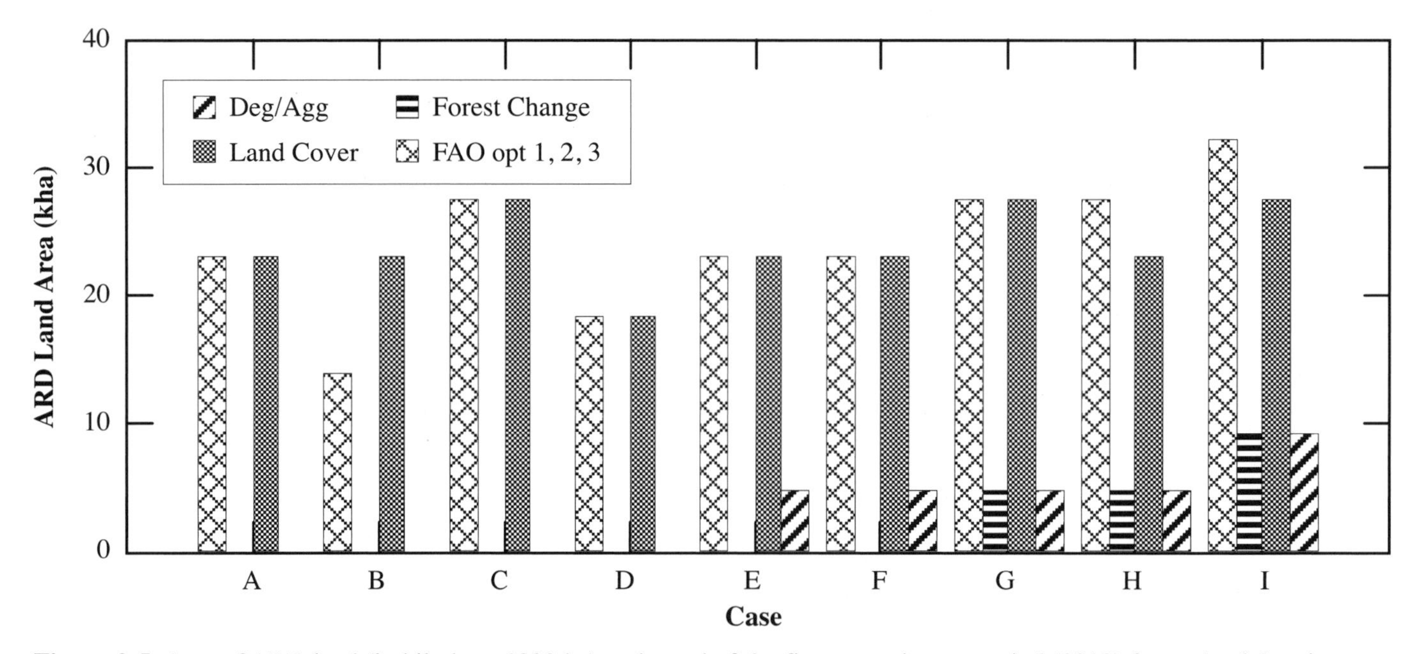

Figure 3-5: Area of ARD land (in kilo ha = 1000 ha) at the end of the first commitment period (2012) for each of the nine cases (A–I) and for various definitional scenarios: FAO, Forest Change [includes all scenarios requiring a forest/non-forest conversion (i.e., IPCC, Land Use, Flexible, and Biome)], Land Cover, and Degradation/Aggradation (Deg/Agg). Missing bars indicate zero area. FAO options 1, 2, 3 refers to the three approaches for accounting carbon stock changes.

only the results of the first commitment period are reported here.

The area of ARD land in each commitment period is a function of the human activities simulated in each of the nine cases, as well as the definitional scenarios that specify which activities create ARD land (see Table 3-10). In all cases, the area of ARD land in the first commitment period is much less than the total area in the landscape. For cases A through F, no ARD land is created for the definitional scenarios that require a forest change (Figure 3-5). For these scenarios, only cases G, H, and I (i.e., those with a true forest/non-forest change) create ARD land at the end of the first commitment period. The Degradation/Aggradation definitional scenario also reports ARD land for cases E and F, in which the potential forest cover at maturity is changed for some areas.

The definitional scenarios in which a harvest/regeneration cycle creates ARD land (i.e., FAO and Land Cover) report large areas of ARD land. In case A, the area of ARD land is 23 kha for both the FAO and the Land Cover scenarios. In case B, the area naturally regenerated in the FAO scenario does not create ARD land. In cases H and I, the afforested area is not included in the Land Cover scenario because the planted areas have not grown past the forest cover threshold at the end of the first commitment period. In case I, all deforested and afforested area contributes 9.2 kha of ARD land at the end of the first commitment period—except in the Land Cover scenario, in which the afforested area again has not yet grown past the forest cover threshold.

Figure 3-6 summarizes the results for the nine cases defined in Table 3-10 for the first commitment period and for three carbon pools: aboveground biomass; dead organic matter, including soil carbon; and total ecosystem carbon, which is the sum of aboveground and below-ground biomass and dead organic matter. Below-ground biomass is not shown separately because it is proportional to aboveground biomass. All carbon stock changes are reported in kilo tons over the 5-year commitment period. The actual stock change is that observed in the hypothetical landscape.

In most combinations of definitional scenarios and simulated cases, Figure 3-6 shows large discrepancies between the actual carbon stock change in the landscape and that reported for the ARD land. In cases A and B, the actual carbon stock change in the landscape is zero. The scenarios that limit the creation of ARD land to only forest change activities report no ARD land, therefore no carbon stock changes for cases A–F. These definitional scenarios fail to account for increases (case D) or decreases (case C) in actual carbon stocks resulting from accelerated or decelerated harvesting rates and from changes in forest productivity (cases E and F). These scenarios do capture the land-use change activities of cases G, H, and I and represent the correct carbon stock increase in case H, as well as the decrease in cases G and I.

Definitional scenarios in which the harvest/regeneration cycle creates ARD land (FAO, Land Cover) report a large decrease in carbon stock in the ARD land, except with an activity-based accounting approach in the FAO scenario. In cases in which the actual landscape carbon stock increases (cases D, F, and H) or decreases (cases C, E, G, and I), the reported stock change becomes somewhat larger or smaller relative to case A, respectively. Despite these small differences between cases, the reported change is always negative, even if the actual carbon stock increases in the landscape (e.g., cases D and F).

The choice between the three carbon accounting approaches (land-based I and II and activity-based) for the FAO definitional scenario does not affect the area of ARD land. The approaches do report different carbon stock changes for the ARD land, however. The stock change in land-based approach I is reported for the 5-year commitment period. For land-based approach II, the stock change on the land is reported from the time of the activity to the end of the commitment period. In the activity-based approach, stock changes that do not result from the reforestation activity but happen on the land after reforestation has taken place are excluded from the accounting.

The impacts of the three accounting approaches differ between carbon pools. For aboveground biomass, the FAO definitional scenario with land-based approach I (FAO opt 1 in Figure 3-6) always yields a large negative carbon stock change because this option indirectly accounts for harvesting during the commitment period. Land-based approach II (FAO opt 2) reports a small positive carbon stock change in aboveground biomass, resulting from the growth of reforested seedlings. This increase in aboveground biomass is more than offset by decreases in dead organic matter carbon stocks. Land-based approach II does not account for the increase in dead organic matter from the input of logging slash and root biomass during harvest. It does account for the decay of all dead organic matter from the reforestation activity to the end of the commitment period. Thus, land-based approach II reports a large negative carbon stock change for the dead organic matter pools. Total ecosystem carbon stock change under land-based approach II is negative—albeit only about one-third of the values reported for land-based approach I. The results for the activity-based approach are the same as for land-based approach II, except that the negative stock change from decaying slash between reforestation and 2012 is excluded from the accounting because it does not result from the reforestation activity. Therefore, the FAO scenario with activity-based accounting results in carbon credits in the first commitment period; that is, only the aboveground biomass increase under FAO (option 2) would be reported.

The definitional scenario that is designed to capture degrading or aggrading carbon stocks (the Degradation/Aggradation scenario) correctly reports no change in carbon stocks in cases A and B. By design, it does not capture carbon stock changes associated with altered harvest rates (cases C and D) because they do not change the potential cover of carbon density at maturity. This scenario fails to accurately report the actual carbon stock changes in cases E and F, which represent degrading and aggrading forest conditions. As before, the changes at the

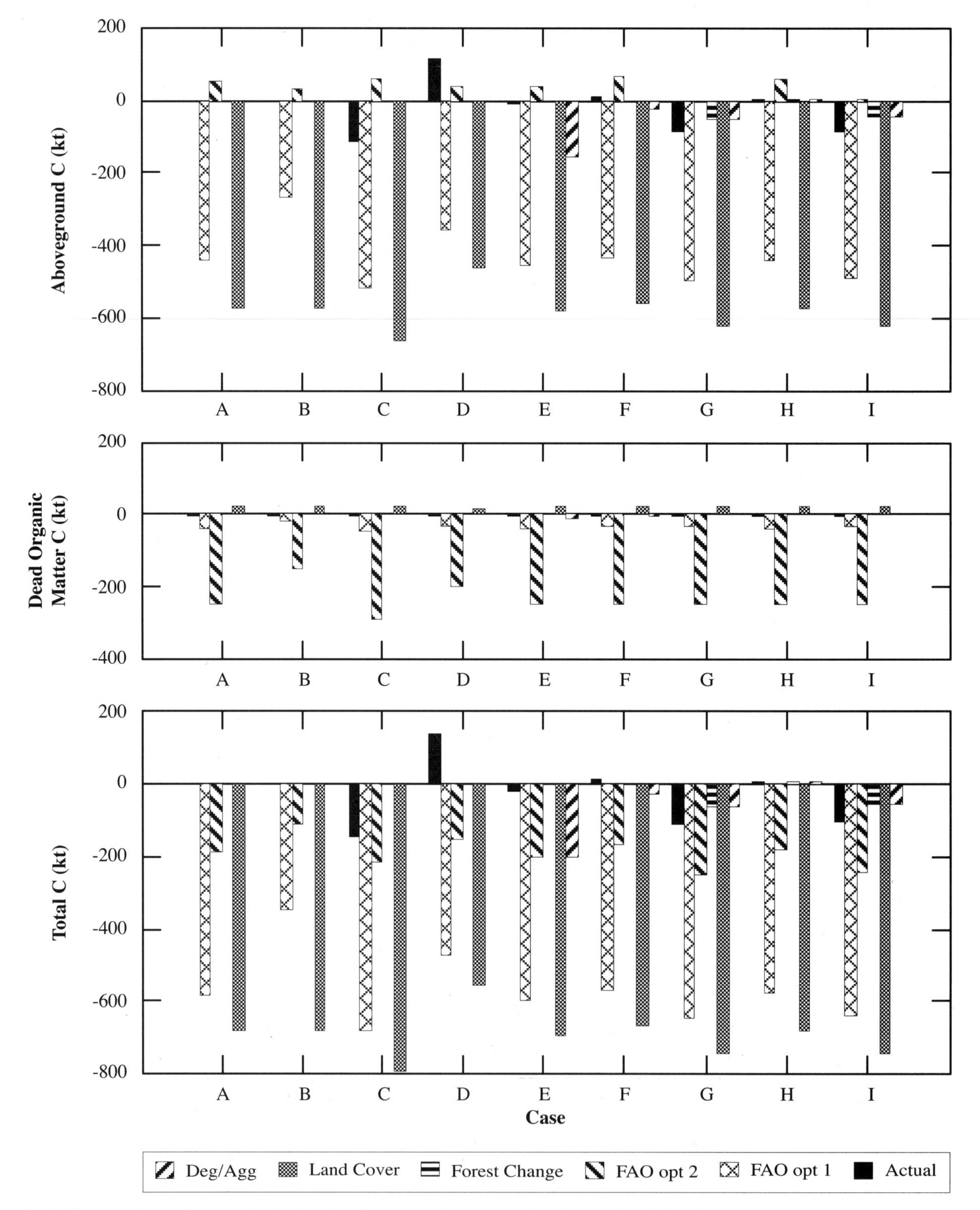

Figure 3-6: Changes in carbon stock during the first commitment period for each of the nine simulated cases of human activities for aboveground (top), dead organic matter including soil (middle), and total ecosystem (bottom) carbon. "Actual" illustrates simulated carbon stock change in the landscape. The other bars represent carbon stock change reported for various definitional scenarios: FAO, Forest Change [includes all scenarios requiring a forest/non-forest conversion (i.e., IPCC, Land Use, Flexible, and Biome), Land Cover, and Degradation/Aggradation (Deg/Agg). Missing bars indicate that no ARD land was created. FAO opt 1 and 2 are the land-based accounting approaches I and II, respectively, described in the text. The results for an activity-based accounting approach used for the FAO scenario are similar to the results reported for FAO opt 2, but reporting only the changes in aboveground biomass.

landscape level are different from those reported for the ARD land. Moreover, even when the carbon stock at the landscape level is increasing (case F), the reported carbon stock change is still negative.

The decision about which pools to include will also affect the differences between the actual carbon stock changes in the landscape and those reported for the ARD land. For aboveground biomass, below-ground biomass (not shown), and total ecosystem carbon, the differences are in the same direction but of different magnitude (Figure 3-6). The differences can be of opposite direction for the dead organic matter pool alone, but it is not likely that carbon stock changes for only this pool will be reported. For definitional scenarios in which the harvest/regeneration cycle creates ARD land, the dead organic matter pool can increase when biomass pools are decreasing because logging slash—including stumps and root biomass—is added to the dead organic matter pool following harvesting.

We have conducted additional simulation experiments as sensitivity analyses. We systematically altered the rates of land-use change and harvest to cause landscape-level carbon stocks to either increase, remain approximately unchanged, or decrease. It can be demonstrated that combining one rate of land-use change with three different rates of harvesting, or *vice versa*, creates very different reported carbon stock changes for the three groups of definitional scenarios.

We drew the following conclusions from the analyses of the simulations:

- The area of ARD land in the first commitment period is much less than the total area in the landscape. The area is greater in definitional scenarios in which forest harvesting creates ARD land.
- Definitional scenarios that limit ARD land to areas with forest/non-forest change and scenarios in which the harvest/regeneration cycle does not create ARD land (IPCC, Land Use, Flexible, and Biome) report very similar amounts of carbon stock changes for the same amount of deforestation. These definitional scenarios are insensitive, however, to actual changes in the landscape-level carbon stock. Thus, although the same carbon stock change is reported for the ARD land area, the actual carbon stock change in the landscape could be positive, unchanged, or negative.
- Definitional scenarios in which a harvest/regeneration cycle creates ARD land (FAO and Land Cover) report changes in the ARD land in the early commitment periods that are much more negative than the actual changes in the landscape because regrowth on lands harvested prior to 1990 is not counted. This conclusion holds for land-based accounting approaches I and II but not for the activity-based approach: In some cases the activity-based approach does report an increase in aboveground biomass even if the actual carbon stock change is zero or negative. In the FAO and Land Cover scenarios, the rates of harvest and deforestation affect the reported carbon stock change. Thus, with the same deforestation rate, the reported change in carbon stocks becomes more negative with increasing harvest rate.
- If the intent is to report carbon stock changes that are similar to the actual carbon stock changes in a landscape, only full carbon accounting can achieve this objective. If the intent is to best account for carbon stock changes in the ARD land area, a definitional scenario that only includes forest/non-forest transitions is best suited.
- Differences between actual and reported carbon stock changes can be reduced further if forest management activities that alter landscape-level carbon stock are included using the mechanisms defined in Article 3.4.

3.5.3. Country- and Regional-Level Studies

Case studies are employed to demonstrate how the mix of ARD activities and the various definitional scenarios interact to affect different countries. The case studies are drawn mainly from the literature, given the time constraints of this report. In the studies available, only IPCC and FAO definitions for ARD activities have been applied, rather than all of the definitional scenarios in Section 3.2.

Available country case studies show that if definitions of ARD include the harvest/regeneration cycle, the reported carbon stock change for reforestation in the commitment period will be negative for many Annex I countries. This negative change results from limiting credits for regrowth to areas reforested or afforested since 1990 (Table 3-12). Carbon debits, though limited to lands cleared during the commitment period, account for the loss of a large portion of the carbon accumulated during the lifetime of the stand. Debits will exceed credits, except in regions with very quickly maturing forests. If the stock change should be calculated from the onset of the activity (i.e., from reforestation), then under the FAO scenario some countries could get credits for biomass from regrowing areas, depending on how dead organic matter and soil carbon pools are handled.

Using the European Union (EU) as an example, woody biomass stock increased by 53 Tg C yr^{-1} in 1990. Reforestation in the FAO scenario (with land-based accounting approach I) provides the EU (Liski *et al.*, 2000) with a source of 5.5 Tg C yr^{-1} because of negative stock change in the reforested area (regrowth of 12 Tg C yr^{-1} and loss of 17.5 Tg C yr^{-1} in clear-cut harvesting). The FAO scenario generates carbon emissions for countries where the afforestation area is relatively small, large areas are reforested, and regrowth is relatively slow (e.g., Austria, Finland, Germany, Italy, Sweden, and Norway). The same scenario produces net carbon sinks in countries where the afforestation area is relatively large and regrowth on reforested areas is relatively fast, or rotation cycles are shorter (e.g., Ireland, Portugal, and the United Kingdom). In Table 3-12, afforestation under the FAO and IPCC scenarios provided a 1–2 percent reduction in a country's 1990 CO_2 emissions—

Table 3-12*: Country and regional ARD case studies applying FAO and IPCC definitional scenarios.*

Country or Region	Potential for ARD Credits and Debits for Selected Annex I Countries[a] in First Commitment Period (Mt C yr^{-1}) FAO Scenario Land-Based	Activity-Based	IPCC Scenario	Remarks	1990 Fossil Fuel CO_2 Emissions[b] (Mt C yr^{-1})	1990 LUCF Net Removal[b] (Mt C yr^{-1})	Reference
Australia				Different scenarios were used to generate possible patterns of AR and D from 1999 to the end of the commitment period	-79	-24	[c] Kirschbaum, 2000 [d] Hamilton and Vellen,1999 (Nabuurs *et al.*, 2000)
AR		*(0.9)*	0.6-7[c] *(0.1)*				
D		*(-13.6)*	-6 to -14[d] *(-13.6)*				
D in 1990 baseline	-24.5[d]						
AR in 1990 baseline	n/a						
Austria				A and R according to IPCC definitions cannot be separated by the Austrian forest inventory	-17	3.6	National data submitted after the Government Review
A	0.34		0.34				
R	-0.92						
D	-0.84		-0.84				
Canada					-126	12	Robinson *et al.*, 1999 (Nabuurs *et al.*, 2000)
A		*(0.14)*	*(0.01)*				
R		*(5.2)*	*(0.08)*				
D	2.6–3.8	2.6-3.8	2.6-3.8				
EU 15				ARD based on past trends	-907	53	Liski *et al.*, 2000 (Nabuurs *et al.*, 2000)
A	3.6	3.6 *(8)*	3.6 *(0.44)*				
R	-5.5	12.0 *(16)*	incl. under 'A'				
D	-3.5	-3.5 *(-2.2)*	-3.5 *(-2.2)*				
Finland				ARD based on past trends	-15	8	[e] Karjalainen *et al.*, 1999 [f] Mäkipää and Tomppo, 1998
A	0.12[e]	0.12[f]	0.12[f]				
R	-2.79[e]	2.30[f]					
D	-0.48[e]	-0.48[f]	-0.48[f]				
Japan					-315	26	(Nabuurs *et al.*, 2000)
A		*(0.24)*	*(0.002)*				
R		*(11.6)*	*(0.16)*				
D		*(-0.3)*	*(-0.3)*				
Netherlands					-46	0.5	(Nabuurs *et al.*, 2000)
A		*(0.04)*	*(0.0)*				
R		*(0.06)*	*(0.004)*				
D		*(-0.02)*	*(-0.02)*				

Table 3-12 (continued)

Country or Region	Potential for ARD Credits and Debits for Selected Annex I Countries[a] in First Commitment Period (Mt C yr^{-1}) FAO Scenario Land-Based	Activity-Based	IPCC Scenario	Remarks	1990 Fossil Fuel CO_2 Emissions[b] (Mt C yr^{-1})	1990 LUCF Net Removal[b] (Mt C yr^{-1})	Reference
New Zealand		n/a		2.7, assuming no new planting after 1997; 7.5, assuming 55 kha yr^{-1} AR; 10.3, assuming 90 kha yr^{-1} AR	-7	6	Ford-Robertson, *et al.*, 1999
A	2.7, 7.5, 10.3						
R	n/a		2.7, 7.5, 10.3				
D	n/a		n/a				
Norway				ARD based on past trends	-10	3	Karjalainen *et al.*, 1999
A	0.04	0.04	0.04				
R	-1.30	0.07					
D	-0.07	-0.07	-0.07				
Russia				It was assumed that in Russia large areas of agricultural lands have been abandoned since 1990, leading to large areas for afforestation	-647	107	Izrael and Avdjushin, 1999 (Nabuurs *et al.*, 2000)
A	1.5	1.5 *(2.3)*	1.5 *(0.12)*				
R	n/a	2.3-2.5 *(14)*	incl. under 'A'				
D	n/a	n/a *(-60.8)*	n/a *(-60.8)*				
Spain					-62		National data submitted after Government Review
A	0.04		0.04				
R	0.09		n/a				
D	—		—				
Sweden				ARD based on past trends	-17	9	Karjalainen *et al.*, 1999 (Nabuurs *et al.*, 2000)
A	0.32	0.32 *(1.7)*	0.32 *(0.006)*				
R	-4.53	4.03 *(3)*	*(0.04)*				
D	-0.43	-0.43 *(-0.02)*	-0.43 *(-0.02)*				
USA					-1,352	117	(Nabuurs *et al.*, 2000)
A		*(36)*	*(0.4)*				
R		*(68)*	*(8)*				
D		*(-12.4)*	*(-12.4)*				

Notes: Projected ARD credits and debits in first commitment period measured as Mt C yr^{-1} in tree biomass. Negative values are emissions, and positive values removals. n/a = not applicable. Values in brackets are from Nabuur *et al.*, 2000.
[a]Major Annex I forest nations (EU treated as one entity).
[b]As reported in the second national communications to UNFCCC. The term "LUCF" is used in the second national communications.

except in New Zealand, where reductions were 40–150 percent because of the low population density and substantial afforestation programs. In contrast, reforestation under the FAO scenario provided an additional debit to 1990 CO_2 emissions of up to 27 percent for countries with large reforested areas and relatively slow regrowth when stock changes are calculated for the entire commitment period. The same countries earned a credit of up to 25 percent of 1990 CO_2 emissions when stock changes were calculated from the reforestation event to the end of the commitment period (i.e., land-based accounting approach II or activity-based accounting approach). These estimates, however, have excluded possible changes in soil and dead organic matter carbon. Including these pools will change the results, as demonstrated by the stand and landscape-level examples in

Sections 3.5.1 and 3.5.2. Deforestation under the FAO and IPCC scenarios contributes debits of up to 3 percent of 1990 CO_2 emissions. Available country-level data indicate the following:

- The choice of definitions for forest and ARD affects the proportion of the land on which carbon stock changes must be reported; if harvest/regeneration cycles are included in ARD lands, the amount could increase greatly.
- Reported changes in carbon stocks for ARD land do not necessarily represent stock changes in the whole country. A country could be losing forest, and thus reporting debits under Article 3.3, even though the managed forests could be a carbon sink.

***Table 3-13**: Anthropogenic CO_2 emissions and removals (Mt C yr^{-1}) from the LUCF sector in 1990 for selected Annex I countries (based on the second national communications to UNFCCC). Removals (sink) are positive and emissions (source) are negative values.*

		Categories				
Country	**Forest Area (Mha)**	A: Changes in Forest and Other Woody Biomass Stocks (Mt C)[a]	B: Forest and Grassland Conversion (Mt C)[b]	C: Abandonment of Managed Lands (Mt C)[c]	D: Other (Mt C)	**Total Net Removal (Mt C)**
Australia	146[f]	6.3	-32.1	NR	2.2[g]	-23.6
Canada[d]	418	13.6	-0.3	0.8	-1.1[h]	12.0[i]
European Community		51.9	-11.6	2.9	7.9	51.1
Germany	10.7	9.2				9.2
Japan	25.3[f]	23.0	-0.2		-0.1	22.7
New Zealand	7.9	6.0	-0.3			5.7
Norway	9.6[f]	j	j	j	j	2.6[j]
Poland	8.8	9.4	-0.2	3.0	NR	12.2
Russian Federation	763.5	106.9				106.9
United Kingdom[e]	2.4	2.6		k	-8.4[l]	-5.8
United States	295					+125[m]/+311.5

Notes: NR = not reported information. The term "LUCF" is used in the second national communications.
[a] Includes changes in stock by forest management in existing commercial forest (harvest, thinning, and restocking) plus managed forests resulting from afforestation and reforestation.
[b] Includes conversion of existing forest and natural grassland into other land use (e.g., agriculture). There is no information on forests alone. These data may be considered a rough approximation for deforestation under all definitional scenarios.
[c] Abandoned cropland and pastures. These data may approximate afforestation and reforestation on cropland and pastures based on natural regeneration and without future management.
[d] Based on GHG Inventory submission, 1999.
[e] Categories 5B and 5C have been combined.
[f] From UN-ECE/FAO Forest Resource Assessment (1999).
[g] Pasture improvements.
[h] Emissions and removals from soil.
[i] Numbers in UNFCCC database match.
[j] Not estimated according to IPCC draft reporting instructions (net removal based on increment, harvesting, and natural decay projections for 1990).
[k] Included elsewhere.
[l] Emissions of CO_2 from wetland drainage and peat extraction.
[m] The first figure is interpolated from forest inventories in 1987 and 1992. The calculation method reflects long-term averages, rather than specific events in any given year. The second figure is from GHG Inventory submission, 1999.

- Even when afforestation and reforestation areas exceed deforestation areas, a country will not necessarily report credits. This finding applies to all scenarios.

Table 3-13 illustrates the magnitude of ARD activities derived from the national communications of some Annex I countries. Further methodological work clearly is needed to assure that reported inventory data are consistent, transparent, and comparable and that the categories requested by the Kyoto Protocol can be distinguished. Chapter 6 provides a detailed description how the IPCC Guidelines fit into the reporting requirements of the Kyoto Protocol.

Information needed to judge ARD activities cannot be derived from national communications prepared before the Kyoto Protocol was adopted. Additional difficulties arise because each nation uses either its own scheme (neglecting IPCC categories) or its own definitions. Most nations do not report deforestation.

***Table 3-14**: Average annual area estimates for regeneration of forest[a], extension of forest, and natural colonization of forest and other wooded land [from TBFRA 2000 (UN, 1999)], and loss to other uses [from FRA 1990 (UN-ECE/FAO, 1992)] for specified Annex I countries. Extension of forest (including afforestation and reforestation of other wooded land) refers to late 1980s and early 1990s, and is used to approximate the area of afforestation and reforestation under the IPCC definitional scenario and the area of afforestation under the FAO definitional scenario. Regeneration refers to late 1980s and early 1990s, and is used to approximate the area of reforestation under the FAO definitional scenario. Loss to other uses refers to 1980s, and is used to approximate the area of deforestation.*

Country	Extension of Forest in Late 1980s and Early 1990s (1000 ha yr^{-1})	Regeneration of Forest in Late 1980s and Early 1990s (1000 ha yr^{-1})	Loss to Other Uses in 1980s (1000 ha yr^{-1})
Finland	21	167	9.5
France	10	107	60
Germany	7	70	4.2[b]
Ireland	17	4	
Norway	31	47	
Poland	10	59	3.4
Sweden	2	204	
United Kingdom	23	15	0.4
Europe	283	1,192	103
EU	143	850	80
Russian Federation	50[c]	2,026[d]	450[e]
Belarus	1	26	9
Ukraine	14	39	4
Canada		693	60[f]
United States	187[g]	4,372	506
Australia			550[h]
Japan		170	
New Zealand	53	20	
Boreal TBFRA[i]	110	3,137	520
Temperate TBFRA[j]	488	5,402	1,163
All TBFRA	598	8,538	1,683

[a] Regeneration defined as the reestablishment of a forest stand by natural or artificial means following the removal of the previous stand by felling or as a result of natural causes (e.g., fire or storm).
[b] For the Federal Republic of Germany.
[c] Annual afforestation in the Russian Federation has heavily declined between 1988 and 1996, and the number given is an average value for this time period (RFFS, 1998).
[d] Of this regeneration, 0.5–0.6 Mha were planted annually until about 1990. However, the amount of forest planted annually has since then decreased to about 0.25–0.3 Mha yr^{-1} (Pisarenko *et al.*, 1992; RFFS, 1998).
[e] A. Shvidenko (pers. comm.).
[f] Robinson *et al.* (1999).
[g] The "Extension of Forest" area for the United States of America given in UN (1999) is 1.8 Mha yr^{-1}. However, this number is incorrect and has been replaced with a corrected number of 0.2 Mha yr^{-1} (U.S. Forest Service, 2000a) in Tables 3-14 and 3-17, and in Table 3 of the SPM. After completion of these tables, notification arrived that this number has been further amended to 0.39 Mha yr^{-1} (U.S. Forest Service, 2000b), but has not been considered in these tables.
[h] The data for Australia are based on NGGI (1999), and give the average over the decade from 1981–90.
[i] Canada, Finland, Iceland, Norway, Russian Federation, and Sweden.
[j] Austria, Belarus, Belgium, Bulgaria, Croatia, Czech Republic, Denmark, Estonia, France, Germany, Greece (no data), Hungary, Ireland, Italy, Latvia, Liechtenstein, Lithuania, Luxembourg (no data), The Netherlands, Poland, Portugal, Romania (no data), Slovakia, Slovenia, Spain (no data), Switzerland, Turkey, Ukraine, United Kingdom, United States of America, Australia (no data), Japan, and New Zealand.

Table 3-15: Pan-tropical area transition matrix for 1980–1990 [FAO, 1996; FRA 1990 (UN-ECE/FAO, 1992); Survey of Tropical Forest Cover and Study of Change Processes, Forestry Paper 130, FAO, Rome].

Land Cover Classes in 1980	Land Cover Classes in 1990 (Mha)										
	Closed Forest	Open Forest	Long Fallow	Fragmented Forest	Shrubs	Short Fallow	Other Land Cover	Water	Plantations	***Total 1980 (Mha)***	***Total 1980 (%)***
Closed forest	1275.91	8.97	9.27	9.17	2.53	21.57	34.79	1.78	3.95	***1367.96***	***44.6***
Open forest	0.86	283.31	1.3	5.18	1.46	2.4	10.18	0.05	0.21	***304.94***	***9.9***
Long fallow	1.1	0.26	48.61	1.08	0.79	2.35	2.27	0.05	0.01	***56.54***	***1.8***
Fragmented forest	0.58	0.63	0.63	159.33	0.45	1.41	11.4	0.25	0.39	***175.06***	***5.7***
Shrubs	0.15	0.2	0.26	0.14	152.6	0.34	19.17	0.19	0.15	***173.3***	***5.6***
Short fallow	0.56	0.29	0.46	0.39	0.16	119.79	7.3	0.19	0.17	***129.31***	***4.2***
Other land cover	0.71	0.7	0.26	1.35	1.94	2.03	834.23	1.58	0.44	***843.26***	***27.5***
Water	0.14	0.02	0.01	0.05	0.01	0.07	1.46	stable	0.02	***1.78***	***0.1***
Plantations	0.05	0.03	0	0	0	0.01	0.11	0	15.68	***15.88***	***0.5***
Total 1990 (Mha)	***1280.06***	***294.41***	***60.81***	***176.69***	***160***	***149.97***	***920.91***	***4.09***	***21.03***	**3068.01**	
Total 1990 (%)	***41.7***	***9.6***	***2***	***5.8***	***5.2***	***4.9***	***30***	***0.1***	***0.7***		**100**

Countries that mention deforestation do not include data because of a high degree of uncertainty. In these cases, only the "regrowth of cleared vegetation" sink is reported.

Regional Potentials for ARD Activities

Potential forest areas under Article 3.3 can be roughly estimated from the TBFRA 2000 (UN, 1999). TBFRA 2000 provides estimates of the area of forest extension and natural colonization of forest and other wooded land. These areas could be regarded as afforestation in all definitional scenarios. TBFRA 2000 also provides areas of regeneration of forest (which could be regarded as reforestation according to the FAO scenario). Estimates for the annual loss of forest to other uses (UN-ECE/FAO, 1992) could be regarded as deforestation in all definitional scenarios (Table 3-14). These estimates show recent trends (deforestation in 1980s, afforestation and reforestation in late 1980s and early 1990s) in ARD activities in Annex I countries. Note that some countries have not reported extension or regeneration of forest or losses to other uses. In addition, the terms and definitions used in the statistics do not fully correspond to the terms and definitions of ARD; thus, they provide some error to the ARD estimates. Annex I countries reported in TBFRA 2000 (see Table 3-14) had regeneration area that was 14 times larger than the area of extension of forest, and loss to other uses was nearly 3 times the extension of forest. In Europe, the area of regeneration of forest was 4.2 times larger than the area of extension of forest (23 times larger in the United States); the area of extension of forest was 2.7 times larger than the area of loss to other uses [only one-third as large in the United States (see footnote g in Table 3-14)].

Potential ARD land in the tropics cannot be estimated from recent data. The FAO forest land-cover change in the tropics between 1980 and 1990 (FAO, 1996) shows that 3068 Mha were transferred from one land-cover class to another (Table 3-15). In addition, 92 Mha of closed forest in 1980 (out of a total of 1368 Mha) were transferred to other land-cover classes (i.e., deforested) by 1990. In contrast, only 4.2 Mha were transferred from other land-cover classes to closed forest (i.e., afforested and/or reforested) during the same period. Note that all of these transfers produced changes in woody biomass and thus carbon, as well as changes in soil carbon. Cells below the diagonal in Table 3-15 have most likely experienced increases in biomass, whereas cells above the diagonal have most likely experienced losses in biomass. Areas in the latter cells are much larger.

3.5.4. Global Assessment

Policymakers require information on potential global carbon amounts that may result from implementation of Article 3.3. Scientific data to estimate the potential amount of carbon involved in ARD activities on a global scale since 1990 are unsatisfactorily sparse. At best, with the few available estimates, one can put coarse bounds on those amounts and refine the bounds with results from isolated regional studies and with mathematical models.

3.5.4.1. How Much Carbon is Global Deforestation Likely to Release and How is It Distributed Spatially?

The effects of deforestation will depend in large part on the amount of land deforested and the amount of carbon in forests at the time of deforestation. Published averages are about 400 t C ha^{-1} above and below ground in boreal forests, 150 t C ha^{-1} in temperate forests, and 250 t C ha^{-1} in tropical forests (Dixon *et al.*, 1994). There is considerable variability to these values. Olson *et al.* (1983) estimate that aboveground biomass of all forest and woodland varies between 40 and 250 t C ha^{-1}, with boreal forest and woodland ranging from 20 to 80 t C ha^{-1} in the north and 60 to 140 t C ha^{-1} in the south. These values for boreal forest do not include the large quantities of soil organic carbon reported from Russian forests (Dixon *et al.*, 1994). Temperate forests also vary widely: from 60 to 140 t C ha^{-1} for closed forest. Tropical regions may contain as little as 20–50 t C ha^{-1} in savannas and dry forests but up to 250 t C ha^{-1} in wet tropical forests (Olson *et al.*, 1983). The global area of forests is about 4.2 billion ha, including 1.4 billion ha in the boreal zone, 1.0 billion ha in temperate forests, and 1.8 billion ha in tropical forests (Dixon *et al.*, 1994). The estimate of global forest area is much more sensitive than that of biomass to minimum tree cover on which the forest definition is based.

The partitioning of organic carbon into living biomass and dead organic matter pools differs greatly between ecological zones (Olson *et al.*, 1983). The proportion of total ecosystem carbon in below-ground biomass and soil organic matter is greatest in boreal forests (84 percent) and smallest in the tropics (50 percent), with temperate forest values falling between those two (63 percent). Therefore, the inclusion of soil carbon into the definition of forest carbon will produce considerable liability for highest-latitude countries when deforestation emissions are considered, with increasingly less liability nearer the equator. Moreover, the decision about which carbon pools to include under the Kyoto Protocol will have differential effects on the timing of the carbon release in all three regions. On average, deforestation activities are likely to release less carbon than the true carbon stock values because not all carbon will be oxidized or exported as wood products. In addition, oxidation of organic residues resulting from deforestation declines as a negative exponential process requiring many years (Harmon *et al.*, 1996; Micales and Skog, 1997). Thus, the emission of carbon from soils in the first decade or so following deforestation will be much more rapid than carbon sequestration during the first decade or so following afforestation and reforestation (Harmon *et al.*, 1986; Kirilenko and Solomon, 1998).

The total amount of carbon that may be released annually to the atmosphere by global deforestation because of agriculture and fuel wood extraction was projected with a spatially explicit model (IMAGE 2.1) by Alcamo *et al.* (1996). Their projections at 5-year intervals (Table 3-16) include carbon releases of 0.94 Gt C yr^{-1} in 1990. Linear interpolation of their 5-year interval data produces values of 1.85 Gt C yr^{-1} in 2008 and 2.09 Gt C yr^{-1} in 2012. More recent simulations with a more detailed version of the model (Leemans *et al.*, 1998) are lower overall, with estimated carbon releases of 1.02 Gt C yr^{-1} in 1990, 1.31 Gt C yr^{-1} in 2008, and 1.31 Gt C yr^{-1} in 2012. Yamagata and Alexandrov (1999) obtained a value of 1.0 Gt C yr^{-1} for 2005–2010. These latter values are considerably lower than the most recent estimates of carbon emissions from deforestation during 1980–1989: 1.7 $\pm$ 0.8 Gt C yr^{-1} (Houghton, 1999; see Section 1.2.1.2).

These estimates are sensitive to the forest definitions used, however. For example, if savannas are considered to be forests (normally they are not)—and conversion of savannas is considered as deforestation—the estimate of Yamagata and Alexandrov rises to 1.8 Gt C yr^{-1}. In contrast, inclusion of a 40 percent tree cover threshold (excluding both savanna and some open woodland) reduces the estimate from 1.8 to 0.9 Gt C yr^{-1}.

3.5.4.2. *How Much Carbon Can Be Sequestered by Global Afforestation and Reforestation?*

As Table 3-17 shows, afforestation and reforestation potentially could achieve annual carbon sequestration rates in aboveground and below-ground biomass of 0.4–1.2 t ha^{-1} yr^{-1} in boreal regions, 1.5–4.5 t ha^{-1} yr^{-1} in temperate regions, and 4–8 t ha^{-1} yr^{-1} in tropical regions (Dixon *et al.*, 1994; Nabuurs and Mohren, 1995; Nilsson and Schopfhauser, 1995; Brown *et al.*, 1996; Yamagata and Alexandrov, 1999). The latter two values assume that there is 0.3 t C m^{-3} of wood in boreal and temperate regions and 0.4 t C m^{-3} wood in tropical regions (Nilsson and Schopfhauser, 1995). The maximum amount of carbon that might be sequestered by global afforestation and reforestation activities for the 55-year period 1995–2050 was estimated at 60–87 Gt C, with about 70 percent in tropical forests, 25 percent in temperate forests, and 5 percent in boreal forests (Brown *et al.*, 1996). Hence, an average maximum potential carbon sequestration rate would be 1.1–1.6 Gt yr^{-1} above and below ground (Brown *et al.*, 1996). Although these maximum values represent about 2 percent of the annual global carbon uptake by the terrestrial biosphere, they are much higher values than would be expected from ARD activities in all but the most broadly defined ARD lands (e.g., FAO or Land Cover definitional scenarios).

Note that the annual carbon sequestration value is not constant; it will vary from year to year with annual weather conditions and would change over longer terms. Nilsson and Schopfhauser (1995) calculated a mean annual global carbon increment of 0.4 Gt C yr^{-1} above ground, with an additional 0.1 Gt C yr^{-1} below ground, from potential afforestation and deforestation 20 years after initiation of an optimum set of global forest

Table 3-16: *Carbon emissions in Gt yr^{-1} from deforestation, including fuelwood and timber decay, projected by IMAGE 2.1 integrated assessment model (Alcamo et al., 1996). Total emissions from deforestation are the sum of direct emissions from deforestation and emissions from deforestation caused by harvesting and burning of fuelwood.*

Year	Deforestation	Use of Fuelwood	Total
1990	0.83	0.11	0.94
1995	1.41	0.12	1.53
2000	1.04	0.12	1.16
2005	1.58	0.13	1.71
2010	1.81	0.14	1.93
2015	2.16	0.15	2.31

***Table 3-17**: Estimate of accounted average annual carbon stock change for ARD activities. The IPCC and FAO Definitional Scenarios and three accounting approaches under the FAO Definitional Scenario have been applied to illustrate with the available data the effect of different accounting approaches. Other Definitional Scenarios described in Table 3-4 have not been included in this analysis. It is assumed that recent area conversion rates ["recent" = for Annex I Parties AR late 1980s/early 1990s and for D 1980s (except for Canada and Russian Federation early 1990s); ARD in other regions 1980s] have applied since 1990 and will continue to do so through the first commitment period. It is also assumed that current uptake rates apply during the first commitment period. The figures and ranges of values in the table are illustrative and provide first-order estimates, and may not encompass the full range of uncertainties. Negative numbers indicate carbon emissions and positive numbers carbon removals.*

Region	Activity	AR Average Rate of Uptake[d] (t C ha^{-1} yr^{-1}); D Average Stock[a,d] (t C ha^{-1})	Area Change (Mha yr^{-1}): Post-Harvest Regeneration	Area Change (Mha yr^{-1}): Conversion between Non-Forest and Forest	Estimated Range of Accounted Average Annual Stock Change 2008–2012 (Mt C yr^{-1})[h] *Includes carbon in aboveground and below-ground biomass, excludes carbon in soils and DOC*: FAO Definitional Scenario, Land-Based I Accounting[b]	FAO Definitional Scenario, Land-Based II Accounting[b]	FAO Definitional Scenario, Activity-Based Accounting[b]	IPCC Definitional Scenario[b]
Boreal Region	AR	0.4 0.8 1.2	3.1[e]	0.1[e]	*R:* -209 -191 -164 *A:* 0.2 0.7 1.6	*R:* -56 -38 -10 *A:* 0.2 0.7 1.6	*R:* 5 21 46 *A:* 0.2 0.7 1.6	*AR:* 0.2 0.7 1.6
Total (=Annex I)[g]	D	35		0.5[e]	-18	-18	-18	-18
	Total ARD				-227 -208 -180	-74 -55 -26	-13 4 30	-18 -17 -16
Temperate Region	AR	1.5 3 4.5	5.4[e]	0.5[e]	*R:* -557 -351 -125 *A:* 7 25 44	*R:* -141 49 259 *A:* 7 25 44	*R:* 74 265 475 *A:* 7 25 44	*AR:* 7 25 44
Annex I[g]	D	60		1.2[e,i]	-72[j]	-72[j]	-72[j]	-72[j]
	Total ARD				-622 -398 -153	-206 2 231	9 218 447	-65 -47 -28
Annex I Total[g]	AR		8.5[e]	0.6[e,i]	*R:* -766 -542 -289 *A:* 7 26 46	*R:* -197 11 249 *A:* 7 26 46	*R:* 80 289 527 *A:* 7 26 46	*AR:* 7 26 46
	D			1.7[e]	-90	-90	-90	-90
	Total ARD				-849 -606 -333	-280 -53 205	-3 225 483	-83 -64 -44
Temperate Region	AR	1.5 3 4.5	n/a	1.9[f,i]	*R:* n/a *A:* 27 93 167	*R:* n/a *A:* 27 93 167	*R:* n/a *A:* 27 93 167	*AR:* 27 93 167
Total	D	60		2.1[c]	-126	-126	-126	-126
	Total ARD				n/a	n/a	n/a	-99 -33 41
Tropical Region	AR	4 6 8	n/a	2.6[f]	*R:* n/a *A:* 170 305 415	*R:* n/a *A:* 170 305 415	*R:* n/a *A:* 170 305 415	*AR:* 170 305 415
Total	D	120		13.7[c]	-1644	-1644	-1644	-1644
	Total ARD				n/a	n/a	n/a	-1474 -1339 -1229
Global Total	AR		n/a	4.6	*R:* n/a *A:* 197 399 584	*R:* n/a *A:* 197 399 584	*R:* n/a *A:* 197 399 584	*AR:* 197 399 584
(summing regional totals)	D			16.3	-1788	-1788	-1788	-1788
	Total ARD				n/a	n/a	n/a	-1591 -1389 -1204

NOTES FOR TABLE 3-17:

n/a = Area of regeneration after harvest not available. In addition, regeneration after selective cutting, as is often used in tropics, is difficult to capture with the FAO Definitional Scenario.

[a] Literature sources do not allow estimation of uncertainty levels of carbon stocks prior to deforestation or of areas deforested.

[b] The values under the three accounting approaches of the FAO Definitional Scenario and in the IPCC Definitional Scenario were calculated with a spreadsheet that includes all relevant cohorts of the landscape subject to ARD activities, based on the following assumptions. (1) Trees grow according to a sigmoidal growth curve with a maximum mean annual increment at the time of harvest ("Average Uptake Rate" in the table). (2) The biomass stock (aboveground and below-ground biomass, but not soil carbon) just before harvest is twice the "D Average Stock" in boreal and temperate regions and 1.33 times in the tropical regions. This, in combination with the growth rates in the table, results in rotation cycles of 175, 88, and 58 years for the boreal region; 80, 40, and 27 years for the temperate region; and 40, 27, and 20 for the tropical region. (3) One-third of the biomass stock at harvest is assumed to be left on the site as slash, litter, and dead roots. This material is assumed to decay at a constant rate in 15 years (boreal), 10 years (temperate), and 5 years (tropical).

The IPCC Definitional Scenario includes transitions between forest and non-forest land uses under Article 3.3. The FAO Definitional Scenario includes the harvest/regeneration cycle, because regeneration is defined as reforestation. Within the FAO Definitional Scenario we distinguish three accounting approaches. Land-based I approach always accounts for stock changes over the full commitment period. Land-based II approach starts the accounting in 2008 or with the activity, whichever is later. Stock changes that do not result from reforestation (e.g., decay of post-harvest slash) are counted. The activity-based approach is the same as land-based II except that stock changes in decaying slash from a preceding harvest are excluded.

The results for R in FAO land-based I approach are negative because the stock changes due to harvest dominate the carbon balance. The stock-change values under FAO land-based II approach are negative in boreal Annex I Parties and range from negative to positive numbers for temperate Annex I Parties depending on how quickly stand growth increases after harvest to offset the loss of carbon from decaying materials. The large positive values shown for FAO activity-based accounting arise because none of the carbon-stock losses are accounted whereas the carbon gains from biomass growth are. In the IPCC Definitional Scenario no initial harvest occurs and therefore there is no difference between the accounting approaches—only one set of numbers is shown. Sigmoidal growth curves as used in the calculations provide lower uptake in the early and late stand development compared to linear growth curves, but higher uptake in between.

Stock changes from deforestation are calculated as the average stock multiplied by the recent annual rate of deforestation. For example, annual deforestation of 1.2 Mha in temperate region (with 60 t C ha^{-1}) results in annual emissions of 72 Mt C yr^{-1}.

[c] FAO (1996) and Brown *et al*. (1996) estimate deforestation in developing countries between 1980 and 1990 at 16.3 Mha yr^{-1} (15.4 Mha yr^{-1} in the tropics). A critique of the methods and database sources with which these estimates were derived can be found at <http://www.fao.org/forestry/for/fra/FO124E/GEP15.HTM>. A more recent number for deforestation between 1990 and 1995 is lower at 13.7 Mha yr^{-1} (FAO, 1999) and has been used here. The older number from FRA 1990 (FAO, 1996) is not comparable with this new number from FAO (1999) because: (i) The two data sets are completely different; (ii) the FRA data sets are updated for new data becoming available after an assessment; and (iii) the data set for the period 1990–1995 (FAO, 1999) still has some gaps (FAO, 2000). It is recommended that definitive conclusions be drawn based on the forthcoming new statistics from the Forest Resource Assessment 2000, which should become available during the year 2000 (FAO, 2000).

[d] Uptake rates are intended to span the range within which the average value for each region is expected to be. These numbers were derived from the sources below, some of which (e.g., Nilsson and Schopfhauser, 1995) give country- or region-specific data, thus allowing to calculate a weighted mean. Uptake rates and carbon stocks given include aboveground and below-ground biomass, but not soil carbon or dead organic matter (DOC). "D Average Stock" is an average over the landscape, assuming that D will equally impact all age classes of the forest estate. All these values include total biomass but exclude carbon in wood products, DOC, and soil carbon (except FAO land-based II approach which factors in post-harvest slash). Average rate of uptake are from Dixon *et al*. (1994), Nabuurs and Mohren (1995), Nilsson and Schopfhauser (1995), Brown *et al*. (1996), and Yamagata and Alexandro (1999); average stock for boreal and temperate regions are from TBFRA 2000 (UN, 1999), for tropics from Dixon *et al*. (1994), supported by Table 1 in the SPM. The number for the tropics may be regarded to be a high estimate, but was used here in absence of additional literature sources.

[e] Areas as given in Table 3-14. Extension of forest is used to approximate the area of afforestation and reforestation under the IPCC Definitional Scenario and the area of afforestation under the FAO Definitional Scenario. Regeneration is used to approximate the area of reforestation under the FAO Definitional Scenario. Loss to other uses is used to approximate the area of deforestation.

Recent AR area estimates for Annex I Parties are data from the TBFRA 2000 (UN, 1999), and D area estimates from FRA 1990 (UN-ECE/FAO, 1992). D for Russian Federation is from A. Shvidenko (pers. comm.). D for Canada is from Robinson *et al*. (1999). Data for Australia are based on NGGI (1999).

[f] FAO (1995). This estimate should be regarded as an upper limit, because some countries may have reported plantations for 1990 but not for 1980, and because some of the plantations may not qualify as resulting from AR activities under the IPCC Definitional Scenario.

[g] Boreal Annex I countries included: Canada, Finland, Iceland, Norway, Russian Federation, and Sweden. Temperate Annex I countries included: Australia, Austria, Belarus, Belgium, Bulgaria, Croatia, Czech Republic, Denmark, Estonia, France, Germany, Hungary, Ireland, Italy, Japan, Latvia, Liechtenstein, Lithuania, The Netherlands, New Zealand, Poland, Portugal, Slovakia, Slovenia, Switzerland, Turkey, Ukraine, United Kingdom, and United States of America. Data for Greece, Luxemburg, and Spain were not available.

[h] Because of the average uptake rates and carbon stocks at harvest that were used in this table, stands planted or regenerated since 1990 are not harvested before the end of the commitment period in the calculations. However, there may be AR forests with a very short rotation that are harvested in the first commitment period. The accounting for such stands may require a deduction of the harvested stock from the reported stock change. See Section 3.3.2.2, item b3.

[i] The regeneration area for the United States of America in UN (1999) is 1.8 Mha yr^{-1}. However, this number is incorrect and has been replaced with a corrected number of 0.2 Mha yr^{-1} (US Forest Service, 2000).

[j] These results do not take into account the effect of the second sentence of Article 3.7 of the Kyoto Protocol.

plantations. Neither the value derived from Brown *et al.* (1996) nor that from Nilsson and Schopfhauser (1995) included carbon losses to deforestation or degradation from tropical fuelwood extraction (Alcamo *et al.*, 1996; Solomon *et al.*, 1996), from deteriorating climate, or from increasing agriculture (Cramer and Solomon, 1993; Alcamo *et al.*, 1996).

The FAO scenario (with the activity-based approach) and the IPCC scenario produce different estimates of potential carbon stock changes. Under the IPCC definitional scenario, if Annex I countries (approximated by the estimates for boreal and temperate region Annex I countries in Table 3-17) maintain recent rates of afforestation and reforestation from 1990 through 2012, the estimated rate of increase in carbon stocks from these activities would be 7–46 Mt C yr^{-1}. Under the same assumptions, this increase would be offset by estimated decreases in carbon stocks from deforestation of 90 Mt C yr^{-1}, producing a net change of -83 to -44 Mt C yr^{-1}. Under the FAO definitional scenario using the activity-based accounting approach (i.e., the stock change accounting does not start before the activity, nor does it include carbon loss from decaying slash), if Annex I countries maintain recent rates of afforestation and reforestation from 1990 through 2012, the estimated rate of increase in carbon stocks from these activities would be from 87–573 Mt C yr^{-1}, with estimated decreases in carbon stocks from deforestation of 90 Mt C yr^{-1}—producing a net change of -3 to 483 Mt C yr^{-1}. Under the FAO scenario using land-based accounting approach I, the estimated stock change from afforestation and reforestation activities continued at recent rates would be -759 to -243 Mt C yr^{-1} and from deforestation activities -90 Mt C yr^{-1}, resulting in an estimated net stock change of -849 to -333 Mt C yr^{-1}. Under the FAO scenario using land-based accounting approach II, the estimated stock change for afforestation and reforestation activities continued at recent rates would be -190 to 295 MtC yr^{-1} and from deforestation activities -90 Mt C yr^{-1}, resulting in an estimated net stock change of -280 to 205 Mt C yr^{-1}.

Under the IPCC definitional scenario, if recent rates of afforestation and reforestation are increased by 20 percent and rates of deforestation decreased by 20 percent in Annex I countries from the year 1990, carbon stocks would increase by 7–49 Mt C yr^{-1} due to afforestation and reforestation activities and decrease by 72 Mt C yr^{-1} due to deforestation activities, resulting in a net change in carbon stocks from these activities of -83 to -23 Mt C yr^{-1} in Annex I countries. For comparison purposes, increases in carbon stocks from afforestation and reforestation globally using the IPCC definitions could result in a stock change of about 197–584 Mt C yr^{-1} and decreases from deforestation of about 1,788 Mt C yr^{-1} if current rates are maintained.

In the IPCC Definitional Scenario and FAO Definitional Scenario with land-based I accounting approach, the accounted carbon stock changes are broadly consistent with the 2008–2012 actual changes in carbon stocks from land under Article 3.3. The IPCC and FAO Definitional Scenarios bring different amounts of land under Article 3.3, hence the estimated carbon stock changes in Table 3-17 differ. In the FAO Definitional Scenario with land-based II and activity-based accounting approaches, the accounted carbon stock change is not consistent with the 2008–2012 actual changes in carbon stocks on land under Article 3.3, except in the case of short rotation cycles. In neither of the two Definitional Scenarios is the accounted carbon stock change consistent with the 2008–2012 actual carbon stock changes, nor with the net exchanges with the atmosphere, at the national and global levels in part because the land under Article 3.3 is small in comparison with the national and global forest area (see Section 3.5.2.5).

3.5.4.3. *What Net Changes in Carbon Stocks from ARD Can Be Expected in the Future?*

To estimate global potential carbon stock changes induced by ARD activities during the commitment period, Yamagata and Alexandrov (1999) used the land-use change scenarios obtained from the IMAGE 2.1 model (Alcamo *et al.*, 1998). The model gives a total carbon stock at maturity of 216 t C ha^{-1}, 142 t C ha^{-1}, and 90 t C ha^{-1} for the low-, mid-, and high-latitude regions respectively (Yamagata and Alexandrov, 1999). These estimates are similar to those made by Dixon *et al.* (1994) for low- and mid-latitude (tropical and temperate forests) but very small compared to the high-latitude (boreal forests) carbon estimates of Dixon *et al.* (1994). The Yamagata and Alexandrov (1999) simulations lack the large soil carbon pools that have accumulated at high latitudes during the past 8,000 to 10,000 years. The simulations yield an emission rate of 1.8 Gt C yr^{-1} during the commitment period if the scenario of land-use changes suggested by Leemans *et al.* (1998) took place.

These estimates for global ARD fluxes would be significantly reduced, however, if the actual tree canopy cover (<<100 percent) of the ARD lands in each region is used. In general, higher thresholds (e.g., 40 percent) capture not only clearing but also forest degradation as a deforestation activity. Higher thresholds also credit more afforestation and reforestation activities on land already covered by trees. On the other hand, higher thresholds are not sensitive to conversions of open forests, giving no credit for establishing open forests and no debit for clearing them.

To illustrate how thresholds change the ARD potential, we simulated the global carbon fluxes from ARD activities during the first commitment period. We combined remotely sensed canopy cover data (Nemani and Running, 1997) with the carbon stock change model of Yamagata and Alexandrov (1999) and the land-use change model of Leemans *et al.* (1998). Each value is predicted assuming the Land Use definitional scenario, with a different canopy cover threshold (Table 3-18).

The result demonstrates several implications for the Annex I countries. First, global carbon removals from ARD activities are as high as 0.2 Gt C yr^{-1}. Second, a lower canopy cover threshold significantly reduces potential carbon removals from

***Table 3-18**: Total estimates of carbon fluxes for Annex I countries during the first commitment period that would be induced by land-use change based ARD activities since 1990, in Gt C yr^{-1}. Estimates for non-Annex I countries are given in parentheses to illustrate the carbon potentials for projects between Annex I and non-Annex I countries. Removals are shown as positive and emissions as negative numbers.*

	Tree Cover Threshold in Forest Definition		**All Forests**
	10% Canopy Cover	40% Canopy Cover	Regardless of Canopy Cover
Removals (Gt C yr^{-1})	0.078 (0.041)	0.170 (0.174)	0.228 (0.206)
Emission (Gt C yr^{-1})	-0.017 (-0.586)	-0.012 (-0.478)	-0.017 (-0.591)
Net flux (Gt C yr^{-1})	0.061 (-0.545)	0.158 (-0.304)	0.211 (-0.385)

afforestation and reforestation activities and increases potential emissions from deforestation activities. Third, net carbon removals from ARD activities decline from 0.16 (40 percent cover threshold) to 0.06 Gt C yr^{-1} (10 percent cover threshold).

3.6. Associated Impacts of ARD Activities

This section briefly considers potentially harmful or beneficial impacts of increased ARD activities on factors other than carbon storage. To what extent might these activities have associated impacts that conflict with, or support, the sustainable development objectives agreed to at the UN Conference on Environment and Development (UNCED) and set out in more detail in other multilateral international agreements, such as the Intergovernmental Panel on Forests and the Convention on Biological Diversity?

The heart of the sustainable development concept is management of resources to maximize the income of an activity while maintaining or increasing the amount and quality of assets or capital affected by that activity. Because resources, assets, and capital all have natural and human dimensions, sustainable development encompasses socioeconomic factors as well as environmental sustainability. Sustainable development requires a balance between social, economic, and environmental considerations to improve current human welfare while ensuring a sound foundation for the welfare of future generations.

Opinions generally differ over the interpretation, identification, and valuation of assets and capital which can make implementation of sustainable development difficult. Section 3.6.1 provides a very brief discussion of specific factors related to ARD activities. These issues are discussed more fully in Section 2.5 and, in relation to project-based activities, in Chapter 5.

3.6.1. Afforestation and Reforestation

Converting non-forest land to forests will typically increase the diversity of flora and fauna, except in situations where biologically diverse non-forest ecosystems are replaced by forests that consist of single or a few species (e.g., plantations of monocultures and especially exotic species).

Where afforestation or reforestation is done to restore degraded lands, it also is likely to have other environmental benefits—such as reducing erosion, controlling salinization, and protecting watersheds. In dry countries, expansion of forested areas can also be viewed as a desertification-reduction activity.

Changing land uses also alters the nature of economic activity. The socioeconomic opportunities provided by the new land use (e.g., forestry) are a benefit of the change, but forgone opportunities in the previous activity (e.g., agriculture) are a cost. Social impacts can include population displacement and loss by some (often disadvantaged) section of society of the use of common property (Fearnside, 1996). The net effect of land-use change on employment, income, and equity cannot be determined *a priori*; it must be evaluated on a case-by-case basis. The social systems in each country will strongly influence the socioeconomic impacts associated with any given activity.

If a definitional scenario under which the harvest/regeneration cycle creates ARD land is adopted, efforts on LULUCF activities under Article 3.3 may concentrate on existing forests rather than creating new forests. As a result, an incentive to enhance the early growth rates of regenerating stands, such as through fertilization, could be put in place. In the FAO scenario combined with land-based accounting approach II, emissions from harvest slash are accounted if they occur during the regeneration phase. This factor could lead to an incentive to burn the harvest slash at the time of harvest. If the definitions of afforestation and reforestation exclude natural establishment of tree cover and restrict themselves to planting, the occurrence of monocultures on afforestation/reforestation land could increase. If the definition of reforestation in the FAO scenario excludes regeneration of trees after a selective cut, clear-cut management might be favored over selective-logging management.

3.6.2. Avoiding Deforestation

Any reduction in the rate of deforestation has the benefit of avoiding a significant source of carbon emissions (especially in the tropics) and reducing other environmental and social problems associated with deforestation. Current rates of tropical deforestation—estimated to be 0.7 percent of the remaining forest area per year (FAO, 1997)—are a primary cause of global biodiversity loss (Heywood, 1995; Stork, 1997). Deforestation and degradation of upland catchments can disrupt hydrological systems, replacing year-round water flows in downstream areas with flood and drought regimes (Myers, 1997). Deforestation can also diminish the social, aesthetic, and spiritual values of forests.

Limiting deforestation forgoes the opportunity to utilize the land for other purposes, such as agriculture or other developed uses, therefore would potentially be subject to the same opportunity costs that might arise with afforestation and reforestation.

Although there are often synergies between increased carbon storage through ARD activities and other desirable associated impacts, no general rules can be applied; impacts must be assessed individually for each specific case. Associated impacts can often be significant, and the overall desirability of specific ARD activities can be greatly affected by their associated impacts.

References

Alcamo, J., E. Kreileman, J. Bollen, G.J. van den Born, R. Gerlagh, M. Krol, S. Toet, and B. de Vries, 1996: Baseline global changes in the 21st century: baseline scenarios of global environmental change. *Global Environmental Change: Human and Policy Dimensions,* **6(4),** 261–303.

Alcamo, J., R. Leemans, and E. Kreileman, 1998: Global change scenarios of the 21st century. Results from the IMAGE 2.1 model. Elsevier Science Publishers, London, United Kingdom, 1998, 287 pp.

Alexandrov, G.A., Y. Yamagata, and T. Oikawa, 1999: Towards a model for projecting net ecosystem production of the world forests. *Ecological Modelling,* **123,** 183–191.

Brown, S., J. Sathaye, M. Cannell, and P. Kauppi, 1996: Management of forests for mitigation of greenhouse gas emissions. In: *Climate Change 1995: Impacts, Adaptations and Mitigation of Climate Change: Scientific-Technical Analyses. Contribution of Working Group II to the Second Assessment Report of the Intergovernmental Panel on Climate Change* [Watson, R.T., M.C. Zinyowera, and R.H. Moss (eds.)], Cambridge University Press, Cambridge, United Kingdom and New York, NY, USA, pp. 773–797.

Cannell, M.G.R., R.C. Dewar, and D.G. Pyatt, 1993: Conifer plantations on drained peatlands in Britain: a net gain or loss of carbon? *Forestry,* **66,** 353–369.

Cramer, W.P. and A.M. Solomon, 1993: Climatic classification and future distribution of global agricultural land. *Climate Research,* **3,** 97–110.

Dixon, R.K., S. Brown, R.A. Houghton, A.M. Solomon, M.C. Trexler, and J. Wisniewski, 1994: Carbon pools and flux of global forest ecosystems. *Science,* **263,** 185–190.

FAO, 1995: Forest Resource Assessment 1990, Tropical Forest Plantation Resources. Forestry Paper 128, Food and Agriculture Organization of the United Nations, Rome, Italy, 90 pp.

FAO, 1996: *Forest Resource Assessment Programme 1990, Survey of Tropical Forest Cover and Study of Change Processes.* Forestry Paper 130, Food and Agriculture Organization of the United Nations, Rome, Italy, 152 pp.

FAO, 1997: *State of the World's Forests.* Food and Agriculture Organization of the United Nations, Rome, Italy, 200 pp.

FAO, 1998: *Forest Resource Assessment Programme.* Working Paper 1, Food and Agriculture Organization of the United Nations, Rome, Italy, 18 pp.

FAO, 1999: *State of the World's Forests.* Food and Agriculture Organization of the United Nations, Rome, Italy, 154 pp.

FAO, 2000: Alberto DelLungo, personal communication, 11 May 2000.

FCCC, 1998a: *FCCC/CP/1998/L.5, Matters Related to the Kyoto Protocol; Matters Related to Decision 1/CP.3, Paragraph 5, Land-Use Change and Forestry.* Available at: http://www.unfccc.de/fccc/docs.

FCCC, 1998b: *FCCC/CP/1998/16 Add. 1, Report of the Conference of the Parties on Its Fourth Session, held at Buenos Aires from 2 to 14 November 1998. Addendum. Part Two: Action Taken by the Conference of the Parties at Its Fourth Session.* Available at: http://www.unfccc.de/resource/docs/cop4/16a01.pdf.

FCCC, 1999: On-line searchable database of GHG inventory data (updated September 1999). Available at: http://www.unfccc.de.

Fearnside, P.M. 1996: Socio-economic factors in the management of tropical forests for carbon. In: *Forest Ecosystems, Forest Management, and the Global Carbon Cycle* [Apps, M.J. and D.T. Price (eds.)]. NATO ASI Series, Subseries I: Global Environmental Change, Vol. 40, Springer-Verlag, Heidelberg, Germany, pp. 349–361.

Ford-Robertson, J., K. Robertson, and P. Maclaren, 1999: Modelling the effect of land-use practices on greenhouse-gas emissions and sinks in New Zealand. *Environmental Science and Policy,* **2,** 135–144.

Harmon, M.E., J.F. Franklin, F.J. Swanson, P. Sollins, S.V. Gregory, J.D. Lattin, N.H. Anderson, S.P. Cline, N.G. Aumen, J.R. Sedell, G.W. Lienkaemper, K. Cromack, and K.W. Cummins, 1986: Ecology of coarse woody debris in temperate ecosystems. *Advances in Ecological Research,* **15,** 133–302.

Harmon, M.E., S.L. Graman, and W.K. Ferrell, 1996: Modeling historical patterns of tree ulitisation in the Pacific Northwest: carbon sequestration implications. *Ecological Applications,* **6,** 641–652.

Hamilton, C. and L. Vellen, 1999: Land-use change in Australia and the Kyoto Protocol. *Environmental Science and Policy,* **2,** 145–152.

Heywood, V.H. (ed.), 1995: *Global Biodiversity Assessment.* Cambridge University Press, Cambridge, United Kingdom, 1140 pp.

Houghton, R.A., 1999: The annual net flux of carbon to the atmosphere from changes in land use 1850–1990. *Tellus,* **50B,** 298–313.

IPCC, 1997: *Revised 1996 IPCC Guidelines for National Greenhouse Gas Inventories* [J.T. Houghton, L.G. Meira Filho, B. Lim, K. Tréanton, I. Mamaty, Y. Bonduki, D.J. Griggs, B.A. Callander (eds.)]. Intergovernmental Panel on Climate Change, Meteorological Office, Bracknell, United Kingdom.
- *Volume 1: Greenhouse Gas Inventory Reporting Instructions.* 130 pp.
- *Volume 2: Greenhouse Gas Inventory Workbook.* 346 pp.
- *Volume 3: Greenhouse Gas Inventory Reference Manual.* 482 pp.

IPCC, 1999: *Evaluating Approaches for Estimating Net Emissions of Carbon Dioxide from Forest Harvesting and Wood Products* [Brown, S., B. Lim, and B. Schlamadinger (eds.)]. IPCC Meeting Report, May 1998, Dakar, Senegal. International Panel on Climate Change/Organization for Ecomonic and Commercial Development/International Energy Agency Programme on National Greenhouse Gas Inventories. Organization for Economic Cooperation and Development, Paris, France, 51 pp.

Izrael, Y.A., S.I. Avdjushin, I.M. Nazarov, A.O. Kokorin, A.I. Nakhutin, A.F. Yakovlev, and Y.A. Anokhin, 1999: *Russian Federation Climate Change Country Study (cooperative agreement DE-FCO2-93PO10118). Task 6. Climate Change Action Plan Report.* Russian Federal Service for Hydrometeorology and Environment Monitoring, Moscow, Russia, 125 pp.

Karjalainen, T., J. Liski, A. Pussinen, and T. Lapveteläinen, 1999: *Sinks in the Kyoto Protocol and Considerations for Ongoing Work in the UNFCCC Work.* The Nordic Council of Ministers, Copenhagen, Denmark, (in press).

Kirilenko, A.P. and A.M. Solomon, 1998: Modeling dynamic vegetation response to rapid climate change using bioclimatic classification. *Climatic Change,* **38,** 15–49.

Kirschbaum, M.U.F., 2000: What contribution can tree plantations make towards meeting Australia's commitments under the Kyoto Protocol? *Environmental Science and Policy,* (in press).

Laine, J., J. Silvola, K. Tolonen, J. Alm, H. Nykänen, H. Vasander, T. Sallantaus, I. Savolainen, J. Sinisalo, and P.J. Martikainen, 1996: Effect of water-level drawdown on global climatic warming:northern peatlands. *Ambio,* **25(3),** 179–184.

Leemans, R., R. E. Kreileman, G. Zuidema, J. Alcamo, M. Berk, G. J. van den Born, M. den Elzen, R. Hootsmans, M. Janssen, M. Schaeffer, S. Toet, and B. de Vries, 1998: *The IMAGE User Support System: Global Change Scenarios from IMAGE 2.1.* RIVM publication 4815006, National Institute of Public Health and the Environment (RIVM), P.O. Box 1, 3720 BA Bilthoven, Netherlands, CD-ROM.

Liski, J., T. Karjalainen, G.J. Nabuurs, A. Pussinen, and P. Kauppi, 2000: Trees as carbon sinks and sources in the European Union. *Environmental Science and Policy,* (in press).

Lund, G., 1999: A 'forest' by any other name.... *Environmental Science and Policy,* **2(2),** 125–134.

Lund, G., 2000: Definitions of Forest, Deforestation, Afforestation, and Reforestation. Available at: http://home.att.net/~gklund/DEFpaper.html

Mäkipää, R. and E. Tomppo, 1998: Suomen metsät ovat hiilinielu—vaikka Kioton ilmastosopimuksen mukaan muulta näyttää. Metsätieteen aikakausikirja. *Folia Forestalia,* **2,** 268–274. (in Finnish).

Mantovani, A.C.M. and A.W. Setzer, 1997. Deforestation detection in the Amazon using an AVHRR-based system. *International Journal of Remote Sensing,* **18(2),** 273–286.

Marland, G. and B. Schlamadinger, 1999: The Kyoto Protocol could make a difference for the optimal forest-based CO_2 mitigation strategy, *Environmental Science and Policy,* **2(2),** 111–124.

Micales, J.A. and K.E. Skog, 1997: The disposition of forest products in landfills. *International Biodeterioration and Degradation,* **39(2-3),** 145–158.

Myers, N. 1997: The world's forests and their ecosystem services. In: *Nature's Services: Societal Dependence on Natural Ecosystems* [Daily, G.C. (ed.)]. Island Press, Washington, DC, USA, pp. 215–235.

Nabuurs, G.J. and G.M.J. Mohren, 1995: Modelling analyses of potential carbon sequestration in selected forest types. *Canadian Journal of Forest Research,* **25,** 1,157–1,172.

Nabuurs, G.J., A.V. Dolman, E. Verkaik, P.J. Kuikman, C.A. van Diepen, A. Whitmore, W. Daamen, O. Oenema, P. Kabat, and G.M.J. Mohren, 2000: Article 3.3 and 3.4. of the Kyoto Protocol: consequences for industrialised countries' commitment, the monitoring needs and possible side effect. *Environmental Science and Policy,* **00(2000): 1-12,** (in press).

Nemani, R. and S. Running, 1997: Land cover characterization using multi-temporal Red, Near-IR and Thermal-IR data from NOAA/AVHRR. *Ecological Applications,* **7(1),** 79–90.

NGGI, 1999: *National Greenhouse Gas Inventory 1997 with Methodology Supplements.* National Greenhouse Gas Inventory Committee, Australian Greenhouse Office, Canberra, Australia, 202 pp.

Nilsson, S. and W. Schopfhauser, 1995: The carbon-sequestration potential of a global afforestation program. *Climatic Change,* **30,** 267–293.

O'Connell, A.M. and K.V. Sankaran, 1997: Organic matter accretion, decomposition and mineralization. In: *Management of Soil, Nutrients and Water in Tropical Plantation Forests* [E.K. Sadanandan Nambiar and A.G. Brown (eds.)]. Australian Centre for International Agricultural Research Monograph No. 43, Commonwealth Scientific and Industrial Research Organization (CSIRO), Canberra, Australia, pp. 443–480.

Olson, J.S., J.A.Watts, and L.J. Allison, 1983: *Carbon in Live Vegetation of Major World Ecosystems.* ORNL-5862, Oak Ridge National Laboratory, Oak Ridge, TN, USA, 152 pp.

Pisarenko, A., G. Redok, and M. Merzlenko, 1992: *Artificial Forests, Vol. 1.* Ministry of Ecology and Natural Resources of the Russian Federation, Moscow, 307 pp. (in Russian)

RFFS, 1998: *Basic Indicators of Activities of the Federal Forest Service of Russia in 1998, 1992–1997.* Federal Forest Service of Russia, Moscow, 233 pp. (in Russian)

Robinson, B.C.E., W.A. Kurz, and C. Pinkman, 1999: *Estimating the Carbon Losses from Deforestation in Canada.* Report prepared by ESSA Technologies Ltd., Vancouver, British Columbia, for the National Climate Change Secretariat, Ottowa, Ontario, Canada, 81 pp.

Ruhiyat, D., 1995: *Estimasi Biomasa Tegakan Hutan Hujan Tropika di Kalimantan Timur. Lokakarya Inventarisasi Emisi dan Rosot Gas Rumah Kaca, Bogor : 4–5 Agustus 1995.* Bogor Agricultural University and Ministry of Environment, Republic of Indonesia, Bogor and Jakarta, Indonesia, 15 pp. (in Indonesian)

Schlamadinger, B. and G. Marland, 1998: The Kyoto Protocol: provisions and unresolved issues relevant to land-use change and forestry. *Environmental Science and Policy,* **1,** 313–327.

Solomon, A.M., N.H. Ravindranath, R.B. Stewart, S. Nilsson, and M. Weber, 1996: Wood production under changing climate and land use. In: *Climate Change 1995: Impacts, Adaptations and Mitigation of Climate Change: Scientific-Technical Analyses. Contribution of Working Group II to the Second Assessment Report of the Intergovernmental Panel on Climate Change* [Watson, R.T., M.C. Zinyowera, and R.H. Moss (eds.)], Cambridge University Press, Cambridge, United Kingdom and New York, NY, USA, pp. 487–510.

Stork, N.E., 1997: Measuring global biodiversity and its decline. In: *Biodiversity II* [Readka-Kudla, M.L., D.E. Wilson, and E.O. Wilson (eds.)]. John Henry Press, Washington, DC, USA, pp. 41–68.

UN, 1999: *Forest Resources of Europe, CIS, North America, Australia, Japan and New Zealand (industrialized temperate/boreal countries), UN-ECE/FAO Contribution to the Global Forest Resources Assessment 2000.* United Nations, New York, NY, USA and Geneva, Switzerland, (in press).

UN-ECE/FAO, 1992: *The Forest Resources of the Temperate Zones. The UN-ECE/FAO 1990 Forest Resource Assessment. Vol. 1, General Forest Resource Information.* United Nations Economic Commission for Europe. Food and Agriculture Organization of the United Nations, Rome, Italy. ECE/TIM/62. UN Publ. No. E.92.II.E.27, 348 pp.

U.S. Forest Service, 2000a: Fax sent by USFS to UN-ECE/FAO (K. Prins), 2 May 2000.

U.S. Forest Service, 2000b: Personal communication with B. Stokes/USFS, 9 May 2000.

Yamagata, Y. and G.A. Alexandrov, 1999: Political implication of defining carbon sinks under the Kyoto Protocol. *World Resource Review,* **11(3),** 346–359.

4

Additional Human-Induced Activities—Article 3.4

R. NEIL SAMPSON (USA) AND ROBERT J. SCHOLES (SOUTH AFRICA)

Lead Authors:
C. Cerri (Brazil), L. Erda (China), D.O. Hall (UK), M. Handa (Japan), P. Hill (USA), M. Howden (Australia), H. Janzen (Canada), J. Kimble (USA), R. Lal (USA), G. Marland (USA), K. Minami (Japan), K. Paustian (USA), P. Read (New Zealand), P.A. Sanchez (Kenya), C. Scoppa (Argentina), B. Solberg (Norway), M.A. Trossero (FAO), S. Trumbore (USA), O. Van Cleemput (Belgium), A. Whitmore (UK), D. Xu (China)

Contributors:
B. Burrows (Australia), R. Conant (USA), G. Liping (China), W. Hall (Australia), W. Kaegi (Switzerland), P. Reyenga (Australia), N. Roulet (Canada), K.E. Skog (USA), G.R. Smith (USA), Y. Wang (China)

Review Editors:
J.W.B. Stewart (Canada) and J. Stone (Canada)

CONTENTS

EXECUTIVE SUMMARY

The chapter addresses the implications of including—as adjustments to assigned amounts under the Kyoto Protocol—the effects of activities related to land use, land-use change, and forestry (LULUCF) other than those covered by Article 3.3. It derives a set of core questions from Article 3.4 of the Protocol and arranges these questions in a sequence designed to help decisionmakers work through issues such as which additional human-induced activities should be included and how those activities should be defined, measured, reported, monitored, and verified.

Although Article 3.4 of the Kyoto Protocol applies only to Parties listed in Annex I, examples and estimates that are relevant to current Annex I and non–Annex I countries are included because the list of countries interested in controlling greenhouse gas (GHG) emissions is likely to grow in the future. The numerous potential practices are summarized in broad activity groups, and "Fact Sheets" presenting the details of individual practices are included after the reference section.

The potential global impact of these additional LULUCF activities in Annex I countries in the first commitment period is estimated to be up to 0.52 Gt C yr^{-1}. The combined Annex I and non–Annex I impact could reach 2.5 Gt C yr^{-1} in 40 years (Table 4-1). These estimates include only on-site carbon stock changes; reductions of CO_2 emissions caused by the increased use of bioenergy or wood products are not included. The estimates are based on reviewed studies and the assumption that economic, social, and technical constraints will limit the land that is actually available for these activities to a small percentage of the land that is theoretically available. The main component of the large uncertainty associated with these estimates relates to where, and to what extent, LULUCF activities may occur. Once those facts are known, the estimates of GHG impacts of a given activity in a given location are far less uncertain.

Change in management within a land use or change in land use to one with a higher potential carbon stock can increase carbon stocks in an ecosystem, leading to a net removal of CO_2 from the atmosphere. The carbon is stored as soil carbon, wood, leaf, root, and litter biomass. These different carbon pools have mean residence times that range from days to millennia. The stable carbon pools remain virtually unchanged over very long periods if management, climate, and disturbance regimes do not change. The rate at which ecosystems can accumulate carbon, the ultimate size of the pools, and the rate at which the carbon can be lost again under altered circumstances all depend on the form of the newly stored carbon, the magnitude of the land-use or management change, the inherent biological productivity of the site, and the type and depth of the soil.

The net effect of LULUCF activities on the atmosphere depends not only on changes in on-site carbon stocks but also on:

- The lifetime of carbon in agricultural and forest products and how they replace other products that require more or less energy to produce and use
- Concomitant changes in the net fluxes of other GHGs (especially CH_4 and N_2O)
- Changes in GHG emissions resulting from changes in the fossil fuel energy needed to maintain the new land-use practices
- Changes in non-GHG-related radiant forcing (such as changes in albedo).

These non-CO_2 effects can add to, or reduce, the effects from CO_2. In several important cases, the non-CO_2 effects are sufficiently large that failure to include them in the accounting procedure would lead to substantially misleading conclusions. For example, changes in CH_4 and N_2O emissions resulting from management changes in wetlands, rice crops, or dryland crops can offset a large portion of changes in the carbon stock.

The capacity to store on-site carbon through LULUCF activities is finite because the land area available for this purpose has competing uses and is limited and because the carbon pools have practical upper limits. Although the rates suggested in Table 4-1 will eventually decline, they are considered to be applicable for a 30- to 50-year period.

The potential contribution of these additional LULUCF activities to the reduction of global radiative forcing is therefore a substantial portion of the total human-induced effect, though it is too small to balance current fossil fuel emissions by itself. Including major LULUCF activities as adjustments to the assigned amounts under the first commitment period of the Kyoto Protocol could potentially reduce the degree to which many countries may need to alter energy use and energy production technology in the short term.

The opportunities for reducing the net flux of GHGs to the atmosphere through the application of these activities is large because of the extensive areas of cropland, agroforests, grazing lands, and forests that exist. The associated impacts—in terms of altered crop, animal, and tree production; water yield; biodiversity; energy use; and socioeconomic effects—are in many cases significant; these impacts can be negative or positive

***Table 4-1**: Potential net carbon storage of additional activities under Article 3.4 of the Kyoto Protocol. Increases in carbon storage may occur via (a) improved management within a land use, (b) conversion of land use to one with higher carbon stocks, or (c) increased carbon storage in harvested products. For (a) and (b), rates of carbon gain will diminish with time, typically approaching zero after 20–40 years. Values shown are average rates during this period of accumulation. Estimates of potential carbon storage are approximations, based on interpretation of available data. For some estimates of potential carbon storage, the uncertainty may be as high as ±50%.*

Activity (Practices)	Group[a]	Area[b] (10^6 ha)	Adoption/ Conversion (% of area) 2010	2040	Rate of Carbon Gain[b] (t C ha^{-1} yr^{-1})	Potential (Mt C yr^{-1}) 2010	2040
a) Improved management within a land use							
Cropland (reduced tillage, rotations and cover crops, fertility management, erosion control, and irrigation management)	AI	589	40	70	0.32	75	132
	NAI	700	20	50	0.36	50	126
Rice paddies[c] (irrigation, chemical and organic fertilizer, and plant residue mgmt.)	AI	4	80	100	0.10	<1	<1
	NAI	149	50	80	0.10	7	12
Agroforestry[d] (better management of trees on croplands)	AI	83	30	40	0.50	12	17
	NAI	317	20	40	0.22	14	28
Grazing land (herd, woody plant, and fire management)	AI	1297	10	20	0.53	69	137
	NAI	2104	10	20	0.80	168	337
Forest land (forest regeneration, fertilization, choice of species, reduced forest degradation)	AI	1898	10	50	0.53	101	503
	NAI	2153	10	30	0.31	69	200
Urban land (tree planting, waste management, wood product management)	AI	50	5	15	0.3	1	2
	NAI	50	5	15	0.3	1	2
b) Land-use change							
Agroforestry (conversion from unproductive cropland and grasslands)	AI	~0	~0	~0	~0	0	0
	NAI	630	20	30	3.1	391	586
Restoring severely degraded land[e] (to crop-, grass-, or forest land)	AI	12	5	15	0.25	<1	1
	NAI	265	5	10	0.25	3	7
Grassland (conversion of cropland to grassland)	AI	602	5	10	0.8	24	48
	NAI	855	2	5	0.8	14	34
Wetland restoration (conversion of drained land back to wetland)	AI	210	5	15	0.4	4	13
	NAI	20	1	10	0.4	0	1
c) Off-site carbon storage							
Forest products	AI	n/a[e]	n/a	n/a	n/a	210	210[e]
	NAI	n/a	n/a	n/a	n/a	90	90
Totals	AI					497	1063
	NAI					805	1422
	Global					*1302*	*2485*

[a] AI = Annex I countries; NAI = non-Annex I countries.
[b] Areas for cropland, grazing land, and forestland were taken from IGBP-DIS global land-cover database derived from classification of AVHRR imagery (Loveland and Belward, 1997). Each land-use/land-cover type was subdivided by the climatic regions defined in Table 4-4, using a global mean climate database (Schimel *et al.*, 1996) of temperature and precipitation, with additional calculations of potential evapotranspiration (Thornthwaite, 1948). Each climatic region was further subdivided by Annex I and non-Annex I countries. Modal rate estimates from Table 4-4 were then weighted by the relative area of each land use by climatic region for Annex I and non-Annex I countries to derive a global area-weighted mean rate for each land use.
[c] Riceland area was subtracted from cropland area.
[d] Of the 400 Mha presently in agroforestry, an estimated 300 Mha are included in the land-cover classification for cropland; the remaining 100 Mha are included in forestland cover. These areas were subtracted from the respective totals for cropland and forestland.
[e] Assumes that severely degraded land is not currently classified as cropland, forestland, or grassland.
[f] Estimates for 2040 are highly uncertain because they will be significantly affected by policy decisions; n/a = not applicable.

(frequently a mixture of both), depending on the specific activity and the environment in which it is applied. In deciding whether to develop policies or programs to encourage these activities, the associated non-climate benefits and tradeoffs of the activities are usually very important factors and may dominate over climate considerations at a local or national scale (Section 4.7).

Decisions about how additional activities will be implemented under the Protocol will significantly affect the cost and difficulty of implementation by the Parties, as well as the choice of reporting methods that will be used (particularly in the initial years). Carbon stocks in agricultural soils and forests can be measured, monitored, and verified by a range of methods with varying precision. The cost of measurement increases with the required precision. The methods continue to improve in accuracy and cost-effectiveness. One impact of including additional activities will be accelerated development and improved cost-effectiveness of new methods.

There are two basic options for inclusion of additional LULUCF activities: include a limited list of activities, or include essentially all LULUCF activities that affect carbon pools and/or GHG fluxes. Choosing to include a very limited list of additional activities may reduce short-term demands on a Party's accounting system, but it may fail to encourage activities and practices that have important impacts on the atmosphere and benefits for environmental quality and sustainable development and may result in displacement of emissions into areas or activities that are not included. A reporting regime in which a large number of small land areas must be monitored and tracked far into the future may make some practices (particularly those that tend to shift from place to place over time) very expensive to monitor and verify operationally. In that case, accounting for the main carbon stock changes over the entire land area of the country—as is done by some current national inventory systems—may be a more cost-effective approach in the long run. Such full-area accounting may simplify accounting procedures by allowing one set of rules to apply everywhere, making the precise definition of different land uses and LULUCF activities less critical and enabling a statistical sampling approach rather than spatially explicit recordkeeping.

For many LULUCF activities—particularly those that involve land-use change or widespread adoption of new management systems—measuring the direct human-induced effects independently of the background of indirect effects and natural variability is difficult or impossible. Where a single or well-defined change in management (such as the addition of fertilizer, a switch to conservation tillage, or a change in forest harvesting technique) can be identified and there are comparable areas nearby where that management is not undertaken, estimating the fraction of the total effect from the activity may be possible through the use of control plots, models, or predicted baselines. The decisions taken on these aspects will be important in affecting measurement and accounting difficulty.

Some agricultural practices (e.g., excessive soil disturbance, nutrient depletion, planting varieties with low biomass production, crop residue removal, inadequate erosion control practices) have caused reductions in soil organic carbon (SOC) estimated at 55 Gt C globally. The application of best-practice crop management—which may include conservation tillage, frequent use of cover crops in the rotation cycle, agroforestry, judicious use of fertilizers and organic amendments, site-specific management, soil water management that involves irrigation and drainage, and improved varieties with high biomass production—can recover a substantial part of this loss over a period of several decades. In rice agriculture, careful water and fertilizer management can lead to increased carbon storage, but calculation of the net effect must consider simultaneous changes in CH_4 and N_2O emissions. In general, carbon-enhancing cropland management will increase the capacity of croplands to feed the world's growing population (Section 4.4.2).

Forest management can create a significant increase in the standing stock of forest biomass and produce positive associated benefits such as improved water quality, reduced soil erosion, increased biodiversity, and enhanced rural income. Forest management includes all of the practices relating to the regeneration, tending, use, and conservation of forests. All aspects that are not included as afforestation, reforestation, and deforestation (ARD) activities under Article 3.3 could be covered by Article 3.4. Examples of the main forestry practices likely to alter carbon stocks include forest regeneration, including sub-practices such as human-induced natural regeneration, enrichment planting, less grazing of savanna grassland, change of tree provenances or species to include short-rotation forestry; forest fertilization; pest management; forest fire management; harvest quantity and timing; low-impact harvesting; and reducing forest degradation (Section 4.4.4).

The carbon impacts of forest management usually can be measured best as a net result of the complete suite of practices applied to a given forest area, as well as the end-use of the forest products. End use is important for two reasons: First, wood products provide a carbon sink, the size of which depends on the lifetime of the product; second, greater use of wood allows reduced use of fossil fuel (either by using the wood directly for energy production or indirectly to replace energy-intensive products such as steel, aluminum, and concrete). Including forest products as an activity under Article 3.4 will require decisions about how to report credits because the products will come from ARD and non-ARD forests (Chapter 3), and their original sources will be hard to trace once they are in use. Such products are widely traded between countries, requiring decisions about where and how to assign credits (Section 2.4.2.2).

Carbon storage rates per hectare in grazing systems are low, but these systems cover very large areas. Net carbon uptake can be achieved through reversion of degrading processes (overgrazing, erosion, acidification and salinization)—a strategy that also tends to increase productivity. Increased carbon storage can also occur through enhanced plant productivity by fertilization of nutrient-limited locations, introduction of better adapted grass and legume species, and irrigation in water-limited situations. Changes in grazing and fire management can also increase soil

***Table 4-2**: Some options and implications associated with selected questions that may arise in deciding how to implement Article 3.4.*

Selected Question	Options and Implications
Which additional activities should be included under Article 3.4? (Section 4.3.1)	*A few selected activities* • Decisions on criteria for inclusion are needed; may require regional flexibility. • Careful definition of each activity is needed; makes regional flexibility challenging. • May reduce cost and difficulty of near-term accounting.
	All activities that affect GHG emissions and sinks • Allows more flexibility in definitions and criteria. • May contribute to full carbon accounting. • May provide more accurate reflection of a Party's total GHG impact. • May have a more significant effect on first-period reports (Section 4.3.4). • If coupled with activity-based reporting (see below), requires development of more activity rate factors.
Should "activities" be broadly or narrowly defined? (Section 4.3.2)	*Broadly (i.e., cropland management or forest management)* • Definitions of different activities can be fairly general and adapted to regional differences. • Double-counting potential is reduced. • Land-based measures of stock change are well-adapted. • Full carbon accounting may be more feasible. • Indirect effects and/or background variability may need to be calculated.
	Narrowly (i.e., conservation tillage or forest fertilization) • Definitions of activities need to be unambiguous, implementable, and verifiable. • May tend to favor activity-based rather than land-based accounting methods (see below). • May make non-CO_2 estimates easier where activities are clearly associated with non-CO_2 emissions.
Should additional activities be measured by land-based or activity-based accounting methods? (Section 4.3.3)	*Land-based (i.e., sampling of C pools at start and end of accounting period)* • Statistical sampling methods for different pools are well-established. Cost varies with degree of precision demanded and frequency of measurement required. Methods can be transparent and results verifiable. • Will not measure non-CO_2 emissions. • If coupled with broad definition of activity, statistical sampling of large areas (e.g., a regional forest or agricultural area) at two points in time could capture net effect of emissions and sinks, eliminating need to track separate activities on individual forest patches or agricultural fields. Offers one possible method of reducing difficulty in tracking Articles 3.3 and 3.4 forest activities separately. • Statistical sampling could capture on-site effects of activities such as biofuel production or agroforestry. • If coupled with a narrow definition of activity, reduces the problem of tracking individual activities from place to place or year to year but increases problem of separating effects of different activities.
	Activity-based (i.e., area of activity times calculated rate of impact) • May be more suited if decisions are made to include only a limited group of narrowly defined additional activities under Article 3.4. • Could be adapted where a particular narrow activity (i.e., conservation tillage) was promoted by incentives and records were kept on amount applied by participating farmers. • Where more than one activity occurs on a particular piece of land, carbon impacts of different activities may be difficult to verify. • Requires development of emission or sink factors for each activity in each region. Some factors may need to be tied to specific land uses or soil types under some conditions.

Table 4-2 (continued)

Selected Question	Options and Implications
Should additional activities be measured by land-based or activity-based accounting methods? (Section 4.3.3) (continued)	*Activity-based (i.e., area of activity times calculated rate of impact) (continued)* • Can be used to estimate non-CO_2 emissions. • Methods can be transparent, but verification of seasonal activities may be difficult or impossible at a later time.
	Some combination of the two methods • May provide an opportunity to better fit an accounting method to different activities and/or available data. • Activity-based methods or default values could provide a transition method while a Party established new institutional capacity. • If land-based measures were used to estimate changes in carbon pools, activity-based measures might be used to calculate associated non-CO_2 effects to achieve more complete accounting. • Could provide a different approach for verification (i.e., land-based approach could measure total effect after several activity-based estimates had been applied to one piece of land or over several years of application). • Where land-based measures were used, an associated activity-based estimate could be one means of adjusting measured carbon stock change to compensate for indirect effects or natural variability.
Should soil carbon stocks be included in the accounting? (Section 4.2)	*Yes* • This pool contains large amounts of carbon, is the major pool in agricultural (cropland and grassland) situations, and soil carbon stocks can be altered (+ or -) by management or land-use change. • Incentives to improve soil carbon also tend to help improve soil quality, productivity, and sustainability of agricultural and forestry systems. • Soil surveys and periodic monitoring needed. Methods are readily available from a scientific viewpoint but require investment and may be currently lacking in some areas, suggesting technology transfer and/or transition strategies to produce adequate data in short term.
	No • Need for soil surveys and soil monitoring is reduced. • Major carbon transfers between atmosphere and biosphere (emissions and sinks) will go unreported. • Opportunities to provide incentives for improving agricultural system sustainability, expanding agroforestry, producing biofuels, or restoring severely degraded land may be lost. • Incentives for development of more cost-effective ways to monitor soil carbon (e.g., remote sensing) are reduced.
Should forest products be included as an additional activity under Article 3.4? (Section 4.5.6)	*Yes* • Contributes to full accounting, if that is a desired goal. • Decisions on accounting methods are needed. • Activity- or flux-based accounting methods appear more feasible for this activity, as opposed to land-based accounting. • A more complete accounting of the total carbon impact of certain forest management strategies (e.g., utilization vs. protection) could result. • Incentives may be increased for using wood in place of higher energy-using substitutes in construction. • Incentives may be increased for recycling and extending product life.
	No • Land-based accounting, if used on forest areas, need not attempt to track fate of forest products after they leave site. • Full carbon implication of forest management options such as extended rotations is not captured.

Table 4-2 (continued)

Selected Question	Options and Implications
Should forest products be included as an additional activity under Article 3.4? (Section 4.5.6) (continued)	*No (continued)* • Fewer incentives will be created to use wood in construction, to recycle, and to extend product life. • Accounting in terms of trade between Parties will not be required.
Should wetland restoration or drainage be included as an additional activity? (Sections 4.4.6 and 4.7.2)	*Yes* • May capture effect of locally important land-use activities. • May require activity-based accounting methods to be developed that can estimate both CO_2 and CH_4 impacts. • In some cases, results may be different than expected in terms of climate-forcing impact. • Associated impacts may be more important than climate impacts in affecting land-use and management decisions.
	No • Reduces complexity of monitoring and accounting systems. • Probably has minor effect on national accounting totals for Parties. • May miss opportunities to encourage wetland management that has other environmental benefits.

and biomass carbon pools. The increases in woody biomass that usually follow a reduction in fire frequency may have adverse impacts on grazing productivity, other production values, and biodiversity. Monitoring the extensive but relatively small per-unit area changes in grassland carbon stocks is technically challenging. The net assessment must include changes in CH_4, N_2O, and ozone-precursor emissions (Section 4.4.3).

Drainage of wetlands can reduce methane emissions, but it results in a rapid emission of carbon dioxide. Re-flooding of previously drained wetlands allows carbon to be stored, but only at a slow rate—which can result in domination of the radiative forcing benefit by the restored methane emission. These offsetting impacts make full and symmetrical accounting of wetland management a difficult and complex problem (Sections 4.4.6, 4.7.2).

The chapter identifies several activities that can occur under different land use and cover conditions. These activities therefore are difficult to separate out under most land-cover mapping schemes (Section 4.3.2) that separate the landscape into areas of cropland, grassland, forest, and so forth. These activities include agroforestry and biofuel production—which are options for growing a different mix of crops on crop, grass, or forest land—and restoration of severely degraded soils.

The use of agricultural and forest crops to produce biofuels reduces the use of fossil fuels—currently by the equivalent of about 1 Gt C yr^{-1} and potentially by up to 10 Gt C yr^{-1}. If biomass crops were deployed at this maximum level, they would be a major land use. The fossil fuel replacement effect is captured under existing reporting rules, but the direct effects of land conversion to biofuel crops (which can be positive or negative) are not. The latter could be included under Article 3.3 (where forest crops were planted on former cropland) and/or under Article 3.4 (where growing grasses or short-rotation woody crops for biofuels were defined as a cropping change). Modern biofuel technology could contribute to sustainable development and GHG mitigation and may reduce GHG levels sooner than other energy sector measures, at low cost.

Agroforestry can increase carbon sequestration while creating positive environmental and social impacts such as improved biodiversity and farm income security. It includes conversion of land recently derived from forest to agroforests rather than to croplands or pastures, restoration of nutrient-depleted lands using trees, and improvement of productivity in current agroforestry systems in tropical and temperate regions. At present, the application of agroforestry is largely limited to non-Annex I countries.

Restoration of severely degraded soils can produce significant increases in soil and biomass carbon per unit area. In general, restoration activities also improve biodiversity, promote sustainable productivity, and reduce air and water pollution. Their adoption will often require some form of additional incentives because the immediate benefits to land managers can be negligible or negative. If relapse is avoided by addressing the causes of the original degradation, the carbon pools formed are effectively permanent. Reclamation, restoration, and removal of the causes of severe land degradation are almost always expensive, so no more than a few percent of the affected area can or will realize this potential. Fully reclaimed land can be put to a variety of uses and offers the opportunity to adopt many of the sequestration methods discussed in this Special Report for increasing the carbon sink still further.

To facilitate the assessment of existing proposals and future proposals as they arise, we have suggested a decision framework that can be adapted to assist in the consideration of issues such as verifiability, additionality, permanence, and additional impacts (see Table 4-2).

4.1. Introduction and Context

The Kyoto Protocol assigns amounts of GHG emissions to the countries listed in Annex B. Under Article 3.4 of the Protocol, Parties may add to or subtract from those assigned amounts GHG emissions by sources and removals by sinks that result from "additional" activities in agricultural soils, land-use change, and forestry. Decisions about which additional activities will be considered under Article 3.4 have not yet been made; these decisions are partly related to definitions adopted under Article 3.3 because in this context the term "additional" means activities that are in addition to ARD. Some key policy choices raised by Article 3.4 are illustrated in Figure 4-1 (note that numbers in parentheses indication the section in this report in which respective questions are most extensively discussed). This chapter presents scientific and technical information that may help inform those choices. There are also choices to be made on accounting rules (options are discussed in Chapter 2), which will affect how the adjustments are to be calculated. Some of the questions posed in Article 3.4 are more fully addressed in other chapters of this Special Report.

This chapter does not limit its consideration to practices or regions that are exclusively of interest to countries that have Kyoto Protocol commitments during the first commitment period. Subsequent periods may include a wider range of countries, and some countries currently outside Annex B may make voluntary commitments or be drawn into land-use activities under Article 12, if that is permitted.

We use the term "land use" to cover the entire range of direct management activities that affect agricultural soils, result in land-use change, alter forest management, or affect the long-term storage of carbon-containing products. All such activities are implicitly "human-induced." It is not clear that all measurable changes in carbon stocks are "directly" human-induced—a qualifier that appears in Article 3.3 but not in Article 3.4. Chapter 2 discusses the implications of the direct versus indirect distinction.

4.2. Processes and Time Scales

This section presents the scientific understanding of the mechanisms by which land-use activities lead to changes in carbon stocks contained in terrestrial ecosystems, as well as changes in the net emissions of non-CO_2 GHGs. It explains why some of the mechanisms are temporary, over what periods they can be expected to operate, and why the carbon pools that are formed differ in their capacity and permanence.

4.2.1. *Land Use, Management, and Ecosystem Carbon Balances*

Ecosystem carbon stock changes are determined by the balance of carbon inputs—via photosynthesis and organic matter imports—and carbon losses through plant, animal, and decomposer respiration; fire; harvest; and other exports.

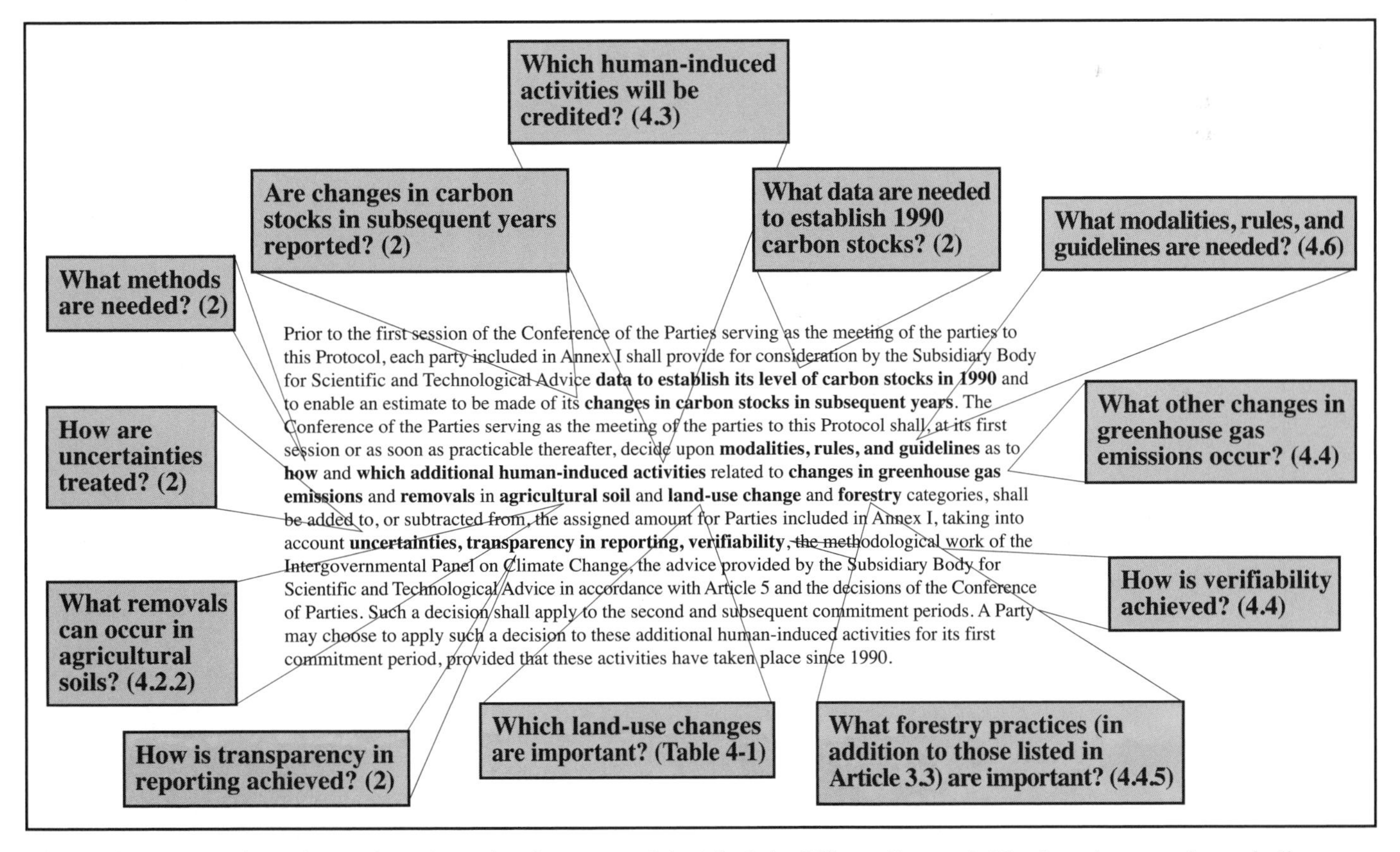

Figure 4-1: Illustration of questions that arise from text of Article 3.4 of Kyoto Protocol. Numbers in parentheses indicate sections of this Special Report where the questions are discussed.

Typically, but not always, intensive human use of an ecosystem leads to a net depletion of carbon storage relative to lightly exploited ecosystems. For example, conversion of grasslands and forests to agriculture typically causes the loss of aboveground and below-ground carbon stocks and accelerates respiration relative to photosynthesis, resulting in a decline of carbon reserves until the rates of inputs and outputs again converge (e.g., Cerri *et al.*, 1991; Davidson and Ackerman, 1993; Balesdent *et al.*, 1998). Conversely, a change in management that favors inputs relative to losses will elicit a gain in carbon storage until losses again equal inputs (see Figure 4-2). Such an increase in stored carbon, relative to the current trajectory, can be achieved by a shift in management or environmental conditions. Broadly, there are two ways of achieving an increase in carbon stocks:

- A change in land use to one with higher carbon stock potential, usually revealed by a change in land cover—for example, conversion of cropland to grassland (e.g., Robles and Burke, 1998; Conant *et al.*, 2000).
- A change in management within a land use that does not lead to a qualitative change in land cover—for example, introduction of more productive species in grasslands (Fisher *et al.*, 1994), reduction in tillage intensity in croplands (Paustian *et al.*, 1997a; Rosell and Galantini, 1998), or forest regeneration (Ravindranath *et al.*, 1999).

The rate of accumulation of additional stored carbon from a change in land use or management cannot be sustained indefinitely. Eventually, input and loss rates come into balance and carbon stocks approach some new, higher plateau (Odum, 1969; Johnson, 1995). Consequently, additional carbon storage in response to any management shift is of finite magnitude and duration (Greenland, 1995; Scholes, 1999). Rates of carbon

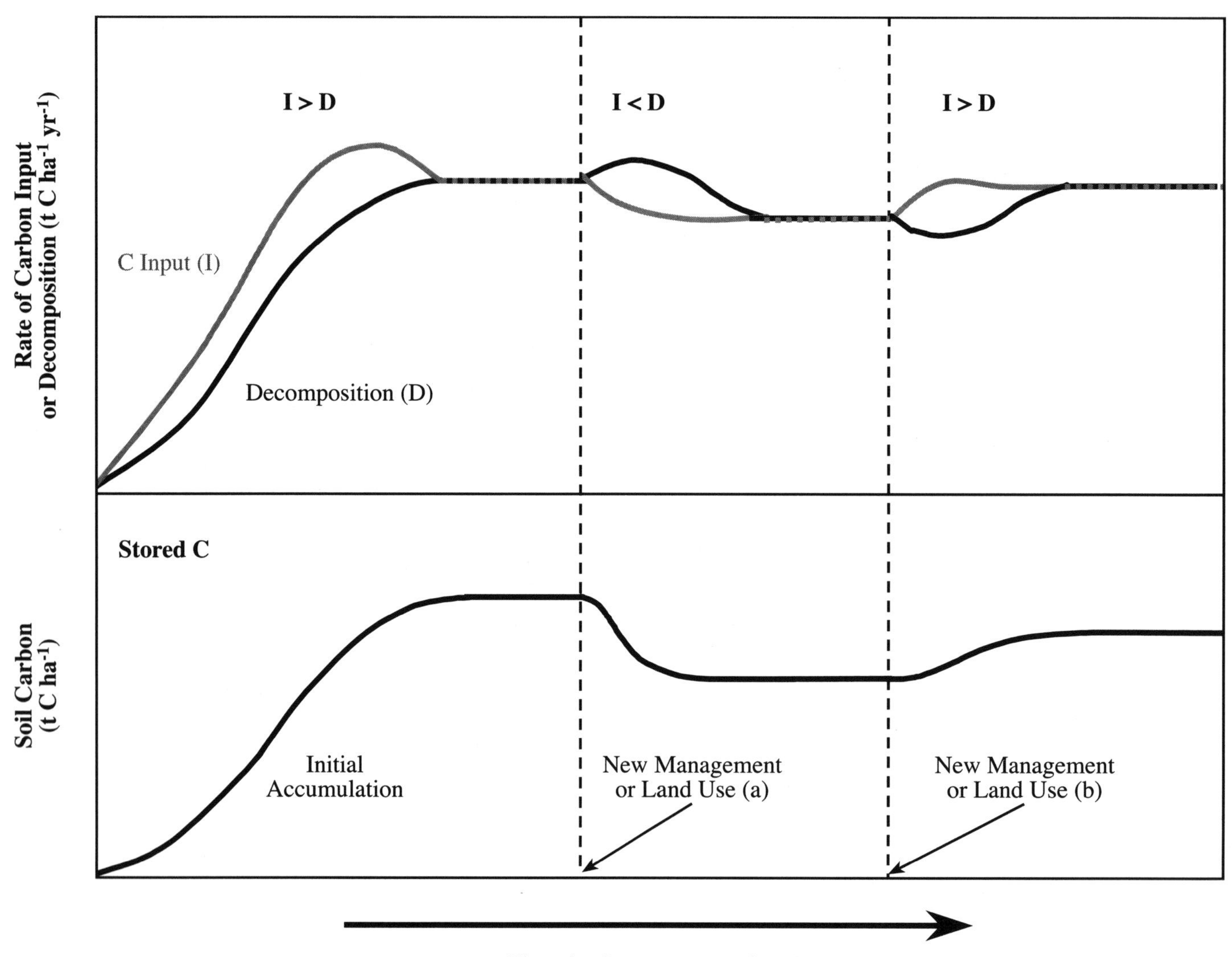

Figure 4-2: Idealized view of changes in ecosystem carbon storage. During early succession, rate of carbon input (I) exceeds rate of decomposition (D), resulting in an accumulation of stored carbon, equivalent to accumulated difference in area between the two curves. Eventually the rates converge and carbon storage approaches a maximum. Adoption of a new land use or management within a land use may alter the relative rates of I and D, resulting in either a loss of stored carbon (a) or a gain in stored stored (b). In either case, rates of I and D eventually converge, and the ecosystem approaches a new "equilibrium" carbon level (based on Odum, 1969).

***Table 4-3**: Possible repositories for additional carbon storage in terrestrial ecosystems or their products, and approximate residence times for each pool. Mean residence time is average time spent by a carbon atom in a given reservoir.*

Repository	Fraction	Examples	Mean Residence Time
Biomass	woody	tree boles	decades to centuries
	non-woody	crop biomass, tree leaves	months to years
Soil organic matter	litter	surface litter, crop residues	months to years
	active	partially decomposed litter; carbon in macro-aggregates	years to decades
	stable	stabilized by clay; chemically recalcitrant carbon; charcoal carbon	centuries to millennia
Products	wood	structural, furniture	decades to centuries
	paper, cloth	paper products, clothing	months to decades
	grains	food and feed grain	weeks to years
	waste	landfill contents	months to decades

gain often are highest soon after adoption of a new practice, but they subside over time. Although the temporal pattern of carbon accumulation varies, the benefits of biological mitigation will likely be most pronounced in the first few decades after adoption of carbon-sequestering land use and management practices (Dumanski *et al.*, 1998; IGBP, 1998; Smith *et al.*, 1998).

The main factors that determine the rate and duration of carbon gain in a given ecosystem after a management change follow:

- *The form the newly stored carbon takes.* Carbon is stored in terrestrial ecosystems in diverse organic forms with a wide range of mean residence times (Balesdent and Mariotti, 1996; Harrison, 1996; Skjemstad *et al.*, 1996; Trumbore and Zheng, 1996). The mean residence time (MRT) is the average time a carbon atom spends in a given pool. Some carbon—for example, much of the carbon in fresh litter or non-woody biomass—is ephemeral and returns quickly to the atmosphere (Stott *et al.* 1983). Other carbon, such as that in woody materials or "active" soil organic matter fractions, may persist for decades (Table 4-3). The most persistent stock of carbon is stable soil carbon (including charcoal) that may have an MRT as high as 1000 years or more because of chemical inertness or interaction with soil minerals (Balesdent and Mariotti, 1996; Paul *et al.*, 1997a). The amount and duration of carbon gain within an ecosystem depend on the temporal dynamics of these different pools: Transient pools may increase rapidly but quickly level off, whereas carbon that is incorporated into more stable pools can produce slow but long-term increases. Consequently, the initial impact of land-use or management change occurs disproportionately in pools with shorter residence times (Cambardella and Elliott, 1992; Huggins *et al.*, 1998a), whereas increases in stable soil pools occur slowly over a much longer time period (see Figure 4-3).

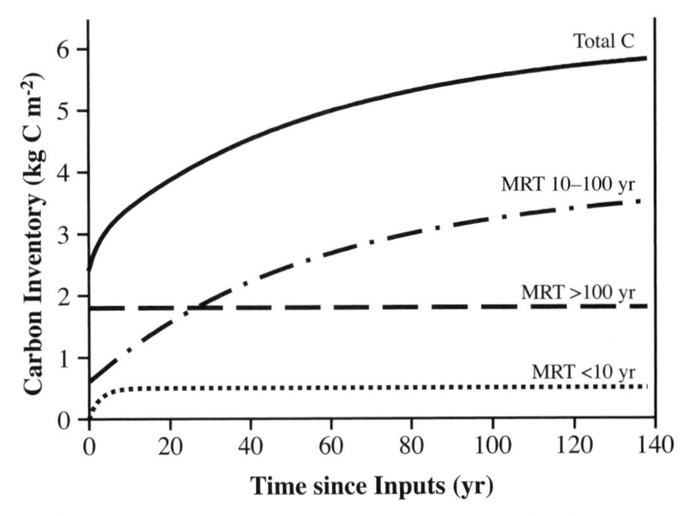

Figure 4-3: Predicted response of different pools of soil organic matter for an agricultural soil converted to forest in northeastern United States of America (Gaudinski *et al.*, 2000). Early response reflects changes in the relatively small pools with mean residence time (MRT) <10 years (leaf and root residues). Pools with intermediate MRT (10–100 years; including humified organics in litter layers) dominate the overall response because this pool contains most organic matter in this soil. Persistent carbon pools (MRT >100 years) do not change appreciably over a 100-year period. MRT = average time spent by a carbon atom in a given reservoir.

- *The degree of land-use or management change*. The potential carbon gain depends on the carbon stock potential of the new practice relative to that of the previous practice (Greenland, 1995). Consequently, a wholesale change from one land use to another (e.g., cropland to forestland) often elicits the largest response in carbon stock change. Within a given land use, potential carbon sequestration depends on how far the current carbon status is below the eventual maximum storage of the newly adopted practice. All else being equal, an ecosystem that has been under effective carbon-conserving management for several years has less potential for further storage than one that has been severely depleted of carbon. In ecosystems in which carbon stocks are already near maximum, the primary focus may be preserving existing stocks.
- *The persistence of current management and climatic regimes*. Continued accumulation of carbon in response to new management or land use depends on continued application of that new practice; a reversal can lead to partial or complete loss of previous gains (Dick *et al*., 1998; Stockfisch *et al*., 1999). The long-term pattern of carbon accumulation also is responsive to changes in climatic conditions. For example, accumulations of stored carbon may be susceptible to loss from accelerated decomposition under higher temperatures in future decades (Jenkinson, 1991; Trumbore *et al*., 1996), although losses may be at least partially offset by the effects of CO_2 fertilization (see Chapter 1). As carbon stocks increase, so does the potential for future losses from regressive management or unfavorable climatic conditions.

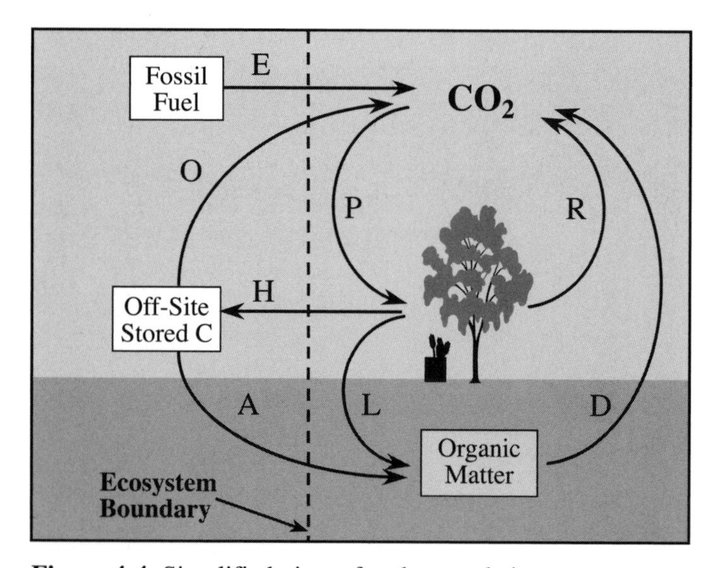

Figure 4-4: Simplified view of carbon cycle in an ecosystem and associated off-site carbon. Designations are as follows: P = photosynthesis; R = plant respiration [includes respiration from herbivory and abiotic respiration (e.g., fire)]; H = harvest; L = litter fall; D = decomposition; E = CO_2-C emission from energy use in the ecosystem; O = oxidation of harvested carbon (e.g., consumption/respiration of food products, burning); A = amendment of ecosystem with off-site organic carbon (e.g., biosolids, wood chips). Net change in stored carbon = P – (R + D + H) – O. Net effect of ecosystem on atmospheric CO_2 = (E + O + D + R) – P. Some practices can reduce net emissions by substitution; for example, biofuels reduce E by substitution with O.

The effect of a new management or land use on atmospheric CO_2 cannot be judged solely on the basis of net carbon storage within the ecosystem (see Figure 4-4). In many "managed" ecosystems, there is significant removal of carbon in harvested product. Some of this harvested carbon may accumulate in long-term repositories (e.g., wood products), and some is quickly returned to the atmosphere via respiration (e.g., agricultural products) (see also Figure 2-2). Thus, the full impact of a new management practice on atmospheric CO_2 can be assessed only by including net changes in off-site carbon stocks.

Energy use—notably that from fossil fuel used to establish and maintain a given land use—also affects the net exchange of carbon between the ecosystem and the atmosphere (Figure 4-4). Consequently, if a management or land-use change affects energy use, the corresponding CO_2 emission affects the net carbon balance. For example, if an effort to increase soil carbon in cropland requires higher fertilization, the CO_2 from energy involved in manufacturing that fertilizer may partially offset soil carbon gains (Flach *et al*., 1997; Janzen *et al*., 1998; Schlesinger, 1999). By replacing fossil fuel, biofuels can reduce the net emission of CO_2 (Cole *et al*., 1997).

Some land management practices may also affect emission of GHGs other than CO_2, thereby augmenting or offsetting CO_2 sources and sinks. For example, wetland restoration may increase methane emissions (see Fact Sheet 4.18); greater use of nitrogen fertilizers to enhance crop productivity (Fact Sheet 4.1) may enhance N_2O emissions; biomass burning emits CH_4 and N_2O; and conversion of arable land to grassland may reduce N_2O emissions (Fact Sheet 4.7). Emissions from these activities are already estimated and reported in a country's GHG inventory under the Revised Guidelines for National Greenhouse Gas Inventories, hereafter referred to as the IPCC Guidelines (IPCC, 1997); nevertheless, these emissions must be considered when a Party is contemplating the adoption of any new carbon-conserving practices.

4.2.2. Soil Carbon Dynamics

Article 3.4 of the Kyoto Protocol refers specifically to agricultural soils. However, the basic principles discussed herein cover soil carbon dynamics in general regardless of the land use and apply to all ecosystems (cropland, forest, grazing lands, and wetlands).

Although most carbon enters ecosystems via leaves, and carbon accumulation is most obvious when it occurs in aboveground biomass, more than half of the assimilated carbon is eventually transported below ground via root growth and turnover, exudation of organic substances from roots, and incorporation of fallen dead leaves and wood (litter) into soil. Soils contain the major proportion of the total ecosystem carbon stock in all ecosystems.

As with total ecosystem carbon stocks, soils tend toward "equilibrium" carbon levels. With a change in carbon input and/or decomposition rates, soil carbon stocks change. This change is most rapid for the active fraction, including structural carbon (i.e., cellulose and hemicellulose) and metabolic carbon components (i.e., proteins, lipids, starches, nucleic acids); the slow fraction (i.e., microbial walls and metabolic components protected by soil clays and aggregates); and the passive soil carbon (i.e., clay-protected humics).

Carbon inputs to soil are determined by the amount and distribution of primary production, the life cycle of the vegetation, and exogenous organic matter additions (e.g., composts, manure). Thus, practices that increase net primary production (NPP) and/or return a greater portion of plant materials to the soil have the potential to increase soil carbon stocks. Organic matter decomposition is influenced by numerous physical, chemical, and biological factors that control the activity of microorganisms and soil fauna (Swift *et al.*, 1979). These factors include the abiotic environment (temperature, water, aeration, pH, mineral nutrients), plant residue quality (i.e., C:N ratio and lignin content), soil texture and mineralogy, and soil disturbance (tillage, traffic, logging, grazing, etc.). The root system, depth distribution, and chemical characteristics of the root biomass also play significant roles in SOC dynamics (Gale and Cambardella, 1999). Practices that reduce the decomposition rate by altering these physical, chemical, or biological controls also lead to carbon storage.

In some soils, changes in soil inorganic carbon (SIC) stocks can be significantly influenced by land use and management (Lal *et al.*, 2000), making SIC either a sink or source of atmospheric CO_2:

- In native ecosystems, the weathering of base-rich silicate minerals releases calcium or magnesium, which can combine with CO_2 and precipitate as secondary carbonates ($CaCO_3$ or $MgCO_3$). The addition of excess calcium or magnesium (e.g., in sewage sludge) to irrigated cropland soils of dry regions may increase rates of accumulation (Gislason *et al.*, 1996; Nordt *et al.*, 2000). In contrast, secondary carbonates formed in soils having calcareous parent materials (e.g., limestone) are neither a sink nor a source of CO_2.
- In humid regions, losses of SIC from naturally calcareous (e.g., limestone-derived) soils can be accelerated through biomass removal and the use of acidifying fertilizers in cropland and grassland (van Bremen and Protz, 1988). The use of carbonate-rich groundwater for irrigation can also result in a net flux of CO_2 from groundwater storage to the atmosphere (Schlesinger, 1999).

4.2.2.1. *Factors Leading to Aggradation of Soil Carbon*

Net accumulation of soil carbon occurs through practices that increase the amount of plant-fixed carbon that is returned to soils in the form of residues (i.e., leaves, stems, and branches) and especially roots (Balesdent and Balabane, 1996) and/or reduce the specific rate of decomposition. Strategies to increase productivity—including plant selection and breeding, plant protection, irrigation, and reducing the frequency of bare fallowing—contribute to higher carbon inputs. In addition, reducing the removal or burning of biomass (in crop, forest, and grazing lands) increases the portion of total productivity that is returned to soil.

Specific rates of decomposition (i.e., CO_2 emissions per unit soil carbon per unit time) can be reduced by creating a less favorable environment for decomposer organisms. The pores and surfaces of the soil are the habitat for decomposer organisms. Soil structure (i.e., the arrangement and stability of primary particles, aggregates, and pores within the soil) controls that habitat and therefore has a large effect on decomposition. Soil structure affects aeration and moisture dynamics and is itself affected by tillage, other disturbances, rooting habit, soil texture, clay mineralogy, and base cation status (Lal, 2000a). Maintaining surface mulches, reducing incorporation of residues, and enhancing plant/forest canopy coverage can reduce soil surface temperatures. Increasing the opportunities for water extraction and atmospheric CO_2 fixation throughout a greater portion of the year (e.g., through crop selection and greater cropping frequency) can cause soil to dry more thoroughly, thereby reducing microbial activity and leading to increased soil organic carbon storage. Anaerobic conditions in suitably adapted ecosystems (i.e., native wetlands, flooded rice) retard decomposition rates and CO_2 emissions but increase emissions of CH_4. Reducing soil disturbance associated with tillage, logging, and roading tends to reduce specific rates of soil organic matter decomposition (Lal and Kimble, 2000; Paustian *et al.*, 2000a).

4.2.2.2. *Factors Leading to Degradation of Soil Carbon*

Reductions in soil carbon stocks are driven by the inverse of the conditions described in Section 4.2.2.1. The carbon content of agricultural soils has generally been depleted by 20–50 percent relative to their original native condition (Mann, 1986; Schlesinger, 1986; Davidson and Ackerman, 1993). Historically, low crop productivity, harvest export, residue removal (including burning), intensive tillage, and bare fallowing have been immediate causes of decreases in soil carbon stocks (Paustian *et al.*, 1997a). In many European countries, removal of forest litter was a common practice that caused severe degradation of forest soils (Kreutzer, 1972). Similarly, where mismanagement leads to significant, long-term degradation of the productive capacity of forest, grasslands, or crops, soil carbon stocks will be depleted. Wetland drainage and conversion to cropland, pasture, and forests results in rapid decomposition and loss of organic carbon from the carbon-rich soils beneath them (Armentano and Menges, 1986). Secondary salinization, leaching and acidification, and other soil degradation processes (i.e., accelerated erosion) have adverse effects on SOC stocks.

Disturbance of the land cover by human activity sometimes leads to soil erosion by wind and water at rates greatly exceeding the rate of new soil formation. This effect has severe adverse impacts on productivity, as well as water and air quality. Erosion tends to selectively displace organic matter-enriched components (particulate and silt- and clay-associated organic matter) of surface soils and thus can significantly reduce soil carbon stocks at a particular location. Erosion processes may also stimulate soil organic matter decomposition rates by dispersing soil aggregates and releasing physically protected organic matter (Lal *et al.*, 1998). Adverse effects of erosion on NPP from loss of fertility and/or soil structural degradation would also have a negative impact on the soil carbon balance. A portion of the eroded material is deposited in lower portions of the landscape (depressions, floodplains, aquatic sediments), and some small fraction finally ends up in marine sediments on the continental shelves. Burial in these typically anaerobic environments may reduce the decomposition rate of organic matter attached to the eroded material, thereby increasing sequestration of this fraction of the eroded carbon (van Noordwijk *et al.*, 1997; Stallard, 1998). SOC loss on cropland occurs because of erosion, tillage-induced soil displacement, and exacerbated rates of mineralization (Govers *et al.*, 1994; Reicosky, 1994; Lobb *et al.*, 1995; Reicosky and Lindstrom, 1995; Lindstrom *et al.*, 1990, 1992, 1998, 1999; Lindstrom, 1997). Soil erosion control is necessary to maintain and enhance the SOC stock.

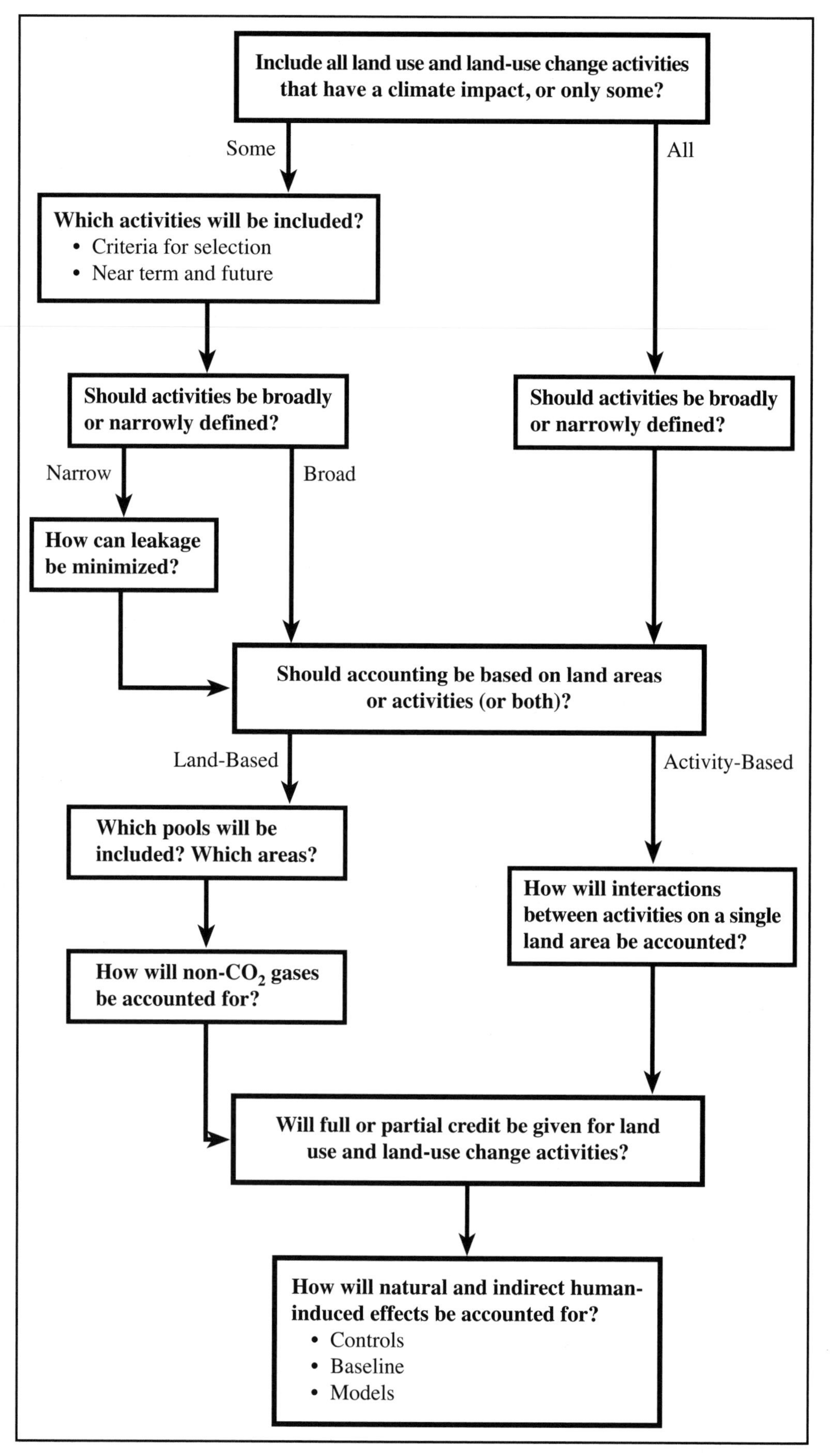

Figure 4-5: Decision tree to assist in determining which additional activities to include under Article 3.4 of Kyoto Protocol.

4.3. Choices Relating to Inclusion and Definition of Activities

The decision about which activities should be included and the decision as to how broadly those activities should be defined are interdependent (see Figure 4-5).

4.3.1. Which Additional Activities are to Be Included?

The Parties have two broad choices with respect to including activities under Article 3.4:

- Include a limited, selected set of activities ("some" in Figure 4-5)
- Include all activities that can be shown to have consequences on the atmospheric concentration of GHGs.

A limited set of activities will be easier to implement and verify; if activities are narrowly defined (see Section 4.3.2), however, only a fraction of the true atmospheric impact of LULUCF will be included within inventories and reported amounts. A narrow definition could also result in displacement of atmospheric impact from activities that are included into activities that are not included. Because specific, narrowly defined activities are typically concentrated within particular geographic areas, a short, limited list may favor some regions over other regions. Precise definition of what is meant by each specific activity would be more critical under this option.

Including all possible activities would encourage Parties to increase the land area that would likely be measured and reported, leading toward full carbon accounting. The potential for undetected leakage would be reduced—as would the consequences of leakage if it does occur because an impact created in another land area would be measured wherever it occurred. For this reason, precision of definition of activities is less critical under this option.

4.3.2. How Broadly or Narrowly are "Activities" Defined?

In this context, the term "broad" is used to denote an activity definition that is essentially land- or area-based, where the net effect of all practices that might be applied to the same area over the same time period are included within the activity. For example, the broad activity of "cropland management" might embrace practices such as conservation tillage, improved fertilization, and irrigation water management. If broad definitions are applied, periodic stock change measurements that statistically sample the area to be monitored could be used to estimate the net effect of altered carbon emission or sink processes. Where statistical sampling methods are used, monitoring larger areas becomes more cost-effective because the number of samples required for a given degree of statistical accuracy does not scale up proportionately with increased area.

The term "narrow" denotes an activity definition that is based on individual practices, such as reduced tillage or irrigation water management. Where individual practices are used as the basis for reporting, associated rates of reduced emissions or increased sinks need to be developed through model estimates or research plots—which in many instances are not yet available for all environmental circumstances. Estimates of annual rates of carbon stock change could then be estimated as a product of the area affected by the practice and the associated rate of emission or sink (perhaps by general soil type, if needed for accuracy). Such estimates could be verified by using other data sources or by a sampling technique that measured soil carbon stock change on the area where the practice had been reported. Non-CO_2 GHG emission estimates could be included where appropriate (on agricultural soils, these emissions may already be reported under the IPCC Guidelines, so care may need to be taken to avoid double-counting).

Some implications regarding definition of activities follow:

A) Broad
   - Activity definitions can be fairly loose.
   - Lends itself to land-based stock change CO_2 flux measurements; helps minimize double-counting or undercounting.
   - Non-CO_2 GHG fluxes must be estimated with practice-based models or other methods.
   - Verifiable by a third party, using different methods and at a later date, using techniques such as statistical sampling or remote sensing.

B) Narrow
   - Activity definitions critical; must be unambiguous, implementable, and verifiable.
   - Lends itself to activity-based model estimates of fluxes. Where multiple activities occur on the same land area, there is a potential for double-counting.
   - CO_2 and non-CO_2 emissions can be estimated with activity-based models. Where more than one activity occurs on a land area, some of which are not included in Article 3.4 accounting, stock change measures would be more difficult.
   - Verifiable by auditing activity statistics and independently checking them using other data sources. Verifying where activities have occurred in prior years may be difficult or impossible if no visible evidence remains.

Both options would require further decisions about the modalities, rules, and guidelines necessary for implementation. Some options are discussed in Section 4.6 and in Chapter 2.

The process of selecting activities would be facilitated and made more transparent and consistent by the development of criteria for activity selection (Section 4.7).

One approach to structuring activities broadly is suggested by Figure 4-6, which indicates how broadly defined activities might be fitted into a land-cover classification system. If such a scheme were used, every activity, no matter how it were defined, would fit in one of the boxes in the matrix. A national inventory based on a sampling scheme that produced statistically reliable estimates for each land cover/land use box in the matrix could produce national estimates of carbon stock change across all land uses and land cover areas. Therefore, if Parties differed somewhat over the definition of an activity, the practices carried out within the activity, or a definition of land cover such as forest, the impact on total accounting would be minimal because the impact would simply shift from one box in the matrix to another. If such an approach were adopted, a full-carbon accounting system (Chapter 2) would be required, wherein each major land area and land use were sampled on a periodic basis. Such a system is in place in some countries (national forest inventories are one example) but may currently be infeasible for some.

The activities reviewed in Section 4.4 are categorized in a way that is consistent with this broadly defined approach—not to prejudge the decisions to be taken by the Conference of Parties (COP) but to keep the presentation to a reasonable length. For more specific information on a few of the practices that might be considered as additional activities under the narrow-definition approach, see the Fact Sheets at the end of the chapter.

4.3.3. Land-Based versus Practice-Based Accounting

The choice of accounting approaches will be important in establishing the difficulty in implementing additional activities under Article 3.4 (see Section 2.3.2.2 for a discussion of these accounting issues and Section 3.3.2 for a discussion of various measurement methods as they apply to Article 3.3 activities). If the Parties decide on land-based measurement for activities under Articles 3.3 and 3.4 and the affected Parties are successful in identifying and tracking individual land parcels in a land-based system, some additional cost and difficulty would result from the extra parcels of land that would need to be tracked, measured, and recorded as a result of additional Article 3.4 activities. This increased cost and difficulty may not pose a significant challenge if a Party's reporting is based solely on project activities. If the Party wishes to report on the implications of a policy or program with national impacts, however, the technical and operational difficulties may become significant. The additional demand for field monitoring resources would essentially scale linearly with the number of separate land areas involved. The incremental effort may be small, however, relative to the large investment of effort needed for the initial establishment of a system for Article 3.3, unless the sheer number of land areas overwhelms the system's capability.

If a land-based approach were chosen, the demand on a Party's measurement and accounting system created by the additional activities selected under Article 3.4 would be significantly affected by the manner in which those activities were defined. Under the broad definition, the monitored land areas might be large regions of similar land use where periodic measurements based on statistical sampling methods were used to document changes in stock over the whole area.

Land Cover ⇩ *from* *to* ⇨	**Cropland**	**Grassland, Desert, Savanna**	**Forest, Woodland**	**Urban, Industrial**	**Wetland, Tundra**
Cropland	Cropland management (3.4)	Cropland conversion (3.4)	Afforestation (3.3)	Development (3.4)	Wetland restoration (3.4)
Grassland	Grassland conversion (3.4)	Grassland management (3.4)	Afforestation (3.3)	Development (3.4)	Wetland restoration (3.4)
Forest	Deforestation (3.3)	Deforestation (3.3)	Forest management (3.4)	Deforestation (3.3) Land-use change w/o deforestation (3.4)	Wetland restoration (3.4)
Urban - Industrial				Urban ecosystem management (3.4) Products (3.4)	
Wetland Tundra	Drainage (3.4)	Drainage (3.4)	Drainage (3.4)	Drainage (3.4)	Peat and rice management

Figure 4-6: Suggested land-cover, land-cover/use change, and forestry matrix that illustrates how activities might be identified with different land cover areas. Numbers in parentheses indicate relevant Article in Kyoto Protocol, where apparent.

If the narrow definition of activities is utilized—and a fairly lengthy list adopted—maintaining a spatially explicit database on each land area where activities were implemented may be beyond the capacity of any existing national environmental inventory system, so the Parties may wish to consider options that do not require spatially explicit recordkeeping. In selecting among these options, the major issues are likely to be cost and feasibility (Figure 4-7). Although all options are technically feasible, they vary widely in their degree of difficulty. Where Parties have strong institutions of private land ownership and management responsibility, establishing a system in which land owners and managers carry out the basic measurements and annual monitoring, with verification by a third party or government, may be feasible. (The issues of accounting methods discussed in Chapters 2 and 3 would be encountered in a similar manner with activities under Article 3.4.)

The impact of these policy choices on monitoring and reporting effort may or may not be significantly affected by selection of a short rather than long list of activities under Option A, if a broad activity definition is adopted. In instances in which a national inventory system is established and the entire land surface is periodically monitored, the addition of further activities would represent no additional effort. If a narrow definition of activity were adopted, the effort would essentially scale up with the number of land areas affected by that activity, and by the number of different activities in the list. The effort would continue to rise, even once the entire land surface of an area was under observation.

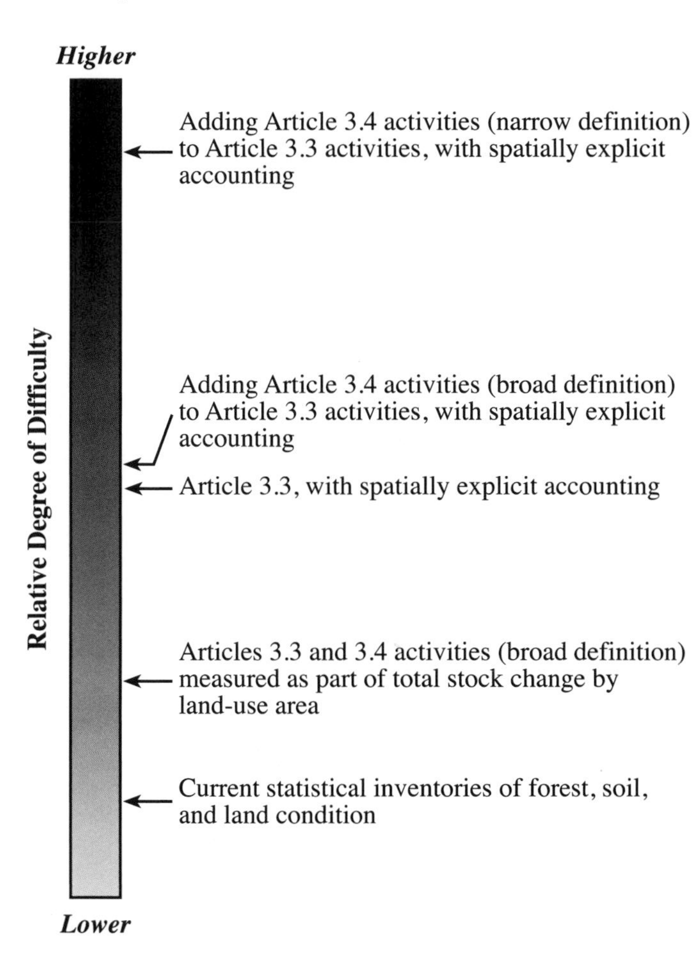

Figure 4-7: Relative costs of measuring and reporting carbon stock changes under different decisions regarding definition of "activities" and requirements for spatially explicit accounting.

One effect of selecting Option B associated with a flexible approach to definitions would be to encourage research and development of new and more efficient methods of measurement, monitoring, and verification. The Parties might choose to make an initial decision that allows for a broad approach during a testing period, with a second review period to take place at a specified future date, as a means of encouraging innovation while protecting against the permanent establishment of methods that might later be regarded as inappropriate.

4.3.4. Measuring "Human-Induced" Effects of Activities

The land management activities discussed in Section 4.4 are all "human-induced" in the sense that they are carried out by people, usually in intentional efforts to produce food, fiber, and other goods and services that people need or desire. There have been discussions, however, about whether changes in carbon stocks to be reported for credit need to be limited to those that were the direct result of that human activity; those questions are discussed elsewhere in this Special Report (e.g., Section 2.3.3.1).

If it is decided to limit reportable amounts to reduced emissions or increased sinks resulting from "direct human-induced" activities (as the term appears, for example, in Article 3.3), separation of natural from direct and indirect human effects can be approximated in some situations by the application of one or more of the following techniques:

- Model-based inference
- Control plots where "no-treatment" regime is monitored
- Specification of agreed natural-plus-indirect human-induced background corrections. One option might be a baseline or "business-as-usual" scenario from which to estimate the net addition attributable to the additional activities undertaken under Article 3.4. The implications of different options for calculating baselines are discussed in Section 4.6 and Chapter 2.

Some of the situations that might be created under potential Article 3.4 activities will provide relatively easy opportunities to separate the direct effects of an activity from those that might be either indirect (i.e., CO_2, N, or S fertilization from anthropogenic atmospheric changes) or the result of natural variability in climate conditions. If the narrow definition of activity is chosen, for example, many activities could be applied next to a control plot where the activity was not applied. The carbon stock difference, if any, measured on the two areas would be an indication of the effect of the activity.

In other situations—such as where land-use change is involved (i.e., conversion of cropland to grass)—field operations will be unable to test how the grass would have impacted soil carbon stocks in the absence of indirect effects or natural variation. If a broad definition of activity is chosen, a wholesale change in management (i.e., an improved grazing system over a large area) may be equally difficult to test for indirect effects because retaining an exact model of the prior management system for comparison will be difficult or impossible.

4.3.5. Applying Decision to the First Commitment Period

Article 3.4 of the Kyoto Protocol does not clearly state whether the decision made by the COP for the second and subsequent commitment periods will apply identically to the Parties who choose to include the additional human-induced activities in their accounts for the first commitment period (see Figure 4-1). A challenge for inclusion of additional activities during the first commitment period would be rapid implementation of systems for monitoring, reporting, and verifying changes in carbon stocks.

If activity is defined broadly enough that the accounting can reflect changes in carbon stock over entire land use areas (such as a Party's entire managed forest area), measurements could be done with a combination of forest inventory and models. This approach would be consistent with the manner in which many countries conduct forest and natural resource inventories today. Establishing a reporting system for additional human-induced activities by 2008–2012 will be a substantial challenge for most Parties. This effort will be particularly daunting if there must be a way to identify and track the specific patches upon which an activity has occurred since 1990. There could also be a need for a way to measure the changes in carbon stocks in a geographically explicit way between 2008 and 2012. This potential requirement will challenge many Parties given the technology, data, resources, and short time available. New techniques for assessing land cover and condition that currently are in the research domain are likely to continue to develop. These techniques should deliver quicker, less labor-intensive, and more cost-effective approaches in the future. Post *et al.* (1999) provide a thorough review of the prospects for monitoring and verifying changes in soil carbon.

4.4. Activities Categorized and Described

In this section, we define activities broadly, primarily to keep the complexity and length of the presentation manageable. Some of the activities listed here, such as agroforestry and restoration of degraded lands, cut across several of the "land cover" boxes of the matrix in Figure 4-6. Key practices are discussed in some detail in several of the Fact Sheets.

4.4.1. Overview of Rates and Duration of Potential Carbon Sequestration

Potential rates of carbon sequestration in response to improved management vary widely as a function of land use, climate, soil, and many other factors (refer to Tables 4-5 through 4-12). Although research to date does not allow definitive evaluation of potential rates of carbon gain for all regions and management options, Table 4-4 presents rough estimates for broad activity and eco-zone groupings. For each grouping, Table 4-4 shows the carbon-conserving practices that are most likely to achieve substantial rates of carbon gain. The rates of carbon gain alongside each practice reflect the best combination of these practices for that activity and eco-zone. For example, estimated rates of potential carbon gain for cropland management might reflect the effects of reduced tillage, better fertilization, and improved crop rotation applied together on the same unit of land.

The rates of carbon gain in Table 4-4 refer to the average accumulation rate from the time a practice is started until carbon storage again reaches a new equilibrium (see Box 4-1). Almost invariably, the rates in Table 4-4 are lower than those from published studies, which are usually measured for time intervals shorter than that needed to reach saturation. As shown, some of the rate values have greater uncertainty than others, largely because of a paucity of studies in many regions. The net effect of the practices on climate forcing is also affected by impacts on other GHGs (Table 4-4). For example, where the proposed practice increases N_2O or CH_4 emission, the rate of carbon sequestration alone overestimates the net benefit of the practice to the atmosphere.

Adoption of a proposed practice seldom depends solely on perceived effects on the atmosphere; indeed, other benefits or disadvantages of the proposed activities will usually outweigh any effect on atmospheric CO_2 in affecting land-use decisions. Table 4-4 shows a few of the key associated impacts for each of the proposed activities.

4.4.2. Cropland Management

The current global arable land area is approximately 1.5 Gha, or 11 percent of total terrestrial area (FAO, 1999). Most of the carbon stocks in croplands are in the soil because of frequent biomass removal during harvest. Soils now contain about 8–10 percent of total global carbon stocks (Cole *et al.*, 1993). For the purposes of this Special Report, arable land is divided into conventional (aerobic) systems and (anaerobic) paddy rice systems. The remainder is classified as agroforestry and degraded lands, which are discussed as separate activities.

The conversion of natural systems to cultivated agriculture results in losses of soil organic carbon on the order of 20–50 percent from the pre-cultivation stocks in the surface 1 m (Davidson and Ackerman, 1993; Cole *et al.*, 1997; Paustian *et al.*, 1997a; Lal and Bruce, 1999). Loss rates are higher during the first years of cultivation and then tend to level off after 20–50 years (Díaz-Raviña *et al.*, 1997; Lal *et al.*, 1998). There are several exceptions to this rule. One is with paddy cultivation (flooded for extended periods), where decomposition rates are retarded, thereby maintaining or increasing carbon stocks (Greenland, 1985; Bronson *et al.*, 1997). In contrast, when organic

Table 4-4: *Summary of potential rates of carbon gain and associated impacts for various activities.*

Activity	Ecozone[a]	Key[b] Practices	Rate[c] ($t\ C\ ha^{-1}\ yr^{-1}$)	Confi-dence[d]	Duration (yr)[e]	Other GHGs[f]	Associated Impacts
Cropland management	Boreal	Ley/perennial forage crops, organic amendments	0.3–0.6 (0.4)	M	40	$+N_2O$	Increased food production, improved soil quality
	Temperate – dry	Reduced tillage, reduced bare fallow, irrigation	0.1–0.3 (0.2)	H	30	$+N_2O$	Increased food production, improved soil quality, reduced erosion, possibly higher pesticide use
	Temperate – wet	Reduced tillage, fertilization, cover crops	0.2–0.6 (0.4)	H	25	$+N_2O$	Increased food production, improved soil quality, reduced erosion, possibly higher pesticide use
	Tropical – dry	Reduced tillage, residue retention	0.1–0.3 (0.2)	L	20	$+N_2O$	Increased food production, improved soil quality, reduced erosion, possibly higher pesticide use
	Tropical – wet	Reduced tillage, improved fallow management, fertilization	0.2–0.8 (0.5)	M	15	$+N_2O$	Increased food production, improved soil quality, reduced erosion, fertilizers often unavailable, possibly higher pesticide use
	Tropical – wet (rice)	Residue management, fertilization, drainage management	0.2–0.8 (0.5)	L	25	$++CH_4$, $+N_2O$	Increased food production
Agroforest management	Tropical	Improved management	0.5–1.8 (1.0)	M	25	$+N_2O$	
Grassland management	Temperate – dry	Grazing management, fertilization, irrigation	0–0.3 (0.1)	M	50	$\pm CH_4$, $+N_2O$	Increased energy use, salinity, higher productivity
	Temperate – wet	Grazing management, species introduction, fertilization	0.4–2.0 (1.0)	M	50	$\pm CH_4$, $++N_2O$	Higher productivity, acidification, erosion, reduced biodiversity
	Tropical – dry	Grazing management, species introduction, fire management	0.1–1.5 (0.9)	L	40	$-CH_4$, $++N_2O$	Reduced soil degradation, higher productivity, woody encroachment (reduced productivity)
	Tropical – wet	Species introduction, fertilization, grazing management	0.2–3.9 (1.2)	L	40	$-CH_4$, $++N_2O$	Increased productivity, reduced biodiversity, acidification
Forestland management	Boreal and Temperate – dry	Forest regeneration, fertilization, plant density, improved species, increased rotation length	0.1–0.8 (0.4)	L	80	$+N_2O$, $+NO_x$	Leakage (rotation length), high cost efficiency

Table 4-4 (continued)

Activity	Ecozone[a]	Key[b] Practices	Rate[c] (t C $ha^{-1} yr^{-1}$)	Confi-dence[d]	Duration (yr)[e]	Other GHGs[f]	Associated Impacts
Forestland management (continued)	Temperate – wet	Forest regeneration, fertilization, species change	0.1–3.0 (1.0)	L	50	$+N_2O$, $+NO_x$	Leakage (rotation length), reduced biodiversity
	Tropical – dry	Forest conservation, reduced degradation	(1.75)	L	40		Ecological improvement, high cost efficiency
	Tropical – wet	Reduced degradation	3.1–4.6 (3.4)	L	40		Environmental improvement
Wetland management	All	Restoration	0.1–1 (0.5)	L	100	$++CH_4$, $\pm N_2O$	Increase in water quality, decrease in flooding, increased biodiversity
Restoration of degraded land	All	Restoration of eroded lands, saline soil reclamation	0.1–7 (0.25)	M	30	$+N_2O$	Increased productivity, may be expensive
Urban land management	All	Tree planting	(0.3)	M	50		Increased biodiversity
Conversion to agroforestry	Tropics	Conversion from cropland or grassland at forest margins	1–5 (3)	L	25		Improved biodiversity, CH_4 sinks, poverty alleviation, food security
Conversion (cropland to grassland)	Temperate – dry	Marginal cropland re-seeded to grassland	0.3–0.8 (0.5)	H	50	$-N_2O$; $-CH_4$	Enhanced biodiversity, reduced erosion
	Temperate – wet	Surplus cropland seeded to grassland	0.5–1.0 (0.8)	M	50	$--N_2O$; $-CH_4$	Enhanced biodiversity, reduced erosion

[a] "Wet" vs. "dry" based on potential evapotranspiration:precipitation ratio. "Wet" <1 and "Dry" >1; "Tropical" = latitude <30°.
[b] List of practices that may yield largest gains in carbon stocks, roughly in order of importance.
[c] Range of carbon increase rates that might reasonably be expected to occur in response to adoption of best-possible complement of key practices. Actual rate will depend on previous management practices (e.g., rates of gain may be higher in a carbon-depleted system), climate, ecosystem properties (e.g., soil carbon gain may be favored by higher clay content), and many other factors. Value in parentheses is default estimate. Rates for tropical forest management to recover carbon stocks on degraded forests apply only to the present area of degraded forest (as of 1990) as reported by FAO (1996)—that is, closed-canopy forest having full biomass stocks are excluded.
[d] Relative reliability of rate estimates. Generally, confidence increases with the number of studies conducted in the activity-ecozone grouping. L = low, M = medium, and H = high.
[e] An estimate of the time required for the system to approach a new steady state after the adoption of the new practices.
[f] Relative magnitude of potential effects on emission of N_2O, CH_4, and other GHGs. "+" denotes increased emission; "-" denotes reduced emission; number of "+" or "-" denotes relative magnitude of possible effects.

soils (Histosols) are drained and brought into cultivation, losses of soil carbon may continue as long as the soil is exposed (Lal *et al.*, 1998). When arid or semi-arid soils are brought under irrigation, increased organic matter input from roots and crop residues may increase soil carbon stocks (Leuking and Schepers, 1985). This increase could be partially offset by a concomitant increase in residue decomposition resulting from added moisture. On a global basis, the cumulative historic loss of carbon from agricultural soils has been estimated at 55 Gt C (IPCC, 1996b)—nearly one third of total carbon loss from soils and vegetation (150 Gt C) (Houghton, 1995).

Studies suggest that most of the world's agricultural soils have not reached saturation of their carbon stocks; therefore, most soils are potential carbon sinks (Kern and Johnson, 1993; Donigian *et al.*, 1995). Future increases in carbon stocks are likely to be smaller in areas that have been highly productive during past decades than in areas that currently are experiencing sharp increases in crop yields, as well as those moving into conservation tillage; the magnitude of these increases is highly dependent on inputs and management, however. Barriers to adoption of soil carbon-enhancing activities include a perceived lack of profitability and land managers' lack of understanding

Box 4-1. The Rate of Carbon Gain

The rate of carbon gain following application of a given practice that stores carbon will decrease over time (see Chapter 1 and Section 4.2). Figure 4-8a shows an approximation of this relationship, which normally is used to describe changes in soil carbon following application of a particular practice that stores carbon. Therefore, rates of carbon change observed during the initial stages of an activity are usually higher than the average rates (Figure 4-8a). Idealized saturating curves such as those in Figure 4-8a require two constraining parameters: the rate of inputs and the time required to reach "saturation." Data for carbon stock increases reported in the literature sometimes do not cover the entire duration; hence, the rates reported are higher than the average rate. We have corrected for this in determining the average rates in Table 4-4. Many management practices are applied over a period of years before saturation can occur. In such cases, the average rate of carbon storage underestimates stock changes that will occur using the approach outlined in Figure 4-8a. In this case, we have estimated the average change in stock once the management has come to an approximate steady-state for carbon (Figure 4-8b).

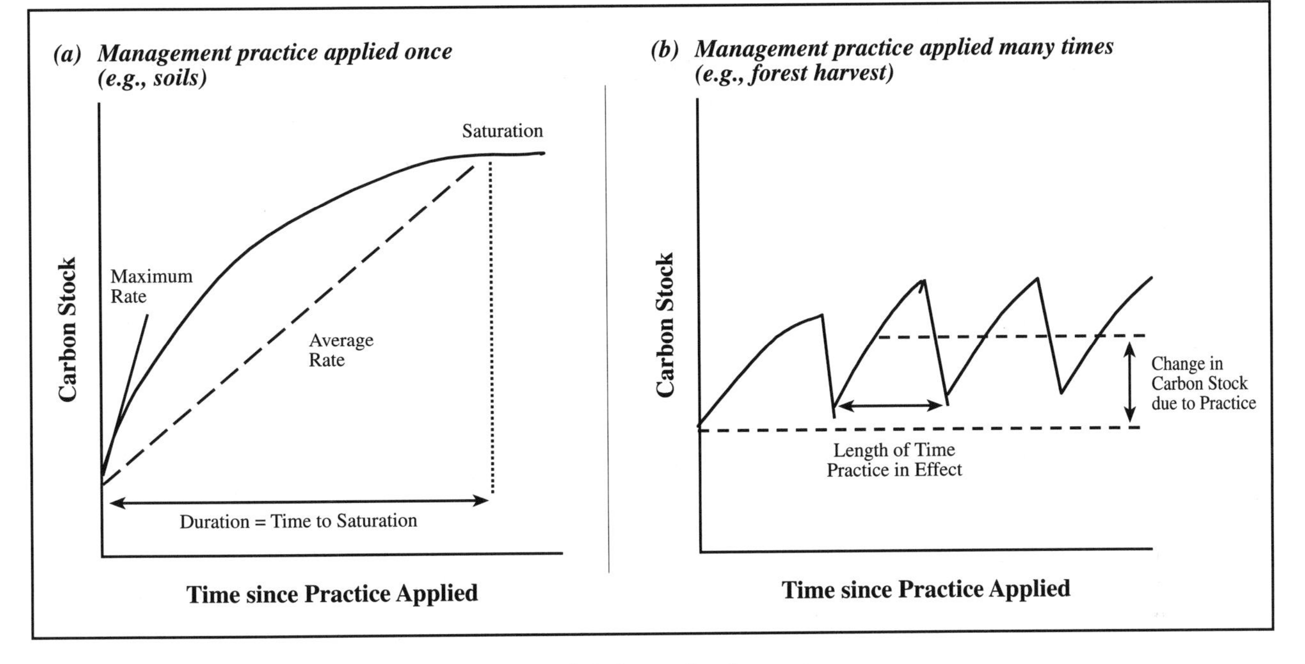

Figure 4-8. Rate of carbon gain.

The magnitude of change in carbon stocks for a given practice depends on three factors: the average rate of carbon stock change per unit area after the practice has been applied, the time required for saturation to occur, and the total area over which the activity is applied. Figure 4-8a shows how we have defined these terms for calculation of magnitudes.

and acceptance of improved techniques for controlling pests and weeds, applying fertilizers, and switching to crop varieties that are matched to different soil moisture and temperature conditions. Because these barriers are addressable by public policies and programs, increasing the adoption of cropland management activities appears to be an opportunity for a broad range of Parties. Virtually everything that occurs in arable lands is "human-induced;" therefore, that criterion, if chosen, should pose no technical difficulty with regard to these activities.

Global warming is expected to increase yields in higher latitude cropland zones by virtue of longer growing seasons and CO_2 fertilization (Cantagallo *et al.*, 1997; Travasso *et al.*, 1999). At the same time, however, global warming may also accelerate decomposition of carbon already stored in soils (Jenkinson, 1991; MacDonald *et al.*, 1999; Niklinska *et al.*, 1999; Scholes *et al.*, 1999). The net effect of these changes on carbon sequestration in croplands is likely to impact some geographic areas more than others, but there are no reliable estimates of those future impacts. Although much work remains to be done in quantifying the CO_2 fertilization effect in cropland, van Ginkel *et al.* (1999) estimate the magnitude of this effect (at current rates of increase of CO_2 in the atmosphere) at 0.036 t C ha^{-1} yr^{-1} in temperate grassland, even after the effect of rising temperature on decomposition is deducted.

Efforts to improve soil quality and raise SOC levels have been grouped into three practices: agricultural intensification, conservation tillage, and erosion reduction. Agricultural intensification (Section 4.4.2.1) involves the use of improved water management, new or higher yielding varieties, integrated pest management, judicious use of organic and inorganic amendments, crop rotations, and other sound technologies that are available now, except for conservation tillage. Conservation tillage (Section 4.4.2.2) includes various forms of reduced tillage and is discussed as a distinct practice because of its specific effects of accumulating soil carbon and reducing soil erosion. Soil erosion control and improved water management are discussed in Section 4.4.2.3. Care has been taken so the areas listed under each practice are not double-counted.

4.4.2.1. Agricultural Intensification

Improved cultivars, irrigation, organic and inorganic fertilization, management of soil acidity, green manure and cover crops in rotations, integrated pest management, double-cropping, and crop rotation (including reduction of bare fallow) are some of the ways to increase crop yields (see Fact Sheet 4.1). Increasing crop yields results in more carbon accumulated in crop biomass or in an alteration of the harvest index. The higher residue inputs associated with those higher yields favor enhanced soil carbon storage (Paustian *et al.*, 1997a). Estimates and experimental data from around the world indicate that the application of management practices to improve agricultural productivity results in increased SOC content (Table 4-5). For example, increases in biomass production resulting from advances in improved crop germplasm and agronomy are estimated to sequester carbon at rates ranging from 0.01–0.7 t C ha^{-1} yr^{-1}, with a mean value of 0.27 t C ha^{-1} yr^{-1} (Lal and Bruce, 1999). This rate could provide a carbon capture of 0.02–0.07 Gt C yr^{-1} in an area of 122–152 Mha.

Irrigated agriculture produces about one-third of the Earth's total crops, including 40 percent of all crops harvested on only about one-sixth of the cultivated cropland (Lal, 2000a). Because most irrigation is located in arid and semi-arid regions, many irrigable soils are inherently low in soil organic carbon in their native state. Converting these dryland soils to irrigated agriculture may increase soil organic carbon content in the soil by 0.05–0.15 t C ha^{-1} yr^{-1}, with a modal rate of 0.10 t C ha^{-1} yr^{-1} (Lal *et al.*, 1998). Irrigation of arid and semi-arid soils can also affect the inorganic soil carbon pool (carbonates) and its dynamics (Suarez, 1998). Although the processes involving inorganic soil carbon dynamics are complex and poorly understood, irrigation of arid and semi-arid soils may result in inorganic soil carbon sequestration rates that are similar to the rates estimated for soil organic carbon (Lal *et al.*, 1998). Irrigated lands are susceptible to high levels of soil erosion and salinization, however—both of which can reduce soil organic carbon levels and increase emissions.

Improved management of drained croplands—through conversion to conservation tillage and/or management of sub-surface drainage to keep soil moisture levels high—can increase soil organic carbon. In some cases, practices that promote higher productivity may entail greater energy use. For example, increased fertilization and expansion of irrigation may lead to higher fossil fuel use (Schlesinger 1999). Soil carbon gains, therefore, may be partially offset by higher CO_2 emissions from energy use.

4.4.2.2. Conservation Tillage

Conservation tillage is any tillage and planting system in which 30 percent or more of the crop residue remains on the soil surface after planting to reduce soil erosion by water. Where soil erosion by wind is the primary concern, conservation tillage is any system that maintains at least 1,000 kg ha^{-1} of flat, small-grain residue equivalent on the surface throughout the critical wind erosion period (CTIC, 1998). Conservation tillage can include specific tillage types such as no-till, ridge-till, mulch-till, zone-till, and strip-till systems that meet the residue requirements (see Fact Sheet 4.3). These tillage systems are chosen by farmers to address soil type, crop grown, machinery available, and local practice. Although these systems were originally developed to address water quality, soil erosion, and agricultural sustainability problems, they also lead to higher soil carbon and increased fuel efficiency (Table 4-5) (Lal, 1989, 1997; Blevins and Frye, 1993; Kern and Johnson, 1993; Dalal and Mayer, 1996).

No-till methods were initiated in North America in the 1950s; by 1996, the estimated area of conservation tillage in the United States was 42 Mha—about 36 percent of planted cropland (CTIC, 1998). Conservation tillage also has been widely adopted in China (Luo, 1991), Brazil (Hebblethwaite and Towery, 1998), Argentina (AAPRESID, 1999), the Pacific, and other parts of the humid and subhumid tropics (Lal, 1989). Yearly estimates provided by CTIC (1997, 1998, 1999) suggest that conservation tillage adoption appears to have reached a plateau because the area under conservation tillage is believed to have decreased in the United States during the past 2 years. Nevertheless, conservation tillage is now achieving widespread adoption around the world. In Argentina, conservation tillage started in 1990; the area under this type of cultivation system was 7.3 Mha in 1998—31 percent of the agricultural lands of the country (AAPRESID, 1999). The area under conservation tillage in Brazil has now reached 12 Mha (CTIC, 1998).

The opportunity for carbon emissions reduction and soil carbon sequestration in the United States has been estimated at 0.012–0.023 Gt C ha^{-1} yr^{-1} (Lal *et al.*, 1998). Globally, conservation tillage can sequester 0.1–1.3 t C ha^{-1} yr^{-1} and could feasibly be adopted on up to 60 percent of arable lands (Table 4-5). These estimates depend on continued use of conservation tillage. Use of intensive tillage or moldboard plowing can negate or offset any gains made in carbon sequestration (Lindstrom *et al.*, 1999).

Soil carbon sequestration can be further increased when cover crops are used in combination with conservation tillage. Cover cropping shares some of the environmental benefits of no-tillage, such as reduced soil erosion and a resulting decrease in

Table 4-5*: Rates of potential carbon gain under selected practices for cropland (including riceland) in various regions of the world.*

Practice	Country/ Region	Rate of Carbon Gain (t C ha^{-1} yr^{-1})	Time[1] (yr)	Other GHGs and Impacts	Notes[2]
Improved crop production and erosion control	Global	0.05–0.76	25	+N_2O	a
– Partial elimination of bare fallow	Canada	0.17–0.76	15–25		b
	USA	0.25–0.37	8	±N_2O	q
– Irrigation water management	USA	0.1–0.3			c
– Fertilization, crop rotation, organic amendments	USA	0.1–0.3		+N_2O	c
– Yield enhancement, reduced bare fallow	Tropical and subtropical China	0.02	10		e
– Amendments (biosolids, manure, or straw)	Europe	0.2–1.0	50–100		f
– Forages in rotation	Norway	0.3	37		d
– Ley-arable farming	Europe	0.54	100		m
Conservation tillage	Global	0.1–1.3	25	±N_2O	a
	UK	0.15	5–10		g
	Australia	0.3	10–13		h
	USA	0.3	6–20		i
		0.24–0.4			c
	Canada	0.2	8–12		r
	USA and Canada	0.2–0.4	20		j
	Europe	0.34	50–100		k
	Southern USA	0.5	10		l
		0.2	10–15		s
Riceland management					
– Organic amendments (straw, manure)		0.25–0.5		++CH_4	n
– Chemical amendments				--CH_4, +N_2O	o
– Irrigation-based strategies				--CH_4	

[1]Time interval to which estimated rate applies. This interval may or may not be time required for ecosystem to reach new equilibrium.

[2]a. Lal and Bruce (1999). Estimates of carbon gain shown represent range of values presented by the authors for various regions throughout the world.

b. Dumanski *et al.* (1998). Estimates presented are for the 0–30 cm layer. Estimated carbon gain for the 0–100 cm layer are twice those for the 0–30 cm layer. Estimated rates of carbon gain are higher for conversion of fallow to forages (0.48–0.76 Mg C ha^{-1} yr^{-1}) than for conversion to cereal crops (0.17–0.52 Mg C ha^{-1} yr^{-1}).

c. Lal *et al.* (1999b).

d. Singh *et al.* (1994). Reported rate is from one long-term study.

e. Li and Zhao (1998). Rate of carbon gain based on total carbon gain (0.7 Tg C yr^{-1} for 10 years) and total cropland area in the region (~40 Mha) reported by the authors.

f. Smith *et al.* [1998, including data from Smith *et al.* (1997a,b) and Powlson *et al.* (1998)]. Carbon gain from manure, sewage sludge, and straw incorporation assumes that carbon in these materials would otherwise all be lost as CO_2. Rates reported here were calculated from annual mitigation potential (Tg C yr^{-1}) and area values in source reference.

g. Mean carbon accumulation rate in four sites sampled after 5–10 years; from literature data compiled and cited by Paustian *et al.* (1997a).

h. Mean carbon accumulation rate in two sites sampled after 10 or 13 years; from literature data compiled and cited by Paustian *et al.* (1997a).

i. Mean carbon accumulation rate in 22 sites sampled after 6–20 years. Includes one site from Canada.

j. Bruce *et al.* (1999). Rates of carbon accumulation assume "best management practices," including no-till.

k. Smith *et al.* (1998). Based on data from 14 sites in UK and Germany, ranging in duration from 2 to 23 years. Rates reported here were calculated from annual mitigation potential (46.6 Tg C yr^{-1}) and area of arable land (135 x 10^6 ha).

l. Franzluebbers *et al.* (1998). Increase in soil carbon in soybean/wheat double crop vs. soybean (averaged across tillage treatments).

m. Smith *et al.* (1997b). Rates reported here were calculated from annual mitigation potential (73 Tg C yr^{-1}) and area of arable land (135 x 10^6 ha).

n. Net local increase in carbon stored in organic matter; likely small net carbon gain regionally, depending on fate of organic amendments if not applied as fertilizer.

o. Addition of sulfate, nitrate, or iron decreases activity of methanogens by providing alternative electron acceptors and restricting availability of substrates in submerged soils (e.g., Hori *et al.*, 1990). Amendments tend to reduce CH_4 emissions by 0–77% in different experiments (Schutz *et al.*, 1989; Lindau *et al.*, 1993; Wassmann *et al.*, 1993; Denier van der Gon and Neue, 1994). Amendments will likely result in net loss of organic carbon, though estimates were not reported.

p. Drainage of field during cropping season. Oxygen availability stimulates CH_4 oxidation and reduces CH_4 emission (Yagi *et al.*, 1997). Reduced CH_4 emission may be offset by increased CO_2 emission (soil carbon loss).

q. Peterson *et al.* (1998); values for increased carbon levels with continuous crop rotations vs. wheat-summer fallow rotations for four experiments in Montana and Colorado.

r. Janzen *et al.* (1998); mean for six experiments with no-till vs. tilled treatments and continuous crop rotations. No apparent increases in soil carbon with no-till were found for wheat-summer fallow rotations.

s. Potter *et al.* (1998); mean for no-till vs. conventional tillage treatments at three sites (11 crop rotations) in Texas.

surface water runoff. The use of cover crops can also aid in reducing the use of herbicides (e.g., smother crops; Pan, 1999) and fertilizers (legumes; Subak, 1999). The magnitude of carbon sequestration from the use of cover crops can be greater than that of conservation tillage (Donigian *et al.*, 1995; Grant *et al.*, 1997; Paustian *et al.*, 1997b; Buyanovsky and Wagner, 1998).

4.4.2.3. *Erosion Reduction*

Soil erosion from wind and water under cropland can be three to four times higher than under grass or trees on similar soils (Lal and Bruce, 1999; see Fact Sheet 4.4). Thus, erosion control is an important strategy to enhance productivity and sequester SOC. Soil loss prevented in the Loess Plateau region of China by soil conservation measures was estimated at 10.6 billion t for 1995 alone (Nie, 1996). Although estimating the total amount of soil erosion from the world's croplands is difficult, it would be much more difficult to estimate the carbon loss from erosion. Lal and Bruce (1999) estimate that the total amount of SOC displaced by soil erosion annually to be in the range of 0.5 Gt, of which 20 percent may be emitted into the atmosphere (the remainder is re-located on the landscape). A recent detailed study, however, indicates that soil erosion losses have been overestimated (Trimble, 1999). This finding suggests that revised estimates of the role of soil erosion control practices beyond conservation tillage are needed.

In addition to conservation tillage, another strategy to reduce erosion is to plant perennial grasses and legumes, either as a regular phase in an arable farming system or in permanent "set-aside" lands. Not only do these practices reduce erosion, they also favor carbon storage because of reduced soil disturbance and greater allocation of carbon below ground (Table 4-5) (Cole *et al.*, 1997; Feller and Beare 1997; Grace *et al.*, 1997; Neill *et al.*, 1997; Paustian *et al.*, 1997b; Smith *et al.*, 1997b, 1998; Carter *et al.*, 1998; Huggins *et al.*, 1998b).

4.4.2.4. *Overall Assessment*

Improved productivity and conservation tillage typically allow increases in soil carbon at a rate of about 0.3 t C ha^{-1} yr^{-1}. If these practices were adopted on 60 percent of the available arable land worldwide, they might result in a capture of about 0.27 Gt C yr^{-1} over the next few decades (Lal, 1997). It is unclear if this rate is sustainable because research shows a relatively rapid increase in carbon sequestration for a period of about 25 years, with a gradual leveling off in about 50 years (Lal *et al.*, 1998). Important secondary benefits of conservation tillage adoption include soil erosion reduction, improvements in water quality, increased fuel efficiency, and increases in crop productivity.

4.4.2.5. *Managing Rice Agriculture*

Rice paddies are an agricultural form of wetland that may replace natural wetlands or be drylands that have been converted through irrigation (see Fact Sheet 4.5). Rice paddies are flooded for long periods. Anaerobic conditions associated with inundation slow decomposition rates and allow accumulation of large stores of carbon over long time scales. Organic matter storage in rice paddy soils may be considerably greater than in adjacent, unflooded arable soils, especially when organic amendments (rice straw and manures) are part of the agricultural inputs. Carbon may be stored or lost to the atmosphere on conversion of wetland or dryland to rice paddy fields, depending on the amount of carbon stored in soils prior to conversion.

Most practices in rice paddies affect CO_2 and CH_4 emissions in opposite ways. Therefore, management of rice agriculture for positive climate impact must consider the combined effects of carbon storage, CH_4 emissions, and N_2O emissions. Table 4-5 provides qualitative rather than quantitative estimates for carbon sequestration and net GHG effects of different management practices. As with managed wetlands (rice paddies) in general, the carbon impacts would constitute the rationale to include this activity under Article 3.4, whereas CH_4 reductions from improved practices are already included under the Protocol. There is a lack of monitoring and measuring of carbon storage of rice paddies in the world. China's 200 years of data show that carbon storage of rice paddies from original dryland can reach levels of 0.2–0.45 t C ha^{-1} yr^{-1}.

Intermittent rice field drainage, which is practiced widely in Asia to improve production, typically decreases CH_4 production and emission. If nitrogen management is not linked to the practice, however, N_2O emission can become quite large (Bronson *et al.*, 1994). Where rice growth is now limited by nitrogen deficiencies, increased deposition of nitrogen associated with intensified production of available nitrogen (Vitousek *et al.*, 1997) may accelerate rice growth—eventually resulting in more soil carbon storage (Wedin and Tilman, 1996; Nadelhoffer *et al.*, 1999). The global impact of nitrogen deposition, however, may be comparatively small. On the other hand, increased soil temperatures associated with atmospheric CO_2 are expected to result in increased soil respiration losses (Woodwell *et al.*, 1998). Higher CO_2 concentrations may also suppress decomposition of stored carbon because C:N ratios in residues may increase and because more carbon may be allocated below ground (Torbert *et al.*, 1997).

Strategies for mitigating CH_4 emission from rice cultivation are based on controlling production, oxidation, or transport processes (Yagi *et al.*, 1997; Fact Sheet 4.5).

4.4.3. *Grazing Lands Management*

Grazing lands (which include grassland, pasture, rangeland, shrubland, savanna, and arid grasslands up to the fringe of hyper-arid deserts—all of which are referred to here as grasslands) occupy 1,900 to 4,400 Mha, depending on definitions (Ojima *et al.*, 1993c); recent global estimates have placed the area of grazing lands at 3,200 Mha (FAO 1999). Grasslands contain 10–30 percent of the world's soil carbon (Anderson, 1991;

Eswaran *et al.*, 1993)—including approximately 200–420 Gt organic C (Ojima *et al.*, 1993c; Scurlock and Hall, 1998; Batjes, 1999) and 470–550 Gt of carbonate C to a depth of 1 m (Batjes, 1999). Management activities can significantly affect carbon storage by reducing carbon loss in degradation processes or increasing carbon inputs and residence times (Table 4-6).

4.4.3.1. Influencing Degradation Processes

Up to 71 percent of the world's grasslands are degraded to some extent (Dregne *et al.*, 1991) as a result of overgrazing, salinization, alkalinization, acidification, and other processes (Oldeman, 1994). Generally, degradation results in reductions

***Table 4-6**: Rates of potential carbon gain under selected practices for grasslands in various regions of the world.*

Practice	Country/ Region	Rate of Carbon Gain ($t\ C\ ha^{-1}\ yr^{-1}$)	Time[1] (yr)	Other GHGs and Impacts	Notes[2]
Reduce degradation	Global	0.5	20	Increases sustainability.	a
	Global	0.024–0.24	50	Generally likely to reduce	b
	Global	0.41	110	methane emissions via	c
– Improve grazing management	Global	0.22	40	reductions in animal	d
	Global	0.7	50+	numbers and increases	e
	Australia	0.24	30	in diet quality. Possible	f
– Protected lands and set-asides	USA CRP	0.52	50	improvement in	g
	China	1.3		biodiversity, particularly in set-aside lands.	h
Increase grassland productivity	Global	0.51		$+N_2O$. Reduced erosion if grazing management appropriate.	d
Fertilization	Global average	0.23	40+	$++N_2O$, off-site nutrient	d
	N. Australia	0.50	10	impacts, acidification.	i
Irrigation	Global average	0.16		Associated fossil fuel emissions, salinization risks.	d
Improved species and legumes				$+N_2O$, $\pm CH_4$. Risk of	
– Legumes	Global	1.09		introduced species	d
– Grasses	Global	3.34		becoming weeds in	d
– Conversion from native pasture	Global	0.36		adjacent areas.	d
	South America	2.8–14.4		Biodiversity loss from native pastures.	j
Fire management	Orinoco Plains, South America	1.4		$-CH_4$, $-N_2O$. Reduced agricultural production,	k
	NE Australia	0.56	50	± biodiversity depending on site.	l

[1]Time interval to which estimated rate applies. This interval may or may not be time required for ecosystem to reach new equilibrium.

[2] a. Glenn *et al.* (1993). 5172 Mh of drylands, halophyte storage over 5 years.
b. Paustian *et al.* (1998a). Assumes potential carbon sequestration of 1–2 kg C m^{-2} on an arbitrary 10–50% of moderately to highly degraded land (1.2 Gha globally; Oldeman *et al.*, 1992).
c. Keller and Goldstein (1998). Revegetation of agricultural and pastoral drylands.
d. Conant *et al.* (2000). Based on literature review.
e. Ojima *et al.* (1993b). Regressive 50% consumption vs. sustainable 30% land management for grassland and rangelands.
f. Ash *et al.* (1996). Average estimate 0–10 cm soil carbon only for transfer from deteriorated condition to sustainable across northern Australia.
g. McConnell and Quinn (1988); Gebhart *et al.* (1994); Barker *et al.* (1995); Burke *et al.* (1995).
h. Li and Zhao (1998). Change in top 20 cm of soil.
i. Dalal and Carter (1999). Phosphorus and sulfur fertilization over 56 Mha in northern Australia.
j. Fisher *et al.* (1994, 1995). Introduced grass/legume pastures with deep, dense root systems.
k. San Jose *et al.* (1998). Extrapolated from one site to 28 Mha region.
l. Burrows *et al.* (1998); Burrows *et al.* (1999). Aboveground and below-ground biomass extrapolated from 47 sites to 60 Mha region.

in perennial plant cover, increased erosion risk, and loss of productive potential. Restoration of vegetation and hence productive capacity results in increases in biomass, litter, and soil carbon pools (Table 4-6). There may also be a reduction in the intensity or frequency of water and wind erosion, which removes carbon (some of which may be oxidized to CO_2).

Wind and water erosion reduce carbon stocks and productivity at the location of soil loss. The discussion in Section 4.2 of the uncertainty relating to the net effect of erosion is relevant here as well. Grazing and burning reduce plant cover, thereby increasing soil erosion rates (Scanlan *et al.*, 1996) and carbon loss (e.g., Ash *et al.*, 1996). Increased erosion can reduce yields (Tenberg *et al.*, 1998) and thus carbon inputs into the soil, which depletes soil carbon levels (e.g., Kelly *et al.*, 1996; Lal, 2000b). Soil conservation procedures that restore landscape patchiness (e.g., by using shrub branches laid across the slope) can rapidly and substantially increase soil carbon (Tongway and Ludwig, 1996). Activities that combat desertification more broadly are also likely to result in substantial carbon storage (Fullen and Mitchell, 1994).

4.4.3.2. *Grazing Management*

Overgrazing is the single greatest cause of degradation in grasslands (Oldeman *et al.*, 1991) and the overriding human-influenced factor in determining their soil carbon levels (Ojima *et al.*, 1993a). Grazing influences carbon storage through removal of biomass and nutrients—much of which are recycled to soil pools but some of which are lost from the grassland (e.g., Haynes and Williams, 1993). Grazing also influences partitioning of carbon to aboveground or below-ground plant organs, changes the temperature and disturbance regimes of the soil, and alters water infiltration and susceptibility to erosion (Follett *et al.*, 2000). Often, extensive heavy grazing practices result in decreases in carbon pools of biomass and soil carbon (e.g., Naeth *et al.*, 1991; Ash *et al.*, 1996; Li *et al.*, 1997; McIntosh *et al.*, 1997a,b). Consequently, in many systems, improved grazing management (e.g., optimizing stock numbers, rotational grazing) will result in substantial increases in carbon pools (e.g., Ash *et al.*, 1996; Eldridge and Robson, 1997; Table 4-6); in some cases, grazing can increase nutrient cycling, animal productivity (e.g., Franzluebbers *et al.*, 2000), and incorporation of litter into the soil—thereby increasing soil carbon (Schuman *et al.*, 1999). Where soil carbon has been lost, however, the rate of recovery of carbon may be considerably slower than the rate of loss (Northup and Brown, 1999). Additional human intervention may be required where degradation is severe. Adoption of more sustainable grazing practices that track climate variability and change is likely to reduce the risk of degradation and hence carbon loss (McKeon *et al.*, 1993).

In some grasslands, changes in species composition under grazing toward those with large and dense root systems (e.g., *Bouteloua gracilis*) can increase carbon levels in the surface soil layers (Dormaar and Willms, 1990; Frank *et al.*, 1995; Manley *et al.*, 1995; Berg *et al.*, 1997) but are associated with reduced grazing utility. Where such species are already dominant, however, heavy grazing will reduce soil carbon levels (Gardner, 1950). In other situations, grazing intensity seems to have little or variable impact on soil carbon levels (Milchunas and Lauenroth, 1993; Basher and Lynn, 1996; Chaneton and Lavado, 1996; Brejida, 1997; Tracy and Frank, 1998; McIntosh and Allen, 1998), provided overgrazing does not occur. Heavy grazing can increase opportunities for establishment of unpalatable woody shrubs, resulting in increased biomass carbon pools but lower grazing utility (e.g., Boutton *et al.*, 1998).

Grazing livestock are significant contributors of methane and nitrous oxide globally (accounted for within the IPCC Guidelines), and their wastes can produce ammonia (a precursor of oxides of nitrogen and subsequent ozone formation). Improvements in the diet of livestock can substantially reduce methane emissions per unit intake (Kurihara *et al.*, 1999), but if livestock are fed grain or other supplements, the emissions embodied in these feeds must be accounted for.

4.4.3.3. *Protected Grasslands and Set-Asides*

Protection of previously intensively grazed land and reversion of cultivated lands to perennial grasslands (e.g., in "set-aside" programs) often increases mean aboveground and below-ground biomass and soil carbon (Tables 4-5 and 4-6). The rate of sequestration will decline with time over a period of about 50 years (McConnell and Quinn, 1988) or longer (Burke *et al.*, 1995). The rate of carbon storage and total carbon storage could be increased with fertilization and in some cases may be nearly doubled (Huggins *et al.*, 1998b); additional sinks may be available if afforestation is promoted in such lands (Barker *et al.*, 1995). If these lands are returned to crop production, there will again be a rapid loss of carbon from the soils (Lindstrom *et al.*, 1999), with the size of losses dependent on the management used (Section 4.4.2). Thus, a continuity of set-aside policies and management purposes is required to maintain or increase soil carbon reliably.

4.4.3.4. *Grassland Productivity*

4.4.3.4.1. *Fertilization*

Where plant growth is limited by soil nutrient availability as well as by water, fertilization can result in large growth responses, as well as increased biomass and soil carbon pools (e.g., Schwab *et al.*, 1990; Haynes and Williams, 1992; Schnabel *et al.*, 2000; Table 4-6). This effect will increase if legumes are introduced in conjunction with fertilizers (Barrow, 1969). The response to increased fertility is generally greater where grazing is either reduced or removed (Cameron and Ross, 1996; McIntosh *et al.*, 1997b). Normal management practice, however, is to increase the harvesting intensity of the additional biomass produced (e.g., Winter *et al.*, 1989), potentially leading to little change in carbon stocks. Fertilizer responses vary significantly between systems and species and can result in undesirable changes such as increased dominance of weedy species. Emissions of oxides of nitrogen, decreases in methane oxidation, and impacts from leaching of nutrients also must be included in the assessment of tradeoffs involved in adopting these practices.

4.4.3.4.2. *Irrigation*

Where water deficiencies limit plant growth, irrigation can increase growth and carbon accumulation in biomass and in the soil—although it can also increase decomposition rates (Table 4-6). Water is usually a scarce commodity in dry regions, so it is more often used for higher value purposes. The arguments presented previously about tradeoffs from fossil fuel emissions from irrigation apply here as well. In some areas where pasture irrigation has been implemented, there have been adverse impacts on production through increased waterlogging and irrigated salinization processes (e.g., Gupta and Abrol, 1990).

4.4.3.4.3. *Species introductions (including legumes and deep-rooted species)*

Introduction of nitrogen-fixing legumes and high-productivity grasses can increase biomass production and soil carbon stocks (Table 4-6). Some of these species have significant potential to become weeds. Replacement of native grass pastures with high-productivity perennial grasses or grass-legume combinations can result in substantial increases in dry matter production and aboveground biomass (e.g., Montes and Masco, 1996) and even greater increases in root biomass and soil carbon (Fisher *et al.*, 1994; Rao *et al.*, 1994, 1998; Guggenberger *et al.*, 1995). The relative increase is sometimes modified by soil type (high in sandy soils, low in clay soils) (Guggenberger *et al.*, 1995). Increases in below-ground carbon pools may result in part from high proportions of resistant root material in these grasses (Urquiaga *et al.*, 1998), which in some systems may cause pasture decline in the longer term from nitrogen immobilization (e.g., Robbins *et al.*, 1989; Robertson *et al.*, 1997). Pasture renovation (light plowing) and re-sowing with high-productivity grasses can temporarily increase productivity but are often associated with short-term carbon losses (e.g., Robbins *et al.*, 1989).

Net gains in ecosystem carbon stocks from replacement of native grass species—such as those observed in Colombia by Fisher *et al.* (1994)—depend heavily on the absence of a need to clear forests to establish the grasses because of the large carbon stores in the forest. Many of these benefits may also depend on concomitant intensification of management, including application of fertilizers—placing constraints in expanding this activity to large areas (Fearnside and Barbosa, 1998).

The effects of these practices (Sections 4.4.3.1 through 4.4.3.4) can be monitored and verified through repeated field sampling of soil and biomass carbon pools over large areas (see Section 2.3). Models can provide an indication of the magnitude of the effect, given verified information on climate, animal types and densities, grassland species composition, irrigation, and fertilizer levels.

4.4.3.5. *Fire Management*

Fire is often an essential management tool in grasslands of the tropics and arid zones for controlling woody weeds, removing dead biomass, clearing land, stimulating grass growth and palatability, hunting, and controlling wildfires and pests (Follett *et al.*, 2000). Fires can also occur naturally through lightning strike. In some places (e.g., Africa and Australia), humans have been setting grassland fires for millennia, and humans can be considered the historically dominant ignition source. In many regions, if managed burns are not implemented, a fire will eventually occur from non-human ignition sources. These two observations illustrate the operational difficulty in attributing pyrogenic emissions to natural or human influence.

Fires transfer a significant portion of aboveground carbon and other material—as well as smaller portions of below-ground materials—into the atmosphere. Emissions of methane, non-methane hydrocarbons, carbon monoxide, oxides of nitrogen, and particles also occur. Burning therefore represents a short-term transfer of carbon from grassland ecosystems to the atmosphere; under an unchanging fire regime, this carbon transfer is balanced by subsequent regrowth of vegetation prior to the next fire (Follett *et al.*, 2000). Increasing fire frequency over time tends to reduce grass biomass production through loss of soil nitrogen (e.g., Scholes and Walker, 1993; Kauffman *et al.*, 1994); in conjunction with enhancement of decomposition processes (Knapp *et al.*, 1998), this effect results in declines in soil carbon pools (Jones *et al.*, 1990). In some grasslands, however, burning frequency appears to have little impact on grass biomass pools (e.g., Senthilkumar *et al.*, 1998), whereas in other grasslands changing the season of burning can substantially alter grassland productivity and biomass pools (Cox and Morton, 1986).

Changes in fire management, often in association with increased grazing pressure, can also influence the vegetation dynamics of grasslands and result in long-term changes in structure and carbon pools (Scholes and Hall, 1996; Scholes and Archer, 1997). Management that removes fire from previously burned landscapes will tend to increase the woody components, resulting in large increases in carbon pools in biomass and soil (e.g., San Jose *et al.*, 1998; Burrows *et al.*, 1999). This increase in woody vegetation cover is occurring in many grasslands throughout the world (reviewed in Archer *et al.*, 1995), largely in response to management change. Other factors, such as climatic variations, are sometimes significant in the establishment and removal of woody species (e.g., Fensham, 1998). Atmospheric CO_2 increases probably have not had major effects to date (Archer *et al.*, 1995; Howden *et al.*, 1999), although there is some dispute over this point (e.g., Polley *et al.*, 1997). The usefulness of areas with increased woody biomass for other purposes (e.g., grazing, thatch production) will often decline as a result of reductions in grass productivity and species changes (e.g., Burrows *et al.*, 1990). Current aspirations to return to more sustainable grazing systems are likely to result in increased burning opportunities to manage woody vegetation (Hall *et al.,* 1998). In many ecosystems, fauna and non-woody flora species are fire-dependent, and removing fires will result in loss of biodiversity (e.g., Edroma, 1986); a reduction in fire frequency will not necessarily lead to biodiversity loss, although it may.

Charcoal generated by fires can constitute 8 g C kg^{-1} soil and represent up to 30 percent of the soil carbon content of some grassland soils (Skjemstad *et al.*, 1996); it probably constitutes a significant part of the inert or passive soil carbon pool (Skjemstad *et al.*, 1996). A small fraction of the biomass consumed by fire is converted to black carbon—a highly condensed substance that is virtually impervious to decay—part of which may eventually end up in oceanic sediments (Verardo and Ruddiman, 1996; Verardo, 1997; Stallard, 1998). Management that reduces or removes fire will result over centuries in reduction of these small but highly stable pools. The net atmospheric consequences are unknown.

Fires are routinely monitored with remote sensing. The consequences of altered fire regimes on emissions and carbon storage can be modeled (Scholes and van der Merwe, 1996; Moore *et al.*, 1997). Verification of the altered carbon stores would require field measurement of woody biomass and soil carbon. Monitoring the impacts of changing fire management may be possible, although attribution of changes in carbon stores will remain difficult given the complexity of interactions between human actions, climate variability, and other environmental factors.

4.4.4. Agroforestry

Agroforestry is a management system that integrates trees on farms and in the agricultural landscape. It leads to a more diversified and sustainable production system than many treeless alternatives and provides increased social, economic, and environmental benefits for land users at all levels (Sanchez, 1995; Leakey, 1996; Fay *et al.*, 1998). Agroforestry is practiced from the Arctic to the south temperate regions, but it is most extensive in the tropics. Approximately 1.2 billion people (20 percent of the world's population) depend directly on agroforestry products and services in rural and urban areas of developing countries (Leakey and Sanchez, 1997). Agroforestry encompass a wide variety of practices, including crop-fallow rotations, complex agroforests, simple agroforests, silvopastoral systems, and urban agroforestry (Steppler and Nair, 1987; Fujisaka *et al.*, 1996; Huxley, 1999).

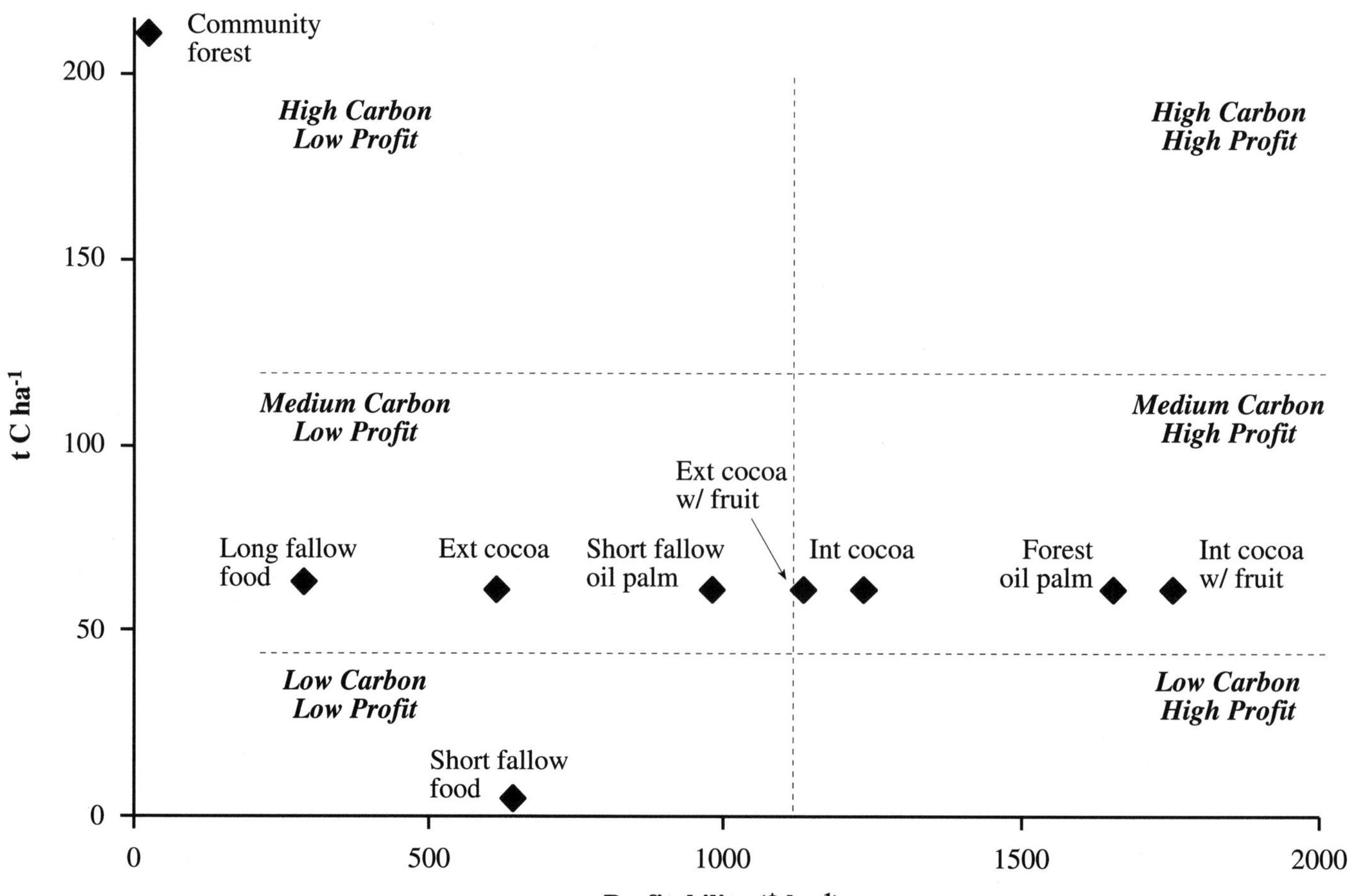

Figure 4-9: Tradeoffs between carbon stocks and social profitability of land-use systems in Cameroon (Gokowski *et al.*, 1999).

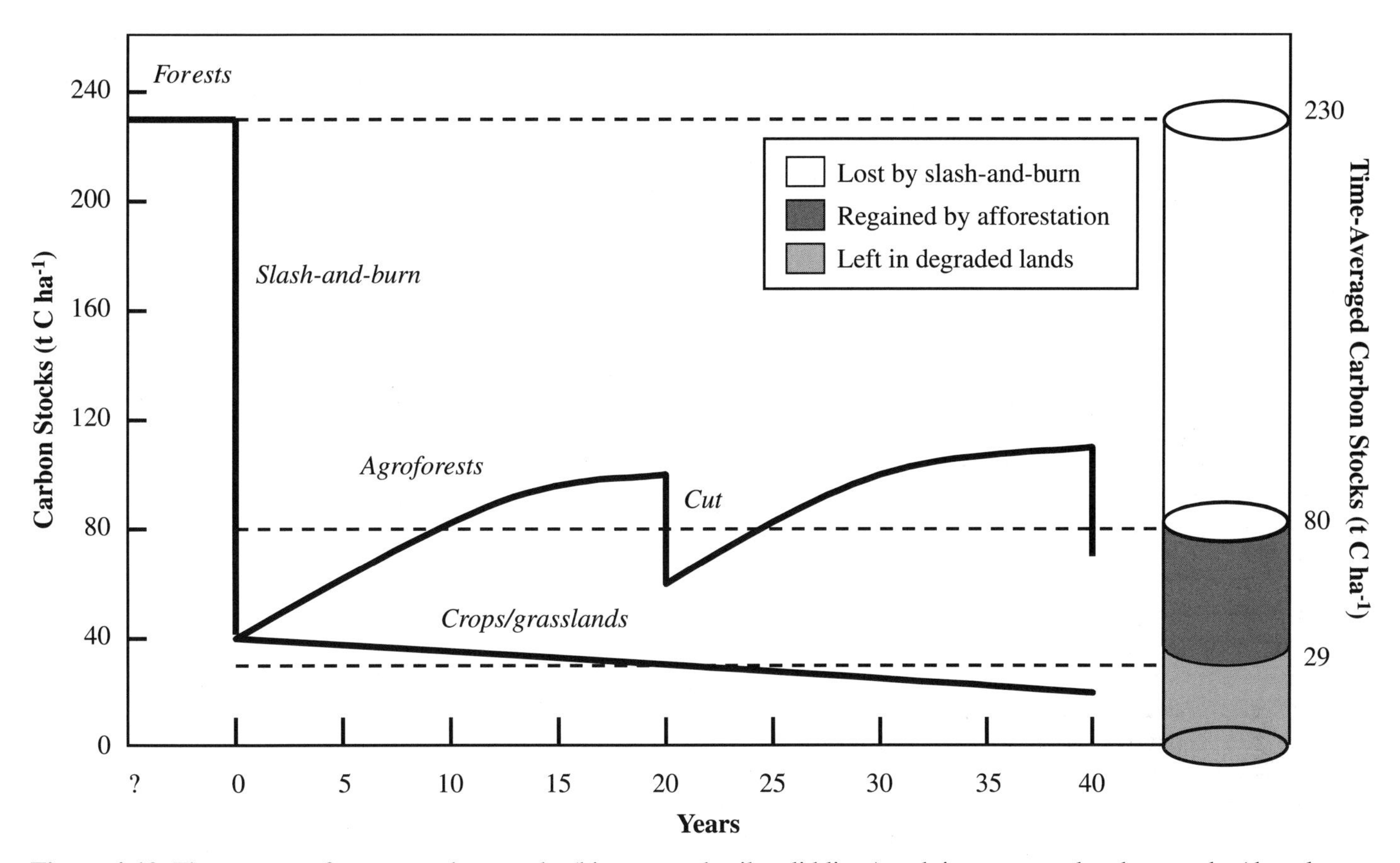

Figure 4-10: Time course of system carbon stocks (biomass and soil, solid lines) and time-averaged carbon stocks (dotted lines) in agroforestry systems vs. crops followed by grasslands at margins of humid tropical forest. Based on Table 4-6 and additional data from Palm *et al.* (2000).

Agroforestry practices in the temperate regions include planting trees at wide spacing in combination with pastures or crops; this practice results in increasing carbon density in Annex I countries (Buck *et al.*, 1999). Examples in Canada include the Permanent Cover Program under the Prairie Farm Rehabilitation Administration. Agroforestry in North America was reviewed by Lassoie and Buck (1999).

Agroforestry systems can be superior to other land uses at the global, regional, watershed, and farm scales because they optimize tradeoffs between increased food production, poverty alleviation, and environmental conservation (Izac and Sanchez, 2000). They can also be inferior to other land uses, particularly when the technology is inappropriate or the accompanying policies are not enabling (Sanchez, 1995). Analysis of tradeoffs between private farmer benefits and global environmental benefits provide a solid basis for partitioning benefits arising from global environmental conventions and protocols. Figure 4-9 illustrates tradeoffs between carbon sequestration and farmer profitability in Cameroon in a range of agroforestry practices.

Incorporation of trees on farms affects carbon stocks differently than cropland or forest management. For example, trees on farms provide tighter coupling of key processes such as nutrient cycling and weed control than in croplands; trees in agroforestry are harvested more frequently than under forest management. One accounting option for agroforestry is the time-averaged carbon sequestration rate (Palm *et al.*, 2000), which takes into account periodic woody biomass harvests based on the "average storage method" of Schroeder (1992). Figures 4-10 and 4-11 show the application of this method for two broad practices. All estimates of changes in carbon stocks and carbon sequestration rates are based on time-averaged carbon stocks.

The potential land area suitable for agroforestry in Africa, Asia, and the Americas is 585–1215 Mha (Dixon, 1996). This estimate is a compilation of several estimates (Unruh *et al.*, 1993; Dixon *et al.*, 1994; Dixon, 1995). The current area in agroforestry is on the order of 400 Mha, of which 300 Mha are "arable land" and 100 Mha are "forest lands" in the FAO database. For example, the 14 Mha of agroforestry in China are classified as agricultural land (Xu, 1999). It is estimated that an additional 630 Mha of current croplands and grasslands could be converted into agroforestry, primarily in the tropics.

This chapter considers two kinds of agroforestry activities: land conversion and improved land use. Land conversion includes transformation of degraded cropland and grasslands, including those from slash-and-burn agriculture, into new agroforests. The potential area of new agroforests from land-use change could be on the order of 400 Mha during the next 25 years. Improved use of current agroforestry systems with interventions that result in increased carbon is analogous to "improved cropland"—but with trees on the farm. Both kinds

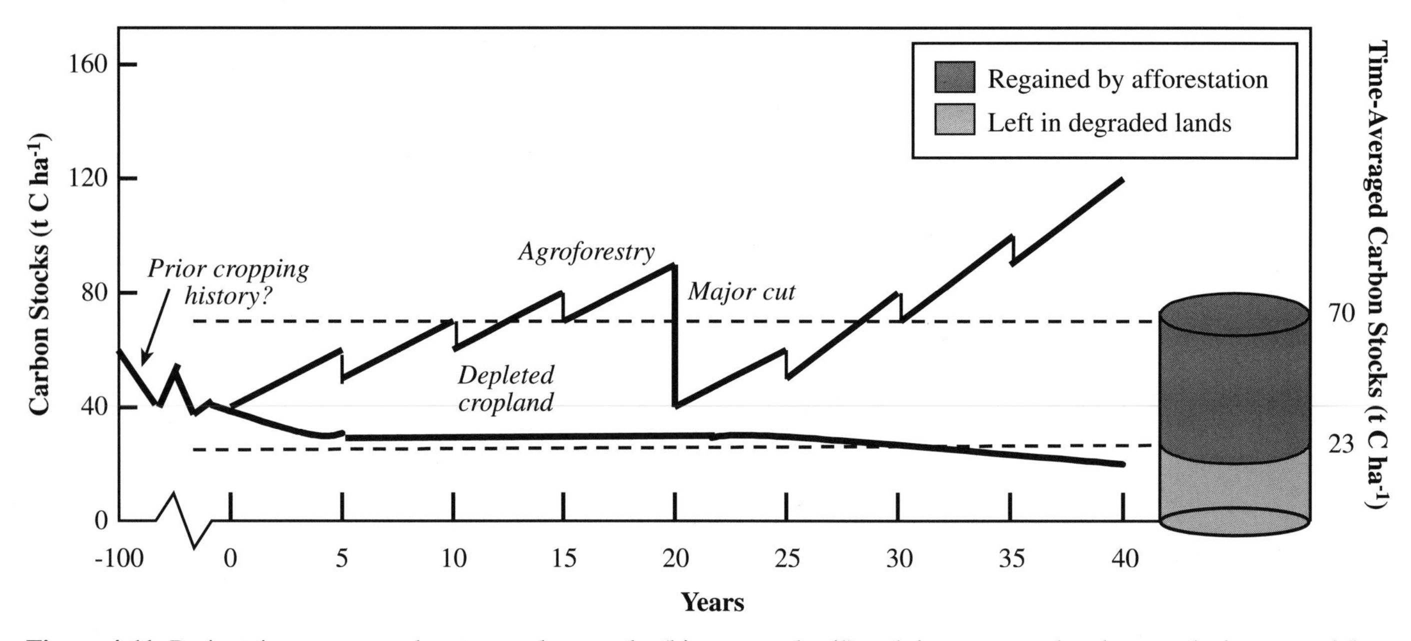

Figure 4-11: Project time course and system carbon stocks (biomass and soil) and time-averaged carbon stocks in sequential agroforestry systems based on soil fertility replenishment and intensification with high-value trees in subhumid tropical Africa. Based on Table 4-7 and additional data from Woomer *et al.* (1997).

of activities increase carbon stocks, and many also prevent carbon losses in adjacent forests and woodlands by avoiding further deforestation or land degradation. Leakage can reverse agroforestry's effect on avoiding deforestation (Fearnside 1995, 1997), however, particularly if new practices are capital-intensive rather than labor-intensive (Kaimowitz and Angelsen, 1998). This section discusses only the effects on carbon stocks in agroforests because Chapter 3 discusses avoidance of further deforestation. Two practices involving land-use change are discussed: one from slash-and-burn agriculture and another that involves converting degraded cropland in Africa.

4.4.4.1. From Forests to Slash-and-Burn to Agroforests after Deforestation

This major land conversion practice takes place mainly at the margins of humid tropical forests. Transformation of the original forest into various types of agroforests results in a smaller decrease in carbon stocks than transformation of forests into cropland, pastures, or degraded grasslands (Palm *et al.*, 2000).

Much of the uncertainty in the values of CO_2 fluxes from the tropics is the result of inadequate estimates of the biomass that is cleared, the fate of the carbon lost, the type of biomass and time course of subsequent land-use systems, and the regrowth rates of vegetation (Houghton, 1997). A project that used standardized methods to compare several such systems in Brazil, Cameroon, Indonesia, and Peru provided data on the foregoing parameters and the carbon sequestration potential of many land-use systems at the margins of the humid tropics (Woomer *et al.*, 1999; Palm *et al.*, 2000; Sanchez, 2000). The time course of the land-use changes is described in the following subsections. An example of one specific practice, complex agroforests, is described in Fact Sheet 4.10.

4.4.4.1.1. Rates

Carbon sequestration rates are highly negative on forest clearance: -92 t C ha^{-1} yr^{-1} during the first 2 years after slash-and-burn (Table 4-7)—a period that is normally under annual cropping or pasture establishment (Neill *et al.*, 1997). Table 4-7 shows that carbon sequestration rates become positive with secondary forest fallows (5–9 t C ha^{-1} yr^{-1}); complex agroforests (2–4 t C ha^{-1} yr^{-1}); and simple agroforests with one dominant species such as oil palm, rubber, or *Albizia falcataria* (7–9 t C ha^{-1} yr^{-1}). The lower carbon sequestration rate of some agroforestry systems in relation to natural secondary succession found by Palm *et al.* (2000) is partly because agroforestry products are removed from the system for family use or for sale. This finding underscores the important tradeoffs between a global public good (carbon) and a private good (economic gain) (Tomich *et al.*, 1998). Croplands, pastures, and degraded grasslands lost carbon at a slow rate or show modest positive rates (-0.4 to +3 t C ha^{-1} yr^{-1}). Land-use systems that include trees, therefore, produce higher carbon sequestration rates than those that are limited to annual crops, pastures, or grasslands (Palm *et al.*, 2000).

4.4.4.1.2. Stocks

Results summarized in Table 4-7 provide a time course of changes in total carbon stocks (aboveground plus below-ground) in a 20-year traditional slash-and-burn sequence (Sanchez, 2000). The area changes from original forest, usually logged (230 t C ha^{-1}), to burned cropland (46 t C ha^{-1}), to bush fallow (34 t C ha^{-1}), and to tall secondary forest fallow (112 t C ha^{-1}). After burning and cropping for an average of 2 years, about 80 percent of the carbon stock is lost. These slash-and-burn systems do not include tillage. Crops are planted by digging holes in the ground with a stick or machete and dibbling seeds, pastures by broadcasting seeds on the surface, and trees by transplanting seedlings in holes. There is essentially no soil disturbance: The surface is covered with partially burned logs, branches, tree stumps, and leaf litter. Most of the carbon stock lost is by biomass burning; some is lost by increasing decomposition of soil carbon because of higher soil temperatures during fires and afterward as a result of increased incident solar radiation.

Crop-short bush fallow rotations, pastures, and *Imperata cylindrica* grasslands that ensue have carbon stocks of about 30 t C ha^{-1}, most in the soil. Therefore, about 88 percent of the carbon stock of the original forest is lost and emitted to the atmosphere in the transformation from forest to croplands or pastures within 4–12 years (Table 4-7). Improved short-term fallow systems or improved pastures do not significantly increase carbon stocks in relation to the current degrading practices.

Below-ground organic carbon stocks averaged 40 t C ha^{-1} at 0–20 cm depth in undisturbed forests (Palm *et al.*, 2000). The following land uses resulted in different proportions of below-ground carbon relative to undisturbed forests:

- Agroforestry systems: 80–100 percent
- Crop/long-term fallow sequence: 90–100 percent
- Pastures: 80 percent
- Crop/short-term fallow: 65 percent
- Degraded *Imperata* grasslands: 50 percent or less.

Except for the crop/short-term fallow and degraded *Imperata* grasslands, the other alternative land uses lost less than 20 percent of topsoil carbon. This finding suggests that the potential for carbon sequestration in the humid tropics is mainly above ground.

4.4.4.1.3. Difference in stocks between agroforestry and common practices

Ten-year-old pastures in Brazil and 13-year-old *Imperata* grasslands in Indonesia (common practices after tropical deforestation) contained 201 t C ha^{-1} less than the original forest that was cleared (Palm *et al*, 2000). In contrast, agroforests that are established immediately after slash-and-burn by planting trees along with food crops contained 150 t C ha^{-1} less than the original forest (Table 4-7). Agroforestry systems can regain 35 percent of the original carbon stock of the forest; croplands and pastures can regain 12 percent. Through the establishment of tree-based systems in degraded pastures, croplands, and grasslands, the time-averaged carbon stock in the vegetation increases by 50 t C ha^{-1} in 20–25 years, whereas that in the soil increases by 7 t C ha^{-1} (Palm *et al.*, 2000). Agroforestry practices therefore permit the sequestration of an additional 57 t C ha^{-1}— three times as much as croplands or grasslands can sequester

***Table 4-7**: Carbon uptake rates and time-averaged system carbon stocks and differences in carbon stocks from land transformation at margins of humid tropics. Summary of 116 sites with different land uses before and after slash-and-burn located in Pedro Peixoto (Acre) and Theobroma (Rondônia), Brazil; Ebolowa, M'Balmayo, and Yaounde, Cameroon; Jambi and Lampung, Sumatra, Indonesia; and Yurimaguas and Pucallpa, Peru.*[a]

Land-Use Practice	Carbon Uptake Rates (t C ha^{-1} yr^{-1})			Duration (yr)	Carbon Stocks (time-averaged) (t C ha^{-1})			Differences in Modal Carbon Stocks (time-averaged) (t C ha^{-1})	
	Low	*Modal*	*High*		*Low*	*Modal*	*High*	*Forest*	*Pasture/ Grasslands*
Primary and logged forest	n/a[b]	n/a[b]	n/a[b]	?	192	230	276	—	-201
Cropping after slash-and-burn	-76	-92	-112	2	39	46	52	-184	+17
Crops/bush fallow	2	3	4	4	32	34	36	-196	+5
Tall secondary forest fallow	5	7	9	23	95	112	142	-118	+83
Complex agroforest	2	3	4	25–40	65	85	118	-145	+56
Simple agroforest	5	7	9	15	65	74	92	-156	+61
Pasture, *Imperata* grassland	-0.2	-0.4	-0.6	4–12	27	29	31	-201	—

[a] Sanchez (2000). Calculated from data of Woomer *et al.* (1999) and Palm *et al.* (2000), assuming the following time-averaged soil carbon stocks (in Mg C ha^{-1}): 40 for primary/logged forest and crops after slash-and-burn, 35 for tall secondary forest fallow and complex agroforest, 30 for bush fallow and simple agroforest, and 25 for pasture and *imperata* grassland.

[b] Not available; likely close to zero.

(Figure 4-10). This finding indicates the key contribution of agroforestry to increasing carbon stocks at the margins of the humid tropics.

The area that could be converted to this practice is 10.5 Mha annually if enabling government policies were in place (Fay *et al.*, 1998; Tomich *et al.*, 1998). The area calculation is based on two assumptions: 20 percent of the 15 Mha that is annually deforested (3 Mha) is put into agroforestry every year, and 3 percent of the 250 Mha of degraded lands at the forest margins (Sanchez *et al.*, 1994) is converted into agroforests every year (7.5 Mha). The difference in time-averaged carbon density from this land-use change is 35–90 t C ha^{-1}, with a modal value of 57 t C ha^{-1}. We further assume that the annual deforestation rate will stay constant for the next 10 years. If this happens, the global contribution of this practice to carbon sequestration would be on the order of 0.105–0.525 Gt C yr^{-1}, with a modal value of 0.315 Gt C yr^{-1}.

4.4.4.2. *From Low-Productivity Croplands to Sequential Agroforestry in Africa*

The second major agroforestry practice is the transformation of unproductive cropland into agroforestry-based crop/tree fallow rotations. Although various expressions of this practice are found throughout the tropics (Buresh and Cooper, 1999), it is illustrated in this Special Report with the recent movement to replenish soil fertility in subhumid areas of tropical Africa. Soil carbon stocks have dramatically decreased in smallholder farms of sub-Saharan Africa because of nutrient depletion, which is increasingly recognized as the fundamental biophysical cause for declining food security in this region (Sanchez *et al.*, 1997a,b). Given the acute poverty and limited access to mineral fertilizers, an ecologically robust approach is being used by tens of thousands of farmers in eastern and southern Africa. This approach consists of bringing natural resources to farmer fields where crops can utilize them: nitrogen from the air by biological nitrogen fixation, phosphorus from indigenous phosphate rock deposits, and nutrient-rich shrub biomass from roadsides and farm hedges (Rao *et al* 1998; Kwesiga *et al.*, 1999; Sanchez, 1999). Details of this practice are described in Fact Sheet 4.11. Replenishment of nitrogen and phosphorus has important effects on changes in carbon stocks.

4.4.4.2.1. *Rates*

Conventional cropping on clayey, oxidic Alfisols in Kabete, Kenya, shows a 28 percent decrease in topsoil carbon, from 36 to 26 t C ha^{-1} during an 18-year period (Kapkiyai *et al.*, 1998). Another long-term trial in Muguga, Kenya, with similar soils shows a total loss of 91 t C ha^{-1} to a depth of 120 cm with 8 years of continuous cultivation without inputs. About half of this loss (48 t C ha^{-1}) took place in the top 15 cm (Woomer *et al*, 1997). The loss of topsoil organic carbon associated with soil nutrient depletion has been estimated at an average rate of -0.22 t C ha^{-1} yr^{-1} (Sanchez *et al.*, 1997b). When soil fertility is replenished, maize grain yields increase from 0.5 to 2 t C ha^{-1} and carbon sequestration rates become positive, averaging 1.5 t C ha^{-1} (Table 4-8). When more trees are planted on field boundaries and as orchards, carbon sequestration rates increase further to 3.5 t C ha^{-1}. Overall carbon sequestration rates may range from 1.2 to 5.1 t C ha^{-1} yr^{-1}, with a modal value of 3.1 t C ha^{-1} yr^{-1}.

4.4.4.2.2. *Stocks*

Nutrient-depleted fields have little biomass carbon stock; a time-averaged modal figure is on the order of 23 t C ha^{-1}, virtually all below ground (Table 4-8). Soil fertility replenishment practices that are based on improved fallows, rock phosphate, and biomass transfers of *Tithonia diversifolia* for 25 years are estimated to result in time-averaged carbon stocks of 35 t C ha^{-1}. Such stocks are virtually all in the soil; crop and fallow accumulation may account for only 1 t C ha^{-1} above ground. The increase in soil carbon at equilibrium (8 t C ha^{-1}) reflects 80 percent replenishment of the lost soil carbon. When trees are incorporated after fertility replenishment, total time-averaged stocks reach 70 t C ha^{-1}, which includes 34 t C ha^{-1} in aboveground biomass and 36 t C ha^{-1} below ground (Table 4-8). It should be noted that all the foregoing figures are estimates rather than hard data as in the case of the humid tropical forest margins.

4.4.4.2.3. *Difference in stocks between agroforestry and common practices*

The transformation of low-productivity croplands to sequential agroforestry in subhumid smallholder Africa can triple carbon stocks (from 23 to 70 t C ha^{-1}) in a 25-year period (Figure 4-11). This transformation consists of a two-stage process: first fertility replenishment, then more trees on the farm. The first stage increases carbon stocks by 9 t C ha^{-1}—all but 1 t C ha^{-1} as soil carbon. The second step increases carbon stocks by an additional 28 t C ha^{-1}—5 t C ha^{-1} as soil carbon and 23 t C ha^{-1} as aboveground biomass. Assuming such increases can take place in 46 percent (37.5 Mha) of the smallholder farms of subhumid tropical Africa (Sanchez *et al.* 1997b), during the next 25 years—after which we assume equilibrium is reached—this practice would provide a global contribution of 0.045–0.191 Gt C yr^{-1}, with a modal value of 0.116 Gt C yr^{-1}, in subhumid tropical Africa alone.

Carbon sequestration in subhumid Africa may therefore be considerable with land conversion to agroforestry systems that involve soil fertility replenishment and intensification and diversification of farming with the use of high-value domesticated trees. Lesser amounts can be expected in semi-arid tropical areas with similar practices. The widespread use of green manure cover crops in subhumid Latin America (Bunch, 1999) and subhumid West Africa (Buckles *et al.*, 1998) suggest considerable carbon sequestration potentials, but the rates have yet to be estimated.

***Table 4-8**: Estimates of carbon uptake rates and time-averaged system carbon stocks and differences in carbon stocks from land transformation from low-productivity cropland to sequential agroforestry in subhumid tropical Africa (Sanchez, 2000).*

Land-Use Practice	Carbon Uptake Rates ($t\ C\ ha^{-1}\ yr^{-1}$) Low	Modal	High	Duration (yr)	Carbon Stocks (time-averaged) ($t\ C\ ha^{-1}$) Low	Modal	High	Differences in Modal Carbon Stocks (time-averaged) ($t\ C\ ha^{-1}$)
Current nutrient-depleted small farms[a]		-0.22		?		23		—
Fertility-replenished farms with maize-tree fallow rotations[b]	1.0	1.5	2.4	25	20	35	49	+12
Fertility-replenished farms (as above) + more trees on farm[c]		3.5		25		70		+47

[a] From Sanchez *et al*. (1997b), based on calculations from Smaling (1993).
[b] Calculations from agronomic data by Kwesiga and Coe (1994), Rao *et al*. (1998), Kwesiga *et al*. (1999), and Jama *et al*. (2000). Improved fallows of *Sesbania sesban* and other species planted for 1 year in eastern Africa and 2 years in southern Africa add 100–200 kg N ha^{-1} to the soil. Tree fallows are followed by one maize crop in eastern Africa and three consecutive maize crops in southern Africa. Where phosphorus is limited (primarily eastern Africa), replenishment also includes basal application of 125–250 kg P ha^{-1} as rock phosphate plus biomass transfer of 1.8 tons dry matter ha^{-1} of *Tithonia diversifolia* to every maize crop (Sanchez *et al*., 1997b).
[c] Same as note b, plus adding trees as orchards and to farm boundaries (Woomer *et al*., 1997).

4.4.4.3. *Improved Agroforests*

Improvements include the incorporation of shelter belts; superior germplasm of tree and crop species; trees planted at a higher density; better nutrient management; integrated pest management; and other agronomic, silvicultural, and silvopastoral techniques in Annex I and non-Annex I countries (Dixon *et al*., 1994; Dixon, 1995; Doran and Turnbull, 1997; Buck *et al*., 1999; Lassoie and Buck, 1999; Xu, 1999).

Rising saline water tables occur in 15 of the 65 Mha (23 percent) of cultivated land in Australia. This slow process is mainly blamed on the gradual removal of trees from the agricultural landscape; because of their deep rooting habits, these trees formerly kept saline water tables well below the surface. Although deep drainage and other standard salinity management practices are recommended, the main vehicle of restoration is to repopulate trees on farms and in the landscape—particularly trees that are deep-rooted and salinity-tolerant. Woody species could provide more stable biomass pools than grasses in locations with variable climate.

Agroforestry improvement practices generally have a lower carbon uptake potential than land conversion to agroforestry practices because existing agroforestry systems have much higher carbon stocks than degraded croplands and grasslands that can be converted into agroforestry. Improving agroforestry land use could sequester carbon at time-averaged rates of 0.02–1.0 t C ha^{-1} yr^{-1}, with a modal value of 0.50 t C ha^{-1} yr^{-1}, in Annex I countries. The estimate for non-Annex I countries (0.08–0.33 t C ha^{-1} yr^{-1}, with a modal value of 0.22 t C ha^{-1} yr^{-1}) is lower because of greater institutional constraints. The foregoing estimates are based on expert opinion that considers technical, social, and institutional constraints, as well as existing policy environments. Policy issues affecting tree species selection in agroforestry projects (e.g., benefits for local biomass energy use, commercial pulping, native vs. non-native species) and requisite institutional and regulatory frameworks affect the success of agroforestry programs. Policy improvements that support secure land tenure, improved road infrastructure, and increased access to credit and markets are likely to contribute as much or more to increasing the productivity of existing agroforests than technological improvements in non-Annex I countries (Sanchez, 1995, 1999).

In temperate areas, the potential carbon storage with agroforestry ranges from 15 to 198 t C ha^{-1} (Dixon *et al*., 1994), with a modal value of 34 t C ha^{-1} (Dixon *et al*., 1993).

4.4.4.4. *Overall Contribution*

The potential contributions of agroforestry systems to carbon sequestration are summarized in Table 4-1 under the headings of improved management within a land use and land-use change. The latter is an order of magnitude higher than the former, given the lower initial levels of carbon stocks. Overall, agroforestry can sequester carbon at time-averaged rates of 0.2–3.1 t C ha^{-1}. In temperate areas, the potential carbon storage with agroforestry ranges from 15 to 198 t C ha^{-1} (Dixon *et al*., 1994), with a modal value of 34 t C ha^{-1} (Dixon *et al*., 1993). The associated impacts of agroforestry include helping to attain food security and secure land tenure in developing countries, increasing farm income, restoring and maintaining aboveground and below-ground biodiversity (including corridors between protected forests), serving as CH_4 sinks, maintaining watershed hydrology, and decreasing soil erosion.

4.4.5. Forest Management

Forest management is the application of biological, physical, quantitative, managerial, social, and policy principles to the regeneration, tending, utilization, and conservation of forests to meet specified goals and objectives while maintaining forest productivity. Management intensity spans the range from wilderness set-asides to short-rotation woody cropping systems. Forest management encompasses the full cycle of regeneration, tending, protection, harvest, utilization, and access.

A vast number and variety of activities exist and are being developed by researchers to manage forests. The actual outcome of forest management (e.g., whether it sequesters carbon, produces industrial wood or wood fuel, or protects biodiversity) usually can be measured only as the integrated outcome of the suite of practices used, not as the outcome of individual practices evaluated alone. The positive impact of any practice may be realized only if applied in concert with one or more other practices—each of which may have a minimal, or even negative, impact. Thus, for forest management, measuring carbon stocks with a broad definition of the activity (Section 4.3.2) and land-based accounting methods (Section 4.3.3) may lead toward full accounting, particularly if wood products are included in the accounting (Section 2.4.2.2).

The end use of wood products is important for two reasons. First, wood products provide a stored carbon stock that depends on the life span of the product. Wood-in-use stocks are growing larger in many countries, and management choices (efforts to extend useful life, recycling, etc.) can contribute to further growth in these stocks. Second, greater utilization of wood allows reduced use of fossil fuel—by utilizing the wood either directly for energy production or to replace energy-intensive products such as steel, aluminum, plaster board, and bricks.

Forest management in this chapter refers to forestry activities that are not ARD activities as defined in Article 3.3 of the Kyoto Protocol. Although there are many possible forest management activities, the following are examples that are likely to alter carbon stocks:

- Forest regeneration
- Forest fertilization
- Pest management
- Forest fire management
- Harvest quantity and timing
- Low-impact harvesting
- Reducing forest degradation.

Brief descriptions of these activities follow (the Fact Sheets provide more detail). The carbon sequestration potential of forestry activities varies considerably between ecosystems, countries, and regions, and few empirical studies exist. Table 4-9 lists some examples of existing studies to illustrate the potential. Finally, it is assumed that forest management practices that are implemented for carbon sequestration purposes will comply with existing multilateral agreements (e.g., the United Nations Convention on Biological Diversity and Ramsar Convention of Wetlands) and the results of the ongoing United Nations Intergovernmental Forum on Forests regarding sustainable forest management.

4.4.5.1. Forest Regeneration

Forest regeneration generally is done promptly after the previous stand or forest has been removed. The method (natural regeneration, artificial planting, seeding), species, and density are chosen to meet landowner goals. Forest regeneration includes activities such as changing tree density through human-assisted natural regeneration; enrichment planting; reduced grazing of savanna woodlands; and better matching of tree provenances, genetic strains, or species to soils and sites. "Human-assisted natural regeneration" is establishment of a forest age class from natural seeding or sprouting after activities such as selection cutting, shelter (or seed-tree) harvest, soil preparation, or restricting the size of a clear-felling stand to secure natural regeneration from surrounding trees. "Enrichment planting" is inter-planting additional trees in an existing forest stand.

Forest regeneration influences on-site carbon stocks by accelerating the return of a growing forest after harvest, altering growth rates of aboveground and below-ground tree biomass through better species selection, and changing the potential mix of final wood products. The impacts on the litter layer and soil vary with many factors. Generally, over the rotation period, the annual increase of carbon in tree biomass is much higher than the additions to soil or litter carbon. Where normal forest management activities (e.g., careful harvest followed by rapid regeneration) are carried out, it is generally thought that soil carbon stocks do not change significantly from rotation to rotation.

The connection between forest regeneration and the end use of wood is an important consideration in full accounting of carbon stocks. For example, a higher tree density may not be the best option for carbon sequestration. High densities may lead to rapid crown closure and early growth, but crowded stands often suffer early onset of mortality and rapid growth decline, so they can become sources of carbon emissions considerably sooner than stands that are managed at lower densities. Additionally, trees grown in less dense conditions generally reach a size that is suitable for solid wood products earlier, leading to long-lived wood products sooner, adding to the stock of sequestered carbon, and substituting for non-wood products that use more fossil fuel in their production.

The accumulation time of aboveground and below-ground biomass ranges from 5 years (for the shortest rotation times in tropical plantations) to 150 years or more on low-potential sites in boreal forests. The tree biomass carbon accumulation process is not difficult to quantify and predict where well-developed forest growth and yield models are available. Conversion factors have been published to convert merchantable stemwood estimates into total tree biomass, including roots (Marklund, 1988; Birdsey, 1996). Soil carbon accumulation processes under forest management are less well researched in many forest regions.

Only a few studies exist on the relationship between forest regeneration and carbon sequestration (e.g., Lunnan *et al.*, 1991; Hoen and Solberg, 1994; Xu, 1995; Row, 1996; Nabuurs *et al.*, 1999; Ravindranath *et al.*, 1999).

4.4.5.2. Forest Fertilization

Fertilization is the addition of nutrient elements to increase growth rate or overcome a nutrient deficiency in the soil. Effective fertilization can entail either an increase in the quantity of fertilizer or an improvement in application through the selection of compounds, timing, and dosage so that as much fertilizer as possible is taken up by trees and less becomes waste to groundwater. Unintentional fertilization is occurring in many forests that are downwind of industrial centers, as a result of the deposition of nitrogen and sulfur from the atmosphere.

Appropriate fertilization leads to higher growth rates of aboveground and below-ground biomass, thereby increasing carbon storage. The process is well understood except for some of the soil processes, and in many countries reliable models are available to predict increased biomass growth. Fertilization may have negative associated environmental impacts because it can increase the emission of N_2O and NO_x to air and water and influence soil processes. Few studies exist on forest fertilization as a method of carbon sequestration (e.g., Lunnan *et al.*, 1991; Hoen and Solberg, 1994; Nabuurs *et al.*, 1999).

***Table 4-9**: Rates of potential carbon gain under selected practices for forestland in various regions of the world.*

Practice	Country/ Region	Rate of Carbon Gain ($t\ C\ ha^{-1}\ yr^{-1}$)	Time[1] (yr)	Other GHGs and Impacts	Notes[2]
Improved Natural Regeneration	India	0.55	30		a
Increased Rotation Length	Canada	0.022	80	Leakage (increased	b
	USA	0.036	80	harvest elsewhere)	b
	The Netherlands	0.035	80		b
Forest Fertilization	Canada	0.03–0.19	20	$+N_2O$, $+NO_x$	b
	USA	0.08–0.48	20	Ecological changes	b
	The Netherlands	0.1–0.6	20		b
	Norway	0.44	20		c
Forest Conservation	India	0.48	30	Environmental improvements	a
Reduced Forest Degradation	Tropical/ Global	1.7–4.6	40	Environmental improvements	h
Several Practices Combined	USA	3.1	50	Ecological changes	d
	Norway	0.12–0.20	20		e
Several Practices Combined,	USA	1.2	40	Ecological changes	f
Lobloly Pine	USA	3.5	25		g
Species Change (Aspen to Red Pine)	USA	0.88	80	Ecological changes	f

[1]Time interval to which estimated rate applies. This interval may or may not be time required for ecosystem to reach new equilibrium.

[2] a. Ravindranath *et al.* (1999).
b. Nabuurs *et al.* (1999).
c. Lunnan *et al.* (1991).
d. Birdsey *et al.* (2000).
e. Hoen and Solberg (1994); assuming harvest volume is kept constant.
f. Row (1996).
g. Albaugh *et al.* (1998); refers to intensive fertilization and irrigation on an infertile drained sandy soil in North Carolina. Rate is an average estimate of 3 years of measurements starting in 8-year-old stands.
h. Based on mean biomass stock differences between non-degraded and degraded tropical forests as reported in FAO (1996). Stock differences are 182, 126, and 70 tons dry matter per hectare for tropical wet, moist, and dry zones, respectively, with carbon content as 50% of dry matter.

4.4.5.3. Fire Management

"Fire management" describes activities that are required for the protection of wildland values from fire, as well as the use of fire to meet land management objectives. Fire management is usually carried out in an attempt to produce less damaging wildfires through lowered fire intensity and severity. Fire management can entail changing the type of fire (i.e., reducing the chances of a lethal crown fire by thinning a stand so that a subsequent fire will be a cooler, surface fire) or increasing the frequency of intentional fires to reduce the chance that fuels will build to levels that will support a severe wildfire. For several types of forests, such activities increase the standing stock and reduce the risk of large, severe wildfires that may cause soil damage, nutrient depletion, and watershed damage (MacDonald *et al.*, 2000). Although prescribed fires produce a variety of GHGs and particulate pollution, if managed properly they may reduce time-averaged emissions from some forest ecosystems (Neuenschwander and Sampson, 2000). In extreme situations, severe wildfires can so degrade a site that forest recovery may be delayed, a totally different ecosystem may emerge, or desertification may begin (Cromack *et al.*, 2000). Fire management is complicated and must be viewed in connection with the long-term health of the particular forest ecosystem involved, as well as harvest and wood utilization practices. Burning aboveground and below-ground biomass, humus, and soil increases the emission of carbon and other GHGs (Goldammer, 1990; Dixon and Krankina, 1993; Price *et al.*, 1998).

Globally, forest burning releases large quantities of CO_2, about 10 percent of annual global CH_4 emissions, and 10–20 percent of global N_2O emissions—a significant effect on atmospheric chemistry (IPCC, 1992). In the United States, effective fire suppression for the past century has dramatically reduced wildfire emissions and atmospheric pollution, but it also has restored high-risk fuel conditions that are now a major policy concern (Leenhouts, 1998; Sampson *et al.*, 2000). Forest ecosystems that developed with fire as a major ecosystem process (i.e., nearly all forests)—even those where fires, though infrequent, generally killed most or all of the trees (i.e., boreal forests)—could be expected to re-assimilate the CO_2 emitted during the growing period between fires, provided that the fire frequency as measured over large regions is relatively unchanging. The latter requirement may no longer hold true in many parts of the world because of human intervention and climate change (Apps *et al.*, 1999a).

4.4.5.4. Pest Management

Pest management is the application of approved strategies to maintain a pest population within tolerable levels. Improved pest management may prevent damage to standing stocks of forests and thus prolong and increase carbon storage. The processes related to carbon storage are well known, whereas the reasons pests occur and how they can be prevented are far less well known. The interactions between pests, fire, and climate change are likely to be important factors in forest cover change in the next century. Methodological and scientific uncertainty will make this practice difficult to measure and monitor if practice-based accounting is used. As part of a broad forest management strategy, good pest management will contribute to improved carbon stocks over time by helping to maintain sustainable forest conditions and higher standing carbon stocks within managed forests.

4.4.5.5. Harvest Quantity and Timing

Harvesting is a procedure by which a forest stand is logged with an emphasis on meeting logging requirements while concurrently attaining silvicultural objectives. Harvest scheduling is a process for allocating cutting and other silvicultural treatments over a forest with an emphasis on which treatments to apply, as well as where and when to apply them. The practice relates to when the harvesting is done (thinnings including pre-commercial thinnings, selection fellings, or clear-fellings) and the timber volume extracted.

Harvest intensity significantly affects the quantity and quality of timber produced. The carbon sequestration impact therefore must be considered in connection with the end use of wood products that are manufactured from a particular harvest. For example, merely increasing the rotation time, lengthening the period between harvest operations, or reducing the volume extracted may lead to reduced growth rates and reduced sequestration opportunities in the forest, as well as less wood for bioenergy or replacement of energy-intensive products such as steel, aluminum, and concrete (Row, 1996).

Optimal thinning and clear-felling times (and quantities) are interlinked and dependent on other forest management measures (e.g., fertilization or plant density) and objectives. Hoen and Solberg (1994) analyzed combinations of thinnings, clear-fellings, and other silviculture measures (e.g., fertilization, types and intensity of regeneration) simultaneously for a boreal forest region, keeping harvesting levels constant and maximizing carbon storage (including end use and decay of wood products) for the region for a long period. The study illustrates that in such a situation, thinning and clear-felling times will be significant and complementary. For example, thinnings on good site classes substitute for clear-fellings on low site classes that have high standing volumes (but low annual growth and carbon sequestration potential if harvested). Boscola *et al.* (1997) show how the length of harvest cycles and minimum cutting diameters can be combined to produce maximum carbon sequestration; they arrive at a cost of US\$1.20 t^{-1} C for an increase in cutting cycles from 40 to 50 years for lowland tropical rain forest in Malaysia. Other studies on this topic include Boscola and Buongiorno (1997), Hoen and Solberg (1997), and Plantinga and Birdsey (1994).

4.4.5.6. Low-Impact Harvesting

Low-impact harvesting methods have been developed and executed to provide minimum disturbance to soil, remaining vegetation, and extracted trees during the logging process. This practice influences the stock of carbon in the soil as well as the growth (and corresponding carbon storage) of new and remaining trees/vegetation. It also influences carbon storage in end products by influencing timber quality and thus the type of utilization of the timber. Among the studies on this topic are Putz and Pinard (1993), Winjum *et al.* (1993), and Pinard and Putz (1996, 1997).

Uncertainties in the quantification of differences in carbon stock changes associated with changed harvesting practices may make this practice difficult to measure on an activity-based approach. If this practice were incorporated within a broader activity that is measured in terms of stock change over time, the impact would be captured.

4.4.5.7. Reducing Forest Degradation

Reducing forest degradation may be a possible practice under Article 3.4. Forest degradation can yield large carbon losses without qualifying as deforestation, depending on the definitions adopted by the Parties for the implementation of Article 3.3 activities. Because reduced forest degradation is concerned with avoiding emissions rather than creating additional sinks, it poses a different set of issues related to baseline and leakage.

The practice is closely linked to deforestation; as such, it is complicated from the natural and social sciences points of view. For example, the question of driving forces behind tropical deforestation/degradation has spawned a huge literature (e.g., Repetto and Gillis, 1988; Mahar, 1989; Anderson, 1990a,b; Ara, 1990; Pearce, 1990; Binswanger, 1991; Amelung and Diehl, 1992; Grainger, 1993; Rietberger, 1993; Brown and Pearce, 1994; Kägi, 1997; Rudel and Roper, 1997; von Amsberg, 1998; Palo, 1999) but little consensus. This issue and others related to forest degradation/deforestation are treated more thoroughly in Chapters 2 and 5.

4.4.5.8. Associated Impacts, Leakage, and Verification

Several associated impacts (positive as well as negative) and leakage effects are possible for most forest management activities. For a more detailed discussion of these impacts, see the Fact Sheets at the end of this chapter and Chapters 2, 3, and 5 of this Special Report.

In principle, all forest management practices can be verified, particularly by using land-based, stock change methods. There are always questions of accuracy and cost, however. Monitoring and verification capacity varies between countries, and combinations of methods often might be applied. For example, if one is interested only in carbon stock changes from enrichment planting, a control plot without the enrichment planting should provide a basis to calculate the difference, as well as help filter out background variability and indirect effects (e.g., airborne nutrients). On the other hand, if a complex and variable suite of activities is applied across a broad forest region, separating direct effects from indirect effects and natural variability will be much more difficult. In this case, statistical sampling may be coupled with forest yield models and inventory data for similar, unmanaged stands to arrive at an estimate.

Many forestry activities that do not involve land-use change or clear-cut forest harvesting are difficult to observe with remote sensing; therefore, the net carbon storage must be estimated through statistical sampling (land-based accounting), from activity reports, or through some combination of approaches (Section 4.3.4).

4.4.6. Wetlands Management

Wetlands are defined as areas of land that are inundated for at least part of the year, leading to physico-chemical and biological conditions characteristic of shallowly flooded systems (IPCC, 1996b). Anaerobic conditions associated with inundation slow decomposition rates and allow accumulation of large stores of carbon over long time scales, even in systems of relatively low productivity. Although wetlands occupy only 4–6 percent of the Earth's land area (~0.53–0.57 Gha) (Matthews and Fung, 1987; Aselmann and Crutzen, 1989), they store an estimated 20–25 percent of the world's soil carbon (350–535 Gt C) (Gorham, 1995). The rates of carbon accumulation in peats (organic soils commonly associated with wetlands) vary with age (Armentano and Menges, 1986; Tolonen and Turunen, 1996) but eventually reach equilibrium when inputs equal losses (slow decomposition rates applied to very large carbon stores) (Clymo, 1984). Most of the wetland area and associated carbon storage is in peatlands in temperate and boreal regions; roughly 10–30 percent is in the tropics.

Decomposition under anaerobic conditions produces methane—a greenhouse gas. Wetlands are the largest natural source of methane to the atmosphere, emitting roughly 0.11 Gt CH_4 yr^{-1} of the total of 0.50–0.54 Gt CH_4 yr^{-1} (Fung *et al.*, 1991). Using a Global Warming Potential (GWP) of 21 for CH_4, emissions of ~1.7 g CH_4 m^{-2} yr^{-1} will offset the CO_2 sink equivalent to a 0.1 Mg C ha^{-1} yr^{-1} accumulation of organic matter. The range of CH_4 emissions from freshwater wetlands ranges from 7 to 40 g CH_4 m^{-2} yr^{-1}; carbon accumulation rates range from small losses up to 0.35 t C ha^{-1} yr^{-1} storage (Gorham, 1995; Tolonen and Turunen, 1996; Bergkamp and Orlando, 1999). Most freshwater wetlands therefore are small net GHG sources to the atmosphere. Two exceptions are forested upland peats, which may actually consume small amounts of methane (Moosavi and Crill, 1997) and coastal wetlands, which do not produce significant amounts of methane (e.g., Magenheimer *et al.*, 1996). Wetlands appear to be relatively small sources of N_2O to the atmosphere, except when they are converted for agricultural use. Although methane emissions from wetlands are now reported as part of a nation's GHG emissions, we compare methane and CO_2 emissions equivalently here to demonstrate the impact of various wetland activities on atmospheric GHGs.

Wetlands are vulnerable to future climate change (see IPCC, 1996b, Chapter 6). Increased decomposition rates in warmer temperatures—if associated with drier conditions—may lead to large carbon losses to the atmosphere, particularly from northern peatlands (Gorham, 1995); warmer temperatures may also lead to enhanced CH_4 emissions. Changes in regional hydrology caused by precipitation changes may cause loss or new growth of wetlands locally. Changes in permafrost extent and depth will alter the extent and dynamics of tundra wetlands. Sea-level rise will impact coastal wetland areas.

Wetlands management takes several forms: conversion to agriculture, drainage for forestry or agriculture, conversion for urban/industrial land uses, creation through construction of dams (energy uses), direct harvesting, and wetlands reconstruction.

Table 4-10 lists major practices that impact wetlands, along with associated processes affecting carbon storage and methane emission. Most practices affect CH_4 and CO_2 emissions in opposite ways. Data for the area of total wetlands, the area of human-impacted wetlands, and effects on GHGs are largely unknown for many regions. Hence, Table 4-10 gives qualitative rather than quantitative estimates for the net GHG effect of different management practices. Drainage of wetlands is associated with potentially large carbon losses as organic matter that has accumulated slowly over centuries to millennia is oxidized. Methane emissions from drained wetlands will be reduced (drained systems may even consume methane), offsetting some of the net GHG emission. For wetlands that do not emit significant methane (coastal wetlands), carbon stock changes will dominate. For many freshwater wetlands, methane emissions roughly balance the effect of carbon stock changes in CO_2-equivalent emissions.

To calculate the effect of wetlands activities on GHG emissions, information on areal conversion by wetland type is needed, along with net changes in GHG emissions. The area of global wetland converted for human use is poorly known, and estimates are mostly unavailable at the country level. Global estimates range from 6 percent (Armentano and Menges, 1986) to 50 percent (Moser *et al.*, 1996), with most conversion in temperate and tropical regions.

Wetland extent and duration of inundation are observable through remote sensing (e.g., Johnston and Barson, 1993). Carbon accumulation and methane emission rates can potentially be modeled, although these models are still in developmental stages for most wetland types. Validation of changes in carbon stores through field sampling is challenging because of the large size of the pools (e.g., the depth of many peaty soils) and the difficulty of physical access. Methane measurements on a wide scale would be difficult and expensive.

Several non-GHG impacts are associated with land practices that affect wetlands. Wetlands are specialized habitats that have distinct and often valuable flora and fauna; their loss is a biodiversity issue. Wetlands sequester many pollutants at the local level; in several countries, wetlands are constructed to treat wastewater. They act as a buffer for rapid changes in hydrology, and removal of wetlands can lead to increased flooding in some areas. Harvested organic matter from organic soils and peatlands is used as a fuel in some regions. Several international negotiations pertain to these aspects of wetlands—in particular, the Convention on Wetlands, the Convention on Biodiversity, and the Marine and Coastal Work Program (Bergkamp and Orlando, 1999).

4.4.6.1. *Wetland Conversion to Agriculture or Forestry*

Drainage of wetlands during conversion for agriculture or forestry results in a loss of carbon, as soil organic matter previously stored under anaerobic conditions is exposed to oxygen in air. In many cases, the organic carbon stores that accumulated slowly over centuries to millennia can be lost in days (in the case of burning) to decades. Rates of carbon loss are often inferred from changes in the surface elevation of the peat layer; careful analysis, however, shows that physical compaction of peat, if unaccounted for, will lead to subsidence without carbon loss (Minkkinen and Laine, 1998). Loss of anaerobic conditions near the wetland surface allows for greater oxidation of produced methane. Drained wetlands decrease methane emissions to zero, in some cases even consuming small amounts of methane from the atmosphere. Roulet and Moore (1995) report, however, that decreases in methane emission from the drained wetlands themselves may be offset (in some cases completely) by increased methane emissions from standing water in ditches used to promote drainage.

Kasimir-Klemedtsson *et al.* (1997) examined the net effect of agriculture on GHG emissions from temperate wetlands in Europe. The conversion of bogs and fens to different cropping types led to five- to 23-fold increases in CO_2-equivalent emissions, with a large increase in CO_2 emissions dominating over a drop in CH_4 emissions. Increases in N_2O emissions have also been observed in drained organic soils (Kasimir-Klemedtsson *et al.*, 1997), although few data are available.

Cutting of trees in wetlands damages the organic layer of soils through physical disruption, leading to loss of organic carbon. Growth of trees in drained wetlands—a practice affecting areas of boreal wetland in Europe—results in increased carbon stores in vegetation, which can offset short-term losses from soils (Minkkinen and Laine, 1998). The water table in forested peatlands is often below the surface, which causes much of the methane to be oxidized before it can be emitted. Hence, methane changes will not be as important to overall GHG balance in forested peatlands as in bogs and fens.

Using areas of wetland drainage from Armentano and Menges (1986) and the rates of carbon loss in Table 4-10, the increase in carbon released by drainage from that stored in natural wetlands is 0.063–0.085 Gt C yr^{-1} from temperate and boreal peatlands (Armentano and Menges, 1986) and 0.058 Gt C yr^{-1} (Armentano and Menges, 1986) to 0.11 Gt C yr^{-1} (Maltby and Immirzi, 1993) in tropical peatlands. For comparison, peat

***Table 4-10**: Rates of potential carbon gain under selected practices for wetland management activity in various regions of the world.*

Practice	Country/ Region	Rate of Carbon Gain ($t\ C\ ha^{-1}\ yr^{-1}$)	Time[1] (yr)	Other GHGs and Impacts	Notes[2]
Conversion to agriculture	Annex B (boreal and temperate)	-1 to -19 (loss)	D	---CH_4 (net effect: generally an increase in GHG emissions, depending on initial CH_4 emission rate and actual rate of CO_2 release); loss of biodiversity, increase in flooding, decrease in water quality, increased availability of food	a
	Non-Annex B (tropical)	-0.4 to -40 (loss)	D		b
Conversion to forestry	Annex B (boreal and temperate)	-0.3 to -2.8 (loss)	D	---CH_4 (net effect: small increase to small decrease in GHG emission become reduction of CH_4 emissions largely offsets loss of CO_2); loss of biodiversity, increase in flooding, decrease in water quality, increased availability of food or harvested products	c
	Non-Annex B (tropical)	-0.4 to -1.9 (loss)	D		d
Conversion for urban and industrial use		Potentially high losses (rates unknown)	D?	---CH_4 (net GHG effect ~0); loss of biodiversity, increase in flooding, decrease in water quality	e
Wetland restoration	Range of reported values	0.1–1.0	>100	+++CH_4 [net GHG effect ~0, where CH_4 emissions are large enough to offset carbon sink to decreased emissions in wetlands where CH_4 emissions are small (especially coastal areas that do not emit significant CH_4)]; increase in water quality, decrease in flooding	f
Creation of new flooded lands		Short term: -0.1 to -2; Long term: 0–0.05	>100	Short-term GHG source (in long term may be small sink through sediment deposition), loss of biodiversity, higher stability of water supply, increased availability of energy that requires no fossil fuel burning	g
Peat harvesting	Boreal and temperate	Unknown	>100	Unknown effect on CH_4	h

[1]Either duration of emissions of stored carbon (which will last as long as carbon is available to decompose; signified by D) or persistence of carbon stored in wetland organic matter as sinks (>100 years).

[2]a. Bergkamp and Orlando (1999).
b. Maltby and Immirzi (1993). Locally, rates of carbon loss may be as high as 150 $t\ C\ ha^{-1}$ when drained fields are burned.
c. Armentano and Menges (1986).
d. Maltby and Immirzi (1993); Sorenson (1993). Burning can locally release 11,000 $t\ C\ ha^{-1}\ yr^{-1}$.
e. Roulet (2000).
f. Tolonen and Turunen (1996).
g. Fearnside (1995, 1997); Galy-Lacaux *et al.* (1997); Dumestre *et al.* (1999); Kelly *et al.* (1999).
h. Armentano and Menges (1986). CO_2 emission through peat burning ~0.03 $Gt\ C\ yr^{-1}$.

combustion for fuel releases an estimated 0.03 Gt C yr^{-1} (Armentano and Menges, 1986). Assuming (as a maximum) that all of the areas converted for forestry, agriculture, and pasture reduced methane emissions to zero, the global methane emissions reduction would be roughly 0.6 Mt C yr^{-1}—equivalent to reducing the CO_2 emissions from decomposing organic matter by only ~0.004 Gt C.

4.4.6.2. *Wetland Conversion to Urban or Industrial Land*

Complete removal or covering of a wetland will limit exchanges of CO_2 and CH_4 with the atmosphere and eliminate plant production. The total effect on GHGs depends on whether the original carbon stock of organic matter is removed and mineralized or buried and subject to slowed decomposition. The potential for carbon release is high. Estimates of wetland land area converted for urban and industrial use are not available, except locally.

4.4.6.3. *Impoundments*

Water impoundments (dams, weirs, reservoirs) are significant sources of CO_2 and CH_4 to the atmosphere (Duchemin *et al.*, 1995, 1999; Fearnside, 1995, 1997; Galy-Lacaux *et al.*, 1997; Dumestre *et al.*, 1999), largely because of enhanced GHG emissions during decomposition of organic matter in flooded soils and wetlands. Kelly *et al.* (1997) report that flooding boreal land to create a reservoir changed it from a small net sink of GHGs (-6.6 g C m^{-2} yr^{-1}) to a large source (130 g C m^{-2} yr^{-1}). Turbines used to generate power lead to more efficient loss of methane (Duchemin *et al.*, 1999). Rates of emission may slow over time as organic matter inundated by the impoundment decomposes, but in some cases emission will be sustained by aquatic plant production in flooded lands (Fearnside, 1995, 1997).

4.4.6.4. *Wetland Restoration*

Wetland restoration (see Fact Sheet 4.18) is increasing as the role of wetlands in water quality and flood management is recognized. Few countries account formally for changes in wetland area, so the global importance of this practice is not known. The net effect on GHG emissions will depend on previous land use, the degree to which the restored wetland functions biogeochemically like the predisturbance wetland, and the degree to which enhanced carbon storage in organic matter is offset by methane emissions.

For the purposes of estimating the area available for wetland restoration (see Fact Sheet 4.18 and Table 4-1), estimates of area conversion are those in Patterson (1999). It was assumed that 50 percent of available wetland has been drained in Annex I countries and 10 percent in non-Annex I countries. The total area of wetlands globally is 570 Mha (Matthews and Fung, 1987), of which 350 Mha are estimated to be in Annex I countries (Maltby and Immirzi, 1993—North America, Europe, and former Soviet Union). Hence, the total area available for restoration of former wetlands (Table 4-1) is estimated at 210 Mha for Annex I countries and 20 Mha for non-Annex I countries. Table 4-1 is based on the assumption that 5 percent of this area will be restored in Annex I countries in the 2010 time frame and 15 percent in the 2040 time frame. Restoration rates in non-Annex I countries are assumed to be lower—1 and 10 percent, respectively. Table 4-4 gives the range of carbon storage rates in restored wetlands. Again, few data are available. A value of 0.4 t C ha^{-1} yr^{-1} is used to estimate net carbon storage from wetlands restoration in Table 4-1.

4.4.7. *Restoration of Severely Degraded Lands*

Restoration of degraded land generally involves revegetation that increases carbon stocks in biomass and soil. It can occur on croplands, grazing lands, forests, or "other" lands (mine spoils, deserts, etc.) As such, many of the practices relating to restoration of partially degraded land are discussed elsewhere in this Special Report under the appropriate land use. This section deals with soils that are so badly degraded that normal practices no longer have an effect, soils that are no longer capable of supporting crops, and soils that must be taken out of agricultural practice before progress can be made (Fact Sheet 4.19). For example, where land is polluted with heavy metals, these pollutants may have to be removed before revegetation can proceed. Other important categories of degraded land are salinized, sodic, desertified, and eroded soils; Section 2.2.5.7 discusses the nature of degradation and lists definitions. Areas of most categories of severely degraded land are increasing. Restoration brings multiple benefits: Not only can carbon be sequestered, the loss of carbon can be arrested and nonproductive land brought back into use. Restored lands may be put in to crops (Section 4.4.2, Fact Sheets 4-1 through 4-4), pasture (Section 4.4.3, Fact Sheets 4-6 through 4-9), or forest (Chapter 3, Section 4.4.4, Fact Sheets 4-12 through 4-17) to sequester still more carbon.

Degraded land has a large potential for sequestration in relation to undisturbed land because it often contains little carbon, but there are almost always factors that limit this potential. Nabuurs *et al.* (1999) cite rates of 0.2–2 t C ha^{-1} yr^{-1}. Rates as large as 7–9 t C ha^{-1} yr^{-1} appear in the literature (Table 4-11), but these rates often involve other measures such as the application of animal manures or deal with special problems. With severely degraded land, it is probably unreasonable to expect the largest rates to apply everywhere. Lal and Bruce (1999) use a conservative rate of 0.25 t C ha^{-1} yr^{-1}, which is used in Table 4-4. The very low carbon levels in most of these soils means that a 1 percent increase in the carbon content of soil mass is feasible. Such an increase would take about 60 years at a rate of 0.25 t C ha^{-1} yr^{-1}; in practice, however, the land is likely to be used for some other purpose, such as agriculture, before that rate and other rates of carbon sequestration will apply (Table 4-4).

Because severely degraded soils require intervention above and beyond revegetation, extra costs inevitably are involved.

Table 4-11: *Rates of potential carbon gain under selected practices for degraded lands in various regions of the world.*

Practice	Country/ Region	Rate of Carbon Gain ($t C ha^{-1} yr^{-1}$)	Time[1] (yr)	Other GHGs and Impacts	Notes[2]
Saline/alkali soils				If grazed, livestock may generate CH_4	
– Saline soil reclamation	India	2			a
– Alkali soil reclamation	India	4	5		b
– Irrigate halophytes with seawater	Australia	1–2	20		c
Polluted soils					
– Reclamation of mineland	USA	1.5–2.0	25		d
		1–7	4		
Eroded soils					
– Rehabilitation practices	Australia	0.1–0.4		Improved grazing with lower variability of production	e
Desertified soils					
– Restorative practices	China	<1	>6		f
	India	0.4–0.3	25		g

[1]Time interval to which estimated rate applies. This interval may or may not be time required for ecosystem to reach new equilibrium.
[2]a. Singh *et al.* (1994); Lal and Bruce (1999).
b. Garg (1998); Sumner and Naidu (1998); Lal and Bruce (1999).
c. Glenn *et al.* (1993).
d. Paustian *et al.* (1997b); Akala and Lal (1999). Where sites are polluted with heavy metals, these pollutants may first need to be removed [e.g., hyperaccumulators (McGrath, 1998)].
e. Tothill and Gillies (1992); Ash *et al.* (1996).
f. Fullen and Mitchell (1994); see also Li and Zhao (1998).
g. Lal and Bruce (1999); see also Gupta and Rao (1994).

Remote sites may impose additional costs for transporting the means of intervention to the site. Although remediation is expensive, additional income from carbon credits may make restoring some land that would otherwise remain abandoned worthwhile.

Estimates of the net global carbon sequestration potential of polluted land are difficult to make. On one hand, these lands might be the last to be restored because of the expense and difficulty. On the other hand, their polluted or damaged nature means that in many cases—especially in developed countries—legislation may compel owners to restore the land. Where land remediation is a legal requirement, a decision might be needed on whether to allow the claiming of carbon credits.

4.4.7.1. Salt-Affected Soils

Saline soils may cover as much as 930 Mha worldwide (Sumner and Naidu, 1998; Lal *et al.*, 1999b). Restoring these soils presents two problems: first, encouraging strength and desirable flocculation in soil particles by replacing sodium in the clay with calcium or magnesium; second, reducing high levels of salt in soil solution to levels tolerated by plants. Applications of manure and gypsum can help, but the cost may be prohibitive if a site is distant from distribution centers.

Glenn *et al.* (1991, 1993) suggest that halophyte (salt-tolerant) shrubs may be suitable for rejuvenating coastal deserts, inland saline soils, or salinized irrigated lands. They suggest further that up to 130 Mha of land may have the potential to sequester 1–2 $t C ha^{-1}$ at a cost of \$44–53 t^{-1} C; the marginal cost may be as little as \$12 t^{-1} C if the halophytes are used to produce agricultural products (Glenn *et al.,* 1993; Table 4-11). There are a range of management issues to overcome, such as the relatively short life span of many halophytes and uncertainties in the longevity of biomass incorporated under the soil. Garg (1998) reports that sodic soils in Northern India have been able to sequester up to 4 $t C ha^{-1} yr^{-1}$ by afforestation with *Prosopis juliflora* (Table 4-11).

An additional risk to the permanence of measures taken on many low-lying salinized soils is posed by the risk of sea-level rise with climate change. In these same areas, salinity may be accompanied by periodic waterlogging, so halophyte production may be poor because of the interacting effects of salinity and poor soil aeration (Barson *et al.*, 1994). Under some situations, the high evapotranspiration rates achieved by trees may provide an effective way of maintaining the productivity of once-inundated land.

4.4.7.2. Badly Eroded and Desertified Soils

Eroded soils have a large potential for carbon sequestration if the erosion can be halted to prevent the further loss of sequestered carbon. In many cases, much fertile topsoil has already been lost, and the soil may need fertilizers to support restoration. Soil structure and stability are of great importance if soil is to avoid further damage; in some cases, the addition of clay or silt may be needed.

Of a total estimated area of 2,000 Mha of degraded land (Oldeman, 1994), badly eroded land constitutes about 1,200 Mha; 250 Mha of this eroded land is considered to be severely eroded (Paustian *et al.*, 1998b). Almost half of the severely eroded land area (112 Mha) is in Africa; 88 Mha are in Asia, and 37 and 13 Mha are in Latin America and Europe, respectively. Lal and Bruce (1999) consider 100 Mha of land worldwide to be so badly degraded that it is unsuitable for agriculture; assuming a sequestration rate of 0.25 t C ha^{-1}, these lands have a global potential to sequester 0.025 Gt C yr^{-1}. Lal *et al.* (1999b) report that 3,500–4,000 Mha of land are susceptible to desertification throughout the world; much of this land is rangeland or degraded tropical forests (see Sections 4.4.3 and 4.4.4). Desertified cropland constitutes 757 Mha.

Activities instigated to combat degradation and desertification include irrigation with silt-rich water; land stabilization and revegetation; enclosures to restrict grazing; and recovery of buried, fertile soils (Fullen and Mitchell, 1994). In the tropical and subtropical regions of China, about 48 Mha are regarded as "wasteland"—of which about 6 percent could be restored as pasture (Li and Zhao, 1998). In India, there are estimated to be more than 100 Mha of land in which organic carbon can be as low as 0.2 percent (Gupta and Rao, 1994). Revegetation of 35 Mha of these lands with suitable grass and multi-purpose trees could sequester up to 0.84 and 1.06 Gt C in vegetation and soil, respectively. Increase in soil carbon following cultivation is considerably enhanced by the introduction of legume trees or shrubs into the perennial pasture (e.g., Prinsloo *et al.*, 1990). The global potential from restoration of desertified lands is estimated at 1.0–1.4 Gt C yr^{-1} (Squires *et al.*, 1995; Lal *et al.*, 1999b); the potentials estimated in Table 4-11 are more conservative because they assume that only a fraction of the affected land is likely to be treated, given economic and other constraints.

The potential for aboveground increases in carbon stocks in degraded lands following afforestation may be about 2–3 t C ha^{-1} yr^{-1}; potential increases in soil carbon are probably on the order of 0.25 t C ha^{-1} yr^{-1} (Lal *et al.*, 1999a). Rehabilitation of severely degraded lands was estimated to have the potential to store 0.14 Gt C in the 0–10 cm soil layer in northeast Australia; such rehabilitation, however, would require higher cost inputs (Ash *et al.*, 1996) that have not yet been assessed for financial viability. The rate of establishment of perennial grasses appears to determine the rate of recovery of soil carbon (Burke *et al.*, 1995), and soil carbon can be lost from these systems—through a combination of inappropriate management and drought years—much more quickly than it can be replaced (Northup and Brown, 1999). Thus, a continuity of management purpose is required to maintain or increase soil carbon reliably.

4.4.7.3. Mine Spoils and Industrially Polluted Sites

Reclamation of polluted land is especially beneficial because land is brought back into production as well as being used to sequester carbon. The downside is that reclamation can be very expensive; because of the cost, current practice often is to carry out minimum remediation to comply with regulations. Given that crop biomass and soil microbial biomass—the means by which all carbon eventually becomes sequestered in soil—have been shown to decline with increasing levels of heavy metal in soil, carbon sequestration will be less than the potential maximum in unpolluted soils unless remediation is complete.

Land that is polluted with heavy metals such as mine-spoil waste or agricultural land that has been degraded by sewage sludge can be remediated with plants that take up abnormal amounts of the pollutants and concentrate it in harvested biomass. Such plants are known as hyperaccumulators (McGrath, 1998). Because the hyperaccumulator must be removed along with the heavy metals it contains, the main increase in carbon stocks on the site will occur when the soil is brought under normal vegetation after initial remediation. Because of the expense and physical limit to the amount of contaminant that can be taken up, the use of hyperaccumulators is usually restricted to lightly polluted land where the aim of restoration is to bring the content of the pollutant below some legal norm.

Unreclaimed lands are more prone to erosion than other land—which emphasizes the need for action. Up to 0.63 Mha of strip-mined land in the United States may require reclamation (Lal *et al.*, 1999a). Bennett (1977) reported an increase in SOC from 3 to 30 t C ha^{-1} during a 4-year study on reclaimed mine land where organic manures were applied. Akala and Lal (1999) measured soil sequestration of 35–37 t C ha^{-1} in 30 cm of soil by reclamation of strip-mined land in Ohio, through afforestation and establishment of improved pasture during a 25-year period (Table 4-11).

Soils that are polluted with organic substances such as fossil fuels are often restored by oxidation. This process will lead to an emission of CO_2. The spill already may have been accounted under fossil fuel emissions at some stage in the past. The oxidation is almost always microbial; in using the organic compounds as a substrate, the microorganisms will add to soil carbon. Thus, accounting of carbon associated with the pollution of land with organic substances is fraught with complications.

Where land remains polluted or is unattractive for other uses, biofuels may be a sensible option—particularly in industrial areas where power stations are nearby (Lal and Bruce, 1999; Fact Sheet 4.21).

4.4.7.4. *Verification*

At a scale greater than a few hectares, land restoration generally can be verified by remote sensing. A change in vegetation or in the density of vegetation is usually apparent; where it can be made quantitative, the effectiveness of the restoration can be assessed. Carbon gains associated with the new vegetation may be modeled with acceptable accuracy at this scale (except, perhaps, in very toxic soils). Confirmation requires field sampling of soil carbon and aboveground biomass.

4.4.8. *Urban and Peri-Urban Land Management*

Improving land management in urban and peri-urban areas can affect stocks of carbon through additions to aboveground and below-ground biomass in areas where grass and trees can be grown. Urban forests—in the form of trees and other greenery—constitute a major amenity feature of urban areas, but they are also elements of the urban ecosystem: They moderate air temperatures and water runoff, clean pollutants from low-level air, and provide wildlife habitat in addition to sequestering carbon (McPherson *et al.*, 1997). In Tokyo, with its high population densities, the proportion of green coverage is 23 percent; in Vienna, a city known for its open areas, the proportion is 50 percent.

Rapid expansion of urban areas is occurring in the United States: Urban expansion of approximately 1.34 Mha yr^{-1} was recorded in the period 1992–1997, of which around 0.5 Mha was converted from forest (USDA-NRCS, 1999). Conversion of forest to urban development may remove only a portion of the forest cover (unless the development includes large buildings or solid hard surface), but there are few data to help estimate the amount of carbon loss that might be associated with the activity for activity-based carbon flux estimates.

Urban areas around the world generally exhibit a "heat island" effect that makes the center of the city warmer than its surrounding rural regions (Akbari *et al.*, 1989, 1992). This effect increases the demand for energy for cooling, which typically is provided by fossil fuels. Urban vegetation, particularly urban trees, is an important factor in reducing urban heat build-up, reducing the energy demand and thus saving fossil fuels. Studies in the United States indicate that the daily electrical usage for air-conditioning could be reduced by 10–50 percent by properly locating trees and shrubs (Akbari *et al.*, 1992). Although these fossil fuel savings will be accounted for elsewhere, the net carbon store in increased urban vegetation is not. A 10-year program aimed at increasing the canopy cover by 10 percent on residential lands and 5–20 percent on other urban lands in the United States could result in sequestration of 3–9 Mt C yr^{-1} in trees and soils and a 7–29 Mt C reduction in emissions as a result of energy conservation from improved shading, increased evapotranspiration, and reduction of the urban heat island, along with wintertime heat savings (Sampson *et al.*, 1992).

Improving urban trees also contributes to the reduction of ozone formation through their effect on lowering urban heat island temperatures. Ozone in the lower atmosphere is an important GHG as well as a threat to human health. Mitigating the heat island effect is important in reducing the formation of ozone in the lower atmosphere. Rosenfeld *et al.* (1996) estimated that, for the city of Los Angeles, a program of tree planting and installing cooler roofing and pavement products could reduce by 12 percent the number of days when ozone levels exceed health standards, with half of the benefit coming from trees and the other half from increased reflection of incoming solar radiation during summer months.

Trees contribute to ozone formation by emitting volatile organic compounds (VOC) in the presence of NO_x (largely from automobile exhaust); they also "scrub" ozone from city air. Because these VOC emissions—which generally constitute less than 10 percent of total VOC emissions in urban areas (Nowak, 1991)—are temperature-dependent, increased urban tree cover is believed to lower overall VOC and therefore reduce ozone formation (Cardelino and Chameides, 1990). Nowak (1994) found that Chicago's urban forest was removing 2,000 t O_3 yr^{-1} from that city's atmosphere.

4.4.8.1. *Practice*

Increasing the number of urban trees, improving their growth rates, and prolonging their life spans are effective ways of increasing urban ecosystem carbon stocks (Table 4-12). Urban spaces that may be covered with trees include parks, green areas, forests, roadsides (alleys and street trees), riversides, and gardens. Consideration also should be given to the tree home gardens that are common in the urban tropics, which often consist mainly of fruit and nut trees. Because urban trees are often planted in dry soils that are disturbed and compacted by

***Table 4-12**: Rates of potential carbon gain under urban land management.*

Practice	Country/ Region	Rate of Carbon Gain (t C ha^{-1} yr^{-1})	Time[1] (yr)	Other GHGs and Impacts	Notes[2]
Planting trees		4.6–4.8		Enhanced biodiversity	a

[1]Time interval to which estimated rate applies. This interval may or may not be time required for ecosystem to reach new equilibrium.
[2]a. Sampson *et al.* (1992). Carbon sequestration opportunities on available urban growing space in the United States of America.

construction, and in small spaces, it is necessary to pay attention to site conditions, including addition of special soil amendments or facilities to improve soil aeration, moisture, and nutrient levels.

Urban tree and other vegetation waste constitutes a significant portion of the waste stream in many urban areas. Local programs to convert this material to energy may provide benefits such as reduced need for landfill space, reduced methane emissions from landfills, and economic benefits that help offset costs. Although the energy production benefits would be accounted for in fossil fuel accounting, reporting reduced methane emissions from improved urban waste management activities may be feasible in some situations.

Prolonging the life span of trees is an effective way to maximize changes in carbon stores and delay carbon emission. Because urban trees are surrounded by air that often contains high levels of pollution, they have short life spans compared to their rural counterparts (Kielbaso and Cotrone, 1990). Programs that improve tree maintenance and care can extend life spans significantly. In urban areas, dead or cut-down trees are seldom left to decompose naturally; they are collected and often burnt. Carbon emission can be delayed somewhat by reusing the wood to produce compost and mulch chips, which slowly decompose in soils and contribute to long-term SOC accumulation.

Important associated impacts of urban forests include contributions to biodiversity, purification of the atmosphere, improvement of scenic beauty, and support for environmental education.

4.4.8.2. *Verification*

The degree of urban greenness can be remotely sensed. Estimating the resultant carbon store will require on-site measurements, given the wide range of forms that an increase in vegetation can take.

4.5. Biofuels and Forest Products

This section focuses on biofuels and forest products and their implications for land use and carbon dynamics. Section 4.4.5 discusses the *in situ* stock of carbon in growing forest; this section deals with carbon in forest products (Section 4.5.6) and the tradeoff between biofuels and *in situ* sequestration, including environmental and socioeconomic impacts (Sections 4.5.2 through 4.5.5). Fact Sheets at the end of this chapter deal with tradeoffs (Fact Sheet 4.20), biofuel from plantations (Fact Sheet 4.21), and biofuel from food and fiber production wastes such as sawdust (Fact Sheet 4.22).

4.5.1. *Introduction*

When biomass displaces fossil fuels, the mitigation is captured as a decrease in fossil fuel use. The change in carbon stored in and on the land during biofuel growth must be accounted for separately. Tradeoffs between carbon storage and displacement of fossil fuels and energy-intensive materials have implications for land use and forest management. Article 3.3 of the Kyoto Protocol clearly distinguishes between biofuels and fossil fuels, establishing that biofuels are part of the cycling of carbon in the biosphere. Distinguishing biofuels from other fuels entails assessing options for managing carbon through the land-use change and forestry sector and looking at effects in the energy sector.

Globally, biofuel contributes about 14 percent of primary energy supply. Most biofuel use is traditional wood fuel in developing countries, but agricultural and forest wastes provide significant industrial feedstocks for energy production in developed economies (see Fact Sheets 4-21 and 4-22). Modern biofuel technology can provide electricity, gases, and transportation fuels and more efficiently support traditional uses of wood fuel, with environmental and social benefits. These benefits include job creation, productive use of surplus agricultural land, avoidance of health hazards from traditional wood burning, reduced urban and agricultural wastes, and nutrient recycling.

Agroforestry systems can provide multiple benefits, including energy, to rural communities, with synergies between sustainable development and GHG mitigation. Large-scale biofuel production raises questions, however, involving land availability and productivity (short- and long-term), species selection and mixtures, environmental sustainability, social and economic feasibility, and ancillary effects. Issues include fertilizer and pesticide requirements, nutrient cycling, energy balances, biodiversity impacts, hydrology and erosion, conflicts with food production, and the level of financial subsidies required.

Three broad questions arise if biofuels are to significantly reduce net CO_2 emissions:

- Is sufficient land available to meet demand for food, fiber, and energy?
- Do adverse environmental and social impacts negate the advantages of biofuels?
- Why aren't modern biofuels more widely used now?

Underlying these questions is the tradeoff between stocking carbon in standing forest and producing a flow of woody biomass that displaces fossil fuels directly as biofuel or by displacing energy-intensive building materials.

The key message of this section is that managing land use for maximum on-site carbon storage may not always result in the most effective mitigation of GHG emissions. Increased carbon storage in the biosphere yields benefits, but over time greater mitigation is possible by managing the entire system—including the production and use of biofuels and other products.

From a policy perspective, the potential for biofuel displacement of fossil fuel is an order of magnitude greater than any other land-use change. It may also impact atmospheric carbon levels earlier and at lower cost than other energy sector measures.

This consideration has significant precautionary potential against the possibility that undesired climate effects may occur at lower levels of atmospheric GHG than have been previously postulated (e.g., the threshold for rapid, nonlinear climate change, which is currently unknown) (Houghton, 1998).

4.5.2. CO_2 Sequestration versus Fossil Fuel Substitution

Since Dyson (1977) and Dyson and Marland (1979), numerous analysts have studied the potential for mitigating GHG levels by sequestering carbon in standing forest. Forest land management can impact GHG levels in four ways: through carbon stored in standing biomass and soils, carbon stored in durable biomass/wood products and landfills, fossil fuel left underground because biofuels are used instead, and forest products and other bio-products displacing fossil fuel-intensive materials. Matthews (1996) discusses the importance of capturing the full system impacts.

Using biomass for energy and other products holds great long-term potential for GHG mitigation but generally means less carbon is stored than would be under a pure sequestration strategy. There is a need to consider the rate and character of carbon flows and their short- and long-term benefits. Read (1996, 1997) suggests that larger stocks of standing timber can increase short-term carbon sequestration while building a "buffer stock" of wood fuel for future biofuel systems, when biofuel market penetration is greater. Short-rotation biofuel production could follow when biofuel-based infrastructure is in place and the initial rotation is harvested.

Managing long-rotation forests for timber is also complementary with some biofuel production. Suppression of fires, insects, and other disturbances may result in large accumulations of fuel components in the forest. Wildfires that begin in this litter can damage living trees (Fact Sheet 4.15), so the long term well-being of the forest, as well as its carbon storage, is enhanced by thinning and removal of potential fuel materials.

Modeling of forestland use for net CO_2 mitigation shows that the merits of different options depend on current use, potential productivity, how biofuel is substituted for fossil fuels, and the time horizon. In terms of CO_2 benefit alone, low-productivity mature forests are best conserved as carbon stores; low-productivity unforested land is best reforested and managed for carbon storage; and more productive land is best forested and managed for biofuel with modern conversion technologies, as well as for products that displace fossil fuel-intensive materials. Longer time horizons tip the balance toward harvesting and replanting (e.g., Hall *et al.*, 1991; Marland and Marland, 1992; Schlamadinger and Marland, 1996; Marland and Schlamadinger, 1997).

Strategies that rely only on sequestering carbon eventually find the reservoirs filled: trees grown to maturity with increased risks of natural disturbances. This prospect is distant, however, with large potential for sequestration before saturation (see Fact Sheet 4.20). In addition, market forces will eventually shift the pattern to biofuel production, with rising costs as reservoir saturation is approached. Ultimately, net CO_2 impacts require consideration of carbon flows in the energy, forest product, and land-use sectors. Optimal answers can be explored through dynamic modeling of these sectors jointly (Read 1998, 1999; Fact Sheet 4.20).

Substitution of biomass products for energy-intensive materials is less researched, but prospects for CO_2 mitigation appear to be promising (Schopfhauser, 1998). Wood used in place of aluminum, concrete, or steel saves fossil fuel used to process these materials. Estimates of the gain must allow for production energy requirements and service lifetimes of alternative products (Marland and Schlamadinger, 1997).

4.5.3. Global Scenarios for Biomass Energy

Recent scenario studies (Hall *et al.*, 2000) show biomass energy contributing 150–200 EJ yr^{-1} by 2050, avoiding CO_2 emissions of ~3.5 Gt C yr^{-1}—more than half of present fossil fuel emissions. Previous global energy scenarios show a rising trend for biofuel use, at small or no additional cost, with Latin America and Africa becoming large net exporters of liquid biofuels. WEC (1993) projects 62 EJ in 2020, plus traditional wood fuel in developing countries; IEA (1998) projects that biomass fuels will grow at 1.2 percent per year to 60 EJ in 2020; Lazarus *et al.* (1993) project 91 EJ in 2030; Dupont-Roc *et al.* (1996) showed a business-as-usual scenario for Shell International with 221 EJ—of which 179 EJ were from fuel plantations—in 2060. The Global Energy Perspectives' high-growth, high-biomass scenario has 316 EJ of biomass by 2100 (Nakicenovic *et al.*, 1998). Without arguing these and other numerical scenarios, the common vision is that there is a large and increasing potential for biofuels (see Fact Sheet 4.21 for land-use implications).

Under its Biofuel Activity Program, the International Energy Agency (IEA) monitors a wide range of commercial and near-commercial processes, many of which use small-scale plants for converting biofuel into heat, light, and transportation fuels (Overend and Chornet, 1999; Rosillo-Calle *et al.*, 2000). Walter (2000) reviews new technology. Large sunk costs in long-lived capital stock and infrastructure impedes market entry for renewable energy. Biofuel is relatively compatible, however, with the fossil fuel-based energy systems (e.g., blending with petroleum products, wood chips with coal at power stations). Modern biofuel is efficient at small scales (e.g., in rural areas and developing countries).

4.5.4. Land Availability

In general, land availability (out of the very large area that might be used) will be influenced by its value (opportunity cost) in the variety of services that land provides, from wilderness through food production to urban occupation, as well as by its biomass productivity. The potential for increased production of biofuels can be accomplished through increased use of existing forest and other land resources, higher rates of plant productivity, and more efficient conversion processes and capture of wastes.

Unmanaged woody species have yields of less than 5 t ha^{-1} yr^{-1} (dry weight biomass). Optimal management and planting of selected species and clones on appropriate soils currently achieves 10–15 t ha^{-1} yr^{-1} in temperate areas and 15–25 t C ha^{-1} yr^{-1} in the tropics; 40 t C ha yr^{-1} has been obtained with *Eucalyptus* in Brazil and Ethiopia. High yields (30–40 dry t ha^{-1} yr^{-1}) are also possible with herbaceous crops such as switchgrass (Hall *et al.*, 1992). In Brazil, the average annual yield of sugar cane has risen from 28 to 39 t ha^{-1} yr^{-1} (dry weight) over 15 years, with more than 70 t ha^{-1} yr^{-1} achieved in Hawaii, southern Africa, and Queensland (FAO, 1999).

There are large areas of deforested and degraded lands in tropical countries that could produce multiple benefits from the establishment of biofuel plantations (Brown, 1998). Conversion of these and other lands to biofuel plantations can provide economic value to the local people. Large-scale biofuel production will require specific energy crops, improved land management, species selection and mixes, genetic engineering, and so forth.

4.5.5. Associated Impacts

There is concern over short- and long-term environmental and socioeconomic effects of large-scale biofuel production. These concerns relate to the energy balance of biofuel production, conversion, and use; soil and water quality effects; poor resilience of monocultural plantations; and the implications of biofuels for biodiversity, sustainability, and amenity (Cook *et al.*, 1991; National Audubon Society, 1991; Brown, 1998; Christian *et al.*, 1998).

Large-scale bioenergy plantations that generate high yields with production systems that resemble intensive agriculture would have adverse impacts in place of natural forest. With the development of conversion technologies that are efficient at small scales, however (Bowman and Lane; 1999; Larson and Jin, 1999; Prabhu and Tiangco, 1999), transportation costs are reduced and large monocultures are unnecessary (see Fact Sheet 4.21). Small-scale plantations on degraded land or abandoned agricultural sites would have environmental benefits. The key is to consider site-specific circumstances rather than make generalizations.

Significant amounts of fossil fuels are used in the production, harvest, and transport of biofuels; the net carbon benefit of biofuels must account for this fossil fuel use. Energy input-output ratios in biofuel production are 1:10 to 1:15 and improve to 1:30 (Turhollow and Perlack, 1991). As with fossil fuels, overall system efficiency is lower because of the characteristics of conversion processes (e.g., Graham *et al.*, 1992; Matthews *et al.*, 1994; Boman, 1996; Mann and Spath, 1997).

Biodiversity concerns relate to plantation species and plantation habitat conditions. Plantations with only a small number of species achieve the highest yields and the greatest efficiency in management and harvest, but good plantation design now includes set-asides for native flora and fauna and blocks with different clones and/or species. The variety of species in biofuel plantations falls between that for natural forests and annual row crops. Research on multi-species plantations and management strategies and thoughtful land-use planning to protect reserves, natural forest patches, and migration corridors can help address these issues.

Concerns regarding food supply and access to land for host communities are addressed through community-scaled plantations that feed small-scale conversion technologies, meet local fuel and timber needs, provide employment with biofuel-powered rural electrification, and export liquid fuel products (Read, 1999; Fact Sheet 4.21). Higher incomes may enable communities to invest in modern food production and replace traditional agriculture, which appears unsustainable with predicted population and climate change. A barrier to community-scaled biofuel systems is a lack of institutional and human capital to ensure biofuel projects that meet local needs rather than foreign investors' carbon credit priorities.

4.5.6. Forest Products

This section presents the opportunities offered by forest products as a recyclable store of carbon and a renewable source of fibers for the mitigation of climate change. Policy options to account for these wood product pools are discussed in Section 6.2.2.

In 1996, the world's forests produced 3.4 billion m^3 of harvested roundwood. About 1.9 billion m^3 (56 percent) of this harvest was fuelwood; the remainder (1.5 billion m^3) was industrial roundwood (e.g., sawlogs and pulpwood). The industrial roundwood corresponds to a harvesting flux of about 0.3 Gt C yr^{-1}. Developed countries account for 70 percent of total world production and consumption of industrial wood products (FAO, 1999).

The timber that is harvested is converted into a wide variety of wood products (Skog and Nicholson, 1998). The carbon in the wood is fixed in products until they decay or are burned (i.e., for energy production) and the carbon is subsequently released back into the atmosphere. Models are available to assist in developing estimates of carbon fate in wood products, depending on their initial size, quality, and industrial utilization (Row and Phelps, 1996; Apps *et al.*, 1999b). With increasing industrial use of wood, the amount of carbon fixed in wood products will raise proportionally. There are several ways to positively influence the carbon balance:

- *Shifting the product mix to a greater proportion of wood products*. Manufacturing and transport of wood products requires less fossil fuel than energy-intensive construction materials such as aluminum, steel, and concrete (Matthews *et al.*, 1996). Recent comparisons show that the production of steel and concrete as building material requires up to two times more energy than wood-based product—with concomitant greater generation of GHGs. Increased use of solid timber, engineered wood and wood-based panels, paper, and fuels displaces energy-intensive materials (cement, steel, bricks, and plastics) in two main ways. On one hand, it increases the amount of carbon stored in wood products, particularly those with long life spans. On the other hand, it reduces emissions during production processes. The use of by-products (wood fuels) for energy generation in production processes has an additional positive impact on the overall carbon balance.

- *Increasing the useful life of products*. Extending the life of wood products implies not only a longer service life but a longer carbon sequestration period and less energy consumption for their replacement through other new materials. The service life of wood products can be extended by using the appropriate timber species for particular end uses, applying constructive or chemical wood protection against fungi and insect attack, and wise use of the products themselves.
- *Increasing product recycling*. Wood and paper products are among the most commonly used materials for recycling into new products and fuels. Utilization of recovered wood in the paper industry and power plants is a good example. In Europe, for example, recovered paper accounts for 40 percent of annual paper production and is predicted to increase to 45 percent in the future. An increase in the number of times a material is recycled and the recycling of more wood and paper products will enhance the storage of carbon and reduce emissions.

4.5.6.1. Need for Research and Technology Transfer

Wood products must be affordable, safe, durable, and more cost-effective to be preferable to other more energy-intensive alternatives. Research plays a key role in developing new applications, improving processing efficiency, and extending product life spans. Policy and its implementation play a crucial role in promoting technology transfer and technological and economic transition, particularly in developing countries.

4.5.6.2. Impacts on Forest Resources

FAO (1999) has projected that the annual production of wood will reach almost 3.7 billion m^3 by 2010. The demand for industrial roundwood is expected to increase by 1.7 percent annually between 1996 and 2010, driven by population growth, market forces, and new policies that address climate change issues.

If these projections are realized, investment in reforestation, afforestation, and forest management activities will have to rise accordingly, and timber supplies may increase. This trend may imply increased carbon stocks in forest products—but also more pressure on forest resources. Climate change mitigation requirements will have to be balanced with other requirements, such as soil and water protection, wildlife habitat, or biodiversity.

4.5.6.3. Economic Impacts

At the global level, forestry is estimated to contribute only 2 percent toward the world's GDP and 3 percent to international trade. The estimated value of wood consumption in 1996 exceeds US$400 billion, with industrial usage accounting for 75 percent. Forest products are the main source of foreign exchange for several countries. Any change in the status of their forests is likely to affect their balance of trade. Measures aimed at mitigating climate change may open new opportunities for the development of the forestry sector, but provision must be made to avoid negative impacts.

4.5.6.4. Construction of a "Closed Carbon Cycle"

Actions and measures for the sequestration and displacement of carbon by means of forest management, forest products, and bioenergy constitute a "closed carbon cycle." These actions are closely linked to other priority areas, such as food production, erosion control, and biodiversity protection. These actions can be tailored to different economic situations, industrial conditions, resources availability, consumption patterns, and social and cultural behaviors. It is important to recognize, however, that they must be carefully planned and monitored at national and local levels to generate a perpetual cycle of tree planting and forest product utilization.

4.6. How to Include Activities under Article 3.4: Modalities, Rules, and Guidelines

Article 3.4 calls for a decision not only on which additional activities will be added to the accounting of national commitments under the Protocol but also on how they will be added. This section discusses technical issues that relate to selected questions about how activities might be treated if they are added under Article 3.4. What modalities, rules, and guidelines (MRGs) might be considered to preserve the spirit, intent, and consistency of the Protocol and the United Nations Framework Convention on Climate Change (UNFCCC)?

The text of Article 3.4 provides two requirements: Activities should be "human-induced," and they should be "related to changes in greenhouse gas emissions by sources and removals by sinks in the agricultural soils and land-use change and forestry categories...." (Figure 4-1). Section 2.3.3.1 discusses the concept of "human-induced" (and whether it differs substantially from "direct human-induced" as in Article 3.3). Articles 5.1 and 5.2 of the Protocol refer specifically to the IPCC Guidelines for estimating and reporting GHG emissions by sources and removals by sinks (see Chapter 6). Article 3.4 also requires that uncertainty, transparency in reporting, and verifiability be taken into account.

The remainder of the Kyoto text includes at least four MRGs that are imposed on some other activities. Although there is no explicit linkage between these portions of the Kyoto Protocol and Article 3.4, they suggest the kinds of MRGs that might be considered for Article 3.4 activities:

- Article 3.3 limits credits to activities "since 1990;" Article 3.4 has the same requirement for the first commitment period.
- Article 3.3 measures net changes in GHG emissions in the land-use change and forestry categories as "changes in carbon stocks in each commitment period."
- Articles 6 and 12 require that any creditable reduction in emissions or enhancement of sinks be "additional to any that would otherwise occur."
- Article 12 permits banking of "certified emission reductions obtained during the period from the year 2000 up to the beginning of the first commitment period."

Other potential MRGs have been discussed, and many more could be suggested. Three that have generated particular discussion follow:

- A requirement that activities be defined or paired to ensure that both increases and decreases in carbon stocks are captured.
- A system to apply an adjustment factor to credits for carbon sequestration—to encourage emissions reductions in other categories (e.g., the energy sector), recognize higher uncertainty in measurement of sinks, or recognize the lower reliability over time of enhanced sinks in the biosphere relative to forgone emissions in some other sectors.
- A system to put a limit on the extent to which a party can meet its commitments through credits for sequestration.

The following sections briefly discuss these seven potential MRGs and the implications of their application to Article 3.4. Clearly, many different combinations or scenarios could be chosen. Moreover, we consider the possibility that MRGs might be different for the first commitment period (for which commitments have already been agreed) than for the subsequent commitment period (for which commitments have yet to be agreed). The appropriate set of MRGs may depend on whether the term "activity" is defined broadly or narrowly. We recognize that there is a linkage between the opportunities available for meeting commitments under the Protocol, the MRGs for calculating credits, and the expectations that the Parties had for these credits when they negotiated their commitments.

4.6.1. "Since 1990"

Article 3.3 allows credit only for ARD activities undertaken since 1990; Article 3.4 allows credit only for activities during the first commitment period. Aside from limiting credits to newly undertaken activities, these requirements have two obvious consequences: They withhold credits from Parties that behaved in the desired way (regardless of motivation) prior to 1990, and they complicate the accounting system by requiring differentiation of activities by date of initiation. If the Protocol were to adopt more comprehensive accounting of carbon in the biosphere, relaxing the "since 1990" requirement for Article 3.4 activities in commitment periods after the first would make accounting simpler. If the intent is to give credits only for a short list of narrowly defined activities that can be identified in time and space, however, retaining the "since 1990" requirement could work toward this objective. Article 3.4 stipulates, in its last sentence, that if the additional activities are used to meet commitments during the first commitment period (a choice offered to each Party), the "since 1990" requirement will apply; it does not impose this requirement on subsequent commitment periods, however. The options are to either relax this MRG or impose another initiation date. Obviously, the later the initiation date, the fewer the projects that will qualify and the more likely that qualifying projects (or the national scale of an activity) were undertaken for purposes that include gaining GHG emissions credits. The earlier the initiation date before 1990 (unless no initiation date is set), the more difficult it will be to retrospectively gather baseline data, if needed. For instance, remote-sensing data on land cover are not available from before the mid-1970s.

4.6.2. "Measured as...changes in carbon stocks in each commitment period"

There are two possible approaches for measuring CO_2 emissions by sources and removal by sinks in the biosphere. In most instances, the IPCC Guidelines and other conventions estimate the actual exchange of a gas between the system of interest and the atmosphere. This approach is used, for example, to measure emissions of CO_2 from fossil fuel combustion and CH_4 from landfills. The Kyoto Protocol has chosen in Article 3.3 to use the other approach: measuring net emissions as the change in carbon stocks in the various reservoirs (see Brown *et al.*, 1996; Apps *et al.*, 1997; Sections 2.3.2.3 and 4.3.3). Having made this land-based choice for Article 3.3 activities, the accounting system may become more complex if an activity-based approach is chosen for Article 3.4 activities. This approach may be needed in the case of some activities for which distinguishing whether products (e.g., forest products) originate from Article 3.3 or 3.4 activities could be difficult once the products are in industrial trade channels. A stock-based approach does not work for non-CO_2 GHGs; these emissions must be treated by activity-based methods.

4.6.3. Reference Year, "Additionality," and Baseline Issues

Reference in Article 3.4 to "changes in greenhouse gas emissions...and removals" imply that credit will be based on a comparison between two points in time or two paths through time. There are several ways this comparison might be done, and the Protocol itself prescribes three alternatives in different places. Any one of the alternatives might be adapted (or slightly modified) for activities under Article 3.4. The choice will have significant implications for the amount of credits available and the data required for accounting. The choice could also affect accounting of the interaction of activities undertaken under Article 3.3 and Article 3.4 and the consistency of national and project accounting (see Chapter 5). These three possibilities—all of which may pose problems in implementation—are discussed in the following sections and illustrated in Figure 4-12.

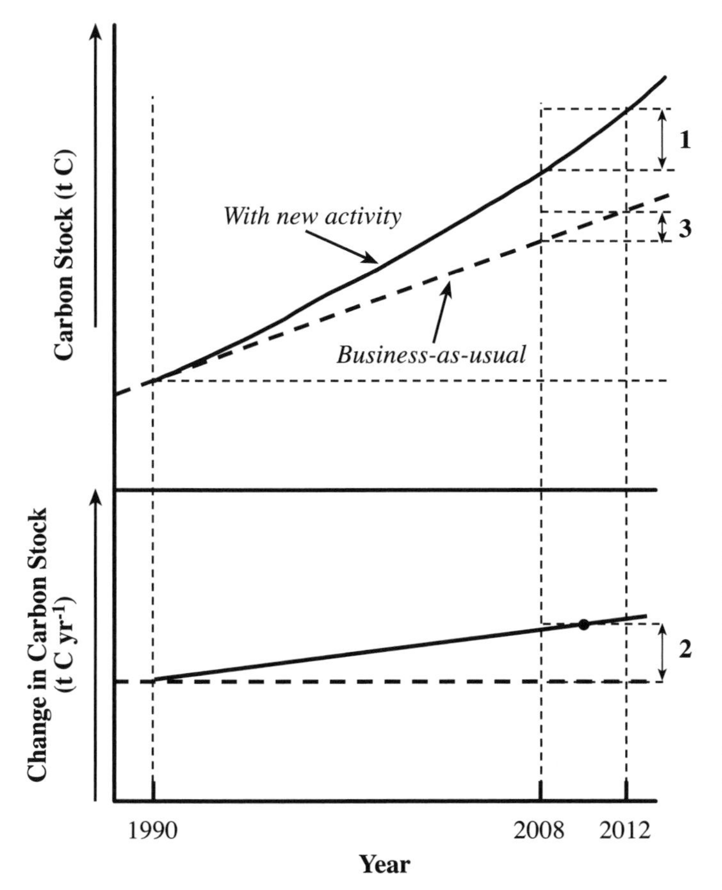

Figure 4-12: Three possibilities for evaluating change in carbon stocks attributable to Article 3.4 activities. Figure shows total carbon stock (above) and annual change (below) for an activity that causes an increase in carbon stocks and the path that would have been followed without the activity. The activity is arbitrarily assumed to have begun in 1990. The example chosen shows an increase in carbon stocks with respect to business-as-usual, but the principles are equally relevant if the activity resulted in a decrease in carbon stocks with respect to business-as-usual. If the stock in 2008 is taken as the reference (Section 4.6.3.2), the credit for the first commitment period will be as shown by arrow 1. If the change in carbon stocks in 1990 is taken as the reference (Section 4.6.3.1), the credit will be as shown by arrow 2. If the business-as-usual scenario is taken as a baseline (Section 4.6.3.3), the credit will be as shown by arrow 1 minus arrow 3.

4.6.3.1. Change in Carbon Stocks in 1990 as the Reference

For CO_2 emissions from fossil fuel combustion and most other categories of emissions from Annex I countries, the Kyoto Protocol would require that Parties reduce emissions by some specified amount below emissions levels in 1990. (We note for completeness that some countries are in fact permitted an increase in emissions and that, for some countries and some GHGs, an alternate reference year is permitted.) In application to Article 3.4, this requirement would suggest that the carbon stock change during the commitment period be compared with the carbon stock change during the base period (1990).

In mathematical terms, the change in stocks over time is equal to net emissions, so net emissions are the first derivative of the stocks (the first derivative of stocks gives the rate at which stocks are changing). The change in net emissions over time is the second derivative of stocks, which indicates the rate at which net emissions are changing. The opening line of Article 3.3 suggests that, in fact, the second derivative of stocks is of interest; this phrase conflicts with later portions of Article 3.3, however, and has now been officially rendered meaningless by a decision at COP4 interpreting Article 3.3 (UNFCCC, 1998b). Nonetheless, some observers believe that our interest should be in the second derivative of stocks. This conflict has been referred to as the "gross-net disparity" (IGBP, 1998).

Use of the second derivative of stocks (i.e., change in stocks in 1990 as a reference) was specifically rejected by negotiators in Kyoto for Article 3.3 but could be reconsidered for Article 3.4 activities. This calculation would require measuring the change in carbon stocks in the base year (1990 or other year chosen) and comparing it with the change in carbon stocks during the commitment period. Article 3.3 measures the net change in GHG emissions for a limited set of activities by making the tacit assumption that the net rate of emissions from approved activities during the base period (1990) was zero. The second derivative of the carbon stocks is equivalent to the first derivative only when the change in carbon stocks in the base period is zero.

For Parties to whom Article 3.7 applies (i.e., Parties for whom land-use change and forestry constituted a net source of GHG emissions in 1990), aggregate emissions and sinks from land-use change in 1990 are to be included in the 1990 reference. Subtraction of base-year emissions from commitment-year emissions then yields a true measure of the change in net GHG emissions from the base period to the commitment period if the same list of activities is included in both calculations—limited by the "since 1990" stipulation. If the change in carbon stocks in 1990 (or other base year) is not taken as the reference for Article 3.4 activities, and if some of the activities brought in under Article 3.4 were significant sources of GHG emissions in the base year, a qualification such as that in Article 3.7 might be considered. If some of these Article 3.4 activities are defined as land-use changes, Article 3.7 may be interpreted to apply.

One of the procedural problems with this approach has to do with year-to-year variability in the natural part of the global carbon cycle, particularly in the terrestrial biosphere (see Chapter 1). Another difficulty is that retrospectively determining the rate of change in some stocks may not be technically feasible—restricting this approach to application only in the future. Although this approach was rejected in Kyoto, it would most nearly treat emissions from the biosphere in the same way that emissions from fossil fuel burning are treated, while retaining the linkage to carbon stocks.

Another issue that has been raised with regard to this approach is the possibility that best efforts could still result in a net rate of removals that shrinks with time. As an activity continues over time, it could use all of the land available (conservation tillage, for example, might become adopted on virtually all cropland) and/or begin to fill up the ecologically available carbon stocks (a soil achieves a general equilibrium at a certain SOC level, or a large region of forests matures and growth rates decline); thus, the rate of change would begin to decline compared to an earlier period. This issue may be more immediately germane to project-level accounting—which is more likely to contain a few soil situations or ecosystem successional phases—but it could become important at the Party level if aggressive implementation of effective activities were continued for several decades. The result is that a Party with a declining rate of carbon sequestration would appear to be a source from an accounting standpoint even though it was a net carbon sink.

4.6.3.2. Stocks in 2008 as the Reference

For LUCF activities undertaken under Article 3.3, net emissions are defined as the change in stocks during the first commitment period, 2008–2012. In essence, this definition specifies that, for qualifying activities, carbon stocks in 2008 are the reference against which to measure carbon sequestration by ARD during the first commitment period. This situation is the no-baseline case of Section 2.3.2.1 and the accounting rule of Table 2-3.

In keeping with the precedent set in Article 3.3, the level of carbon stocks in 2008 (or the first year of any subsequent commitment period) could be used as the reference point for that commitment period for activities under Article 3.4. This approach would give countries time to prepare the reference measure and provide a uniform and well-defined base against which to evaluate changes.

If implementation of Article 3.4 (Figure 4-5) permits credits for only a small number of narrowly defined activities that are specifically identified in time and space, this approach could have little impact on negotiated commitments. If a more comprehensive approach to carbon accounting and admission of credits for a broad range of activities (Figure 4-5) is chosen, this approach could have a significant effect on negotiated commitments.

For example, if there is a large and continuing sink for carbon in the Northern Hemisphere terrestrial biosphere—as is now generally believed (Ciais *et al.*, 1995; Keeling *et al.*, 1996)—and if the driving force behind this sink can be characterized as "human induced," this reference measure could open up the possibility for large numbers of carbon credits that lead to significantly less stringency in many national commitments. This concern was expressed in some national submissions prior to and following the drafting of the Kyoto Protocol. One way to address this issue would be to ensure that credits for additional activities apply only in commitment periods after the first and that negotiated commitments for subsequent periods are agreed to with clear understanding of the implications of the selected reference. Such a decision would need to interpret the last sentence of Article 3.4, which allows Parties to choose whether to use these additional activities to meet commitments during the first commitment period in a way that would permit different accounting rules in the first and subsequent commitment periods.

The Kyoto Protocol acknowledges that using stocks in 2008 as the reference could cause Parties for whom land-use change and forestry activities were a large source of emissions in 1990 to have a net debit that continues into the commitment period even though they have reduced emissions. For this reason, Article 3.7 permits some Parties to include aggregate emissions by sources less removals by sinks from land-use changes in their reference-year (1990) accounts (see also Section 4.6.3.1).

4.6.3.3. Business-as-Usual Baseline

Activities undertaken under Articles 6 and 12 of the Protocol are eligible for credits toward national commitments if they result in net emissions reductions that are additional to any that would have occurred in the absence of the activity. In essence, this provision specifies that, for activities undertaken under Joint Implementation (JI) and the Clean Development Mechanism (CDM), "business-as-usual" emissions act as a baseline, and credits accrue when net emissions fall below this baseline.

This same approach—that is, defining a business-as-usual baseline and awarding credits only when the actual path of emissions falls below this baseline—could be used for Article 3.4 activities to provide rewards for improved behavior. Defining a meaningful business-as-usual path of emissions—a measurement of a counterfactual path that never exists—presents significant problems (see Section 5.2.3). The last sentence of Article 3.4 suggests the possibility of using different reference systems for the first and subsequent commitment periods. A business-as-usual baseline might be used when activities are used to meet commitments during the first commitment period, but a different reference system might be applied at the beginning of subsequent periods for which commitments have yet to be negotiated. For carbon sequestration, credits would accrue when the change in stocks over the commitment period is greater than the change in stocks for the business-as-usual scenario.

Business-as-usual is only one of a variety of baselines that might be chosen. Other possibilities include giving credit for any practice that improves on existing practice or for continuation of current levels of adoption of a mitigation practice that improves on some practices still in use; giving credit for any practice that improves on "standard management practice;" or using a performance standard so that any practice that exceeds the standard could receive credit. These baselines are not unique to the land-use change and forestry sector; they are listed (though not elaborated) here to emphasize that if a baseline is desired, there are several ways it might be defined.

4.6.3.4. Additional Implications of the Choice of Baseline or Reference Measure

Article 3.4 establishes that any decision on incorporation of additional activities "shall apply" in the second and subsequent commitment periods. It states that a Party "may choose to apply" this decision in the first commitment period. It does not stipulate that the activities will be governed by the same MRGs in the first commitment period as in subsequent periods; in fact, it implies the contrary by establishing that the "since 1990" rule will apply during the first commitment period but not necessarily during subsequent periods. Thus, in arriving at a decision regarding Article 3.4 activities, Parties may be able to prescribe different selection criteria, different activities, and different MRGs for individual commitment periods. It seems possible, then, to phase in over commitment periods either the breadth of activities included or the breadth of the accounting rules. This phasing-in could accommodate changes in available data, development of methodologies, or compatibility between commitments and opportunities for emission reductions.

Decisions yet to be made by the Parties will determine which specific forestry practices will qualify for inclusion under Article 3.3 (see Chapter 3). All practices not included under Article 3.3 will be eligible for incorporation under Article 3.4. Under the broadest definition of the term "activities," most practices that are excluded from Article 3.3 are likely to fall into the "forest management" activity. If "activity" is defined more narrowly, care will be required to ensure that practices that are defined out of Article 3.3 will be considered under Article 3.4. Even if there is a seamless fit between activities included under Article 3.3 and those included under Article 3.4, however, there will be an accounting misfit unless both activities are governed by the same MRGs. If a business-as-usual baseline is adopted for Article 3.4, for example, activities under Article 3.4 would likely generate fewer credits than under Article 3.3 for the same total change in carbon stocks, and it would be necessary to continue to distinguish between the two for accounting purposes.

4.6.4. Banking of Emissions Offsets

Article 12.10 of the Kyoto Protocol permits any certified emissions reduction from the year 2000 (8 years prior to the first commitment period) to be used to achieve compliance during the first commitment period under the CDM. This provision could have the effect of encouraging participation by non-Annex B countries and stimulating early implementation of activities that reduce emissions. Recognizing the slow rate at which biomass activities sequester carbon compared to the high rate at which they can discharge carbon, as well as the associated advantages of early implementation of improved land-management activities, the Parties could choose to permit this sort of banking for early credits under Article 3.4 activities. This strategy might encourage early adoption of sustainable land-management practices; it would also create additional credits that would permit Annex B countries to more easily meet their commitments in the first commitment period.

4.6.5. Capturing Both Increases and Decreases in Carbon Stocks

Article 3.4 permits Parties to choose whether to include these additional activities in meeting commitments during the first commitment period. Without appropriate guidance or the adoption of a full and symmetrical accounting system (see Chapter 2 and Figure 4-5), a Party might be able to select projects or activities that provide a net sink for carbon without accounting for those that are a net source. This type of selection could encourage adoption of desirable land-management activities in some places with no net gain for the atmosphere if countervailing choices are made elsewhere.

A partial solution to this potential problem would be to adopt any additional activities allowed for credit under Article 3.4 through a single decision that required acceptance of all activities or none in one package. This approach could ensure that a Party that chose to apply the decision to the first commitment period would have to account for the net effect of all of the additional activities. A full accounting approach for all changes in carbon stocks would also avoid this problem.

4.6.6. Adjusting Credits from Changes in Carbon Stocks

The Kyoto Protocol currently treats all GHG flows equally, in CO_2-equivalent units: A ton of carbon is a ton of carbon, whether it is from fossil fuel combustion or is sequestered by reforestation. An alternative approach would be to treat different flows of carbon differently by giving partial, limited, or exaggerated credit for some carbon flows with respect to others or by putting a cap on the credits or debits available from a particular sink or source. Four reasons have been suggested for choosing to treat different flows of carbon differently:

- Encouraging emissions reductions in other sectors
- Compensating for uncertainty in measurement
- Compensating for differences in permanence
- Compensating for leakage.

One could choose an adjustment factor that is greater or less than 1 to encourage sequestration in the biosphere or to discourage it with respect to other mitigation strategies. Such an approach could be applied to provide differential encouragement for different activities. A system might be designed to reward improvements with respect to uncertainty, permanence, or leakage. For example, providing credits at the 90-percent confidence level for estimates of changes in carbon stocks would reward improved accuracy in measurement with additional credits (Canada, 1998). There is a risk that valuing different carbon flows differently could lead to outcomes that are not economically efficient or environmentally optimal.

4.7. Technical Issues Related to Implementation of Article 3.4

4.7.1. Assessing Activities

If the Parties decide to include only some activities under Article 3.4, they will need to address two generic problems:

- How to weigh the large number of factors relevant to the decision
- How to be consistent across widely different practices, some of which probably have not even been suggested yet.

One way to deal with both of these issues is to adopt a decision framework that allows practices now and in the future to be screened in a consistent way and simplifies a complex problem by dividing it into simpler steps and following a logical procedure. The framework outlined in Figure 4-13 is one example of such a procedure, which could be modified by the Parties. It works through multiple criteria to reach an endpoint where an activity falls into one of the following categories:

- Clearly a priority candidate for inclusion
- Clearly not a candidate
- May need to be considered in terms of reduced radiative forcing of the atmosphere traded off against non-climate impacts.

This approach can be made semi-quantitative up to the point at which climate impacts need to be compared to non-climate impacts, which unavoidably involves value judgments. The procedure can be applied at all scales (project, national, or global), although the importance-weighting of different criteria is likely to vary with scale. For example, the cost-efficiency of the activity (especially the monitoring cost) is very important at the project level, and non-climate impact is very important at the national level. At the global level, the net atmospheric impact (discounted for risk and uncertainty) and its verifiability may be the key criteria.

The total global magnitudes for carbon stored through various activities (Table 4-1) are relevant to the discussion of whether activities should be included as adjustments to Kyoto Protocol targets or whether they are so insignificant that the additional complexity of accounting is not justified. From the perspective of individual nations or projects, the local rate (the per-hectare rate multiplied by the area under consideration) may be of highest concern. Even activities that make minor global contributions can be important in the GHG inventories of individual nations.

Arguments about the cost of implementing an activity or the cost of measurement and reporting an activity are not directly relevant to whether an activity should be permitted under Article 3.4. Such arguments, however, will strongly affect the degree to which permitted activities will actually take place or be reported by individual Parties. Most of the additional activities analyzed in this chapter are undertaken primarily for non-climate reasons, so only a portion of the cost would be allocated to climate impact reduction. For activities whose cost-effectiveness is already demonstrated by their partial adoption, the costs of broader application may consist only of the marginal transaction cost of inventorying and reporting. Individual countries and projects will make cost-effectiveness decisions on the basis of their own circumstances, which vary greatly, and in relation to the cost of alternative land-use activities and energy sector options. Inventory and reporting costs are important at the national level if spatially comprehensive (and especially spatially explicit) accounting is required.

The sample assessment procedure begins by estimating the carbon storage rate for the activity in a given area (Figure 4-13). It then successively reduces this value, taking into consideration the following factors:

- The confidence with which the storage rate can be estimated
- The degree to which the observed storage is likely to result from direct human-induced activity
- The security of the carbon pools formed (given the risks to which they are exposed).

The example deliberately refrains from being prescriptive regarding cutoff criteria or formulae to be used because these judgments are decisions for the policy process. The weighting factors conceptually range between zero and one: In the example, "fully weighted" means a weighting of 1.0, "highly weighted" approximately 0.7, "low weighted" approximately 0.4, and "very low weighted" approximately 0.2.

4.7.2. Examples

Several examples are given in the following sections. These examples have been selected for no special reason other than that they illustrate different criteria coming into play, leading to different outcomes (see Figure 4-14). The details underpinning the examples can be found in relevant Fact Sheets at the end of this chapter.

4.7.2.1. Restoration of Wetlands

This example illustrates the tradeoffs between the carbon stock increase that can be achieved by reflooding areas of drained wetlands and the reestablishment of methane emissions that

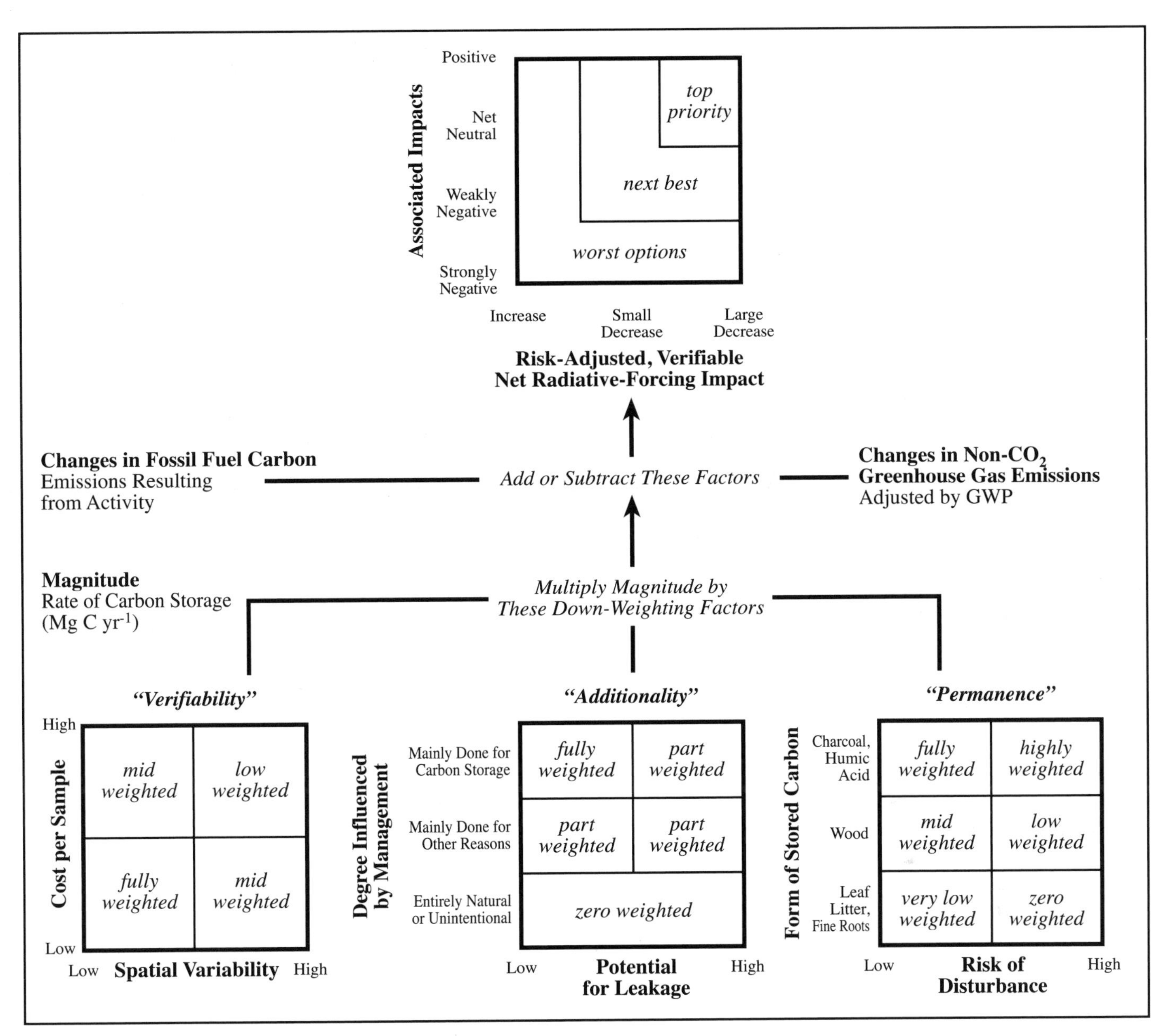

Figure 4-13: One possible framework for systematically considering a variety of factors that have a bearing on the suitability of an activity for inclusion under Article 3.4 of Kyoto Protocol. The entry point is an estimate of the magnitude of the carbon stored by the activity (lefthand side, lower middle). This estimate is then progressively down-weighted by considerations such as how easy it is to verify, the degree to which it is an intended consequence of a management action, and how likely it is to be lost through disturbance (note that this is an example list; the Parties have yet to decide which criteria will be employed). The estimate then must be adjusted for changes in non-CO_2 GHG emissions and changes in fossil fuel consumption resulting from the activity. Finally, this adjusted estimate must be weighted up in relation to the non-climate benefits or disbenefits it may cause.

will result (see Fact Sheet 4.18). The average annual carbon sequestration is less than 10 t C ha^{-1} yr^{-1}, once spatial variability has been taken into account (but note that where restoration prevents an ongoing emission of carbon, the effective carbon storage rate relative to no intervention may be higher). The carbon that is formed is relatively stable because much of it is in the form of humus and is dependent only on the continuation of the flooding regime. In the process, methane emissions are reestablished. Given a 100-year GWP of 21 for methane relative to carbon dioxide (i.e., a GWP of 5.7 relative to carbon) and an annual methane emission that can exceed 2 t CH_4 ha^{-1} yr^{-1}, the net effect can range from slightly reduced to a slightly increased radiative forcing. The associated impacts—principally on biodiversity and hydrology—are generally positive.

4.7.2.2. *Drainage of Wetlands*

Despite the fact that restoration of wetlands could lead in some cases to a net increase in radiative forcing, drainage of wetlands will not necessarily reduce global warming. The rate at which carbon is lost from wetlands following drainage is much faster

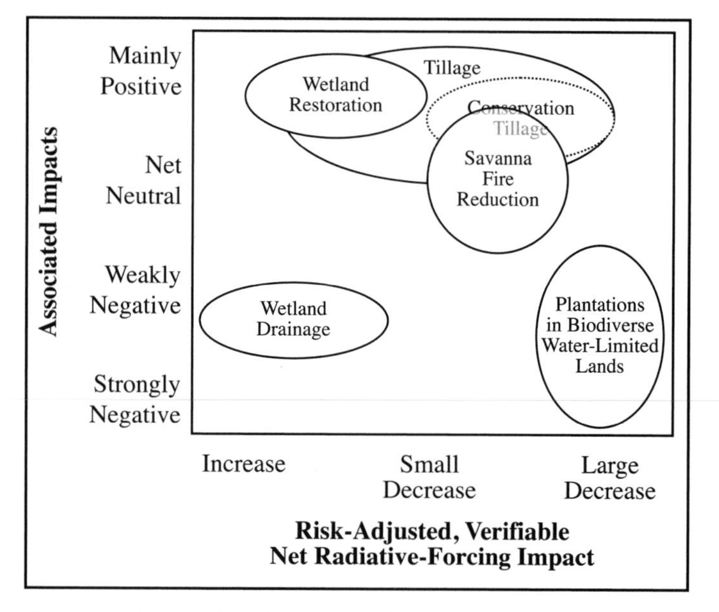

Figure 4-14: Application of decision framework described in Section 4.7.1. Cases are elaborated in Section 4.7.2 and are chosen simply to demonstrate different outcomes, not to promote or discourage any particular activity. The most favorable activities would fall in the top right corner of the graphic. Those in the left third would be hard to justify in purely climate terms; those in the bottom quarter may be disqualified by unacceptably high associated impacts.

than the rate at which it accumulates when they are flooded; in general, therefore, the increased radiative forcing caused by carbon dioxide outweighs the reduced forcing resulting from lower methane emissions. Where the drained wetlands are planted with trees, increased carbon inputs from fine roots may slow or reverse carbon loss by respiration (Finér and Laine, 1998; Minkkinen and Laine, 1998). In addition, there are negative impacts on biodiversity and water quality that must be balanced against non-environmental benefits such as food production or disease prevention.

4.7.2.3. Fire Management in Savannas

Reduction in the frequency and/or intensity of fires in savannas is technically possible, leading to a modest reduction in the time- and space-averaged emission of methane and ozone precursor gases from fires, as well as a gradual accumulation of wood and soil carbon. Smoke particles have a small radiative cooling effect, and some carbon is stored in the long term as charcoal, but the net effect of fire reduction is reduced radiative forcing (Moore *et al.*, 1997). The change in fire frequency is monitored relatively easily with remote sensing; if such monitoring is conducted over large areas and multiple years, the change in fire frequency can be compared to historical or contemporary baselines. The increase in carbon stocks is sufficiently large to be quantifiable through a combination of ground or aerial survey. The decrease in fire emissions is difficult to measure directly in an operational system, but it can be reliably modeled given fire frequency and fuel biomass. There is little risk of leakage: Suppressing fires in one savanna region does not encourage them elsewhere because there is no demand for fires *per se*.

The avoided non-CO_2 emissions can be considered permanent, whereas the increase in carbon stocks is somewhat vulnerable to future fires. The associated impacts are a mixture of positive and negative: On the positive side are an increased stock of timber resources, reductions in air pollution, and a reduced risk of injury or asset loss from fire; negative impacts are mostly related to potentially reduced biodiversity and reduced grass production for grazing use. The breadth of outcomes on the associated impacts axis reflects the varying degrees to which benefits arising from grazing are important in savannas, as well as the varying risks of loss of biodiversity (higher carbon gain is associated with greater chance of biodiversity changes).

4.7.2.4. Afforestation on Biodiverse, Water-Limited Lands

Large areas that could be considered for afforestation in non-Annex I countries in sub-humid regions are currently covered with native vegetation. Planting fast-growing, usually non-indigenous, tree crops into grassland or savanna landscapes has the potential to substantially increase the carbon sequestered (e.g., Christie and Scholes, 1995)—although this effect depends on effective ongoing plantation management, which sometimes is not socioeconomically viable (e.g., Carrere and Lohmann, 1996). The area planted and the biomass per hectare are relatively easily verified. This activity is clearly human-induced, and it has little leakage potential (although it does displace the current land use, which may be reestablished somewhere else with negative GHG consequences).

The carbon is stored as leaf, litter, wood, and soil carbon; the first three are susceptible to rapid loss through fire—a risk that can be reduced through established fire control practices. Where plantations replace biodiverse native vegetation in water-limited lands, there are significant negative environmental impacts: The water yield from the catchment is significantly reduced (van Wilgen *et al.*, 1998), as is biodiversity. Positive socioeconomic benefits, such as wealth or job creation, must be balanced by the loss of welfare resulting from reductions in available water, grazing, natural resources, and agricultural land. Afforestation of previously eroded or otherwise degraded land may have a net positive environmental impact; in catchments where the water yield is large or is not heavily used, the streamflow reduction may not be critical.

4.7.2.5. Conservation Tillage

Conventional tillage generally leads to a gradual reduction in carbon stocks, whereas conservation tillage (including increased crop residue return) on areas formerly under conventional tillage generally results in a slow recovery of soil carbon towards the original, untilled level (Fact Sheet 4.3). Spatial variability is high because tillage is patchily applied in time and space by a large number of individual farmers, each working on a small area. The cost per sample is low to moderate, but a large number of samples are needed. Leakage is low because crop yields are similar to those achieved with conventional tillage, and the activity is clearly human-induced. The carbon gained is easily lost again if the land use reverts to conventional tillage even briefly (Lindstrom *et al.*, 1999). There generally is a reduction in fossil fuel usage (through less machinery use), and there may be minor increases in CH_4 uptake and N_2O emission by the soil.

The net increased emission of GHGs resulting from conventional tillage (excluding emissions resulting from the initial land-use conversion) is small on a per-hectare basis, as is the net uptake resulting from conservation tillage. The total global area to which these practices could be applied is very large, however. The associated environmental impacts of conventional tillage (erosion, declining soil native fertility, loss of biodiversity) are generally negative, and the associated impacts of conservation tillage are therefore positive relative to conventional tillage; on the negative side is possibly increased use of biocides and an increase in soil water—which in some cases may increase drainage that can affect salinization in downslope soils.

The socioeconomic benefits of tillage are large, of course (it is the basis of global food and fiber production), and can be maintained under conservation tillage. Conservation tillage currently is not more widely applied largely because the economic benefits are marginal (and may even be marginally negative), and the added organic material at the surface may create some additional disease problems in some situations. At a project or national level, the decision to use conservation tillage as a climate change reducing activity would probably be strongly influenced by the balance of reporting costs relative to carbon stored.

4.7.3. Additional Technical Considerations for Implementing Article 3.4

If a real reduction in radiative forcing is to be achieved through land use-related activities, the Parties will need to address the following technical considerations:

- *Either* the land area on which changes in carbon stocks are to be monitored—and the pools to be measured—must be defined exactly and spatially explicitly, *or* the major management practices leading to carbon storage must be defined in terms that are verifiable and correlated with net changes in the flux of GHGs resulting from their application. The change in carbon stocks approach is applicable only to carbon dioxide (which is the dominant land use-related GHG at a global level), whereas the activity-based approach applies to all gases. Note that both approaches will probably make use of models, and both apply to specific land areas. The land-based and activity-based approaches are not mutually exclusive; they could be applied simultaneously. This consideration allows for an internal check and is analogous to the simultaneous use of the "top-down" and "bottom-up" accounting procedures for fossil fuels.
- Direct measurements of changes in GHG fluxes resulting from changes in land-use practices currently (and for the first commitment period) are not feasible on an operational basis at the national scale. Therefore, estimation of fluxes based on activities requires the use of models. Very simple models (such as default emission factor per hectare) do not capture the range of variation that typically occurs in land use-related activities. Models require more information than simply the degree of implementation of the practice to have acceptable accuracy; they may need to incorporate information about the environment in which the practice is applied, the land-use history, the timing of actions within the practice, and the intensity with which it is applied. The details vary from practice to practice.
- *If* the land-based approach is followed *and* the area thus defined is less than the total land area of each country reporting stock changes for purposes of adjustment of Kyoto Protocol targets, *or if* activities are defined that are substantively less than the full range of activities that are applied to a given land use, *then* there is a significant risk of displacement of the putative gains to other areas (leakage) or their erosion by other activities. Such effects may not be preventable, but they could be accounted. Effectively, this accounting would require attention to full and symmetrical accounting of land area or activities, or both.
- *If* the land-based approach is used, the reporting interval for changes in carbon stocks probably should not be less than about 5 years because changes over shorter periods will be difficult to detect reliably in many pools. There is no theoretical upper limit to the period, but the Parties may need to consider some practical upper limit for managing the global atmosphere via the international treaty process.
- Reporting intervals for carbon stock changes that result from Article 3.4 activities may need to be contiguous for the accounting to remain accurate because some of the carbon pools are sufficiently volatile that substantial losses to the atmosphere could occur within a period of a few years.
- The pools to which stock-change calculations are to be applied must be defined consistently over time and space, and full accounting requires that all pools that are substantially affected by the land use and land-use change activities be counted.
- Adjustments to commitments that are based on Article 3.4 activities should include the effects of all major GHGs (CO_2, CH_4, and N_2O) because in many cases the net effect is strongly affected by two or more gases. In some cases, non-GHG climate impacts (such as changes in albedo) also may need to be considered. Where other GHGs are already accounted for in the inventory (for example, methane emissions from rice paddies), care will be needed to avoid double-counting.

- Failure to distinguish between changes in land-based GHG fluxes that are a direct consequence of human-induced land-use activities and those that result from natural processes or indirectly from human activities (e.g., from CO_2 fertilization or deposition of atmospheric nitrogen) could allow Parties to claim a large portion of the current terrestrial carbon sink as an adjustment to the Kyoto targets. Measuring which portion of the change in fluxes is directly human induced is not technically feasible, but inferring this fraction is possible, by using models, agreed baselines, or control areas.

References

AAPRESID, 1999: Siembra Directa: mas del 30% de la superficie agrícola se cultiva en SD. *Gacetilla Informática de la Asociación Argentina de Productores de Siembra Directa,* **9(49),** 1–3.

Abel, N., 1997: Mis-management of the productivity and sustainability of African communal rangelands: a case study and some principles from Botswana. *Ecological Economics,* **23,** 113–133.

Akala, V. and R. Lal, 1999: Mineland reclamation and soil carbon sequestration in Ohio. *Land Degradation and Rehabilitation,* (in press).

Akbari, H., A.H. Rosenfeld, and H.G. Taha, 1989: Recent development in heat island studies. In: *Controlling Summer Heat Islands* [Akbari, H., K. Garbesi, and P. Martien (eds.)]. Proceedings of Workshop on Saving Energy and Reducing Atmospheric Pollution by Controlling Summer Heat Islands, February 1989. U.S. Department of Energy, Berkeley, CA, USA,pp. 14–30.

Akbari, H., S. Davis, S. Dorsano, J. Huang, and S. Winnett (eds.), 1992: *Cooling Our Communities: A Guidebook on Tree Planting and Light-Colored Surfacing.* U.S. Government Printing Office, Washington, DC, USA, 217 pp.

Albaugh, T.J, H.L. Allen, P.M. Dougherty, L.W. Kress, and J.S. King: Leaf area and above- and below-ground growth responses of Loblolly pine to nutrient and water additions. *Forest Science,* **44(2),** 317–328.

Amelung, T. and M. Diehl, 1992: *Deforestation of Tropical Rain Forests—Economic Causes and Impact on Development.* Kieler Studien No. 241, Institut für Weltwirtschaft, Kiel, Germany.

Anderson, A.B., 1990a: Deforestation in Amazonia: Dynamics, causes and alternatives. In: *Alternatives to Deforestation: Steps Toward Sustainable Use of the Amazon Rain Forest* [Anderson, A.B. (ed.)]. Columbia University Press, New York, NY, USA, pp. 3–23.

Anderson, A.B. (ed.), 1990b: *Alternatives to Deforestation: Steps Toward Sustainable Use of the Amazon Rain Forest.* Columbia University Press, New York, NY, USA, 281 pp.

Anderson, J.M., 1991: The effects of climate change on decomposition processes in grassland and coniferous forests. *Ecological Applications,* **1,** 326–347.

Apps, M., T. Karjalainen, G. Marland, and B. Schlamadinger, 1997: Accounting system considerations: CO_2 emissions from forests, forest-products, and land-use change, a statement from Edmonton. Available at: http://www.joanneum.ac.at/IEA-Bioenergy-TaskXV.

Apps, M., J.S. Bhatti, D.H. Halliwell, H. Jiang, and C.H. Peng, 1999a: Simulated carbon dynamics in the boreal forest of central Canada under uniform and random disturbance regimes. *Advances in Soil Science,* (in press).

Apps, M., W.A. Kurz, S.J. Beukema, and J.S. Bhatti, 1999b: Carbon budget of the Canadian forest product sector. *Environmental Science and Policy,* **2,** 25–41.

Ara, (ed.), 1990: *Naturerbe Regenwald—Strategien und Visionen zum Schutz der tropischen Regenwälder.* Focus-Verlag, Giessen, Germany.

Archer, S., 1995: Tree-grass dynamics in a *Prosopis*—thornscrub savanna parkland: Reconstructing the past and predicting the future. *Ecoscience,* **2,** 83–99.

Archer, S., 1989: Have southern Texas savannas been converted to woodlands in recent history? *American Naturalist,* **134,** 545–561.

Archer, S., 1994: Woody plant encroachment into southwestern grasslands and savannas: rates, patterns and proximate causes. In: *Ecological Implications of Livestock Herbivory in the West* [Vavra, M., W.A. Laycock, and R.D. Pieper (eds.)]. Society for Range Management, Denver, CO, USA, pp. 13–68.

Archer, S., D.S. Schimel, and E.A. Holland, 1995: Mechanisms of shrubland expansion: land use, climate or CO_2? *Climate Change,* **29,** 91–99.

Armentano, T.V. and E.S. MenGes, 1986: Patterns of change in the carbon balance of organic soil—wetlands of the temperate zone. *Journal of Ecology,* **74,** 755–774.

Aselmann, I. and P.J. Crutzen, 1989: Global distribution of natural freshwater wetlands and rice paddies, their net primary productivity, seasonality, and possible methane emissions. *Journal of Atmospheric Chemistry,* **8,** 307–358.

Ash, A.J., S.M. Howden, J.G. McIvor, and N.E. West, 1996: Improved rangeland management and its implications for carbon sequestration. In: *Proceedings of the Fifth International Rangeland Congress* [West, N.E. (ed.)]. Salt Lake City, UT, USA, 23–28 July 1995, Vol. I. Society for Range Management, Denver, CO, USA, pp. 19–20.

Aumonier, S., 1996: The greenhouse gas consequences of waste management—identifying preferred options. *Energy Conservation Management,* **37(6-8),** 1117–1122.

Austin, M.P. and P.C. Heyligers, 1989: Vegetation survey design for conservation: gradsect sampling of forests in north-eastern New South Wales. *Biological Conservation,* **50,** 13–32.

Back, P.V., E.R. Anderson, W.H. Burrows, M.K.K. Kennedy, and J.O. Carter, 1997: *TRAPS—Transect Recording and Processing System: Field Guide and Software Manual.* Department of Primary Industries, Rockhampton, Queensland, Australia.

Balesdent, J. and M. Balabane, 1996: Major contribution of roots to soil carbon storage inferred from maize cultivated soils. *Soil Biological Biochemistry,* **28(9),** 1261–1263.

Balesdent, J., E. Besnard, D. Arrouays, and C. Chenu, 1998: The dynamics of carbon in particle-size fractions of soil in a forest-cultivation sequence. *Plant and Soil,* **201,** 49–57.

Balesdent, J. and A. Mariotti, 1996: Measurement of soil organic matter turnover using C natural abundance. In: *Mass Spectrometry of Soils* [Boutton, T.W. and S. Yamasaki (eds.)]. Marcel Dekker, Inc., New York, NY, USA, pp. 83–111.

Barker, J.R., G.A. Baumgardner, D.P. Turner, and J. Lee, 1995: Potential carbon benefits of the Conservation Reserve Program in the United States. *Journal of Biogeography,* **22,** 743–751.

Barrow, N.J., 1969: The accumulation of soil organic matter under pasture and its effect on soil properties. *Australian Journal of Experimental Agriculture and Animal Husbandry,* **9,** 437–444.

Barson, M.M., B. Abraham, and C.V. Malcolm, 1994: Improving the productivity of saline discharge areas: an assessment of the potential role of saltbush in the Murray-Darling Basin. *Australian Journal of Experimental Agriculture,* **34,** 1143–1154.

Basher, L.R. and I.H. Lynn, 1996: Soil changes associated with cessation of sheep grazing in the Canterbury high country, New Zealand. *New Zealand Journal of Ecology,* **20,** 179–189.

Bastin, G.N., R.W. Tynan, and V.H. Chewings, 1998: Implementing satellite-based grazing gradient methods for rangeland assessment in South Australia. *Rangeland Journal,* **20(1),** 61–76.

Bationo, A., S.P. Wani, C.L. Bielders, P.L.G. Velk, and A.U. Mokwunye, 2000: Crop residues and fertilizer management to improve soil organic carbon content, soil quality and productivity in the desert margins of West Africa. In: *Global Cimate Change and Tropical Ecosystems* [Lal, R., J.M. Kimble, and B.A. Stewart (eds.)]. CRC-Lewis Publishers, Boca Raton, FL, USA, pp 117–146.

Batjes, N.H., 1999: Management options for reducing CO_2-concentrations in the atmosphere by increasing the carbon sequestration in soil. ISRIC Technical Paper 30, Wageningen, The Netherlands.

Beaumont, P., 1996: Agricultural and environmental changes in the upper Euphrates catchment of Turkey and Syria and their political and economic impacts. **16(2),** 137–157.

Bennett, O.L., 1977: Stripmining new solutions to an old but growing problem. *Crops and Soils,* 12–14 January 1997.

Berg, W.A., J.A. Bradford, and P.L. Sims, 1997: Long-term soil nitrogen and vegetation change on sandhill rangeland. *Journal of Range Management,* **50,** 482–486.

Bergkamp, G. and B. Orlando, 1999: *Wetlands and Climate Change—Exploring Collaboration Between the Convention on Wetlands and the UNFCCC.* Ramsar Bureau, Geneva, Switzerland.

Binswanger, H.P., 1991: Brazilian policies that encourage deforestation in the Amazon. *World Development,* **19,** 821–829.

Birdsey, Richard A., 1996: Carbon storage for major forest types and regions in the conterminous United States. In: *Forests and Global Change, Volume 2: Forest Management Opportunities for Mitigating Carbon Emissions* [Sampson, R.N. and D. Hair (eds.)]. American Forests, Washington, DC, USA, pp. 1–26, 261–372.

Birdsey, R.A., R. Alig, and D. Adams, 2000: Mitigation activities in the forest sector to reduce emissions and enhance sinks of greenhouse gases. In: *The Impact of Climate Change on America's Forests* [Joyce, L. and R. Birdsey (eds.)]. U.S. Department of Agriculture, Rocky Mountain Research Station, Ft. Collins, CO, USA, (in press).

Blevins, R.L. and W.W. Frye, 1993: Conservation tillage: An ecological approach to soil management. *Advances in Agronomy,* **51,** 33–78.

Boman, U., 1996: *Carbon Dioxide from Integrated Biomass Energy Systems—Examples from Case Studies in the USA.* Project Number 38704, Vattenfall Support AB, Vallingby, Sweden.

Boscolo, M., J. Buongiorno, and T. Panayotou, 1997: Simulating options for carbon sequestration through improved management of a lowland tropical rainforest. *Environmental and Development Economics,* **2,** 241–263.

Boscolo, M. and J. Buongiorno, 1997: Managing a tropical rainforest for timber, carbon storage and tree diversity. *Commonwealth Forestry Review,* **76(4),** 246–254.

Bouamrane, M., 1996: The season of gold—putting a value on harvests from Indonesian agroforests. *Agroforestry Today,* **8(1),** 8–10.

Boutton, R.W., S.R. Archer, A.J. Midwood, S.F. Zitzer, and R. Bol, 1998: ^{13}C values of soil organic carbon and their use in documenting vegetation change in a subtropical savanna ecosystem. *Geoderma,* **82,** 5–41.

Bowman, L. and N.W. Lane, 1999: Micro-scale biomass power. In: *Proceedings of the 4th Biomass Conference of the Americas, Oakland, CA* [Overend, R.P. and E. Chornet (eds.)]. Elsevier Science Publishers, Oxford, United Kingdom, pp. 1445–1452.

Brejida, J.J., 1997: Soil changes following 18 years of protection from grazing in Arizona chaparral. *Southwestern Naturalist,* **42,** 478–487.

Bridge, B.J., J.J. Mott, and R.J. Hartigan, 1983: The formation of degraded areas in the dry savanna woodlands of northern Australia. *Australian Journal of Soil Research,* **21,** 91–104.

Bronson, K.F., K.G. Cassman, R. Wassmann, D.C. Olk, M. van Noordwijk, and D.P. Garrity, 1997: Soil carbon dynamics in different cropping systems in principal ecoregions of Asia. In: *Management of Carbon Sequestration in Soil* [Lal, R., J.M. Kimble, R.F. Follett, and B.A. Stewart (eds.)]. CRC Press, Boca Raton, FL, USA, pp. 35–37.

Bronson, K.F. and A.R. Mosier, 1991: Effect of encapsulated calcium carbide on dinitrogen, nitrous oxide, methane and carbon dioxide emissions from flooded rice. *Biology and Fertility of Soils,* **11,** 116–120.

Brookman-Amissah, J., J.B. Hall, M.D. Swaine, and J.Y. Attakorah, 1980: A re-assessment of a fire protection experiment in north-eastern Ghana savanna. *Journal of Applied Ecology,* **17,** 85–99.

Brown, S., 1997: *Estimating Biomass and Biomass Change of Tropical Forests: A Primer.* Food and Agriculture Organization of the United Nations, Rome, Italy, 69 pp.

Brown, S., A.Z. Shvidenko, W. Galinski, R.A. Houghton, E.S. Kassischke, P. Kauppi, W.A. Kerz, I.A. Nalder, and V.A. Rojkov, 1996: Forests and the global carbon cycle; past, present, and future role. In: *The Role of Forest Ecosystems and Forest Management in the Global Carbon Cycle* [Apps, M. and D. Price (eds.)]. NATO ARW Series, Springer-Verlag, New York, NY, USA.

Brown, K. and D.W. Pearce (eds.), 1994: *The Causes of Tropical Deforestation.* UCL Press, London, United Kingdom.

Brown, P., 1998: *Climate, Biodiversity, and Forests: Issues and Opportunities Emerging from the Kyoto Protocol.* World Resources Institute, Washington, DC, USA, 36 pp.

Bruce, J.P., M. Frome, E. Haites, H. Janzen, R. Lal, and K. Paustian, 1999: Carbon sequestration in soils. *Journal of Soil and Water Conservation,* **54,** 381389.

Buck, L.E., J.P. Lassoie, and E.C.M. Fernandes (eds.), 1999: *Agroforestry in Sustainable Agricultural Systems.* CRC Press, Boca Raton, FL, USA, 416 pp.

Buckles, D., A. Eteka, O. Osiname, M. Galiba, and G. Galiano, (eds.), 1998: *Cover Crops in West Africa: Contributing to Sustainable Agriculture.* International Development Research Centre, Ottawa, Canada.

Bunch, R., 1999: Greener fields with greener technology: Case studies of sustainable low-input agricultural development in Central America. *Environment, Development and Sustainability,* (in press).

Buresh, R.J. and P.J.M. Cooper (eds.), 1999: The science and practice of short-term improved fallows. *Agroforestry Systems,* **47,** 1356.

Burke, I.C., W.K. Lauenroth, and D.P. Coffin, 1995: Soil organic matter recovery in semiarid grasslands: implications for the Conservation Reserve Program. *Ecological Applications,* **5(3),** 793–801.

Burrows, W.H., J.O. Carter, J.C. Scanlan, and E.R. Anderson, 1990: Management of savannas for livestock production in north-east Australia: contrasts across the tree-grass continuum. *Journal of Biogeography,* **17,** 503–512.

Burrows, W.H., J.F. Compton, and M.B. Hoffmann, 1998: Vegetation thickening and carbon sinks in the grazed woodlands of north-east Australia. In: *Proceedings of the Australian Forest Growers Conference.* Lismore, New South Wales, Australia, pp. 305–316.

Burrows, W.H., M.B. Hoffmann, J.F. Compton, P.V. Back, and L.J. Tait,. 2000: Allometric relationships and community biomass estimates for some dominant eucalypts and associated species in Central Queensland woodlands. *Australian Journal of Botany,* (in press).

Burschel, P., E. Kürsten, and B.C. Larson, 1993: Die Rolle von Wald und Forstwirtschaft in Kohlensoffhaushalt—eine Betrachtnung fur die Bundesrepublik Deutschland. In: *Forstliche Forschungsberichte,* Munich, Germany, p. 126.

Buyanovsky, G.A. and G.H. Wagner, 1998: Carbon cycling in cultivated land and its global significance. *Global Change Biology,* **4,** 131–141.

Cai, Z.C., H. Xu, H.H. Zhang, and J.S. Jin, 1994: Estimate of methane emission from rice paddy fields in Taihu region, China. *Pedosphere,* **4,** 297–306.

Cambardella, C.A. and E.T. Elliott, 1992: Particulate soil organic matter changes across a grassland cultivation sequence. *Soil Science Society of America Journal,* **56,** 777–783.

Cameron, A.G., and B.J. Ross, 1996: Fertiliser responses of established grass pastures in the Northern Territory. *Tropical Grasslands,* **30,** 426–429.

Canada, 1998: *Additional Submissions by Parties.* Report FCCC/SBSTA/1998/Misc.6/ADD1, UNFCCC Subsidiary Body for Scientific and Technological Advice. Available at: http://www.unfccc.de.

Cantagallo, J.E., C.A. Chimenti, and A.J. Hall, 1997: Number of seeds per unit area in sunflower correlates well with a photothermal quotient. *Crop Science,* **37,** 178–1786.

Cardelino, C.L. and W.L. Chameides, 1990: Natural hydrocarbons, urbanization, and urban ozone. *Journal of Geophysical Research,* **95(D9),** 13,971–13,979.

Carrere, R. and L. Lohmann, 1996: *Pulping the South: Industrial Tree Plantations and the World Paper Economy.* Zed Books, London, United Kingdom.

Carter, M.R., E.G. Gregorich, D.A. Angers, R.G. Donald, and M.A. Bolinder, 1998: Organic C and N storage, and organic C fractions, in adjacent cultivated and forested soils of eastern Canada. *Soil and Tillage Research,* **47,** 253–261.

Cerri, C.C., B. Volkoff, and F. Andreaux, 1991: Nature and behavior of organic matter in soils under natural forest, and after deforestation, burning and cultivation near Manaus. *Forest Ecology and Management,* **38,** 247–257.

Chadwick, O.A., E.F. Kelly, D.M. Merritts, and R.G. Amundson, 1994: Carbon dioxide consumption during soil development. *Biogeochemistry,* **24,** 115–127.

Chaneton, E.J. and R.S. Lavado, 1996: Soil nutrients and salinity after long-term grazing exclusion in a flooding pampa grassland. *Journal of Range Management,* **49(2),** 182–187.

Chen, Z., L. Li, K. Shao, and B. Wang, 1993: Features of CH_4 emission from rice paddy fields in Beijing and Nanjing. *Chemosphere,* **26,** 239–245.

Christian, D.P., W. Hoffman, J.M. Hanowski, G.J. Niemi, and J. Beyea, 1998: Bird and mammal diversity on woody biomass plantations in North America. *Biomass and Bioenergy,* **14,** 395–402.

Christie, S.I. and R.J. Scholes, 1995: Carbon storage in eucalyptus and pine plantations in South Africa. *Environmental Monitoring and Assessment,* **38,** 231–241.

Ciais, P., P.P. Tans, M. Trolier, J.W.C. White, and R.J. Francey, 1995: A large northern hemisphere terrestrial CO_2 sink indicated by 13C/12C ration of atmospheric CO_2. *Science,* **269,** 1098–1102.

Clymo, R.S., 1984: The limits to peat bog growth. *Philosophical Transactions of the Royal Society of London,* **303,** 605–654.

Cole, C.V., J. Duxbury, J. Freney, O. Heinemeyer, K. Minami, A. Mosier, K. Paustian, N. Rosenburg, N. Sampson, D. Sauerbeck, and Q. Zhao, 1997: Global estimates of potential mitigation of greenhouse gas emissions by agriculture. *Nutrient Cycling in Agroecosystems,* **49,** 221–228.

Cole, C.V., K. Flach, J. Lee, D. Sauerbeck, and B. Stewart, 1993: Agricultural sources and sinks of carbon. *Water, Air, and Soil Pollution,* **70,** 111–122.

Conant, R.T., K. Paustian, and E.T. Elliot, 2000: Grassland management and conversion into grassland: Effects on soil carbon. *Ecological Application,* (in press).

Cook, J.H., J. Beyea, and K.H. Keeler, 1991: Potential impacts of biomass production in the United States on biological diversity. *Annual Reviews of Energy and Environment,* **16,** 401–431.

Cox, J.R., and H.L. Morton, 1986: The effects of seasonal fire on live biomass and standing crop in a big sacaton (*Sporobolus wrightii*) grassland. In: *Rangelands: A Resource Under Siege. Proceedings of the Second International Rangeland Congress, Adelaide, Australia.* Australian Academy of Science, Canberra, Australia, pp. 596–597.

Critchley, C.N.R. and S.M.C. Poulton, 1998: A method to optimize precision and scale in grassland monitoring. *Journal of Vegetation Science,* **9,** 337–346.

Cromack, Jr., K., J.D. Landsberg, R. Everett, R. Zeleny, C. Giardina, T.D. Anderson, R. Averill, and R. Smyrski, 2000: Assessing the impacts of severe fire on forest ecosystem recovery. *Journal of Sustainable Forestry,* **11(1-2),** 177–227.

Crush, J.R., G.C. Waghorn, and M.P. Rolston, 1992: Greenhouse gas emissions from pasture and arable crops grown on a Kairanga soil in the Manuwatu, North Island, New Zealand. *New Zealand Journal of Agricultural Research,* **35,** 253–257.

CTIC, 1998: *17th Annual Crop Residue Management Survey Report.* Conservation Technology Information Center, West Lafayette, IN, USA. Available at: http://www.ctic.purdue.edu.

CTIC, 1997: *16th Annual Crop Residue Management Survey Report.* Conservation Technology Information Center, West Lafayette, IN, USA. Available at: http://www.ctic.purdue.edu.

CTIC, 1999: *Crop Residue Management: United States.* Available at: http://www.ctic.purdue.edu/cgi-bin/CRMMap.

Cunningham, A.B. and F.T. Mbenkum, 1993: *Sustainability of Harvesting* Prunus africana *Bark in Cameroon. A Medicinal Plant in the International Trade.* People and Plants Working Paper 2, United Nations Educational, Scientific and Cultural Organization, Paris, France, 28 pp.

Current, D., E. Lutz, and S. Scherr, 1998: Costs, benefits and farmer adoption of agroforestry. In: *Agriculture and the Environment. Perspectives on Sustainable Rural Development* [Lutz, E. (ed.)]. World Bank, Washington, DC, USA, pp. 323–344.

Dalal, R.C. and J.O. Carter, 2000: Soil organic matter dynamics and carbon sequestration in Australian tropical soils. *Global Climate Change and Terrestrial Ecosystems,* 283–314.

Dalal, R.C., and R.J. Mayer, 1996: Long-term trends in fertilization of soils under continuous cultivation and cereal cropping in Southern Queensland. II. Total organic carbon and its rate of loss from the soil profile. *Australian Journal of Soil Research,* **24,** 281–292.

Danaher, T.J., G.R. Wedderburn-Bisshop, L.E. Kastanis, and J.O. Carter, 1998: The Statewide Landcover and Trees Study (SLATS)—monitoring land cover change and greenhouse gas emissions in Queensland. In: *9th Australasian Remote Sensing and Photogrammetry Conference.* Volume 1, Sydney, Australia, July 1998, pp. 1301–1341.

Davidson, E.A. and I.L. Ackerman, 1993: Changes in soil carbon inventories following cultivation of previously untilled soils. *Biogeochemistry,* **20,** 161–164.

Davidson, E.A., D.C. Nepstad, C. Klink, S. Trumbore, M.J. Fisher, I.M. Rao, C.E. Lascano, J.I. Sanz, R.J. Thomas, and R.R. Vera, 1994: Pasture soils as a carbon sink. *Nature,* **376,** 472–473.

Díaz-Raviña, R.A., G.O Magrin, M.I. Travasso, and R.O. Rodríguez, 1997: Climate change and its impact on the properties of agricultural soils in the Argentinean Rolling Pampas. *Climate Research,* **9,** 25–30.

De Foresta, H. and G. Michon, 1994: Agroforestry in Sumatra—where ecology meets economy. *Agroforestry Today,* **6(4),** 12–13.

Denier van der Gon, H.A.C. and H.U. Neue, 1994: Impact of gypsum application on methane emission from wetland rice field. *Global Biogeochemical Cycles,* **8,** 127–134.

Dick, W.A., R.L. Blevins, W.W. Frye, S.E. Peters, D.R. Christenson, F.J. Pierce, and M.L. Vitosh, 1998: Impacts of agricultural management practices on C sequestration in forest-derived soils of the eastern Corn Belt. *Soil and Tillage Research,* **47,** 235–244.

Dixon, R.K. and O.N. Krankina, 1993: Forest fires in Russia: carbon dioxide emissions to the atmosphere. *Canadian Journal of Forest Research,* **23,** 700–705.

Dixon, R.K., J.K. Winjum, and P.E. Schroeder, 1993: Conservation and sequestration of carbon: the potential of forest and agroforest management practices. *Global Environmental Change,* **3(2),** 159–173.

Dixon, R.K., J.K. Wimjum, K.J. Andrasko, J.J. Lee, and P.E. Schroeder, 1994: Integrated systems: assessing of promising agroforest and alternative land use practices to enhance carbon conservation and sequestration. *Climate Change,* **30,** 1–23.

Dixon, R.K., 1996: Agroforestry systems and greenhouse gases. *Agroforestry Today,* **8(1),** 11–14.

Dixon, R.K., 1995: Agroforestry systems: sources or sinks of greenhouse gases? *Agoforestry Systems,* **31,** 99–116.

Donigian, A.S. A.S. Partwardhan, R.V. Jackson, T.O. Barnwell, K.B. Weinrich, and A.L. Rowell, 1995: Modeling the impacts of agricultural management practices on soil carbon in the central U.S. In: *Soil Management and Greenhouse Effect* [Lal, R., J. Kimble, E. Levine, and B.A. Stewart (eds.)]. CRC-Lewis Publishers, Boca Raton, FL, USA, pp. 121–135.

Doran, J.C. and J.W. Turnbull (eds.), 1997: Australian trees and shrubs: Species for land rehabilitation and farm planting in the tropics. ACIAR Monograph No. 24, Australian Centre for International Agricultural Research, Canberra, Australia, 384 pp.

Dormaar, J.F.and S. Smoliak, 1985: Recovery of vegetative cover and soil organic matter during revegetation of abandoned farmland in a semiarid climate. *Journal of Range Management,* **38(6),** 487–491.

Dormaar, J.F. and W.D. Willms, 1990: Effects of grazing and cultivation on some chemical properties of soils in the mixed prairie. *Journal of Range Management,* **43(5),** 456–460.

Dregne, H., M. Kassa, and B. Rzanov, 1991: A new assessment of the world status of desertification. *Desertification Control Bulletin,* **20,** 6–18.

Duchemin, E., M. Lucotte, R. Canuel, and A. Chamberland, 1995: Production of the greenhouse gases CH_4 and CO_2 by hydroelectric reservoirs of the boreal region. *Global Biogeochemical Cycles,* **9(4),** 529–540.

Duchemin, E., M. Lucotte, A.G. Queiroz, R. Canuel, H.C.P. da Silva, D.C. Almeida, J. Dezincourt, and L.E. Ribeiro, 1999: Greenhouse gas emissions from an old tropical reservoir in Amazonia: Curuá-Una reservoir. *Verhandlungen International Verein Limnologie,* (in press).

Dumanski, J., R.L. Desjardins, C. Tarnocai, C. Monreal, E.G. Gregorich, V. Kirkwood, and C.A. Campbell, 1998: Possibilities for future carbon sequestration in Canadian agriculture in relation to land use changes. *Climate Change,* **40,** 81–103.

Dumestre, J.F., J. Guezennec, C. Galy-Lacaux, R. Delmas, S. Richard, and L. Labroue, 1999: Influence of light intensity on methanotrophic bacterial activity in Petit Saut Reservoir, French Guiana. *Applied and Environmental Microbiology,* **65(2),** 534–539.

Dupont-Roc, G., A. Khor, and C. Anastasi, 1996: *The Evolution of the World's Energy System.* Shell International Limited, London, United Kingdom.

Dyson, F.J., and G. Marland, 1979: Technical fixes for the climatic effects of CO_2. In: *Workshop on the Global Effects of Carbon Dioxide from Fossil Fuels, Miami Beach, FL, March 7–11, 1977* [Elliott, W.P. and L. Machta (eds.)]. CONF-770385, U.S. Department of Energy, Washington, DC, USA, pp. 111–118.

Dyson, F.J., 1977: Can we control carbon dioxide in the atmosphere? *Energy,* **2,** 287–291.

Edroma, E.L., 1986: Effects of fire on the primary productivity of grasslands in Queen Elizabeth National Park, Uganda. In: *Rangelands: A Resource Under Siege. Proceedings of the Second International Rangeland Congress.* Australian Academy of Science, Adelaide, Australia, pp. 599–600.

Eldridge, D.J. and A.D. Robson, 1997: Bladeploughing and exclosure influence soil properties in a semi arid Australian woodland. *Journal of Range Management,* **50,** 191–198.

Eswaran, H., E. Van den Berg, and P. Reich, 1993: Organic carbon in soils of the world. *Journal of the Soil Science Society of America,* **57,** 192–194.

Eve, M., K. Paustian, R. Follett, and E.T. Elliott, 2000: An inventory of carbon emissions and sequestration in U.S. cropland soils. In: *Soil Management for Enhancing Carbon Sequestration* [Lal, R. and K. McSweeney (eds.)]. SSSA Special Publication, Madison, WI, USA (in press).

Fan, S., M. Gloor, J. Mahlman, S. Pacala, J. Sarmiento, T. Takahashi, and P. Tans, 1998: A large terrestrial carbon sink in North America implied by atmospheric and oceanic carbon dioxide data and models. *Science,* **282,** 442–446.

FAO, 1996: *Forest Resources Assessment 1990. Survey of Tropical Forest Cover and Study of Change Processes.* FAO Report 130, Food and Agriculture Organization of the United Nations, Rome, Italy, 152 pp.

FAO, 1999: FAOSTAT Agricultural Production Database. Food and Agriculture Organization of the United Nations, Rome, Italy.

Fay, C., H. de Foresta, M. Sarait, and T.P. Tomich, 1998: A policy breakthrough for Indonesian farmers in the Krui damar agroforests. *Agroforestry Today,* **10(2),** 25–26.

Fearnside, P.M. and R.I. Barbosa, 1998: Soil carbon changes from conversion of forest to pasture in Brazilian Amazonia. *Forest Ecology and Management,* **108(1-2),** 147–166.

Fearnside, P.M., 1999: Forests and global warming mitigation in Brazil: opportunities in the Brazilian forest sector for responses to global warming under the "Clean Development Mechanisms." *Biomass and Energy,* **16,** 171–189.

Fearnside, P.M., 1997: Transmigration in Indonesia: lessons from its environmental and social impacts. *Environmental Management,* **21(4),** 553–570.

Fearnside, P.M., 1995: Agroforestry in Brazil's Amazonian development policy: The role and limits of a potential use for degraded land. In: *Brazilian Perspectives on Sustainable Development of the Amazon Region* [Clusener-Godt, M. and I. Sachs (eds.)]. United Nations Educational, Scientific and Cultural Organization, Paris, France, pp. 125–148.

Feller, C., and M.H. Beare, 1997: Physical control of soil organic matter dynamics in the tropics. *Geoderma,* **79,** 69–116.

Feller, C., 1993: Organic inputs, soil organic matter and functional soil organic compartments in low-activity clay soils in tropical zones. In: *Soil Organic Matter Dynamics and Sustainability of Tropical Agriculture* [Mulongoy, K. and R. Merckx (eds.)]. John Wiley and Sons, Chichester, United Kingdom, pp. 77–88.

Fensham, R.J., 1998: The influence of cattle grazing on tree mortality after drought in savanna woodland in north Queensland. *Australian Journal of Ecology,* **23,** 405–407.

Finér, L. and J. Laine, 1998: Fine root dynamics at drained peatland sites of different fertility in southern Finland. The net impact may be enhanced sequestration rate. *Plant and Soil,* **201,** 27–36.

Finnveden, G. and T. Ekvall, 1998: Life-cycle assessment as a decision-support tool—the case of recycling versus incineration of paper. *Resources, Conservation and Recycling,* **24,** 235–256.

Fisher, M.J., R.J. Thomas, and I.M. Rao, 1997: Management of tropical pastures in acid-soil savannas of South America for carbon sequestration in the soil. In: *Management of Carbon Sequestration in Soil* [Lal, R., J.M. Kimble, R.F. Follett, and B.A. Stewart (eds.)]. CRC Press, Boca Raton, FL, USA, pp. 405–420.

Fisher, M.J., I.M. Rao, C.E. Lascano, M.A. Ayarza, J.I. Sanz, R.J. Thomas, and R.R. Vera, 1995: Pasture soils as carbon sink. *Nature,* **376(10),** 473.

Fisher, M.J., I.M. Rao, M.A. Ayarza, C.E. Lascano, J.I. Sanz, R.J. Thomas, and R.R. Vera, 1994: Carbon storage by introduced deep-rooted grasses in the South American savannas. *Nature,* **371(15),** 236–238.

Flach, K.W., T.O. Barnwell, and P. Crosson, 1997: Impacts of agriculture on atmospheric carbon dioxide. In: *Soil Organic Matter in Temperate Agroecosystems: Long-Term Experiments in North America* [Paul, E.A., E.T. Elliott, K. Paustian, and C.V. Cole (eds.)]. CRC Press, Boca Raton, FL, USA, pp. 3–13.

Follett, R.F., J.M. Kimble, and R. Lal (eds.), 2000: *Potential for Carbon Sequestration in U.S. Grazingland Soils.* Ann Arbor Press, Chelsea, MI, USA, (in press).

Frank, A.B., D.L. Tanaka, L. Hofmann, and R.F. Follett, 1995: Soil carbon and nitrogen of Northern Great Plains grasslands as influenced by long-term grazing. *Journal of Range Management,* **48,** 470–474.

Franzluebbers, A.J., J.A. Stuedemann, H.H. Schomberg, and S.R. Wilkinson, 2000: Soil organic C and N pools under long-term pasture management in the Southern Piedmont USA. *Soil Biology and Biochemistry,* (in press).

Franzluebbers, A.J., F.M. Hons, and D.A. Zuberer, 1998: In situ and potential CO_2 evolution from a Fluventic Ustochrept in southcentral Texas as affected by tillage and cropping intensity. *Soil and Tillage Research,* **47,** 303–308.

Friday, K.S., M.E. Drilling, and D.P. Garrity, 1999: *Imperata* grasslands rehabilitation using agroforestry and assisted natural regeneration. International Centre for Agroforestry, Bogor, Indonesia. 167 pp.

Fujisaka, S., L. Hurtado, and R. Uribe, 1996: A working classification of slash-and-burn agricultural systems. *Agroforestry Systems,* **34(2),** 151–169.

Fullen, M.A. and D.J. Mitchell, 1994: Desertification and reclamation in North-Central China. *Ambio,* **23(2),** 131–135.

Fung, I., J. John, J. Lerner, E. Matthews, M. Prather, L.P. Steele, and P.J. Fraser, 1991: 3-dimensional model synthesis of the global methane cycle. *Journal of Geophysical Research-Atmospheres,* **96(7),** 13033–13065.

Gachengo, C.N., C.A. Palm, B. Jama, and C. Othieno, 1999: *Tithonia* and *Senna* green manures and inorganic fertilizers as phosphorus sources for maize in Western Kenya. *Agroforestry Systems,* **44,** 21–36.

Gale, W.J. and C.A. Cambardella, 1999: Carbon dynamics of surface residue and root-derived organic matter under simulated no till. *Journal of the Soil Science Society of America,* (in press).

Galy-Lacaux, C., R. Delmas, C. Jambert, J.-F. Dumestre, L. Labroue, S. Richard, and P. Gosse, 1997: Gaseous emissions and oxygen consumption in hydroelectric dams: A case study in French Guyana. *Global Biogeochemical Cycles,* **11,** 471–483.

Gardner, J.L., 1950: Effects of thirty years of protection from grazing in desert grassland. *Ecology,* **31,** 44–50.

Garg, V.K., 1998: Interaction of tree crops with a sodic soil environment: Potential for rehabilitation of degraded environments. *Land Degradation Development,* **9,** 81–93.

Garrity, D.P. (ed.), 1997, Agroforestry innovations for *Imperata* grassland rehabilitation. *Agroforestry Systems,* **36,** 1–277.

Gaudinski, J.B., S.E. Trumbore, E.A. Davidson, and S. Zheng, 2000: Soil carbon cycling in a temperate forest: radiocarbon-based estimates of residence times, sequestration rates, and partitioning of fluxes. *Biogeochemistry,* (in press).

Gebhart, D.L., H.B. Johnson, H.S. Mayeux, and H.W. Polley, 1994: The CRP increases soil organic carbon. *Journal of Soil and Water Conservation,* **49(5),** 488–492.

Gillies, M.D. and D.G. Leckie, 1996: Forest inventory update in Canada. *The Forestry Chronicle,* **72,** 138–156.

Gillison, A.N. (coordinator), 1999: *Above-Ground Biodiversity Assessment Working Group Summary Report 1996–1998. Alternatives to Slash and Burn Project. Phase II.* International Centre for Agroforestry, Nairobi, Kenya, pp. 4–13, (in press)

Gislason, S.R., S. Srnorsson, and H. Armannson, 1996: Chemical weathering of basin in SW Iceland: Effects of runoff, age of rocks and vegetative/glacial cover. *American Journal of Science,* **296,** 837–907.

Glenn, E., V. Squires, M. Olsen, and R. Frye, 1993: Potential for carbon sequestration in the drylands. *Water, Air and Soil Pollution,* **70,** 341–355.

Glenn, E.P., J.W. O'Leary, M.C. Watson, T.L. Thompson, and R.O. Kuehl, 1991: *Salicornia bigelovii* Torr.: An oilseed halophyte for seawater irrigation. *Science,* **251,** 1065–1067.

Gokowski, J., B. Nkamelu, and K. Wendt, 1999: Implications of resource use intensification for environment and sustainable technology systems in the Central African rainforest. In: *Agricultural Intensification and the Environment* [Barrett, C.R. and D.R. Lee (eds.)]. Blackwell Science Publishers, Oxon, United Kingdom, (in press).

Goldammer, J.G. (ed.), 1990: *Fire in the Tropical Biota: Ecosystem Processes and Global Challenges*. Ecological Studies 84, Springer-Verlag, Berlin, Germany.

Gorham, E., 1995: The biogeochemistry of northern peatlands and its possible responses to global warming. In: *Biotic Feedbacks in the Global Climatic System* [Woodwell, G.M. and F.T. MacKenzie (eds.)]. Oxford University Press, New York. pp. 169–187.

Govers, G., K. Vandale, P.J.J. Desment, J. Poesen, and K. Bunte, 1994: The role of tillage in soil distribution on hillslopes. *European Journal of Soil Science,* **45,** 469–478.

Grace, P.R., W.M. Post, D.C. Godwin, K.P. Bryceson, M.A. Truscott, and K.J. Hennessy, 1997: Soil carbon dynamics in relation to soil surface management and cropping systems in Australian agroecosystems. In: *Management of Carbon Sequestration in Soil* [Lal, R., J.M. Kimble, R.F. Follett, and B.A. Stewart (eds.)]. CRC Press, Boca Raton, FL, USA, pp. 175–193.

Graham, R.L., L.L. Wright, and A.F. Turhollow, 1992: The potential for short-rotation woody crops to reduce U.S. CO_2 emissions. *Climatic Change,* **22,** 223–238.

Grainger, A., 1993: *Controlling Tropical Deforestation.* Earthscan Publications, London, United Kingdom.

Grant, R.F., R.C. Izaurralde, M. Nyborg, S.S. Malhi, E.D. Soberg, and D. Jans-Hammermeister, 1997: In: *Soil Processes and the Carbon Cycle* [Lal, R., J.M. Kimble, R.F. Follett, and B.A. Stewart (eds.)]. CRC Press, Boca Raton, FL, USA, pp. 527–547.

Greenland, D.J., 1995: Land use and soil carbon in different agroecological zones. In: *Soil Management and Greenhouse Effect* [Lal, R., J. Kimble, E. Levine, and B.A. Stewart (eds.)]. Lewis Publishers, Boca Raton, FL, USA, pp. 9–24.

Greenland, D.J., 1985: Nitrogen and food production in the tropics: Contributions from fertilizer nitrogen and biological nitrogen fixation. In: *Nitrogen Management in Farming Systems in Humid and Sub-Humid Tropics* [Kang, B.T. and J. van der Heide (eds.)]. Proceedings of an International Institute of Tropical Agriculture (IITA) Symposium, Ibadan, Nigeria, pp. 9–38.

Greenpeace, 1993: *Towards a Fossil Free Energy Future: The Next Energy Transition.* A technical analysis for Greenpeace International by the Stockholm Environmental Institute Boston Center, Boston, MA, USA.

Guggenberger, G., W. Zech, and R.J. Thomas, 1995: Lignin and carbohydrate alteration in particle-size separates of an oxisol under tropical pastures following native savanna. *Soil Biology and Biochemistry,* **27(12),** 1629–1638.

Gupta, R.K., and I.P. Abrol, 1990: Salt-affected soils: their reclamation and management. In: *Soil Degradation* [Lal, R. and B.A. Stewart (eds.)]. *Advanced Soil Science,* **5,** 223–287.

Gupta, R.K. and D.L.N. Rao, 1994: Potential of wastelands for sequestering carbon by reforestation. *Current Science,* **66,** 378–380.

Hall, D.O., 1991: Biomass energy. *Energy Policy,* **19,** 711–731.

Hall, D.O., J.I. House, and J.I. Scrase, 2000: Introduction: Overview of biomass energy. In: *Industrial Uses of Biomass Energy* [Rosillo-Calle, F., S. Bajay, and H. Rotman (eds.)]. Taylor and Francis, London, UK, (in press).

Hall, D.O., H.E. Mynick, and R.H. Williams, 1991: Cooling the greenhouse with bioenergy. *Nature,* **353,** 11–12.

Hall, D.O., F. Rosillo-Calle, R.H. Williams, and J. Woods, 1992: Biomass for energy: supply prospects. In: *Renewable Energy: Sources for Fuels and Electricity* [Johansson, T.B., H. Kelly, A.K.N. Reddy, and R.H. Williams, (eds.)]. Island Press, Washington, DC., USA, pp. 593–651.

Hall, W.B., G.M. McKeon, J.O. Carter, K.A. Day, S.M. Howden, and J.C. Scanlan, 1998: Climate change in Queenslands grazing lands: II. An assessment of the impact on animal production from native pastures. *Rangeland Journal,* **20,** 177–205.

Haque, A.K.E., P. Read, and M.E. Ali, 1999: *The Bangladesh MSP Pilot Project Proposal for GEF Funding of Capacity Building for Country Driven Projects.* Working Paper, Institute of Development, Environment, and Strategic Studies (IDESS), North-South University, Dhaka, Bangladesh.

Harrison, K.G., 1996: Using bulk soil radiocarbon measurements to estimate soil organic matter turnover times: Implications for atmospheric CO_2 levels. *Radiocarbon,* **38,** 181–190.

Hassett, R.C., H.L. Wood, J.O. Carter, and T.J. Danaher, 1999: Statewide ground-truthing of a spatial model. In: *People and Rangelands: Building the Future. Proceedings of the Sixth International Rangeland Congress* [Eldridge, D. and D. Freeudenberger (eds.)]. Australian Academy of Science, Townsville, Australia, July 1999, Vol. II, pp. 763–764.

Haynes, R.J. and P.H. Williams, 1993: Nutrient cycling and soil fertility in grazed pasture ecosystems. *Advances in Agronomy,* **49,** 119–199.

Haynes, R.J. and P.H. Williams, 1992: Accumulation of soil organic matter and the forms, mineralization potential and plant-availability of accumulated organic sulfur: effects of pasture improvement and intensive cultivation. *Soil Biology and Biochemistry,* **24,** 209–217.

Hebblethwaite, J. and D. Towery, 1998: World wide trends in no-till farming—competing with the competition. Information transfer document Conservation Tillage Information Center (CTIC), West Lafayette, IN, USA.

Helyar, K.R., B.R. Cullis, K. Furniss, G.D. Kohn, and A.C. Taylor, 1997: Changes in the acidity and fertlity of a red earth soil under wheat-annual pasture rotations. *Australian Journal of Agricultural Research,* **48,** 561–586.

Hillel, D., 1997: *Small-Scale Irrigation for Arid Zones: Principles and Options.* FAO Development Series 2, Food and Agriculture Organization of the United Nations, Rome, Italy, 66 pp.

Hobbelstad, K., 1999: *Accuracy of the Norwegian Forest Inventory. Note.* Norway's Institute of Land Use Surveys. (in Norwegian)

Hoen, H.F. and B. Solberg, 1994: Potential and economic efficiency of carbon seqestration in forest biomass through silvicultural management. *Forest Science,* **40,** 429–451.

Hoen, H.F. and B. Solberg, 1997: CO_2-taxing, timber rotations, and market implications. *Economics of Carbon Sequestration in Forestry. Critical Reviews in Environmental Science and Technology,* **27,** 151–162.

Hori, K., K. Inubushi, S. Matsumoto, and H. Wada, 1990: Competition for acetic acid between methane formation and sulfate reduction in the paddy soil. *Japanese Journal of Soil Science and Plant Nutrition,* **61,** 572–578. (in Japanese with English summary)

Houghton, J., 1998: Introduction. In: *Workshop Report.* IPCC Workshop on Rapid Non-Linear Climate Change, Noordwijkerhout, Netherlands, March 1997, Hadley Centre, Meterological Office, Bracknell, United Kingdom, pp. 3–4

Houghton, R.A., 1995: Changes in the storage of terrestrial carbon since 1850. In: *Soils and Global Change* [Lal, R., J. Kimble, E. Levine, and B.A. Steward (eds.)]. CRC-Lewis Publishers, Boca Raton, FL, pp. 45–65.

Houghton, R.A., 1997: Terrestrial carbon storage: Global lessons from Amazonian research. *Ciencia e Cultura,* **49,** 58–72.

Houghton, R.A., 1996: Land-use change and terrestrial carbon: the temporal record. In: *Forest Ecosystems, Forest Management and the Global Carbon Cycle* [Apps, M.J., and D.T. Price (eds.)]. NATO ASI Series, Vol. I-40. Springer Group 1998, Berlin, Germany, pp. 117–134.

Howden, S.M., D.H. White, G.M. McKeon, J.C. Scanlan, and J.O. Carter, 1994: Methods for exploring management options to reduce greenhouse gas emissions from tropical pastures. *Climatic Change,* **30,** 49–70.

Howden, S.M., J.L. Moore, G.M. McKeon, P.J. Reyenga, J.O. Carter, and J.C. Scanlan, 1999: Dynamics of mulga woodlands in south-west Queensland: global change impacts and adaptation. In: *Modsim '99 International Congress on Modelling and Simulation Proceedings,* December 6–9, 1999, Hamilton, New Zealand, pp. 637–642.

Huggins, D.R., G.A. Buyanovsky, G.H. Wagner, J.R. Brown, R.G. Darmody, T.R. Peck, G.W. Lesoing, M.B. Vanotti, and L.G. Bundy, 1998a: Soil organic C in the tallgrass prairie-derived region of the corn belt: effects of long-term crop management. *Soil and Tillage Research,* **47,** 219–234.

Huggins, D.R., D.L. Allan, J.C. Gardner, D.L. Karlen, D.F. Bezdicek, M.J. Rosek, M.J. Alms, M. Flock, and M.L. Staben, 1998b: Enhancing carbon sequestration in CRP-managed lands. In: *Management of Carbon Sequestration in Soil* [Lal, R., J. Kimble, R.F. Follet, and B.A. Stewart (eds.)]. CRC Press, Boca Raton, FL, USA, pp. 323–334.

Huxley, P., 1999: *Tropical Agroforestry.* Blackwell Science Publishers, London, United Kingdom, 371 pp.

IEA, 1998: *World Energy Outlook: 1998 Edition*. International Energy Agency/Organization for Economic and Commercial Development, Paris, France.

IGBP Terrestrial Carbon Working Group, 1998: The terrestrial carbon cycle: Implications for the Kyoto Protocol. *Science,* **280,** 1393–1394.

IPCC, 1992: Climate Change 1992: *The Supplementary Report to the IPCC Scientific Assessment* [Houghton, J.T. (ed.)]. Cambridge University Press, New York, NY, USA, pp. 25–47.

IPCC, 1994: *Radiative Forcing of Climate Change, The 1994 Report of the Scientific Assessment Working Group of IPCC, Summary for Policymakers*. World Meteorological Organization/United Nations Environmental Programme, Geneva, Switzerland, 28 pp.

IPCC, 1996a: *Climate Change 1995: The Science of Climate Change. Contribution of Working Group I to the Second Assessment Report of the Intergovernmental Panel on Climate Change* [Houghton, J.T., L.G. Meira Filho, B.A. Callander, N. Harris, A. Kattenberg, and K. Maskell (eds.)]. Cambridge University Press, Cambridge, United Kingdom and New York, NY, USA, 572 pp.

IPCC, 1996b: Climate change impacts on forests. In: *Climate Change 1995: Impacts, Adaptations and Mitigation of Climate Change: Scientific-Technical Analyses. Contribution of Working Group II to the Second Assessment Report of the Intergovernmental Panel on Climate Change* [Watson, R.T., M.C. Zinyowera, and R.H. Moss (eds.)]. Cambridge University Press, Cambridge, United Kingdom, and New York, NY, USA, 879 pp.

IPCC, 1997: *Revised 1996 IPCC Guidelines for National Greenhouse Gas Inventories* [J.T. Houghton, L.G. Meira Filho, B. Lim, K. Tréanton, I. Mamaty, Y. Bonduki, D.J. Griggs, and B.A. Callander (eds.)]. Intergovernmental Panel on Climate Change, Meteorological Office, Bracknell, United Kingdom.
- *Volume 1: Greenhouse Gas Inventory Reporting Instructions*. 130 pp.
- *Volume 2: Greenhouse Gas Inventory Workbook*. 346 pp.
- *Volume 3: Greenhouse Gas Inventory Reference Manual*. 482 pp.

Izac, A.-M.N. and P.A. Sanchez, 2000: Towards a natural resource management research paradigm: an example of agroforestry research. *Agricultural Systems,* (in press).

Jama, B., C.A. Palm, R.J. Buresh, A.I. Niang, C. Gachengo, G. Nziguheba, and B. Amadalo, 1999a: *Tithonia diversifolia* green manure for improvement of soil fertility: a review from western Kenya. *Agroforestry Systems,* (in press).

Jama, B., R.J. Buresh, and F.M. Place, 1999b: Sesbania tree fallows in phosphorus deficient sites: Maize yield and financial benefit. *Agronomy Journal,* **90,** 717–726.

Janzen, H.H., C.A. Campbell, R.C. Izaurralde, B.H. Ellert, N. Juma, W.B. McGill, and R.P. Zentner, 1998: Management effects on soil C storage on the Canadian prairies. *Soil Tillage Research,* **47,** 181–195.

Jenkinson, D.S., 1991: The Rothamsted long-term experiments: are they still of use? *Journal of Agronomy,* **83,** 2–12.

Johnson, M.G., 1995: The role of soil management in sequestering soil carbon. In: *Soil Management and Greenhouse Effect* [Lal, R., J. Kimble, E. Levine, and B.A. Stewart (eds.)]. Lewis Publishers, Boca Raton, FL, USA, pp. 351–363.

Johnston, R. M. and M.M. Barson, 1993: Remote sensing of wetlands. *Australian Journal of Marine and Freshwater Research,* **44,** 235–252.

Jones, C.L., N.L. Smithers, M.C. Scholes, and R.C. Scholes, 1990: The effect of fire frequency on the organic components of a baslatic soil in the Kruger National Park. *South African Journal of Plant and Soil,* **7(4),** 236–238.

Justice, C.O., J.D. Kendall, P.R. Dowty, and R.J. Scholes, 1996: Satellite remote sensing of fires during SAFARI 92 campaign using NOAA advanced very high resolution radiometer data. *Journal of Geophysical Research,* **101,** 23851–23864.

Kägi, W., 1997: Carbon store protection: how to determine the value of JI projects. In: *Greenhouse Gas Mitigation: Technologies for Activities Implemented Jointly* [Riemer, P.W.F., A.Y Smith, and K.V. Thambimuthu (eds.)]. Elsevier Science Publishers, Amsterdam, The Netherlands, pp. 371–378.

Kaimowitz, D., and A. Anglesen, 1998: *Economic Models of Tropical Deforestation: A Review*. Center for International Forestry Research, Bogor, Indonesia, 139 pp.

Kapkiyai, J.J., N.K. Karanja, P. Woomer, and J.N. Qureshi, 1998: Soil organic carbon fractions in a long-term experiment and the potential for their use as a diagnostic assay in highland farming systems of central Kenya. *African Crop Science Journal,* **6(1),** 19–28.

Kasimir-Klemedtsson, A., L. Klemedtsson, K. Bergelund, P. Martikained, J. Silvola, and O. Ocenema, 1997: Greenhouse gas emissions from farmed organic soils: a review. *Soil Use and Management,* **13,** 245–250.

Kauffman, J.B., D.L. Cummings, and D.E. Ward, 1994: Relationships of fire, biomass and nutrient dynamics along a vegetation gradient in the Brazilian cerrado. *Journal of Ecology,* **82,** 519–531.

Keeling, R.F., S.C. Piper, and M. Heimann, 1996: Global and hemispheric CO_2 sinks deduced from changes in atmospheric O_2 concentration. *Nature,* **381,** 218–221.

Keller, A.A. and R.A. Goldstein, 1998: Impact of carbon storage through restoration of drylands on the global carbon cycle. *Environmental Management,* **22(5),** 757.

Kelly, R.H., I.C. Burke, and W.K. Lauenroth, 1996: Soil organic matter and nutrient availability responses to reduced plant inputs in shortgrass steppe. *Ecology,* **77(8),** 2516–2527.

Kelly, C.A., J.W.M. Rudd, R.A. Bodaly, N.P. Roulet, V.L. St. Louis, A. Heyes, T.R. Moore, S. Schiff, R. Aravena, and K.J. Scott, 1997: Increases in fluxes of greenhouse gases and methyl mercury following flooding of an experimental reservoir. *Environmental Science Technology,* **31,** 1334–1344.

Kern, J.S. and M.G. Johnson, 1993: Conservation tillage impacts on national soil and atmosperic carbon levels. *Journal of the Soil Society of America,* **57,** 200–210.

Kielbaso, J.J. and V. Cotrone, 1990: The state of the urban forest, In: *Make Our Cities Safe for Trees: Proceedings of the Fourth Urban Forestry Conference* [Rodbell, P.D. (ed.)]. American Forests, Washington, DC, USA, pp. 11–18.

Kimura, M., J. Tanaka, H. Wada, and Y. Takai, 1980: Quantitative estimation of decomposition process of plant debris in the paddy field (part 1). *Japanese Journal of Soil Science,* **51,** 169–174. (in Japanese with English summary)

Kitada, K., Y. Ozaki, Y. Akiyama, and K. Yagi, 1993: Effects of high content of NO_3–N in irrigation water and rice straw application on CH_4 emission from paddy fields. *Japanese Journal of Soil Science and Plant Nutrition,* **64,** 49–54. (in Japanese with English summary)

Knapp, A.K., S.L. Conrad, and J.M. Blair, 1998: Determinants of soil CO_2 flux from a sub-humid grassland: effect of fire and fire history. *Ecological Applications,* **8(3),** 760–770.

Koizumi, H., T., Kibe, T. Nakadai, Y. Bekku, Y. Tang, S. Nishimura, H. Kawashima, K. Kobayashi, and S. Mariko, 1998: Carbon dynamics and budgets in upland and paddy agricultural ecosystems in monsoon east Asia. *INTECOL,* **7,** 232.

Krankina, O.N., M.E. Harmon, and J.W. Winjum, 1996: Carbon storage and sequestration in the Russian forest sector. *Ambio,* **25(4),** 284–288.

Krankina, O.N. and M.E. Harmon, 1994: The impact of intensive forest management on carbon stores in forest ecosystems. *World Resource Review,* **6(2),** 161–177.

Kreutzer, K., 1972: Über den Einfluß der Streunutzung auf den Stickstoffhaushalt von Kieferbeständen. *Forstw. Centralblatt,* 263–270.

Kurihara, M., T. Magner, R.A. Hunter, and G.J. McCrabb, 1999: Methane production and energy partitioning of cattle in the tropics. *British Journal of Nutrition,* **81,** 263–272.

Kurz, W.A., M. Apps, T.M. Webb, and P.J. McNamee, 1992: *The Carbon Budget of the Canadian Forest Sector: Phase 1: Information Report*. Technical Report NOR-X-326, Forestry Canada, Northern Forestry Centre, Edmonton, Alberta, Canada, 93 pp.

Kwesiga, F. and R. Coe, 1994: The effect of short rotation *Sesbania sesban* planted fallows on maize yields. *Forest Ecology and Management,* **64,** 199–208.

Kwesiga, F.R., S. Franzel, F. Place, D. Phiri, and C.P. Simwanza, 1999: *Sesbania sesban* improved fallows in Eastern Zambia: Their inception, development and farmer enthusiasm. *Agroforestry Systems,* **47,** 49–66.

Lal, R., 1989: Conservation tillage for sustainable agriculture. *Advances in Agronomy,* **42,** 85–197.

Lal, R., 1997: Residue management conservation tillage and soil restoration for mitigating greenhouse effect by CO_2-enrichment. *Soil and Tillage Research,* **43,** 81–107.

Lal, R., 2000a: Soil aggregation and carbon sequestration. In: *Global Climate Change and Tropical Ecosystems* [Lal, R., J.M. Kimble, and B.A. Stewart (eds.)]. CRC-Lewis Publishers, Boca Raton, FL, USA, pp. 317–329.

Lal, R., 2000b: Soil management in the developing countries. *Soil Science,* (in press).

Lal, R. and J.P. Bruce, 1999: The potential of world cropland soils to sequester C and mitigate the greenhouse effect. *Environmental Science and Policy,* **2,** 177–185.

Lal, R. and J.M. Kimble, 2000: Tropical ecosystems and the global carbon cycle. In: *Global Climate Change and Tropical Ecosystems* [Lal, R., J.M. Kimble, and B.A. Stewart (eds.)]. CRC-Lewis Publishers, Boca Raton, FL, USA, pp. 3–32.

Lal, R., J. Kimble, R. Follett, and C.V. Cole, 1998: *Potential of U.S. Cropland for Carbon Sequestration and Greenhouse Effect Mitigation.* Sleeping Bear Press, Chelsea, MI, USA, 128 pp.

Lal, R., H.M. Hassan, and J. Dumanski, 1999a: Desertification control to sequester C and mitigate the greenhouse effect. In: *Carbon Sequestration Soils, Science Monitoring and Beyond* [Rosenberg, N.J., R.C. Izaurralde, and E.L. Malone (eds.)]. Battelle Press, Columbus, OH, USA, pp. 83–130.

Lal, R., J.M. Kimble, and R.F. Follett, 1999b: Agricultural practices and policies for carbon sequestration in soil. *Recommendation and Conclusions of the International Symposium,* 19–23 July 1999, Columbus, OH, USA, 12 pp.

Lal, R., R.F. Follett, J. Kimble, and C.V. Cole, 1999c: Managing U.S. cropland to sequester carbon in soil. *Journal of Soil and Water Conservation,* **54,** 374–381.

Lal, R., J.M. Kimble, H. Eswaran, and B.A. Stewart (eds.), 2000: *Global Climate Change and Pedogenic Carbonates.* CRC-Lewis Publishers, Boca Raton, FL, USA, 305 pp.

Larson, E.D. and H. Jin, 1999: Biomass conversion to Fischer-Tropsch liquids: Preliminary energy balances. In: *Proceedings of the 4th Biomass Conference of the Americas, Oakland, CA.* Elsevier Science Publishers, Oxford, United Kingdom.

Lassoie, J.P. and L.E. Buck (eds.), 1999: Special issues: Exploring the opportunities for agroforestry in changing rural landscapes. *Agroforestry Systems,* **44,** 105–353.

Leakey, R.R.B. and P.A. Sanchez, 1997: How many people use agroforestry products? *Agroforestry Today,* **9(3),** 4–5.

Leakey, R.R.B., 1996: Definition of agroforestry revisited. *Agroforestry Today,* **8(1),** 5–7.

Leakey, R.R.B., A.B. Temu, M. Melnyk, and P. Vantomme (eds.), 1996: Domestication and commercialization of non-timber forest products in agroforestry systems. In: *Non-Wood Forest Products,* vol. 9, Food and Agriculture Organization of the United Nations, Rome, Italy.

Leenhouts, B., 1998: Assessment of biomass burning in the conterminous United States. *Conservation Ecology,* **2(1),** 1–24. (also available at: http://www.consecol.org.

Leuking, M.A. and J.S. Schepers, 1985: Changes in soil carbon and nitrogen due to irrigation development in Nebraska's sand hill soils. *Journal of the Soil Science Society of America,* **49,** 626–630.

Li, L.H., Z.Z. Chen, Q.B. Wang, X.H. Liu, and Y.H. Li, 1997: Changes in soil carbon storage due to over-grazing in *Leymus chinensis* steppe in the Xilin river basin of Inner Mongolia. *Journal of Environmental Science,* **9,** 486–490.

Li, Z. and Q. Zhao, 1998: Carbon dioxide fluxes and potential mitigation in agriculture and forestry of tropical and subtropical China. *Climatic Change,* **40,** 119–133.

Lindau, C.W., P.K. Bollich, R.D. DeLaune, A.R. Mosier, and K.F. Bronson, 1993: Methane mitigation in flooded Louisiana rice fields. *Biology and Fertility of Soils,* **15,** 174-178.

Lindstrom, M.J., T.E. Schumacher, N.P. Cogo, and M.L. Blecha, 1998: Tillage effects on water runoff and soil erosion after sod. *Journal of Soil and Water Conservation,* **53(1),** 59–63.

Lindstrom, M.J., W.W. Nelson, T.E. Schumacher, and G.D. Lemme, 1990: Soil movement by tillage as affect by slope. *Soil and Tillage Research,* **17,** 225–264.

Lindstrom, M.J., W.W. Nelson, and T.E. Schumacher, 1992: Quantifying tillage erosion rates due to moldboard plowing. *Soil and Tillage Research,* **24,** 243–255.

Lindstrom, M.J., 1997: Soil movement by tillage as influenced by variations in slope gradients. *Journal of Soil and Water Conservation,* **52(4),** 304.

Lindstrom, M.J., T.E. Schumacher, D.C. Reicosky, and D.L. Beck, 1999: Soil quality: Post conservation reserve changes with tillage and cropping. In: *Soil Quality and Soil Erosion* [Lal, R. (ed.)]. CRC Press, Boca Raton, FL, USA, pp. 143–151.

Lobb, D.A., R.G. Kachanoski, and M.H. Miller, 1995: Tillage translocation and tillage erosion on shoulder slope landscape positions measured using 137 cesium as a tracer. *Canadian Journal of Soil Science,* **75,** 211–218.

Loveland, T.R., and A.S. Belward, 1997: The IGBP-DIS global 1 km land cover data set, DISCover: first results. *International Journal of Remote Sensing,* **18,** 3289–3295.

Lugo, A.E., M.J. Sanchez, and S. Brown, 1986: Land use and organic carbon content of some subtropical soils. *Plant and Soil,* **96,** 185–196.

Lugo, A.E. and S. Brown, 1996: Management of tropical soils as sinks or sources of atmospheric carbon. *Plant and Soil,* **149,** 27–41.

Lunnan, A., S. Navrud, P.K. Rorstad, K. Simonsen, and B. Solberg, 1991: Forestry and forest production in Norway as a measure against CO_2-accumulation in the atmosphere. *Aktuelt fra Skogforsk,* no. 6-1991, Institutt for skogfag, NLH, Norway, 67 pp. + appendices. (in Norwegian)

Luo, Y., 1991: Review and outlook on the the extensive application and development of reduced tillage and zero-tillage technologies in China. *Chinese Journal of Tillage and Cultivation,* **11(2),** 1–7. (in Chinese)

MacDicken, K.G., 1997: *A Guide to Monitoring Carbon Storage in Forestry and Agroforestry Projects.* Winrock International Institute for Agricultural Development, Arlington, VA, USA, 87 pp.

MacDonald, L.H., R. Sampson, D. Brady, L. Juarros, and D. Martin, 2000: Predicting post-fire erosion and sedimentation risk on a landscape scale: A case study from Colorado. *Journal of Sustainable Forestry,* **11(1-2),** 57–88.

MacDonald, N.W., D.L. Randlett, and D.R. Zak, 1999: Soil warming and carbon loss from a Lake States Spodosol. *Soil Science Society of America Journal,* **63,** 211–218.

Magenheimer, J.F., T.R. Moore, G.L. Chmura, and R.J. Daoust, 1996: Methane and carbon dioxide flux from a macrotidal salt marsh. Bay of Fundy, New Brunswick. *Estuaries,* **19,** 139–145.

Mahar, D.J., 1989: *Government Policies and Deforestation in Brazil's Tropical Rain Forest.* Kiel Working Paper No. 565, The Kiel Institute of World Economics, Kiel, Germany.

Maltby, E. and P. Immirzi, 1993: Carbon dynamics in peatlands and other wetland soils: regional and global perspectives. *Chemosphere,* **27,** 999–1023.

Manley, J.T., G.E. Schuman, J.D. Reeder, and R.H. Hart, 1995: Rangeland soil carbon and nitrogen responses to grazing. *Journal of Soil and Water Conservation,* **50(3),** 294–298.

Mann, M.K., and P.L. Spath, 1997: *Life Cycle Assessment of a Biomass Gasification Combined-Cycle System.* Technical Report NREL/TP-430-23076, National Renewable Energy Laboratory, Golden, CO, USA.

Mann, L.K., 1986: Changes in soil carbon storage after cultivation. *Soil Science,* **142,** 279–288.

Marklund, L.G., 1988: *Biomass Functions for Pine, Spruce, and Birch in Sweden.* Department of Forest Survey Report 45, Swedish University of Agricultural Sciences, Umeå, Sweden, 15 pp. + appendices (in Swedish with English summary).

Marland, G. and B. Schlamadinger, 1997: Forests for carbon sequestration or fossil fuel substitution? A sensitivity analysis. *Biomass and Bioenergy,* **13,** 389–397.

Marland, G. and S. Marland, 1992: Should we store carbon in trees? *Water, Air, and Soil Pollution,* **64,** 181–195.

Matthews, E. and I. Fung, 1987: Methane emission from natural wetlands: global distribution, area, and environmental characteristics of sources. *Global Biogeochemical Cycles,* **1,** 61–86.

Matthews, R., 1996: The influence of carbon budget methodology on assessments of the impact of forest management on the carbon balance. In: *Forest Ecosystems, Forest Management and the Global Carbon Cycle* [Apps, M.J. and D.T. Price (eds.)]. NATO ASI Series, Vol. I, No. 40, Springer-Verlag, Berlin, Germany, pp. 233–244.

Matthews, R.W., G.-J. Nabuurs, V. Alexeyev, R.A. Birdsey, A. Fischlin, J.P. Maclaren, G. Marland, and D.T. Price, 1996: Evaluating the role of Forest Management and Forest Products in the Carbon Cycle. In: *Forest Ecosystems and the Global Carbon Cycle* [Apps, M.J. and D.T. Price (eds.)]. NATO ASI Series, Vol. I-40, Springer-Verlag, Berlin and Heidelberg, Germany.

Matthews, R., R. Robinson, S. Abbott, and N. Fearis, 1994: *Modeling of Carbon and Energy Budgets of Wood Fuel Coppice Systems*. Technical Report ETSU B/W5/00337/REP, Mensuration Branch, Forestry Commission, for the Energy Technology Support Unit, Department of Trade and Industry, London, United Kingdom.

McConnell, S.G. and M.L. Quinn, 1988: Soil productivity of four land use systems in southeastern Montana. *Soil Science Society of America Journal,* **52,** 500–506.

McCown, R.L., 1996: Being realistic about no-tillage, legume ley farming for the Australian semi-arid tropics. *Australian Journal of Experimental Agriculture,* **36,** 1069–1080.

McGrath, S.P., 1998: Phytoextraction for soil remediation. In: *Plants that Hyperaccumulate Heavy Metals* [Brooks, R.R. (ed.)]. CAB International, Wallingford, United Kingdom, pp. 261–287.

McIntosh, P.D., A.E. Hewitt, K. Giddens, and M.D. Taylor, 1997a: Benchmark sites for assessing the chemical impacts of pastoral farming on loessial soils in southern New Zealand. *Agriculture Ecosystems and Environment,* **65,** 267–280.

McIntosh, P.D., R.B. Allen, and N. Scott, 1997b: Effects of exclosure and management on biomass and soil nutrient pools in seasonally dry high country, New Zealand. *Journal of Environmental Management,* **51,** 169–186.

McIntosh, P.D. and R.B. Allen, 1998: Effect of exclosure on soils, biomass, plant nutrients, and vegetation, on unfertilised steeplands, Upper Waitaki District, South Island, New Zealand. *New Zealand Journal of Ecology,* **22,** 209–217.

McKeon, G.M., S.M. Howden, N.O.J. Abel, and J.M. King, 1993: Climate change: adapting tropical and subtropical grasslands. In: *Grasslands for Our World* [Barker, M.J., (ed.)]. SIR Publishing, Wellington, New Zealand, pp. 426–435.

McPherson, E.G., D. Nowak, G. Heisler, S. Grimmond, C. Souch, R. Grant, and R. Rowntree: 1997: Quantifying urban forest structure, function, and value: The Chicago Urban Forest Climate Project. *Urban Ecosystems,* **1,** 49–61.

Michon, G. and H. de Foresta, 1996: Agroforests as an alternative to pure plantations for the domestication and commercialization of NTFPs. In: *Domestication and Commercialization of Non-Timber Forest Products for Agroforestry* [Leakey, R.R.B., A.B. Temu, and M. Melnyk (eds.)]. Non-Wood Forest Products, vol. 9, Food and Agriculture Organization of the United Nations, Rome, Italy, pp. 160–175.

Milchunas, D.G. and W.K. Laurenroth, 1993: Quantitative effects of grazing on vegetation and soils over a global range or environments. *Ecological Monographs,* **63(4),** 327–366.

Minami, K., 1994: Methane from rice production. *Fertilizer Research,* **37,** 167–180.

Minami, K., J. Goudriaan, E.A. Lantinga, and T. Kimura, 1993: The significance of grasslands in emission and absorption of greenhouse gases. In: *Proceedings of the International Grasslands Congress*. pp. 1231–1238.

Minami, K., A. Mosier, and R. Sass (eds.), 1994: *CH_4 and N_2O: Global Emissions and Controls from Rice Fields and Other Agricultural and Industrial Sources*. NIAES Series 2, YOKENDO Publishers, Tokyo, 234 pp.

Minkkinen, K. and J. Laine, 1998: Long-term effect of forest drainage on the peat carbon stores of pine mires in Finland. *Canadian Journal of Forest Research,* **28,** 1267–1275.

Mitsuchi, M., 1974: Characters of humus formed under rice cultivation. *Soil Science and Plant Nutrition*, 249–259.

Montes, L. and M. Masco, 1996: Alternative species to raise productivity and extend the grazing season in the tussock grasslands of south Patagonia. In: *Proceedings of the Fifth International Rangeland Congress* [West, N.E. (ed.)]. Vol. I. Society for Range Management, Denver, CO, and, Salt Lake City, UT, USA, pp. 383–384.

Moore, J.L., S.M. Howden, G.M. McKeon, J.O. Carter, and J.C. Scanlan, 1997: A method to evaluate greenhouse gas emissions from sheep grazed rangelands in south west Queensland. In: *Modsim '97 International Congress on Modelling and Simulation Proceedings, 8–11 December, University of Tasmania, Hobart* [McDonald, D.A. and M. McAleer (eds.)]. Modelling and Simulation Society of Australia, Canberra, Australia, pp. 137–142.

Moosavi, S.C. and P.M. Crill, 1997: Controls on CH_4 and CO_2 emissions along two moisture gradients in the Canadian boreal zone. *Journal of Geophysical Research,* **102,** 29, 261–229, 277.

Moser, M., C. Prentice, and S. Frazier, 1996: A global overview of wetland loss and degradation. In: *Technical Session B of the 6th Ramsar Conference of Parties (COP)*. UNFCCC, New York, NY, USA and Brisbane, Australia.

Mosier, A.R., J.M. Duxbury, J.R. Freney, O. Heinemeyer, K. Minami, and D.E. Johnson, 1998: Mitigating agricultural emissions of methane. *Climatic Change,* **40,** 39–84.

Murdiyarso, D., K. Hariah, and M. van Noordwijk (eds.), 1994: *Modeling and Measuring Soil Organic Dynamics and Greenhouse Gas Emissions After Forest Conversion*. ASB-Indonesia Report No. 1, International Centre for Agroforestry Southeast Asia Programme, Bogor, Indonesia, 118 pp.

Nabuurs, G.J., A.J. Dolman, E. Verkaik, A.P. Whitmore, W.P. Daamen, O. Oenema, P. Kabat, and G.M.J. Mohren, 1999: *Resolving the Issues on Terrestrial Biospheric Sinks in the Kyoto Protocol*. Report no. 410 200 030, Dutch National Research Programme on Global Air Pollution and Climate Change, Wageningen, Netherlands, 100 pp.

Nadelhoffer, K.J, B.A. Emmett, P. Gundersen, O.J. Kjønaas, C.J. Koopmans, P. Schleppi, A. Tietema, and R.F. Wright, 1999: Nitrogen deposition makes a minor contribution to carbon sequestration in temperate forests. *Nature,* **398,** 145–148.

Naeth, M.A., A.W. Bailey, D.J. Pluth, D.S. Chanasyk, and R.T. Hardin, 1991: Grazing impacts on litter and soil organic matter in mixed prairie and fescue grassland ecosystems of Alberta. *Journal of Range Management,* **44,** 7–12.

Nakicenovic, N., A. Grubler, and A. McDonald (eds.), 1998: *Global Energy Perspectives*. WEC-IIASA, Cambridge University Press, Cambridge, United Kingdom.

National Audubon Society, 1991: *Toward Ecological Guidelines for Large-Scale Biomass Energy Development*. National Audubon Society, New York, NY and Princeton University Center for Energy and Environmental Studies, Princeton, NJ, USA.

Neill, C., J.M. Melillo, P.A. Steudler, C.C Cerri, J.F.L. Moraes, M.C. Piccolo, and M. Brito, 1997: Soil carbon and nitrogen stocks following forest clearing for pasture in the southwestern Brazilian Amazon. *Ecological Applications,* **7(4),** 1216–1225.

Neuenschwander, L.F. and R.N. Sampson, 2000: A wildfire and emissions policy model for the Boise National Forest. *Journal of Sustainable Forestry,* **11(1-2),** 289–309.

Neuenschwander, L.F., J.P. Menakis, M. Miller, R.N. Sampson, C. Hardy, B. Averill, and R. Mask, 2000: Indexing Colorado watersheds to risk of wildfire. *Journal of Sustainable Forestry,* **11(1-2),** 35–55.

Nie, X., 1996: The water and soil conservation and Loess Plateau farming. *Journal of the Science and Technique for Shangxi Water and Soil Conservation*, 6-8. (in Chinese)

Niklinska, M., M. Maryanski, and R. Laskowski, 1999: Effect of temperature on humus respiration rate and nitrogen mineralization: Implications for global climate change. *Biogeochemistry,* **44,** 239–257.

Noble, A.D., M. Cannon, and D. Muller, 1997: Evidence of accelerated soil acidication under *Stylosanthes*-dominated pastures. *Australian Journal of Soil Research,* **35,** 1309–1322.

Nordt, L.C., L.P. Wilding, and L.R. Drees, 2000: Pedogenic carbonate transformations in leaching soil systems: Implications for the global carbon cycle. In: *Global Climate Change and Pedogenic Carbonates* [Lal, R., J.M. Kimble, H. Eswaran, and B.A. Stewart (eds.)]. CRC-Lewis Publishers, Boca Raton, FL, USA, pp. 43–64.

Northup, B.K. and J.R. Brown, 1999: Spatial distribution of soil carbon in grazed woodlands of dry tropical Australia: tussock and inter-tussock scales. In: *People and Rangelands: Building the Future. Proceedings of the Sixth International Rangeland Congress* [Eldridge, D. and D. Freeudenberger (eds.)]. Australian Academy of Science, Townsville, Australia, July 1999, Vol. II, pp. 120–121.

Nowak, D.J., 1991: Urban forest structure and the functions of hydrocarbon emissions and carbon storage. In: *Proceedings of the Fifth National Urban Forest Conference, Los Angeles, CA* [Rodbell, P.D. (ed.)]. American Forests, Washington, DC, USA, pp. 48–51.

Nowak, D.J., 1994: Air pollution removal by Chicago's urban forest. In: *Chicago's Urban Forest Ecosystem: Results of the Chicago Urban Forest Climate Project* [McPherson, E.G., D.J. Nowak, and R.A. Rowntree (eds.)]. General Technical Report No. NE-186, USDA Forest Service, Northeastern Forest Experiment Station, Radnor, PA, USA, pp. 63–82.

Nyasimi, M., A. Niang, B. Amadalo, and E. Obonyo, 1997: *Using the Wild Sunflower,* Tithonia, *in Kenya*. International Centre for Agroforestry, KARI, and KEFRI, Nairobi, Kenya, 12 pp., (available in English and Kiswahili).

Nziguheba, G., C.A. Palm, R.J. Buresh, and P.J. Smithson, 1998: Soil phosphorus fractions and adsorption as affected by organic and inorganic sources. *Plant and Soil,* **198,** 159–168.

Odum, E.P., 1969: The strategy of ecosystem development. *Science,* **164,** 262–270.

Ojima, D.S., W.J. Parton, D.S. Schimel, J.M.O. Scurlock, and T.G.F. Kittel, 1993a: Modeling the effects of climatic and CO_2 changes on grassland storage of soil C. *Water, Air and Soil Pollution,* **70,** 643–657.

Ojima, D.S., B.O.M. Dirks, E.P. Glenn, C.E. Owensby, and J.O. Scurlock, 1993b: Assessment of C budget for grasslands and drylands of the world. *Water, Air and Soil Pollution,* **70,** 95–109.

Oldeman, L.R., 1994: The global extent of soil degradation. In: *Soil Resilience and Sustainable Land Use* [Greenland, D.J. and I. Szabolcs (eds.)]. CAB International, Wallingford, United Kingdom, pp. 99–117.

Oldeman, L.R., V.W.P. van Engelen, and J.H.M. Pulles, 1991: The extent of human-induced soils degradation. In: *World Map of the Status of Human-Induced Soil Degradation: An Explanatory Note*.

Overend, R.P. and E. Chornet (eds.), 1999: *Proceedings of the 4th Biomass Conference of the Americas, Oakland, CA*. Elsevier Science Publishers, Oxford, United Kingdom.

Palm, C.A., R.J.K. Myers, and S.M. Nandwa, 1997: Combined use of organic and inorganic nutrient sources for soil fertility maintenance and replenishment. In: *Replenishing Soil Fertility in Africa* [Buresh, R.J., P.A. Sanchez, and F. Calhoun (eds.)]. SSSA Special Publication 51, Soil Science Society of America, Madison, WI, USA, pp. 193–218.

Palm, C.A., P.L. Woomer, J. Alegre, L. Arevalo, C. Castilla, D.G. Cordeiro, B. Feigl, K. Hariah, J. Kotto-Same, A. Mendes, A. Moukam, D. Murdiyarso, R. Njomgamg, W.J. Parton, A. Ricse, V. Rodrigues, S.M. Sitompul, and M. van Noordwijk, 2000: *Carbon Sequestration and Trace Gas Emissions in Slash and Burn and Alternative Land Uses in the Humid Tropics. ASB Climate Change Working Group Report. Final Report, Phase II*. International Centre for Agroforestry, Nairobi, Kenya, 37 pp.

Palo, M., 1999: No end to deforestation? In: *World Forests, Society and the Environment* [Palo, M. and J. Uusivuori (eds.)]. Kluwer Academic Publishers, Dordrecht, Netherlands, pp. 65–77.

Pan, W., 1999: Personal communication, Washington State University, Pullman, WA, USA.

Patterson, J., 1999: *Wetlands and Climate Change: Feasibility Investigation of Giving Credit for Conserving Wetlands as Carbon Sinks*. Wetlands international, Ottawa, Canada.

Paul, E.A., R.F. Follett, S.W. Leavitt, A.D. Halvorson, G.A. Peterson, and D.J. Lyon, 1997a: Radiocarbon dating for determination of soil organic matter pool sizes and dynamics. *Journal of the Soil Science Society of America,* **61,** 1058–1067.

Paul, E.A., K. Paustian, E.T. Elliott, and C.V. Cole (eds.), 1997b: *Soil Organic Matter in Temperate Agroecosystems: Long-Term Experiments in North America*. CRC Press, Boca Raton, FL, USA, 414 pp.

Paustian, K., O. Andren, H.H. Janzen, R. Lal, P. Smith, G. Tian, H. Tiessen, M. Van Noordwijk, and P.L. Woomer, 1997a: Agricultural soils as a sink to mitigate carbon dioxide emissions. *Soil Use and Management,* **13(4),** 230–244.

Paustian, K., H.P. Collins, and E.A. Paul, 1997b: Management controls on soil carbon. In: *Soil Organic Matter in Temperatate Agroecosystems: Long-Term Experiments in North America* [Paul, E.A., K. Paustian, E.T. Elliott, and C.V. Cole (eds.)]. CRC Press, Boca Raton, FL, USA, pp. 15–49.

Paustian, K., C.V. Cole, D. Sauerbeck, and N. Sampson, 1998a: CO_2 mitigation by agriculture: an overview. *Climatic Change,* **40,** 135–162.

Paustian, K., E.T. Elliot, and K. Killian, 1998b: Modeling soil carbon in relation to management and climate change in some egroecosystems in Central North America. In: *Soil Processes and the Carbon Cycle* [Lal, R., J.M. Kimble, R.F. Follet, and B.A. Stewart (eds.)]. CRC Press, London, United Kingdom.

Paustian, K., J. Six, E.T. Elliott, and H.W. Hunt, 2000a: Management options for reducing CO_2 emissions from agricultural soils. *Biogeochemistry,* **48,** 147–163.

Paustian, K., E.T. Elliott, K. Killian, J. Cipra, G. Bluhm, and J.L. Smith, 2000b: Modeling and regional assessment of soil carbon: A case study of the Conservation Reserve Program. In: *Soil Management for Enhancing Carbon Sequestration* [Lal, R. and K. McSweeney (eds.)]. SSSA Special Publication, Soil Science Society of America, Madison, WI, USA, (in press).

Peterson, G.A., A.D. Halvorson, J.L. Havlin, O.R. Jones, D.J. Lyon, and D.L. Tanaka, 1998: Reduced tillage and increasing cropping intensity in the Great Plains conserves soil C. *Soil and Tillage Research,* **47,** 207–218.

Pearce, D., 1990: *An Economic Approach to Saving the Tropical Forests*. LEEC Paper 90-06, [Who is the publisher?], London, United Kingdom.

Pinard, M.A. and F.E. Putz, 1996: Retaining forest biomass by reducing logging damage. *Biotropica,* **28,** 278–295.

Pinard, M.A. and F.E. Putz, 1997: Monitoring carbon sequestration benefits associated with a reduced-impact logging project in Malaysia. *Mitigation and Adaption Strategies for Global Change,* **2,** 203–215.

Pingoud, K., A. Lehtila, and I. Savolainen, 1999: Bioenergy and the forest industry in Finland after the adoption of the Kyoto Protocol. *Environmental Science and Policy,* **2,** 153–163.

Plantinga, A.J. and R.A. Birdsey, 1994: Optimal forest stand management when benefits are derived from carbon. *Natural Resource Modeling,* **8(4),** 373–387.

Polley, H.W., H.S. Mayeaux, H.B. Johnson, and C.R. Tischler, 1997: Atmospheric CO_2, soil water, and shrub/grass ratios on rangelands. *Journal of Range Management,* **50,** 278–284.

Post, W.M., R.C. Izaurralde, L.K. Mann, and N. Bliss, 1999: Monitoring and verifying soil organic carbon sequestration. In: *Carbon Sequestration in Soils: Science, Monitoring, and Beyond* [Rosenberg, N.J., R.C. Izaurralde, and E.L. Malone (eds.)]. Battelle Press, Columbus, Ohio, USA, pp. 41–66.

Potter, K.N., H.A. Torbert, O.R. Jones, J.E. Matocha, J.E. Morrison, Jr., and P.W. Unger, 1998: Distribution and amount of soil organic C in long-term management systems in Texas. *Soil and Tillage Research,* **47,** 309–321.

Powlson, D.S., P. Smith, and J.U. Smith (eds.), 1996: *Evaluation of Soil Organic Matter Models Using Existing Long-Term Datasets*. NATO ASI Series I, Vol. 38, Springer-Verlag, Heidlelberg, Germany, 429 pp.

Powlson, D.S., P. Smith, K. Coleman, J.U. Smith, M.J. Glendining, M. Korschens, and U. Franko, 1998: A European network on long-term sites for studies on soil organic matter. *Soil and Tillage Research,* **47,** 263–274.

Prabhu, E. and V. Tiangco, 1999: The Flex-Microturbine™ for biomass gases. In: *Proceedings of the 4th Biomass Conference of the Americas, Oakland, CA* [Overend, R.P. and E. Chornet (eds.)]. Elsevier Science Publishers, Oxford, United Kingdom, pp. 1439–1444.

Price, T.D., M.J. Apps, and W.A. Kurz, 1998: Past and possible future carbon dynamics of Canada's boreal forest ecosystems. In: *Carbon Dioxide in Forestry and Wood Industry* [Kohlmaier, G.K., M. Weber, and R.A. Houghton (eds.)]. Springer-Verlag, Berlin, Germany, pp. 63–88.

Prinsloo, M.A., G.H. Wiltshire, and C.C. du Preez, 1990: Loss of nitrogen fertility and its restoration in some Orange Free State soils. *South African Journal of Plant and Soil,* **7(1),** 55–61.

Putz, F.E. and M.A. Pinard, 1993: Reduced-impact logging as a carbon-offset method. *Conservation Biology,* **7,** 755–757.

Rao, M.R., A. Niang, F. Kwesiga, B. Duguma, S. Franzel, B. Jama, and R. Buresh, 1998: Soil fertility replenishment in sub-saharan Africa. New techniques and the spread of their use on farms. *Agroforestry Today,* **10(2),** 3–8.

Rao, I.M., M.A. Ayarza, and R.J. Thomas, 1994: The use of carbon isotope ratios to evaluate legume contribution to soil enhancement in tropical pastures. *Plant and Soil,* **162,** 177–182.

Ravindranath, N.H., P. Sudha, and R. Sandhya, 1999: *CDM Opportunities in Forestry Sector in India*. Technical Report, Indian Institute of Science, Bangalore, India.

Read, P., 1996: Forestry as a medium term buffer stock of carbon. In: *Proceedings [of the] World Renewable Energy Conference, 1996* [Sayigh, A.A.N. (ed.)]. Elsevier Science Publishers Ltd., Oxford, United Kingdom, pp. iii, v, 984–988.

Read, P., 1997: Food fuel fiber and faces to feed. *Ecological Economics,* **23,** 81–93.

Read, P., 1998: Dynamic interaction of short rotations and conventional forestry in meeting demand for bioenergy in the least cost mitigation strategy. *Biomass and Bioenergy,* **15,** 7–15.

Read, P., 1999: *Carbon Sequestration in Forests: Supply Curves for Carbon Absorption*. Paper presented at a U.S. Department of Energy/International Energy Agency workshop on promising technologies for mitigating greenhouse gases, Washington, DC, USA, May 1999. Available at: http://www.iea.org/workshop/engecon/.

Reicosky, D.C. and M.J. Lindstrom, 1995: Impact of fall tillage on short-term carbon dioxide flux. In: *Soils and Global Change* [Lal, R., J. Kimble, E. Levine, and B.A. Stewart (eds.)]. Lewis Publishers, Boca Raton, FL, USA, pp. 177–188.

Reicosky, D.C., 1994: Crop residue management: Soil, crop, climate interaction. Special series. In: *Advances in Soil Science, Crop Residue Management* [Hatfield, J.L. and B.A. Stewart (eds.)]. Lewis Publishers, Chelsea, MI, USA, pp. 191–214.

Repetto, R. and M. Gillis (eds.), 1988: *Public Policies and the Misuse of Forest Resources*. Cambridge University Press, Cambridge, United Kingdom, 432 pp.

Resck, D.V.S., C.A. Vasconcellos, L. Vilela, and M.C.M. Macedo, 2000: Impact of conversion of Brazilian cerrados to cropland and pastureland on soil carbon pool and dynamics. In: *Global Climate Change and Tropical Ecosystems* [Lal, R., J.M. Kimble, and B.A. Stewart (eds.)]. CRC-Lewis Publishers, Boca Raton, FL, USA, pp. 169–195.

Rietberger, S. (ed.), 1993: *Tropical Forestry*. Earthscan Publications Ltd., London, United Kingdom.

Robbins, G.B., J.J. Bushell, and G.M. McKeon, 1989: Nitrogen immobilisation in decomposing litter contributes to productivity decline in aging pastures of green panic (*Panicum maximum vari trichoglume*). *Journal of Agricultural Science (Cambridge),* **113,** 401–406.

Robertson, F.A., R.J. Myers, and P.G. Saffigna, 1997: Nitrogen cycling in brigalow clay soils under pasture and cropping. *Australian Journal of Soil Research,* **35,** 1323–1339.

Robles, M.D. and I.C. Burke, 1998: Soil organic matter recovery on Conservation Reserve Program fields in southeastern Wyoming. *Soil Science Society of America Journal,* **62,** 725–730.

Rosell, R.A. and J.A. Galantini, 1998: Soil organic carbon dynamics in native and cultivated ecosystems of South America. In: *Management of Carbon Sequestration in Soil* [Lal, R., J. Kimble, R. Follett, and B.A. Stewart (eds.)]. CRC Press, Boca Raton, FL, USA, pp. 11–33.

Rosenfeld, A.H., J.J. Romm, H. Akbari, and M. Pomerantz, 1996: Policies to reduce heat islands: benefits and implementation strategies. *Energy and Buildings (Special Issue on Heat Islands)*.

Rosillo-Calle, F., S. Bajay, and H. Rothman (eds.), 2000: *Industrial Uses of Biomass Energy*. Taylor and Francis, London, United Kingdom, (in press).

Roulet, N.T., 2000: Peatlands, carbon storage and flow, and the Kyoto Protocol: prospects and significance for Canada. *Wetlands,* (in press).

Roulet, N.T. and T.R. Moore, 1995: The effect of forestry drainage practices on the emission of methane from northern peatlands. *Canadian Journal of Forestry Research,* **25,** 491–499.

Row, C., 1996: Effects of selected forest management options on carbon storage. In: *Forests and Global Change. Volume 2: Forest Management Opportunities for Mitigating Carbon Emissions* [Sampson, N. and D. Hair (eds.)]. American Forests, Washington, DC, USA, pp. 59–90.

Row, C. and R.B. Phelps, 1996: Wood carbon flows and storage after timber harvest. In: *Forests and Global Change. Volume 2: Forest Management Opportunities for Mitigating Carbon Emissions* [Sampson, N. and D. Hair (eds.)]. American Forests, Washington, DC, USA, pp. 27–58.

Rudel, T. and J. Roper, 1997: The Paths to Rain Forest Destruction: Crossnational Patterns of Tropical Deforestation, 1975–1990. *World Development,* **(25/1),** 53–65.

Russell, J.S., 1960: Soil fertility changes in the long-term experimental plots at Kybybolite, South Australia. *Australian Journal of Agricultural Research,* **11,** 902–926.

Sampson, R.N., G.A. Moll, and J.J. Kielbaso, 1992: Opportunities to increase urban forests and the potential impacts on Carbon storage and conservation, In: *Forests and Global Change, Volume 1: Opportunities for Increasing Forest Cover* [Sampson, R.N. and D. Hair (eds.)]. American Forests, Washington, DC, USA, pp. 51–72.

Sampson, R.N., L.L. Wright, J.K. Winjum, J.D. Kinsman, J. Benneman, E. Kirsten, and J.M.O. Scurlock, 1993: Biomass management and energy. *Water, Air and Soil Pollution,* **70(1-4),** 139–159.

Sampson, R.N., R.D. Atkinson, and J.W. Lewis, 2000: Indexing resource data for forest health decisionmaking. *Journal of Sustainable Forestry,* **11(1-2),** 1–14.

San Jose, J.J., R.A. Montes, and M.R. Farinas, 1998: Carbon stocks and fluxes in a temporal scaling from a savanna to a semi-deciduous forest. *Forest Ecology and Management,* **105,** 251–262.

Sanchez, P.A., 1995: Science in agroforestry. *Agroforestry Systems,* **30,** 5–55.

Sanchez, P.A., P.L. Woomer, and C.A. Palm, 1994: Agroforestry approaches for rehabilitating degraded lands after tropical deforestation. In: *JIRCAS International Symposium Series, vol. 1*. Tsukuba, Japan, pp. 108–119.

Sanchez, P.A., K.D. Shepherd, M.J. Soule, F.M. Place, R.J. Buresh, A.-M.N. Izac, U. Mokwunye, F.R. Kwesiga, C.G. Ndiritu, and P.L. Woomer, 1997a: Soil fertility replenishment in Africa: an investment in natural resource capital. In: *Replenishing Soil Fertility in Africa* [Buresh, R.J., P.A. Sanchez, and F. Calhoun (eds.)]. SSSA Special Publication 51, Soil Science Society of America, Madison, WI, USA, pp. 1–46.

Sanchez, P.A., R.J. Buresh, and R.R.B. Leakey, 1997b: Trees, soils and food security. *Philosophical Transactions of the Royal Society of London, Series B,* **353,** 949–961.

Sanchez, P.A., 1999: Delivering on the promise of agroforestry. *Environment, Development and Sustainability,* (in press).

Sanchez, P.A., 2000: Linking climate change research with food security and poverty reduction in the tropics. *Agriculture, Ecosystems and Environment,* (submitted).

Sass, R.L., F.M. Fisher, Y.B. Wang, F.T. Turner, and M.F. Jund, 1992: Methane emission from rice fields; the effect of floodwater management. *Global Biogeochemical Cycles,* **6,** 249–262.

Scanlan, J.C., A.J. Pressland, and D.J. Myles, 1996: Run-off and soil movement on mid-slopes in north-east Queensland [Australia] grazed woodlands. *Rangeland Journal,* **18(1),** 33–46.

Scanlan, J.C., 1992: A model of woody-herbaceous biomass relationships in eucalypt and mesquite communities. *Journal of Range Management,* **45,** 75–80.

Scheumann, W., 1993: New irrigation schemes in southeastern Anatolia and in northern Syria: more competition and conflict over the Euphrates? *Quarterly Journal of International Agriculture,* **32,** 240–259.

Schimel, D.S., B.H. Braswell, R. McKeown, D.S. Ojima, W.J. Parton, and W. Pulliam, 1996: Climate and nitrogen controls on the geography and timescales of terrestrial biogeochemical cycling. *Global Biogeochemical Cycles,* **10,** 677–692.

Schlamadinger, B. and G. Marland, 1996: The role of forest and bioenergy strategies in the global carbon cycle, *Biomass and Bioenergy,* **10,** 275–300.

Schlesinger, W.H., 1999: Carbon sequestration in soils. *Science,* **284,** 2095.

Schlesinger, W.H., 1986: Changes in soil carbon storage and associated properties with disturbance and recovery. In: *The Changing Carbon Cycle. A Global Analysis* [Trabalka, J.R. and D.E. Reichle (eds.)]. Springer-Verlag, New York, NY, USA, pp. 194–220.

Schnabel, R.R., A.J. Franzluebbers, W.L. Stout, M.A. Sanderson, and J.A. Stuedemann, 2000: Pasture management effects on soil carbon sequestration. In: *Carbon Sequestration Potential of U.S. Grazinglands* [Follett, R.F., J.M. Kimble, and R. Lal (eds.)]. CRC Press, Boca Raton, FL, USA, (in press).

Scholes, R.J. and S.R. Archer, 1997: Tree-grass interactions in savannas. *Annual Review of Ecology and Systematics,* **28**: 517–544.

Scholes, R.J. and M.R. van der Merwe, 1996: Sequestration of carbon in savannas and woodlands. *The Environmental Professional,* **18,** 96–103.

Scholes, R.J., 1999: Will the terrestrial carbon sink saturate soon? *Global Change Newsletter,* **37,** 2–3.

Scholes, R.J., E.-D. Schulze, L.F. Pitelka, and D.O. Hall, 1999: Biogeochemistry of terrestrial ecosystems. In: *The Terrestrial Biosphere and Global Change* [Walker, B., W. Steffen, J. Canadell, and J. Ingram (eds.)]. Cambridge University Press, Cambridge, United Kingdom, pp. 271–303.

Scholes, R.J. and B.H. Walker, 1993: An African savanna: synthesis of the Nylsvley study. Cambridge University Press, Cambridge, United Kingdom, 293 pp.

Scholes, R.J. and D.O. Hall, 1996: The carbon budget of tropical savannas, woodlands and grasslands. In: *Global Change: Effects on Coniferous Forests and Grasslands* [Breymeyer, A., D.O. Hall, J.M. Melillo, and G.I. Agren (eds.)]. SCOPE 56, John Wiley and Sons, Chichester, United Kingdom, pp. 69–100.

Schopfhauser, W., 1998: World forests: The area for afforestation and their potential for fossil carbon sequestration and substitution. In: *Carbon Dioxide Mitigation in Forestry and Wood Industry,* [Kohlmaier, G.H., M. Weber, and R.A. Houghton (eds.)]. Springer-Verlag, Berlin, Germany, pp. 185–203.

Schroeder, P.E., 1992: Carbon storage potential of short rotation tropical tree plantations. *Forest Ecology and Management,* **50,** 31–41.

Schuman, G.E., J.D. Reeder, J.T. Manley, R.H. Hart, and W.A. Manley, 1999: Impact of grazing management on the carbon and nitrogen balance of a mixed-grass rangeland. *Ecological Applications,* **9,** 65–71.

Schwab, A.P., C.E. Owensby, and S. Kulyingyong, 1990: Changes in soil chemical properties due to 40 years of fertilisation. *Soil Science,* **149(1),** 35–43.

Schutz, H., A. Holzapfel-Pschorn, R. Conrad, H. Rennenberg, and W. Seiler, 1989: A 3-year continuous record on the influence of day-time, season, and fertilizer treatment on methane emission rates from an Italian rice paddy. *Journal of Geophysical Research,* **94,** 16405–16416.

Scurlock, J.M.and D.O. Hall, 1998: The global carbon sink: a grassland perspective. *Global Change Biology,* **4,** 229–233.

Sen, A.K., 1981: *Poverty and Famines: An Essay on Entitlement and Deprivation.* Clarendon Press, Oxford, United Kingdom.

Senthilkumar, K., S. Manian, K. Udaiyan, and S. Paulsamy, 1998: Elevated biomass production in burned grasslands in southern India. *Tropical Grasslands,* **32,** 50–63.

Simons, A.J., I.K. Dawson, B. Duguma, and Z. Tchoundjeu, 1998: Passing problems: prostate and prunus. *HerbalGram,* **43,** 49–53.

Singh, G., N.T. Singh, and I.P. Abrol, 1994: Agroforestry techniques for the rehabilitation of degraded salt-affected land. *Land Degradation Development,* **5,** 223–242.

Skjemstad, J.O., P. Clarke, J.A. Taylor, J.M. Oades, and S.G. McClure, 1996: The chemistry and nature of protected carbon in soil. *Australian Journal of Soil Research,* **34,** 251–271.

Skog, K.E. and G. Nicholson, 1998: Carbon cycling through wood products: the role of wood and paper products in carbon sequestration. *Forest Products Journal,* **48(7-8),** 75–83.

Smaling, E.M.A., 1993: *An Agroecological Framework for Integrating Nutrient Management, with Special Reference to Kenya.* Ph.D. thesis Agricultural University, Wageningen, The Netherlands.

Smith, P., D.S. Powlson, M.J. Glendining, and J.U. Smith, 1998: Preliminary estimates of the potential for carbon mitigation in European soils through no-till farming. *Global Change Biology,* **4,** 679–685.

Smith, P., D.S. Powlson, M.J. Glendining, and J.U. Smith, 1997a: Potential for carbon sequestration in European soils: preliminary estimates for five scenarios using results from long-term experiments. *Global Change Biology,* **3,** 67–79.

Smith, P., D.S. Powlson, M.J. Glendining, and J.U. Smith, 1997b: Opportunities and limitations for C sequestration in European agricultural soils through changes in management. In: *Management of Carbon Sequestration in Soil* [Lal, R., J.M. Kimble, R.F. Follett, and B.A. Stewart (eds).]. CRC Press, Boca Raton, FL, USA, pp. 143–152.

Sorenson, K.W., 1993: Indonesian peat swamp forests and their role as a carbon sink. *Chemosphere,* **27,** 1065–1082.

Squires, V.R., E.P. Glenn, and A.T. Ayoub (eds.), 1995: *Combating Global Climate Change by Combating Land Degradation.* Proceedings of a workshop held in Nairobi, Kenya, 4–8 September 1995, United Nations Environmental Programme, Nairobi, Kenya, 348 pp.

Stafford Smith, M., G.M. McKeon, R. Buxton, and J. Breen, 1999: The integrated impacts of price, policy and productivity changes on land use in northern Australia. In: *People and Rangelands: Building the Future. Proceedings of the Sixth International Rangeland Congress* [Eldridge, D. and D. Freeudenberger (eds.)]. Australian Academy of Science, Townsville, Australia, Vol. II, pp. 864–866.

Stallard, R.F., 1998: Terrestrial sedimentation and the carbon cycle: Coupling weather and erosion to carbon burial. *Global Biogeochemical Cycles,* **12,** 231–257.

Steppler, H.A. and P.K.R. Nair (eds.), 1987: *Agroforestry—a Decade of Development.* International Centre for Agroforestry, Nairobi, Kenya, 335 pp.

Stockfisch, N., T. Forstreuter, and W. Ehlers, 1999: Ploughing effects on soil organic matter after twenty years of conservation tillage in Lower Saxony, Germany. *Soil and Tillage Research,* **52,** 91–101.

Stott, D.E., G. Kassim, J.P. Martin, and K. Haider, 1983: Stabilization and incorporation into biomass of specific plant carbons during biodegradation in soil. *Plant and Soil,* **70,** 15–26.

Strich, S., 1998: Carbon mitigation potential of German forestry considering competing forms of land use. In: *Carbon Dioxide in Forestry and Wood Industry* [Kohlmaier, G.K., M. Weber, and R.A. Houghton (eds.)]. Springer-Verlag, Berlin, Germany, 125–135.

Suarez, D., 1998: Impact of agriculture on CO_2 as affected by changes in inorganic carbon. In: *Global Climate Change and Pedogenic Carbonates* [Lal, R., J. Kimble, H. Eswaran, and B.A. Stewart (eds.)]. CRC Press, Boca Raton, FL, USA, (in press).

Subak, S., 1999: *Agricultural Soil Carbon Accumulation in North America: Considerations for Climate Policy.* Natural Resourcs Defense Council, Washington, DC, USA, 25 pp.

Sumner, M.E. and R. Naidu (eds.), 1998: *Sodic Soils: Distribution, Properties, Management and Environmental Consequences.* Oxford University Press, New York, NY, USA.

Swarup, A., M.C. Manna, and G.B. Singh, 2000: Impact of land use and management practices on organic carbon dynamics in soils of India. In: *Global Climate Change and Tropical Ecosystems* [R. Lal, J.M. Kimble, and B.A. Stewart (eds.)]. CRC-Lewis Publishers, Boca Raton, FL, USA, 261–282.

Swift, M.J., O.W. Heal, and J.M. Anderson, 1979: *Decomposition in Terrestrial Ecosystems.* Blackwell Science Publishers, Oxford, United Kingdom, 372 pp.

Swift, M.J. (coordinator), 1999: *Below-Ground Biodiversity Assessment Working Group Summary Report 1996–1998. Alternatives to Slash and Burn Project. Phase II.* International Centre for Agroforestry, Nairobi, Kenya, (in press).

Tang, C., L. Barton, and C. Raphael, 1998: Pasture legume species differ in their capacity to acidify soil. *Australian Journal of Agricultural Research,* **49(1),** 53–58.

Tenberg, A., M. Da Veiga, S.C.F. Dechen, and M.A. Stocking, 1998: Modelling the impact of erosion on soil productivity: a comparative evaluation of approaches on data from southern Brazuk. *Experimental Agriculture,* **34,** 55–71.

Thornthwaite, C.W., 1948: An approach toward a rational classification of climate. *Geographical Review,* **38,** 55–94.

Togtohyn, C., D. Ojima, J. Luvsandorjiin, J. Dodd, S. Williams, and N.E. West, 1996: Simulation studies of grazing in the Mongolian steppe. In: *Proceedings of the Fifth International Rangeland Congress* [West, N.E. (ed.)]. Society for Range Management, Denver, CO, and Salt Lake City, UT, USA, 23–28 July 1995, Vol. I, pp. 561–562.

Tolonen, K., and J. Turunen, 1996: Accumulation rates of carbon in mires in Finland and implications for climate change. *The Holocene,* **6,** 171–178.

Tomich, T.P., M. van Noordwijk, S. Vosti, and J. Witcover, 1998: Agricultural development with rainforest conservation: Methods for seeking best-bet alternatives to slash-and-burn, with applications to Brazil and Indonesia. *Agricultural Economics,* **19,** 159–174.

Tomich, T.P., M. van Noordwijk , S. Budidarsono, A. Gillison, T. Kusumanto, D. Murdiyarso, F. Stolle, and A.M. Fagi, 1999: Agricultural intensification, deforestation and the environment: assessing tradeoffs in Sumatra, Indonesia. In: *Agricultural Intensification and the Environment* [Barrett, C.R. and D.R. Lee (eds.)] Blackwell Science Publishers, Oxon, United Kingdom.

Tongway, D.J. and J.A. Ludwig, 1996: Rehabilitation of semiarid landscapes in Australia. I. Restoring productive soil patches. *Restoration Ecology,* **4(4),** 388–397.

Torbert, H.A., H.H. Rogers, S.A. Prior, W.H. Schlesinger, and G.B. Runion, 1997: Effects of elevated atmospheric CO_2 in agro-ecosystems on soil carbon storage. *Global Change Biology,* **3,** 513–521.

Torquebiau, E., 1984: Man-made dipterocarp forests in Sumatra. *Agroforestry Systems,* **2,** 103–128.

Tothill, J.C. and C.G. Gillies, 1992: *The Pasture Lands of Northern-Australia*. Occasional Publication No. 5, Tropical Grasslands Society, Brisbane, Australia, 106 pp.

Tothill, J.C., J.N.G. Hargreaves, R.M. Jones, and C.K. McDonald, 1992: *BOTANAL—a Comprehensive Sampling and Computing Procedure for Estimating Pasture Yield and Composition. 1. Field Sampling*. Tropical Agronomy Technical Memorandum No. 78, Commonwealth Scientific and Industrial Research Organization (CSIRO) Tropical Crops and Pastures, Brisbane, Australia, 24 pp.

Townsend, A.R., M.T. Sykes, and M.J. Apps, 1996: WG1 summary: natural and anthropogenically-induced variations in terrestrial carbon balance. In: *Forest Ecosystems, Forest Management and the Global Carbon Cycle* [Apps, M.J. and D.T. Price (eds.)]. NATO ASI Series I: *Global Environmental Change,* **40,** 97–107.

Thorsen, B.J. and F. Helles, 1998: Optimal stand management with endogenous risk of sudden destruction. *Forest Ecology and Management,* **108,** 287–299.

Tracy, B.F. and D.A. Frank, 1998: Herbivore influence on soil microbial biomass and nitrogen mineralization in a northern grassland ecosystem: Yellowstone National Park. *Oecologia,* **114,** 556–562.

Travasso, M.I., G.O. Magrin, G.R. Rodríguez, and D.R. Boullón, 1999: Climate Change assessment in Argentina: II. Adaptation strategies for agriculture. Accepted in *Food and Forestry: Global Change and Global Challenge*. GCTE Focus 3 Conference. Reading, United Kingdom, September 1999.

Trimble, S.W., 1999: Decreased rates of alluvial sediment storage in the Coon Creek basin, Wisconsin, 1975–93. *Science,* **285,** 1244–1246.

Trumbore, S.E. and S. Zheng, 1996: Comparison of fractionation methods for soil organic matter C analysis. *Radiocarbon,* **38,** 219–229.

Trumbore, S.E., O.A. Chadwick, and R. Amundson, 1996: Rapid exchange between soil carbon and atmospheric carbon dioxide driven by temperature change. *Science,* **272,** 393–396.

Turhollow, A.H. and R.D. Perlack, 1991: Emissions of CO_2 from energy crop production. *Biomass and Bioenergy,* **1,** 129–135.

United Nations Framework Convention on Climate Change (UNFCCC), 1998: Methodological issues, land-use change and forestry. Division 1/CP.3 Paragraph 5(a). Document FCCC/SBSTA, 1998/CRP.3, 8th Session of the Subsidiary Body for Scientific and Technological Advice, Bonn, Germany, June 2–12.

Unruh, J.D., R.A. Houghton, and P.A. Lefebvre, 1993: Carbon storage in agroforestry: an estimate for sub-Saharan Africa. *Climate Research,* **3,** 39–52.

Urquiaga, S., G. Cadisch, B.J.R. Alves, R.M. Boddley, and K.E. Giller, 1998: Influence of decomposition of roots of tropical forage species on the availability of soil nitrogen. *Soil Biology and Biochemistry,* **30(14),** 2099–2106.

USDA-NRCS, 1999: *Summary Report: 1997 National Resources Inventory*. US Department of Agriculture Natural Resources Conservation Service, Washington, DC, USA, 83 pp.

van Bremen, N. and R. Protz, 1988: Rates of calcium carbonate removal from soils. *Canadian Journal of Soil Science,* **68,** 449–454.

van Ginkel, J.H., A.P. Whitmore, and A. Gorissen, 1999: *Lolium perene* grasslands may function as a sink for atmospheric carbon dioxide. *Journal of Environmental Quality,* **28,** 1580–1584.

van Noordwijk, M., C. Cerri, P.L. Woomer, K. Nugroho, and M. Bernoux, 1997: Soil carbon dynamics in the humid tropical forest zone. In: *The Management of Carbon in Tropical Soils Under Global Change: Science, Practice and Policy* [Elliott, E.T., J. Kimble, and M.J. Swift (eds.)]. *Geoderma,* **79,** 187–225.

van Wilgen, B.W., R.M. Cowling, and D.C. leMaitre, 1998: Ecosystem services, efficiency, sustainability an equity: South Africa's working for water programme. *Trends in Ecology and Evolution,* **13,** 378.

Veldkamp, E., M. Keller, and M. Nunez, 1998: Effects of pasture management on N_2O and NO emissions from soils in the humid tropics of Costa Rica. *Global Biogeochemical Cycles,* **12(1),** 71–79.

Verardo, D.J., 1997: Charcoal analysis in marine sediments. *Limnology and Oceanography,* **42,** 192–197.

Verardo, D.J. and W.F. Ruddiman, 1996: Late Pleistocene charcoal in tropical Atlantic deep-sea sediments: climatic and geochemical significance. *Geology,* **24,** 855–857.

Vine, E., J. Sathaye, and W. Makundi, 1999: *Guidelines for the Monitoring, Evaluation, Reporting, Verification and Certification of Forestry Projects for Climate Change Mitigation*. Lawrence Berkeley National Laboratory, Berkeley, CA, USA.

Vitousek, P.M., H.A. Mooney, J. Lubchenco, and J.M. Melillo, 1997: Human domination of earth's ecosystems. *Science,* **277,** 494–499.

von Amsberg, J., 1998: Economic parameters of deforestation. *The World Bank Economic Review,* **12/1,** 133–153.

Vosti, S.A., J. Witcover, C.L. Carpentier, and E. do Amaral, 1999: Intensifying small-scale agriculture in the western Brazilian Amazon: Issues, implications and implementation. In: *Agricultural Intensification and the Environment* [Barrett, C.R. and D.R. Lee (eds.)]. Blackwell Science Publishers, Oxon, United Kingdom, (in press).

Wada, H., 1984: Nonvolatile residues in the anaerobic decomposition of organic matter. In: *Organic Matter and Rice*. IRRI, Los Banos, Philippines, pp. 345–359.

Wada, H., K. Inubushi, Y. Uehara, and Y. Takai, 1981: Relationship between total nitrogen and mineralizable nitrogen. *Japanese Journal of Soil Science and Plant Nutrition,* **52,** 246–252 (in Japanese with English summary)

Waite, R.B., 1994: The application of visual estimation procedures for monitoring pasture yield and composition in exclosures and small plots. *Tropical Grasslands,* **28,** 38–42.

Walter, A., 2000: New technologies for modern biomass energy carriers. In: *Industrial Uses of Biomass Energy* [Rosillo-Calle, F., S. Bajay, and H. Rothman (eds.)]. pp. 200–253.

Wassman, R., H. Pappen, and H. Rennenberg, 1993: Methane emission from rice paddies and possible mitigation strategies. *Chemosphere,* **26,** 201–217 .

WEC, 1993: *Energy for Tomorrow's World*. Kogan Page, London, United Kingdom.

Wedin, D.A. and D. Tilman, 1996: Influence of nitrogen loading and species composition on the carbon balance of grasslands. *Science,* **274,** 1720–1723.

Wilding, L.P., 1999: Comments to paper by Lal, Hassan and Dumanski. In: *Carbon Sequestration in Soil: Science, Monitoring and Beyond,* [Rosenberg, N.J., R.C. Izaurralde, and E.L. Malone (eds.)]. Battelle Press, Columbus, OH, USA, 146–149.

Williams, C.H., and C.M. Donald, 1957: Changes in organic matter and pH in a podzolic soil as influenced by subterranean clover and superphosphate. *Australian Journal of Agricultural Research,* **8,** 179–189.

Wills, B.J., J.S. Sheppard, and K.T. Trainor, 1996: A review of the use of chenopod shrubs for dryland revegetation and forage in the South Island, New Zealand. In: *Proceedings of the Fifth International Rangeland Congress* [West, N.E. (ed.)]. Society for Range Management, Denver, CO, and Salt Lake City, UT, USA, 23–28 July 1995, Vol. I, pp. 612–613.

Winjum, J., S. Brown, and B. Schlamadinger, 1998: Forest harvests and wood products; sources and sinks of atmospheric carbon dioxide. *Forest Science,* **44,** 272–284.

Winjum, J.K., R.K. Dixon, and P.E. Schroeder, 1993: Forest management and carbon storage: an analysis of 12 key forest nations. *Water, Air and Soil Pollution,* **70,** 239–257.

Winter, W.H., J.J. Mott, R.W. McLean, and D. Ratcliff, 1989: Evaluation of management options for increasing the productivity of tropical savanna pastures 1. Fertiliser. *Australian Journal of Experimental Agriculture,* **29,** 613–622.

Woods, J., and D.O. Hall, 1994: *Bioenergy for Development:Technical and Environmental Dimensions*. FAO Energy and Development Paper No. 13, Food and Agriculture Organization of the United Nations, Rome, Italy.

Woodwell, G.M., F.T. MacKenzie, R.A. Houghton, M. Apps, E. Gorham, and E. Davidson, 1998: Biotic feedbacks in the warming of the earth. *Climatic Change,* **40,** 495–518.

Woomer, P.L., C.A. Palm, J. Alegre, C. Castilla, D.G. Cordeiro, K. Hairiah, J. Kotto-Same, A. Moukam, A. Ricse, V. Rodrigues, and M. van Noordwijk, 2000: Slash-and-burn effects on carbon stocks in the humid tropics. In: *Global Climate Change and Tropical Ecosystems* [Lal, R., J.M. Kimble, and B.A. Stewart (eds.)]. CRC Press, Boca Raton, FL, USA, pp. 99–115.

Woomer, P.L., C.A. Palm, J.N. Qureshi, and J. Kotto-Same, 1997: Carbon sequestration and organic resource management in African smallholder agriculture. In: *Management of Carbon Sequestration in Soil* [Lal, R., J.M. Kimble, R.A. Follett, and B.A. Stewart (eds.)]. CRC Press, Boca Raton, FL, USA, pp. 153–173.

Xu, D.Y., 1995: The potential for reducing atmospheric carbon by large-scale afforestation in China and related cost/benefit analysis. *Biomass and Bioenergy,* **8(5),** 337–344.

Xu, D., 1999: Forestry and land use change assessment for China. In: *Forestry and Land Use Change Assessment*. Asian Development Bank, Manila, Philippines, pp. 73–97.

Yagi, K. and K. Minami, 1990: Effect of organic application on methane emission from some Japanese paddy fields. *Soil Science and Plant Nutrition,* **36,** 599–610.

Yagi, K., H. Tsuruta, K. Kanda, and K. Minami, 1996: Effect of water management on methane emission from Japanese rice paddy field; antomated methane monitoring. *Global Biogeochemical Cycles*, **10,** 255–267.

Yagi, K., H. Tsuruta, and K. Minami, 1997: Possible options for mitigating methane emission from rice cultivation, nutrient cycling. *Agroeconomics,* **49,** 213–220.

Fact Sheet 4.1. Agricultural Intensification and Carbon Inputs

Farming practices that enhance production and the input of plant-derived residues to soil include crop rotations, reduced bare fallow, cover crops, high-yielding varieties, integrated pest management, adequate fertilization, organic amendments, irrigation, water table management, site-specific management, and other proper management practices. These practices are referred to collectively as agriculture intensification (Lal *et al.*, 1999b; Bationo *et al.*, 2000; Resck *et al.*, 2000; Swarup *et al.*, 2000). For more detail, see Section 4.4.2.1.

Use and Potential
Increasing global demand for food will drive continued agriculture intensification. Intensification can be applied to all cropping systems, with varying degrees of constraints because of economics and the availability of labor and technology. Rates of residue return to soil are also influenced by potential alternative uses as fodder and fuel. Intensification of systems with previously low use of purchased inputs (e.g., fertilizer, improved varieties, pesticides) may involve increased use of these inputs and/or intensive management using biological inputs (e.g., crop rotations, cover crops, manures). Where the use of purchased inputs is already high, intensification implies increased efficiency (and potentially reduced use) of fertilizer, pesticide, and other inputs. The principal means by which intensification influences soil carbon changes are through the amount and quality of carbon returned to soil (via roots, crop residues, and manures) and through water and nutrient influences on decomposition (Paustian *et al.*, 2000a). Agricultural intensification can occur on all or nearly all of the world's existing cropland (1.6 Bha).

Current Knowledge and Scientific Uncertainties
The rates of SOC sequestration by agriculture intensification differ among soils and ecoregions. The influence of these practices on productivity and soil properties, including organic matter dynamics, has been studied for many decades; there is an extensive body of research results and many well-documented, long-term field studies around the world (Powlson *et al.*, 1998). Uncertainties remain, however, regarding the interactions between different practices (e.g., crop rotations, water table management, fertilization) for different soil and climate conditions.

Methods
Rates of soil carbon sequestration can be established for predominant cropping/management systems on the basis of long-term benchmark experiments, on-site sampling, and modeling (e.g., Powlson *et al.*, 1996; Eve *et al.*, 2000). Annual statistics on cropland area and crop production are available globally at the country level (e.g., FAO, IGBP-DIS), and more detailed data on crop production and the extent and distribution of the practices described above exist to varying degrees for all Annex I countries and many non-Annex I countries. Several generalized models of carbon cycling in agricultural systems exist (see Chapter 2). Quantification of soil carbon changes can be estimated, using models, from the distribution of major cropping systems, data on production and residue returns, and associated soil and climate information, and/or with soil sampling designs. Scaling from local to regional to national levels can be done by using a combination of climate and soil maps, management and yield data, modeling, and geographic information systems (GIS).

Time Scale
These practices can increase soil carbon stocks for 25–50 years or until saturation is reached.

Monitoring, Verifiability, and Transparency
The amount of new carbon sequestration and its residence time (turnover rate) can be verified through ground truthing (on-site sampling) and well-calibrated models. Periodic monitoring can be done by using benchmark sites where SOC content and bulk density can be measured once every 5–10 years to a depth of 1 m. Because of the stratification of SOC, soil samples need to be taken in small depth increments in the surface layers. The practices to be used are well characterized.

Removals
Reversion to conventional agriculture practices (i.e., plowing, residue removal or burning, inappropriate irrigation, improper fertilizer use) can cause the loss of sequestered carbon.

Permanence
Most carbon in agricultural systems is in the soil and has residence times of years to centuries (see Section 4.2).

Associated Impacts
Agriculture intensification has numerous ancillary benefits—the most important of which is the increase and maintenance of food production. Environmental benefits can include erosion control, water conservation, improved water quality, and reduced siltation of reservoirs and waterways. Soil and water quality is adversely affected by indiscriminate use of agriculture inputs and irrigation water. Where intensification involves increased use of nitrogen fertilizers, fossil energy use will increase, as may N_2O emissions.

Relationship to IPCC Guidelines

These practices relate directly to the "Input Factors" used in the IPCC Guidelines for estimates of changes in soil carbon stocks. Default values for three levels of plant residue production and addition to soil are provided in the Workbook, with examples describing the types of management systems that each level would correspond to.

Fact Sheet 4.2. Irrigation Water Management

Productivity of cropland and grazing land can be enhanced by supplemental irrigation in drought-prone ecosystems and by water table management in seasonally wet soils.

Use and Potential

Irrigated crops are grown on about 255 Mha (FAO, 1999). The potentially irrigable land area in sub-Saharan Africa is estimated at 39 Mha (Hillel, 1997), and there is additional potential in Asia and South America. Most irrigation is in areas with low levels of SOC in the native state. Therefore, there is a large potential for carbon sequestration by the use of irrigation (Lal *et al.*, 1998; Conant *et al.*, 2000). Global expansion of irrigation requires successful resolution of socioeconomic and political issues (Scheumann, 1993; Beaumont, 1996; McCown, 1996). Although drainage of wet soils (organic and mineral soils) enhances crop and animal productivity, it reduces biodiversity. Drainage decreases methane emissions but leads to loss of SOC stock. Therefore, adopting judicious methods of water table management, including sub-irrigation and water recycling, is necessary to SOC sequestration.

Current Knowledge and Scientific Uncertainties

Experimental data on soil carbon (SOC and SIC) dynamics in irrigated soils is scanty. Application of water to drylands influences biomass productivity, increases the amount of residue returned, changes mineralization rates, and alters the carbonate balance. Experimental rates of soil carbon sequestration range from 0.05 to 0.15 t C ha^{-1} yr^{-1} for SOC (Lal *et al.*, 1998; Conant *et al.*, 2000) and 0.05 to 0.10 t C ha^{-1} yr^{-1} for SIC (Wilding, 1999; Nordt *et al.*, 2000). Excessive irrigation, lack of proper drainage, and use of poor-quality irrigation water accentuate the risks of soil salinization. Use of proper irrigation methods and improved cropping systems is therefore essential to reap the benefits of irrigation in enhancing productivity and soil carbon sequestration.

Methods

At the plot or field level, soils can be sampled and SOC changes determined by laboratory analysis. Rates can be developed from selected fields by the use of models along with soil maps and other digital databases to expand localized measurements to the watershed or ecoregions and to national scales. Remote sensing can be used to determine the areas under improved water management. (Determining the area under irrigation is very easy with remote sensing, but assessing areas to which improved water table management is being applied is more difficult.)

Time Scale

The soils in arid and semi-arid regions are inherently low in SOC and would take 50–100 years to reach a steady state. With proper water management, dissolved organic carbon (DOC) would continue to move downward in the soil profile, with the potential to bind with soil minerals and form stable organo-mineral complexes.

Monitoring, Verifiability, and Transparency

Monitoring entails using direct sampling and modeling to quantify changes in SOC levels and using remote sensing to determine the area being irrigated. In areas with low initial SOC levels, changes may be relatively easy to measure over short time periods (e.g., 5-year intervals). This practice is very transparent and can be used, monitored, and verified in all areas.

Removals

If the irrigation were stopped, the sequestration would stop. If the soil is tilled, the stored carbon would be lost.

Permanence

The bulk of sequestered carbon will be in the soil, with residence times of years to centuries.

Associated Impacts

Irrigation enhances biomass productivity in water-limited agricultural systems. In addition to increasing the risk of salinization, irrigation also requires energy input for pumping and distributing water.

Relationship to IPCC Guidelines

Irrigation and water management practices are not explicitly included in the calculations for soil carbon. The effects of irrigation on residue carbon inputs can be expressed through the choice of "Input Factors." The influence of irrigation on carbon turnover rates and inorganic (carbonate) carbon changes, however, are not dealt with and would require appropriate revisions to the IPCC Guidelines.

Fact Sheet 4.3. Conservation Tillage

Conservation tillage is a generic term that includes a wide range of tillage practices, including chisel plow, ridge till, strip till, mulch till, and no till (CTIC, 1998). For more detail, see Section 4.4.2.2.

Use and Potential

The practice came into use during the 1950s for row crop production on erodible land in the midwestern United States. In 1998, about 37 percent of the row crops grown in the United States were sown with a conservation tillage system (CTIC, 1998). The upland area managed with conservation tillage is 12 Mha in Brazil and 4 Mha in Argentina. The majority of the area in conservation tillage in Brazil and Argentina is continuous no-till; this is not the case in the United States. There is potential for expansion of cropland under conservation tillage in Asia, Australia, Africa, and Europe.

Current Knowledge and Scientific Uncertainties

The rate of SOC sequestration by conversion from conventional to conservation tillage in North America has been found to differ among soils, cropping systems, and ecoregions and may range from 0.05 to 1.3 t C ha^{-1} yr^{-1}, with a mean of 0.3 t C ha^{-1} yr^{-1}. The rate of sequestration for principal soils and ecoregions must be established through monitoring of carbon dynamics on long-term experiments in different ecoregions. The net rate of sequestration must be assessed by taking into consideration the carbon used in herbicide production and application, which differs among tillage systems. The amount of carbon residue returned to the soil is an important factor because such residue is often removed for use as fodder and fuel.

Methods

Rates of SOC sequestration for specific types of conservation tillage can be established for predominant cropping systems on the basis of long-term benchmark experiments, on-site sampling, and modeling. The rates differ depending on the amount of soil disturbance and the quality and quantity of crop residue returned (Lal, 1997; Paustian *et al.*, 1997a). The rate of carbon sequestration can be quantified on the basis of ground cover, residue returned, and cropping systems determined through remotely sensed data. Scaling from local to regional and national levels can be done by using a combination of soil maps, cropping reports, yield data, modeling, and GIS.

Time Scale

This practice can increase the SOC stock in the soil profile for 25–50 years or until saturation is reached. The rate of carbon sequestration may be highest in the initial 5–20 years.

Monitoring, Verifiability, and Transparency

The amount of new carbon sequestration and its residence time (turnover rate) can be verified through ground truthing (on-site sampling). SOC content and bulk density can be measured at the same location over a period of time (e.g., 3- to 10-year interval) to a depth of 1 m. Because of the stratification of SOC, soil samples must be taken in small depth increments in the surface layers. For a few sites, the rate and magnitude of newly sequestered carbon can be determined by soil sampling and measurements of residue returned. Monitoring/verification of tillage practices can be carried out through ground surveys and potentially through the use of remote-sensing techniques to assess surface residue coverage. The modus operandi of conservation tillage is well known. The rate and type of herbicide use may differ among soils and ecoregions. Effective weed control and use of proper seeding equipment to ensure a good crop stand are important to ensure successful adaptation.

Removals

Reversion to more intensive tillage can cause loss of sequestered carbon.

Associated Impacts

SOC sequestration through conservation tillage depends on continued use. Reversion to conventional methods (high degree of disturbance) can cause loss of sequestered SOC. Policy measures must be in effect to ensure that conservation tillage is adopted on a continued basis. Adoption of conservation tillage has numerous ancillary benefits. Important among these benefits are control of water and wind erosion, water conservation, increased water-holding capacity, reduced compaction, increased soil resilience to chemical inputs, increased soil and air quality, enhanced soil biodiversity, reduced energy use, improved water quality, reduced siltation of reservoirs and waterways, and possible double-cropping. In some areas (e.g., Australia), increased leaching from greater water retention with conservation tillage can cause downslope salinization. In wet years, planting may be delayed in no-till systems, potentially resulting in a yield reduction.

Relationship to IPCC Guidelines

Tillage effects on soil carbon stock changes are included in the IPCC Guidelines (Reference Manual), and default values for three levels of tillage intensity are provided.

Fact Sheet 4.4. Erosion-Control Practices

Safe disposal of surplus runoff at low velocity involves the use of some land-forming and engineering techniques, including terraces, waterways, diversion channels, drop structures, chutes, and so forth. Similarly, vegetative strips are used as filter strips for riparian zone management and as shelter belts for wind erosion control. Development of more sustainable grazing systems leads to a reduction of soil erosion rates (Ash *et al.*, 1996; Scanlan *et al.*, 1996; Tenberg *et al.*, 1998). For more detail, see Sections 4.4.2.3 and 4.4.3.3.

Use and Potential
Terracing and other engineering structures are widely used on sloping lands all over the world. Runoff management to control soil erosion by water and shelter belts to control wind erosion are important strategies to decrease the risk of soil erosion. There are several soils and ecoregions where conservation tillage is not applicable and adoption of these erosion control measures is essential. Many of these techniques are also used in conjunction with conservation tillage. Erosion control enhances the productivity of these lands.

Current Knowledge and Scientific Uncertainties
Potential SOC sequestration through erosion-control measures depends on an increase in biomass productivity through conservation of water and efficient use of fertilizer and farm chemicals. Initial energy input that is required for the installation of engineering techniques is offset by long-term benefits of erosion control and enhanced productivity. Although the productivity benefits of erosion control measures are known, improvements in the SOC pool are not widely established. Improved grazing management systems, which reduce erosion, also lead to increases in the SOC stock (Kelly *et al.*, 1996; Tongway and Ludwig, 1996).

Methods
Methods for the measurement of SOC sequestration through erosion control involve soil sampling for periodic assessment of SOC stocks and estimation of the reduction in depletion of the SOC resulting from decreased soil erosion. The SOC input can also be measured by using established empirical relations for specific soil types and farming systems within an ecoregion.

Time Scale
The effects of erosion-control measures are cumulative and occur over a long time, so changes in SOC stocks need to be measured over a period of 25–50 years through periodic assessment. The rate of SOC sequestration may vary among ecoregions and can be high for shelter belts and contour hedgerows.

Monitoring, Verifiability, and Transparency
Monitoring and verification of SOC sequestration must be carried out by using ground truthing through periodic measurements of SOC stock in benchmark sites. Adoption of these measures can be verified through remote-sensing techniques and ground truthing through local extension organizations. Installations of terraces, waterways, shelter belts, riparian protection zones, and other engineering devices are conspicuous, easily checked, and verified through a combination of measures that involve remote sensing and ground truthing. The empirical relations developed for estimating the rates of SOC sequestration can be verified through soil sampling and analysis for benchmark soils in principal ecoregions.

Removals
Maintenance of engineering devices is critical to their performance and effectiveness. Defective installation and poor maintenance can exacerbate soil erosion. Therefore, regular maintenance of these installations is critical to ensure the benefits of engineering devices.

Associated Impacts
There are numerous ancillary benefits and associated impacts. Important among these impacts are increased productivity; improved water quality; reduced use of fertilizers, especially nitrates; decreased siltation of waterways; reduced methane emissions; associated reductions in risks of flooding; and increased biodiversity in shelter belts and riparian zones.

Relationship to IPCC Guidelines
The effects of erosion-control practices (other than for reduced tillage) on changes in soil carbon stocks are not included in the Guidelines.

Fact Sheet 4.5. Management of Rice Cultivation

Management strategies for rice include irrigation, fertilization, and crop residue management. Rice agriculture is an important source of methane globally; hence, changes in methane emissions likely dominate the overall GHG effect of riceland agriculture on short time horizons (<100 years). Little information is available on carbon stock changes associated with rice paddy management. Rice agriculture tends to increase soil carbon stocks in comparison with adjacent areas without rice. Most practices that reduce methane emissions will likely also reduce the rate of carbon storage in rice paddy soils, however.

Use and Potential

Changes in Soil Carbon Stocks under Rice Management

Increases in organic matter have been observed in paddy field soils that have been in cultivation for 30–100 years in Japan and China. Organic matter may increase in the plow layer by 40–50 percent through rice cultivation and triple in other paddy soils compared with adjacent unflooded arable soils. In deeper soil horizons, increases have been shown to be small (Mitsuchi, 1974). Measurements of annual carbon balance made between May 1991 and April 1997 showed uptake of 0.27–0.32 t C ha^{-1} for an upland single-cropping field, 0.16–0.27 C ha^{-1} for an upland double-cropping field, and only 0.02 t C ha^{-1} for a paddy rice single-cropping field in Japan (Koizumi *et al.*, 1998). These rates agree with long-term rates observed over decades to a century of rice cultivation.

Addition of fertilizers, including manure (Wada *et al.*, 1981; Wada, 1984), plant residues, and chemical amendments (Kimura *et al.*, 1980) increase carbon storage. Storage of 7–26 t C ha^{-1} was observed for 28 to 53 years of manure application in central and southern Japan (average of 0.25–0.5 t C ha^{-1} yr^{-1}). Many researchers have demonstrated that incorporation of rice straw and green manure into rice paddy soils dramatically increases methane emissions. Yagi *et al.* (1997) showed that incorporation of rice straw in soil at rates of 600–900 g m^{-2} after previous harvest increased methane emission rates up to 3.5-fold in Japanese rice paddy fields; application of rice straw compost slightly increased methane emissions (Yagi and Minami, 1990).

Strategies to Reduce Methane Emissions

Strategies to reduce methane emissions from rice cultivation include changes in water management, fertilizer application, and chemical additions. Water management strategies include midseason drainage (Yagi and Minami, 1990; Yagi *et al.*, 1996) and intermittent irrigation (Sass *et al.*, 1992; Chen *et al.*, 1993; Cai *et al.*, 1994). Chemical additions (e.g., sulfate or iron) decrease the activity of methanogens by providing alternative electron acceptors and restricting the availability of substrates in submerged soils (e.g., Hori *et al.*, 1990). Treatments with sulfate have reduced overall methane emissions by 20–77 percent in different experiments (Schutz *et al.*, 1989; Lindau *et al.*, 1993; Denier van der Gon and Neue, 1994). The effect may depend on the amount applied, however: Wassmann *et al.* (1993) reported no effect from sulfate addition to fields in China. Other chemical amendments have included nitrate (Kitada *et al.*, 1993—reduced emissions 23 percent), thiourea (Cai *et al.*, 1994—no effect), and calcium carbide (Bronson and Mosier, 1991—large reduction). Other options for reducing methane emissions include changes in tillage and selection of rice cultivars that are associated with lower methane emissions (Yagi *et al.*, 1997).

Mosier *et al.* (1998) estimate the potential to reduce methane emissions by 8–35 Mt CH_4 yr^{-1} [total emissions estimated at 10–113 Mt CH_4 yr^{-1} (Minami, 1994)] if practices were applied in all amenable areas. This reduction is equivalent to reducing carbon emissions by 0.04–0.2 Gt C. By comparison, the global potential for GHG reduction through carbon storage in rice paddy soils is small: Storage at a rate of 0.25 t C ha^{-1} yr^{-1} over 30 percent of the total rice area (140 Mha) would remove 0.01 Gt C yr^{-1}—equivalent to a reduction of about 1 Mt CH_4 yr^{-1} globally. Carbon storage likely would actually decrease under most of the management practices for methane reduction, decreasing the net GHG effect from that determined from methane emissions reduction alone.

Scientific Uncertainties

There are large uncertainties regarding the area amenable to various rice cultivation practices, and very few data exist on the rates of carbon accumulation or loss and methane emission changes under these practices. The large range in estimates of global methane sources associated with rice cultivation illustrate the large uncertainties associated with extrapolation of methane emissions data over larger land areas.

Time Scales

Data from areas where rice agriculture has been practiced continuously for a century or more show that gains in soil carbon are long term (>100 years). Increased drainage to decrease methane emissions may increase decomposition rates dramatically, however, if the soil changes from largely anaerobic to aerobic conditions.

Monitoring, Verifiability, and Transparency

Measurement of methane fluxes is technically challenging and expensive, although several models now attempt to predict methane emissions from rice. Because methane fluxes are highly variable in space and time, monitoring of methane emissions involves significant effort and cost. Changes in carbon storage may be monitored as changes in bulk density and percent carbon, as discussed in Chapter 3.

Permanence

Carbon storage depends on the degree to which the soil remains anaerobic as opposed to aerobic. Permanence of carbon storage therefore depends on the duration of the cultivation practice. On long time horizons (>100 years), carbon storage changes will dominate changes in methane because of the short atmospheric lifetime of methane.

Associated Impacts

Rice is a major world food crop. The impact of management strategies on costs to farmers and on rice yield and sustainability has yet to be assessed. Very few data are available on nitrous oxide emissions and how they will be affected by various management strategies.

Fact Sheet 4.6. Grazing Management

Grazing management alters the amount and consumption of biomass by domestic animals and wildlife to achieve production and other goals. This technique requires management of intensity, frequency, and seasonality of grazing and animal distribution.

Use and Potential

Grazing influences carbon and nutrient cycling, as well as many other properties of grassland ecosystems (e.g., species composition, light interception, soil compaction). Grazing results in some of the plant carbon being routed through the digestive tracts of animals—where some is converted to weight gain, some is emitted as CO_2 and CH_4, and 25–50 percent is returned in wastes to the grassland. The response of soil carbon stocks to grazing intensity varies for different grasslands. In general, where grazing is managed to maintain or increase plant productivity, soil carbon stocks can be maintained or increased (Conant *et al.*, 2000). Overgrazing—the definition of which is determined by social and economic values, as well as ecosystem function (Abel, 1997)—to the extent of significantly decreasing primary productivity and stimulating erosion decreases soil carbon. For example, Li *et al.* (1997) reported that overgrazing during a 40-year period in Inner Mongolia resulted in a 12.4 percent loss of soil carbon.

In northern Australia, 40 percent of grazing lands have been degraded, resulting in increased incidence of annual grasses and reduced cover of perennial grasses (Tothill and Gillies, 1992). Adoption of reduced stocking rates to increase perennial grasses could sequester 315 Mt C in the top 10 cm of soil over 30 years (Ash *et al.*, 1996); this management change often does not significantly reduce farm income (Stafford Smith *et al.*, 1999). The 0–10 cm layer contains about 16 percent of the profile soil carbon (Dalal and Carter, 1999); thus, total storage may be much greater. The rate of establishment of perennial grasses appears to determine the rate of recovery of soil carbon (Burke *et al.*, 1995). Soil carbon can be lost from these systems—through a combination of inappropriate management and drought—much more quickly than it can be replaced (Bridge *et al.*, 1983; Northup and Brown, 1999). Thus, a continuity of management purpose is required to maintain or increase soil carbon reliably.

Developing countries face additional problems in that the change from nomadic/semi-nomadic grazing management to permanent settlements may result in overgrazing and increased soil carbon losses (Togtohyn *et al.*, 1996). In tropical pastures established after deforestation, overgrazing and nutrient deficiencies have resulted in soil erosion and soil carbon losses, weed invasion, and land degradation (Feller, 1993; Woomer *et al.*, 1997). In contrast, appropriate pasture management can lead to increased soil carbon levels compared with the native forest (Lugo *et al.*, 1986; Cerri *et al.*, 1991; Lugo and Brown, 1996); Neill *et al.* (1997) estimate a potential soil carbon increase of 12–18 t C ha^{-1} in the top 30 cm of soil.

Current Knowledge and Scientific Uncertainties

Data exist for many locations, predominately in temperate countries but increasingly in tropical regions. Most studies use paired-site measurements, which have higher uncertainties than time series-based measurements. The mechanisms by which grazing influences ecosystem carbon storage are relatively well understood, but the magnitude of carbon change as a function of grazing intensity is uncertain for many grasslands. Rates of increase of soil carbon are likely to be higher in more mesic environments than in drier ones (Table 4-4); woodlands and native grasslands show the greatest increases (Conant *et al.*, 2000). Environmental changes such as climate variability, climate change, and atmospheric CO_2 increases are likely to impact carbon stocks.

Methods

Where direct sampling is used, soil ideally should be measured throughout the profile (1 m or more) because of deep carbon storage with some species. In at least some environments, surface soil carbon is correlated with pasture condition (Ash *et al.*, 1996), which could be assessed using field measurements and/or remote sensing (Bastin *et al.*, 1998)—providing the potential for regular, large-scale, geographically referenced assessments. Rapid, direct assessments of aboveground pasture biomass at site (Tothill *et al.*, 1992; Waite, 1994) and regional scales (Hassett *et al.*, 1999) are well established. Allometric relationships allow estimation of root biomass.

Time Scale

Carbon accumulation from improved grazing could extend for 25–50 years or more depending on the rate of plant productivity response. Low rates of change in soil carbon (e.g., because of slow plant recovery from overgrazing) will be associated with long recovery times (Burke *et al.*, 1995); conversely, rapid increases in soil carbon will tend to diminish more rapidly over time. Soil carbon content and bulk density can be measured at intervals of 5 years or more to depths of 1 m or more, depending on species and soil conditions (see Chapter 2).

Monitoring, Verifiability, and Transparency
Rates of change could be verified through repeated field measurements (soil and vegetation) over time, for representative grassland types and grazing regimes. Existing grassland carbon models that incorporate grazing, appropriately parameterized, could be used in verification. Conventional vegetation mapping and/or remote sensing can be used in conjunction with country statistics to verify the areal extent of grazing lands of different types. Statistics or surveys of animal stocking rates could be used for rough estimates of past and current grazing intensity.

Removals
Biomass is removed by livestock and other herbivores, by fire, and through detachment and decomposition. Soil carbon is removed mainly by decomposition and erosion. Both processes are strongly influenced by management.

Permanence
If continuity of management purpose is maintained, storage of carbon in soils and root biomass can provide pools that persist for years to centuries.

Associated Impacts
Improved land conditions where overgrazing is addressed reduces erosion; reduces methane emissions by reducing animal numbers and improving intake quality; and is likely to reduce nitrous oxide emissions as a result of lower levels of excreted nitrogen (Howden *et al.*, 1994) and probably greater uptake by vegetation.

Relationship to IPCC Guidelines
Procedures for estimating grazing effects on carbon stocks are not explicitly defined in the Guidelines. For soil carbon, however, grazing effects will be a function largely of changes in productivity and carbon additions to soils, which can be addressed through the selection of input factors (Reference Manual). In addition, default values are provided for improved, unimproved, and degraded pasture soils (Workbook).

Fact Sheet 4.7. Protected Grasslands and Set-Asides

Changing land use from cropping or degraded lands to perennial grasslands in response to government policies can increase aboveground and below-ground biomass and soil carbon stocks.

Use and Potential

Set-aside of marginal and/or degraded cropland to grassland is likely to be most predominant in countries with agricultural surpluses, but opportunities for set-asides for environmental protection reasons are possible in all countries. The Conservation Reserve Program (CRP) in the United States has resulted in the set-aside of more than 17 Mha of erodable and environmentally sensitive croplands, most of which have been planted to perennial grasses and legumes. Several estimates of the potential carbon storage on CRP lands have been made, ranging from 12 to 18 Mt C yr^{-1} (Gebhart *et al.*, 1994; Barker *et al.*, 1995; Follett *et al.*, 2000; Paustian *et al.*, 2000b). Comparable per-unit area increases have been reported for studies of conversions to grasslands in other temperate regions (Paustian *et al.*, 1998b), although low or non-significant increases have been found in some studies in semi-arid climates (Burke *et al.*, 1995; Robles and Burk,e 1998). Globally, estimates of the potential area of cropland that could be placed into set-asides are on the order of 100 Mha (IPCC, 1996b).

Current Knowledge and Scientific Uncertainties

The processes involved in carbon storage with conversion of cultivated land to perennial grasslands are relatively well understood. Rates vary, however, as a function of many site-specific factors, including stand composition and establishment, fertilization, and nutrient availability (Huggins *et al.*, 1998a). Many rate measurements in the literature are based on paired-site comparisons, which increases levels of uncertainty (Huggins *et al.*, 1998a; Robles and Burke, 1998; Follett *et al.*, 2000).

Methods

Where direct sampling is used, soil carbon should be measured to sufficient depth (e.g., 1 m) to represent the full rooting zone because of deep carbon storage with some species. Carbon stock changes can be directly measured through repeated sampling at intervals of 3–5 years or more; the length of the re-measurement interval depends on the initial carbon level and the productivity of the grassland conversion. Scaling up can be implemented with the use of models and information on the location of the pertinent areas; time since conversion; and climate, soils, and vegetation composition.

Time Scales

Carbon accumulation to levels comparable to native grasslands may take up to 50 years (McConnell and Quinn, 1988) or longer (Dormaar and Smoliak, 1985; Burke *et al.*, 1995) in temperate regions, but recovery rates will vary according to site and management variables. Rapid rates of carbon accumulation are most likely in highly productive grassland conversions in mesic environments.

Removals

Aboveground biomass is subject to removal by livestock and other herbivores, as well as by fire. In the absence of significant erosion or degradation of vegetation, sequestered carbon will be maintained in the soil.

Permanence

If these lands are returned to crop production, there will be a rapid loss of carbon from the soils (Barker *et al.*, 1995), although re-conversion to no-till cropping may reduce these losses. If continuity of management purpose is maintained, storage of carbon in soils and root biomass can provide pools that persist for years to centuries.

Associated Impacts

Associated impacts include reduced crop production, increased animal production (if the land is grazed), increased biodiversity of native grass ecosystems (if they are reestablished), increased wildlife habitat, reduced erosion, improved water quality, and reduced dryland salinity downslope (in landscapes with this hazard). If the land is grazed, methane and nitrous oxide emissions may more than offset the sink provided by increasing carbon pools (Crush *et al.*, 1992).

Relationship to IPCC Guidelines

Grassland set-asides are included in the Reference Manual (which refers to "abandonment of managed lands"); the Workbook provides default values of carbon stock changes for set-asides that are less than 20 years old and older than 20 years.

Fact Sheet 4.8. Grassland Productivity Improvements

The productivity of many pastoral lands, particularly in the tropics and arid zones, is restricted by nitrogen and other nutrient limitations and the unsuitability of some native species to high-intensity grazing. Introduction of nitrogen-fixing legumes and high-productivity grasses or additions of fertilizer can increase biomass production and soil carbon pools.

Use and Potential

Native grasses in grasslands, woodland, and shrublands in some regions are sensitive to heavy grazing, and growth is often nutrient-limited. Introduction of grass and legume species to overcome these limitations has a long history. Such introductions, where successful, can significantly increase primary production by up to a factor of three or more (Montes and Masco, 1996), thereby increasing soil and biomass carbon. Legume and grass introductions increase soil carbon by about 0.08 and 3.97 t C ha^{-1} yr^{-1}, respectively (Conant *et al*., 2000; Tables 4-4 and 4-6), although the value for grasses is significantly higher in a study by Fisher *et al*. (1997) that introduced African grasses to South America.

The area of grassland is variously estimated as 1900–4400 Mha, with a median value of 3220 Mha derived from the IGBP-DIS database (Loveland and Belward, 1997). In developed nations, much of this land may be improved, but in these and other regions there is likely to be considerable further potential. For example, Fisher *et al*. (1994) suggest that there may be 35 Mha of suitable land in South America that could store 100–507 Mt C yr^{-1} by using deep-rooted African grasses. This estimate is thought to be too high because of economic and management constraints (Fearnside and Barbosa, 1998); carbon storage will likely plateau as decomposition rates match increased carbon inputs (Davidson *et al*., 1994).

Species characteristics such as resistant root material appear to influence carbon storage (Urquiaga *et al*., 1998). In some environments, productivity declines over time because of nitrogen immobilization in the root mass (e.g., Robbins *et al*., 1989; Robertson *et al*., 1997). Nitrogen can be released by soil disturbance, which can also release soil carbon in the short term. Wills *et al*. (1996) suggest that there is potential for the establishment of chenopod shrubs over about 6 Mha in New Zealand, with growth of 2–3 t C ha^{-1}. Similar opportunities exist to grow halophyte shrubs in coastal deserts, inland saline soils, or salinized irrigated lands (Table 4-5).

Plant growth on many pastoral lands is limited by lack of nutrients. Additions (particularly phosphorus, nitrogen, and sulfur) can result in large growth responses—which lead to increases in biomass and soil carbon pools, especially in more mesic environments. Normal management practice, however, is to increase the harvesting intensity of the additional biomass produced (e.g., Winter *et al*., 1989), leading to potentially little change in carbon stocks. In the absence of grazing, increases in aboveground biomass carbon are maximized. In New Zealand, for example, fertilizer application to pasture increased total biomass and soil carbon (0–25 cm) by 7.6 t C ha^{-1} in ungrazed grasslands, compared with 6.3 t C ha^{-1} in grazed pasture (McIntosh *et al*., 1997). Even with grazing, fertilizer application in the semi-arid tropics increased aboveground grass biomass from 0.88 to about 3 t C ha^{-1} (Cameron and Ross, 1996). In prairie vegetation, nitrogen fertilization (0.224 t N ha^{-1} yr^{-1}) over 40 years resulted in an increase of about 22 t SOC ha^{-1} compared with unfertilized plots (Schwab *et al*., 1990).

Addition of superphosphate to phosphate-poor soil in conjunction with planting of legumes has been shown to more than double soil organic matter levels (Williams and Donald, 1957; Russell, 1960; Barrow, 1969). Dalal and Carter (1999) suggest that phosphorus and sulfur fertilization over 56 Mha of northern Australia could increase carbon sequestration by 28 Mt C yr^{-1} (280 Mt C over 10 years). Haynes and Williams (1992) reported that SOC (0–4 cm) levels of 37-year-old grazed pastures that received 0, 188, and 376 kg of superphosphate per year were greater than or equal to SOC levels at their "wilderness" site.

Curent Knowledge and Scientific Uncertainties

The basic processes that change carbon stocks under this activity are well understood. The extent of the potential area for this activity and the proportion of potential carbon storage benefits that will result are uncertain. Rates of increase of soil carbon are generally higher in more mesic environments than in drier ones (Table 4-4); these rates vary between ecosystems, with woodlands and native grasslands exhibiting the greatest increases (Conant *et al*., 2000). Background environmental changes such as climate variability, climate change, and atmospheric CO_2 increases may impact on carbon stocks.

Methods

Where direct sampling is used, soil carbon pools must be measured throughout the profile down to at least 1 m because of deep carbon storage with some species. In at least some environments, surface soil carbon is correlated with pasture condition (Ash *et al*., 1996), which could be assessed as described in Fact Sheet 4.6. There is some capacity to model carbon changes, if information about climate, soils, species characteristics, fertilizer applications, and livestock usage is sufficiently well known.

Monitoring, Verifiability, and Transparency
Techniques for repeat direct sampling exist for the biomass and soil components (Chapter 2). There may be opportunities to develop approaches to scale up these results by using the approaches that are briefly described in Fact Sheet 4.6. Agricultural statistics on areas of improved pastures and fertilizer use and associated information on livestock density and characteristics may be needed to make evaluations of likely impacts on system carbon stores.

Time Scales
Soil carbon can continue to accumulate for periods of greater than 40 years (Conant *et al.*, 2000). For example, linear increases in soil carbon over 40- to 50-year periods have been recorded (e.g., Russell, 1960). Carbon accumulation will plateau at some point, however, and this time is likely to vary substantially between systems and with specific activities.

Removal
Biomass is removed by livestock and other herbivores, by fire, and through detachment and decomposition. Soil carbon is removed mainly by decomposition and erosion. Both processes are strongly influenced by management.

Permanence
Cessation of these practices will tend to result in restrictions of further carbon storage. Soil carbon is likely to have varying residence times, extending to millennia. Biomass carbon is vulnerable to disturbance but will recover rapidly in most circumstances, and mean levels of carbon are likely to be higher than in the absence of the activity given suitable grazing regimes.

Associated Impacts
Increased agricultural productivity is likely, as is some loss of biodiversity from native grass ecosystems. Increased legume components are likely to increase acidification rates in tropical (Noble *et al.*, 1997) and temperate (Helyar *et al.*, 1997) pastures, through increased leaching of nitrate and increased productivity. Productivity may fall if pH is lowered too far. Scanning accessions for lower rates of acid excretion may lead to introduction of new varieties that impose less acidification risk (Tang *et al.*, 1998). Optimization of fertilizer application rates can reduce these risks and reduce off-site impacts from nutrient leaching and pollution of waterways and groundwater.

Increases in legumes may result in more nitrous oxide emissions than from native grass pastures (Veldkamp *et al.*, 1998). Increased digestibility and protein content of improved pastures, however, may reduce livestock methane emissions substantially (Kurihara *et al.*, 1999). The radiative forcing of methane emissions is about an order of magnitude larger than that of nitrous oxide emissions from tropical grasslands (Howden *et al.*, 1994). Addition of nitrogenous fertilizer is associated with nitrous oxide emissions. These emissions and methane emissions from livestock are accounted under the current IPCC Guidelines.

Relationship to IPCC Guidelines
The Reference Manual includes procedures for estimating changes in soil carbon stocks associated with pasture management as it affects productivity. Individual practices (e.g., species replacement, fertilization), however, are not dealt with, and general default values are provided in the Workbook only for improved, unimproved, and degraded pasture soils.

Fact Sheet 4.9. Fire Management in Grasslands

Fire management entails changing burning regimes to alter carbon pools in the landscape.

Use and Potential

Fire often is an essential tool in pastoral lands for controlling woody weeds, removing dead biomass, stimulating regrowth, hunting, controlling pests, and clearing land. In many areas, fire regimes are strongly influenced by human actions such as controlled burning, back-burning, firebreaks, and rapid response. Reduced fire frequency or fire prevention tends to increase mean soil, biomass, and litter carbon levels (Jones *et al.*, 1990). In particular, fire management increases the density of woody species in many landscapes (Archer, 1994; Archer *et al.*, 1995; Scholes and Archer, 1997).

The magnitude of carbon storage associated with woody growth can be large. In the Orinoco Llanos, for example, protection from fires for 25 years increased the total carbon pool by a mean of 1.4 t C ha^{-1} yr^{-1}; carbon stocks increased by 5.6 and 29 t C ha^{-1} in the vegetation and soil pools, respectively. If open forest is allowed to form, up to 5.69 Gt C may accumulate over 51 years, if extrapolated over the full area of the Orinoco Plains (28 Mha) (San Jose *et al.*, 1998). Similarly, in 60 Mha of savanna lands in northeastern Australia, Burrows *et al.* (1998, 2000) report that management and environmental changes (predominantly decreased fire frequency) are increasing carbon pools by 30 Mt yr^{-1} in aboveground woody biomass and 10 Mt C yr^{-1} in below-ground woody biomass; aboveground biomass pools could increase 2–5 t C ha^{-1} (open grassland) to 15–75 t C ha^{-1} (closed woodland), and similar changes in below-ground biomass stocks can be expected (Burrows *et al.*, 1998). Similar changes have been found in Africa (northern Guinea savanna): Protection from burning for 26 years resulted in large increases in tree density and basal area compared with burned plots, as well as increases in soil carbon concentrations (Brookman-Amissah *et al.*, 1980). Scholes and Hall (1996) suggest that increased tree cover in savannas could be contributing a worldwide sink of 2 Gt C yr^{-1}. This potential could be limited in parts of Africa and South America, however, because of population pressures on land use and in other regions because of changes toward more sustainable stocking rates that will increase burning opportunities for woody vegetation management (Hall *et al.*, 1998).

Optimizing fire timing may increase biomass in some systems while increasing productivity. Conventional spring and autumn burning in some perennial grasslands, for example, have a long-term negative effect on live biomass and standing crop; midsummer burns facilitate more effective recovery of the grasses (Cox and Morton, 1986). In some ecosystems, burning has little effect on aboveground or below-ground biomass (e.g., Senthilkumar *et al.*, 1998).

Charcoal generated by fires can constitute 8 g C kg^{-1} soil and represent up to 30 percent of the soil carbon content of some Australian soils (Skjemstad *et al.*, 1996); it probably constitutes a significant part of the inert or passive soil carbon pool. Reduction or removal of fire will result over long periods (several centuries) in reduction of this pool.

Methods

Direct, repeated measurements of basal area increase in woody species can be made cost-effectively to high levels of accuracy (e.g., Back *et al.*, 1997; Brown, 1997; Vine *et al.*, 1999); these measurements can be combined with allometric equations for the species involved (e.g., Burrows *et al.*, 2000) to calculate aboveground and below-ground biomass carbon change. Associated soil carbon sampling (see Chapter 2) can be carried out to calculate total system carbon change. The results can be scaled up to regional levels by using statistical sampling methods (Austin and Heyligers, 1989) or via remote sensing (Danaher *et al.*, 1998). For areas where woody components are not a large part of the carbon fluxes, sampling regimes described in Fact Sheet 4.6 can be used.

Sampling for charcoal pools is a unique feature related to this activity. The slow rate of change of these relatively inert pools create uncertainty about whether including this pool is appropriate to the short time frame of the Kyoto Protocol. Furthermore, analysis is likely to be expensive. The size of the pool and the relatively poorly known dynamics, however, suggest that research is needed to determine the significance of this pool under reduced-fire regimes.

Current Knowledge and Scientific Uncertainties

There is copious documentation of the increase in woody density in savannas and other woodlands (e.g., Archer, 1994; Archer *et al.*, 1995), supported by analyses such as soil C^{13}/C^{12} profiles (e.g., Boutton *et al.*, 1998; Burrows *et al.*, 1998). Management appears to be a more significant factor in these changes than environmental factors such as CO_2, climate, and increased nitrogen deposition (Archer *et al.*, 1995). There is difficulty in definitively attributing the proportion of change from each factor, which will vary by location. There is significant variation in rates of accumulation of woody biomass by location (e.g., 0.25–2.5 t ha^{-1} yr^{-1} for South African savannas) (Scholes and van der Merwe, 1996), and differences in potential pool sizes are likely. This situation requires some spatial disaggregation to meet uncertainty levels specified for reporting.

Monitoring, Verifiability, and Transparency

Monitoring to detect change can be carried out by repeat sampling procedures (Back *et al.*, 1997; Critchley and Poulton, 1998). Detailed guidelines for establishing a monitoring network and ensuring its representativeness are given by Brown (1997), MacDicken (1997), and Vine *et al.* (1999). At the individual plot level, the combination of allometry plus measured stem growth increment (including the use of dendrometers) is a powerful and accurate indicator of aboveground biomass flux. A verification and auditing team could evaluate satellite imagery to confirm that the integrity of registered sites was maintained, and auditing could be undertaken of a subset of these sites.

Time Scales

Biomass carbon increases are likely to continue for at least 50 years, though at reduced rates with time (Scholes and van der Merwe, 1996; Burrows *et al.*, 1998). In some systems, the period of accumulation may be 100 years or longer (Archer, 1989).

Removals

Burning transfers a large proportion (up to 90 percent) of the aboveground carbon and nitrogen pools in some grasslands, and 3 percent of the total nitrogen pool to 10 cm in the soil (e.g., Kauffman *et al.*, 1994), into the atmosphere as CO_2, CO, CH_4, N_2O, NO_x, and particulates. The mix of the gases depends on the material burned and the conditions of burning. Burning results in greater soil temperature, which increases soil CO_2 fluxes (Knapp *et al.*, 1998). Thus, burning results in a short-term loss of carbon from ecosystems. Replacement of this carbon, however, generally occurs within 1 to 3 years for grasslands (somewhat longer for woody plants). Where demands for fuel wood or agricultural products dictate or where population density is high (e.g., West Africa), increase in woody plant density will most likely be kept in check by ongoing management.

GHG emissions directly from burning (CO_2, CO, CH_4, N_2O, NO_x) and associated grazing activities (CH_4, N_2O) cause most productive savannas to be net sources of greenhouse emissions (0.06–0.2 t CO_2-eq ha^{-1} yr^{-1} for semi-arid grasslands in Australia; Moore *et al.*, 1997). Elimination of burning along with subsequent increases in woody plants and reductions in grazing livestock numbers can cause these systems to become net sinks for GHGs (about 1 t CO_2-eq ha^{-1} yr^{-1}; Moore *et al.*, 1997).

Permanence

In the absence of human intervention, catastrophic fire is the major threat to carbon storage in savannas. This situation may be impossible to prevent in the long term (50–100 years) (Scholes and van der Merwe, 1996), but because this vegetation type evolved under a regular burning regime, recovery after fire to the pre-fire structure is usually rapid. Extended droughts can also cause mortality (Fensham, 1998). Where mature woody plants die, they are likely to be replaced, provided that fire frequency remains low. Dead trees may remain standing, undergoing slow decomposition and providing a continuing but decreasing carbon store for periods of up to decades. Tree clearing and thinning may occur on dense stands to improve agricultural productivity.

Associated Impacts

For much of the world's broad-leaved savannas, as woody plant density increases, potential livestock carrying capacity declines (e.g., Scanla, 1992)—as does production for other purposes (e.g., Brookman-Amissah *et al.*, 1980). In many ecosystems, fauna and flora species are fire-dependent, so removing or reducing fires may result in localized extinction or decline (e.g., Edroma, 1986). Increasing woody biomass may also reduce environmental services such as catchment water yield. Reducing burning will reduce atmospheric loads of particulates and various other forms of pollution.

Relationship to IPCC Guidelines

The Guidelines deal with savanna burning only in terms of non-CO_2 gases; it is assumed that there are no net losses of carbon, and the system is assumed to be in balance on average.

Fact Sheet 4.10. Agroforests at the Margins of the Humid Tropics

The complex agroforests of Indonesia are indigenous systems invented over generations by local people living at the margins of tropical rainforests in Sumatra (Torquebiau, 1984). After slash-and-burn of a primary forest, food crops are planted along with coffee, pepper, fruit trees (*Lansium domesticum*—duku, *Durio zibethinus*—durian), and the resin-producing damar tree (*Shorea javanica*). The trees eventually shade out the crops, occupy different strata, and produce high-value products such as fruits, resins, medicinals and high-grade timber (de Foresta and Michon, 1994). Biophysical scientists have studied the productivity and ecological dimensions of these systems (Michon and de Foresta, 1996). Villagers in Krui, Lampung Province, who live off these complex agroforests obviously have a higher standard of living than those that grow only crops (Bouamrane, 1996). Detailed explanations of other best-bet agroforestry activities are described elsewhere, including activities that start from secondary forest fallows, reclamation of abandoned *Imperata* grasslands (Garrity, 1997; Friday *et al.*, 1999), or degraded pastures in the Amazon.

Use and Potential

The greatest potential area for expanding agroforestry practices and other forms of land-use intensification is in areas considered "degraded" at the margins of the humid tropics, such as many secondary forest fallows, *Imperata* grasslands, and degraded pastures. These areas amount to about 250 Mha, or 42 percent of the total deforested area of the humid tropics (Sanchez *et al.*, 1994). A major advantage of such lands is their proximity to roads and urban areas because they often were the first ones to be cleared. We assume that 3 percent of these lands (7.5 Mha) plus 20 percent of the 15 Mha annually deforested (3 Mha)—a total of 10.5 Mha—can be put into agroforestry yearly with enabling government policies such as those described by Fay *et al.* (1998) and Tomich *et al.* (1998).

Methods

Changes in time-averaged aboveground and soil carbon can be measured as described elsewhere in this Special Report. Newly developed allometric equations for estimating biomass of shrubs and small trees provide an important tool (Palm *et al.*, 2000).

Current Knowledge and Scientific Uncertainties

Data by Palm *et al.* (2000) can be used to calculate the difference between carbon lost when moist tropical forests are transformed into agroforests and carbon lost when forests are transformed into croplands or pastures. In this case, the estimate is 46 t C ha^{-1}. The largest uncertainty is the area that will benefit from such practices. Other human-induced activities at the forest margins that do not include trees (e.g., active slash-and-burn cropping phase, pastures, and grasslands) are sources rather than sinks of carbon.

Monitoring, Verifiability, and Transparency

A 10-year period is recommended to assess the impact on soil carbon stocks. Direct measurement (time-averaged) of aboveground and soil stocks should be used for monitoring. Combining present algorithms for estimating biomass in shrubs and small trees (Palm *et al.*, 2000), standard soil carbon sampling, and GIS techniques appears feasible. The level of carbon stock change can be readily verified through land-based techniques described above. Assumptions and methodologies associated with this practice can be explained clearly to facilitate replication and assessment. Scientific methods are open to review and replicable over time.

Permanence

Aboveground carbon stocks can be rapidly eliminated by shifting from this practice to slash-and-burn, followed by cropland or pasture. About half of the carbon stored in the soil is likely to have a turnover rate of >20 t C yr^{-1}. Stopping the activity would lead to an estimated loss of soil carbon of 50 percent in about 5 years.

Associated Impacts

Agroforestry is an economic activity that helps to reduce or eliminate poverty at the forest margins if high-value products (fruits, resins, medicines, and high-grade timber) are produced. Agroforestry also facilitates land tenure because it encourages settled farming systems (Fay *et al.*, 1998). Tree-based systems can serve as a methane sink (Murdiyarso *et al.*, 1994; Palm *et al.*, 2000). Methane emitted by 1 ha of paddy rice can be absorbed by 24 ha of agroforests if the patches are close enough in a landscape mosaic. There seems to be no difference in N_2O emissions between the original forest, agroforests, cropland, or grassland at the humid tropical forest margins (Palm *et al.*, 2000).

Agroforestry systems are more biodiverse above ground than crops, grasslands, and secondary forest fallows in the humid tropics (Gillison, 1999). Differences in below-ground biodiversity seem less important (Swift, 1999). Plant diversity in mature complex agroforests of Indonesia is on the order of 300 species per hectare, which approximates that of adjacent undisturbed forests (420 plant species per hectare). The diversity of bird species in these agroforests is approximately half that of the original rainforest, and almost all mammal species are present in the agroforest (de Foresta and Michon, 1994). This biodiversity is possible because such agroforests, which are composed of hundreds of small plots that are managed by individual families, occupy contiguous areas of several thousand hectares in Sumatra. Such agroforestry "corridors" are an important tool for avian biodiversity conservation in Central America (Current *et al.*, 1998).

In view of human migrations to the forest margins, the optimal tradeoffs between carbon capture and economic and social benefits are an important policy determination. Examples of such tradeoffs are described by Gokowski *et al.* (1999), Vosti *et al.* (1999), and Tomich *et al.* (1998, 1999).

Relationship to IPCC Guidelines

Changes in woody biomass stocks associated with agroforestry practices can be included by using the procedures for estimating "changes in forest and other woody biomass stocks." The Guidelines do not provide specific default values for agroforest systems, however. Changes in soil carbon stocks can also be estimated using the Guidelines, although specific examples and default values for agroforest systems are not provided. Soil carbon stock change estimates would require values for "Input Factors," and potentially "Tillage Factors," related to the level of management of agroforests. For land-use conversions to agroforestry, "Base Factors" exist in the Workbook for conversions from forest and from cultivated land.

Fact Sheet 4.11. Replenishing Soil Fertility through Agroforestry in Subhumid Tropical Africa

Leguminous fallows of *Sesbania sesban*, *Tephrosia vogelii*, *Gliricidia sepium*, *Crotalaria grahamiana*, and *Cajanus cajan* accumulate 0.1–0.2 t N ha^{-1} in their leaves and roots in 1–2 years. These large amounts of nitrogen are uncommon in the organic farming literature; they are equivalent to mineral fertilizer input levels in modern agriculture. Upon incorporation of leguminous biomass into the soil and subsequent mineralization, these improved fallows provide sufficient nitrogen for one to three subsequent maize crops—doubling to quadrupling maize yields at the farm scale (Rao *et al.*, 1998; Kwesiga *et al.*, 1999). There are no transport costs involved because all of the nitrogen is fixed in the same fields where crops are grown in rotation.

In phosphorus-deficient soils, farmers are beginning to use phosphate rock applications of 0.125–0.25 t P ha^{-1} as a capital investment, with an expected residual effect of 5 years. In addition, biome transfers from hedges of the wild sunflower tithonia *(Tithonia diversifolia)* have shown large yield increases of maize and high-value crops such as vegetables in western Kenya (Jama *et al.*, 1999a,b). Tithonia leaves contain high nutrient concentrations (3 percent N, 0.3 percent P, 3 percent K) and decompose rapidly in the soil, providing a source of soluble carbon that enhances nutrient cycling (Gachengo *et al.*, 1999). Combinations of tithonia biomass with phosphorus fertilizers have been particularly effective (Palm *et al.*, 1997; Nziguheba *et al.*, 1998; Rao *et al.*,1998). Farmers incorporate leguminous fallows, tithonia, and phosphate rock into their farming systems in a variety of ways. Food security has been effectively achieved with these practices. Economic analysis shows high net present values for these technologies (Sanchez *et al.*, 1997b).

The next step envisioned is planting vegetables that produce high-value products as a way to increase small farmer income and reduce poverty. Some farmers have reported increases in their net profits from US$91–1665 yr^{-1} when they have shifted from maize to vegetables in their now-fertile soils (Nyasimi *et al.*, 1997). A further step will be the switch to newly domesticated tree crops that produce high-value products. These "Cinderella" species—so called because their value has been largely overlooked by science although they are appreciated by local people—include indigenous fruit trees and other plants that provide medicinal products, ornamentals, or high-grade timber (Leakey *et al.*, 1996). One example is *Prunus africana*, a timber tree that is indigenous to montane regions of Africa. A substance extracted from its bark to treat prostate gland-related diseases has an annual market value of US$220 million (Cunningham and Mbenkum, 1993; Simons *et al.*, 1998). Because these trees are cut and killed in indigenous forests and the bark shipped to Europe, *Prunus africana* is now in the CITES Appendix II list of endangered species. With domestication, this tree is now being turned into a crop, as researchers select superior ecotypes, ways to harvest the bark sustainably, and eventually the development of extraction industries located in nearby rural areas (Simons *et al.*, 1998).

Use and Potential

These practices are quite new, having been tested in research in the 1990s; about 20,000 smallholder farmers currently practice them on roughly 20,000 ha, primarily in western Kenya and eastern Zambia. The total potential area could become very large, assuming enabling policies in 10 percent of smallholder farms in subhumid Africa (8.1 Mha) and 25 Mha in the subhumid tropics of Latin America and non-paddy rice areas of Asia, all during the next 20 years.

Methods

Changes in time-averaged aboveground and soil carbon can be measured via methods described elsewhere in this Special Report.

Current Knowledge and Scientific Uncertainties

Estimates cited are very preliminary. Hard data are now being developed at ICRAF. Starting from soils that are 40–60 percent depleted in carbon and have very little aboveground biomass, measurement of differences can be made and modeled. The largest uncertainty is the area that will benefit from this technology.

Time Scale

A 10-year period is recommended to assess impact on soil carbon stocks.

Monitoring, Verifiability, and Transparency

Direct measurement (time-averaged) of aboveground and soil stocks should be used for monitoring. Combining present algorithms for estimating biomass in shrubs and small trees, standard soil carbon sampling, and GIS techniques appears feasible. The level of additional sequestered carbon can be readily estimated by the techniques described above. Assumptions and methodologies associated with this practice can be explained clearly to facilitate replication and assessment. Scientific methods are open to review and are replicable over time.

Permanence

About half of the carbon stored in the soil is likely to have a turnover rate of >50 years. Cessation of the activity would lead to a loss of soil carbon, which has been estimated to be 40–60 percent in about 20 years.

Associated Impacts

Major increases in food security and poverty reduction seem assured. Spillover effects could occur in developing rural industries and employment. There would be less dependence on nitrogen fertilizer, the manufacture of which entails major consumption of fossil fuels (Schlesinger, 1999). Using biologically fixed nitrogen would lead to potential savings in N_2O emissions. These practices also would reduce dependence on superphosphates, whose manufacture is also highly fossil fuel-intensive with high risk of pollution. Increases in soil conservation and below-ground biodiversity are likely. Extinction of the endangered tree species *Prunus africana* would be less likely.

Relationship to IPCC Guidelines

See Fact Sheet 4.10.

Fact Sheet 4.12. Forest Regeneration

Forest regeneration is the act of renewing tree cover by establishing young trees naturally or artificially—generally, promptly after the previous stand or forest has been removed. The method, species, and density are chosen to meet the goal of the landowner. Forest regeneration includes practices such as changes in tree plant density through human-assisted natural regeneration, enrichment planting, reduced grazing of forested savannas, and changes in tree provenances/genetics or tree species. "Human-assisted natural regeneration" means establishment of a forest age class from natural seeding or sprouting after harvesting through selection cutting, shelter (or seed-tree) harvest, soil preparation, or restricting the size of a clear-cut stand to secure natural regeneration from surrounding trees. "Enrichment planting" means increasing the planting density (i.e., the numbers of plants per hectare) in an already growing forest stand.

Use and Potential

This activity influences carbon storage through changes in the growth of aboveground and below-ground tree biomass and changes in wood end use. The impacts on the litter layer and soil vary with many factors (see other chapters in this Special Report). Generally, over the rotation period, annual growth in carbon storage in tree biomass in most cases is much higher than soil/litter carbon storage. Regarding carbon sequestration, the connection between forest regeneration and the end use of wood is important. For example, higher planting density is not generally preferable for carbon sequestration. High densities can lead to rapid crown closure and early growth, but such stands reach maximum increment early and may suffer onset of mortality and rapid growth decline—thereby potentially becoming sources of carbon considerably faster than stands that are managed at lower densities. Additionally, trees that are grown in less dense conditions generally reach a suitable size for solid wood products earlier; as a result harvest and conversion to long-lived wood products occurs sooner, adding to the stock of sequestered carbon and substituting for non-wood products that may use more fossil fuel in their production.

All of these activities are used today, with varying intensities, in many countries without consideration of carbon sequestration, largely on the basis of decisions about present costs and expected future benefits from timber and other forest values. If carbon management is introduced, these activities could be effective in sequestering considerably more carbon than is occurring today (e.g., Lunnan *et al.*, 1991; Hoen and Solberg, 1994; Xu, 1995; Row, 1996; Nabuurs *et al.*, 1999; Ravindranath *et al.*, 1999).

No published estimate of the global carbon sequestration potential of these practices is available.

Methods and Uncertainty

At least two methods can be used to quantify changes in carbon stocks from these practices: existing forest growth yield tables and forest inventories that measure standing and incremental aboveground stem volume. The second method can be done as accurately as one wants, though with increasing costs. The first method is less accurate but would be good enough in some cases, at least in the initial phases, and could later be checked by more accurate inventories to secure adequate precision for verification.

Tree biomass growth (and correspondingly carbon accumulation) processes are well known. Soil carbon accumulation generally is less certain.

Mortality caused by wind, fire, pest, rot, or insect damages can lead to a loss of carbon pools for all of these activities. Most yield tables include estimates of natural plus mortality rates (for example, mortality is estimated to be 0.4 percent of living trees per year in Norway). Regarding accidental mortality, fewer estimates exist. Thorsen and Helles (1998) estimate the probability of total damage caused by strong winds for a *Picea abies* stand in Denmark to be about 1.5–3.0 percent per year if the stand were thinned no more than a year previously (the probability declines strongly with time after thinning). Climate changes may increase the risks of tree loss—for example, by more frequent winds or increased insect attacks.

The accuracy of national forest inventories varies considerably. Hobbelstad (1999) reports that the present national inventory of Norway gives estimates of total standing volume and annual yield for the country as a whole at an accuracy of 1.6 percent as standard deviation. This level of accuracy is based on 8000 permanent sample plots, of which 20 percent are measured each year. At a regional level, the standard deviation is 3.2 percent (the country is divided into four regions). Countries such as Sweden and Finland have the same accuracy in their forest inventories.

In addition to national inventories, Norway conducts a county inventory, which covers one-third of the counties every 5 years. This inventory provides a county-level accuracy that corresponds to a standard deviation of 3–4 percent (Norway is composed of 20 counties). The costs for the national inventory and county inventory are about US\$0.17 ha^{-1} yr^{-1}, covering a total productive forest area of about 7.5 Mha.

Time Scale and Monitoring

The accumulation time for aboveground and below-ground biomass ranges from 5 years (for the shortest rotation times in tropical plantations) to 150 years or more on low-potential sites in boreal forests. The tree biomass carbon accumulation process is not difficult to quantify and predict, particularly where well-developed forest growth and yield models exist. Allometric studies provide factors that can be used to estimate total biomass (aboveground and below-ground) from the timber yield tables (Marklund, 1988; Birdsey, 1996). Soil carbon accumulation processes are generally less confidently predicted, but increasingly there are research results to guide these estimates.

The duration of the carbon biomass stored in forests or forest products depends on factors such as the following:

- Forest rotation length (or harvest intensity over time for selection felling systems)
- Thinning intervals and intensity
- Decaying time of timber not used (roots, branches, stumps, logging residues)
- Average lifetime of end use of wood and decay time of end-use product after its use.

These times vary widely; the best estimates for 1992 come from Norway (Hoen and Solberg, 1994), as tabulated below.

End-Use Category	**Anthropogenic Time** (years from felling until decay starts)	**Decay Time** (years until all fiber has decayed)
Bark in land fillings	0	8
Bark for burning	0	1
Needles	0	7–11
Branches, stumps, stems in forest	0	12
Root system after felling	0	100
Construction material	80	80
Furniture and interiors	20	50
Impregnated lumber	40	70
Pallets	2	23
Losses	0	1
Composites, plywood	17	33
Sawdust	1	2
Pulp/paper	1	2
Fuelwood	0	1

Verifiability

In principle, all of these activities can be verified, at varying accuracy and costs. The capacity varies between countries, and combinations of methods might be applied. To estimate the carbon impact from enrichment planting, for example, one would measure a control plot and take the difference as the estimated impact of the activity. Where several activities are combined, land-based measures will probably be required. These estimates can be made from yield models (if available), historical inventory data for similar stands, or a combination of these methods.

Transparency

The assumptions and methodologies associated with this activity can be explained clearly to facilitate replication and assessment of carbon impacts. The scientific and technical methods are open to review and are replicable over time.

Permanence

Carbon will be stored in a forest as long as the forest is not harvested or damaged by natural events. Where the harvested timber is used for bioenergy or forest industry production, for example, the degree of permanency will depend on the end use of the timber extracted and the carbon substitution impact of these products.

Associated Impacts

Improved natural regeneration would result in most cases in increased biodiversity and recreational/landscape improvements. These effects could also result from increased mixed-species stands and higher tree density in savanna woodlands. Regarding environmental damage, tree planting and change of tree species could result in decreased biodiversity and reduced recreational benefits, particularly if monoculture stands are emphasized. All activities will produce more jobs and income in the establishment phase, as well as at harvesting and end-use activities, especially in rural areas. The potential is probably highest in tropical countries; as such, developing countries may benefit more than developed countries. The costs and benefits of associated impacts are difficult to quantify. The economic benefit from increased timber production, by comparison, is easy to estimate by using market prices. Leakages through market dislocations may occur. For example, increased investment in forest management for increased carbon sequestration may increase the long-term timber supply, implying lower future timber prices and thereby reducing total forest management investments. These leakages, however, are probably not higher than for those occurring for other GHG mitigation options in other sectors of the economy (see Chapters 2 and 5 for more discussion of leakage).

Relationship to IPCC Guidelines

All forest management practices that affect the rate of biomass increment and biomass losses through harvesting or other removals are implicitly included in the Reference Manual under the calculations for "Changes in Forest and Other Woody Biomass Stocks." Changes in soil carbon, litter, and below-ground biomass stocks as affected by forest management practices are not included in the Workbook.

Fact Sheet 4.13. Forest Fertilization

Fertilization is the addition of nutrient elements to increase growth rates or overcome a nutrient deficiency in the soil. Fertilization can be divided into two sub-activities: increasing the quantity of fertilizer and improved fertilizing quality (i.e., timing and dosage) so that as much nutrient as possible is taken up by the trees and correspondingly less becomes waste to groundwater. Unintentional fertilization is occurring in many forests downwind of industrial centers, as a result of the deposition of nitrogen and sulfur from the atmosphere.

Fertilization leads to higher growth of aboveground and below-ground biomass, thus increasing carbon storage. The process is well understood except for some of the soil processes, and reliable models are available in many countries to predict the increased biomass growth. Among the few studies on forest fertilization and carbon sequestration are Hoen and Solberg (1994), Lunnan *et al.* (1991), and Nabuurs *et al.* (1999).

Use and Potential

This practice is used in most plantation management systems around the world, with varying intensity. In capital intensity forestry in Scandinavia, it represents one of the most profitable investments in forestry on low to medium site classes.

Lunnan *et al.* (1991) report the effects of carbon storage on boreal forests in Norway with (i) two applications of fertilizer, each with 173 kg N (as NH_4NO_3) ha^{-1}; (ii) 10 and 15 years between the two fertilizations; and (iii) 30 years before clear-fellings. The table provides estimated carbon storage and costs in forest biomass (only in biomass of stem, branches, and root—not in soil and humus) by fertilizing stands of *Picea abies* and *Pinus sylvestris* in boreal forests.

Factor	Species*	
	Picea abies	*Pinus sylvestris*
Carbon storage (tC ha^{-1} yr^{-1} for 10 years)	0.79	0.65
Costs per fertilization (1999 US$ ha^{-1})	250	250
Cost efficiency (1999 US$ t^{-1} C)	5.5–18.0	5.5–29.3

*On relatively low site classes; costs and carbon fluxes (including end-use decays) discounted with 7-percent real annual rate of interest.

The rate of the carbon storage varies with many factors—such as species, site productivity, climate, soil conditions, the degree to which nutrition is the limiting growth factor, and fertilization amounts.

There are no global or regional statistics available regarding the total area to which forest fertilization may apply. This practice might be feasible at the country level, however. For example, Lunnan *et al.* (1991) estimate that the potential area for Norway per year is between 6 and 20 percent of the total productive forest area of the country.

Methods and Uncertainty

The same factors are valid here as those described in Fact Sheet 4.12. If yield tables or models are used, one needs to know the dose/response relationship between fertilization amount and stem volume increase. The ecological impacts of forest fertilization may not be fully understood yet for some ecosystems.

Monitoring, Verifiability, Transparency, and Permanence

The situation is similar to that described in Fact Sheet 4.12.

Associated Impacts

Associated positive environmental benefits are unlikely to result from this activity. In some areas, however, it may have several negative environmental impacts. The use of fertilization may increase the leakage/emission of N_2O and NO_x to air, ground, and water and influence soil processes (see other chapters in this Special Report). Impacts on jobs and income are about the same as described in Fact Sheet 4.13. The main barriers today for this activity are relatively high costs and the possibility of negative environmental impacts.

Relationship to IPCC Guidelines

See Fact Sheet 4.12.

Fact Sheet 4.14. Forest Fire Management

This activity includes efforts to regulate the recycling of forest biomass from fires, maintain healthy forest ecosystems, and reduce total emissions of CO_2 and other GHGs.

Fire management practices vary greatly between forest ecosystems, depending on the historical or natural fire regime for the ecosystem, the current condition of the forest, and the management objectives of the landowner. In many cases, mechanical vegetative manipulation or planned prescribed fires can be utilized to maintain conditions approximating historical ranges. In some cases, where past fire suppression has resulted in fuel buildups, the ecological risk of burning—either in a prescribed or wildfire—is now so great that mechanical means of fuel removal may be a precursor to any management scheme designed to restore fire's role in the ecosystem (Neuenschwander and Sampson, 2000). Where such fuel removal is carried out, wildfire ignitions are less likely to result—and when they happen, they will often burn at lowered severities, with reduced fuel consumption, heat production, and GHG emissions. Because fire management is an integral part of forest management, it must be viewed in connection with other management practices, including harvest and wood utilization, to evaluate its full carbon flux effect.

Use and Potential

Destruction of forest biomass by burning releases large quantities of CO_2 and is estimated to create 10 percent of annual global methane emissions as well as 10–20 percent of global N_2O emissions. Thus, fire can have a significant effect on atmospheric chemistry (IPCC, 1992). The process is well known in terms of general effects, but it has many uncertain parameters in relation to specific fire events because fire effects are related to fuel amounts, arrangements, and conditions as well as weather conditions at the time of combustion—all of which can be highly variable or unpredictable (Goldammer, 1990; Dixon and Krankina, 1993; Price *et al.*, 1998; Neuenschwander *et al.*, 2000).

Monitoring, Verifiability, Transparency, and Permanence

These factors are very difficult to achieve in relation to wildfires, which are such highly stochastic events that any estimate of the effect of management on actual changes in wildfire dynamics is likely to be speculative. Post-event monitoring, however, has begun to provide estimates that can be used to build predictive models (Neuenschwander and Sampson, 2000). Quantification of the carbon stock impact can be estimated by comparing data on historical forest fire frequency and fuel consumption with experience in a reporting period. Forecasting can use techniques that take local weather history and use probability analysis to predict the occurrence of future weather combinations that will support large wildfires (Neuenschwander and Sampson, 2000). Such forecasting relies on the assumption that future climate will not alter fire weather occurrence to a great extent and that it will not contribute to large areas of forest die-off that could create hazardous fuel levels; neither assumption may be fully warranted at this time (Apps *et al.*, 1999a).

Associated Impacts

The associated environmental impacts of this activity are difficult to generalize because some ecosystems need fire to be sustainable. Restoring near-historical fire regimes may be an important component of sustainable forestry but may also require access (road construction) and other practices that may create other environmental effects that are known to be deleterious. Wildland fire in areas near human habitation creates major hazards to life and property, so intentional restoration of historical fire regimes is a risky management activity that is associated with serious social and economic considerations that often restrict its application. Jobs and income may increase if the timber saved from fire is used in sustainable forest harvesting. The costs of fire prevention in remote forest regions and the risks of accidental fire escape during prescribed fires are probably the most important reasons that this activity is not more widespread.

Relationship to IPCC Guidelines

Emissions of CO_2 and other GHGs from forest fires following deforestation and conversion to other land uses are included in the Guidelines. CO_2 emissions from fires within managed forests are not included, but non-CO_2 emissions are.

Fact Sheet 4.15. Pest Management

Pest management is the application of approved strategies to maintain a pest's population within tolerable levels. Improved pest management may prevent damage and tree mortality in forests and thus increase carbon stocks. Processes related to tree health are well known, whereas the reasons pests occur and how they can be prevented are less well known. Interactions between climate change, pest populations, and wildfire (see Fact Sheet 4.14) are likely to become more important in affecting forest carbon stocks in managed and unmanaged forests in the next century.

Use and Potential

Nabuurs *et al.* (1999) (based on Kurz *et al.*, 1992; Townsend *et al.*, 1996; Gillies and Leckie, 1996) estimate the area affected by insect attacks in Canada to be roughly equivalent to the area burned. They assume that reducing the area affected by insect attacks by 50 percent would save the same amount of carbon as fire protection (about 2.3 Gt C yr^{-1}). These phenomena are not disconnected, however. Large insect epidemics can create areas of dead and dying trees that provide a fuel source for very large wildfires in the right ignition and weather conditions. As a result, pest management may be one important parameter of fire management and long-term forest ecosystem health.

Methods and Scientific Uncertainties

This activity is very uncertain. Even in areas where access is possible, effective methods of predicting and preventing pest outbreaks may be lacking. In remote areas, mitigation or treatment may be impractical or impossible. Few empirical studies exist to help managers affect why and how pest populations move from endemic to epidemic.

Time Scale, Monitoring, Variability, and Transparency

There seem to be few practical ways in which this activity, by itself, could be linked directly to changes in carbon stocks. As part of a broad forest management activity, it could contribute to measured changes in carbon stocks achieved by land-based methods. Separating the carbon impacts of this single activity would seem to be difficult or impossible.

Associated Impacts

Where biocides are used to control pests, this activity may result in reduced biodiversity and lower landscape/recreational benefits. On the other hand, where it prevents large-scale forest die-off, it may dramatically increase landscape, recreational, watershed, and other benefits. High costs and uncertainty about the effectiveness of various mitigation measures are among the reasons that this activity is limited in many areas.

Relationship to IPCC Guidelines

See Fact Sheet 4.12.

Fact Sheet 4.16. Forest Harvest Quantity and Timing

Harvesting is the procedure by which a forest stand is logged, with an emphasis on meeting logging requirements and attaining silvicultural objectives. Harvest scheduling is a process for allocating cutting and other silvicultural treatments over a forest, with the emphasis on which treatments to apply as well as where and when to apply them. The practice relates to when and how harvesting is done (e.g., thinnings, including pre-commercial thinnings; selection; or clear-cut harvests) and the timber volume extracted.

Use and Potential

Harvest intensity affects the quantity and quality of timber produced. The carbon impact is directly connected to the end use of the wood products. For example, increasing the rotation time, lengthening the period between harvest operations, or reducing the volume extracted may lead to reduced growth rates and reduced carbon sequestration in the forest, as well as producing less wood for bioenergy or for replacing energy-intensive products such as steel, aluminum, plaster board, and concrete.

Burschel *et al.* (1993) calculated that longer rotation periods could increase carbon stocks in Germany by 0.7–1.8 Mt C yr^{-1} in the first 20 years but would have adverse effects (on industrial roundwood productivity and carbon sequestration) if further extended because of higher risk of natural damages. For forests in Russia and the northwestern United States, intensive management reduces forest carbon stocks below the levels found in native forests, whereas increasing rotation lengths, retaining live trees through harvests, and decreasing site disturbance related to harvest and regeneration can substantially increase forest carbon stock (Krankina and Harmon, 1994; Krankina *et al.*, 1996). These two studies did not consider the end-use aspects, however. Several authors have studied the impact on carbon sequestration (e.g., Plantinga and Birdsey, 1994; Boscola and Buongiorno, 1997; Boscola *et al.*, 1997; Hoen and Solberg, 1997); the conclusion is that including the benefit of carbon will encourage the rotation age to increase.

Forest growth studies have shown that the impact of different thinning practices on total growth (and carbon sequestration) is insignificant in Germany (Strich, 1998). Row (1996) found that a thinning regime produces lower total carbon stocks in a 50-year Loblolly pine rotation than thinned stands; Lunnan *et al.* (1991) came to the same conclusion regarding lengthening the rotation periods when studied at the stand level. In addition, the decrease in timber supply caused by increased rotation length may shift the demand for timber to other stands that will be harvested instead—creating a high probability of leakage.

Optimal thinning and clear-felling times (and quantities) are interlinked and depend on other forest management measures (e.g., fertilization or plant density) and objectives. Hoen and Solberg (1994) carried out one of the few studies that analyzes thinnings, clear-fellings, and other silviculture measures (e.g., fertilization, types and intensity of regeneration) simultaneously for a boreal forest region, keeping harvest levels constant and maximizing carbon storage (including the end use and decay of wood products) for the region over a long period. The study illustrates that thinning and clear-felling times will be significant and will complement each other. For example, thinnings on good site classes substitute for clear-fellings on low site classes that have high standing volumes (but low annual growth and carbon sequestration potential if harvested). Boscola *et al.* (1997) show that the combination of harvest cycles and minimum cutting diameters can maximize carbon sequestration, at costs of US\$1.2 t^{-1} C sequestered, for an increase in cutting cycles from 40 to 50 years in lowland tropical rain forest in Malaysia.

Methods, Uncertainty, Time Scale, and Monitoring

Yield tables or ordinary inventories seem sufficient for measuring changes in carbon stocks with confidence.

Verifiability, Transparency, and Permanence

See Fact Sheet 4.12.

Associated Impacts

This practice could have positive and negative environmental benefits regarding biodiversity, recreation, and landscape management, depending on local circumstances. The main barriers are the lack of incentives, including the risk of negative economic incentives.

Relationship to IPCC Guidelines

See Fact Sheet 4.12.

Fact Sheet 4.17. Low-Impact Forest Harvesting

Low-impact harvesting entails harvesting methods that are developed and executed to provide minimum disturbance to soil, remaining vegetation, and extracted trees. The practice influences the carbon stock change from trees left in the forest after harvest, as well as the growth (and corresponding carbon storage) of new trees and vegetation. It also influences carbon storage in end products by influencing timber quality and affecting the type of utilization that can be made of the timber.

Use and Potential

Winjum *et al.* (1998) estimate regional and global harvest volumes and corresponding carbon storage and emission changes. Putz and Pinard (1993) and Pinard and Putz (1996, 1997) demonstrate that reduced-impact logging is an activity that substantially decreases the emission of carbon in tropical forests because of reduced damage to soil and remaining trees.

Scientific Uncertainties

Uncertainties associated with this practice relate to quantification of differences between carbon stock changes associated with better practices compared to those normally applied in a country or region. Estimates are needed of damage to soil, remaining trees, and other vegetation and the consequences of these damages for carbon stocks. In addition, regarding damage on extracted trees, estimates of the decay time of end products into which harvested timber is manufactured are needed. None of these measurements are included in the present IPCC methodology. Compiling results from several measurements in different forests may make possible the development of benchmarks for typical high-impact logging in different forest types and economies.

Monitoring, Verifiability, and Transparency

If the uncertainty factors are adequately met, monitoring and verifying this practice should be possible. Assumptions and methodologies associated with this practice are easily explained for replication and assessment of impacts.

Associated Impacts

In most cases this practice will have positive environmental benefits regarding biodiversity, recreation, and landscape management; no associated environmental damage is likely. In addition, this practice is likely to increase the economic value of remaining trees, as well as logged trees. Boscolo *et al.* (1997) estimate the cost efficiency of reduced-impact harvesting in a lowland tropical rainforest in Malaysia as \$5.5 t^{-1} C. No conflicts are likely. The equity implications are small. The main barrier that prevents these activities from being implemented is lack of economic incentives. Without some incentives based on carbon impacts, costs are likely to remain higher than benefits.

Relationship to IPCC Guidelines

See Fact Sheet 4.12.

Fact Sheet 4.18. Restoration of Former Wetlands

This practice entails restoration of wetlands that formerly were used for agriculture, forestry, or urban/industrial uses through plugging of drain ditches, restoration of prior hydrological conditions, or artificial water diversions.

Use and Potential

Conversion involves inundation of previously drained areas. This inundation will increase CO_2 storage as organic matter because of reduced oxygen in soils, but it may also increase CH_4 emission, depending on whether the new water table is close to the soil surface. Carbon storage rates will be rapid initially, then slow over time (few estimates are available in the literature, however). Overall GHG emissions changes will likely entail small sinks—or even sources, if CH_4 emissions dominate over CO_2 sinks in stored organic matter. The area of wetland that has been converted since 1900 to agriculture, forest, and urban land (mostly in temperate and tropical regions) is poorly known, with estimates ranging from 6 percent (Armentano and Menges, 1986) to as much as 50 percent (Moser *et al.*, 1996) . Most of the loss of wetland area has been associated with drainage for agriculture in temperate regions, though pressures on tropical wetlands have increased in recent decades (Moser *et al.*, 1996). The maximum area available for restoration is in the range of 30–250 Mha (based on the global wetland area of 570 Mha reported by Matthews *et al.*, 1996).

Current Knowledge and Scientific Uncertainties

Carbon storage results from decreased decomposition rates associated with inundation. Methane emissions will increase, however. Rates of carbon accumulation in new wetlands can be high—0.1–1 t C ha^{-1} yr^{-1} (Tolonen and Turunen, 1996)—but slow over time as carbon accumulates and decomposition losses begin to offset additions. The CO_2 sink that is created by reestablishing the wetland may be offset by increased methane emissions because inundation will cause some of the decomposed carbon to be released as methane. Completely offsetting carbon storage at rates of 0.1–1 t C ha^{-1} yr^{-1} would require CH_4 emissions of ~2–20 g CH_4 m^{-2} yr^{-1}. These rates are in accord with those measured in natural freshwater wetlands (which range from ~7 to 40 g CH_4 m^{-1} yr^{-1}; Bergkamp and Orlando, 1999); the actual amount of emitted methane will depend, however, on the degree of inundation (methane can be oxidized before emission if the water table is below the surface) and on whether other oxidants are present. For example, carbon storage effects will dominate the net GHG budget in coastal wetlands, which do not emit significant amounts of methane.

There are large uncertainties about the actual areas of wetland converted, some of which result from differing definitions of wetlands and differing interpretations of whether rice agriculture is included as wetland. Uncertainties exist about carbon storage rates and methane emission changes (likely 10–30 percent of reported values) because very few published studies exist for these factors in restored wetlands.

Time Scales

Carbon storage is long term; decomposition rates of organic matter stored under anaerobic conditions are slow (decades to millennia, depending on the degree of inundation and the type of vegetation in the new wetland).

Monitoring, Verifiability, and Transparency

Methane fluxes must be monitored. Because methane fluxes are highly variable in space and time, monitoring of methane emissions involves significant effort and cost. New carbon storage may be determined by monitoring changes in bulk density and percentage of carbon. Wetland area may be verified by remote sensing (regionally) or repeated surveys (locally).

Permanence

Carbon storage depends on continued inundation. Stored organic matter is susceptible to rapid decomposition if the wetland dries out and exposes it to oxic conditions. A wetland may dry because of climate change or because of a reversion to drainage.

Associated Impacts

Wetlands have positive impacts on water quality, provide protection against local flooding, help control soil and coastal erosion, and increase biodiversity. Several international negotiations pertain to these aspects of wetlands—in particular, the Convention on Wetlands, the Convention on Biodiversity, and the Marine and Coastal Work Programme (Bergkamp and Orlando, 1999).

Relationship to IPCC Guidelines

The Guidelines do not explicitly mention wetland restoration, although land flooding is included in the Reference Manual in relation to methane emissions. Conceptually, associated changes in soil carbon stocks could be estimated similarly to the procedures for conversions to wetlands, with the need for appropriate values for net carbon uptake rates.

Fact Sheet 4.19. Restoration of Severely Degraded Land

Severely degraded land is understood in this context to mean land that will not revert to its former state through good agricultural, rangeland management, or forestry practice alone. Practices include restoration of severely eroded land and land polluted with heavy metals or mine spoils (Bennett, 1977; Akala and Lal, 1999), as well as reclamation of deserts (Lal *et al*., 1999b), saline soils (Glenn *et al*., 1993), and alkaline soils (Garg, 1998; Sumner and Naidu, 1998). Section 3.2.5.7 discusses definitions and databases; Section 4.6.1.4 includes a small discussion in relation to degraded forest soils. Other relevant categories include acidified soils; soils polluted with hydrocarbons; compacted, sealed, or crusted soils; and waterlogged and slumped soils.

Use and Potential

The specific techniques needed depend on the nature of the degradation. The potential is large, but there is always a reason why land has become degraded; unless this cause is removed, restoration and carbon sequestration will not occur. Globally, very large areas are affected, although there is much variability in estimates of the extent of the problem. Most authors (Dregne *et al*., 1991; Glenn *et al*., 1993; Gupta and Rao, 1994; Sumner and Raidu, 1998; Lal and Bruce, 1999; Lal *et al* 1999b) estimate, however, that no more than 20–40 percent of badly degraded land is restorable at a reasonable cost.

Current Knowledge and Scientific Uncertainties

Eroded soil may have become unstable and susceptible to further erosion. Remaining soil may be highly weathered and lacking basic fertility to permit restoration by growing crops, in particular hyperaccumulators or soil microorganisms. Where land has become degraded—for example, by overgrazing—the underlying cause must be addressed, or the overgrazing is likely to happen again once vegetation is restored. Certain soils are always going to be susceptible simply because of their constitution; stabilizing these soils or achieving anything other than low amounts of carbon stocks may be impossible. Remaining fertility may be extremely variable across a site, making restoration patchy. Resistance to further erosion depends partly on establishing full vegetative cover to provide protection from the action of wind or water.

Time Scales

Highly degraded soils often start from a low carbon base in relation to undisturbed conditions. Given this difficulty, full restoration of the soil may take many years. The combination of large potential and low rates, however, means that some degraded lands could function as a sink for very many years—perhaps more than half a century.

Monitoring, Verifiability, and Transparency

Monitoring depends on the sequestration/revegetation method chosen. The extent or effectiveness of vegetation cover can be monitored by using satellites, and carbon can be estimated in the usual way. Soil carbon can be estimated using models and verified by taking samples on-site. If a soil is still susceptible to erosion, there is a danger that erosion may continue during the monitoring period. The height and location of soil may need to be referenced. Removal of organic pollutants requires special treatment. This practice is as verifiable as any other means of changing soil carbon stocks. Organic pollutants may cause problems that require special attention. Transparency depends on the method of analysis chosen. Uncertainty may be greatest if erosion continues to remove soil or if the original pollutant is organic.

Permanence

If degraded soils start from a low carbon stock, the permanence of additional carbon is probably high, but adequate measures must be taken to prevent repeat degradation.

Associated Impacts

This practice may lead to an increase in aboveground biomass on soil that previously could not support cropping; increased fertility and profitability; reduced erosion and environmental problems such as landslip and silting of water courses; and "recycling" of land where land is scarce.

Relationship to IPCC Guidelines

The Guidelines do not explicitly deal with restoration of severely degraded lands. Changes in soil carbon stocks could be incorporated similarly to what is done with conversions to set-asides, provided appropriate time-dependent "Base Factors" were determined. Changes in non-woody biomass stocks for restoration of degraded lands are not included in the Guidelines.

Fact Sheet 4.20. Sequestration/Displacement Tradeoff

This Fact Sheet illustrates the sequestration/displacement tradeoff discussed in Section 4.5.2.

Use and Potential

Most of the projects under the Activities Implemented Jointly (AIJ) phase have involved sequestration rather than displacement. This situation reflects the reality that the pattern of response is influenced by the time horizon: Displacement of fossil fuel emerges as more effective only in the long run (Marland and Schlamadinger, 1997). Projects such as these provide a future opportunity for biofuel and forest products in the long term. This future potential does not mean that different (additional) land is needed later. The tradeoff is not a once-for-all choice but a dynamic process; the pattern will shift from initial long-rotation sequestration to eventual short rotation-based displacement of fossil fuel (Figure 4-15). In Figure 4-15, Read (1999) has modeled a 70-year projection of the impact on GHG levels, relative to two reference scenarios—business-as-usual (IPCC, 1992) and fossil-free (Greenpeace, 1993)—on two patterns of land-use changes: enhanced biofuel alone and "buffer stock" sequestration plus enhanced biofuel. The latter leads, by about 2040, to ~4 Gt C annual absorption and a cumulative reduction of ~40 ppm CO_2 in the atmosphere. This effort implies an ambitious program of sustainable development pursued consistently over several decades and involving a large number of community-scaled plantations (Read, 1999). It has only a modest effect in 2008–2012.

Biomass growth absorbs CO_2 that is returned to atmosphere after felling—either immediately when it is used as biofuel or more slowly when it is used in conventional forest products, from paper to timber. Long-term removals from atmospheric CO_2 result from substitution in the commercial energy and forest product system, with fossil fuel left underground or biodiverse natural forests—which might otherwise be lost—left standing.

Current Knowledge and Scientific Uncertainties

Modeling of land-use change dynamics is in its infancy and has so far been performed only in global (one region) simulations. Multi-region simulations are expected to show improved outcomes when the buffer stock is focused on industrialized regions, where sunk costs in energy sector infrastructure investments are very large (and where populations are largely urbanized, with surplus agricultural land and only minor rural unemployment).

Time Scale

The tradeoff involves a several decade dynamic process of land-use change that lasts several decades.

Monitoring, Verifiability, and Transparency

Because monitoring processes are required under the CDM, verification in developing countries would be covered by arrangements for project-related carbon credits. Verification of these essentially commercial activities in industrialized countries would also be covered by commercial accounting procedures, subject to disclosure requirements.

Permanence

Although sequestration is of limited duration, the related CO_2 mitigation is not necessarily impermanent because carbon credit creation is a commercial activity that will not leave mature timber from the sequestration phase to rot on felling. It will be used

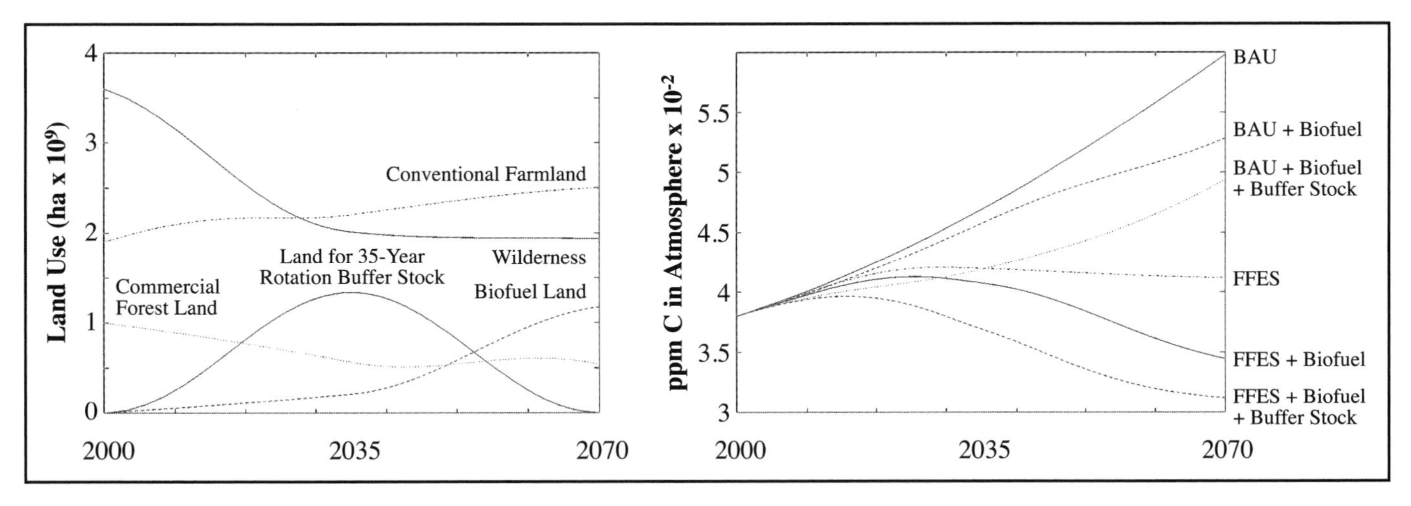

Figure 4-15: CO_2 mitigation with dynamic land-use policy (FLAMES model). Impact on GHG levels over 70 years comparing two reference scenarios [business-as-usual (BAU) and fossil-free energy scenario (FFES)] with two land-use scenarios (enhanced biofuel and enhanced biofuel plus "buffer stock") (Read, 1999).

either in lieu of timber from biodiverse natural forest or as biofuel or in lieu of fossil fuel-intensive materials. Thus, permanence resides in fossil fuel left underground or natural forest left undisturbed.

Associated Impacts

Investing in "buffer stock" forestry is a low-cost, low-risk option, precautionary against climate science finding a low threshold GHG level for a dangerous "rapid non-linear climate change event" (Houghton, 1998). Carbon stored in new forest provides greater flexibility, enabling possible responses to bad news from lower CO_2 levels and providing raw material for a rapid shift to "back-stop" biofuel technology, if needed. If no threshold is revealed, the new forests can be left to grow to maturity to meet demands for timber, avoiding depletion of biodiverse natural forests.

Fact Sheet 4.21. Biofuel Produced from Planted Land

This Fact Sheet details the land-use implications of an expanding role for biofuel.

Use and Potential

Industrial-scale ethanol production exists in Brazil (from sugar cane) and in the United States from corn. There is little fuelwood plantation project experience from the AIJ phase (see Table 5-1). No significant electric power supply based on plantation biofuel exists.

This practice's potential depends on success in enlisting and retaining local support for biofuel production on land from which communities have drawn their livelihood in other ways. Short-rotation forestry, corn, sugar cane, herbaceous plants, and grasses have been proposed, but the focus here is forestry.

The biofuel scenarios captured here project a rise in use from around 60 EJ (~10 EJ from waste) in 2020 to 300 EJ (~50 EJ from waste) in 2100, with land-use implications that depend on plantation productivity (see Section 4.5.3). Biofuel usage would rise from 10 oven dry tons of wood (~5 t C) (~200 GJ) ha^{-1} yr^{-1} to 25 oven dry tons of wood (~12.5 t C) (~500 GJ) ha^{-1} yr^{-1} over a century, leading to land usage that would rise from 250 Mha in 2020 to 500Mha in 2100. If these changes are realized, the potential fossil fuel offset in ~2040 would be as tabulated below.

	Area[b]	Percent Used[c]	Avg. C-Capture Range[d]	Annual Capture[e]
Biofuel production[a]	6.2 Gha	10% 7 t C ha^{-1} yr^{-1}	<2.5–20 t C ha^{-1} yr^{-1}	4.4 Gt C yr^{-1}

[a] Community-scaled production for small-scale gas turbine electricity generation and conversion to transport fuels (e.g., liquid-phase Fischer Tropsch processing) (5 percent) combined with agroforestry meeting local needs.
[b] Cropland, grazing land, degraded land plus forest area vulnerable to predicted climate change. Of this area, 5 percent is in concentrated (a few km in size) biofuel plantations; an additional 10 percent is in 50-percent cover agroforestry, located in settlements in the locality of the plantations. In countries with developed energy supply systems and urbanized populations, less agroforestry is envisioned, with biomass initially accumulated in a long-rotation "buffer stock" awaiting renewal of existing capital stock.
[c] Global average predicted after several decades of technological progress and management experience. A moderately conservative figure is used because species selection and management practices are assumed to be driven by multi-purpose sustainable development criteria.
[d] Low figure = current, for conventional forestry; high figure = current small-plot experience in good growing conditions.
[e] Subject to carbon content of displaced fossil fuel, which depends on fuel mix in power generation and on refinery balances in alternative fossil fuel supply system.

Removal from Atmosphere

Biomass growth absorbs CO_2 that is returned to the atmosphere when it is used as biofuel, providing a renewable fuel system that can be based on sustainable cultivation practices. Removals from atmospheric CO_2 result from substitution in the commercial energy system, leaving fossil fuel underground. With traditional wood fuel, the removal results from leaving standing natural forests that would otherwise be lost.

Scientific and Socioeconomic Uncertainties

Technological uncertainties arise in relation to the community-scale application of existing and near-term future biofuel conversion processes. The area covered by individual plantations, the transportation cost of supply, and the environmental and socioeconomic impacts depend on the scale of application technology. Recent advances in reduced-scale electricity generation and liquid fuel production from gas feedstocks (e.g., gasified biofuel) suggest that community-scaled plantations no greater than about 10-km diameter can support cost-effective production of commercial energy products (Read, 1999).

Economic uncertainties, for export-led growth based on the production of liquid biofuels, relate to future oil price trends, future credits for CO_2 mitigation, and possible support for diversifying the liquid fuel resource base, reducing strategic dependence on dwindling low-cost oil supplies.

Sociocultural uncertainties arise with regard to sustaining commitment to biofuel-based employment and wealth-creating projects that are designed to match the cultural traditions and micro-economic interests of the communities concerned. This sustained commitment requires major capacity-building to enable sustainable technology transfer through country-driven projects that reflect the needs of local communities and host country priorities. This capacity-building entails a career structure for "Project Champions" qualifying at a rate of ~3000 p.a. in ~200 institutions in developing countries (Haque *et al.*, 1999).

Time Scale

Even if the incentives provided by carbon credits and the potential energy productivity of suitable land are sufficient to make reversal of deforestation trend driven by economic pressures of the past few decades possible, this reversal cannot be a short-term process. Rates of policy-driven land-use change that have been modeled are broadly in line with that proposed in the Nordwijk Ministerial Declaration of 1989.

Monitoring, Verifiability, and Transparency

Where products are commercially traded, market statistics and biofuel conversion technology data provide an accurate basis for carbon absorption measurements, as with fossil fuel emissions. Where products are used traditionally, "best practice" project monitoring procedures and benchmark default estimates would be needed.

Permanence

Where emissions savings come through retention of existing stocks of carbon underground and in natural forest, permanence is no different from emissions reduction and forest preservation measures, respectively. With absorption in standing plantation timber, permanence depends on perpetuating community involvement and the incentives that underpin project initiation. For environmental effectiveness, insurance against natural hazards must take the form of additional planting on lands in diverse locations.

Associated Impacts

Community-scaled plantations can transform lifestyles and fund investments in sustainable food systems (e.g., based on agroforestry concepts) in the community. Negative socioeconomic and environmental impacts are avoidable through good project design. Rural electricity and fuelwood used in modern appliances provide rural employment and reduced health risks from smoke inhalation. Environmental benefits include cleaner air with sulfur-free liquid fuels; reduced soil degradation, water runoff, and downstream siltation; capture of polluting agricultural runoff; and utilization of wastes for plantation fertilization, avoiding landfill (Woods and Hall, 1994). Famine is caused by poverty, not land shortage (Sen, 1981), so carbon credit funding could help raise rural living standards and agricultural productivity.

Relationship to IPCC Guidelines

The treatment of biofuels in the IPCC Guidelines is discussed in Chapter 6.

Fact Sheet 4.22. Energy By-Products from Food and Fiber Wastes

Use and Potential

Biomass waste products arising from traditional agricultural and forest production, and processing contributes significantly to the energy mix (Hall, 1991; Pingoud *et al.,* 1999; Hall *et al.,* 2000). Resources include rice husks, corn cobs, cereal straw, bagasse, wood process residues (bark, sawdust, off-cuts), black liquor, nut shells, animal manures, and municipal solid wastes. A wide range of energy conversion routes exist to produce electricity, heat, or fuels (landfill gas, biogas, producer gas, ethanol, methanol, briquetted pellets, charcoal, etc.). Currently, a significant proportion of biomass waste at food and fiber processing plants is either burned to waste or dumped into landfills for disposal. Where this dumping entails a disposal cost, use of the resource for energy production may be economically feasible and reduce the possibility of local air and water pollution.

Monitoring and Verification

The resource is widely distributed, highly variable, and difficult to assess with any degree of accuracy. It tends to have a low energy density and high moisture content. In countries with extensive forest industries—such as Sweden, Finland, and Austria—energy from woody biomass by-products can be monitored relatively easily and can supply up to 30 percent of the country's primary energy. In developing countries with dispersed rural populations, use of biomass wastes for cooking and heating can only be estimated.

Removals

Agricultural and forest production wastes are traditionally left in the field after harvest to decompose; hence, these wastes return organic matter and nutrients to the soil and carbon to the atmosphere. Sustainable production methods would need to be carefully evaluated for each site and soil type if this biomass were to be removed from the site along with the traditional food and fiber products. Conversely, removal of these waste by-products (straw, logging slash, etc.) may have benefits—such as ease of cultivation and replanting, disease control, and avoiding methane production during natural decomposition. Because of the importance of organic residues to the maintenance of soil quality, it has been suggested that only about 50 percent of agricultural residues can be removed from fields without affecting future crop productivity (Sampson *et al.*, 1993).

Scientific Uncertainties

Waste-to-energy plants that produce biofuels or generate heat and electricity may create local emission problems relating to heavy metals, dioxins, and so forth, depending on their design and operation, as well as the nature of the biomass. For example, considerable debate continues about the benefits of incineration of municipal solid waste and sewage sludge (e.g., Aumonier, 1996) versus disposal to properly designed landfills after increased diversion of the organic component to composting facilities or anaerobic digestors (e.g., Finnveden and Ekvall, 1998). If GHG emissions reduction is the primary objective, incineration with energy recovery may be preferred. When other economic and environmental factors are also considered, there may be no general solution.

Collection and transport of biomass by-products such as cereal straw and forest residues to a central processing plant are energy-intensive. Full life-cycle analyses need to be undertaken and energy ratios evaluated.

Time Scales

Many commercial waste-to-energy plants are already being operated successfully. Bagasse is used for on-site co-generation and electricity exports in Australia, South Africa, and Hawaii. Cereal straw is used for district heating in Denmark, Germany, and the United Kingdom. Wood processing residues are used in the United States, Australasia, and northern Europe. Landfill gas plants are widely distributed in most developed countries. Community-scale biogas plants are gaining in popularity in Denmark, India, and China. Several power plants that burn chicken litter are operating in the United Kingdom. Development of further plants depends on local energy prices and government strategies for waste avoidance and carbon trading.

5

Project-Based Activities

SANDRA BROWN (USA), OMAR MASERA (MEXICO),
AND JAYANT SATHAYE (USA)

Lead Authors:
K. Andrasko (USA), P. Brown (USA), P. Frumhoff (USA), R. Lasco (Philippines), G. Leach (UK), P. Moura-Costa (Brazil), S. Mwakifwamba (Tanzania), G. Phillips (UK), P. Read (New Zealand), P. Sudha (India), R. Tipper (UK)

Contributors:
A. Riedacker (France), M. Pinard (USA), M. Stuart (USA), C. Wilson (UK)

Review Editor:
M. Trexler (USA)

CONTENTS

EXECUTIVE SUMMARY

Land use, land-use change, and forestry (LULUCF) activities aimed at mitigating greenhouse gas emissions are often organized as projects. An LULUCF project may integrate one or more activities aimed at reducing greenhouse gas emissions or enhancing greenhouse gas sinks in terrestrial ecosystems and related sectors. LULUCF projects are confined to a specific geographic location, time period, and institutional framework to allow changes in carbon stocks or greenhouse gas emissions to be monitored and verified. There are three broad types of LULUCF projects: (i) avoiding emissions via conservation of existing carbon stocks, (ii) increasing carbon storage by sequestration, and (iii) substituting carbon for fossil fuel and energy-intensive products. Each of these types of project has a variety of subtypes. Integrated multi-component projects may combine many of these subtypes.

LULUCF projects have raised specific concerns regarding duration, additionality, leakage, risks, accounting, measuring and monitoring, and verification of greenhouse gas benefits. These concerns include the ability to construct reasonable, empirically based, without-project baselines; the ability to quantify and reduce potential leakage of greenhouse gases across project borders to other areas or markets; and the ability to cope with natural or human-induced risks that may reduce or eliminate accrued greenhouse gas benefits. Many of these issues are also applicable to climate mitigation projects in other sectors. There are further questions about the degree to which projects can be designed to contribute to sustainable development and improved rural livelihoods. This chapter addresses each of these concerns.

Assessment of the experience of LULUCF projects is constrained by the small number of such projects, their limited activity and geographic scope, and the short period of field operations since the first greenhouse gas mitigation project began in 1988. About 3.5 Mha of land are currently included in 27 LULUCF greenhouse gas mitigation projects being implemented in 19 countries. In addition, LULUCF project experience to date has focused only on mitigating carbon (as carbon dioxide) emissions.

Because no internationally agreed set of guidelines or methods yet exists to quantify carbon benefits, costs, and the carbon and financial efficiency of project activities, projects have used a wide range of methods to estimate changes in carbon stocks or greenhouse gas emissions and financial indicators. Few of the results of these projects have been independently verified, which makes comparative assessments difficult. Using data reported by projects that have been reviewed, average carbon sequestration or emissions avoidance per unit area ranges from about 4–440 t C ha^{-1}; there is wide variation across regions and specific project types. The cost of greenhouse gas mitigation effects in these projects ranges from \$0.1–28 per t C, based on dividing the total financial commitment by the estimated long-term greenhouse gas mitigation effect.

A fundamental component of project assessment is to determine whether changes in carbon stocks or greenhouse gas emissions associated with a project are "additional" to "business as usual." The first step in determining additionality has been to develop a without-project (baseline) scenario against which carbon stocks in the project can be compared. Currently there is no standard method for developing baselines. Approaches for developing and applying baselines include: *project specific*, established through a case-by-case exercise, or *generic*—based on regional, national, or sectoral aggregated data. These baselines may remain fixed throughout the duration of a project, or they may be periodically adjusted in light of new data or evidence. Methods to quantify (or estimate) carbon stocks in the baseline scenario include the use of models to project the fate of land in the project area in combination with data on carbon stocks from proxy or control areas or from the literature.

Experience shows that reducing access to food or fiber resources without offering alternatives or substituting for the activity leading to greenhouse gas emissions may result in project leakage as people move elsewhere to find needed supplies. A few pilot projects to date have been designed to reduce leakage by explicitly incorporating components that supply the resource needs of local communities (e.g., planting fuelwood plantations to reduce pressures on other forests) and provide socioeconomic benefits that create incentives to maintain the project.

Project accounting and monitoring methods could be matched with project conditions to address leakage issues. For example, if flows of LULUCF products or people across project boundaries are negligible, leakage is likely to be small, and the monitoring area can be roughly equal to the project area. Conversely, where flows are significant and leakage is likely to be large, the monitoring area will need to be expanded beyond the project area to account for the leakage. Alternative approaches for accounting and monitoring leakage may be required where monitoring and project areas cannot be easily matched. Potential options include national or regional LULUCF sectoral benchmarks (empirically derived values that relate leakage levels to activities and/or regions) that could capture and report leakage outside the project area, and standard risk coefficients developed by project or activity type and region, with adjustments to project greenhouse gas benefits made accordingly. However, the effectiveness of these approaches is untested.

Implementation of projects in countries without assigned amounts for national emissions presents specific concerns regarding baselines, greenhouse gas accounting, leakage, and monitoring. Unlike Annex I countries, non-Annex I countries are not required to account for emissions on a national level. Therefore, leakage and emissions arising after the project has been completed will not be detected.

Several approaches have been used to account for changes in carbon stocks or greenhouse gas emissions over the lifetimes of LULUCF projects. One method is based on calculating the difference in carbon stocks between a project and its baseline at a given point in time—the *carbon stock method*. The values provided by this method vary depending on the decision of when to account for the project's benefits. The *average storage method* has been used to account for dynamic systems in which planting, harvesting, and replanting operations take place. The advantage of this method is that it accounts for the dynamics of carbon storage over the whole project duration, not only at the times chosen for accounting. Another approach is to credit only a fraction of the total changes in carbon stocks or greenhouse gas emissions for each year that the project is maintained—the *ton-year* method. A variety of methods have been proposed for establishing an equivalency factor by analogy to Global Warming Potentials (GWPs). Depending on the accounting method used, the year-to-year distribution of changes in carbon stocks or greenhouse gas emissions over the project lifetime varies.

The Kyoto Protocol requires that LULUCF projects result in long-term impacts on carbon dioxide concentrations in the atmosphere. The definition of "long-term" varies, however, and there is no consensus on minimum time frames for project duration. Different approaches have been proposed to define the duration of projects. According to one view, the changes in carbon stocks or greenhouse gas emissions must be maintained in perpetuity. This argument is based on the assumption that "reversal" of changes in carbon stocks or greenhouse gas emissions of a project at any point in time would invalidate a project. A second view is that the changes in carbon stocks or greenhouse gas emissions must be maintained for a period of 100 years to be consistent with the time frames adopted in the Kyoto Protocol for the calculation of GWP values. Under a third view, the changes in carbon stocks or greenhouse gas emissions must be maintained until they counteract the effect of an equivalent amount of greenhouse gases emitted to the atmosphere. A fourth view holds that the changes in carbon stocks or greenhouse gas emissions may vary over different time frames, acknowledging that different projects may have different operational time frames; this approach has been adopted during the Activities Implemented Jointly (AIJ) Pilot Phase. Eventually, guidelines will be needed on how to calculate the changes in carbon stocks or greenhouse gas emissions of projects that are conducted over different lifetimes.

Quantification of greenhouse gas emissions or removals in LULUCF projects is subject to a variety of risks and uncertainties. Some of these factors (such as fires, pest and disease, storms) are inherent to certain land-use activities, particularly forestry; others (such as political and economic factors) may be generic and applicable to any greenhouse gas mitigation project in LULUCF and other sectors. These risks and uncertainties could be estimated and the changes in carbon stocks or greenhouse gas emissions adjusted or mitigated through project design, diversification of project portfolios, or insurance methods.

The changes in carbon stocks or greenhouse gas emissions associated with individual LULUCF projects are likely to be more readily quantified and monitored to desired precision levels than national inventories of greenhouse gas emissions and removals because of the clearly defined boundaries of project activities, the ease of stratification of project area, sampling efficiency, and measurement of only a selection of carbon pools. Techniques and methods for measuring carbon in vegetation and soils in LULUCF projects to relatively high levels of precision exist. These techniques have not been universally applied to all projects, however, and methods for accounting of the changes in carbon stocks have not been standardized. A selective accounting system could be used to choose which carbon pools to measure; the choice must include all pools that are expected to decrease and a selection of pools that are expected to increase as a result of the project. The requirements for verifiability in the Protocol suggest that only carbon pools that can be measured and monitored could be claimed.

The costs of measuring and monitoring carbon pools in LULUCF projects are mainly related to the desired precision level, which varies by project type, size of project, distribution of project lands (contiguous or dispersed), and natural variations within the various carbon pools. Different levels of sampling intensity can be used to balance the costs of estimating, monitoring, and verifying the change in carbon stocks. In a few forestry projects in tropical countries, project developers in the early stages of project implementation have measured and monitored relevant aboveground and below-ground carbon pools to precision levels of about 10 percent of the mean at a cost of about US\$1–5 ha^{-1} and US\$0.10–0.50 per t C. The attainable accuracy and precision of carbon measurements and monitoring is likely to be similar among LULUCF project types, but differing measuring and monitoring costs will result from decisions about which particular carbon pools are to be measured and monitored, as well as their variability.

Qualified independent third-party verification plays an essential role in ensuring unbiased monitoring. Although there is growing experience in verification of baseline and project design, there is no experience with verification of monitored data. Guidelines are needed to help establish a procedure and institutional structure for verification.

LULUCF projects may provide significant socioeconomic and environmental benefits to host countries and local communities, though some types of projects pose significant risk of negative impacts. Experience from many pilot projects to date indicates that the involvement of local stakeholders in the design and management of project activities is often critical for success. Critical factors affecting the capacity of projects to provide greenhouse gas and other benefits include consistency with nationally defined sustainable development goals, institutional and technical capacity to develop and implement project guidelines and safeguards, and the extent and effectiveness of local community participation in project development and implementation.

5.1 Introduction

5.1.1. Scope of the Chapter

Projects that are based on land use, land-use change, and forestry are important means of mitigating greenhouse gas (GHG) emissions. These projects are the required approach for putting some parts of the Kyoto Protocol into practice. In this context, they have special features and raise issues that differ sharply from those relating to GHG accounting at the national level (see Chapters 2, 3, and 4).

Although experience has shown that many types of LULUCF projects can mitigate GHG emissions in a cost-effective, measurable, and verifiable manner, there have been questions about the practicality of including LULUCF projects generally within the Kyoto Protocol. These concerns center on the permanence, additionality, leakage, measuring and monitoring, and risks of project-based changes in carbon stocks or GHG emissions. There are also questions about the degree to which LULUCF projects can meet tests for sustainable development and compatibility with national development priorities.

This chapter reviews these project-related issues with two aims in mind. The first goal is to provided policymakers and others with broad guidance about the nature of LULUCF projects. What is their potential for meeting national emission reductions commitments, and with what costs? Are some types of projects more or less efficient in producing GHG and other socioeconomic and environmental benefits? How accurately can carbon be measured and monitored, and with what tradeoffs between accuracy and cost? Will the compliance costs of LULUCF projects deter potential investors or create biases for large projects at the expense of small ones? How do LULUCF projects differ from projects in other sectors, such as energy, with respect to key issues such as additionality, leakage, duration, and risks? The GHG mitigation effect of LULUCF projects results from "*changes in carbon stocks or GHG emissions*;" in this chapter, the terms *GHG benefits* or *carbon benefits* are used as shorthand for this phrase.

Answers to many of the foregoing questions depend on rules and guidelines that remain to be agreed. The second aim of this chapter, therefore, is to provide information to help policymakers develop internationally agreed rules or guidelines concerning a variety of challenging project-specific issues. The chapter presents and discusses these issues together with relevant scientific information, alternative options, and the implications of these options.

5.1.2. Characteristics of Projects

A LULUCF project can be defined as a planned set of activities within a specific geographic location that is implemented by a specific set of subnational or, occasionally, national institutions. These activities may relate to Articles 3.3, 3.4, or 6 of the Kyoto Protocol—and possibly to Article 12, should LULUCF activities be included for certified emissions reductions (CERs) in the Clean Development Mechanism (CDM). There are important differences, however, between the status of LULUCF projects, activities, and enabling policies under these Articles, hence between countries with and without assigned amounts (see Figure 5-1):

- Annex I Parties have taken on commitments to reach assigned amounts of GHG emissions by the end of the first commitment period, thus will have national GHG inventories and accounting systems in place to meet these commitments. Articles 3.3 and 3.4 impose limitations on which LULUCF activities are eligible (see Chapters 2, 3, and 4); a project-based approach is possible under Article 3.4 (Chapter 4). The national assigned-amount commitment may allow Annex I Parties to account for emissions reduction or sequestration across Articles 3.3, 3.4, and 6 as lands or activities move among the Articles—potentially minimizing the risk of leakage of GHGs (see Section 5.3.3).
- Policies by governments, the private sector, or non-governmental organizations (NGOs) can facilitate or hinder the socioeconomic and policy conditions that are likely to encourage the diffusion of LULUCF activities or projects. For example, land tenure, agricultural subsidy, and timber concession or taxation policies have a strong impact on the financial and practical feasibility of many forest or agricultural activities that could generate GHG benefits, such as rates of deforestation or afforestation (e.g., Repetto and Gillis, 1988). These policies are not likely to directly produce emissions reductions or sequestration under Articles 3.3, 3.4, 6, or 12, but they may produce enabling conditions.
- Under Article 6, emissions reduction units (ERUs) in Annex I Parties can be generated only by LULUCF activities that are organized as projects; under Article 12, certified emissions reductions can be generated only by projects, which *may* include LULUCF activities.
- Thus, LULUCF activities that are not implemented as projects are likely to be excluded under Articles 6 and 12. Dispersed, individual actions of land users and beneficial policy changes that are not instituted as projects and may have dispersed GHG impacts but cannot be readily measured and verified are not likely to be included in these Articles unless they are specifically organized as projects.

Many potential LULUCF project activities, taken together, can reduce net emissions of a wide range of GHGs. Project experience to date, however, has been limited mostly to reductions in carbon emissions and enhancement of carbon stocks, as well as forestry operations. This chapter therefore concentrates on carbon and forestry, although it refers to other gases and types of projects where pertinent and where information is available.

There are three broad categories of LULUCF projects, each with a variety of subtypes:

- *Emissions reduction through conservation of existing carbon stocks:* For example, avoidance of deforestation

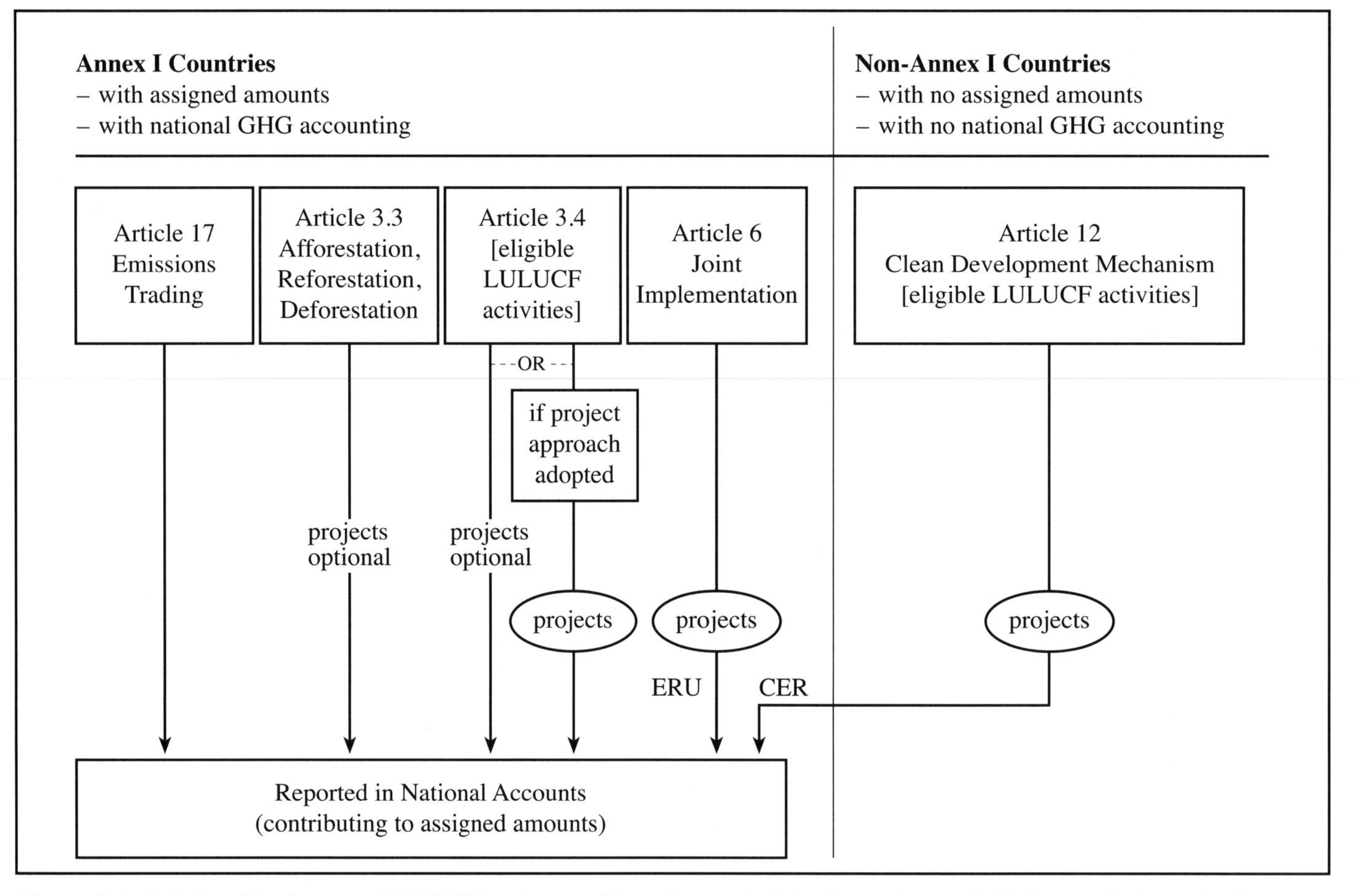

Figure 5-1: Relationships between LULUCF projects and key elements of the Kyoto Protocol (ERU = emission reduction units; CER = certified emission reduction). GHG benefits must be additional to the no-project case.

or improved forest management—including alternative harvest practices such as reduced-impact logging or fire and pest protection.

- *Carbon sequestration by the increase of carbon stocks:* For example, afforestation, reforestation, agroforestry, enhanced natural regeneration, revegetation of degraded lands, reduced soil tillage and other agricultural practices to increase soil carbon, or extended lifetimes of wood products.
- *Carbon substitution:* For example, use of sustainably grown biofuels to replace fossil fuels or biomass to replace energy-intensive materials such as bricks, cement, steel, and plastic.

The eligibility of these different types of LULUCF projects under the Kyoto Protocol and many of the rules that apply to them still have to be decided and formulated. The outcome of this policymaking process will have a large bearing on the potential—and costs—of LULUCF projects as a means of mitigating GHG emissions while contributing to sustainable development.

The very concepts of LULUCF mitigation projects generally and joint implementation (JI) projects specifically—projects that mitigate GHG emissions by Annex I countries in non–Annex I countries, established under the United Nations Framework Convention on Climate Change (UNFCCC)—have been challenged in a growing body of literature. Critics raise three general sets of questions (e.g., Maya and Gupta, 1996; Mulongoy *et al.*, 1998; Lashof and Hare, 1999; Smith *et al.*, 1999). First, do LULUCF projects provide measurable, verifiable, long-term GHG emissions avoidance or reductions? This concern relates to projects' ability to construct reasonable, empirically based, without-project baselines and to quantify leakage of GHGs across project borders to other areas or markets. Second, can LULUCF projects meet tests for sustainable development and be compatible with national sustainable development priorities? Should other policy tests be required for their use under Articles 6 and 12? Third, under what policy circumstances might LULUCF projects be used in the Kyoto Protocol to provide Annex B certified emissions reductions? Should they be limited to fostering energy sector emissions reductions? The technical issues raised are addressed in the following sections.

5.2. Magnitude and Experience of Project-Based Activities

This section reviews the experience of LULUCF projects that generate GHG and other benefits and are being implemented on the ground at least partially. It summarizes estimated GHG benefits for 27 such projects and several portfolios of projects, reviews estimated project costs, and assesses the limited estimates of the potential magnitude of project activities under Article 6 and possibly Article 12 of the Kyoto Protocol.

Some of the key questions addressed in this section follow:

- What is the experience of the voluntary, non-credit, AIJ Pilot Phase established by the UNFCCC in terms of the number, type, and technical issues surrounding LULUCF projects?
- What are the cost estimates of such projects, and how do they vary by project type and location?
- What is the likely supply and cost of projects that might help Annex I countries meet their emissions reduction commitments under Articles 6 and 12?
- Are such projects likely to be implemented at significant scales by 2012?

5.2.1. ***Quantifying Project Activities: Issues and Methods***

By far, most LULUCF projects being implemented within countries or funded internationally are designed to promote economic and social development, without regard for their potential GHG benefits. Such projects provide timber or fuelwood supply, community woodlots, agroforestry crops, soil conservation, biodiversity or watershed protection, and socioeconomic development. The GHG implications of these projects generally have not been estimated or reported.

Six representative case studies of LULUCF projects that are being implemented illustrate the diversity of project types, locations, and estimated GHG benefits and costs (Box 5-1). These cases provide an introduction to the kinds of activities, the projected socioeconomic benefits, and the environmental impacts associated with LULUCF projects described throughout this chapter. These cases are representative of two major categories of LULUCF projects (see Section 5.1): carbon sequestration through increases in carbon stocks, and emissions avoidance through conservation of existing carbon stocks.

About 3.5 Mha of land are currently included in LULUCF GHG mitigation projects being implemented in 19 countries (see Section 5.2.2). Assessment of the experience of LULUCF mitigation projects is constrained by the small number of these projects, the limited range of project types, uneven geographic distribution, and the short period of field operations to date. The first publicized LULUCF mitigation or carbon offset project began in 1988: the CARE-AES Guatemala community forestry project (Trexler *et al.*, 1989; Faeth *et al.*, 1994).

Most reviews of LULUCF climate change mitigation project experience to date are simply summaries of information reported by individual projects or AIJ programs (e.g., Dixon *et al.*, 1993; Stuart and Moura-Costa, 1998; EPA/USIJI, 1998; FACE, 1998; UNFCCC, 1998). A few studies review or analyze several project case studies (e.g., Faeth *et al.*, 1994; Brown *et al.*, 1996; Brown *et al.*, 1997; Goldberg, 1998; Imaz *et al.*,1998; Witthoeft-Muehlmann, 1999). Several projects have been well documented. The Rio Bravo conservation and alternative forest management project in Belize, for example, has produced a set of operational protocols (Programme for Belize, 1997b). These protocols include descriptions of the project's reference case, leakage assessment, a sustainable forest management strategy (including boundary security and fire management), estimates of GHG benefits, baseline, and monitoring plan. These protocols have been filed with the U.S. Initiative for Joint Implementation (USIJI) program, along with other project documents (EPA/USIJI, 1998).

No internationally agreed set of guidelines and methods—comparable to the Organisation for Economic Cooperation and Development/Intergovernmental Panel on Climate Change (OECD/IPCC) *Revised 1996 IPCC Guidelines for National Greenhouse Gas Inventories*—exists to quantify GHG emissions and sequestration, baselines, socioeconomic and environmental impact assessment, and reporting of project activities (Andrasko *et al*, 1996; Swisher, 1997). Such guidelines and methods are urgently needed if projects are to be reported consistently and credibly under several Articles of the Kyoto Protocol (see Chapter 6).

Project data reported in the literature use a wide range of methods and, for the most part, have not been independently verified. Thus, comparing data across projects is difficult. Evaluations of project GHG accounting by different analysts are likely to produce estimated GHG benefits that are different from the estimates of project developers, because GHG accounting methods have not yet been standardized. Analysts and project developers are building on early experience to alter the design of projects; they are beginning to produce data-driven baselines in some cases and revise project estimates of sequestration or avoided emissions.

Table 5-1 compares the initial baseline and net GHG benefits estimated by project developers during the planning phase and reported to the USIJI program (see Section 5.2.2) for two large projects, along with later evaluations by other entities. These projects were conceived in the voluntary, non-credit, exploratory AIJ phase, where steep learning curves were experienced. As Table 5-1 illustrates, estimated GHG benefits have tended to decrease over time as methods and initial assumptions have been refined and applied to a given project (e.g., Busch *et al.*, 1999; Brown *et al.*, 2000). If standardized methods are introduced, estimates should tend to vary mainly as changes occur in project conditions or land uses or with the availability of new data. Over the next 5 years or so, early projects will begin measuring and monitoring their performance, replacing earlier estimates of project baselines and GHG benefits with field data collected for the purpose of monitoring. Reported GHG benefits could change as well if verification of GHG reductions occurs and the results are significantly different from previous estimates.

Box 5-1. LUCF Carbon Mitigation Projects: Selected Case Studies

Krknose and Sumava National Parks, Czech Republic

Objectives and Activities
Temperate forest rehabilitation and reforestation.

Land Area and Type
14,000 ha broadleaf and conifer species forest.

Description of Activities
Designed to restore mountain forest degradation or mortality from air pollution and acid deposition in unique Norway spruce (*Picea abies*) stands, perhaps from coal-fired power plants in Poland. Studies indicate natural regeneration of stands would be difficult, and grasslands would dominate. About 7,000 ha were completely reforested, and mortality tree gaps were planted on another 7,000 ha. Project implementation will be 17 years and its lifetime 99 years.

GHG Estimated Benefits and Methods
Cumulative net sequestration of 2,682,030 t C, using CO2FIX model. Average carbon benefit is 192 t C ha^{-1}, at 6.4 t C ha^{-1} yr^{-1}.

Projected Socioeconomic Benefits
Increase in recreational and tourism value; 200 laborers/farmers involved.

Projected Environmental Impacts
May reduce air pollution levels, conserve and increase forest biodiversity, and improve hydrologic system of the area.

Status of Project
Started in 1992; 5,400 ha planted to date.

Cost Estimate and Efficiency
$38.2 million total cost, public and private financing. Efficiency calculated to be $14.25 per t C.

References
FACE Foundation (1998), Witthoeft-Muehlmann (1998).

Saskatchewan Soil Enhancement Project, Canada

Objectives and Activities
Assess carbon sequestration benefits of farmers changing from conventional soil tillage to reduced-tillage systems.

Land Area and Type
150 experimental sites in Saskatchewan prairie grassland agricultural cropping systems.

Partners
Regional soil conservation agencies, GEMCO (private consortium of Canadian companies initiating GHG reduction pilot projects), and farmers.

Description of Activities
Establishment of experimental research sites to develop baseline soils data and precise measurement locations for a network of fields representing dominant soil types and land uses across Saskatchewan, to measure future changes in soil carbon. Data will be used to calibrate soil carbon simulation models, such as CENTURY, to predict changes in soil carbon across the province, for potential generation of carbon credits for emissions trading in Canada or internationally.

GHG Estimated Benefits and Methods
CENTURY model.

Box 5-1. LUCF Carbon Mitigation Projects: Selected Case Studies (continued)

Saskatchewan Soil Enhancement Project, Canada (continued)

Projected Socioeconomic Benefits
Generation of new carbon commodity for farmers; reduced farm energy costs of alternative tillage systems.

Projected Environmental Impacts
Enhanced soil tilth and carbon sequestration, reduced erosion of soil and carbon off-field, improved local air and water quality.

Status of Project
Several years of data collection from established sites; technical papers and modeling.

Cost Estimate and Efficiency
Not applicable.

References
Padbury (1999).

Rehabilitation of Mt. Elgon and Kibale National Parks, Uganda

Objectives and Activities
Tropical forest rehabilitation and reforestation.

Land Area and Type
27,000 ha broadleaf and conifer species forest.

Partners
FACE Foundation, Netherlands, and Uganda national parks agency.

Description of Activities
Restoration of natural forest and reforestation using native tree species, and fire control, within two national parks. Cooperation with the IUCN-sponsored project in adjacent communities to support sustainable land use in buffer zones around the parks, to reduce forest degradation and forest conversion to agriculture from fire and logging within them. Activities to be implemented over 17 years; project lifetime is 99 years.

GHG Estimated Benefits and Methods
707,000 t C total over lifetime, using CO2FIX model. Average carbon benefit is 26 t C ha^{-1}, at 0.9 t C ha^{-1} yr.

Projected Socioeconomic Benefits
Reforestation planting jobs.

Projected Environmental Impacts
Reduced soil erosion, improved water quality, reduced logging and fire degradation of park, conserving biodiversity.

Status of Project
Began in 1994; rehabilitation or replanting has occurred on 5,700 ha to date.

Cost Estimate and Efficiency
$2.6 million total cost, with private financing. Efficiency calculated to be $27.80 per t C.

References
FACE Foundation (1998), Witthoeft-Muehlmann (1998).

Box 5-1. LUCF Carbon Mitigation Projects: Selected Case Studies (continued)

Rio Bravo Conservation and Management Area (RBCMA) Carbon Sequestration Pilot Project, Belize

Objectives and Activities
Tropical forest sustainable forest management and protection.

Land Area and Type
14,327 ha of tropical forest acquired and protected; 46,406 ha put under sustainable forest management.

Partners
Programme for Belize (private conservation and development organization) and The Nature Conservancy. Financed by U.S. electric utilities (Wisconsin Power Company, CINergy, Detroit Edison, PacifiCorp, EEI Utilitree Carbon Company) and Suncor Energy, Inc.

Description of Activities
Component A involves the purchase of endangered forest land, thereby expanding RBCMA's existing protected forest areas. If not protected, this property is highly likely to be converted to agricultural use, permanently dividing the RBCMA ecosystem. Component B involves the development of a sustainable forestry management program that will increase the level and rate of carbon sequestered within half of the RBCMA, including the purchased parcel. The remaining RBCMA lands will be left undisturbed for experimental control areas, conservation, and research.

GHG Estimated Benefits and Methods
2.4 million t C total over 40-yr lifetime, using Winrock GHG estimation and monitoring software. Average carbon benefit is 31 t C ha^{-1}, at 0.8 t C ha^{-1} yr^{-1}.

Projected Socioeconomic Benefits
Create jobs in newly established forest products industry; enhance commercial value of forests; transfer forest monitoring and management technologies; protect Mayan archaeological sites.

Projected Environmental Impacts
Conserve biodiversity from agricultural conversion; enhance water quality via reduced soil erosion.

Status of Project
In year 5 of a 40-yr project.

Cost Estimate and Efficiency
$2.6 million total cost, with private financing. Efficiency calculated to be $3 per t C.

References
EPA/USIJI (1998), Witthoeft-Muehlmann (1998), TNC (1999).

This chapter surveys projects that were in early or later stages of implementation by 1999 (i.e., projects that have been at least partially funded and have begun activities on the ground that will generate GHG benefits). It focuses on LULUCF projects formally reported to the UNFCCC AIJ Pilot Phase program (17 projects, as of late 1998), as well as more than a dozen other projects (Trexler *et al.*, 1999; UNFCCC, 1999b). The AIJ program was established in 1995 by Decision 5/CP.1 of the first Conference of the Parties (COP) to the UNFCCC as a voluntary program to experiment with concepts of joint implementation that evolved during the negotiation of the UNFCCC (UNFCCC, 1995). Many projects have not been reported to the voluntary AIJ program, which precludes transfer of emissions reduction or avoidance credits to Parties. Unreported projects often began prior to the AIJ program and faced reluctance by host countries to grant formal acceptance, as well as a lack of incentives for investors or developers to report. One review identified 18 offset projects underway in 14 countries that have not been reported to the AIJ program (Trexler *et al.*,1999).

5.2.2. *Experience in LULUCF Project-Based Activities: Estimates of Sequestration, Emissions Avoidance, Substitution, and Land Areas Involved*

Table 5-2 summarizes a representative set of LULUCF projects currently underway that have been reported to provide carbon sequestration or emissions reduction benefits. The projects are divided into six subcategories: (i) reforestation, afforestation,

Box 5-1. LUCF Carbon Mitigation Projects: Selected Case Studies (continued)

Scolel Te Pilot Project for Community Forestry and Carbon Sequestration, Chiapas, Mexico

Objectives and Activities
Tropical forest and highland conifer forest reforestation and community agroforestry on individual farmers' small plots.

Land Area and Type
2,000 ha within 13,200 ha (project size depends on funding received).

Partners
Wide range, including local credit unions, regional research institute (ECOSUR), University of Edinburgh, UK-Overseas Development Administration, International Automobile Federation, among others.

Description of Activities
Consultations with local village farmers identified preferred candidate reforestation, forest management, and agroforestry practices. The project designed a system of technical assistance to farmers by producing plans for each parcel, calculating carbon benefits, and developing a monitoring protocol. International Automobile Federation funded the first implemented management plans. Project is designed to reduce degradation and conversion of remnant forest, and to enhance village land-use sustainability and financial returns.

GHG Estimated Benefits and Methods
Cumulative net sequestration of 15,000–333,000 t C total over lifetime, using CO2FIX model. Average carbon benefit is 26 t C ha^{-1}, at 0.9 t C ha^{-1} yr^{-1}.

Projected Socioeconomic Benefits
Build local economy through sustainable agroforestry; improve welfare of women and villagers.

Projected Environmental Impacts
Conserve and increase forest biodiversity, reduce forest fragmentation and soil erosion, serve as buffer zone by slowing in-migration to the forest.

Status of Project
50 ha funded for initial implementation. Detailed studies at community and regional scale completed. Management, research, and financial institutions established.

Cost Estimate and Efficiency
$3.4 million projected total cost, with initial phase at $0.5 million, and public and private financing. Efficiency calculated to be $10 per t C.

References
EPA/USIJI (1998), Tipper and de Jong (1998), Witthoeft-Muehlmann (1998).

and restoration; (ii) soil carbon management; (iii) forest conservation; (iv) forest management and alternative harvest practices; (v) agroforestry; and (vi) multi-component or community forestry projects that combine several of these activities. The projects listed in Table 5-2 are predominately forestry projects because experience to date has been most influenced by electric utility companies and conservation NGOs seeking projects likely to produce credible GHG benefits at costs that are lower than their emissions reduction options in their home territories, as well as conservation, biodiversity, and community development benefits. Many soil management, bioenergy, and other LULUCF management projects exist, but few have estimated and reported changes in carbon stocks or greenhouse gas emissions, so they are underrepresented in Table 5-2.

The 3.5 Mha of projects currently being implemented could eventually total 6.4 Mha if the projects are fully funded. Most of these 3.5 Mha (2.9 Mha, or 83 percent) are in forest land protection or conservation, potentially avoiding emissions or sequestering about 41–48 Mt C if the projects are fully financed and implemented (Table 5-2). Another 92,000 ha (3 percent) are in projects primarily undertaking afforestation, reforestation, or forest restoration, potentially generating an estimated 10 Mt C. Projects involving forest management and alternative silvicultural or harvesting practices occupy about

Box 5-1. LUCF Carbon Mitigation Projects: Selected Case Studies (continued)

INFAPRO: Innoprise-FACE Foundation Project, Sabah, Malaysia

Objectives and Activities
Enrichment planting and forest rehabilitation in previously harvested mature forest.

Land Area and Type
25,000 ha of selectively logged dipterocarp lowland tropical forest concession lands.

Partners
Innoprise Corporation (forestry arm of Sabah Foundation, Sabah, Malaysia) and FACE (Forests Absorbing CO_2 Emissions) Foundation of Dutch Electricity Board, The Netherlands.

Description of Activities
The project estimates it will sequester approximately 4.3 million t C over the course of 60 years, largely using literature data for the estimate. Permanent sample plots to measure stem growth are established; necromass, understory, and soil data collected for the ICSB-NEP RIL project are being used (see Table 5-2).

GHG Estimated Benefits and Methods
707,000 t C total over lifetime, using CO2FIX model. Average carbon benefit is 26 t C ha^{-1}, at 0.9 t C ha^{-1} yr^{-1}.

Projected Socioeconomic Benefits
Generate US$ 800 million in timber, which will revert to the social programs of the Sabah Foundation. Build capacity through technical training, at all levels of project staff, and with local, regional, and international organizations. Direct employment of more than 150 people.

Projected Environmental Impacts
Improve at least 25,000 ha of degraded logged forests.

Status of Project
The project is in the seventh year of its implementation phase, which is planned to last 25 years; project lifetime is 99 years. If CDM guidelines and crediting are not in place soon, implementation of this project may be halted.

Cost Estimate and Efficiency
$15 million total cost, with private financing. Efficiency calculated to be $3.50 per t C.

References
FACE Foundation (1998), Stuart and Moura-Costa (1998), Witthoeft-Muehlmann (1998).

NOTE: For all case studies described, GHG and cost estimates are provisional. These estimates are genearally provided by project developers and do not use consistent, comparable methods.

60,000 ha (less than 2 percent) and may generate about 5.3 Mt C. Multi-component community forestry or agroforestry system projects cover at least 530,000 ha (15 percent) and may provide about 20 Mt C in benefits. Only a few very small projects currently exist for soil carbon management (see Chapter 4).

Carbon sequestration or emissions avoidance per unit area over the reported lifetime of the projects varies by project type from an average of about 110 t C ha^{-1} for afforestation and reforestation projects, to 88 t C ha^{-1} for forest management projects, to 40 t C ha^{-1} for community forestry and agroforestry projects, to a low of 16 t C ha^{-1} for forest protection projects (mainly from avoided logging), with very large ranges within and across project types (Table 5-2). These averages reflect project designs to date and vary across design, site condition, and implementation conditions.

Emissions avoidance per hectare of forest protection projects, in particular, is highly sensitive to the total project area involved and the activity avoided (e.g., avoided deforestation or avoided logging). These projects generally conserve a large area of forest considered under threat of deforestation at rates of about 1–5 percent of total forest area per year. In the Noel Kempff Climate Action Project (NKCAP), for example, areas where deforestation is expected to be avoided are estimated to generate about 143 t C ha^{-1} over the life of the project and areas where logging is avoided about 12 t C ha^{-1}; the project overall is estimated to generate about 7 t C ha^{-1} (because the total project area is large) (Brown *et al.*, 2000). For project components designed solely to avoid deforestation, typical emissions avoidance values are likely to range from 28–80 t C ha^{-1} for boreal forest to about 30–140 t C ha^{-1} for temperate and 100–175 t C ha-1 for tropical forests (Brown *et al.*, 1996).

Several models for the design and funding of projects are already being used in many of the projects reviewed in Box 5-1 and Table 5-2:

- Project funding is provided by investors who are committed to offset their carbon emissions, irrespective of the status of the international climate change negotiations. Monies are provided to a central office which seeks out, designs, and implements projects meeting investor criteria.
- Entities (e.g., electric utilities) that consider themselves likely to face emissions reduction mandates in the future are implementing their own projects.
- Project proponents identify and design projects on the basis of expected GHG and non-GHG benefits, then seek funding from donor sources. These projects are developed primarily to mobilize resources for non-climate services (e.g., biodiversity protection by a land management NGO) and to gain experience in project implementation (often reporting under the AIJ pilot program).

Other models are likely to develop as entities seeking certified emissions reductions organize their investments to spread liabilities and risks. One potential trend may be the emergence of flexible derivatives involving brokers, traders, and insurers who trade various attributes of the potential emissions reductions of bundles of projects. Experience using the aforementioned models in the early stages of pilot project implementation has helped produce several advances, including quantifying and monitoring the GHG benefits of a range of project types using the Winrock estimation and monitoring methodology (MacDicken, 1997a); reviewing and refining without-project baseline assumptions in an independent review of the Protected Areas Project (PAP) in Costa Rica (Busch *et al.*, 1999); and addressing ways to minimize leakage in the design and implementation of the NKCAP (Brown *et al.*, 2000).

***Table 5-1**: Examples of the effect of non-standardized methods and improved data on greenhouse gas accounting and baseline development, for two projects: original proposals, developer estimates, and reviews by third parties.*

Project and Description of Revisions	Estimated Baseline (MMTC)	Estimated Lifetime Project GHG Benefits (Project Case – Baseline)	Source of Estimates; Reference (MMTC)
Costa Rica Protected Area Project (PAP)—Original proposal to USIJI, 1997, using remote sensing for 1979–92 to estimate deforestation rate	15.7 (422,800 ha)	15.7 (422,800 ha)	Project, report to USIJI; EPA/USIJI (1998)
PAP—Adjusted baseline and GHG estimate (year 1 actions, benefits over 20 years) certified by third party entity, 1998	3.5 (first 30,000 ha only)	1.7 (first 30,000 ha only)	SGS, as certifier of project baseline for project; SGS (1998)
PAP—Independent review, using remote sensing for 1986–97 for deforestation estimate, 1999	8.9 (medium scenario)	4.6–29.9 (medium scenario = 8.9)	LBNL, for USEPA; Busch *et al.* (1999)
Noel Kempff Climate Action Project (NKCAP), Bolivia—Original project proposal to USIJI, 1996	18.8	18.8	Witthoeft-Muehlmann (1998)
NKCAP—Project report to USIJI and UNAIJ, 1998	14.7	14.7	Project report to USIJI; EPA/USIJI (1998)
NKCAP—Revised estimates for project developer, 2000	6.5–8.5	6.5–8.5	Brown *et al.* (2000)[a]

[a] Noel Kempff project revised estimates in Brown *et al.* (2000) are preliminary results from field work in 1999, and are in external review prior to expected publication in 2000.

Table 5-2: Overview of selected LULUCF AIJ pilot program and other projects, in at least early stages of implementation.

Project and Host Country	Dominant Activity	Project Information[a]	Area (ha)	Estimated Lifetime CO_2 Benefits (000 t C)	Estimated CO_2 Benefits per Hectare (t C ha^{-1})[b]
Carbon Sequestration through Increase in Carbon Stocks: Aforestation, Reforestation, and Restoration Projects					
FACE Foundation Kroknose and Sumava National Parks, Czech Republic	Reforestation, regeneration	99; 1992; The Netherlands	14,000	2,682	191
RUSAFOR, Russian Federation	Afforestation plantation	40 (2 sites), 60 (2 sites); 1993; USA	900 EPA, AWM	80	89
Klinki Forestry, Costa Rica	Agroforestry, afforestation	46; 1997; USA	Phase I: 100 Total: 6,000	1,970	328
INFAPRO: FACE Foundation, Malaysia	Enrichment planting	25 implement, 99 total; 1992; The Netherlands	14,000	3,000	170
FACE Netherlands, The Netherlands	Urban forest afforestation	1992; The Netherlands	5,000	885	177
FACE Elgon/Kibale, Uganda	Forest rehabilitation	1994; The Netherlands	27,000	707	26
Bottomland Hardwood Restoration, UtiliTree, Louisiana, USA	Reforestation of marginal riparian farmland	70; 1996; USA	32	12.8	400
Western Oregon Carbon Sequestration Project, UtiliTree, USA	Afforestation, sequestration in wood products	65; 1997; USA	127	54.5	440
Salt Lake City Urban Tree, PacifiCorp, USA	Urban forestry	1995; USA	NA	5	NA
UNSO Arid Savanna Protection, Benin	Woody savanna protection, live fences	1993; U.N. Sudano-Sahelian Office	25,000	660–1,000	33
Subtotal Range (or Average)		*61*	*92,059*	*10,056–10,400*	*26–440*
Carbon Sequestration through Increase in Carbon Stocks: Soil Carbon Management					
Project Salicornia, Mexico	Halophyte planting, soil carbon	59; 1996; USA	30 (Phase I)	0.89	18

Table 5-2 (continued)

Project and Host Country	Dominant Activity	Project Information[a]	Area (ha)	Estimated Lifetime CO_2 Benefits (000 t C)	Estimated CO_2 Benefits per Hectare (t C ha^{-1})[b]
Carbon Sequestration through Increase in Carbon Stocks: Soil Carbon Management (continued)					
Saskatchewan Soil Enhancement Project, GEMCO, Canada	Soil carbon management	5; 1995; Canada	NA	NA	NA
Subtotal Range (or Average)		*32*	*30*	*0.89*	*18*
Emissions Avoidance through Conservation of Existing Stocks: Forest Management and Alternative Harvest Practices					
ICSB-NEP 1, Malaysia	Reduced-impact logging	40; 1992; USA	1,400	58	41
ICSB-NEP 2, UtiliTree, Malaysia	Reduced-impact logging	40; 1997; USA	1,012	104	102
Olafo Project-Peten, Guatemala	Sustainable timber, sustainable agriculture	40; 1995; Denmark, Norway, Sweden	57,800	4,920	85
Pacific Forest Stewardship, Oregon, USA	Improved forest management, conservation easements	1995; USA	NA	242	NA
Subtotal Range (or Average)		*40*	*60,212*	*5,324*	*41–102*
Emissions Avoidance through Conservation of Existing Stocks: Forest Conservation—Protection					
Amazon Basin, AES/Oxfam, Ecuador, Bolivia, Peru	Protection, land tenure	1992; USA	1,500,000	15,000	10
Paraguay Forest Protection, AES, Paraguay	Protection	1992; USA	58,000	14,600	252
ECOLAND, Costa Rica	Protection	16; 1995; USA	2,500	366	146
Rio Bravo, Belize	Protection, forest management	40; 1994; USA	14,000 protection; 46,406 forest management	2,400	39

Table 5-2 (continued)

Project and Host Country	Dominant Activity	Project Information[a]	Area (ha)	Estimated Lifetime CO_2 Benefits (000 t C)	Estimated CO_2 Benefits per Hectare (t C ha^{-1})[b]
Emissions Avoidance through Conservation of Existing Stocks: Forest Conservation—Protection (continued)					
Noel Kempff, Bolivia	Protection from logging and deforestation	30; 1996; USA	~696,000	4,000–6,000	7
Protected Area Project, Costa Rica	Preservation via purchase and land title enhancement	25; 1997; USA	530,000	4,600–8,900	17
Virilla Basin Project, Costa Rica	Protection, reforestation	25; 1997; Norway	52,000	231	4
Subtotal Range (or Average)		*27*	*2,852,500*	41,200–47,500	4–252
Multi-Component Community Forest					
FACE Profafor, Ecuador	Small farmer plantations	1993; The Netherlands	75,000	9,660	129
Sustainable Energy Management, Burkina Faso	Community forest management (component II)	30; 1997; Norway	270,000	67	0.2
Subtotal Range (or Average)		*30*	*345,000*	*9,700*	*0.2–129*
Agroforestry					
AES CARE, Guatemala	Agroforestry, woodlots	35; 1989; USA	186,000	10,500	56
Scolel Te, Mexico	Agroforestry, reforestation, sustainable harvesting	30; 1997; UK, France	Phase I: 50 Total: 2,000 within 13,000 area	Phase I: 15 Total 330	26
Subtotal Range (or Average)		*32*	*186,000–188,000*	*10,500–10,800*	*26–56*
Grand Total		***41***	***3,535,000–3,537,000***	***76,780–83,725***	***23***

[a] Project lifetime (in years); date initiated; investor country.

[b] Estimated CO_2 benefits per hectare and totals for projects are generally reported by project developers, do not use standardized or consistent GHG accounting methods, generally only report CO_2 (not other GHGs), and have not been independently reviewed. The wide range of estimates for conservation/protection projects results from the type of activity (e.g., avoided logging or avoided deforestation) and from a large project area with only a fraction affected by the activity per year (see Section 5.2.2).

Major References: Brown *et al.* (1997), EPA/USIJI (1998), FACE Foundation (1998), Stuart and Moura-Costa (1998), Witthoeft-Muehlmann (1998), Moura-Costa and Stuart (2000).

Several portfolios of projects have been assembled by national, NGO, or private JI or AIJ pilot programs. For example, the FACE Foundation—founded in 1990 by the Dutch Electricity Generating Board—has targeted 150,000 ha of new forest planting in five projects in six countries, to absorb the lifetime CO_2 emissions of a coal-fired 600 MW power station. About 40,000 ha had been planted by 1999. The projected carbon benefits are 75 Mt C over the lifetime of the projects. Total estimated, undiscounted costs are $100 million, of which $30 million has been committed, with an estimated unit cost of $8 per t C (FACE Foundation, 1998; Verweij and Emmer, 1998).

The USIJI began in 1993. It has accepted 14 forestry projects and one soil carbon management project as of February 2000. Estimated total carbon benefits over the lifetimes of eight projects in at least initial implementation stages are about 13 Mt C—rising to 25.5 Mt C if the projects are fully funded and implemented—on 1.27 Mha. Total funding committed to date is about $17 million, at an estimated carbon cost of $3.90 per t C (EPA/USIJI, 1998; Table 5-2).

Some projects have been designed that could expand across whole regions. The Scolel Te project in southern Mexico has initiated agroforestry activities on about 150 small farms. If an incentive rate of $15 per t C were available, it could supply 150–200 Mt C over 40 years (de Jong *et al*, 1997; Tipper *et al.*, 1998).

Projects offer varying rates of carbon benefits over time. Projects summarized in Table 5-2 have reported project lifetimes ranging from 16 to 99 years, averaging 41 years. Forest conservation projects designed to slow deforestation are highly sensitive to estimated baseline assumptions about non-project forest loss rates (see Section 5.3; Busch *et al.*, 1999). These projects appear to deliver carbon benefits quickly relative to other project types, however, by annually avoiding high losses of carbon stocks per hectare of mature forest. Conversely, soil carbon management and afforestation or reforestation projects in boreal forest deliver carbon benefits slowly because carbon sequestration rates in both systems are generally less than 1 t C ha^{-1} yr^{-1}.

5.2.3. *Financial Analysis of LULUCF Project Activities*

Financial analyses of GHG reductions by projects are rarely comparable because no standard method of evaluation has emerged and come into wide use. Financial analysis of direct, indirect, initial, and recurring costs, as well as the stream of revenues, varies across projects. Available cost estimates for LULUCF projects often include direct costs incurred by the project developers: land purchase or rental costs, if necessary; land clearing and site preparation; initial planting or other activity costs; annual, recurring costs of project maintenance and management—including, for example, periodic thinning or other stand improvement or weed control in agricultural soil management; and sometimes the establishment of monitoring data collection and evaluation systems.

Opportunity costs of land (i.e., the present value of alternative opportunities or uses of the land, at the margin) are often not included in financial analyses of projects. Other costs often overlooked are infrastructure costs (e.g., road development), monitoring data collection and interpretation costs, and maintenance or other recurring costs that will be incurred in the future (Mulongoy *et al.*, 1998; Witthoeft-Muehlmann, 1998). The stream of revenues is not widely reported for projects to date, in part because few revenues have accrued in their early stages of implementation. Revenues may be generated by the sale of logs or value-added products from timber harvest, sale of fuelwood or non-timber products such as medicinal plants, usage fees for access, government or NGO grants for subsidies, in-kind contributions, and sale of emissions reductions.

Project-level financial analysis methods are widely used and fairly standardized in development assistance and private investment projects. They have yet to be consistently applied to and reported for LULUCF projects, however—in part because of the highly varied expertise of early actors in such projects (Mulongoy *et al.*, 1998). A standard approach for comparing the economic attractiveness of different projects would compare the time flow of revenues—including the sale of emissions reductions and crediting rules applying to them—with the time flow of expenditures, applying appropriate discount rates. Detailed financial data are not available for most LULUCF projects, however, so the economic indicators often are obtained simply by dividing a project's total carbon sequestration or emissions avoidance over time by total expenditures (e.g., Witthoeft-Muehlmann, 1998). A further complication relates to how emissions reductions are allocated between the sellers and investors. The unit cost of reduction will vary directly with the percentage of total reductions that accrue to the investor.

Cost and investment estimates are available for virtually all of the projects in Table 5-2; because of the different methods used in the estimates, however, only summary ranges are reported in Table 5-3. The costs of GHG benefits in these projects range from $0.1–28 per t C, simply dividing project costs by their total reported carbon benefits. Most of the cost estimates are in the range of $1–15 per t C, with a higher range for reforestation and afforestation projects (reflecting the inclusion of temperate and boreal projects). Mulgonoy *et al.* (1998) reviewed cost estimates for LULUCF carbon projects and found that most estimates for the tropics fall in the range of $2–25 per t C. Two other reviews reported costs of sets of projects in temperate and tropical biomes ranging from $4–26 (Swisher and Masters, 1992) and $2–12 per t C (Witthoeft-Muehlmann, 1998). Other studies are consistent with these results (Dixon *et al.*, 1993; Brown *et al.*, 1996; Stuart and Moura-Costa, 1998).

Other methodological issues include the absence of discounting in most of the available cost estimates, to reflect the time value of the investment and the production of GHG benefits. The choice of accounting approach also is important. If the ton-year approach (Section 5.4.2.) were used, these costs would tend to rise from about 50 percent to several times that, because fewer GHG benefits could be credited over a similar time frame.

Table 5-3: Undiscounted cost and carbon mitigation over project lifetime of selected AIJ Pilot Phase and other LULUCF projects in some level of implementation.[a]

Project Type (number of projects)	Land Area (Mha)	Total Carbon Mitigation (Mt C)	Costs (\$/t C)	Total Carbon Mitigation per Unit Area (t C ha^{-1})
Emissions Avoidance via Conservation:				
Forest Protection (7)	2.8	41–48	0.1–15	4–252
Forest Management (3)	0.06	5.3	0.3–8	41–102
Carbon Sequestration				
Reforestation and Afforestation (7)	0.10	10–10.4	1–28	26–328
Agroforestry (2)	0.2	10.5–10.8	0.2–10	26–56
Multi-Component and Community Forestry (2)	0.35	9.7	0.2–15	0.2–129

[a] Figures taken from project reports and published project reviews. Cost and carbon mitigation figures have been estimated using different methodologies, may not be comparable, and have not been independently reviewed. Cost values are estimated by dividing undiscounted costs and investment by estimated total carbon mitigation.

Estimated total investment committed to date in projects in Table 5-3 is about \$160 million; this amount could grow to about \$330 million if these projects were fully funded and implemented, although these estimates are provisional (EPA/USIJI, 1998; Stuart and Moura-Costa, 1998; Witthoeft-Muehlmann, 1998). Developers' project costs per ton of carbon are likely to change from these initial estimates over time. The price and supply of certified emissions reductions will be revealed if a market for them develops and as the eventual eligibility and requirements for various articles of the Kyoto Protocol become known. Costs may tend to decrease if economies of scale and technology transfer become widely available, potentially via development of portfolios of projects by entities transferring common, state-of-the-art methods to countries and projects. The Costa Rican Government's PAP, for example, undertook land-use data collection, baseline development, and establishment of monitoring systems for virtually all public lands in the country. The parallel Private Forest Program (PFP) provided some of the same services for private forest lands. In both cases, the goal was to reduce barriers to investment for carbon benefits (Tattenbach, 1996; Subak, 2000).

5.2.4. *Potential Magnitude of LULUCF Projects*

The form and magnitude of the eventual markets for ERUs under Article 6 (i.e., Annex I JI) and CERs under Article 12 (the CDM) are difficult to estimate. Key policy decisions have not been made by the Parties. This discussion of the potential for LULUCF activity in the CDM makes no judgment about the policy issue of whether the CDM includes specific LULUCF activities.

No credible, detailed estimates of the magnitude of the potential for LULUCF activities in Annex I and in non–Annex I countries are available for the first commitment period, 2008–2012 (see Chapter 4). To date, macroeconomic model assessments of supply and demand for emissions reductions by Annex I countries using the Kyoto Protocol flexible mechanisms (e.g., emissions trading, JI, or the CDM) have not separated LULUCF activities or projects. These analyses generally do not reflect policy or technical tests and guidance likely to be included in the operationalization of Articles 6 and 12. Analyses are needed of the potential supply, cost, and demand for LULUCF project-based activities, especially at the national level, under realistic scenarios for operating conditions under Articles 6 and 12.

The best approximations of Annex I project-level activity would be some small fraction—as yet unknown—of estimated country-level Article 3.3 and possibly Article 3.4 activities (depending on additional activities the Parties decide to include under Article 3.4, if any). Prospective activity levels under these two articles are reviewed in Chapter 3 and 4 (see also Nabuurs *et al.*, 2000). Project-level activities under Article 6 likely would be a small subset of these activities, which otherwise have been widely assumed to be reported nationally under the two articles, not as projects. No estimates of the demand for LULUCF project ERUs under Article 6 have been widely reviewed and reported for individual countries or for Annex I as a whole.

Global economic general equilibrium models have been used to project GHG target levels for 2010 for Annex I countries, to estimate the percentage of emissions reductions and total financial flows that might occur under Annex I JI or the CDM. Austin and Faeth (2000) summarized the results of four independent modeling teams. These models have been used to estimate where emissions reductions could occur at the lowest cost, largely on the basis of fossil fuel CO_2 emissions. Modeling results project that estimated emissions reductions by Annex I countries in the year 2010 (for that year) would come predominately from domestic reductions (15–45 percent), Annex I trading (6–10 percent), and "hot air" (8–41 percent). (*Hot air* is a term that describes ERUs predicted to be generated by country emissions during the first commitment period below countries' assigned amounts, as an artifact of macroeconomic and political changes in economies in transition, such as the Russian Federation, Ukraine, and Poland.) Some limited set of JI projects under Article 6 might be developed, although such projects would compete directly with these emissions reduction alternatives. Reductions in developing countries were estimated at 33–55 percent (another estimate is 19–57 percent; Vrolijk, 1999) of the demand for reductions by Annex I countries.

These models are neither designed to assess LULUCF activities, nor JI or the CDM. Project-oriented mechanisms are not likely to deliver the same stream of least-cost GHG abatement activities as an efficient emissions trading system, carbon tax, or other economic instrument (Austin and Faeth, 2000). Projects under Articles 6 and 12 may require certification and reporting costs, and costs for projects under Article 12 may also include charges for an adaptation fund and administration expenses of the CDM, as well as sustainable development considerations. These constraints may reduce the economic efficiency of JI and the CDM relative to emissions trading, according to reviews by some economists (e.g., Manne and Richels, 1999).

5.2.5. *Factors that May Affect the Realized Magnitude of Projects*

Studies and pilot project experience indicate that the net costs per ton of carbon of LULUCF mitigation activities in developing countries can be relatively modest or even negative (i.e., such projects may be profitable) in some projects and conditions (e.g., Makundi and Okiting'ati, 1995; Masera *et al* 1995; Ravindranath and Somashekhar, 1995; Wangwacharakul and Bowonwiwat, 1995; Xu, 1995). Annex I country estimates of LULUCF activities are generally found to be relatively higher per ton of carbon, but a substantial supply of sequestration or GHG reductions may be available at less than $20 per t C (Brown *et al*., 1996).

Only a limited number of potential projects are likely to be funded and implemented, however, as a result of community, investor, and national government priorities and cost-effectiveness (Mulongoy *et al*., 1998; Smith *et al*., 1999). The cost-effectiveness of LULUCF project activities will compete with the costs of achieving emissions reductions in other sectors—domestically within each country and internationally—under continual technological innovation in the energy sector, as well as the development of the GHG emissions reduction market.

Pilot projects in Annex I and non–Annex I countries commonly face high transaction costs (e.g., for implementing, monitoring, and reporting project activities) (World Bank, 1997; UNFCCC, 1999a). One key uncertainty is how transaction costs will be affected by the implementation of any eventual standardized guidelines for monitoring and verifying project emissions reductions and associated impacts on sustainable development. Transaction costs and risk may decline as carbon markets develop and standard financial techniques to spread risk and reduce uncertainty evolve (e.g., diversified portfolios, futures options contracts, and project performance insurance) (Frumhoff *et al*., 1998; Smith *et al*., 1999).

The types of future projects financed may not reflect patterns to date because economies of scale may favor larger scale activities with low costs (Smith *et al*., 1999). Investors with substantial near-term carbon liabilities may have a strong incentive to invest in projects that have the potential to provide carbon credits quickly though at a net cost (such as forest conservation). By contrast, investors with relatively modest near-term liabilities may have a strong incentive to invest in projects that provide carbon credits relatively slowly, but at a net profit—such as managed plantations (Frumhoff *et al*., 1998; Smith *et al*., 1999).

An example of how mixed incentives for LULUCF activities could occur has been raised by critics of the Kyoto Protocol. Non–Annex I countries would not have commitments to meet assigned amounts of GHG emissions, hence would not have emissions from deforestation or forest degradation counted against their assigned amounts. Financial incentives might exist to harvest or degrade forest lands to receive revenues from the timber products produced and the CERs generated if such lands were eligible for reforestation as project-based activities (Greenpeace, 1998; Chomitz, 2000). This situation could produce tensions for Parties between objectives of the UNFCCC and the Biodiversity Convention. At least two options exist to address this concern. First, the definition of reforestation activities selected by the Parties could limit reforestation to lands deforested prior to the commencement of non–Annex I project-based activities (Chapter 3 discusses this approach for Article 3.3 reforestation). Second, individual Parties could use the sustainable development conditionality of Article 12 to preclude eligibility for projects that reforest recently deforested lands—on biological diversity, conservation, or other grounds. The economic benefits to the host country of large-scale projects could be a disincentive for countries to limit LULUCF investments in any way, however, eroding their ability to manage such investments and their associated socioeconomic and environmental impacts (Smith *et al*., 1999).

Integrated projects or portfolios may offer potential synergies that address several technical issues. A sequestration component could provide sustainably managed forest products and reduce leakage from a conservation component, and a bioenergy component could provide jobs and low-cost power that is important to the sustainable development priorities of host countries, as well as enhanced profitability for investors (Niles and Schwarze, 2000). This approach has not been widely experimented with to date.

The public policy environment for the agriculture, forestry, and industrial sectors varies across countries and may facilitate or inhibit the penetration rate of LULUCF projects. Such policies could address tax incentives or subsides for afforestation, reforestation, or deforestation; land conversion to agriculture or alternative agricultural practices; land tenure; agrarian reform; and sustainable development more generally (Repetto and Gillis, 1988; Smith *et al.*, 1999). A review of the feasibility of significant levels of project-based LULUCF activity in non–Annex I countries under the Kyoto Protocol argues that the removal of distortionary national policies that promote forest degradation and land-use change may be a prerequisite for projects in some developing countries (Smith *et al.*, 1999).

A major potential limitation on LULUCF project penetration into the market for CERs and ERUs is the perception that LULUCF projects are less likely to produce credible, real, additional reductions. Two major perceptions are often advanced: the perceived difficulty of establishing the additionality of project benefits versus baselines and the claim that LULUCF projects are more difficult to measure and monitor and have greater leakage of GHG benefits than energy sector projects (Greenpeace, 1998; Trexler and Associates, 1998). A review of projects in the energy and LULUCF sectors (Chomitz, 2000) assessed five critical technical issues: additionality, baseline and systems boundary issues (including leakage), measurement, duration, and local social and environmental impacts. This assessment found that LULUCF and energy projects face parallel, comparable issues in measurement and in ensuring social and environmental benefits. In general, it is not possible to assert that energy projects are superior as a class to LULUCF projects on these grounds. The one significant difference identified between projects in these two sectors is the issue of project duration: LULUCF activities can be halted or their emissions reductions emitted. Similarly, a review of eight commonly raised technical issues in 12 CDM-like projects or activities in Brazil, India, Mexico, and South Africa (including seven LULUCF projects) found that about half of the concerns were minor or well managed by the project developers. Additionality, host country institution capacity, and baselines and leakage were the main concerns that needed more effort to be adequately addressed (Sathaye *et al.*, 1999).

5.3. Issues Arising from the Implementation of Projects

5.3.1. Project Boundary

Adequate determination of the physical and conceptual project boundaries is one of the critical steps in project design and implementation. The choice of accounting boundary influences the carbon credit that can be assigned to a project. It can also raise carbon accounting problems, particularly with regard to the relationship between project and national accounting.

Estimates of project impacts on carbon stocks may be limited at one extreme to aboveground vegetation within the geographical area of the project. At the other extreme, "total carbon" accounting may be used to include not only below-ground vegetation and soils on the project site but also the effects of wood products, fossil fuel substitution, and other changes at the national or even the international level. Because of these accounting problems, assessments of project impact should provide explicit details about the spatial, temporal, and conceptual boundaries used. Examples of carbon stocks and emission sources that may not always be captured within project boundaries include the following:

- Emissions associated with preparation of land prior to the official start of a project
- Emissions or removals of GHGs associated with the use of harvested timber
- Emissions associated with project development (car and air transport, machinery use, etc.)
- Fossil fuel emissions avoided through the use of biomass fuels as substitutes for energy production.

Decisions will have to be made about the level of standardization required for boundary-setting in projects. The cost implications of extending project boundaries to include many secondary effects could be significant.

5.3.2. Baselines and Additionality

A fundamental component of project assessment under the AIJ program has been the determination of the extent to which project interventions lead to GHG benefits that are *additional* to "business as usual" (UNFCCC, 1995; UNCCCS, 1997; Baumert, 1999). The concern about additionality also appears in Articles 6 and 12 of the Kyoto Protocol. Although additionality arguments have several different components and are based on multiple sources of information, most additionality problems apply equally to projects in the energy sector as to those in LULUCF (Chomitz, 2000).

The first step in determining a project's additional GHG benefits (*GHG emissions additionality*) is the elaboration of a without-project baseline scenario against which changes in carbon stocks occurring in the project can be compared (see Section 5.3.2.1). It is then necessary to demonstrate that the purported GHG benefits are truly additional, not simply the result of incidental or non-project factors such as new legislation, market changes, or environmental change (see Section 5.3.2.2).

Establishing the baseline scenario therefore requires knowledge of historical series of conventional practices in the affected area, the local socioeconomic situation, wider (national, regional, or even global) economic trends that may affect the conventional outputs of a project, and other relevant policy parameters. The baseline is established by projecting these past trends and current situations into the future. Consequently, baseline scenarios are necessarily based on a range of assumptions.

Currently, there is no standard method for determining baselines and additionality (Puhl, 1998; Matsuo, 1999). This section describes approaches used or proposed to date.

5.3.2.1. *Alternative Approaches Proposed for Establishing Baselines*

The main choices to be considered when deciding on how to establish a baseline are as follows:

- *Project specific versus generic:* Should baselines be developed as a case-by-case, project-specific exercise, or could they be based on generic data that are aggregated in a "top-down" approach? Should baselines be developed by project proponents or by independent bodies (regional, national, or international institutions)?
- *Fixed or adjustable:* Should baselines established at the start of the project be maintained for the project's lifetime or be periodically adjusted?
- *Simple or complex models:* Should baselines be derived by simple extrapolation of past trends in the use of land, or should they be derived from models that attempt to simulate the driving forces of change?

These options are discussed below. Table 5-4 provides examples of how baselines of different pilot projects have been constructed.

5.3.2.1.1. *Project-specific versus generic*

Most projects developed under the AIJ Pilot Phase have used project-specific, bottom-up baselines determined by project developers (Moura-Costa *et al*., 2000; see also Table 5-4). The attractions of this approach are that analysis focuses on specific areas and activities relating to the project and that developers may have a better knowledge of local conditions. Because land-use practices and change processes are often spatially and temporally variable, a detailed project-specific study arguably is likely to yield a more accurate prediction of emissions than a broader regional or sectoral assessment. Giving project developers the task of developing baselines also introduces the risk, however, that they may choose scenarios that maximize their perceived benefits (Tipper and de Jong, 1998). Moreover, ensuring consistency between assessments may be difficult if different teams develop many baselines. Allowing *ad hoc* project baselines could lead to inconsistent approaches among similar projects and increase the risk that project baselines would be set strategically to maximize the potential to generate credits.

Generic methods that have been proposed but not yet tested include benchmarking models similar to those being assessed for the industrial and energy sectors (Center for Clean Air Policy, 1998; Hargrave *et al*., 1998; Baumert, 1999; Ellis and Bosi, 1999; Friedman, 1999; Jepma, 1999; Michaelowa, 1999). For example, certain practices could be considered "standard management practice," and baselines might be set to reflect the level of carbon sequestration or emissions avoidance that would occur if these practices were universally applied. Credit would then be available only to the extent that a project improved on the results that would be obtained by simply applying these standard practices. Because the development of credible baseline scenarios represents a significant capital cost, the use of generic baselines for sectors, technologies, or regions could provide economies of scale (Baumert, 1999). If such baselines were set by an organization independent from project developers, they could also provide transparency and reduce the potential for discrepancies between projects. The applicability of this approach to the LULUCF sector is unclear; no project to date has used a benchmarking approach. Generic baselines set by a coordinating body have been used in a few cases (e.g., the Protected Areas Project in Costa Rica, SGS, 1998; the Profafor project in Ecuador, FACE Foundation, 1998).

Another proposed approach involves minimum performance benchmarks (Brown, 1998). Minimum baselines or benchmarks could help to avoid rewarding countries or investors with poor practices or policies by paying for improvements over an exceedingly low baseline (Brown, 1998). If countries hosting LULUCF projects have policies that encourage carbon-emitting activities, such as subsidies for deforestation, LULUCF projects may only be mitigating the impact of poor policies. For instance, if project baselines are influenced by the threat that a particular area will be deforested in the absence of the project, this situation could create an incentive to "demonstrate" the threat of deforestation—by building roads through isolated areas, for example.

5.3.2.1.2. *Approaches for determining baselines*

Most projects to date have adopted a two-step approach to determine baselines. First, the likely fate of terrestrial ecosystems within the project boundary is predicted. Second, changes in carbon stocks that would occur as a result of this scenario are estimated.

Specification of the without-project scenario for the project area usually have been based on projections of past trends of land use into the future. These predictions have taken into consideration events that are expected to alter current behavior (changes in legislation related to land use and tenure, changes in market preferences or prices, changes in environmental awareness, etc.). Even a thoroughly investigated without-project baseline, however, is prone to the risk that unexpected social or policy changes will confound predictions over the longer time frame. For example, the baseline for a reduced-impact logging project could change radically if national policy dictated adoption of this practice in all forest concessions. Key factors used in projecting the baselines have included planned land-use decisions of landowners/stakeholders, designation of land by national authorities, and historical patterns of land-use change in the local area.

Different approaches likely would be required, however, for different types of projects operating in different circumstances:

- Afforestation projects might use simple models that predict zero uptake/emissions without intervention.

Table 5-4*: Approaches used by different projects to establish baselines.*

Type of Project	Project-Specific or Generic	Approach Used to Establish Baseline	Fixed or Adjustable	**Reference**
	Baseline Issues			
Avoided Emissions				
Noel Kempf Climate Action Project, Bolivia	Project-specific	Simple logical argument based on adjusting observed trends; quantification of baseline carbon done in proxy areas	Adjustable, based on changes of demand for timber, changes in marketable species, forest law, and rates of deforestation	Brown *et al.* (2000)
Reduced Impact Logging, Sabah, Malaysia	Project-specific	Simple logical argument, based on assuming continuation of business-as-usual trends; quantification of baseline carbon done in control plots	Fixed	Pinard and Putz (1997)
Rio Bravo Carbon Sequestration Project, Belize	Project-specific	Simple logical argument, based on assuming continuation of business-as-usual trends	Fixed	Programme for Belize (1997)
Sequestration				
Farm Forestry Scolel Te Pilot Project, Chiapas, Mexico	Mixed approach	Regional land-use model with community-specific adjustments based on land-use needs	Fixed	Tipper *et al.* (1998)
INFAPRO Rainforest Rehabilitation, Malaysia	Project-specific	Simple logical argument, based on assuming continuation of business-as-usual trends; quantification of baseline carbon done in control plots	Fixed	Moura-Costa *et al.* (1996)
Multi-Component (avoided emissions and sequestration)				
Protected Areas Project and Private Forests Project, Costa Rica	Generic, set by Costa Rican Office for Joint Implementation (OCIC)	Simple logical argument, based on adjusting current land-use trends; quantification of baseline carbon based on literature values	No adjustments planned	SGS (1998)
Guaraqueçaba Climate Action Project, Brazil	Project-specific	Spatial land-use models incorporating socioeconomic factors	Adjustable, to recalibrate model at frequent intervals	Brown *et al.* (1999a)

- Projects to conserve forests used by small farmers are likely to need models that reflect local demands for agricultural land, firewood, and timber.
- Projects aiming to reduce emissions through better forest management may need models that compare technological alternatives.

Different approaches for data collection have been used, including compilations of national/regional statistics, satellite imagery, and interviews with relevant authorities and key stakeholders. There is debate about the level of detail required and the weight given to different criteria (historical trends, available technology, population pressure, etc.) (Busch *et al.*, 1999).

Several approaches have been proposed and/or used during the AIJ Pilot Phase for deciding how to carry out baseline projections. These approaches vary with regard to data requirements and treatment:

- *Simple, logical arguments* do not use quantitative methods for predicting changes in current trends (or use simple ones). For example: "Without intervention, the forest concerned will be sold for agricultural development" [Rio Bravo project (Programme for Belize, 1997a)] or "without intervention, loss of aboveground carbon stocks within the area will continue at approximately 1.5 percent per year" [Scolel Té pilot project (Tipper *et al.*, 1998); see also Box 5-2]. Variations of this approach have been used by most projects during the AIJ Pilot Phase [e.g., the NKCAP in Bolivia (Brown *et al.*, 2000); the RIL project in Sabah, Malaysia (Pinard and Putz, 1997); the PAP in Costa Rica (SGS, 1998)].
- *Spatial or social-economic models* simulate land-use change processes on the basis of factors such as proximity of towns, roads, and agricultural frontiers; population growth; food requirements; and the productivity of local agricultural technology [e.g., LUCS model (Faeth *et al.*, 1994); Ludeke, 1990; Jepma, 1995). This approach is being used in The Nature Conservancy's project in Guaraqueçaba, Brazil (Brown *et al.*, 1999a,b).
- *Econometric models* use an econometric treatment to data factors such as historical series of productivity, price, costs, and so forth. This approach has not been used in the AIJ Pilot Phase, but it has been discussed in a few publications (e.g., Chomitz, 1998).

Simple, logical arguments are not necessarily less accurate in terms of predictive ability. Their applicability will probably be

Box 5-2. Historical and Projected Carbon Storage in an Area of ~300,000 ha in the Highlands of Chiapas, Mexico, based on a Series of Multi-Spectral Scanner (MSS) Images

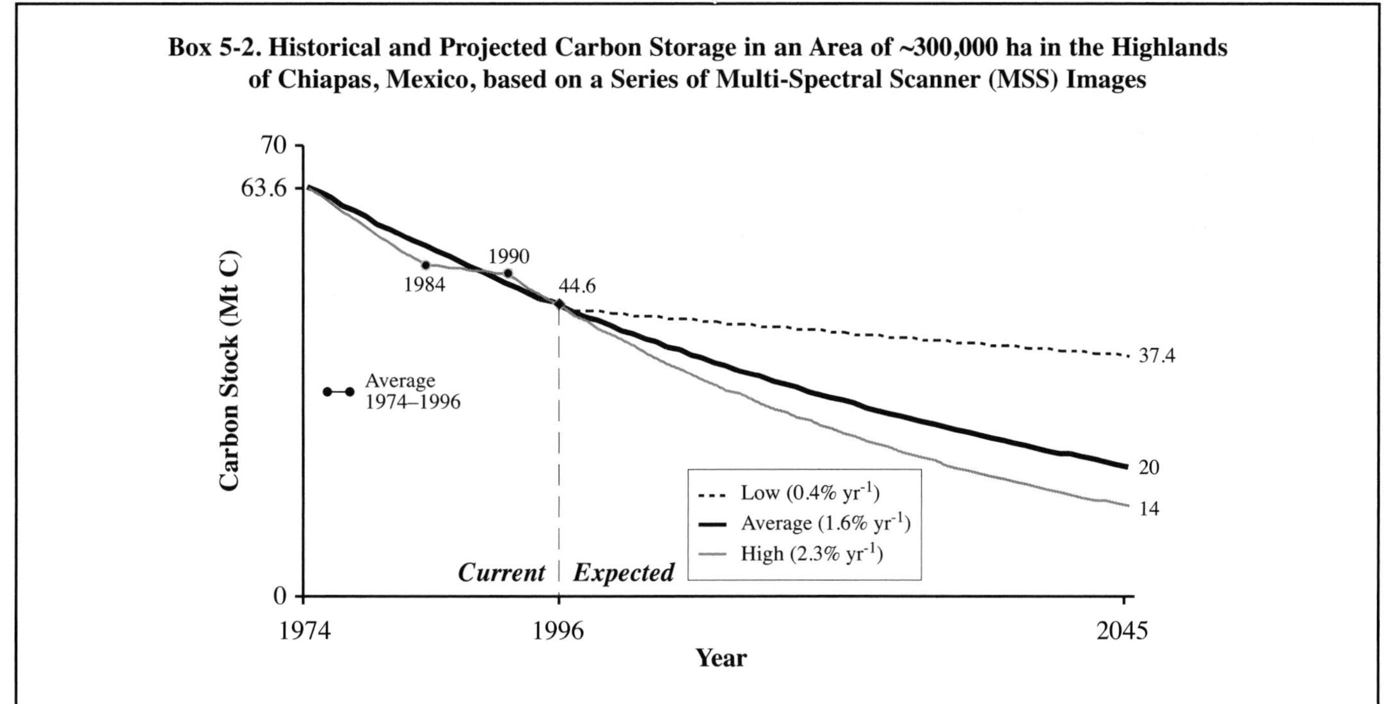

A series of satellite images, from 1974 to 1996, was used to estimate changes in land use between 12 categories of vegetation/land use for an area of approximately 300,000 ha in the highlands of Chiapas. Measurements of the biomass of each vegetation type were then used to derive an estimate of the change in carbon stocks. Extrapolation of the historical changes of carbon stocks into the future can be used as a basis for without-project baselines. Because the rate of land-use change varied considerably over the 1974 to 1996 period, however, so did the baseline rate of carbon loss over the time period chosen as the reference. The spatial frame of reference used to derive estimates of land-use change is also important. Deforestation activity is often concentrated in particular areas (e.g., along roads and river valleys). The historical rate of change may therefore vary considerably according to the geographical coordinates of the reference area.

limited, however, to specific areas and contexts. Increasing model complexity is likely to be required to attempt credible predictions across a range of land uses. Such models, however, generally require large amounts of input data and may still be poor predictors of specific local changes. Requirements for complex baseline models could represent a serious barrier to small-scale projects or initiatives in poorer countries unless "umbrella" approaches are adopted (Bass *et al.*, 2000). Procedures for selection or approval of models and a program for model testing and improvement to ensure some degree of consistency and quality would have to be considered.

Once a baseline scenario for land-use and ecosystem changes has been developed, changes in carbon stocks associated with this scenario must be estimated. Different approaches have been used or proposed during the AIJ Pilot Phase (see examples in Table 5-4), including the following:

- Quantification of carbon in proxy areas [e.g., the NKCAP (Brown *et al.*, 2000)]
- Control plots where project activities are not applied, which are set aside for measurement of carbon stocks in the absence of the project intervention [e.g., the RIL project in Sabah, Malaysia (Pinard and Putz, 1997)]
- Modeling [e.g., the PAP in Costa Rica (SGS, 1998)]
- Combinations of the foregoing approaches.

5.3.2.1.3. Fixed or adjustable baselines?

Baselines could be fixed for the lifetime of the project or adjusted following periodic reviews or the occurrence of unexpected events. A preliminary report to the Secretariat to the Convention (UNCCCS, 1997) argued that baselines for AIJ projects should not be revised because such revisions would increase the uncertainty associated with any investment and entail significant additional costs. The central argument for revising the baseline over the length of the project is that such revisions may ensure more realistic offsets. A key counter-argument is that continuous revision of baselines could have a significant impact on the economic value of the project, introducing another source of risk. Moreover, disassociating changes observed after the implementation of the project from the impact of the project itself also is difficult. Detailed discussion of methods for adjusting baselines and their implications appears in Michaelowa (1998, 1999) and Ellis and Bosi (1999).

5.3.2.2. Additionality Tests

After a project baseline is determined, it may then be necessary to demonstrate that the purported GHG benefits of the project are truly additional (environmental additionality). Several additionality tests have been devised to assess the eligibility of projects to enter the AIJ program. Tests applied by the USIJI (USIJI, 1997a) included the following:

- *Technological tests:* Where activities have resulted from the introduction of new technologies or the removal of technological barriers. Evidence would include comparison of current practices and technologies with those to be adopted by the project (Carter, 1997b).
- *Institutional or program tests*: Where activities go beyond the scope of the programs of the institutions involved in the development of the project. Evidence would include removal of institutional constraints or implementation of measures in excess of current activities and regulatory requirements.
- *Financial tests:* Demonstration that the project incurred higher costs (or has higher risks) compared with those of comparable baseline activities. Evidence could include an assessment of the potential for commercial finance, as well as cost-benefit analyses.

Projects may demonstrate additionality with one or more (but not necessarily all) of the foregoing tests. According to the USIJI experience, additionality criteria are difficult to evaluate objectively on a project-by-project basis (Carter, 1997a). As with other screening programs, two types of errors exist: approval of non-additional projects and exclusion of valid projects (Chomitz, 1998). The concept itself is complicated because it requires assessment of hypothetical future scenarios in the absence of the project.

For projects implemented under the AIJ modality, additionality has not only been required in terms of expected GHG benefits but also with regard to funding. The first COP ruled that "the financing of AIJ shall be additional to the financial obligations of Parties included in Annex II to the Convention within the framework of the financial mechanism as well as to current international development assistance flows." This requirement applies to country-level Official Development Assistance (ODA) transfers, funding mechanisms under the UNFCCC, and various multilateral development bank and development agency activities.

5.3.3. Leakage

Leakage is defined as the unanticipated decrease or increase in GHG benefits outside of the project's accounting boundary (the boundary defined for the purposes of estimating the project's net GHG impact) as a result of project activities. For example, conserving forests that otherwise would have been deforested for agricultural land may displace farmers to an area outside of the project's boundaries. There, the displaced farmers may engage in deforestation—and the resulting carbon emissions are referred to as leakage.

Projects may also yield greater GHG benefits than anticipated—positive leakage or "spillover." For example, if a project introduced a new land management approach or technology—such as increased use of agroforestry or cover crops or increased saw mill efficiency—and this technology was more widely adopted outside the project's boundaries, the net GHG benefits would be larger than initially estimated.

5.3.3.1. Assessing Leakage

Leakage has been divided into various effects. This section discusses leakage effects that are most relevant to forest and land-use projects.

Market effects occur when project activities change the supply/demand equilibrium, such as if demand is unmet because a project reduces supply or because it unexpectedly increases demand. For example, large-scale plantation projects may depress the local price of wood products, causing nearby plantations to be replaced with pasture or other low-biomass land uses (Fearnside, 1995). Activity-shifting occurs when the activity that causes carbon loss in the project area is displaced outside the project boundary. For example, prevention of deforestation in the project area may displace the GHG-emitting activity.

Although project experience to date is limited, case studies have indicated that landscape dynamics may signal if the project has no or low potential for leakage or a moderate to high risk for leakage.

No/Low Leakage Potential: Experience to date indicates that projects implemented on land that has few or no competing uses are unlikely to impact areas outside of project activities, and leakage potential is minimal. For example, the Krkonose project in the Czech Republic (see Table 5-2 and Box 5-1) is situated in a protected area with virtually no danger of encroachment or displacement because the park had protected status for many years (Brown *et al.*, 1997).

Moderate/High Leakage Potential: Where land has competing uses or in dynamic settings where factors such as population growth, logging or agricultural production for export, subsistence agriculture, fuelwood needs, and concerns about deforestation interact, a project's impact may extend beyond the area of direct project activities (Brown, 1998). If net GHG benefits estimated and monitored fail to account for emissions that arise because of the project outside the area of direct activities, leakage is an issue. For example, a project that stops the conversion of forest to agricultural land or ends timber harvest by effectively "putting a fence around the forest" will face leakage problems because if an economic activity in the forest is stopped with no alternative taking its place, people will shift the activity to a surrounding area.

Changes in national or international policies also can lead to leakage. For example, when a government changes its policy to lower the country's overall emissions, the emissions may be displaced to other countries (see Section 2.1.1).

5.3.3.2. Methods for Monitoring Leakage

To date, two approaches have been used and proposed to monitor leakage. One approach involves determining the appropriate spatial area in which to monitor project effects; the other involves identifying key indicators of leakage on the basis of demand that drives land-use change and management.

Monitoring by Area: Leakage may be monitored by expanding the project's boundary. The monitoring area may be larger than the area on which project activities are implemented (Brown, 1997; Trexler and Associates, 1998). Potential monitoring boundaries for leakage are at the project level, the local/regional level, or the global level:

- *Project activity boundaries:* Projects implemented on land that has few or no competing uses may have to consider only the area of direct project activities because the project's impact is unlikely to extend beyond its immediate boundaries. For example, the RUSAFOR project (see Table 5-2) has no competing land uses, and the timber will not be harvested.
- *Regional/local boundaries:* Where land has competing uses or in dynamic settings where factors such as emigration, population growth, and fuelwood gathering are important, a project's impact is likely to extend beyond its immediate boundary and may extend to the local area or region (Brown, 1998). For example, monitoring can be expanded to include the local cattle, timber, or food market (Chomitz, 2000).
- *Global market boundaries:* Still other projects—notably, those involving timber harvesting or agricultural production for export—may be operating in a global market. In these cases, if the project causes a restriction on the goods produced, leakage may occur because the project will be unable to affect global market demand. For example, a logging project in which the timber is for a global market could monitor regional wood product production surveys, regional or national wood product flows, or survey mills (Brown *et al.*, 2000).

Monitoring by Key Indicators: Alternately, it has been proposed that leakage be monitored by determining key indicators for demand that drives land-use patterns or management that leads to carbon emissions (such as demand for timber, fuelwood, or agricultural land) (Brown *et al.*, 1997). The key indicator is the output of the product demanded. A project that reduces output or access to resources without offering alternatives is likely to result in leakage because people within the project area will move elsewhere to find other sources of resource supply. A review at the project level has suggested that leakage indicators can be developed by determining whether the project has displaced activities that lead to carbon emissions, rather than replacing or substituting for them (Brown *et al.*, 1997). For example, to monitor leakage potential for a project that seeks to replace conventional logging with reduced-impact logging, timber output would be the key indicator to be monitored. If timber output from the project area decreases while prices and demand for wood products remain the same, the project could have leakage. The assumption would be that additional areas would be logged to compensate for the timber loss (Brown *et al.*, 1997). Under this method, it would be unnecessary to track global markets or the harvest intensity of nearby timber concessions; instead, the key indicator of output would be used to monitor the leakage potential. Similarly where demand for agricultural land is driving land-use change: If conversion of forest to agricultural land is halted but agricultural productivity is not increased on existing lands, the project is likely to result in leakage.

Several projects have developed leakage indicators. In the Noel Kempff Mercado project, for example, the government of Bolivia used carbon mitigation funds to compensate forest concessionaires for giving up logging rights on government-owned forest lands and expand the park boundaries (Box 5-3). A legally binding "leakage agreement" was signed by the logging companies, obliging them not to invest the funds they received in logging elsewhere. The key indicators are the use of received carbon funds and the harvesting rates in the concessions. The concessionaires will be monitored to ensure that they do not increase production elsewhere because of the project funding (Brown *et al.*, 2000).

Table 5-5 presents indicators of leakage for LULUCF project activities on the basis of whether the project has addressed demands that drive carbon emissions from the project area (Brown *et al.*, 1997). The underlying concept is that decreasing output or access to needed resources will prevent a project from meeting its carbon benefit goals. The extent of the unmet demand determines the potential magnitude of leakage caused by project activities. Multi-component projects are missing from the table, but potential management strategies point to adding activities, particularly to conservation projects (Chomitz, 2000).

5.3.3.3. *Options for Responding to Leakage*

To date, two approaches have been used /proposed to address leakage; these two approaches may be employed independently or simultaneously. One approach involves addressing leakage at the project level through either project design or re-estimating of net GHG benefits. Some people question whether project-level approaches can adequately ensure that leakage will be addressed, however. As a response, macro-level approaches to address leakage have been proposed that would involve developing regional or national baselines or establishing risk coefficients by project type or characteristic.

Project-level approaches: Leakage potential may be identified at the front end of project design and additional activities incorporated if the project appears vulnerable to leakage. If evidence of leakage emerges after project implementation has begun, project implementers may undertake additional activities to mitigate leakage or to monitor it and subsequently revise net GHG estimates.

Project design elements incorporated in projects: Although experience to date is limited, several elements have emerged that may help avoid leakage, depending on the socioeconomic and physical context of the project. Project design strategies that have been used to avoid leakage include providing socioeconomic benefits to local people that create incentives to maintain the project and its GHG benefits because of these associated benefits and using replicable or transferable technologies that can help avoid leakage because they allow project benefits to be duplicated outside project boundaries, so that social benefits are not restricted to a limited area. Incorporating these elements can help avoid leakage.

Multi-component projects may also help avoid leakage because they can combine project activities to fully address demands that drive land-use change (Chomitz, 2000). For example, the Costa Rican PAP generated carbon offsets by avoiding carbon emissions and through carbon sequestration. The PAP is consolidating approximately 570,000 ha of primary, secondary, and pasture lands within the National Parks and Biological Reserves of Costa Rica (Tattenbach, 1996; Stuart and Moura-Costa, 1998). The PAP plans to reduce deforestation of primary forest, thereby reducing carbon emissions resulting from deforestation. The PAP also plans to allow secondary forest and pasture to regenerate, thereby sequestering carbon through tree growth and accumulation of woody biomass. Concurrently, Costa Rica has also developed a parallel program, the PFP, which provides financial incentives for land owners outside the PAP area to opt for forestry-related land uses as opposed to agriculture—thereby generating a series of environmental services, such as CO_2 fixation, maintenance of water quality, biodiversity, and landscape beauty (Forestry Law N. 7575, April 1996) (Stuart and Moura-Costa, 1998). The PFP is also expected to offset the effects of decreasing timber harvest in the project area, reducing possible leakage effects.

Another example of a multi-component project is the CARE/ Guatemala project, which increased fuelwood availability and agricultural productivity by encouraging agroforestry. The project also protected some forest areas, allowing degraded areas to regenerate. The CARE/Guatemala project began in 1988, and persisted through years of political strife and high demand for agricultural land because the project combined elements of forest protection with agricultural extension that provided social benefits that gave local people a stake in the project's success (Brown *et al.*, 1997).

Re-estimation of net GHG benefits: Leakage cannot always be avoided at the outset or mitigated with additional activities. In some cases, GHG estimates can be recalculated. If project implementers can quantify the shortfall in output from the project, they can quantify the amount of leakage (Brown *et al.*, 1997). To recalculate the original net GHG benefits, the project evaluator must determine approximately how much area must be logged or converted to agriculture to compensate for the decrease in output.

For example, the RIL project in Malaysia (see Table 5-2) was originally estimated to avoid 38,700 t of carbon emissions. However, the project may have resulted in carbon leakage because on 450 ha of the 1,400 ha project, timber production was

Box 5-3. Carbon Inventorying and Monitoring of the Noel Kempff Climate Action Project (NKCAP), Department of Santa Cruz, Bolivia

The project area of approximately 634,000 ha is located within the newly expanded western region of the Noel Kempff Mercado National Park. Prior to the initiation of the NKCAP, much of the forest in the expansion area had been high-graded over a period of about 15 years. In addition to logging, this area was also under pressure for conversion to agriculture (for further details, see Brown *et al.*, 2000). The forests in the expansion area were divided into six strata for sampling: tall evergreen, liana, tall inundated, short inundated, mixed liana, and burned forest.

The project design for inventorying and monitoring the carbon pools in the with-project case is based on the methodology and protocols in MacDicken (1997a). The carbon inventory of the area was based on data collected from a network of 625 permanent plots; the number of plots sampled in a given strata was based on the variance of an initial sample of plots in each strata and the desired precision level (±10 percent) with 95-percent confidence. A fixed-area, nested-plot design was used, and carbon stocks were measured or calculated for each of the following pools in each plot: all trees with diameter at breast height >5 cm, understory, fine litter standing stock, standing dead wood, and soil to 30-cm depth. Root biomass was estimated from root-to-shoot ratios given in Cairns *et al.* (1997). The total amount of carbon in the park expansion area was about 115 million t C, most of which was in aboveground biomass of trees (60 percent), followed by soil to 30-cm depth (18 percent), roots (12 percent), and dead wood (7 percent); the understory and fine litter accounted for about 3 percent of the total. The 95-percent confidence interval of the total carbon stock was ±4 percent, based on sampling error only; regression and measurement error were not included.

Averted Logging

The carbon benefits from this activity result from halting the removal of commercial timber and eliminating damage to the residual stand. Estimates of changes in major carbon pools from logging and projections of timber extraction if logging had been allowed to continue over the project life were assessed to generate the without-project baseline. The main carbon pools considered in this activity are aboveground tree biomass, dead biomass, and wood products. Bolivia recently enacted a new forestry law and developed new regulations for forest harvesting. This information is used to predict how much forest area in the project area would have been harvested in a given year for each year over the length of the project. From data provided by logging concessionaires, and analysis of concessionaire management plans in areas nearby, the likely quantity of wood (in cubic meters per hectare) extracted per year is also estimated.

The change in carbon stocks from logging activities is measured in a nearby proxy forest concession. Permanent plots are established to measure the amount of dead biomass produced during the felling of a tree and associated activities, as well as the rate of regrowth after harvesting. Dead biomass results from the crown and stump of the felled timber tree and damage to other trees. Total production of dead biomass carbon per unit of harvested biomass carbon is determined from these plots.

C benefits from averted logging = Δlive biomass C + Δdead biomass C + Δwood product C

where Δ is the difference in carbon stocks between the with- and the without-project case. The annual benefits are calculated from a carbon accounting model that tracks all of the changes in these pools from a scenario that is based on the annual area logged, log extraction rates, and logging damage.

Δlive biomass C = (biomass C from logging damage + C in timber extracted) x growth factor

To estimate the change in live biomass, one could measure the live biomass in the proxy concession before an area was logged and then again after it was logged; the difference would give the change in the live biomass C. One main problem with this approach, however, is that two large carbon stocks are being subtracted; although the error on each stock could be small, the error on the difference, expressed as a percentage, will be much larger. To overcome this problem, the change in live biomass was measured directly. The change in live biomass between the with- and without-project cases is a result of the extraction of timber and damage to residual trees from logging activities (the quantity in parentheses). The quantity in parentheses, expressed on an area basis, multiplied by the area logged per year gives the total change in live biomass without adjustment for logging effects on the growth of the residual stand (the growth factor). It is not clear if

Box 5-3. Carbon Inventorying and Monitoring of the Noel Kempff Climate Action Project (NKCAP), Department of Santa Cruz, Bolivia (continued)

harvesting stimulates or reduces regrowth in recently logged areas. The logging of large trees and damage to residual trees may be enough to actually reduce net biomass growth of the stand per unit area for several years after logging rather than stimulate it. For projects that prevent or modify logging, this effect of logging on the growth of the residual trees must be determined. Monitoring of paired permanent plots in logged and unlogged areas of the proxy concession is under way to establish the sign and magnitude of the growth factor over the length of the project.

Δdead biomass C = (dead biomass from logging damage x decomposition factor)

In projects that are related to preventing or reducing logging, dead wood cannot be ignored because it is a long-lived pool, and logging increases the size of this pool. Thus, stopping logging has the effect of reducing the dead biomass carbon stock, and the dead biomass carbon in the with-project case is less than in the without-project case. The change in the dead biomass pool has to be corrected for decomposition, however. Estimates of the decomposition correction factor are taken from the literature (Delaney *et al.*, 1998), but field measurements are under way to improve this factor.

Δwood products C = (timber extracted x proportion converted to long-lived products)

Stopping logging reduces the long-term wood product pool because the input of new products is reduced; thus, the change in the wood products pool is negative. The harvested timber in the Santa Cruz area is from a small number of speciality tree species; a reduction in their supply may not be supplied from elsewhere. In the NKCAP, the proportion of harvested roundwood that goes into long-term wood products was obtained from literature sources for Brazil (Winjum *et al.*, 1998). The project assumed that wood waste generated at each stage of the conversion of timber to products (50 percent was converted to sawdust in the first milling stage) was oxidized in the year of harvest.

The difference between the with- and the without-project case is that the with-project case has more carbon in the live biomass pool and less carbon in the dead biomass and wood product pools than the without-project case.

Averted Conversion to Agriculture

The carbon benefits from this activity result from the elimination of carbon loss in forest biomass and soil. The without-project baseline for this component was established by using projected human demographics in areas adjacent to the project area. The two factors affecting conversion of forestlands to agriculture in the area surrounding the NKCAP are increasing human populations and the resulting demand for farmland. In constructing the deforestation scenario, it was assumed that migration into the area will fuel continued demand for agricultural land, as in other areas nearby the NKCAP.

C benefits from averted forest conversion = Δtotal biomass C + Δsoil C

Carbon loss from a change in biomass is calculated as the product of the projected area cleared and the difference between carbon in forest biomass (the sum of trees, understory, litter, dead wood, and roots) and agriculture crop biomass. Changes in soil carbon are estimated as the product of area cleared, weighted average forest soil carbon, and an average soil oxidation rate for converted tropical forest soils obtained from Detwiler (1986).

decreased by approximately 49 m^3 ha^{-1} relative to conventional logging. Total timber shortfall was 450 ha x 49 m^3 ha^{-1} = 22,050 m^3 of reduced timber output. To quantify the amount of potential leakage, it is possible to estimate the additional area that must be logged to make up for the deficit. The leakage potential could be roughly determined by estimating the amount of emissions resulting from logging to compensate for the 22,050 m^3 of reduced output. Assuming that RIL makes up for the shortfall, leakage could be estimated as follows: RIL emits 108 t C ha^{-1} and yields 103 m^3 of timber per ha (Pinard and Putz 1997); therefore, harvesting 214 ha using RIL methods would make up for the reduced output. Leakage then equals 23,112 t C emitted (214 ha x 108 t C ha^{-1}). Thus, the estimated net carbon benefit is 38,700 t C (the original amount) - 23,112 t C (the leakage) = 15,558 t C.

These estimates are approximate, and they represent only one harvest cycle. They illustrate one means of quantifying leakage. In this example, RIL still results in a net carbon gain—which might or might not be the case for all projects. In addition, RIL projects are designed to increase output over time because there is less damage to young trees. In the long run, RIL sites may produce greater output than conventionally logged sites.

***Table 5-5**: Factors contributing to leakage and potential options.*

Project Components	Activity Being Replaced	Conditions Signaling Leakage	Leakage Potential	Management Strategies
Emission Avoidance				
Forest preservation	Conventional timber harvest practices	Decrease or halt in timber output	High	Develop alternative timber sources such as plantations on marginal land; introduce sustainable harvest in buffer areas; reestimate project's GHG benefits
	Conversion to agriculture	Decrease in agricultural output	High	Create alternative income source such as sustainable forestry; add agricultural productivity component
Sustainable forestry, reduced-impact logging, natural forest management	Conventional timber harvest practices	Decrease in short-term output, but increase over long term	Moderate	Reestimate GHG benefits over short term; develop alternative timber sources
		Decrease in timber output	High	Reestimate GHG benefits
Carbon Sequestration				
Agroforestry, improved soil conservation, woodlots, windbreaks	Current agricultural practices	Increase in output, but free resources for development on adjacent lands	Moderate	Protect adjacent forests; implement sustainable forestry
	Fuelwood gathering	Common property resource; off-site market demand	Moderate	Employ transferable technology
Increased agricultural productivity	Current agricultural practices	Free resources for development on adjacent lands	Moderate	Protect adjacent forests; implement sustainable forestry

The components of the table are as follows:

- Project Components—Activities employed in LULUCF projects to date.
- Activity Being Replaced—These activities typically produce agricultural goods, fuelwood, and timber. The underlying concept is that decreases in output or access to needed resources resulting from these activities may result in leakage.
- Conditions Signaling Leakage—Conditions under which components may become vulnerable to leakage. A project that reduces access to resources, without offering alternatives, is likely to result in leakage. If a project expands, has a neutral impact on output of the resource, or provides a substitute, it is likely to avoid leakage.
- Leakage Potential—Offers an assessment of a project's potential for leakage (moderate or high) during the short or long term. Because the index is qualitative, there is no strict interpretation for these designations. A *moderate* designation means that the amount of leakage, as well as its presence or absence, depends on individual site conditions. A *high* designation means that, unless there are mitigation strategies, leakage will occur. Where timber is the primary resource demanded, leakage may be of short- or long-term duration. For example, although sustainable forestry projects may reduce timber output in the short term, in the long term the project sites are more productive than their conventionally logged ones because fewer young trees are damaged.
- Management Strategies—Suggests potential strategies for avoiding or mitigating leakage (or in some cases reestimating project impacts) that have been implemented in ongoing carbon sequestration projects or have been proposed for such projects.

Macro-level approaches: Alternatives to project-based approaches have been proposed, including estimation of empirically based sectoral, national, or regional baselines that can potentially capture leakage, and development of adjustment coefficients for leakage risk and adjustment of net GHG estimates accordingly:

- *National and regional baselines:* Adopting sectoral or regional baselines on LULUCF emissions and sequestration is one alternative to the project-based approach. If a project attempted to reduce the rate or area of deforestation, the project-level effect would need to be demonstrated in subsequent monitoring and baseline estimates improved. By encompassing a large geographic area, leakage could be internalized. One proposal involves developing national, regional, or sectoral baselines on land-use change and management; this approach is based on the concept of tradable development rights. As above, a regional baseline deforestation rate would be determined by using land-use trends. A certain percentage of the forest would be protected, and the remaining forest could be developed (Chomitz, 2000). Allowance or development rights to the forest would be distributed; the owners of these development allowances could then in turn sell the development rights. In areas where development was allowed, the carbon sequestration services of the forest could be sold instead of developing the forest (Chomitz, 2000).
- *Risk premiums and adjustment coeffiecients:* To overcome the complexity of quantifying leakage, another approach is assigning specific leakage coefficients (Trexler and Kosloff, 1998). Project estimates could then be adjusted by this coefficient (Gustavsson *et al.*, 1999). These coefficients could be developed at a regional or national level for different project types. The effect of risk premiums or adjustment coefficients is that the projects can claim only a portion of the estimated GHG benefits. A percentage of the net GHG benefit is retained in a buffer to cover the risk of leakage—the goal being to protect the atmosphere from added carbon emissions. For example, the PAP in Costa Rica has a reserve or buffer of carbon sequestered to insure against various risks, including leakage. The PAP assumes that some of the subsistence-based farmers who move from the forest may squat on new land, resulting in leakage. The PAP estimates a low risk of leakage because only 22,223 ha of the total of 530,498 ha is currently in private hands. Therefore, if 25 percent of all private owners choose to buy or occupy new areas of land and deforest them, the carbon offsets arising from over than 1 percent of the total project area would be negated. As a response, in its first year the PAP will offer only half of the estimated emission reductions for sale, with the rest serving as an insurance buffer for this and other estimated risks (Chomitz *et al.*, 1999).

5.3.4. Project Duration

5.3.4.1. How Long Do Projects Have To Be Run?

The Kyoto Protocol requires that LULUCF projects result in long-term changes in terrestrial carbon storage and CO_2 concentrations in the atmosphere. The definition of "long-term" varies substantially, however, and there is no consensus regarding a minimum time frame for project duration.

During the AIJ Pilot Phase, projects have been conducted for a variety of time frames: from 20 years [e.g., the PAP in Costa Rica (Trines, 1998a)] to 99 years [e.g., the FACE Foundation projects (Verweij and Emmer, 1998)]. Most projects state that their GHG benefits are expected to be maintained beyond the project time frame [see the list of AIJ projects on the UNFCCC Web site (UNFCCC, 1999b)], although their contractual arrangements are finite. This lack of definition has caused uncertainty to all parties involved—from regulatory bodies to project developers and investors.

There is a need, therefore, to agree on what time frame should be used as the basis for quantification of GHG benefits of a project. Different time frames or approaches have been proposed to define the duration of projects:

- *Perpetuity*; Under this approach the environmental benefits of projects must be maintained forever. This argument is based on the assumption that the reversal of GHG benefits from a project at any point in time would totally invalidate the project (Carbon Storage Trust, 1998; Maclaren, 1999) and that only maintenance of carbon stocks in perpetuity could counter the environmental effects of GHG emissions from fossil fuel sources. It is also argued that this approach is the only one that is compatible with the stock change method currently used by the IPCC for national GHG inventories (Houghton *et al.*, 1997). Criticisms of this approach argue that it is impossible to guarantee that a project will be run in perpetuity; that maintenance of projects in perpetuity may create conflicts with other land uses in the long term; and that because of the decay pattern of GHGs in the atmosphere, there is no need for mitigation effects to be perpetual.
- *100 years*: Under this approach, the GHG benefits of a project must be maintained for a period of 100 years to be consistent with the Kyoto Protocol's adoption of the IPCC's GWPs (Article 5.3) and the Protocol's 100-year reference time frame (Addendum to the Protocol, Decision 2/CP.3, para. 3) for calculation of the AGWP for CO_2. Although this concept has limitations (IPCC, 1996), it has been adopted for use in the Kyoto Protocol to account for total emissions of GHGs on a CO_2-equivalent basis.
- *Equivalence based*: Under this approach, the GHG benefits of LULUCF mitigation projects must be maintained until they counteract the effect of an equivalent amount of GHGs emitted to the atmosphere,

estimated on the basis of the cumulative radiative forcing effect of a pulse emission of CO_2 during its residence in the atmosphere (i.e., its AGWP) (IPCC, 1992). Variations of this concept have been developed that proposed minimum time frames of 55 years (Moura-Costa and Wilson, 2000) or 100 years (Fearnside *et al.*, 2000) (see Chapter 2).
- *Variable:* This approach acknowledges that different projects may have different operational time frames. Given the wide range of time frames of projects carried out to date, it can be inferred that this approach has been adopted during the AIJ Pilot Phase.

Adoption of a standard definition of the minimum required time frame for project duration would greatly facilitate consistency in accounting for the GHG benefits of different projects. It would also reduce the uncertainty of all parties involved in project development (project developers, investors, certifiers, regulatory bodies, and the general public).

5.3.4.2. *How Should Projects with Shorter Time Frames Be Treated?*

Once the minimum project duration has been defined, it is also important to decide how to treat projects that have a shorter duration than the minimum required time frame. The options can be divided into two main approaches:

- *Full liability:* In the event of reversal of GHG benefits, projects should return an amount of credits equal to the total amount of GHGs released. This approach is consistent with the stock change method, which consists of giving credits to projects as carbon is fixed and removing credits if stocks of carbon diminish. In essence, this approach does not recognize the temporal value of carbon storage. This is the only method possible if it is decided that projects have to be run in perpetuity.
- *Proportional liability:* Projects should be debited an amount of credits proportional to the difference between the minimum required time frame and the actual project duration (the "period of noncompliance'). This method is applicable only if a finite minimum project duration is adopted. If a minimum time frame of 100 years is adopted, for instance, a plantation project that is harvested at 60 years (assuming that all carbon is released to the atmosphere) would be liable for not maintaining carbon stocks for the last 40 years of the required time frame. Different methods have been proposed for calculating this proportional liability:
  - *Linearly*—Dividing the period of noncompliance by the required time frame. In the foregoing example, the project would have to return 40 percent of the credits it earned/claimed.
  - *Ton-year based*—Calculating the liability based on the ton-year approach (Fearnside *et al.*, 2000; Moura-Costa and Wilson, 2000).
  - *Adjusted for time preference*—Using any of the methods described above but applying discount rates to reflect time preference (see Chapter 2).

The choice of method for dealing with liability is linked with methods chosen for accounting for GHG benefits and when credits are given to projects (see Section 5.4).

5.3.5. *Risks*

Quantification of GHG emissions or removals in LULUCF projects is subject to a variety of risks and uncertainties. Some of these risks and uncertainties are inherent to certain land-use activities (particularly forestry); others may be generic and applicable to any GHG mitigation project in the energy and LULUCF sectors.

Risks refer to events that negatively affect the expected GHG benefits of the project. Land-use projects are exposed to a series of risks, such as natural risks (e.g., rainfall, sunlight, pests and diseases, reductions in growth rates, fire, climate change); anthropogenic factors (e.g., encroachment, fires, theft); political risks (e.g., non-enforcement of legally binding contracts between project partners, noncompliance with guarantees, expropriation, uncertain property rights, policy changes); economic risks [e.g., exchange rate and interest rate fluctuations (Shapiro, 1996), changes in the prices of the relevant factor and product markets (Janssen, 1997), changes in the opportunity costs of land]; financial risks; institutional risks (e.g., land tenure); and market risks. Not all of these risks are exclusive to land-use activities. Because land-use activities have strong social implications; rely on a land base; and depend on natural factors such as rainfall, sunlight, pollinators, and exposure to natural and anthropogenic factors, however, such activities are particularly exposed to these risks.

Risks of project failure because of fire, climatic variations (e.g., drought or storms), and pests also entail potential negative environmental and social impacts associated with failed projects. Implementation of large-scale teak plantation projects in India, for instance, may have led to cultivation of monocultures, that are susceptible to pest infestation, loss of timber affecting local timber markets, and associated release of sequestered carbon (Ravindranath *et al.*, 1998). Because carbon mitigation projects also have to address issues of sustainable forest management, the risks associated with these new endeavors—where there is less experience and infrastructure to draw on—may not realize the full potential of co-benefits. In the Salicornia project (Box 5-1), for instance, a new concept is being tested to evaluate the cultivation possibilities and commercial uses of a previously uncultivated crop. The entry of Salicornia straw to wood markets could lower the price of wood, reducing the incentive for forest plantations locally (Imaz *et al.*, 1998). Project developers will have to establish procedures to deal with extra costs in the event of such impacts. For example, the Costa Rican government has committed to find replacement farmers if targets are not met in the PFP project. Another example is the contractual obligations required by the FACE Foundation, which require project implementers to replant any forests that are lost during the project's time frame (Verweij and Emmer, 1998). Alternatively, in the context of a growing trend in trading in carbon credits, management can be expected to seek to lay off these risks in conventional insurance and reinsurance markets.

Risk mitigation can be accomplished through a variety of internal and external mechanisms to the project. *Internal* methods include the following:

- Introducing good practice management systems to control the occurrence of damaging events.
- Project design, aiming at diversification of activities within a project and spreading of projects in different areas, reducing the risks that damage (e.g., fire, pests and diseases, flood) will spread.
- Maintaining self-insurance reserves or keeping a portion of the project's benefits as a reserve to insure against any shortfalls. This reserve could be financial or in-kind (GHG benefits). This approach was used by the national program of the Costa Rican Office for Joint Implementation, which placed about 40 percent of credits derived from the PFP project in a self-insurance buffer reserve (SGS, 1998). If damage does not occur, this reserve can be used at the end of the project lifetime.
- Diversification of sources of funding, reducing financial dependency on a single source.
- Involvement of a wide range of stakeholders, through consultation and participatory management.
- Creation of positive local side effects from hosting the project, such as transferring needed technologies, fostering local social developments (e.g., job creation), or creating positive side effects on other local or regional environmental goals in the host country (Janssen, 1997).
- Project auditing and external verification, which may serve as a way to highlight project risks early on.
- Timed allocation of GHG benefits. If GHG benefits are credited to project partners only after they are fully realized, there will be less need for long-term guarantees and a lower perception of risk. This allocation could be accomplished by staggering sequestration and crediting or by only allowing crediting according to a ton-year factor calculated according to an equivalence factor between CO_2 sequestration and emissions (Moura-Costa and Wilson, 2000).

External methods include the following:

- Cross-project insurance through direct arrangements in which projects would guarantee each other.
- Regional carbon pools—a similar approach in which "carbon banks" are established with contributions from a diversified pool of projects to insure contributing projects.
- Financial insurance. Some insurance companies are already offering services related to risk mitigation for carbon offset projects. It is important to note that a series of project risks are common to non-GHG specific activities and traditionally have been covered by standard insurance schemes (e.g., crop or timber insurance).
- Portfolio diversification in terms of placing different projects in different locations [e.g., the FACE Foundation's portfolio (Verweij and Emmer, 1998)].

There are still issues related to liability, such as allocation of responsibilities for ensuring compliance and deliverables. The UN Conference on Trade and Development's Emissions Trading Forum has raised issues of responsibility such as "buyer beware"—in which buyers are responsible to ensure that offsets are valid—or "seller beware," in which an exporting country would have the entire transaction invalidated if projects do not deliver (Tietenberg *et al.*, 1998). This approach has different implications for countries with and without emissions limitation caps. Additional issues raised during the meetings of the Ad Hoc Working Group on CDM included allocation of liabilities between nations, individuals, and certifiers (Stuart, 1998; Stewart *et al.*, 1999).

5.4. Measuring, Accounting, Monitoring, and Verifying GHG Benefits

Many pilot projects have been developed (see Tables 5-1 and 5-2), and much experience has been gained—particularly at the early stages of project implementation. On the basis of this experience, this section presents an assessment of the nature of measuring, monitoring, accounting, verifying, and reporting GHG benefits. Some key questions that guide this section are the following: With what accuracy and precision can GHG benefits be measured and monitored in LULUCF projects? Does the accuracy and precision of the measuring and monitoring of GHG benefits vary across project types? What are the tradeoffs between cost and precision in measuring and monitoring GHG benefits? What effects do different accounting methods have on the GHG benefits accruing to a project? How long should monitoring, verification, and reporting be pursued? How can verification costs be managed? What alternative formats are available for reporting project-level GHG benefits?

5.4.1. Methods for Quantification of Project GHG Benefits

A key aspect of implementing LULUCF projects to mitigate GHG emissions and trading is accurate and precise quantification of project-level GHG benefits. In LULUCF projects, the main focus is on carbon (as CO_2) benefits, but other gases should be included as appropriate. Table 5-6 presents typical examples of generic projects, some of which could include carbon only or both carbon and non-CO_2 GHG benefits (see also Chapter 4). For instance, a project designed to stop deforestation typically would include carbon benefits, but it could also include nitrous oxide and carbon monoxide benefits that would result from stopping the burning of biomass during forest clearing. Soil

Table 5-6: *Examples of LUCF projects and their corresponding without-project case. Some of these projects could generate carbon benefits only, whereas others could generate both carbon and non-CO_2 GHG benefits.*

With-Project Case	Without-Project Case
Stop deforestation	Forest clearing for agriculture or pasture
Stop forest logging	Forest with continued logging
Reduced impact logging	Traditional logging with high damage
Improved forest management (e.g., longer rotations, wide buffers, shelterwood, less-intensive site preparation)	Traditional management
Plantation establishment	Degraded or marginal non-forested lands
Agroforesty:	
– Trees and crops	Annual crops
– Shade coffee	Sun coffee
Soil management:	
– No to low till	Traditional till
– Manure, fertilize, crop rotation	Traditional practice
– Restore degraded soils	Continued soil degradation
Biofuel tree plantations	Marginal lands and fossil fuel use

and agricultural projects could include non-CO_2 GHGs such as nitrous oxide and methane as well. Whereas carbon benefits are generally measured as changes in carbon stocks, however, non-CO_2 GHGs are measured as fluxes, and the methods are less well developed (See Chapters 1 and 2; Houghton *et al.*, 1997); thus, the following discussion focuses on carbon (as CO_2). Moreover, the example projects in Table 5-6 could vary in size and distribution; they could be contiguous, extending over hundreds to thousands of hectares, or a "bundle" of small scattered landowners each of whose total area could be hundreds of hectares.

This section discusses which pools need to be quantified, how they can be accurately measured to a known level of precision, and techniques to monitor the carbon benefits over the length of the project. The initial carbon inventory is distinguished from subsequent monitoring: In the initial inventory, the relevant major pools or fluxes are quantified; in subsequent monitoring, only selected pools or fluxes may be measured, and even indicators could be used, depending upon the type of LULUCF project.

5.4.1.1. Identification of Carbon Pools

Possible criteria affecting the selection of carbon pools to inventory and monitor are the type of project; the size of the pool, its rate of change, and its direction of change; the availability of appropriate methods; the cost to measure; and attainable accuracy and precision (MacDicken, 1997a,b). A selective or partial accounting system can be used; such a system must include all pools expected to decrease and a choice of pools expected to increase as a result of the project (Hamburg, 2000). Only measured (or estimated from a measured parameter) and monitored pools are incorporated into the calculation of GHG benefits. Carbon benefits are calculated as the net differences between selected pools for the with- and without-project baseline conditions on the same piece of land over a specified time period.

The major carbon pools in LULUCF projects are live biomass, dead biomass, soil, and wood products; each of these pools can be subdivided further (e.g., live biomass may include leaves, twigs, branches, stems, coarse and fine roots of trees, herbaceous plants, shrubs, and vines—see Chapter 2 for further details). Table 5-7 illustrates how decisions about which pools to choose for quantification and monitoring may be made for different types of LULUCF projects. Accurately and precisely measuring soil carbon pools presents several challenges; of the projects listed in Table 5-7, however, the soil carbon pool need be measured in only two cases (Y).

Table 5-8 includes a selection of projects and their measured or estimated carbon pools (for other project details, see Table 5-1 and Box 5-1). Although soil carbon is measured in two of the emission avoidance projects, using these data for calculating the carbon benefits could be problematic. In the NKCAP, for example, the soil carbon benefits from averted deforestation could be calculated as the difference between the soil carbon in the project area and soil carbon in a nearby reference area.

***Table 5-7:** Decision matrix of main carbon pools for examples of land-use and forestry projects to illustrate selection of pools to quantify and monitor. Y = yes, indicating that the change in this pool is likely to be large and should be measured. R = recommended, indicating that the change in the pool could be significant but measuring costs to achieve desired levels of precision could be high. N = no, indicating that the change is likely small to none thus not necessary to measure this pool. M indicates that the change in this pool may need to be measured depending on forest type and/or management intensity of the project.*

	Carbon Pools						
	Live Biomass			**Dead Biomass**			
Project Type	Trees	Herbaceous	Roots	Fine	Coarse	**Soil**	**Wood Products**
Avoid Emissions							
– Stop deforestation	Y	M	R	M	Y	R	M
– Reduced-impact logging	Y	M	R	M	Y	M	M
– Improved forest management	Y	M	R	M	Y	M	Y
Sequester Carbon							
– Plantations	Y	N	R	M	M	R	Y
– Agroforestry	Y	Y	M	N	N	R	M
– Soil carbon management	N	N	M	M	N	Y	N
Carbon Substitution							
– Short-rotation energy plantations	Y	N	M	N	N	Y	*

*Stores carbon in unburned fossil fuels.

Without careful selection of the reference site, its average soil carbon could be higher or lower than the average of the project area solely because of variability in soil characteristics rather than human management. Thus, simply subtracting the forest soil carbon from the agriculture soil carbon would give erroneous carbon offsets.

5.4.1.2. Measurement of Carbon Benefits

Land-use and forestry projects generally are easier to quantify and monitor than national inventories because of the clearly defined boundaries for project activities, the relative ease of stratification of the project area, and the choice of carbon pools to measure (Section 5.4.1.1). Techniques and methods for sampling design and for accurately and precisely measuring individual carbon pools in LULUCF projects are based on commonly accepted principles of forest inventory, soil sampling, and ecological surveys (Pinard and Putz, 1996, 1997; MacDicken, 1997a,b; Post *et al.*, 1999; Winrock International, 1999; Hamburg, 2000). For example, there is a wealth of experience with well-developed and accepted methods to inventory forests for merchantable volume and growth; these methods can be and are being readily adopted to inventory forest biomass carbon. Likewise with regard to measuring soil carbon, standardized techniques are well established. Further descriptions of methods for estimating the carbon pool in live tree biomass, understory and herbaceous plants, roots, fine and coarse litter, and soil are described in Chapter 2. Standard methods have not been universally applied to all projects, however, and methods of accounting for carbon benefits have not been standardized—resulting in some difficulties in comparing results across different LULUCF projects.

For most LULUCF projects, it would be necessary to measure non-project reference or control sites as well. These sites must be sufficiently similar to the project area to serve as valid proxies under the assumption that the project was not implemented (Vine *et al.*, 1999). To help overcome the difficulty of establishing proxy areas, non-project reference sites could be identified during the project design phase. The location of proxy sites as close as possible to the project would be the most desirable situation. For example, in projects in which many small landowners convert to no-till agriculture, proxy areas would be farmers in the area who do not practice no-till agriculture. Box 5-3 illustrates the types of measurements being taken to estimate with- and without-project cases and the resulting carbon benefits.

The total carbon stock has been measured to <10 percent of the mean with 95-percent confidence in several pilot LULUCF projects (e.g., Programme for Belize, 1997a; Hamburg, 2000; NKCAP—see Box 5-3). Although techniques and tools exist to measure carbon stocks in project areas to a high degree of precision, the same level of precision for carbon benefits may not be achieved. The carbon benefit per unit area of land is the difference between the carbon stocks in the with-project case—which is high if, for example, the project is conserving carbon in existing forests through an avoided deforestation project—and the carbon pools in the without-project case, which is low if the baseline is agricultural or degraded lands. In this case, the estimated carbon benefit is likely to be high (a small carbon stock subtracted from a large carbon stock), and the error estimate, expressed as a percentage of the mean difference, likely to be small and similar to that obtained for carbon stocks in the forests. As the difference between the with- and without-project cases decreases as, for example, in reduced impact logging projects, however, the percentage error of the carbon benefit increases. To reduce this error, monitoring can be designed to measure the change in carbon stocks directly, as in the NKCAP (see Box 5-3). Difficulties in establishing baselines and leakage effects also affect the precision of the carbon benefits.

5.4.2. Accounting

5.4.2.1. Carbon Accounting Methods

Various methods have been used to account for GHG mitigation effectiveness of LULUCF projects. Some of these methods are based on absolute measurements at a point in time; others take into account the time dimension of carbon sequestration and storage. These methods are discussed below.

5.4.2.1.1. Stock change method

The method most commonly used to express carbon storage is based on calculating the difference in carbon stocks between a project and its baseline at a given point in time. This method is referred to as the *stock change method* [previously the *flow summation method* (Richards and Stokes, 1994)]; measurements are usually expressed in tons of carbon per hectare. This method is limited, however, insofar as it provides only a "snapshot" of the carbon fixed: The values will vary depending on the often arbitrary decision of when to account for the project's benefits. Furthermore, this method does not differentiate between projects that earn credits earlier rather than later. For these reasons, this method does not provide a useful tool for comparison between projects.

For example, Figure 5-2 illustrates a projection of carbon stored in two hypothetical tree plantation projects with different growth rates. The arrows illustrate that stock change measurements carried out at time *t1* would provide different results between the two projects, whereas the same result would be reached if measurements were carried out at time *t2*. If measurements were carried out at time *t3*—after harvesting—an entirely different result would be reached for both projects in relation to measurements at *t2*.

5.4.2.1.2. Average storage method

To account for dynamic systems—such as afforestation projects, in which planting, harvesting, and replanting operations take

Table 5-8: *Main carbon pools measured in a selection of forest-based pilot carbon-offset projects. Y means the pool was measured, N means it was not measured, and E means it was estimated from field measurements and literature data.*

	Carbon Pools						
Project Type	Trees	Herbaceous/ Understory	Roots	Dead Fine	Coarse	Soil	Wood Products
Avoid Emissions							
– Noel Kempff Climate Action Project, Bolivia[a]	Y	Y	E	Y	Y	Y	E
– Reduced-Impact Logging, Sabah, Malaysia[b]	Y	Y	Y	Y	Y	N	N
– Rio Bravo Carbon Sequestration Project[c]	Y	Y	E	Y	N	Y	N
Sequester Carbon							
– Farm Forestry Scolel Te Pilot Project, Chiapas, Mexico[d]	Y	N	N	N	N	Y	N
– FACE, Malaysia[e]	Y	Y	E	Y	Y	N	N
Multi-Component							
– PAP/PFP, Costa Rica[f]	E	E	E	N	N	N	N
– Guaraqueçaba Climate Action Project, Brazil[g]	Y	Y	E	Y	Y	N	N

[a]Brown *et al*. (2000).
[b]Pinard and Putz (1997).
[c]Programme for Belize (1997).
[d]de Jong *et al*. (1997).
[e]Moura-Costa (1993 1996a).
[f]SGS (1998).
[g]Brown *et al*. (1999b).

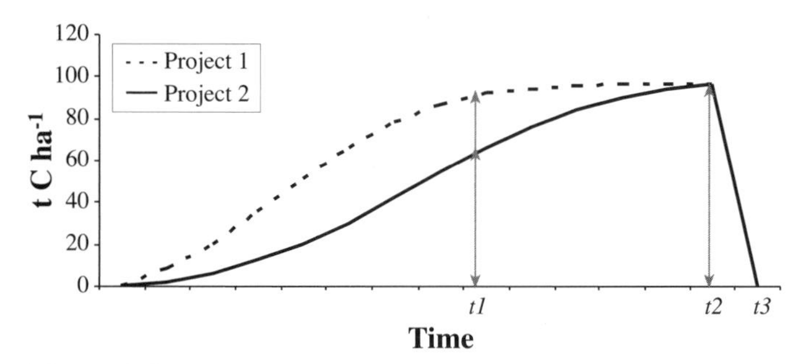

Figure 5-2: Projection of carbon stored in two tree plantation projects with different growth rates. For simplicity, it is assumed that the baseline is zero and that harvesting leads to an immediate release of all carbon stored. Arrows illustrate the net carbon storage of the projects at different points in time, calculated by the stock change method.

place—an alternative approach has been used (e.g., Dixon *et al.*, 1991, 1994; Masera, 1995): the *average storage method* (Schroeder, 1992). This method entails averaging the amount of carbon stored in a site over the long term according to the following equation:

$$\textit{Average net carbon storage (t C)} = \frac{\sum_{t=0}^{t=n} (\textit{carbon stored in project} - \textit{carbon stored in baseline}), \textit{in t C}}{n \textit{ (years)}}$$

where t is time, n is the project time frame (years), and measurements are expressed in tons of carbon per hectare. The advantage of this method is that it accounts for the dynamics of carbon storage over the whole project duration, not only at the times chosen for accounting. This method is also useful for comparing different projects with different growth patterns. As Figure 5-3 shows, the average storage over three rotations of Project 1 is higher than that of Project 2. A weakness of this method, however, relates to the still subjective time frame, n, chosen for running the analysis. In the case of Figure 5-3, for example, the average net carbon storage in either project would

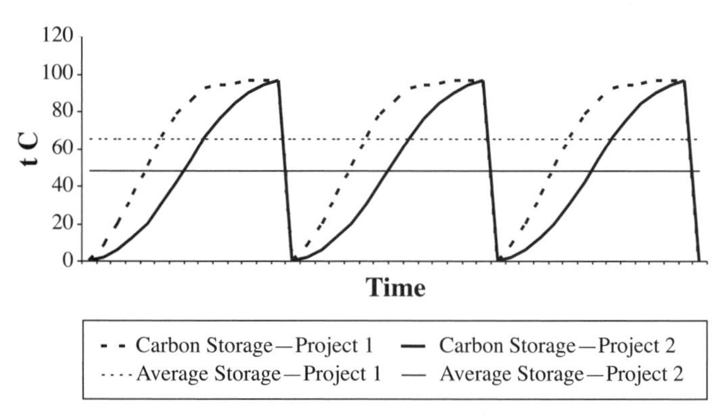

Figure 5-3: Projection of carbon stored in two tree plantation projects over three rotations. For simplicity, it is assumed that the baseline is zero, that harvesting leads to an immediate release of all carbon stored, and that equilibrium of carbon pools is reached in the first rotation cycle. The curves illustrate carbon storage over time; the straight horizontal lines show the average storage calculated for the two projects.

be equal whether the calculation was performed for one, two, or an infinite number of rotations, as long as the denominator chosen for the foregoing equation coincided with the last year of a rotation.

5.4.2.1.3. Alternative approaches

Alternative approaches have been proposed to better address the temporal dimension of carbon storage. Most of these approaches are based on adopting a two-dimensional measurement unit that reflects storage and time—namely, the ton-year. Many authors have proposed the concept of a ton-year unit (Moura-Costa, 1996a,b; Fearnside, 1997; Greenhouse Challenge Office, 1997; Chomitz, 1998; Dobes *et al.*, 1998; Tipper and de Jong, 1998; Fearnside *et al.*, 2000; Moura-Costa and Wilson, 2000). The general concept of the ton-year approach is in the application of a factor to convert the climatic effect of temporal carbon storage to an equivalent amount of avoided emissions (the remainder of this section refers to this factor as the *equivalence factor*, or E_f); this factor varies from 0.007 to 0.02 (Dobes *et al.*, 1998; Tipper and de Jong, 1998; Moura-Costa and Wilson, 2000). This factor is derived from the "equivalence time" concept (referred to as *Te* in the remainder of this section): the length of time that CO_2 must be stored as carbon in biomass or soil to prevent the cumulative radiative forcing effect exerted by a similar amount of CO_2 during its residence in the atmosphere (Moura-Costa and Wilson, 2000). Chapter 2 describes the theory and methods used for determining E_f.

Irrespective of the method used to calculate the equivalence factors, they could be useful for the accounting of GHG benefits of LULUCF projects. Different applications have been proposed (Moura-Costa and Wilson, 2000); in practice, a combination of approaches can be used, as follows:

- *Equivalence-adjusted average storage,* using *Te* as the denominator of the *average storage* equation. This method could be used to standardize the way in which the average storage method is currently used.
- *Stock change crediting with ton-year liability adjustment,* giving projects credits according to the stock change method but using ton-years to calculate the amount of credits to be removed in the case of any noncompliance (in the case of occurrence of risk-related events).
- *Equivalence-factor yearly crediting (ton-years),* by which a project is credited yearly with a fraction of its total GHG benefit; this fraction is determined by the amount of carbon stored each year, converted using E_f (Figure 5-4). This approach would greatly discourage the implementation of LULUCF projects.
- *Equivalence-delayed full crediting,* recognizing the full benefits of carbon sequestration only after storage for a time period *Te* (Figure 5-5). This delayed crediting likely would discourage the implementation of LULUCF projects.
- *Ex-ante ton-year crediting,* giving projects an amount of credits at the beginning of the project, according to

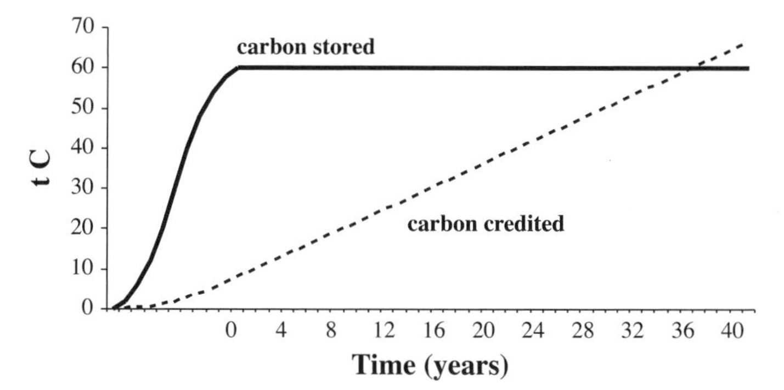

Figure 5-4: Projection of carbon stored in an afforestation project (with baseline assumed to be zero), illustrating the concept of *equivalence-factor yearly crediting (ton-years)*. The project receives yearly credits calculated as the total amount of carbon stored in any given year, multiplied by an equivalence factor, E_f. Alternatively (in the case of *stock change crediting with ton-year liability adjustment*), credits could be given as carbon is stored (solid line); in case of any event leading to the release of carbon stored, the amount of credits to be returned would be calculated as the difference between the solid line and the dotted line at that point in time.

the planned project duration, using the ton-year approach. This approach would reduce the disadvantages that delayed crediting would create for project developers.

If an equivalence factor ton-year approach is used, carbon storage could be credited according to the time frame over which storage takes place. Such a crediting system would reduce the need for long-term guarantees, hence the risks associated with long time frames. If forests storing this carbon pool suffer any damage, the proportion of carbon credits lost could be easily calculated. This method also allows for comparisons between projects. The main disadvantage of this method is that there is still a great deal of uncertainty in relation to the permanence of CO_2 in the atmosphere—consequently the values of the equivalence parameters *Te* and E_f. Depending on the manner in which ton-year accounting is used, there may also be disadvantages in relation to when crediting occurs,

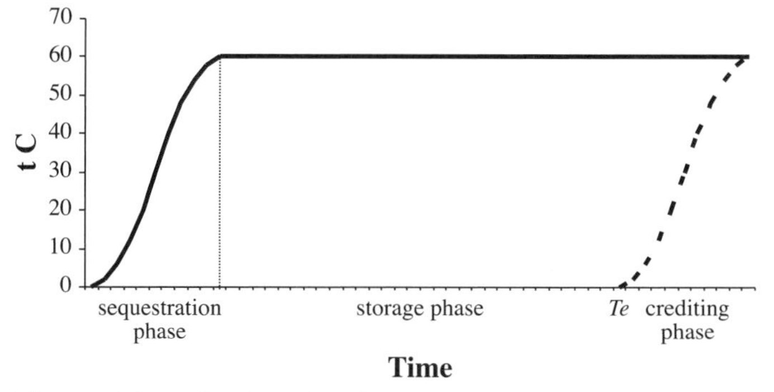

Figure 5-5: Projection of carbon stored in an afforestation project (with baseline assumed to be zero), illustrating the concept of *equivalence-delayed full crediting*. In this example, the project receives credits only after planted trees have grown and been kept for a period of time, *Te*.

discouraging the implementation of LULUCF GHG mitigation projects (particularly in the case of the equivalence factor yearly crediting and equivalence-delayed crediting approaches). Table 5-9 provides a comparison of the GHG benefits of each method. Whichever method is chosen, it would need to be made compatible with UNFCCC reporting requirements (Chapter 6).

5.4.2.1.4. Comparison of methods

Table 5-9 shows a comparison of the GHG benefits attributed to the sequestration project illustrated in Figure 5-3. The example assumes the following:

- The project is run for three rotations of 18 years each.
- At the end of each rotation, the carbon stock in the forest reaches 140 t C ha^{-1}.
- Harvesting reduces carbon stock to zero, and the baseline is zero.

Calculations were conducted assuming a minimum required project duration of 55 years [based on the *Te* of 55 years (Moura-Costa and Wilson, 2000)] and 100 years [based on the equivalence time of 100 years; see Chapter 2 (Fearnside *et al.*, 2000)]. It is clear from this example that, depending on the accounting method used, different amounts of carbon benefits accrue to the project, as is shown by the following results:

- According to the stock change method, this project would receive 140 t C ha^{-1} during the sequestration phase of each rotation and would need to return an equivalent amount after each harvest.
- The average storage calculated for the duration of this project is 84 t C ha^{-1} (using the traditional average storage method, without a fixed minimum project duration), which is reached before the end of the first rotation and remains the same irrespective of the duration of the project. If a specified time frame is adopted for the calculation of the average storage (i.e., with a predetermined denominator in the average storage equation), the GHG benefits of a project would increase proportionally to the time frame under which the project is conducted.
- If a minimum project duration of 55 years were required, the equivalence-adjusted average storage of this project (which is conducted for 54 years) would be 83 t C ha^{-1}, whereas if the minimum time frame required were 100 years, the equivalence-adjusted average storage would be 45 t C ha^{-1}. Furthermore, if this project were conducted for only one rotation, the project's benefits would be lower (see values in parentheses in Table 5-9).
- Another accounting option (the stock change crediting with ton-year liability adjustment method) is to use the stock change method to calculate the benefits of the projects during the sequestration phase and to use ton-years to calculate the "loss" of benefits when

Table 5-9: *Comparison of GHG benefits attributed to a sequestration project at different points in time, according to different carbon accounting methodologies (t C ha^{-1}).*

Method	Year 20	Year 20 after Harvest	Year 60	Year 60 after Harvest	Balance
Stock change	140	-140	140	-140	0
Average storage, with the end of each rotation as denominator	84 (84)	0 (84)	0 (84)	0 (84)	84
Equivalence-adjusted average storage, with minimum required project duration of 55 years (*Te* = 55)[a]	83 (28)	0 (28)	0 (83)	0 (83)	83
Equivalence-adjusted average storage, with minimum required project duration of 100 years (*Te* = 100)	45 (15)	0 (15)	0 (45)	0 (45)	45
Stock change crediting with tonne-year liability (*Te* = 55)	140	-112	140	-57	110
Stock change crediting with tonne-year liability (*Te* = 100)	140	-136	140	-100	44
Ton-year yearly crediting (*Te* = 55; E_f = 0.0182)[b]	28	28	83	83	83
Ton-year yearly crediting (*Te* = 100; E_f = 0.010)[b]	3	4	38	40	40

Notes: Positive values denote GHG benefits (crediting between parentheses), and negative values mean "reversal" of benefits (removal of credits). Calculations are based on an example of an afforestation project conducted for three rotations of 18 years each. It is assumed that at the end of each rotation, the carbon stock in the forest reaches 140 t C ha^{-1}, and that harvesting reduces carbon stocks to zero. Forests are not replanted after the third rotation. For simplicity, it also is assumed that the baseline is zero. Figures in parentheses refer to GHG benefits accumulated until that point in time, in case the project was terminated at that time.

[a]Minimum project duration values were chosen based on different proposed *equivalence time* factors (*Te*, the length of time CO_2 must be stored as carbon in biomass or soil for it to prevent the cumulative radiative forcing effect exerted by a similar amount of CO_2 during its residence in the atmosphere). Moura-Costa and Wilson (2000) propose a *Te* = 55 years, and Fearnside *et al.* (2000) propose a *Te* = 100 years (see Chapter 2).

[b]In both cases, E_f (the *equivalence factor* used to determine the GHG mitigation benefit of a tonne-year of storage) is calculated linearly by 1/*Te*.

emission takes place. Using this approach, the calculated GHG benefits of the project at the end of the first rotation would be 140 t C ha^{-1} (the same as in the stock change method); when emissions take place after harvesting, however, the calculated GHG benefits "lost" are either 112 t C ha^{-1} (if a ton-year equivalence factor E_f = 0.0182 is chosen, based on *Te* = 55) or 136 t C ha^{-1} (if a ton-year equivalence factor E_f= 0.010 is chosen, based on *Te* = 100). The longer the project duration, the smaller the amount of GHG benefits "lost" after harvesting.

- If the GHG benefits of the project are calculated using the equivalence-factor yearly crediting method (ton-year accounting), the GHG benefit attributed to the project would increase gradually as the project is conducted for a longer time frame. Because this method assumes that the ton-year equivalence factor reflects the GHG benefit to the atmosphere derived from temporary storage, no loss of benefits is assumed when emissions take place.

5.4.2.2. *Accounting for Risks and Uncertainty*

Projects have dealt with risks and uncertainty in different ways, depending on the type of uncertainty (see also Section 5.3.4). Mensuration error can be dealt with by the following methods:

- *Error acceptance:* Acknowledging that measurement error is inevitable and listing a range of acceptable errors for different pools.
- *Error minimization:* By setting acceptable errors at a low level, this method forces projects to engage in more effective inventorying and monitoring exercises by increasing the number of samples, the sample size, and the frequency of sampling (see Section 5.4.3). This approach may affect the eligibility of certain types of projects that present mensuration difficulties.
- *Error deduction:* This method consists of deducting the error from a carbon estimate. This approach has the advantage that it allows the project to decide what

is more cost-effective: data gathering or carbon claims (see Section 5.4.3). This approach was used by the international certification company SGS in the certification of the Costa Rican national carbon offset program (SGS, 1998; Moura-Costa *et al.*, 2000).

Methods to account for baseline uncertainty include estimation of the effect of different uncertainty assumptions on the baseline adopted and deduction of the claims. In the case of quantifiable risks, these uncertainties can be accounted for by keeping a portion of the project's GHG benefits as a reserve to insure against any shortfalls. This reserve could be financial or in-kind (GHG benefits), as in the Costa Rican PAP (SGS, 1998). If damage does not occur, this reserve may be used at the end of the project lifetime.

5.4.2.3. *Accounting for Time (Discounting)*

The time frame of project benefits can affect their attractiveness. Projects that bring benefits at an earlier stage may be favored by some planners, which raises the issue of *time preference*. Time preference relates to society's preference for benefits that accrue at an earlier rather than a later stage. In the context of climate change, time preference can be used to introduce a sense of urgency in relation to GHG emission mitigation measures. Not using it implies an endorsement of the assumption that a GHG mitigation activity can be postponed indefinitely without any effect on the overall objective of reducing the impacts of GHG concentrations in the atmosphere.

To account for the value of time and include the concept of time preference, the *discounting method* has been proposed (Richards and Stokes, 1994; Fearnside, 1995). It consists of using a discount rate to calculate the present value of the total amount of carbon stored over the lifetime of a project, according to the following equation:

$$\text{Present value of carbon storage (t C)} = \frac{\sum_{t=0}^{t=n} \text{carbon stored by a project (t C)}}{(1+i)^t}$$

where i is the discount rate and n is the project's time frame (usually in years).

One problem in using discounting, however, relates to the selection of an appropriate discount rate to reflect financial (interest rates), economic, or social degrees of time preference attached to the carbon mitigation benefits of a project. High rates favor short-term projects, discouraging long-term sustainability and forest maintenance. Rates that are two low discourage efficiency and approaches that promote more rapid results. Discounting, however, favors activities that prevent the release of carbon, such as conservation or reduced-impact logging, instead of activities that actively remove carbon from the atmosphere over a longer period (e.g., forest establishment). This dynamic is obtained because conservation activities internalize large amounts of carbon at the beginning of the project cycle, so they suffer less from the effects of discounting.

5.4.3. *Monitoring*

Monitoring relates to the periodic measurement of carbon pools in the project area and in proxy or reference non-project areas. Permanent sample plots, which often are used in the initial carbon inventory (e.g., Box 5-3), are generally considered to be statistically superior means for evaluating changes in forest carbon pools. Methods are well established and tested for determining the number, size, and distribution of permanent plots (i.e., sampling design) in several LULUCF projects to maximize the precision for a given fixed monitoring cost (MacDicken, 1997a; Winrock International, 1999). The use of permanent plots allows for efficient assessments of changes in carbon stocks over time and for cost- and time-efficient verification of the project's reported carbon benefits (MacDicken, 1997a). Moreover, a random selection of permanent plots may be measured only as part of the ongoing monitoring program. In addition, not all of the initial carbon pools need be measured at every interval in some projects; judicious selection of some pools could enable them to serve as indicators that the project is following the expected trajectory. For example, projects that are designed to avoid emissions by arresting deforestation or logging need only establish that no trees are removed or clearings made over the course of the project. In projects that are designed to sequester carbon, changes in vegetation carbon or soil carbon pools need to be re-measured periodically.

Remote sensing can provide a useful means to monitor LULUCF projects (see Chapter 2). A variety of remote data collection technologies are now widely available, ranging from satellite imagery to aerial photographs from low-flying airplanes. A new advance in this area couples dual-camera videography with a pulse laser profiler, data recorders, and differential global positioning system (GPS) mounted on a single-engine airplane (Department of Forestry and Conservation Management, University of Massachusetts, 1999). This system can produce indices of crown density, number of trees per unit area, and tree height; it also can identify the extent of gaps, which will be especially useful for projects that are related to arresting or modifying logging, as well as monitoring for small-scale human disturbance in protected forests.

In some circumstances, models (parameterized for project conditions) can be used to project changes in carbon pools over short time periods for which direct measurements fall below easily detectable levels, followed by direct measurements over longer time intervals to verify model projections (Post *et al.*, 1999; Vine *et al.*, 1999). Process-based models are particularly useful in projecting slowly occurring changes in soil carbon pools (Paustian *et al.*, 1997; Post *et al.*, 1999). Likewise, models exist for plantations and agroforestry systems (e.g., Maclaren, 1996; Schlamadinger and Marland, 1996; Mohren *et al.*, 1999; ICRAF, n.d.) that could be used in conjunction with direct field measurements to estimate changes in carbon pools over shorter time frames.

5.4.4. *Precision and Costs*

Field methods to accurately quantify carbon pools exist, but the level of precision can vary by pool. The total error in measuring a given carbon pool is based on sampling error (variation among sampling units—e.g., the number of plots, within the population of interest), measurement error (error in measuring the parameter of interest—e.g., stem diameter and soil carbon) and regression error, when appropriate (e.g., error resulting from conversion of tree diameter to biomass based on a regression equation). Sampling error is usually the largest source of error (Phillips *et al.*, 2000), and increased precision generally comes at increased costs of inventorying because of the time and cost involved in establishing the appropriate number and distribution of permanent plots. Carbon inventory in forests can be more complicated than traditional forest inventories because each carbon pool will have a different variance. The sample size for each pool can be calculated individually; based on resources available for monitoring the project and the information in Table 5-7, informed decisions can be made about which pools to measure and count. Such information can be used at the design stage to select pools to be included in the project, with significant implications for the total cost of the project and the measurement and monitoring costs per ton of carbon.

The costs of measuring and monitoring carbon offsets are a function mainly of the desired level of precision—which may vary by the type of project activities, the size of the project (areal extent, as well as whether the project area is contiguous or a dispersed bundle of small landowners), and the natural variation within the various carbon pools. For example, an increase in the coefficient of variation (a measure of the variation around the mean) within a forest stratum of about 160 percent would increase the cost of measurement by about 280 percent to maintain the same level of precision (Figure 5-6a). Stratification of the project area into more or less homogeneous units—based on vegetation type, soil type, topography or management practice—can increase the precision of carbon measurements without increasing the cost unduly by lowering the amount of variation around the mean, thereby requiring fewer plots to be within acceptable levels of precision.

A few data provide some preliminary estimates of costs for measuring and monitoring of carbon in LULUCF projects in three tropical countries (Powell, 1999; Subak, 2000; Box 5-4). For the first inventory of the NKCAP, the total fixed operational costs (including human resource costs, project management, mapping, and so forth) were estimated to be about $196,000; variable costs (including labor, equipment, transport, and so forth) ranged between $230 and $281 per plot and totaled about $154,000 (625 plots). The grand total cost was about $350,000 (Powell, 1999). The precision of the inventory, based on sampling error only, was ±4 percent with 95-percent confidence (see Box 5-3). The variable costs dropped rapidly from about $108,000 for a precision level of ±5 percent to $1,000 for a level of ±30 percent; fixed costs would be the same for all levels of precision (Figure 5-6b). Estimates of the revised carbon benefits from this project for its duration based on additional measurements and data collection (Brown *et al.*, 2000) and the additional cost to collect this information result in an estimate of about $0.10 per t C. Estimating future monitoring costs based on the first inventory is difficult, but they are likely to be less than those for the initial inventory because different sampling intensities will be used, project implementers can build on previous experience, and advances in technology will be available (see, e.g., Section 5.4.3).

The organization responsible for monitoring carbon sequestration in Costa Rica's Private Forestry Project and for acquiring remote-sensing information has an annual budget of $200,000 (Subak, 2000). Additional costs associated with the PFP relate to the costs of monitoring forests and plantations. The implementing organizations do not absorb all of the monitoring costs: They charge landowners at a rate that varies in different regions. In the Central Volcanic Range (including the upper Virilla), for example, landowners pay implementing organizations 10 percent of their annual environmental services payment for monitoring

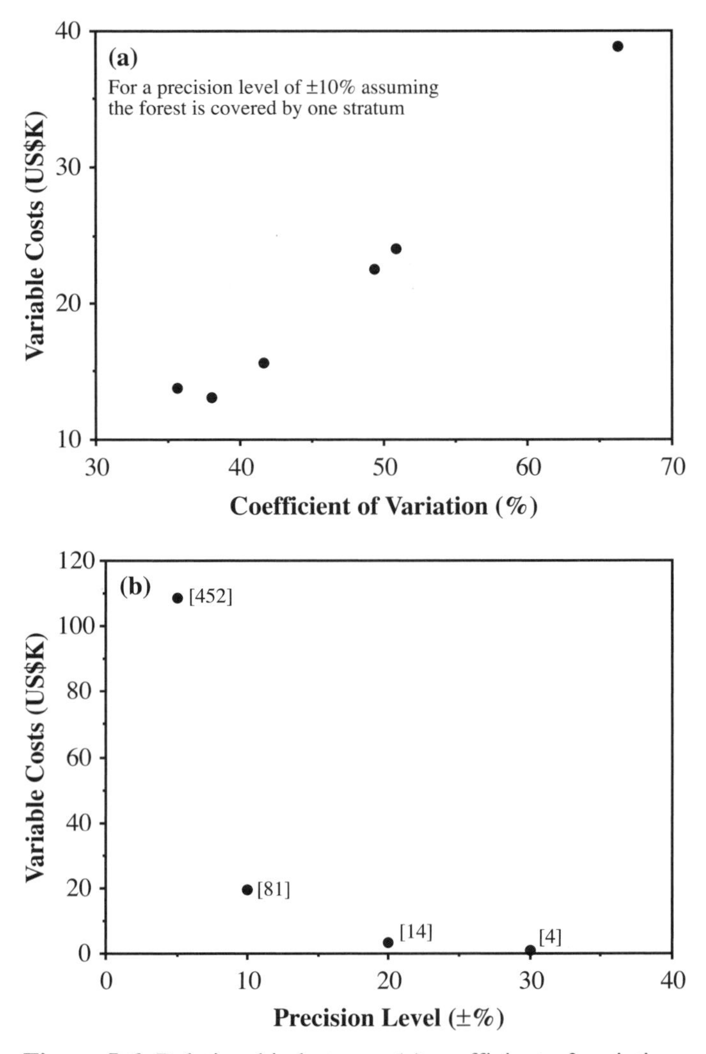

Figure 5-6: Relationship between (a) coefficient of variation and (b) level of precision and total variable costs for the first inventory phase of the Noel Kempff Climate Action Project in Bolivia (data from Powell, 1999).

Box 5-4. Cost of Monitoring and Verification of a Forest-Based Project in the Western Ghats, India

The dominant activity in this project is to reforest degraded lands. The specifics are to perform enrichment planting of trees in partially degraded forests and establish multi-purpose tree plantations on fully degraded lands (Ravindranath and Bhat, 1997). The carbon benefits from this project are from carbon conservation of biomass in native forest by substituting with wood from tree plantations, carbon sequestration in trees and soil in the enrichment planting of partially degraded forests (logging is banned in this area), and enhancing soils carbon in the badly degraded lands. The total area to be reforested is 42,000 ha, with a total budget of US$11.7 million over a period of six years starting in 1991. The cost of reforesting with the dominant multi-purpose tree plantations is US$609 ha^{-1}. The allocation for monitoring and research of the reforested lands (survival and growth rate) is about US$1 million, which accounts for about 9 percent of the total budget allocated for this project (Ravindranath and Bhat, 1997). The annual cost for monitoring of the project lands is about US$5 ha^{-1}.

of forest protection and slightly more for monitoring of forest management but do not pay anything for monitoring of plantations (Subak, 2000). The implementing agency's unit costs will tend to be higher when monitoring of smaller landholdings, and although a stated objective of the PFP is to compensate small and medium-sized landowners, some implementers may favor larger parcels. Monitoring of the PFP is supposed to include site visits by forest engineers as well as more detailed audits of some sites. The annual visits involve making a report on the size, density, and health of the trees on the land; the more detailed audits are to assess management as well as the conditions of the trees and soil. The intention is to audit as few as 5 percent of the PFP sites. The labor costs for auditing are estimated to be $10 ha^{-1} yr^{-1}—compared to $1 ha^{-1} yr^{-1} for monitoring and $2 ha^{-1} yr^{-1} for certification. The aggregate costs of project development, recruiting, and auditing are significant, but they have not been judged to be excessive or to reverse the cost-effectiveness of the PFP as an LULUCF project.

Currently, there are no guidelines regarding the level of precision to which pools should be measured and monitored. Setting such a level would facilitate comparison of projects and could encourage project developers to measure projects more precisely if the price of carbon was high. If the total average carbon benefit was 5 million t C with a ±30-percent confidence interval, for example, the lower bound carbon benefit would be 3.5 million t C. If LULUCF projects could only claim benefits for a lower bound of the confidence interval and if carbon was worth $10 per t C, the "loss" of carbon benefits would be 1.5 million tons—equivalent to $15 million, a value likely to greatly exceed the cost of monitoring to a ±5-percent precision level. Thus, project developers would probably chose high precision levels for their monitoring.

5.4.5. *Verification*

Verification by third-party institutions offers a way to provide confidence to governments, investors, project developers, NGOs, and the public at large regarding the validity of the carbon benefits claimed by a project. Third-party verification could be based on an assessment of the project's compliance with defined eligibility criteria. A single set of internationally accepted eligibility criteria would facilitate direct comparison of projects, whereas a variety of such criteria would result in projects and GHG benefits of differing quality.

Verification activities may include review of data or documentation (e.g., procedures, methodologies, analyses, reports), including interviews with project personnel; inspection or calibration of measurement and analytical tools and methods; repeat sampling and measurements; assessment of the quality and comprehensiveness of the data used in calculating the project baseline and offsets, therefore the confidence in the final claims; assessment of risks associated with the project as well as the carbon benefits; and the presence or absence of non-GHG externalities such as environmental and social impacts. Existing programs describe alternative ways that verification could be accomplished. Notable elements of these alternative programs include periodic verification of project performance against defined criteria (EcoSecurities, 1997; Trines, 1998b; Moura-Costa *et al.*, 2000); an external evaluation panel, site visits, and third-party inspections (USIJI, 1996); and designation of verifiers by the proposer (World Business Council for a Sustainable Development, 1997).

Unlike projects in other sectors, the carbon stocks of LULUCF projects may require verification and monitoring beyond the project time horizon. The verification period will depend on the method chosen for accounting of carbon stocks (see Section 5.4.2). The carbon stock method may require verification until the end of the project; the average net carbon storage method may need verification in perpetuity. The ton-year approach may require verification for periods ranging from the project lifetime to some specified time period beyond the project lifetime, depending on the specifics of the accounting method chosen.

There has been little experience to date with third-party verification of the carbon stock of projects (Moura-Costa *et al.*, 2000). The Forest Stewardship Council (FSC) offers a model, however, for how verification might be accomplished and how verifiers might be accredited by an independent accreditation body. The FSC accredits organizations that inspect forest operations and grants labels certifying that the timber has been produced from well-managed forests. The FSC is funded by organizations other than the industries it monitors. Other institutions, such as SGS, are establishing certification councils with similar responsibilities.

The costs of verification by third parties can be alleviated by taking several steps:

- A single set of eligibility criteria accompanied by standardized accounting and reporting methodologies will reduce the costs of developing such services.
- Definition of acceptable confidence intervals will enable project developers to maximize their sampling efficiency and enable verifiers to minimize their costs.
- Development of "group verification programs," which have been successful in other sectors, can make verification available to small-scale projects.

5.4.6. *Reporting*

The purpose of reporting is to provide information about a project's measured GHG and non-GHG benefits to government and/or intergovernmental entities, so that they can establish GHG credits that might be used for offsetting an Annex I country's commitments during the budget period (see Chapter 6). Reporting guidelines for each of the Kyoto Protocol's flexibility mechanisms are to be developed by the COP. This section discusses the types of data that may be required for reporting, as well as issues relating to multiple reporting of project activities.

The UNFCCC's Subsidiary Body for Scientific and Technological Advice (SBSTA) developed a Uniform Reporting Format (URF) for activities implemented jointly under a pilot program. The format was approved by SBSTA as part of the implementation of the UNFCCC (UNFCCC, 1997). In completing the URF, project proposers are to estimate the projected emissions for their without-project baseline scenario and with-project activity scenario. They are to estimate cumulative effects for carbon dioxide, methane, nitrous oxide, and other GHGs. The URF also contains a section on environmental and socioeconomic benefits. Project developers are to describe how their project is compatible with and supportive of national economic development and socioeconomic and environmental priorities and strategies. Furthermore, the URF requests information on "practical experience gained or technical difficulties, effects, impacts or other obstacles encountered." As of October 1998, 95 AIJ projects had reported the foregoing information using the URF format (UNFCCC, 1999a). Other programs, such as the USIJI, have reporting requirements as well.

Proposed improvements to the URF format (Vine *et al.*, 1999) include basic project contact information; a description of the project; projected and actual changes in carbon stock; net changes in carbon stock; information on the precision of results; data collection and analysis methods used in calculating changes in carbon stock; estimates of project leakage (negative and positive); and market transformation (where calculated). Finally, information on environmental and socioeconomic impacts and an indication of whether there is consistency between environmental laws, environmental impact statements, and expected environmental impacts could be included.

Unlike projects in other sectors, the time period over which reporting needs to occur will depend on the method chosen for accounting of carbon stocks of a project. The project developer or some other organization will need to be designated to report on changes in the carbon stock, should the accounting method require continued monitoring and verification after the end of the project. Governments may need to establish a procedure and set rules for post-project reporting, if needed.

Several types of reporting might occur in forestry projects: The impacts of a particular project could be reported at the project and/or program level (where a program consists of two or more projects); the impacts of a particular project could be reported at the project level and at the entity level (e.g., a utility company reports on the impacts of all of its projects); and the impacts of a particular project could be reported by two or more organizations or by two or more countries as part of a joint venture (partnership). To reduce any problems that might occur in multiple reporting, project-level reporters would need to indicate whether other entities might be reporting on the same activity and, if so, who. Establishment of a clearinghouse with an inventory of stakeholders and projects might solve this problem. For example, in comments on an international emissions trading regime, Canada (on behalf of Australia, Iceland, Japan, New Zealand, Norway, the Russian Federation, Ukraine, and the United States) has proposed a national recording system to record ownership and transfers of assigned amount units (i.e., carbon offsets) at the national level (UNFCCC, 1998). A project synthesis report could confirm, at an aggregate level, that bookkeeping was correct, reducing the possibility of discrepancies among Parties' reports on emissions trading activity.

5.5. Associated Impacts (Benefits and Costs) of LULUCF Projects

Several authors have noted that LULUCF projects to reduce or offset GHG emissions can also provide significant environmental and socioeconomic "co-benefits" to host countries and local communities (Makundi, 1997; Brown, 1998; Frumhoff *et al.*, 1998; Trexler and Associates, 1998; Klooster and Masera, 2000; Lasco and Pulhin, 2000; Losos, 2000; Reid, 2000). Because the scale of such projects is prospectively large (Section 5.1), they may have substantial potential to help countries meet multiple sustainable development objectives. Some authors have also expressed concern, however, that some types of LULUCF projects pose significant risk of negative environmental and socioeconomic impacts (e.g., Cullet and Kameri-Mbote, 1998; German Advisory Council on Global Change, 1998).

This section follows on the general assessment of sustainable development aspects of LULUCF measures in Section 2.5 to address the following project-specific questions: What are the environmental and socioeconomic implications of different LULUCF project types? Do any of these projects pose inherently negative or positive impacts?

Representative data on the socioeconomic and environmental impacts of several LULUCF projects carried out under the AIJ Pilot Phase are provided in Box 5-1. Relatively few AIJ LULUCF projects to date have provided detailed quantification of observed and expected local socioeconomic impacts (Witthoeft-Muehlmann, 1998). The assessment in this section draws on available pilot project data and information from similar LULUCF projects in its evaluation of associated impacts.

5.5.1. *Associated Impacts of Project Activities that Avoid Emissions*

Pilot LULUCF projects that are designed to avoid emissions by reducing deforestation and forest degradation have produced marked environmental and socioeconomic co-benefits, including biodiversity conservation, protection of watershed and water resources, improved forest management and local capacity building, and employment in local enterprises. Substantial biodiversity benefits, for example, have been realized in the Rio Bravo project in northwestern Belize (Box 5-1) and the AES Barbers Point carbon-offset project in Paraguay (Dixon *et al.*, 1993), where protection of 56,800 ha of tropical forest can conserve existing biodiversity and restore native flora lost from logging activities.

Although any LULUCF project that slows deforestation or degradation will help to conserve biodiversity, successful projects in threatened forests that contain assemblages of species that are unusually rich, globally rare, or unique to that region can provide the greatest biodiversity co-benefits (Dinerstein *et al.*, 1995; Olson and Dinerstein, 1998). One example is the Noel Kempff Mercado carbon-offset project in Bolivia: In a region of globally outstanding biological distinctiveness, a 634,000 ha timber concession has been converted into an extension of a national park (Dinerstein *et al.*, 1995; USIJI, 1997b; Box 5-3).

Projects that are designed to protect natural forests from land conversion or degradation could pose significant costs to some stakeholders if they restrict options for alternative land uses, such as crop production. Such costs might be mitigated, however, by siting projects in regions where conservation measures are consistent with regional land-use policies and by promoting sustainable agricultural intensification on associated non-forested lands. Indeed, forest conservation projects in areas where policies encourage agricultural expansion are unlikely to be successful. Critical to shaping project success in meeting carbon mitigation and sustainable development goals is effective participation by local communities affected by project activities (Section 5.6). In the Noel Kempff Mercado project, this local participation includes community-run revolving funds financed by the project that provide loans for local sustainable development enterprises such as ecotourism, bakeries, and hearts-of-palm production (Brown *et al.*, 2000).

LULUCF projects that protect forests from land conversion or degradation in key watersheds have substantial potential to slow soil erosion, protect water resources for rural communities and municipalities (Reid, 2000), and conserve biodiversity (Hardner, 1996; Frumhoff *et al.*, 1998; Hardner *et al.*, 2000). Benefits can also include reduced risk of flood damage and reduced siltation of rivers; the latter can protect fisheries and investments in hydroelectric power generation facilities (Chomitz and Kumari, 1998). One AIJ pilot project that is designed to provide these benefits is Costa Rica's PFP (Subak, 2000).

Several AIJ pilot carbon offset projects include measures to reduce the impacts of logging and more generally improve the sustainability of forest management (Brown, 1998). As evidenced by the RIL project in Sabah, Malaysia, such projects can combine reduced carbon emissions with reductions in the environmental impacts of commercial logging, as well as socioeconomic development through technical training (Putz and Pinard, 1993; Pinard and Putz, 1996, 1997).

The carbon benefits and the associated environmental benefits of reduced-impact logging are captured only in forest sites that otherwise would have been logged by conventional methods or converted to agriculture; such benefits would not be gained in forests that otherwise would have been unlogged. In developing countries, such projects also might slow deforestation under some conditions by making long-term timber production more profitable than forest clearing for low-productivity agriculture or pasture (e.g., Boscolo *et al.*, 1997).

Projects that are designed to promote reduced-impact logging as a carbon offset may produce fewer biodiversity co-benefits than forest protection, but they provide larger socioeconomic benefits for local owners (Kurz *et al.*, 1997; Marland *et al.*, 1997; Bawa and Seidler, 1998; Frumhoff and Losos, 1998; Klooster and Masera, 2000). Policymakers may wish to identify and consider prospective tradeoffs between meeting these objectives on a national basis and meeting them on a project-by-project basis.

5.5.2. *Associated Impacts of Projects that Sequester Carbon*

Under a carbon market, projects that promote afforestation through plantation forestry may be attractive to many prospective investors, given their potential to generate profitable financial returns in addition to carbon credits (Frumhoff *et al.*, 1998). The potential impacts of projects that are designed to promote afforestation through plantation forestry will vary significantly with location, scale, use of native versus exotic tree species, and intensity of management. Intensively managed plantations, for example, can help maintain and improve soil properties, particularly if understory vegetation and leaf litter is not cleared (Chomitz and Kumari, 1998), as well as providing a source for biomass fuels and other wood products. They can have highly variable impacts on water resources (Section 2.5.1). Plantations typically do not appear to reduce pressure on natural forests in the humid tropics (Kanowski *et al.*, 1992; Johns, 1997) because these forests are not generally cleared for the sawn wood, pulpwood, or other products that plantations provide. Kanowski *et al.* (1992) suggest that fuelwood plantations might help reduce pressure on natural woodlands in relatively arid regions. Thus, they might help to stem desertification in some settings.

Plantation projects would have negative impacts on biodiversity if they replace native grassland or woodland habitat or if permanent plantations of exotic species were planted in sites where natural or assisted restoration of indigenous forests is feasible. Many grassland ecosystems, for example, are rich in endemic species; in the Mpumalanga province of South Africa, the expansion of commercial plantations (*Eucalyptus* spp. and *Pinus* spp.) has led to significant declines in several endemic and threatened species of grassland birds (Allan *et al.*, 1997).

In contrast, nonpermanent plantations of exotic or native species can be designed to enhance biodiversity co-benefits by jump-starting the process of restoring natural forests (Keenan *et al.*, 1997; Lugo, 1997; Parrotta *et al.*, 1997a,b). Commercial forestry plantations can also increase biodiversity co-benefits by adopting longer rotation times, reducing or eliminating measures to clear understory vegetation, using native tree species, and minimizing chemical inputs (e.g., Allen *et al.*, 1995a,b; Da Silva *et al.*, 1995).

Afforestation or reforestation measures could have positive or negative impacts on local communities. Negative impacts can result if projects are implemented on land for which communities have alternative priorities, such as agricultural production, and if communities are not effectively engaged in all phases of project design and implementation (Cullet and Kameri-Mbote, 1998; Section 6.6.3). In urban or peri-urban areas, they can also produce significant local socioeconomic benefits through improvements in air quality (McPherson, 1994).

Some observers have expressed concern that carbon-offset financing for reforestation projects in non–Annex I countries could promote deforestation by financing the expansion of plantations that replace natural forests whose associated emissions would not be constrained by a national cap (German Advisory Council on Global Change, 1998). Section 2.5.2.2 discusses possible options, should Parties wish to constrain such projects.

Agroforestry activities can sequester carbon and produce a range of environmental and socioeconomic benefits. For example, trees in agroforestry farms improve soil fertility through control of soil erosion, maintenance of soil organic matter and physical properties, increased nutrient inputs through nitrogen fixation and uptake from deep soil horizons, and promotion of more closed nutrient cycling (Young, 1997). Thus, agroforestry systems that incorporate trees on farms can improve and conserve soil properties (Nair, 1989; MacDicken and Vergara, 1990), as is the case in the AES Thames Guatemala project (Dixon *et al.*, 1993). Agroforestry projects also may provide local economic benefits, with farmers gaining higher income from timber, fruits, medicinals, and extractives than they would from alternative agricultural practices (Cooper *et al.*, 1996).

Poorly planned and implemented agroforestry projects, however, can fail to benefit or have negative impacts on local farmers. For example, the introduction of labor-intensive agroforestry technologies can lead to labor competition between agroforestry practices and traditional farming (Laquihon, 1989; Repollo and Castillo, 1989). Poorly planned projects can also lead to excessive light and water competition between crops and trees, as well as reducing the area available for food crops.

The associated environmental benefits of project activities that promote assisted regeneration of natural forests are similar to those of forest conservation. As the forest matures, key benefits may include protection of watersheds, soil fertility, and biodiversity. As with forest conservation or plantation forestry, assisted forest regeneration could lead to negative social impacts if communities are prevented from changing to preferred land uses in the future. This negative impact also can be reduced by ensuring that the designation of areas for reforestation is consistent with long-term regional land-use plans and that community development priorities are effectively incorporated during project development and implementation (Section 5.6).

There is very limited experience of LULUCF pilot projects that sequester carbon or reduce carbon emissions from agricultural soils. There are vast areas of degraded and desertified land in developed and developing countries, however, where well-designed projects can add carbon to the soil while increasing agricultural productivity and sustainability.

5.5.3. Associated Impacts of Carbon Substitution Projects

Projects that use short-rotation tree plantations as woody biomass energy sources have equivalent associated impacts to the managed plantation projects described in Section 5.5.2. There are also a broad range of prospective environmental and socioeconomic impacts associated with the production of biomass energy from agricultural crops, such as sugarcane and corn, and oil crops such as soybeans. The impacts of substitution projects can occur on-site (where projects are located) or off-site (where electricity or fuel supply is offset). On-site impacts include local environmental and socioeconomic benefits of the forestry and energy generation components of a bioenergy project. The environmental impacts can include reclamation of degraded lands; potential promotion of biodiversity, provided part of the plantation area is left for natural regeneration (Carpentieri *et al.*, 1993); and reduction of pressure on primary forests to the extent that fuelwood derived from such sources is replaced by other energy sources. Rural bioenergy programs can also help local communities achieve self-reliance and decentralize political power by giving control of resources to the local community (Ravindranath and Hall, 1995).

Provision of small-scale bioenergy in place of wood may often directly benefit women more than men. The foregoing options will decrease the labor and time needed to gather wood and reduce indoor air pollution from smoke (a recognized health hazard). The success of rural projects depends on equitable distribution of benefits that community involvement in rural energy projects can provide (Agarwal and Narain, 1989). On-site energy generation can increase the production of local pollutants. Well-designed projects, however, can offset another more-polluting local source—as in the Bio-Gen Biomass Power Generation Project in Honduras. There, emission control technologies are used to produce fewer pollutants than would have been emitted in the non-project case, with the continued uncontrolled burning of sawmill and logging residues. Giampietro *et al.* (1997) provide a more general discussion of the environmental impacts of biofuel production.

In conclusion, LULUCF GHG mitigation projects are neither inherently good nor inherently bad in terms of potential environmental and socioeconomic co-benefits. Adequately designed and implemented, projects in each major category can provide significant socioeconomic and environmental benefits to host countries and local communities, though projects of all types pose some risk of negative impacts. Section 5.6 addresses how the sustainable development contributions of these projects can be strengthened and the negative impacts mitigated.

5.6. Factors Affecting the Sustainable Development Contributions of LULUCF Projects

Six factors that are critical to strengthening the sustainable development contributions of LULUCF GHG mitigation projects have been identified:

- The consistency of project activities with international principles and criteria of sustainable development
- The consistency of project activities with nationally defined sustainable development and/or national development goals, objectives, and policies
- The availability of sufficient institutional and technical capacity to develop and implement project guidelines and safeguards
- The extent and effectiveness of local community participation in project development and implementation
- The transfer and local adaptation of technology (including hardware and software)
- The application of sound environmental and social assessment methodologies to assess sustainable development implications.

Chapter 2 highlights the international principles and criteria of sustainable development that may facilitate more successful implementation of LULUCF projects. It also discusses the application of social and environmental assessment methodologies. This section discusses in more detail the other four factors.

5.6.1. Consistency with Nationally Defined Sustainable Development and/or National Development Goals

Prospective investors in LULUCF projects and host countries may have different priorities in selecting projects. From the investors' perspective, criteria such as land availability and the suitability of the country to undertake the project, the estimated GHG benefits, project cost-effectiveness, risk, and other environmental effects are some of the major concerns. From the host country's perspective, projects that more specifically consider regional or local land-use priorities and significantly strengthen sustainable development contributions will be favored. Some observers have also expressed concern that selecting only the cheapest projects will be detrimental to non–Annex I countries if they subsequently take on GHG emissions (Lee *et al.*, 1997; Brown, 1998). For LULUCF projects to be designed, conceived, and implemented successfully to provide economic and environmental benefits, however, the support of different stakeholders of the project—project investors, host countries, and local communities—is crucial.

The voluntary nature of host country participation in climate mitigation projects increases the prospects that only projects that satisfy investor and host country interests will be implemented. Moreover, host countries can take steps to ensure that the goals of accepted projects are consistent with national and local development and natural resource protection priorities (Intarapravich, 1995; Michaelowa and Schmidt, 1997; Hardner *et al.*, 2000). Dutch and Costa Rican criteria for approval of AIJ projects, for example, state that projects should be compatible with and supportive of sustainable development priorities of each country; fulfill the obligations of various conventions; and enhance income opportunities and quality of life for rural people and members of certain vulnerable groups, including cultural minorities (Andrasko *et al.*, 1996; Ministry of Housing, Spatial Planning and the Environment, 1996; Subak, 2000).

One way to ensure that a mitigation project is consistent with the host country's developmental goals is for the host country to set up a simple approval process for accepting projects and to list criteria that are based on national and local needs. Projects may not satisfy all criteria, but it is important to ensure that they adhere to all applicable laws and/or regulations of the host country. To meet national or regional sustainable development priorities, project transaction costs should be kept low. High costs can reduce investor interest in financing LULUCF climate mitigation projects (Section 5.2) and reduce the proportion of funding available to promote and monitor environmental and social aspects of implemented projects.

To achieve consistency with national and regional environmental and development goals, it is also important to ensure that policies and programs support rather than undermine project objectives. Changes in key policies that may affect project sustainability either positively or negatively include financial subsidies for forestry or agriculture, land tenure, policies to expand agricultural production, import-export policies, and paper recycling programs (World Bank, 1997). For example, Brazil's government-subsidized program to produce ethanol vehicle fuel from sugarcane withered in the face of low gasoline prices (La Rovere, 1998). Therefore, incorporating projects that minimize conflicts or institutional changes relative to existing land-use policies in the host country may be essential.

5.6.2. *Availability of Sufficient Institutional and Technical Capacity to Develop and Implement Project Guidelines and Safeguards*

In industrialized countries, relatively good expertise exists to understand the technical issues involved in the preparation and implementation of LULUCF projects. In many developing countries, however, there is not enough technical capacity to design, implement, monitor, and evaluate LULUCF projects; this deficit raises the issue of capacity-building needs.

As suggested by decisions at the fourth COP, capacity building for country-driven projects must be greatly enlarged. If forestry and biofuel options are to play key roles in least-cost and early (precautionary) GHG reductions, there is a need for experts to initiate and implement projects (Haque *et al*., 1999). Furthermore, the twin objectives of carbon mitigation and sustainable development present additional technical challenges to monitoring and verification (Andrasko, 1997), which are vital to the commercial credibility of LULUCF projects (Fearnside, 1997; MacDicken 1997a).

The capacity to implement LULUCF projects can be developed through investment in training in information programs, demonstration projects, training and outreach, and general capacity building (Swisher, 1997). For instance, Australia and New Zealand have developed capacity-building programs to facilitate strong awareness of modalities that govern projects in developing countries (Read, 1999; UNFCCC, 1999a; Warrick *et al*., 1999). In Africa, capacity building is regarded as an equity issue (Sokona *et al*., 1999). Costa Rica integrated several NGOs into its AIJ program from the beginning; these NGOs provided technical and operational support to Costa Rica's Office for Joint Implementation (OCIC) (MINAE, 1996).

At different stages of a project, appropriate meetings, information workshops, formal hearings, government-supervised notices, consultation, access to documents and reports, employment of members of the public, use of public third-party auditors, and complaint and dispute resolution forms of participation may be most appropriate (Environmental Law Institute, 1996). Training in gathering a conjunction of stakeholders to obtain mutual benefits is a crucial aspect of general capacity building (Haque *et al*., 1999).

5.6.3. *Extent and Effectiveness of Local Community Participation in Project Development and Implementation*

The involvement of local communities that directly depend on forest resources is a precondition for the success of community-based projects. Local communities can be involved by designing a project to develop local skills, create employment in the project, and promote equity, all of which will lead to the long-term sustainability of the project activity. In the Scolel Te project, for instance, local communities and their agroforestry traditions are included in the project design process (Imaz *et al*., 1998). On the other hand, the ECOLAND project in Costa Rica has caused discontent among local residents who did not sell their lands and now face hardships caused by the inclusion of their lands in a national park (Goldberg, 1998). It is also important that the host country and the project designers recognize the land titles and legal rights of indigenous people to ensure their effective participation in the project (see Box 5-5 for a case involving the social forestry program in India).

In bioenergy projects, local people could be trained in the operation and maintenance of biogas plants; this training could lead to the creation of new jobs in rural areas and reduce migration to urban centers, thereby achieving equitable development between rural and urban areas (Ravindranath and Hall, 1995). It would also promote the sustainability of the project by providing financial, social, and environmental benefits even after the investors have withdrawn.

The success of community management projects also depends on equitable discussion, participation, and distribution of benefits, which is crucial for the development of rural areas (Sokona *et al*., 1999). It is important to have institutional arrangements to ensure land tenure and product ownership by local communities or to meaningfully involve local participants in decision-making processes regarding species choice, mode of production, harvesting, and benefit sharing that encourages them to commit themselves to the protection and management of LULUCF projects.

5.6.4. *Transfer and Local Adaptation of Technology*

For LULUCF projects, technology adaptation, diffusion, and transfer requires a broad definition. Such transfer may include sustainable forest management practices; forest conservation and protected area management systems; silvicultural practices for afforestation and reforestation programs; genetically superior planting material; efficient harvesting, processing, and end-use technologies; indigenous knowledge of forest conservation; and low-tillage agriculture and ruminant management practices (Ravindranath *et al*., 2000).

Most LULUCF projects require the transfer of such technology. The absence of these technologies may frustrate delivery of the mitigation and developmental benefits associated with them

Box 5-5. Social Forestry Program in India

Several developmental projects in the forestry sector have been implemented in the tropics that could be sources for understanding the possible implications of future LULUCF projects. One such afforestation program implemented in India was funded by several donor agencies during the 1980s. In terms of number of trees planted (18,876 million trees in 1980–87, Chambers *et al.*, 1989), the project was a success. The lessons learned from the program are briefly described below (Saxena, 1997).

Social forestry projects were implemented by the Forest Department in India with the goal of meeting the demands of rural people and reducing the burden on production forestry. The species planted in the village commons and revenue lands were mainly monocultures of *Eucalyptus*, *Casuarina*, and *Acacia* sp. Tree planting and management was carried out by the Forest Department in the initial years and later handed over to the Panchayat (village governing body).

Local Participation
The selection of species reflected the choice of the Forest Department rather than the local preferences. Participation was limited to a few members of the village elite. Community involvement was limited to handing over the common land for plantation and participation as wage labor. In designing the project, foresters and foreign experts did not fully grasp the complexity of the rural power structure and assumed that the village Panchayats represented the interests of all concerned in the village (SIDA, 1992). Thus, a large portion of the benefit from the project went to the urban areas, industries, and retailers—defeating the purpose of the project.

Land Tenure
Throughout the social forestry phase, it was not clear whether village land belonged to the Forest Department, the Revenue Department, or the village body. Such uncertainty about ownership and legal rights impeded community action. Non-forest laws often conflicted with the social forestry projects.

Technical Issues
Species selection, spacing, and other silvicultural issues were not properly examined and implemented. Benefits that could flow to poor villagers from species yielding intermediate products were not properly appreciated. The production of grass, legumes, leaf fodder, fruits, and non-timber forest products was neglected. Close spacing was prescribed to avoid intermediate management options, reduce plantation costs, and cut down on staff supervision time. As a consequence, thinning and pruning, which could have produced intermediate yields of grass and tree products for the people, were not undertaken (Saxena, 1997). Because of the close spacing, grass production was affected. Because projects were designed around the ultimate felling of the planted trees, degradation often set in after the trees were harvested.

Policy Issues
The failure to define, establish, and publicize the rights for marketing and allocating benefits to the community led to the failure of their participation. Rights to trees and a distribution policy that were not official preoccupations in the early stages of the tree planting led to inequitable distribution later.

Equity Issues
A government review found that only 20 percent of the respondents knew about the woodlots during the planning stage, only 14 percent of the people participated in the meetings, and about 83 percent of the low-status people were adversely affected by the closure of the community land. The landless farmers and artisans depend on the village commons to graze their animals and collect fuelwood.

Capacity-Building
The funding projects provided the Forest Department with vehicles and foreign training, but little emphasis was given to building the capacity of the Forest Department.

Multiplicity of Donors
Multiplicity of donors with different priorities within single provinces resulted in conflicting policies being followed.

(Sathaye *et al.*, 1999). Poorly designed LULUCF projects may lead to the importation of inadequate or inappropriate technologies into recipient countries. In agroforestry projects, for example, inappropriate selection of species and crop timbering processes or machinery may fail to bring out the full potential of associated co-benefits, which depend on local biophysical, social, cultural, and organizational factors (Lemaster, 1995).

Current and emerging pathways and mechanisms for technology transfer through LULUCF GHG mitigation projects have several limitations—namely, limited financial resources, inadequate information on costs and potential benefits of projects, limited host country technical capacity, absence of policies and institutions to process and evaluate mitigation projects, and long gestation periods. In addition, the forest sector faces land-use regulation and other policies that favor conversion to other land uses, such as agriculture and cattle ranching. Insecure land tenure and subsidies that favor agriculture or livestock are among the most important barriers to ensuring sustainable forest management and sustainability of GHG mitigation (Ravindranath *et al.*, 2000).

LULUCF projects have surmounted some of these barriers through means that include extensive capacity building and establishing institutions at the local level (e.g., NKCAP, Bolivia; Scolel-Te, Mexico); development of improved forest management systems and joint ventures between private companies and local organizations (RIL, Malaysia); and systems of financial incentives that directly benefit farmers by increasing the relative cost-effectiveness of forestry options (Costa Rica Joint Implementation Program) (see Box 5-1). LULUCF projects in non–Annex I countries have the potential to fund improved technologies that can yield environmental benefits by raising agricultural productivity through the transfer of irrigation or management practices; increasing milling efficiency; improving silvicultural practices; promoting sustainable forest management (Brown, 1998), as in Senegal (Box 5-6); or—where LULUCF projects involve biofuel production—supporting energy sector development that "leap-frogs" the fossil fuel stage, moving directly to sustainable energy development (Read, 1999).

5.7. Implications of Project-Based Activities for Countries with and without Assigned Amounts of Emissions

All of the major issues—project permanence, additionality, and potential leakage and risks—present different implications for countries with national assigned amounts than for countries without assigned amounts. Some of these issues also show specific characteristics by project type (Table 5-10). The implications for carbon accounting, as well as associated socioeconomic and capacity-building components, are also different depending on whether the countries currently have or do not have national assigned amounts (Table 5-10).

The fate of GHG benefits when the project ends, the risks associated with projects, leakage, and additionality are all major issues for countries without national assigned amounts—particularly for emissions avoidance and carbon sequestration projects—because these countries are not required to capture project activities in national greenhouse gas inventories. For the same reasons, the choice of accounting methods and the control of leakage are also critical. This last issue may be addressed voluntarily and reported on national communications (Table 5-10).

For countries with assigned amounts, project duration is important if the project does not fall under Articles 3.3 or 3.4, if liability for post–project period emissions is not clear, or if the commitment periods are not contiguous. Determination of adequate baselines and establishment of project additionality are required for projects that fall under Article 6 (and perhaps for projects falling under Article 12). Concerns regarding methods for GHG accounting at the project level are not as critical because all countries—including those with assigned amounts—are required to prepare a national GHG inventory. Double-counting could be an issue, however, if project activities cannot be captured in national inventories. Potential transnational leakage between countries with and without assigned amounts is important to consider because such leakage is not captured by the emissions limitation of Annex I countries (Gustavsson *et al.*, 1999).

Box 5-6. Technology Transfer and Capacity-Building in an Agroforestry Project in Senegal

Enda Syspro, an international institution, has developed an ecologically sustainable agroforestry practice in Senegal. This system involves planting hedges in the boundary of the fields with drip irrigation to produce various crops and vegetables for local markets and exports. This type of project improves food security which has been considered the primary concern of African countries at the 1999 Abidjan, Ivory Coast, meeting for climate change. The agroforestry project not only reduces GHG emissions (by avoiding deforestation, sequestering carbon in hedges and soils, and replacing fossil fuels with sustainably harvested firewood), it also improves soil fertility and reduces soil erosion. The project maintains biodiversity by reducing deforestation and fragmentation of the landscape. By reducing the need for water for irrigation, it helps to reduce vulnerability to climate change in the Sahelian countries. A training center has been set up for Senegal and Sahelian countries to replicate such farming systems. Various high-technology agricultural activities, including biotechnology transfer, have been developed in Enda Spyro to improve food security and to measure carbon sequestration. Today, more than 1,000 ha of such agroforestry systems have been established in Senegal.

***Table 5-10**: Key implications of LULUCF projects for countries with and without national assigned amounts.*

Technical Issue	Countries with National Assigned Amounts	Countries without National Assigned Amounts
Permanence and risks (fate of carbon at end of project cycle)	Fate is not an issue if on Article 3.3 Kyoto lands, but fate is an issue if project is not on Article 3.3 or 3.4 lands. Article 3.4 activities (if projects): Fate is not an issue, if commitment periods are contiguous.	Fate is an issue for emissions avoidance and carbon sequestration projects because the country has no assigned amounts or binding baseline, and project and product lifetimes are relatively short.
Baselines and additionality	Not an issue under Article 3.3. Under Article 3.4, depending on the accounting system, it could be an issue. Under Article 6, establishing baselines and additionality is a requirement.	Major issue for all categories of projects. Required under Article 12. Approaches still being tested.
Leakage of GHGs across project boundary	Minor issue because leakage should be picked up in national GHG accounting. Potential transnational leakage (e.g., if projects change net C stocks and wood imports from non-Annex I countries increase).	Leakage is a major issue for emission avoidance and sequestration projects, but not as important for substitution projects. Leakage control would be useful, but without a baseline or assigned amount, a country need not account for it. Might be voluntarily counted in a national communication.
Accounting		
– Double-counting of GHG benefits	Only an issue if project cannot be identified and separately tracked in national baseline and accounting.	Not an issue because only projects are counted.
– Accounting methods	Minor issue because of national assigned amounts.	Major issue for emission avoidance and sequestration projects. Important to count both carbon credits and debits, or to establish value of delayed emissions.
Associated impacts		
– Sustainable development	A general goal stated in Article 2 of the Kyoto Protocol is to promote sustainable development.	Applies to all categories of projects. Project contribution to sustainable development is a stated purpose of Article 12.
– Capacity-building and technology transfer	Issue for some Annex II countries. Important issue for some Annex I countries.	Major issue. Applies to all categories of projects. Several decisions of COP3 and COP4 have emphasized this.

References

Agarwal, A. and S. Narain, 1989: *Toward Green Villages: A Strategy for Environmentally Sound and Participatory Rural Development.* Center for Science and Environment, New Delhi, India, 52 pp.

Allen, D.G., J.A. Harrison, R.A. Navarro, B.W. van Wilgen, and M.W. Thompson, 1997: The impact of commercial afforestation on bird populations in Mpumalanga Province, South Africa—insights from bird atlas data. *Biological Conservation,* **79,** 173–185.

Allen, R.K., K. Platt, and S. Wiser, 1995a: Biodiversity in New Zealand plantation. *New Zealand Forestry,* **February,** 26–29.

Allen, R.K., K.H. Platt, and R.E.J. Coker, 1995b: Understory species composition patterns in a *Pinus radiata* plantation on the central North Island volcanic plateau, New Zealand. *New Zealand Journal of Forestry Science,* **25,** 301–317.

Andrasko, K. 1997: Forest management for greenhouse gas benefits: Resolving monitoring issues across project and national boundaries. *Mitigation and Adaptation Strategies for Global Change,* **2,** 117–132.

Andrasko, K., L. Carter, and W. van der Gaast, 1996: *Technical Issues in JI/AIJ Projects: A Survey and Potential Responses.* Prepared for United Nations Environmental Programme Activities Implemented Jointly Conference: New Partnerships to Reduce the Buildup of Greenhouse Gases, San Jose, Costa Rica, pp. 13–48.

Austin, D. and Faeth, P. (eds.), 2000: *Opportunities for financing sustainable development via the Clean Development Mechanism.* World Resources Institute, Washington, DC, USA. 113 pp. (in press).

Bass, S., J. Ford, O. Dubois, P. Moura-Costa, C. Wilson, M. Pinard, and R. Tipper, 2000: *Rural Livelihoods and Carbon Management: An Issues Paper.* International Institute for Environment and Development, London, United Kingdom, (in press).

Baumert, K.A., 1999: The Clean Development Mechanism: understanding additionality. In: *Promoting Development while Limiting Greenhouse Gas Emissions: Trends and Baselines.* United Nations Development Programme (UNDP) and World Resources Institute, United Nations Publications, New York, New York, USA, pp. pp. 135–145.

Bawa, K.S. and R. Seidler, 1998: Natural forest management and conservation of biodiversity in tropical forests. *Conservation Biology,* **12,** 46–55.

Boscolo, M., J. Buongiorno, and T. Panayotou, 1997: *Environment and Development Economics,* **2,** 241–263.

Brown, P., 1998: *Climate, Biodiversity and Forests.* World Resources Institute, Washington, DC, USA, 35 pp.

Brown, P., B. Cabarle, and R. Livernash, 1997: *Carbon Counts: Estimating Climate Change Mitigation in Forestry Projects.* World Resources Institute, Washington, DC, USA, 25 pp.

Brown, S., 1997: *Estimating Biomass and Biomass Change of Tropical Forests: a Primer.* FAO Forestry Paper 134, Food and Agriculture Organization of the United Nations, Rome, Italy, 55 pp.

Brown, S., J. Sathaye, M. Cannell, and P. Kauppi, 1996: Management of forests for mitigation of greenhouse gas emissions. Chapter 24. In: *Climate Change 1995: Impacts, Adaptations and Mitigation of Climate Change: Scientific-Technical Analyses. Contribution of Working Group II to the Second Assessment Report of the Intergovernmental Panel on Climate Change* [Watson, R.T., M.C. Zinyowera, and R.H. Moss (eds.)]. Cambridge University Press, Cambridge, United Kingdom and New York, NY, USA, pp. 773–797.

Brown, S., M. Burnham, M. Delany, R. Vaca, M. Powell, and A. Moreno, 2000: Issues and challenges for forest-based carbon-offset Projects: A case study of the Noel Kempff Climate Action Project in Bolivia. *Mitigation and Adaptation Strategies for Global Change,* (in press).

Brown, S., M. Calmon, and M. Delaney, 1999a: *Development of a Deforestation and Forest Degradation Trend Model for the Guaraquecaba Climate Action Project.* Winrock International, Carbon Monitoring Program, Arlington, VA, USA, 14 pp.

Brown, S., M. Calmon, and M. Delaney, 1999b: *Carbon Inventory and Monitoring Plan for the Guaraqueçaba Climate Action Project.* Winrock International, Carbon Monitoring Program, Arlington, VA, USA, 26 pp.

Busch, C., J. Sathaye, and G. Sanchez-Azofeifa, 1999: *Lessons for Greenhouse Gas Accounting: A Case Study of Costa Rica's Protected Areas Project.* Report LBNL-42289, Lawrence Berkeley National Laboratory, Berkeley, CA, USA, 102 pp.

Cairns, M.A., S. Brown, E.H. Helmer, and G.A. Baumgardner, 1997: Root biomass allocation in the world's upland forests. *Oecologia,* **111,** 1–11.

Carbon Storage Trust, 1998: *The Carbon Storage Trust and Climate Care—a Detailed Analysis.* The Carbon Storage Trust, Oxford, United Kingdom, 9 pp.

Carpentieri, A., E. Larson, and J. Woods, 1993: Future biomass based electricity supply in Brazil. *Biomass and Bioenergy,* **4,** 149–176.

Carter, L., 1997a: *Additionality: The USIJI Experience.* Paper presented at the workshop on environmental benefits of Activities Implemented Jointly, 9–10 September 1997, International Energy Agency, Paris, France, 35 pp.

Carter, L., 1997b: Modalities for the operationalization of additionality. In: *Proceedings of the Activities Implemented Jointly Workshop, March 1997* [United Nations Environmental Programme and German Federal Ministry of Environment (eds.)]. German Federal Ministry of Environment, Leipzig, Germany, pp. 79–92.

Center for Clean Air Policy, 1998: *Top-down Baselines to Simplify Setting of Project Emission Baselines for JI and the CDM.* Washington, DC, USA, 6 pp.

Chambers, R., N.C. Saxena, and T. Shah, 1989: *To the Hands of the Poor: Water and Trees.* Oxford University Press and IBH, New Delhi and Intermediate Technology, London, United Kingdom, 273 pp.

Chomitz, K. 2000: *Evaluating Carbon Offsets from Forestry and Energy Projects: How Do They Compare?* Development Research Group, The World Bank, Washington, DC, USA, 25 pp.

Chomitz, K., 1998: *Baselines for Greenhouse Gas Reductions: Problems, Precedents, Solutions.* Carbon Offsets Unit, Development Research Group, World Bank, Washington, DC, USA, 62 pp.

Chomitz, K.M. and K. Kumari, 1998: The domestic benefits of tropical forests: A critical review. *The World Bank Research Observer,* **13(1),** 13–35.

Chomitz, K.M., E. Brenes, and L. Constantino, 1999: Financing environmental services: The Costa Rican Experience and its implications. *The Science of the Total Environment,* **240,** 157–169.

Cooper, P.J.M., R.R.B. Leakey, M.R. Rao, and I. Reynolds, 1996: Agroforestry and the mitigation of land degradation in the humid and sub-humid tropics of Africa. *Experimental Agriculture,* **32,** 235–290.

Cullet, P. and A.P. Kameri-Mbote, 1998: Joint implementation and forestry projects: conceptual and operational fallacies. *International Affairs,* **74(2),** 393–408.

Da Silva, Jr., F., S. Rubio, and F. de Souza, 1995: Regeneration of an Atlantic forest formation in the understory of a *Eucalyptus grandis* plantation in south-eastern Brazil. *Journal of Tropical Ecology,* **11,** 147–152.

de Jong, B.H., R. Tipper, and J. Taylor, 1997: A framework for Monitoring and evaluation of carbon mitigation by farm forestry projects: example of a demonstration project in Chiapas, Mexico. *Mitigation and Adaptation Strategies for Global Change,* **2,** 231–246.

Delaney, M., S. Brown, A.E. Lugo, A. Torres-Lezama, and N. Bello Quintero, 1998: The quantity and turnover of dead wood in permanent forest plots in six life zones of Venezuela. *Biotropica,* **30,** 2–11.

Department of Forestry and Conservation Management, University of Massachusetts, 1999: *Final Report and Results on Assessing Dual Camera Videography and 3D Terrain Reconstruction as Tools to Estimate Carbon Sequestration in Forests, Crooksville, Ohio Test Site.* Report to Winrock International, Morrilton, AR, USA, 20 pp. + appendices.

Detwiler, R.P., 1986: Land use change and the global carbon cycle: the role of tropical soils. *Biogeochemistry,* **2,** 67–93.

Dinerstein, E., D.M. Olson, D.J. Graham, A.L. Webster, S.A. Primm, M.P. Bookbinder, and G. Ledec, 1995: *A Conservation Assessment of the Terrestrial Ecoregions of Latin America and the Caribbean.* The World Wildlife Fund, The World Bank, Washington, DC, USA, 129 pp.

Dixon, R.K., J.K. Winjum, K.J. Andrasko, and P.E. Schroeder, 1994: Integrated land-use systems: assessment of promising agroforest and alternative land-use practices to enhance carbon conservation and sequestration. *Climate Change,* **30,** 1–23.

Dixon, R.K., K.J. Andrasko, F.G. Sussman, M.A. Lavinson, M.C. Trexler, and T.S. Vinson, 1993: Forest sector carbon offset projects: near-term opportunities to mitigate greenhouse gas emissions. *Water, Air and Soil Pollution,* **70,** 561–577.

Dixon, R.K., P.E. Schroeder, and J. Winjum (eds), 1991: *Assessment of Promising Forest Management Practices and Technologies for Enhancing the Conservation and Sequestration of Atmospheric Carbon and Their Costs at the Site Level.* Report of the US Environmental Protection Agency No. EPA/600/3-91/067. Environmental Research Laboratory, Corvallis, OR, USA, 138 pp.

Dobes, L., I. Enting, and C. Mitchell, 1998: Accounting for carbon sinks: the problem of time. In *Trading Greenhouse Emissions: Some Australian Perspectives* [Dobes, L. (ed.)]. Occasional papers No. 115, Bureau of Transport Economics, Canberra, Australia, pp. 1–15.

EcoSecurities Ltd., 1997: *SGS Forestry Carbon Offset Verification Services.* Draft paper. SGS Forestry, Oxford, United Kingdom, 45 pp.

Ellis, J. and Bosi, M., 1999: *Options for Project Emission Baselines.* Organization for Economic and Commerical Development and International Energy Information Paper. Organization for Commercial and Economic Development, Paris, France, 60 pp.

Environmental Law Institute, 1996: *Incorporating Public Participation in Joint Implementation of the Framework Convention on Climate Change,* Washington, DC, USA, 58 pp.

Environmental Protection Agency and US Initiative on Joint Implementation (EPA/USIJI), 1998: *Activities Implemented Jointly: Third Report to the Secretariat of the United Nations Framework Convention on Climate Change.* 2 volumes. EPA report 236-R-98-004. US Environmental Protection Agency, Washington, DC, USA, 19 (vol. I) and 607 (vol. II) pp.

FACE, 1998: *Annual Report 1997.* FACE Foundation, Arnhem, Netherlands. 28 pp.

Faeth, P., C. Cort, and R. Livernash, 1994: *Evaluating the Carbon Sequestration Benefits of Forestry Projects in Developing Countries.* World Resource Institute, Washington, DC, USA, 96 pp.

Fearnside, P., 1997: Monitoring needs to transform Amazonian forest maintenance into a global warming-mitigation option. *Mitigation and Adaptation Strategies for Global Change,* **2,** 285–302.

Fearnside, P.M., 1995: Global warming response options in Brazil's forest sector: Comparison of project-level costs and benefits. *Biomass and Bioenergy,* **8(5),** 309–322.

Fearnside, P.M., D.A. Lashof, and P. Moura-Costa, 2000: Accounting for time in mitigating global warming. *Mitigation and Adaptation Strategies for Global Change,* (in press).

Friedman, S., 1999: The use of benchmarks to determine emissions additionality in the Clean Development Mechanism. In: *Proceeding of the Global Industrial and Social Progress Research Institute (GISPRI) Baseline Workshop, 25–26 February.* GISPRI, Tokyo, Japan, pp. 157–165.

Frumhoff, P.C. and E.C. Losos, 1998: *Setting Priorities for Conserving Biological Diversity in Tropical Timber Production Forests.* Union of Concerned Scientists, Cambridge, MA, 14 pp.

Frumhoff, P.C., D.C. Goetze, and J.J. Hardner, 1998: *Linking Solutions to Climate Change and Biodiversity Loss Through the Kyoto Protocol's Clean Development Mechanism.* Union of Concerned Scientists, Cambridge, MA, 14 pp.

German Advisory Council on Global Change (WBGU), 1998: *The Accounting of Biological Sinks and Sources Under the Kyoto Protocol—A Step Forwards or Backwards for Global Environmental Protection?* WBGU, Bremerhaven, Germany, 75 pp.

Giampietro, M., S. Ulgiati, and D. Pimentel, 1997: Feasibility of large-scale biofuel production. *BioScience,* **47(9),** 587–600.

Greenpeace, 1998: *Making the Clean Development Mechanism Clean and Green.* Greenpeace Position Paper, Fourth Conference of the Parties to the United Nations Framework Convention on Climate Change, Amsterdam, Netherlands, Greenpeace International, 12 pp.

Goldberg, D.M., 1998: *Carbon Conservation Climate Change, Forests and the Clean Development Mechanism.* Center for International Environmental Law, Washington, DC, USA, 55 pp.

Greenhouse Challenge Office, 1997: *Greenhouse Challenge Carbon Sinks Workbook: A Discussion Paper.* Greenhouse Challenge Office, Canberra, Australia, 93 pp.

Gustavsson, L., T. Karjalainen, G. Marland, B. Savolainen, B. Schlamadinger, and M. Apps, 1999: *Accounting System Considerations: CO_2 Emissions from Forests, Forest Products, and Land-Use Change.* Available at: http://www.joanneum.ac.at/iea-bioenergy-task25/publication/fpubl.htm.

Hamburg, S.P., 2000: Simple rules for measuring changes in ecosystem carbon in forestry-offset projects. *Mitigation and Adaptation Strategies for Climate Change,* (in press).

Haque, A.K.E., P. Read, and M.E. Ali, 1999: *The Bangladesh MSP Pilot Project Proposal for GEF Funding of Capacity building for Country Driven Projects.* Working Paper, Institute of Development, Environment, and Strategic Studies (IDESS), North-South University, Dhaka, Bangladesh, 15 pp.

Hardner, J., 1996: *Forest Conservation and Watershed Protection in Ilheus, Bahia: An Avoided Cost Approach.* Prepared for Conservation International and Instituto de Estudos Socio-Ambientais do Sul da Bahia, Washington, DC, USA, 30 pp.

Hardner, J.J., P.C. Frumhoff, and D.C. Goetze, 2000: Prospects for mitigating carbon, conserving biodiversity, and promoting socioeconomic development objectives through the Clean Development Mechanism. *Mitigation and Adaptation Strategies for Global Change,* (in press).

Hargrave, T., N. Helme, and I. Puhl, 1998: *Options for Simplifying Baseline Setting for Joint Implementation and Clean Development Mechanism Projects.* Center for Clean Air Policy, Washington, DC, USA, 11 pp.

Houghton, J.T., L.G. Meira Filho, B. Lim, K. Tréanton, I. Mamaty, Y. Bonduki, D.J. Griggs, and B.A. Callander (eds.), 1997: *Revised 1996 Guidelines for National Greenhouse Gas Inventories: Volume 3: Greenhouse Gas Inventory Reference Manual.* Intergovernmental Panel on Climate Change, Meterological Office, Bracknell, United Kingdom, 482 pp.

Imaz, M., C. Gay, R. Friedmann, and B. Goldberg, 1998: *Mexico Joins the Venture: Joint Implementation and Greenhouse Gas Emissions Reduction.* Research Paper LBNL-42000, Berkeley National Laboratory, Berkeley, CA, USA, 36 pp.

Intarapravich, D., 1995: *Joint Implementation: Thailand Environment Institute's Perspective.* Paper presented at Southeast Asian Regional Workshop on International Prospects for Joint Implementation, Bangkok, Thailand, pp. 2–6.

International Centre for Research in Agroforestry (ICRAF), (n.d.): *Soil Changes Under Agroforestry (SCUAF).* Gigiri, Nairobi, Kenya, n.d.

IPCC, 1992: *Climate Change 1992: The Supplementary Report to the IPCC Scientific Assessment.* Prepared by IPCC Working Group I [Houghton, J.T., B.A. Callander, and S.K. Varney (eds.)]. and World Meteorological Association/United Nations Environmental Programme. Cambridge University Press, Cambridge, United Kingdom and New York, NY, USA, 200 pp.

IPCC, 1996: *Climate Change 1995: The Science of Climate Change. Contribution of Working Group I to the Second Assessment Report of the Intergovernmental Panel on Climate Change* [Houghton, J.T., L.G. Meira Filho, B.A. Callander, N. Harris, A. Kattenberg, and K. Maskell (eds.)]. Cambridge University Press, Cambridge, United Kingdom and New York, NY, USA, 572 pp.

Janssen, J., 1997: Strategies for risk management of joint implementation investments. In: *Greenhouse Gas Mitigation. Technologies for Activities Implemented Jointly* [Riermer, P.W.F., A.Y. Smith, and K.V. Thambimuthu (eds.)]. Proceedings of Technologies for the Activities Implemented Jointly Conference, Vancouver, Canada, May 1997. Elsevier Science Publishers, Oxford, United Kingdom, pp. 357–365.

Jepma, C., 1995: *Tropical Deforestation: a Socio-Economic Approach.* Earthscan, London, United Kingdom, 380 pp.

Jepma, C., 1999: Determining a baseline for project cooperation under the Kyoto Protocol: a general overview. In: n: *Proceeding of the Global Industrial and Social Progress Research Institute (GISPRI) Baseline Workshop, 25–26 February.* GISPRI, Tokyo, Japan, pp. 191–199.

Johns, A.G., 1997: *Timber Production and Biodiversity Conservation in Tropical Rain Forests.* Cambridge University Press, Cambridge, United Kingdom. 225 pp.

Kanowski, P.J., P.S. Savill, P.G. Adlard, J. Burley, J. Evans, J.R. Palmer, and P.J. Wood, 1992: Plantation forestry. In: Sharma, N.P. (ed.), *Managing the World's Forests.* Kendall-Hunt, Dubuque, Iowa, USA, pp. 375–401.

Keenan, R., D. Lamb, O. Woldring, T. Irvine, and R. Jensen, 1997: Restoration of plant biodiversity beneath tropical tree plantations in Northern Australia. *Forest Ecology and Management,* **99,** 117–131.

Klooster, D. and O.R. Masera, 2000. Community forest management in Mexico: Making carbon sequestration a by-product of sustainable rural development. *Global Environmental Change,* (in press).

Kurz, W.A., S.J. Beukema, and M.J. Apps, 1997: Carbon budget implications of the transition from natural to managed disturbance regimes in forest landscapes. *Mitigation and Adaptation Strategies for Global Change,* **2,** 405–421.

La Rovere, E.L., 1998: *The Challenge of Limiting Greenhouse Gas Emissions Through Activities Implemented Jointly in Developing Countries: A Brazilian Perspective.* Research Paper LBNL-41998, Berkeley National Laboratory, Berkeley, CA, USA, 43 pp.

Laquihon, W.A., 1989: Some key determinants of SALT adoption in the Philippines: viewpoints of farmer cooperators. In: *Social Forestry in Asia* [Vergara, N.T. and R.A. Fernandez (eds.)]. SEARCA, Los Baños, Philippines, pp. 79–116.

Lasco, R.D. and F.B. Pulhin, 2000: Forest land use change in the Philippines and climate change mitigation. *Mitigation and Adaptation Strategies for Global Change,* (in press).

Lashof, D. and B. Hare, 1999: The role of biotic carbon stocks in stabilizing greenhouse gas concentrations at safe levels. *Environmental Science and Policy,* **2(2),** 101–110.

Lee, R., J.R. Kahn, G. Marland, M. Russell, and K. Shallcross, 1997: *Understanding Concerns About Joint Implementation.* Joint Institute for Energy and Environment, Knoxville, TN, USA, 96 pp.

Lemaster, L., 1995: The relationship between environmental barriers and modes of technology transfer: A study of United States companies with operations in Mexico. *Journal of International Business Studies,* **26(3),** 690–691.

Losos, E., 2000: Can forestry carbon offset projects play a significant role in conserving forest wildlife and their habitats? In: *Conserving Wildlife in Managed Tropical Forests* [Fimbel, R., A. Grajal, and J. Robinson (eds.)]. Columbia University Press, New York, NY, USA, (in press).

Ludeke, A.K., 1990: An analysis of anthropogenic deforestation using logistic regression and GIS. *Journal of Environmental Management,* **31,** 247–259.

Lugo, A., 1997: The apparent paradox of reestablishing species richness on degraded lands with tree monocultures. *Forest Ecology and Management,* **99,** 9–19.

MacDicken, K., 1997a: *A Guide to Monitoring Carbon Storage in Forestry and Agroforestry Projects*. Winrock International, Arlington, VA, USA, 87 pp.

MacDicken, K., 1997b: Project specific monitoring and verification: state of the art and challenges. *Mitigation and Adaptation Strategies for Global Change,* **2,** 191–202.

MacDicken, K.G. and N.T. Vergara, 1990: Introduction to agroforestry. In: *Agroforestry: Classification and Management* [MacDicken, K.G. and N.T. Vergara (eds.)]. John Wiley and Sons, New York, NY, USA, pp. 1–30.

Maclaren, J.P., 1996: Plantation forestry—its role as a carbon sink: conclusions from calculations based on New Zealand's planted forest estate. In: *Forest Ecosystems, Forest Management, and the Global Carbon Cycle* [Apps, M.J. and D.T. Price (eds.)]. Springer-Verlag, Berlin, Germany, pp. 257–270.

Maclaren, P., 1999: *Carbon Accounting Methodologies—a Comparison of Real-time, Tonne-years, and One-off Stock Change Approaches*. Unpublished manuscript, Christchurch, New Zealand, 12 pp.

Makundi, W. and Okiting'ati, A., 1995: Carbon flows and economic evaluation of mitigation options in Tanzania's forest sector. *Biomass and Bioenergy,* **8(5),** 381–393.

Makundi, W.P., 1997: Global climate change mitigation and sustainable forest management—the challenge of monitoring and verification. *Mitigation and Adaptation Strategies for Global Change,* **2,** 133–155.

Marland, G., B. Schlamadinger, and L. Canella, 1997: Forest management for mitigation of CO_2 emissions: How much mitigation and who gets the credits? *Mitigation and Adaptation Strategies for Global Change,* **2,** 303–318.

Manne, A. and R. Richels, 1999. The Kyoto Protocol: a cost–effective strategy for meeting environmental objectives? In: The costs of the Kyoto Protocol: A multi-model evaluation. *Energy Journal,* **Special Issue,** 1–24.

Masera, O.R., M. Bellon, and G. Segura, 1995: Forest management options for sequestering carbon in Mexico. *Biomass and Bioenergy,* **8(5),** 357–368.

Masera, O.R., 1995: Carbon mitigation scenarios for Mexican forests: Methodological considerations and results. *Interciencia,* **20,** 388–395.

Matsuo, N., 1999: Baselines as the critical issue of CDM—possible pathways to standardization. In: *Proceeding of the Global Industrial and Social Progress Research Institute (GISPRI) Baseline Workshop, 25–26 February*. GISPRI, Tokyo, Japan, pp. 9–21.

Maya, R.S. and J. Gupta (eds.), 1996: *Joint Implementation: Carbon Colonies or Business Opportunities?* Southern Centre for Energy and Environment, Harare, Zimbabwe, 164 pp.

McPherson, G., 1994: *Chicago's Urban Forest Ecosystem: Results of the Chicago Urban Forest Climate Project*. General Technical Report NE-186, U.S. Department of Agriculture Forest Service, Northeastern Forest Experiment Station, Radnor, PA, USA, 46 pp.

Michaelowa, A. and H. Schimdt, 1997: A dynamic crediting regime for joint implementation to foster innovation in the long term. *Mitigation and Adaptation Strategies for Global Change,* **2,** 45–56.

Michaelowa, A., 1998: Joint Implementation—the baseline issue. *Global Environmental Change,* **8,** 81–92.

Michaelowa, A., 1999: *Baseline Methodologies for the CDM—Which Road to Take*. Paper presented at the Institute for Global and Environmental Strategies (IGES) meeting, 23 June 1999, Tokyo, Japan, 12 pp.

Ministerio del Ambiente y Energia (MINAE), 1996: Sistema Nacional de Areas de Conservacion, Situación de Tenencia de la Tierra en las Areas Silvestres Protegidas del Pais, Proyectos Gruas. San Jose, Costa Rica, 35 pp.

Ministry of Housing, Spatial Planning, and the Environment, Directorate General for Environmental Protection, Climate Change Department, 1996: *Activities Implemented Jointly: The Netherlands Pilot Phase Programme*. The Hague, Netherlands.

Mohren, G.M.J., J.F. Garza Caligaris, O. Masera, M. Kanninen, T. Karjalainen, and G.J. Nabuurs, 1999: *CO_2FIX for Windows: A Dynamic Model of the CO_2 Fixation in Forest Stands*. Institute for Forestry and Nature Research, Netherlands; Instituto de Ecologia, National University of Mexico (UNAM), Mexico; Centro Agronomico Tropicalde Investigacion y Enseñanza, Costa Rica; and European Forest Institute, Finland, 27 pp.

Moura-Costa, P.H., 1996a: Tropical forestry practices for carbon sequestration: A review and case study from Southeast Asia. *Ambio,* **25,** 279–283.

Moura-Costa, P.H., 1996b: Tropical forestry practices for carbon sequestration. In: *Dipterocarp Forest Ecosystems—Towards Sustainable Management* [Schulte, A. and D. Schone (eds.)]. World Scientific, Singapore, pp. 308–334.

Moura-Costa, P.H., 1993: Large scale enrichment planting with dipterocarps, methods and preliminary results. In: *Proceedings of the Yogyakarta Workshop, BIO-REFOR/IUFRO/SPDC: Bio-reforestation in Asia-Pacific Region* [Suzuki, K., S. Sakurai, and K. Ishii (eds.)]. pp. 72–77.

Moura-Costa, P. and M. Stuart, 1998: Forestry based greenhouse gas mitigation: a story of market evolution. *Commonwealth Forestry Review,* **77,** 191–202.

Moura-Costa, P., M. Stuart, M. Pinard, and G. Phillips, 2000: Issues related to monitoring, verification and certification of forestry-based carbon offset projects. *Mitigation and Adaptation Strategies for Global Change*, (in press).

Moura-Costa, P.H. and C. Wilson, 2000: An equivalence factor between CO_2 avoided emissions and sequestration—description and applications in forestry. *Mitigation and Adaptation Strategies for Global Change*, (in press).

Moura-Costa, P.H., S.W. Yap, C.L. Ong, A. Ganing, R. Nussbaum, and T. Mojiun, 1996: Large scale enrichment planting with dipterocarps as an alternative for carbon offset—methods and preliminary results. In: *Proceedings of the 5th Round Table Conference on Dipterocarps. Chiang Mai, Thailand, November 1994* [Appanah, S. and K.C. Khoo (eds.)]. Forest Research Institute of Malaysia (FRIM), Kepong, Thailand. pp. 386–396.

Mulongoy, K., J. Smith, P. Alirol, and A. Witthoeft-Muehlmann, 1998: *Are Joint Implementation and the Clean Development Mechanism Opportunities for Forest Sustainable Management Through Carbon Sequestration Projects?* Academy of the Environment, Background paper 1, Climate Change in the Global Economy Programme, Geneva, Switzerland, 36 pp.

Nabuurs, G.J., A. Dolman, E.Verkaik, P. Kuikman, C. van Diepen, A. Whitmore, W. Daamen, O. Oenema, P. Kabat, and G. Mohren, 2000. Article 3.3 and 3.4 of the Kyoto Protocol: consequences for industrialised countries' commitment, the monitoring needs, and possible side effects. *Environmental Science and Policy*, (in press).

Nair, P.K.R., 1989: The role of trees in soil productivity and protection. In: *Agroforestry Systems in the Tropics* [Nair, P.K.R. (ed.)]. Kluwer Academic Publishers, Dordrecht, Netherlands, pp. 567–589.

Niles, J. and R. Schwarze, 2000: Long-term forest sector emission reductions under the Kyoto Protocol's Article 12. In: *International Energy Agency (IEA) Bioenergy Task 25, Proceedings of the Workshop on Bioenergy for Mitigation of CO_2 Emissions: the Power, Transportation, and Industrial Sectors, 27–30 September 1999* [Schlamadinger, B. and K. Robertson (eds.)]. Medienfabrik Graz, Graz, Austria, pp. 145–153.

Olson, D.M. and E. Dinerstein, 1998: *The Global 200: A Representation Approach to Conserving the Earth's Distinctive Ecoregions*. World Wildlife Fund, Washington, DC, USA, 152 pp.

Padbury, G., 1999. *Soil Carbon Initiatives on Prairies*. Available at: http://res.agr.ca/clm/padbury.htm.

Parrotta, J., J. Turnbull, and N. Jones., 1997a: Catalyzing native forest regeneration on degraded tropical lands. *Forest Ecology and Management,* **99,** 1–7.

Parrotta, J., O. Knowles and J. Wunderle, Jr., 1997b: Development of floristic diversity in 10-year-old restoration forests on a bauxite mined site in Amazonia. *Forest Ecology and Management,* **99,** 21–42.

Paustian, K., E. Levine, W.M. Post, and I.R. Ryzhova, 1997: The use of models to integrate information and understanding of soil carbon at the regional scale. *Geoderma,* **79,** 227–260.

Phillips, D.L., S.L. Brown, P.E. Schroeder and R.A. Birdsey, 2000: Toward error analysis of large-scale forest carbon budgets. *Global Ecology and Bioegeography,* (in press).

Pinard, M. and F. Putz, 1997: Monitoring carbon sequestration benefits associated with reduced-impact logging project in Malaysia. *Mitigation and Adaptation Strategies for Global Change,* **2,** 203–215.

Pinard, M.A. and F.E. Putz, 1996: Retaining forest biomass by reduced impact logging damage. *Biotropica,* **28,** 278–295.

Post, W.M., R.C. Izaurralde, L.K. Mann, and N. Bliss, 1999: Monitoring and verification of soil organic carbon sequestration. In: *Symposium—Carbon Sequestration in Soils: Science, Monitoring, and Beyond, 3–5 December 1999* [Rosenberg, N.J., R.C. Izaurralde, and E.L. Malone (eds.)]. Batelle Press, Columbus, OH, USA, pp. 41–66.

Powell, M.H., 1999: *Effects of Inventory Precision and Variance on the Estimated Number of Sample Plots and Inventory Variable Cost: the Noel Kempff Climate Action Project.* Winrock International, Morrilton, AR, USA, 6 pp.

Programme for Belize, 1997a: *Rio Bravo Carbon Sequestration Pilot Project. Offsets Attributable to Project Actions for Project Year 2 (1996).* Report to the US Initiative on Joint Implementation and the Government of Belize. Programme for Belize, Belize City, Belize, 15 pp.

Programme for Belize, 1997b: *Rio Bravo Carbon Sequestration Pilot Project: Operational Protocol 3.* Report to the US Initiative on Joint Implementation and the Government of Belize. Programme for Belize, Belize City, Belize, 33 pp.

Puhl, I., 1998: *Status of Research on Project Baselines Under the UNFCCC and the Kyoto Protocol.* Organization for Economic and Commercial Development and International Energy Agency Information Paper. Organization for Economic and Commercial Development, Paris, France, 15 pp.

Putz, F.E. and M.A. Pinard, 1993: Reduced impact logging as a carbon-offset project. *Conservation Biology,* **7(4),** 755–757.

Ravindranath, N. and D. Hall, 1995: *Biomass, Energy, and Environment, A Developing Country Perspective from India.* Oxford University Press, Oxford, United Kingdom, 376 pp.

Ravindranath, N.H., N. Byron, R. Dixon, P. Fearnside, K. MacDicken, W. Makundi, O. Masera, A. DiNicola, and N. Mongia, 2000: *Intergovernmental Panel on Climate Change: Special Report on Technology Transfer: Technology Transfer in the Forestry Sector.* Cambridge University Press, Cambridge, United Kingdom and New York, NY, USA, (in press).

Ravindranath, N. and Somashekhar, B., 1995: Potential and economics of forestry options for carbon sequestration in India. *Biomass and Bioenergy,* **8(5),** 323–336.

Ravindranath, N.H. and P.R. Bhat, 1997: Monitoring of carbon abatement in forestry projects-case study of Western Ghat Project. *Mitigation and Adaptation Strategies for Global Change,* **2,** 217–230.

Ravindranath, N.H., A. Meili, and R. Anita, 1998: *AIJ in the Non-Energy Sector in India: Opportunities and Concerns.* Research Paper LBNL-41999, Berkeley National Laboratory, Berkeley, CA, USA, 39 pp.

Read, P., 1999: Cooperative implementation after Kyoto: Joint Implementation and the need for commercialized offsets trading. In: *On the Compatibility of Flexible Instruments* [Jepma, C. and W. van der Gaast (eds.)]. Kluwer Academic Press, Dordrecht, The Netherlands, pp. 151–163.

Reid, W.V., 2000: Capturing the value of ecosystem services to protect biodiversity. In: *Managing Human-Dominated Ecosystems.* Island Press, Washington, DC and Missouri Botanical Gardens, St. Louis, MO, USA, (in press).

Reppeto, R. and M. Gillis, 1988. *Public Policies and the Misuse of Forest Resources.* Cambridge University Press, Cambridge, United Kingdom and New York, NY, USA, 432 pp.

Repollo, A.Q., and E.R. Castillo, 1989: Agroforestry technology in hillyland households: factors affecting its adoption. In: *Social Forestry in Asia* [Vergara, N.T. and R.A. Fernandez (eds.)]. Southeast Asian Ministers of Education Organization (SEAMEO), Regional Center for Graduate Study in Research in Agriculture (SEARCA), Los Baños, Philippines, pp. 117–132.

Richards, K.R. and C. Stokes, 1994: *Regional Studies of Carbon Sequestration: A Review and Critique.* Paper written for the US Department of Energy, Washington, DC, USA, Contract DE-AC06-76RLO 1830. 40 pp.

Sathaye, J., K. Andrasko, W. Makundi, E.L. La Rovere, N.H. Ravindranath, A. Melli, A. Rangachari, M. Imaz, C. Gay, R. Friedmann, B. Goldberg, C. Van Horen, G. Simmonds, and G. Parker, 1999: Concerns about climate change mitigation projects; summary of findings from case studies in Brazil, India, Mexico and South Africa. *Environmental Science and Policy,* **2(2),** 187–198.

Saxena, N.C., 1997: *The Saga of Participatory Forest Management in India.* Center for International Forestry Research Special Publication, Bogor, Indonesia, 214 pp.

Schlamadinger, B. and G. Marland, 1996: Carbon implications of forest management strategies. In: *Forest Ecosystems, Forest Management, and the Global Carbon Cycle* [Apps, M.J. and D.T. Price (eds.)]. Springer-Verlag, Berlin, Germany, pp. 217–229..

Schroeder, P., 1992: Carbon storage potential of short rotation tropical tree plantations. *Forest Ecology and Management,* **50,** 31–41.

Shapiro, A.C., 1996: *Multinational Financial Management, 5th edition.* Prentice Hall, Upper Saddle River, NJ, USA, 200 pp.

Smith, J., K. Mulongoy, R. Persson and J. Sayer, 1999: *Harnessing Carbon Markets for Tropical Forest Conservation: Towards a More Realistic Assessment.* Center for International Forestry Research, Bogor, Indonesia, and International Academy of the Environment, Geneva, Switzerland, 27 pp.

Société Générale de Surveillance (SGS), 1998: *Final Report of the Assessment of project design and schedule of emission reduction units for the Protected Areas Project of the Costa Rican Office for Joint Implementation.* SGS, Oxford, United Kingdom, 133 pp.

Sokona, Y., S. Humphreys, and J.-P. Thomas, 1999: *The Clean Development Mechanism: What Prospects for Africa?* Environment et Dévelopement du Tiers Monde (ENDA-TM), Dakar, Senegal, 25 pp.

Stewart, R., D. Anderson, M.A. Aslam, C. Eyre, G. Jones, P. Sands, M. Stuart, and F. Yamin, 1999: *The Clean Development Mechanism: Building International Public-Private Partnerships. A Preliminary Examination of Technical, Financial & Institutional Issues.* United Nations Conference on Trade and Development, Geneva, Switzerland, 95 pp.

Stuart, M.D. and P.H. Moura-Costa, 1998: Greenhouse gas mitigation: A review of international policies and initiatives. In: *Policies that Work for People, Series No. 8.* International Institute of Environment and Development, London, United Kingdom, 27–32 pp.

Stuart, M.D., 1998: Financial tools and the Clean Development Mechanism. In: *Design and Implementation of the Clean Development Mechanism: a Concept Paper for the International Working Group on the Clean Development Mechanism,* United Nations Conference on Trade and Development, Geneva, Switzerland.

Subak, S., 2000: Costa Rica's Private Forestry Project: evaluation of a Clean Development Mechanism prototype. *Environmental Management,* (in press).

Swedish International Development Agency (SIDA), 1992: *An Evaluation of the SIDA Supported Social Forestry Project in Tamil Nadu and Orissa.* Swedish International Development Agency. New Delhi, India, 86 pp.

Swisher, J. and Masters, G., 1992. A mechanism to reconcile equity and efficiency in global climate protection: international carbon emissions offsets. *Ambio,* **21(2),** 154–159.

Swisher, J.N. 1997: Joint implementation under the UN Framework Convention on Climate Change: Technical and institutional challenges. *Mitigation and Adaptation Strategies for Global Change,* **2,** 57–80.

Tattenbach, F., 1996: Certifiable, tradeable offsets in Costa Rica. *Joint Implementation Quarterly,* **2(2),** 2.

The Nature Conservancy (TNC), 1999: *Written Information for and Comments on IPCC LUCF.* Special Report draft 1 for Expert Review, August 1999.

Tietenberg, T., M. Grubb, A. Michaelowa, B. Swift, and X.Z. Zhong, 1998: *International Rules for Greenhouse Gas Emissions Trading. Defining the Principles, Modalities, Rules and Guidelines for Verification, Reporting and Accountability.* United Nations Conference on Trade and Development, Geneva, Switzerland, 93 pp.

Tipper, R. and B.H. de Jong, 1998: Quantification and regulation of carbon offsets from forestry: comparison of alternative methodologies, with special reference to Chiapas, Mexico. *Commonwealth Forestry Review,* **77,** 219–228.

Tipper, R., B.H. de Jong, S. Ochoa-Gaona, M.L Soto-Pinto, M.A. Castillo-Santiago, G. Montoya-Gómez, and I. March-Mifsut, 1998: *Assessment of the Cost of Large Scale Forestry for CO_2 Sequestration: Evidence from Chiapas, Mexico.* Report PH12, International Energy Agency Greenhouse Gas R&D Programme, Cheltenham, Gloucester, United Kingdom, 87 pp.

Trexler and Associates, Inc., 1998: *Final Report of the Biotic Offsets Assessment Workshop, Baltimore, Maryland Sept. 5–7 1997.* Prepared for US Environmental Protection Agency, Washington, DC, USA, 107 pp.

Trexler, M., L. Kosloff, and R. Gibbons, 1999: Forestry and land-use change in the AIJ pilot phase: The evolution of issues and methods to address them. In: *The UN Framework Convention on Climate Change Activities Implemented Jointly Pilot: Experiences and Lessons Learned* [Dixon, R. (ed.)]. Kluwer Academic Publishers, Dordrecht, Netherlands, pp. 121–165.

Trexler, M., P. Faeth, and J. Kramer, 1989: *Forestry as a Response to Global Warming: An Analysis of the Guatemala Agroforestry and Carbon Sequestration Project.* World Resources Institute, Washington, DC, USA, 66 pp.

Trexler, M.C. and L.H. Kosloff, 1998: The 1997 Kyoto Protocol: What does it mean for project-based climate change mitigation? *Mitigation and Adaptation Strategies for Global Change,* **3,** 1–58.

Trines, E., 1998a: Assessing and monitoring carbon offset projects: the Costa Rican case. *Commonwealth Forestry Review,* **77,** 214–218.

Trines, E. 1998b: SGS's carbon offset verification service. *Commonwealth Forestry Review,* **77,** 209–213.

United Nations Climate Change Convention Secretariat (UNCCCS), 1997: *UNFCCC AIJ Methodological Issues.* Available at: www.unfccc.de/fccc/ccinfo/inf3.htm.

United Nations Framework Convention on Climate Change (UNFCCC), 1995: *Decision 5/CP.1* from the Conference of the Parties on its First Session Held at Berlin from 28 March to 7 April 1995 Addendum (FCCC(CP/1995/7/Add.1).

UNFCCC, 1997: *Report of the Subsidiary Body for Scientific and Technological Advice on the Work of its Fifth Session, Bonn, Germany, 25–28 February 1997. Annex III. Uniform Reporting Format: Activities Implemented Jointly under the Pilot Phase.* FCCC/SBSTA/1997/Fifth 4.

UNFCCC, 1998: *Activities Implemented Jointly: Review of Progress Under the Pilot.* FCCC/ CP/ 1998/ 2 2, October.

UNFCCC, 1999a: *Views on the Review Process of Activities Implemented Jointly Under the Pilot Phase and Information on Experience Gained and Lessons Learned, Including on the Uniform Reporting Format.* FCCC/SB/1999/MISC.1. Available at: http://www.unfccc.de.

UNFCCC, 1999b: *UNFCCC-CC: Info /AIJ—List of AIJ Projects.* Available at: http://www.unfccc.de/fccc/ccinfo/aijproj.htm

United States Initiative on Joint Implementation (USIJI), 1996: *Guidelines for a USIJI Project Proposal.* US Initiative on Joint Implementation, Washington, DC, USA, 8 pp.

USIJI, 1997a: *USIJI Project Criteria.* US Initiative on Joint Implementation, Washington, DC, USA, 28 pp.

USIJI, 1997b: *Activities Implemented Jointly: 2nd Report to the UN Framework Convention on Climate Change, Vol. 2.* US Environmental Protection Agency, Office of Policy, Planning and Evaluation, Washington, DC, USA, 512 pp.

Verweij, H.J.A. and I.M. Emmer, 1998: Implementing carbon sequestration projects in two contrasting areas: the Czech Republic and Uganda. *The Commonwealth Forestry Review,* **77(3),** 203–208.

Vine, E., J. Sathaye, and W. Makundi, 1999: *Guidelines for the Monitoring, Evaluation, Reporting, Verification, and Certification of Forestry Projects for Climate Change Mitigation.* Paper LBNL-41877, Energy Analysis Department, Environmental Energy Technologies Division, Lawrence Berkeley National Laboratory, Berkeley, CA, USA, 79 pp. + appendices.

Vrolijk, C., 1999: *The Potential Size of the Clean Development Mechanism.* Paper presented at Second International Conference, Emerging Markets for Emissions Trading, April 26–27, 1999, London, United Kingdom, 9 pp.

Wangwacharakul, V. and R. Bowonwiwat, 1995: Economic evaluation of CO_2 response options in the forestry sector: the case of Thailand. *Biomass and Bioenergy,* **8(5),** 293–308.

Warrick, R.A., G.K. Kenny, G.C. Sims, W. Ye, and N. de Wet, 1999: The CC:TRAIN/PICCAP training course on climate change vulnerability and adaptation assessment for Pacific islands. In: *Proceedings of Heads of Forestry Meeting, 21–25 September 1998, Nadi, Fiji* [Tang, H.T., S. Bulai, and B. Masianini (eds.)]. Pacific Islands Forests and Trees Support Programme, RAS/97/330 Field Document No. 1, pp. 197–203.

Winjum, J.K., S. Brown, and B. Schlamadinger, 1998: Forest harvests and wood products: sources and sinks of atmospheric carbon dioxide. *Forest Science,* **44,** 272–284.

Winrock International, 1999: *Field Tests for Carbon Monitoring Methods in Forestry Projects.* Forest Carbon Monitoring Program, Winrock International, Arlington, VA, USA, 65 pp.

Witthoeft-Muehlmann, A., 1998: *Carbon sequestration and sustainable forestry: an overview from ongoing AIJ-forestry projects.* W-75 Working Paper, International Academy of the Environment, Geneva, Switzerland, 37 pp.

World Bank, 1997: *Guidelines for Climate Change Global Overlays.* Paper No. 047, Environment Department Papers, Climate Change Series. World Bank, Washington, DC, USA, 45 pp.

World Business Council for a Sustainable Development, 1997: *Climate Change Projects: Guidelines for Completing Proposals.* Available at: www.wbcsd.climatechange.com/home.html.

Xu, D., 1995: The potential for reducing atmospheric carbon by large-scale afforestation in China and related cost/benefit analysis. *Biomass and Bioenergy,* **8(5),** 323–336.

Young, A., 1997: *Agroforestry for Soil Management (2nd ed).* CAB International, Oxford, United Kingdom. 320 pp.

6

Implications of the Kyoto Protocol for the Reporting Guidelines

BO LIM (UK/UNDP), GRAHAM FARQUHAR (AUSTRALIA),
AND N.H. RAVINDRANATH (INDIA)

Lead Authors:
J. Ford-Robertson (New Zealand), Y. Sokona (Mali), E. Vine (USA)

Contributors:
K. Paustian (USA), S. Calman (New Zealand), P. Frost (Zimbabwe), A. Kokorin (Russian Federation), G. Inoue (Japan)

Review Editor:
K. Ramakrishna (India)

CONTENTS

EXECUTIVE SUMMARY

Under Article 5.2 of the Kyoto Protocol (see Box 6-1), the Revised 1996 IPCC Guidelines for National Greenhouse Gas Inventories (Guidelines) provide the basis for estimating and reporting anthropogenic emissions by sources and removals by sinks of greenhouse gases (GHGs).

Annex A of the Protocol lists all sectors from the Guidelines that may be used to meet commitments (Article 3.1), with the exception of land-use change and forestry (LUCF). The LUCF sector is excluded from Annex A because the accounting rules for this sector are yet to be defined by Parties. However, Article 3 refers to the potential use of agricultural soils and LUCF.

The Guidelines were developed to report national greenhouse gas inventories under the United Nations Framework Convention on Climate Change (UNFCCC), not the Protocol. Depending on decisions made by Parties, it is likely that improvements to the Guidelines will be needed to estimate and report GHGs, stock changes, and associated activities. This chapter suggests options for improving the Guidelines in relation to the verifiability and transparency of reported data.

The IPCC Guidelines consist of three volumes: the Reference Manual, the Workbook, and the Reporting Instructions. To analyze their adequacy for reporting under the Protocol, there is a need to distinguish between them. The desired principles of these inventory guidelines include completeness, comparability, transparency, and accuracy. The UNFCCC has developed reporting guidelines for National Communications of Annex I Parties, which are based on the IPCC Guidelines.

The Workbook contributes to comparability and transparency by virtue of a common, simplified approach. It is also helpful when there are limitations of expertise and data. To improve accuracy, however, country-specific data and more elaborate methods, such as those described in the Reference Manual, may be required. The Workbook encourages the use of national data but provides default data because the methods were designed for use by all Parties. The use of national data is likely to increase the accuracy of estimates of emissions and removals of GHGs, changes in carbon stocks, and their associated activities. This increased accuracy could be important for compliance under the Protocol.

The Reference Manual provides a comprehensive approach to carbon accounting (by covering all of the main land-use change and forestry activities). The Reference Manual mentions all pools and encourages countries to undertake comprehensive accounting of carbon pools affected by anthropogenic activities in relation to LUCF. The Guidelines do not differentiate, however, between direct human-induced activities and indirect human activities (e.g., atmospheric CO_2 fertilization or nutrient deposition).

With the Reference Manual, it is feasible to estimate national changes in aboveground and below-ground biomass, soil, soil surface litter, and harvested wood products for all forests, shrublands, grasslands, and agricultural lands. For countries in which the change in stocks of forest products is significant, the Reference Manual recommends that the products pool be reported.

The underlying principles of national accounting for GHG emissions and removals in the Reference Manual are the same as those applied in the Workbook. The Workbook contains accounting methods and default data for a *subset* of carbon pools, such as aboveground biomass and carbon in the top 0.3 m of the soil. The Workbook does not give explicit methods for other pools, including below-ground biomass, harvested wood products, or deep soil carbon. For the treatment of wood products, the Workbook assumes that the stock of forest products is not increasing significantly in most countries. Hence, carbon accounting for wood products is not required, although the amount of wood harvested for biofuels is necessary for countries to estimate non-CO_2 gases from combustion and report these emissions in the energy sector. The accounting methods in the Workbook may need to be improved to directly link changes in soil carbon and other pools with the activities that cause these changes.

The Reporting Instructions provide a range of definitions for anthropogenic activities, a glossary of terms, and tables to report anthropogenic emissions and removals of GHGs. These tables are flexible enough to allow for reporting of all carbon pools, even if the accounting methods and default data are not provided by the Guidelines. This flexibility in reporting contributes to inventory completeness, though not necessarily comparability.

Under the Protocol, if emissions and removals of GHGs resulting from land use, land-use change, and forestry (LULUCF) are to assist Annex B Parties in meeting their commitments, the Guidelines must provide a basis for transparent and verifiable reporting. In addition, there are other potential requirements:

- Reporting of changes in carbon stocks resulting from afforestation, reforestation, and deforestation (ARD) activities separately (Article 3.3)

- Reporting of changes in carbon stocks and emissions and removals of GHGs resulting from additional human-induced activities, yet to be decided by the Parties (e.g., Article 3.4)
- Reporting of net GHG emissions and removals associated with LULUCF projects (e.g., Article 6).

Many of the issues discussed in this chapter are not unique to the LUCF sector, including uncertainties, transparency, and verifiability. Generally, the level of uncertainty may be inherently higher for all GHG emissions associated with biological systems than the level of uncertainty for emissions associated with energy and industrial systems. These uncertainties can be reduced, however, through scientific and technological developments, as well as improvements in data collection and analysis. Parties would need to consider the tradeoffs between improved accuracy of data, transparency, verifiability, and costs. An underlying principle of the Guidelines, however, is the consistent treatment and reporting of uncertainties of GHGs among all sectors.

Afforestation, Reforestation, and Deforestation

The Guidelines include a definition of each of the three ARD activities. Afforestation is defined as the planting of new forests on lands that historically have not contained forests. Reforestation is defined as the planting of forests on lands that have previously contained forests but have been converted to some other use or some other species. Deforestation is defined as conversion of forest. At one place in the Guidelines, afforestation and reforestation are described differently. There is no distinction, however, between afforestation and reforestation for accounting purposes within the Guidelines; the same method is used for both.

The Guidelines also provide examples of different categories of trees. Variability in such categories allows flexibility in accounting; therefore, the Guidelines do not by themselves ensure consistency in reporting among Parties. The choice of definitions has implications for wooded lands that are included in an inventory (see Chapters 2 and 3). If the Parties adopt definitions for forests that differ from the IPCC definitions, the Guidelines would have to be improved to account for them. Changing these definitions by forest ecosystem or crown cover is unlikely to have significant implications for the accounting approach in the Guidelines. The approach for estimating changes in aboveground biomass is expected to remain similar. If Parties adopted different definitions of ARD activities, however, more significant improvements to the Guidelines might be required—depending on how the regeneration and harvest cycle were treated, for example. Accounting methods for additional carbon pools also might be requested by Parties.

For all three ARD activities, there would be implications if Parties adopt an extension beyond the Workbook. For the Workbook, carbon pools are not only limited to changes in aboveground biomass and soil carbon; the calculations do not explicitly link the two pools for a specific activity. To overcome this issue, the reporting of ARD for aboveground and belowground biomass could be geographically explicit (mapped) or linked through modeling. The same principle would apply to the linking of other carbon pools.

For improved transparency, the reporting of ARD could also be made geographically explicit at some appropriate scale, with geo-referencing of stocks according to species, forest type, and age of stand. The same principle applies with regard to verifying that a forest qualifies as ARD land and verifying changes in these stocks. Geo-referencing in itself does not provide transparency, however. A well-designed national system, together with the application of good practice, may be needed to ensure the verifiability and transparency of the reported data.

For reporting of ARD, the activities or areas subject to these activities since 1 January 1990 need to be identified. The Guidelines do not currently apply this "since 1990" time clause.

Additional Activities

The accounting approach in the Guidelines could be adapted to capture most additional activities, as broadly defined in Chapter 4. Examples of additional activities may include management of cropland, grasslands, and wetlands; if Article 3.3 were to exclude forest management activities, such activities also would include management of forest (aggradation and degradation). Some narrow definitions of activities can introduce land-use categories, however, as well as additional carbon pools that are not currently covered by the Workbook. Default data may not be provided for some of these activities (see Fact Sheets in Chapter 4 for a full discussion).

Although most of these activities could be accounted by the Guidelines to varying degrees, many of the same issues noted with regard to ARD apply. One issue is integrating changes in aboveground biomass, and possibly other carbon pools, with the associated activities. As with Article 3.3, a decision would need to be made regarding which activities after 1 January 1990 would be reported, and how. Depending on if and how baselines and additionality are treated under Article 3.4, additional accounting methods and reporting tables may needed (see Chapters 2 and 4).

Projects

The Guidelines do not address projects, although many of the accounting methods can be applied to estimate emissions and removals of GHGs, changes in carbon stock, and associated activities. Development of guidelines for project activities will depend greatly on the level of detail required and future decisions by the Conference of Parties (COP). The degree of improvements to the Guidelines depends on their intended use. Within each of the options below, Parties could choose their own methods, or internationally agreed methods could be stipulated. There are three main options, of increasing complexity:

Option 1. If the Guidelines are intended to show impacts to a Party's assigned amount, a summary of changes in net GHG emissions and removals from all projects may be all that is required. The first step would involve the transfer of GHG emissions and removals from the project to the national inventory; in the second step, the GHGs from projects would be subtracted from the national inventory.

- *Option 1a*—This option would allow Parties to choose a different reporting format for projects approved by the Parties. An example would be a modification of the Uniform Reporting Format (URF) used in the Activities Implemented Jointly Pilot Project. This option is the simplest to operationalize, but it lacks transparency in methods and data.
- *Option 1b*—This option could stipulate the use of internationally agreed methods and perhaps involve the use of guidance for the URF.

Option 2. If the Guidelines are intended to enable Parties to generate a complete and transparent record of GHG emissions and removals of each project, including a breakdown of all activities, baselines, leakage, risks, and so forth (see methods in Chapter 5), considerable changes will be required.

Option 3. Finally, there may be a need to report all changes in net GHG emissions and removals and environmental and socioeconomic impacts, especially for Clean Development Mechanism (CDM) projects (Article 12).

For option 1, a reporting format in the Guidelines could be developed with minimal effort. For options 2 and 3, considerable effort may be required. The period involved in developing these options would depend on the decisions of the Parties.

For projects, many of the data and reporting needs are similar to those of Articles 3.3 and 3.4. There is a range of additional features to consider, however, including project locations and boundaries, additionality, leakage, permanence, biodiversity, and impacts on sustainable development and the environment. The Guidelines were not designed for reporting of these features. They do provide a basis, however, for reporting changes in emissions and removals of GHGs for projects in a single year. To avoid double-counting, projects and national inventories should be reported in a consistent way in space and time.

Next Steps

Article 5.2 of the Protocol provides a mechanism for Parties to initiate a process to improve the accounting methods and default data in the Guidelines. The time required to make such improvements depends on the decisions of the Parties—including decisions on methods for reporting on projects, which are not currently covered by the Guidelines.

Of importance in this context is a decision by the Subsidiary Body for Scientific and Technological Advice (SBSTA) at its eighth session to encourage the IPCC Inventories Program to "give high priority to completing its work on uncertainty, as well as to prepare a report on good practice in inventory management." In addition, the "SBSTA invited the IPCC to develop a work plan addressing methodological issues raised in the IPCC special report in the context of the IPCC 1996 Revised Guidelines for National Greenhouse Gas Inventories in the areas of agriculture and land-use change and forestry, to commence as soon as practicable following completion of the special report." This work plan may include good practice guidance in LUCF to be initiated in 2000.

Following decisions by the Parties, possible improvements to the Guidelines include the following:

- Reporting of defined activities and changes in carbon stocks associated with them
- Methods for accounting for carbon stocks not currently provided in the Guidelines
- Methods for data collection and evaluation of data quality (uncertainties)
- Methods to integrate changes in aboveground biomass, soil carbon, and other pools, as well as reporting of these changes spatially and by activity (geo-referencing)
- Reporting for projects that takes into account baselines, system boundaries, leakage, and additionality, as well as socioeconomic and non-GHG environmental impacts
- Reporting tables for inventories and projects that differentiate between the base year, 1990, and the commitment period. Projects that qualify under Article 6—and potentially under Article 12—would have to be distinguished, reported, and treated differently in the national GHG inventories of Annex I and non–Annex I Parties.

6.1. Introduction

This chapter analyses the extent to which the Guidelines can be used to report changes in carbon stocks, GHG emissions, and their associated activities under the Kyoto Protocol. This analysis covers the full set of the Guidelines but does not include an analysis of options identified in preceding chapters. This chapter indicates where improvements to the Guidelines may be needed.

Many of the issues discussed in this chapter are not unique to the LUCF sector, including uncertainties, transparency, and verifiability. Others issues to be considered are leakage, baselines, and additionality. Generally, the level of uncertainty may be inherently higher for all GHG emissions associated with biological systems than for those associated with energy and industrial systems. However, these uncertainties can be reduced through scientific and technological developments, as well as improvements in data collection and analysis. Parties would need to consider the tradeoffs between improved accuracy of data, transparency, verifiability, and costs. An underlying principle of the Guidelines, however, is the consistent treatment and reporting of uncertainties of GHG fluxes among all sectors.

The term "reporting" is used to include definitions of forests and activities, accounting methods, default and national data, and reporting tables. As an approximation, the term "accuracy" is used when "reliability" and "confidence of estimates" may be more precise.

The remainder of Section 6.1 reviews the background to the Guidelines. Section 6.2 reviews issues in the Guidelines that are relevant for the reporting of LULUCF (FCCC, 1998b, Decision 9/CP.4). Section 6.3 discusses the adequacy of the Guidelines for the reporting of national activities as a result of ARD (UNFCCC, 1997, Article 3.3), as well as additional human-induced activities (UNFCCC, 1997, Article 3.4). Section 6.4 deals with issues related to projects that potentially fall under Articles 3.4, 6, and 12 (UNFCCC, 1997). Finally, Section 6.5 suggests options for improving the Guidelines.

6.1.1. The IPCC Guidelines

The IPCC Guidelines consist of three volumes: the Reference Manual, the Workbook, and the Reporting Instructions. To analyze their adequacy for reporting under the Protocol, there is a need to distinguish between them. The desired principles of these inventory guidelines include completeness, comparability, transparency, and accuracy (FCCC, 1999). The SBSTA has developed revised UNFCCC reporting guidelines for national communications for Annex I Parties (FCCC, 1999, 2000) that include GHG inventories. These UNFCCC guidelines are based on the IPCC Guidelines but are more detailed.

The Reference Manual provides a comprehensive approach to carbon accounting (by covering all of the main land-use change and forestry activities) (see Figure 6-1). The Reference Manual mentions all pools and encourages countries to undertake comprehensive accounting of carbon pools affected by anthropogenic activities in relation to LUCF. The Guidelines apply the principle of balanced accounting for all changes in carbon stocks for all reported pools on a land area affected by a given activity; it quantifies sources and removals of GHGs. The Guidelines do not differentiate, however, between direct human-induced activities and indirect human activities (e.g., atmospheric CO_2 fertilization or nutrient deposition).

Using the Reference Manual, it is feasible to estimate national changes in aboveground and below-ground biomass, soil, soil surface litter, and harvested wood products for all forests, shrublands, grasslands, and agricultural lands (see Figure 6-1). For countries in which the change in stocks of forest products is significant, the Reference Manual recommends that the products pool be reported.

To improve accuracy, however, country-specific data and more elaborate methods—such as those described in the Reference Manual—may be required. The Guidelines encourage the use of national data, but because the Guidelines were designed for use by all Parties, a range of default data is provided in the Workbook. Although the use of a range of default data does not mean that inventories are not comparable (Ravindranath and Sudha, 1999; Houghton and Ramakrishna, 2000), the Guidelines state that the use of default data "is unlikely to be considered credible for any country which has significant emissions or activities in this area" (IPCC, 1997, Vol. 3, p. 5.15). In other words, the use of national data is likely to increase the accuracy of estimates in GHGs, changes in carbon stocks, and their associated activities. This increased accuracy could be important for compliance under the Protocol.

The underlying principles of national accounting for GHG emissions and removals in the Reference Manual are the same as those applied in the Workbook. The Reference Manual contributes to comparability and transparency by virtue of a common, simplified approach. It is also helpful when there are limitations of expertise and data.

Unlike the Reference Manual, however, the Workbook considers changes in only two carbon pools, such as aboveground biomass and carbon in the top 0.3 m of the soil (see Figure 6-1). The soil carbon method accounts for cultivation of soils, land-use change involving soils, and liming of agricultural soils. The Workbook does not provide explicit methods for other pools, including below-ground biomass, harvested wood products, and deep soil carbon. The treatment of wood products in the Workbook differs from that of the Reference Manual. The Workbook assumes that stocks of forest products are not increasing significantly in most countries. Hence, carbon accounting for wood products is not required, although the amount of wood harvested for biofuels is necessary for estimating non-CO_2 gases and reporting these emissions in the energy sector. The Workbook does not account for the effects of forest fires and other disturbances.

The Reporting Instructions provide a range of definitions for anthropogenic activities, a glossary of terms, and tables to report anthropogenic emissions and removals of GHGs. The tables are flexible enough to allow for reporting of all carbon pools, even if the Guidelines do not provide accounting methods and default data. This flexibility in reporting contributes to completeness by facilitating the reporting of all carbon pools, though it does not necessarily contribute to inventory comparability.

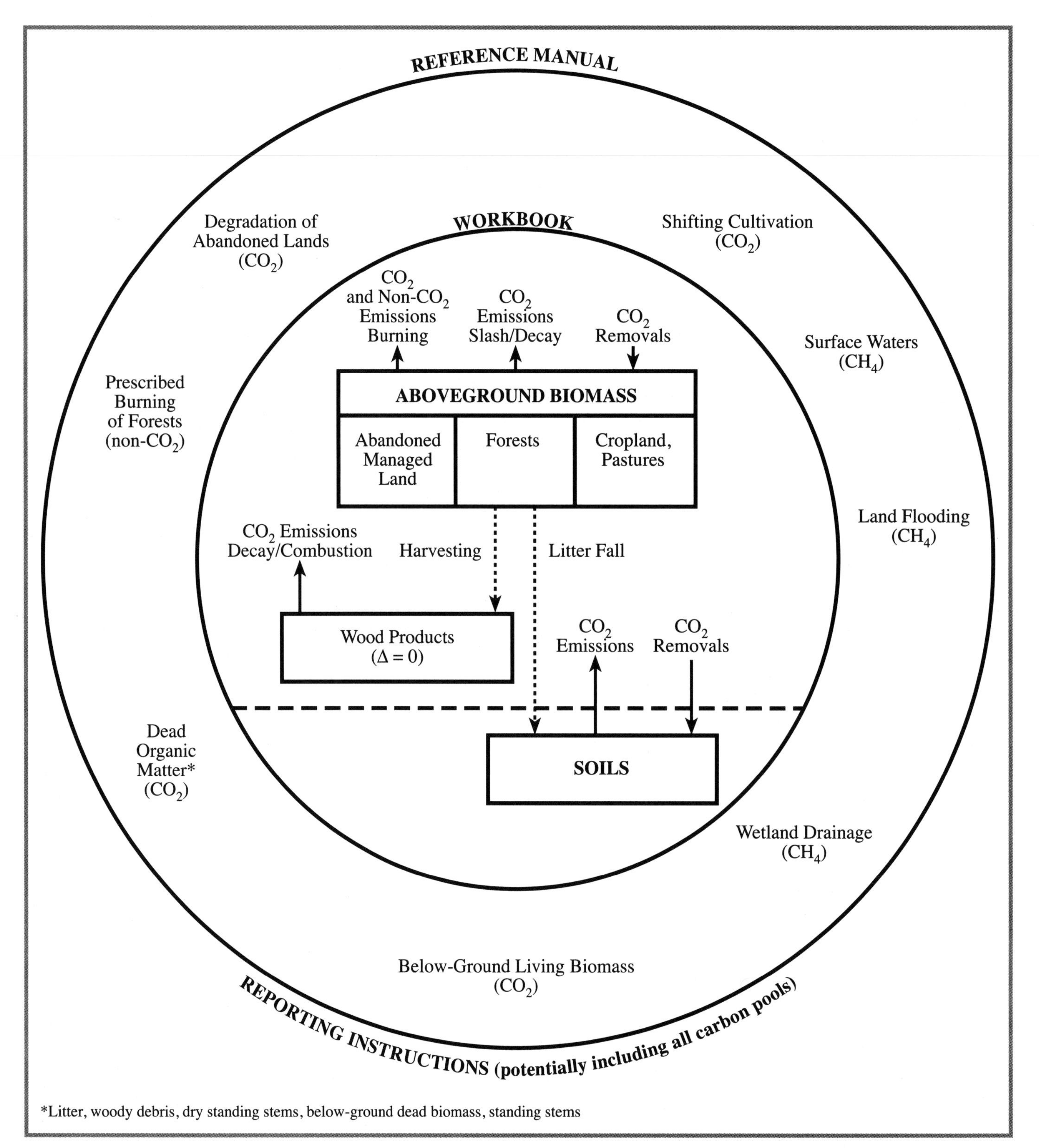

Figure 6-1: Structure in the IPCC Guidelines to account for national changes in carbon pools. The Reference Manual describes all activities within the outer and inner circles. The Workbook accounts for all changes in pools due to activities within the inner circle only. Emissions and removals of greenhouse gases can be reported within the Reporting Instructions.

6.1.2. The IPCC Guidelines and the UNFCCC

The IPCC developed its Guidelines for National Greenhouse Gas Inventories in 1995 to enable all Parties to report national inventories of anthropogenic emissions of all GHGs by sources and removals by sinks under Articles 4.1a and 12.1a of the UNFCCC. The IPCC recognizes that improving the Guidelines is an ongoing process, consistent with Article 7.2 of the UNFCCC.

The IPCC improved the 1995 guidelines and published the Revised 1996 IPCC Guidelines for National Greenhouse Gas Inventories in 1997; the SBSTA recommended that Parties use these Revised Guidelines.

The process for improving the Guidelines can be initiated by a request from the SBSTA to the IPCC. Currently, there is no mandate for the IPCC to improve the Guidelines or develop guidelines for projects. Any improvement to the IPCC Guidelines, however, may remain under the remit of the IPCC.

The IPCC is now developing good practice guidance (GPG) for managing uncertainty of GHG inventories for many sectors (IPCC, 1999), including LUCF in 2000. The combined use of the Guidelines and the GPG can enhance the accuracy and comparability of national GHG inventories.

Box 6-1. Article 5.2 of the Kyoto Protocol

Methodologies for estimating anthropogenic emissions by sources and removals by sinks of all greenhouse gases not controlled by the Montreal Protocol shall be those accepted by the Intergovernmental Panel on Climate Change and agreed upon by the Conference of the Parties at its third session. Where such methodologies are not used, appropriate adjustments shall be applied according to methodologies agreed upon by the Conference of the Parties serving as the meeting of the Parties to this Protocol at its first session. Based on the work of, inter alia, the Intergovernmental Panel on Climate Change and advice provided by the Subsidiary Body for Scientific and Technological Advice, the Conference of the Parties serving as the meeting of the Parties to this Protocol shall regularly review and, as appropriate, revise such methodologies and adjustments, taking fully into account any relevant decisions by the Conference of the Parties. Any revision to methodologies or adjustments shall be used only for the purposes of ascertaining compliance with commitments under Article 3 in respect of any commitment period adopted subsequent to that revision.

6.1.3. The IPCC Guidelines and the Kyoto Protocol

Under Article 5.2 of the Kyoto Protocol (see Box 6-1), the Revised 1996 IPCC Guidelines for National Greenhouse Gas Inventories provide the basis for estimating and reporting of anthropogenic emissions by sources and removals by sinks of GHGs. The reporting requirements of the Protocol are likely to differ from those of the UNFCCC, however. A new development may be the use of national GHG inventories to assess compliance with emission limitation and reduction commitments of Annex I Parties, as described under Article 3.1 of the Protocol.

To meet commitments, Parties may use emissions and removals of GHGs from the sectors listed in Annex A of the Kyoto Protocol. Annex A of the Protocol details all of the sectors from the Guidelines that may be used, with the exception of LUCF. The LUCF sector is excluded from Annex A because Parties have yet to define the accounting rules for this sector. Articles 3.3, 3.4, and 3.7, however, refer to the potential use of agricultural soils and LUCF, as explored elsewhere in this Special Report. There is a further requirement to report changes in carbon stocks in a way that is transparent and verifiable (Articles 3.3 and 3.4).

The inclusion of projects under the Protocol may also introduce new criteria for reporting. The Protocol includes two project-based mechanisms. The first allows for the joint implementation of projects between Annex I Parties (UNFCCC, 1997, Article 6). The second, the CDM, allows for certified emission reduction projects in countries of non-Annex I Parties to assist Annex I Parties in meeting their commitments (UNFCCC, 1997, Article 12). In addition, the CDM will assist non-Annex I Parties in achieving sustainable development and contribute toward the ultimate objective of the FCCC. The Guidelines are intended for reporting of national inventories of emissions and removals of GHGs. The principles of the Guidelines could be applied, however, to estimate changes in carbon stock in projects, although they were not designed to gauge the non-carbon impacts of projects or to address sustainable development.

6.2. Relevant issues in the IPCC Guidelines for Reporting under the Kyoto Protocol

This section reviews the Guidelines and summarizes issues relevant to Articles 3.3 and 3.4 of the Protocol.

6.2.1. Relevance of the IPCC Guidelines to the Protocol

Two linked themes underpin the accounting methods in the LUCF chapters of the Guidelines. First, net emissions and removals of atmospheric carbon (as CO_2) by the land are assumed to be equal to changes in biomass carbon and soil carbon stocks. Second, annual changes in carbon stocks are calculated from rates of change in land use and the human activities that bring about these changes.

The estimated emissions and removals of CO_2 are reported in the following categories in the LUCF chapter:

- Changes in forest and other woody biomass stocks (e.g., afforestation, reforestation)
- Decay and on- and off-site burning of aboveground biomass from conversion of forest and grassland to other land-cover types (e.g., deforestation), together with non-CO_2 trace gas emissions from on-site burning of biomass
- Regrowth of biomass on abandoned managed lands
- Changes in soil carbon stocks.

All three ARD activities are described in the Guidelines. Deforestation (IPCC, 1997, Vol. 3, p. 5.6) is referred to as conversion of forest, often accompanied by burning. Afforestation is defined as planting of new forests on lands that historically have not contained forests (IPCC, 1997, Vol. 1, p.1, Glossary). Reforestation is defined as planting of forests on lands that have previously contained forests but have been converted to some other use or some other species (IPCC, 1997, Vol. 1, Glossary).

Afforestation and reforestation activities are described differently in a footnote of the Reference Manual (Vol. 3, p. 5.14), which states that "plantations are forest stands that have been established artificially, to produce a forest product 'crop.' They are either on lands that previously have not supported forests for more than 50 years (*afforestation*), or on lands that previously supported forests within the last 50 years and where the original crop has been replaced with a different one (*reforestation*) (Brown *et al*., 1986)."

Consistent with the concept of comprehensive accounting, all trees can be potentially included in the national inventory. The default biomass data for forests are based on the Food and Agriculture Organization (FAO), at least for the tropics (IPCC, 1997, Vol. 3, p. 5.7). According to the Guidelines, however, percentage crown cover can also be used to characterize open and closed forests (IPCC, 1997, Vol. 3, p. 5.7). The IPCC categories of trees are potentially wide-ranging and can be extended to include those that are not traditionally considered "forests," such as village and farm trees (IPCC, 1997, Vol. 3, p. 5.7) and urban trees (IPCC, 1997, Vol. 1, p. 1.14). Chapters 2 and 3 discuss other definitions of forests and their implications.

Users of the Guidelines are required to select appropriate input data and assumptions for the methods chosen. Users must decide on forest definitions, land-use classifications, activities, land areas subject to ARD and other activities, accounting methods, and the number of carbon pools to be included in the inventory. The Guidelines provide definitions and guidance for choosing appropriate definitions, but they are not prescriptive. As noted in Chapters 2 and 3, an inventory for LULUCF is sensitive to varying definitions. The amount of land that is brought into the inventory, for example, depends critically on the definitions applied. The Guidelines allow considerable flexibility in their interpretation and do not, in themselves, ensure consistency in reporting among Parties (see Section 6.2.3).

For Parties that wish to move away from defaults and collect national data, the Guidelines offer only a limited range of methods—though this situation is not unique to LULUCF. Currently, the Guidelines do not provide a framework for reporting the variability and uncertainty of national data; again, this situation is not unique to LULUCF.

6.2.2. *Reporting of Greenhouse Gas Emissions among IPCC Sectors*

The IPCC Common Reporting Framework in the Reporting Instructions describes how GHGs are reported among the IPCC sectors. GHG emissions associated with one activity may be reported in different sectors. This framework is important in understanding how GHG emissions from a given activity might show up in national accounts.

For the LUCF sector, the reporting of CO_2 is straightforward. Emissions and removals of CO_2 resulting from changes in biomass stocks and soil carbon are reported in the LUCF sector only. On the other hand, non-CO_2 emissions from LULUCF can be reported in different sectors.

For example, non-CO_2 GHG emissions from the combustion of biofuels appear in the energy sector, whereas net CO_2 emissions and removals associated with the production of biofuels are reported in LUCF. Emissions of non-CO_2 GHGs from most activities are reported in the agriculture sector, but emissions for a few activities are reported in the LUCF sector. This latter reporting applies to the management of wetlands and to biomass burning in relation to land clearing. Non-CO_2 GHGs from savanna burning are reported in agriculture, however. For soils, non-CO_2 GHGs are reported in agriculture, whereas CO_2 emissions and removals appear in LUCF.

This reporting framework has several implications. Under the UNFCCC, care must be taken to avoid double-counting or underreporting of GHGs, particularly in the energy, agriculture, and LUCF sectors. Under the Kyoto Protocol, there is a potential for double-counting because the agriculture sector mentioned in Article 3.4 is already listed in Annex A. This chapter focuses on LUCF and hence on CO_2 emissions and removals from soils, rather than on emissions and removals of trace gases from soils. Chapters 2 and 3 discuss non-CO_2 emissions and removals resulting from ARD.

6.2.3. *Methodological Issues for National Reporting under the UNFCCC*

Parties have identified several issues in applying the Guidelines. Some of these issues have been the subject of continuing debate; several remain unresolved and relevant for the Protocol:

Carbon pools. Because the Guidelines allow flexible reporting of carbon pools, Parties do not always report the same set of carbon pools. Some Parties report changes in aboveground biomass and soil carbon only; others include below-ground biomass, soil carbon, and wood products (FCCC, 1997a).

Managed land and forests. All managed land and forests should be included in national inventories. Natural, unmanaged (for wood products) forests are not considered to be subject to anthropogenic activities (IPCC, 1997, Vol. 3, p 5.11) and are therefore excluded from national inventories. Some Parties, however, have reported that the distinction between "managed" and "unmanaged" is problematic (Lim *et al*., 1997). Ambiguity often arises in the treatment of conservation reserves or where land is managed to enhance the growth of natural forests. As a result, the definition of "managed" land and forest is not consistently applied among Parties—which has implications for the land area that is included in a national inventory. Several options could be applied to ease reporting: Parties could report the definition of "managed" used for the inventory; the existing definitions of "managed" in the Guidelines could be clarified; or a range of categories for "managed" could be developed for the Guidelines.

Forest fires. Only GHGs from anthropogenic fires should be included in the inventory. Separation of natural from anthropogenic forest fires is technically very difficult, however (Chapter 2). Furthermore, the Guidelines do not treat biomass burning in a consistent way. They do not provide accounting methods to estimate CO_2 emissions from anthropogenic forest fires and other disturbances, nor for the uptake of CO_2 through regrowth following disturbance; the Guidelines assume that emissions and regrowth balance on a national scale when averaged over several years (Lim *et al*., 1997). Only emissions of trace gases from prescribed burning and forest clearing are included. The foregoing issues have implications for the comparability of national inventories. Chapter 2 notes that even with clear definitions, however, distinguishing between changes of carbon stock resulting from human activity and those caused by natural events may be difficult.

Uncertainty. The uncertainty of estimates in changes in carbon stock is often cited as variable (IPCC, 1997). Much of this imprecision is caused by uncertainty in emission factors and activity data or the use of inappropriate default data and assumptions. Uncertainty also arises from the differing interpretation of the IPCC categories of sinks and sources provided in the Guidelines. These reported uncertainties—especially in the category of changes in forest and other woody biomass stocks—may not be larger than those for CH_4 and N_2O emissions from other sectors (FCCC, 1997a). Parties can report levels of confidence in their estimates of GHGs with the Guidelines. Although the GPG do not specifically address the LUCF sector, they provide information on the generic treatment of uncertainties for all sectors.

Transparency and verifiability. In the Guidelines, Parties are requested to provide reporting tables, worksheets, and supporting documentation for transparency. The worksheets provide a higher degree of transparency than the summary tables, although they are not sufficient for transparency. The UNFCCC Guidelines on Reporting and Review (FCCC, 1999, 2000) provide more rigorous reporting tables for Annex I Parties and refer to the use of the GPG for verification of national GHG inventories.

Frequency of measurements. Changes in growing stock tend to be very slow in the boreal region and in part of the temperate region (Chapter 2). Because few countries have measurements of annual growth increments, detecting changes in carbon stock over this time interval may be difficult. Chapter 2 presents a range of methods for detecting changes in stock and flux and indicates the strengths and weaknesses of each method. Although remote sensing is promising, one constraint is that there are no systems in place to collect adequate data on forest clearing or regrowth during the first commitment period. Furthermore, in some regions the data from satellite imagery are often taken several years apart and may not coincide with the desired (annual or 5-year) time frame.

6.2.4. Issues for Consideration in National and Project Reporting

There are two types of issues for consideration: issues that affect changes in carbon stocks, and those that relate to the quality of the stock estimates. The UNFCCC has developed terms to describe quality of inventories as transparency, consistency, completeness, and accuracy (FCCC, 1999). A third category of issues relates to project and national reporting. These latter issues tend to be scale-dependent and project-specific. The issues are grouped as follows:

- Issues related to changes in carbon stock at the national level
- Issues related to data quality at the national level
- Other issues related to reporting at the national and project levels.

These issues are summarized in Tables 6-1 to 6-3.

6.3. Adequacy of the IPCC Guidelines for Reporting National Activities under the Kyoto Protocol

This section assesses the adequacy of the Guidelines in relation to activities under Articles 3.3 and 3.4. It begins with a discussion of reporting issues common to ARD, then addresses issues that are specific to deforestation.

6.3.1. Afforestation, Reforestation, and Deforestation under Article 3.3

6.3.1.1. Generic Issues on Afforestation, Reforestation, and Deforestation

In the Guidelines, stock changes arising from afforestation and reforestation activities are reported under the same category,

***Table 6-1**: Issues related to changes in carbon stocks at the national level.*

Issue	Question	National Reporting
Changes in carbon stocks	How do the Guidelines report changes in carbon stocks?	The Workbook method estimates the rate of change of land use and aboveground biomass in forests, other woody biomass stocks, and soil carbon. Changes in carbon stock are used to derive emissions and removals of CO_2. There are no glossary definitions for carbon flows, fluxes, stocks, emissions, and removals.
ARD (Article 3.3)	How do the Guidelines capture ARD?	Glossary definitions exist for afforestation and reforestation; deforestation is referred to as forest conversion. For accounting purposes, there is no distinction between afforestation and reforestation for changes in aboveground biomass. The same module is used for afforestation and reforestation. All three activities are potentially covered by the methods.
Additional activities (Article 3.4)	How might the Guidelines potentially capture additional activities?	In principle, the accounting framework in the Reference Manual could be adapted to capture most additional activities by changing the input and tillage factors where changes in soil changes are expected. For some practices, these factors are not provided.
Full carbon accounting	Do the Guidelines account for all changes in carbon pools, such as aboveground biomass, roots, forest litter, soil carbon, and wood products?	The Reference Manual allows for carbon accounting in more carbon pools than the methods in the Workbook. There are no explicit methods for below-ground biomass, deep soil litter, surface soil litter, and wood products. Additional methods, data, and more detailed reporting tables would be required for more complete carbon accounting.
Direct and indirect human activities	Do the Guidelines differentiate clearly between direct and indirect human activities?	Currently not well-defined. The emphasis is on "anthropogenic." There is no method to separate the impacts of direct from indirect effects on observed stocks.

"Changes in forest and woody biomass stocks." For accounting purposes, there is no distinction between afforestation and reforestation; the same method is used for both. Deforestation can be dealt with separately, according to the principles for accounting of "Forest and Grassland Conversion." Chapter 2 explores a range of possible definitions of ARD. Chapter 3 further explores a wide range of definitional scenarios, including that of the FAO. If Parties choose the IPCC definitions, afforestation and reforestation can be dealt with using the principles in the Guidelines. For discussion only, this section assumes that the principles of afforestation and reforestation are the same as those in the Guidelines.

There would be implications for all three ARD activities if Parties were to adopt an extension beyond the Workbook. Carbon pools are limited to changes in aboveground biomass and soil carbon in the Workbook. Additional carbon pools are discussed in and can be accounted for using the approach laid out in the Reference Manual. According to the rules, modalities, and guidelines agreed to under Articles 3.3 and 5.2, however, Parties may decide that accounting methods for these additional carbon pools may need to be developed and provided in the Workbook.

The calculations in the Workbook do not explicitly link the aboveground biomass and soil carbon pools for a specific activity. To resolve this issue, reporting of ARD for aboveground and below-ground biomass could be geographically explicit (mapped) or linked through modeling, or the accounting methods for these pools could be slightly modified. The same principle would apply to the linking of other carbon pools, such as more complete treatment of surface soil litter, below-ground biomass, and wood products (see Section 6.3.3). Geographically explicit reporting and modification of accounting methods are not expected to be difficult to implement, although data can be expensive to obtain on a large scale, however, even for Parties with a good deal of scientific infrastructure (see Chapter 3). Furthermore, the data would have to be reported at appropriate

***Table 6-2**: Issues related to data quality at the national level.*

Issue	Question	National Reporting
Measurements and data	Are measured data available for estimating changes in carbon stock?	Collection of national data is outside the scope of the Guidelines. However, the reliance on default data can be a constraint to obtaining reliable estimates.
Uncertainty, accuracy, and precision	Can the uncertainty, accuracy, and precision of estimates of changes in carbon stock be calculated and reported?	Basic statistical methods for estimating uncertainty are provided in the Reporting Instructions. The confidence level of emissions estimates is reported in Table 8A, often qualitatively, by countries. Methods of reporting uncertainty may not be adequate.
Transparency	Do the Guidelines allow for transparent reporting of stock changes resulting from ARD and additional activities?	The Guidelines allow for transparency reporting, but only if countries provide worksheets and additional supporting information.
Verifiability	Are there methods in the Guidelines for verifying changes in carbon stock resulting from ARD and additional activities?	Basic verification steps are outlined but are not explicit. There are no reporting sheets for verification.

time intervals in all cases and with consideration of the "since 1990" time clause.

All three ARD activities cause changes in aboveground biomass. Closely linked to the issue of ARD is the definition of forests, which has implications for lands that are included in an inventory (see Chapters 2 and 3). If Parties adopt definitions for forests other than the existing ones, the Guidelines may have to be improved to account for the new definitions. Changing these definitions by forest ecosystem or crown cover is unlikely to have significant implications on the Guidelines; the approach for estimating changes in aboveground biomass is expected to remain similar. If Parties adopted definitions of ARD activities that differed from the IPCC definitions, however, more significant improvements to the Guidelines might be required—depending on how the regeneration and harvest cycle were treated, for example.

If a land-based accounting system (Chapter 2) were to apply, a generic issue for ARD is that there may be a need to separately identify land into ARD categories and report land categorization in 1990, as well as in the commitment period. The optimum size of the landscape unit at which ARD activities may be detected is another consideration (Chapter 3).

If an activity-based accounting system (Chapter 3) were to apply, another solution might be possible. Differentiation among categories can be made at the beginning of a commitment period (in 2008, in the first instance) on the basis of activities implemented prior to 2008. In this system, the impact of an activity on carbon stocks could be tracked and summed per unit area. Carbon stocks would be more difficult to verify under this system, however, than those tracked under a land-based accounting system (see Chapter 2).

For reporting of ARD under both the land-based or activity-based systems, the activities and the areas subject to these activities since 1 January 1990 would have to be identified. The Guidelines do not currently apply this time clause to such land areas. The Guidelines require the use of land areas and ARD activities to estimate changes in carbon stock, however; hence, either a land-based or activity-based system could be used to track aboveground biomass and soil carbon.

To more accurately account for changes in carbon stocks under ARD, a finer level of geographic detail and subcategories may be appropriate. Stocks estimates could be improved by reflecting forest management type, species, soil type, and so forth, as well as by substituting default assumptions with national or region-specific data and assumptions from local sources (IPCC, 1997, Vol. 3, p 5.15).

On the other hand, if changes in stocks associated with ARD activities were reported at aggregated levels, reporting could be made geographically explicit at an appropriate scale, with geo-referencing of stocks according to species, forest type, and age of stand. The same principle applies with regard to verifying that a forest qualifies as ARD land and verifying changes in these stocks. Geo-referencing in itself does not provide transparency. The Guidelines would not be adequate for this purpose. A well-designed national system, together with the application of good practice, may be needed to ensure verifiability and transparency of reported data.

In relation to ARD, examples of data or assumptions that may be reported include the following:

- Forest inventories at the required level of accuracy (IPCC, 1997, Vol. 3, p. 5.16). New methods based on

Table 6-3: *Other issues related to reporting at the national and project levels.*

Issue	Questions	National Reporting	Project Reporting
Scale dependency	Are the methods in the Guidelines scale dependent? Could they be applied at different scales, including at the project level?	The Guidelines are designed for national level reporting.	The Guidelines are not designed for project-level reporting. A well-defined system boundary should be part of any reporting guidelines for projects.
Additionality and baselines	Do the Guidelines provide methods for reporting baselines and additionality?	Not possible to report baselines and additionality for a given year. At the reporting level, activities that reduce emissions are not distinguished from those that increase carbon storage.	The Guidelines are not designed for reporting additionality and baselines. New methods for reporting system boundaries, additionality, and baselines and for projections may be required.
Leakage	Do the Guidelines provide a basis for reporting leakage?	The Guidelines are not designed to detect leakage at the national level.	The Guidelines are not designed to detect leakage at the project level. The reporting of all projects within a system boundary would be required for detecting leakage.
Periodicity of data and time scales of processes	What is the periodicity of the input data in the Guidelines? How does this periodicity relate to the time scales of the processes, the inventory base year (1990), and the commitment period (2008–2012)?	The periodicity of the input data varies from 3-yr averages to 20 years or more. For delayed processes, historical data over 10- and 20-yr periods and longer are needed. The full impacts of an activity may not occur, or be detected, in the commitment period if the effects are delayed.	The frequency of data collection depends on the nature of the project.
Frequency of data measurements and reporting	What is the frequency of measurements and reporting in the Guidelines? What are the implications of the frequency of measurements for assessing annual carbon stocks as currently required under the Protocol?	The Guidelines are designed for annual reporting. However, annual measurements are not always available. The implications of how stocks will be reported during the commitment period (2008–2012) are not clear.	Project-level measurements and reporting may follow the national reporting.
Cost of data collection	Is the cost of different methods for data collection provided in the Guidelines?	Costs of data collection are not discussed.	Costs of data collection are built in, and fully considered, when determining the feasibility of a project.

satellite and aerial photography are also available for large-scale measurements, but only of area. Generic treatment for data quality is covered in the GPG.

- Conversion factors appropriate to categories of forests (IPCC, 1997, Vol. 3, p. 5.18) for estimating tons of biomass per unit volume of commercial timber and tons of carbon per ton of dry biomass.
- Biomass densities appropriate to the mix of species in the growing stock of a particular region could be used, based on literature data or field measurements (IPCC, 1997, Vol. 3, p. 5.22). Any default values would have to be age-class dependent, especially because stands established since 1990 will be very young during the commitment period, and expansion factors tend to be significantly higher for young stands.
- The same principles apply to below-ground biomass (coarse woody roots, etc.) and to forest litter (dead roots, slash, etc.).

All three ARD activities can affect the level of soil carbon. Although the Guidelines can be readily applied to account for ARD, they focus on agricultural land use and management, rather than ARD. For comprehensive carbon accounting, soil carbon components of soil surface litter could be included. Consideration would have to be given to soil depth and rates of change in carbon, as appropriate. In cropped soils, however, most changes occur in the upper 30 cm (Chapter 2), as assumed in the Workbook. In some situations, consideration of deep soil carbon and more types of soil humus, with differing rates of decomposition, might be necessary for ARD activities. Significant changes may occur at 30–50 cm depth—for example, in deep tropical soils (see Chapters 2 and 4), where changes in soil carbon are generally faster than in boreal areas. Consistency in the accounting rules may need to be established, especially if crediting for soil carbon is allowed.

With respect to soil carbon, other refinements to the Guidelines may include the following:

- Use of soil carbon models or methods with realistic time dependencies that reflect the soil type, moisture, and temperature conditions. In the IPCC default method, changes are assumed to be linear in time, so a straight averaging of past rates of change of land use is used. In the Protocol context, changes in stock resulting from increase or decay of soil carbon could be calculated using a time course (e.g., exponential) appropriate to the region and the commitment period in light of local conditions. This approach will give greater weight to more recent events.
- Use of improved soil carbon data from surveys, field studies, and long-term agricultural experiments to replace default values.
- Use of geo-referencing to relate data on base factors, tillage factors, and input factors and initial soil carbon content (IPCC, 1997, Vol. 2, p. 5.25) that is transparently reported with aboveground biomass changes. The base factor refers to the change in soil organic matter that is associated with the conversion of native vegetation to agriculture (IPCC, 1997, Vol. 3, p. 5.47). The term "initial" is used here rather than "native" to take into account that deforestation may be of forest that had been disturbed sometime in the historical past.
- As discussed in Section 6.3.3, changes in carbon wood products could be considered (Winjum *et al.*, 1998; Brown *et al.*, 1999).

A key question is whether averaging periods for data are appropriate. Changes in biomass are linked to climatic fluctuations, so changes in stocks between 2008 and 2012 could be depressed or boosted by unusual weather patterns. Such fluctuations may affect stock changes over the commitment period for afforestation and reforestation or regrowth after deforestation. Separating the observed stock change that is directly human-induced from the influence of indirect causes and natural variability is generally difficult. One possibility is the use of default data to smooth out the effects of natural variability and extreme climatic events. This use of default data would deviate, however, from the principles of the Guidelines.

Because the Guidelines do not fully address the issue of uncertainties (Lim *et al.*, 1999b), the reporting of ARD activities could be modified to include treatment of uncertainties and methods for error propagation—using, for example, Monte Carlo analyses or standardized calculation methods. Uncertainties in measurements of stocks at two points in time (e.g., 2008 and 2012) may cancel out for systematic errors. Under the gross-net approach of Article 3.3, however, the uncertainties would not cancel out for 1990 and the commitment period: For 1990, the uncertainties would comprise the non-LUCF sectors (gross) only, whereas for the commitment period the uncertainties would be the net of the uncertainties of estimates in non-LUCF and LUCF sectors.

6.3.1.2. *Specific Issues on Deforestation*

In deforestation, an extra component is needed. Not only are estimates of below-ground biomass at the time of deforestation and any replacement vegetation needed, so are estimates of the decomposition rates of remaining roots (see Chapter 2, Table 2-3). The Guidelines contain principles for a general approach to the problem. The approach used for estimating soil carbon is related to typical changes in organic matter content per area and averaged over the time (20 years) during which the changes occur. The Reference Manual (IPCC, 1997, Vol. 3, p. 5.23, footnote 15) focuses on the conversion of tropical forests to pasture and cropland because this conversion accounts for the largest share of emissions from forest clearing at the global level.

Under the Kyoto Protocol, all forest-clearing activities could be accounted for. Improvements to the Guidelines, however, might include the following, although they may be difficult to implement because of lack of data:

- In the IPCC default approach, the average loss from decay of litter per year is taken as (linear) 10 percent, so the average rate of clearing over the previous 10 years is used in the calculation of emissions (IPCC, 1997, Vol. 3, p. 5.31). In the Protocol context, changes in stock resulting from decay could be calculated by using a method analogous to that described above for soil carbon.
- Calculation of the net change in aboveground biomass (biomass before clearing minus biomass that regrows on the land—any new crop or pasture—plus any original biomass that was not completely cleared) could be carried out for each relevant forest type and, if appropriate, by region within a country. Again, these records could be geo-referenced. To assess carbon stocks prior to deforestation, data from that area before deforestation could be used. If the estimation is made after deforestation occurs, control plots could be used.
- The fate and amount of below-ground biomass (coarse woody roots, etc.) (Kurz *et al.*, 1996; Cairns *et al.*, 1997) could be taken into account (IPCC, 1997, Vol. 3, pp. 5.12). Such below-ground biomass could be treated as slash, perhaps with a longer decay time (IPCC, 1997,

Vol. 3, pp. 5.53). The Guidelines provide estimates of root-to-shoot mass ratios for such calculations.
- Given variations in burning practices among regions, users could provide their own information on the fate of biomass that is cleared to reflect practices in the country or region (IPCC, 1997, Vol. 3, pp. 5.30).

6.3.2. *Additional Human-Induced Activities under Article 3.4*

6.3.2.1. *Generic Issues on Additional Activities*

This section considers the adequacy of the Guidelines to account for changes in carbon stocks arising from additional human-induced activities in agricultural soils and LUCF (Chapter 4). Many of the additional activities under consideration are described in the Guidelines, particularly in the Reference Manual.

The key issue that emerges from Article 3.4 is the question of which activities will be included for the purpose of meeting commitments. Chapter 4 notes that the Parties have two broad choices with respect to including activities under Article 3.4:

- Include a limited, selected set of activities (narrow definition)
- Include all activities that can be shown to have consequences on the atmospheric concentration of GHGs (broad definition).

Chapter 4 examines effects in reporting, verifiability, and other issues related to the choice between the narrowly and broadly defined set of additional human-induced activities that could be included under Article 3.4. Chapter 4 identifies the following categories of additional human-induced activities that could be included under Article 3.4:

- Cropland management (including agricultural intensification, conservation tillage, erosion reduction, management of rice agriculture)
- Grazing lands management (including influencing degradation processes, grazing management, protected grasslands and set-asides, grassland productivity, fire management)
- Agroforestry (including conversion from forests to slash-and-burn to agroforests after deforestation, conversion from low-productivity croplands to sequential agroforestry in Africa, improved agroforests)
- Forest management (including forest regeneration, forest fertilization, fire management, pest management, harvest quantity and timing, low-impact harvesting, reducing forest degradation)
- Wetlands management (including wetland conversion to agriculture or forestry, wetland conversion to urban or industrial land, impoundments, wetland restoration)
- Restoration of severely degraded lands (including salt-affected soils, badly eroded and desertified soils, mine spoils, and industrially polluted sites)
- Urban and peri-urban land management.

Many of the foregoing land-use and management practices affect the storage of carbon below ground. Such practices include reduced and no tillage, livestock grazing, shifting of cultivation, fallow rotation, improved and degraded pastures, wetland rice, and agroforestry. The Guidelines provide the framework for such treatment of changes in soil carbon caused by agricultural practices. In addition to changes associated with clearing native vegetation (described earlier for deforestation), Section 5.4.2 of the Reference Manual describes the treatment of the effects of land abandonment, shifting cultivation, differing residue addition levels, differing tillage systems, and agricultural use of organic soils. The underlying principles can then be applied to other activities.

6.3.2.2. *Specific Issues on Additional Activities*

Some of the specific practices within the activities are explicitly included in the Workbook, some could be included with minor modification (see Section 6.3.2.1), and some could be included only with extensive revision. For example, in the cropland management activity, tillage practices are explicitly included, whereas erosion reduction practices are not currently dealt with. The Fact Sheets in Appendix A of Chapter 4 provide a brief assessment of the extent to which the Guidelines can be used to account for the individual activities.

Chapter 4 also identifies the potential carbon net storage of improved management within a land use, land-use change, and off-site carbon storage. Activities with the greatest potential change in the aboveground and below-ground biomass are discussed below.

Cropland management. The following practices are aimed at increasing soil carbon or reducing losses of soil carbon: cropland management on non-flooded soils, conservation tillage, increased crop productivity (i.e., carbon inputs), control of soil water, and erosion reduction. This last activity is a source, so the issues raised above about changes in stocks in 1990 are relevant. The default assumptions in the Guidelines can be adjusted to account for changes in cropland management. For intensive agriculture, these assumptions would be the input factors for carbon inputs. For conservation tillage, the appropriate factors are the tillage assumptions. There are no specific factors to deal with soil erosion or irrigation, except through the use of input factors because these values determine the carbon residues.

Grazing lands management. Increases of stocks of aboveground and below-ground carbon could be accounted for in a manner analogous to that for ARD. Activities that enhance uptake by sinks include those that increase tree numbers; activities that reduce emissions from sources are those that reduce the intensity of grazing. Non-CO_2 emissions would be reported under the agriculture sector. Because of the extensive nature of grasslands, the costs of measurement and verification may need to be considered. Changes in fire management would require classification of "management" because experience has shown that the current guidelines are ambiguous. Although the Guidelines do not specify grazing lands, again the input factors could be adjusted to reflect changes in this management practice. Any grazing land under set-aside management would result in uptake of aboveground biomass, which could be accounted for under the "abandonment of managed lands" section in the Guidelines.

Forest management. The build-up of carbon in protected areas could be accounted for if a definition of human-induced activity were to include the act of protection. Agroforestry and forest management (fertilization, thinning, etc.) could be reported in a manner analogous to ARD. The Guidelines can be used for estimating changes in aboveground biomass. As above, changes in carbon stock resulting from changes in forest management could be accounted for under the calculation in the Guidelines for soil carbon and aboveground biomass. Specific data for the assumptions—such as input and tillage factors—would have to be generated, however, as would assumptions for aboveground biomass.

Wood products. If an activity that brings about changes in wood products is included under Article 3.4, the approach in the Guidelines would not be adequate to estimate such changes.

6.3.2.3. Fossil Fuel Substitution

The Guidelines capture reduced emissions resulting from fossil fuel substitution in the energy sector. Biomass fuels include wood, wood waste, charcoal, bio-alcohol, biogas, and so forth (IPCC, 1997, Vol. 1, p 1.20). The Guidelines state that "Biomass fuels are included in the national energy and carbon dioxide emissions accounts for information only. Within the energy module biomass consumption is assumed to equal its regrowth. Any departures from this hypothesis are counted within the Land Use Change and Forestry module" (IPCC, 1997, Vol. 2, p. 1.3). Specifically, under the UNFCCC, the accounting of biomass changes occurs in the "changes in forest and other woody biomass stocks" category of the LUCF module. Hence, fossil fuel substitution by reduced carbon dioxide emissions in the energy sector is already "rewarded."

To avoid underreporting, therefore, any changes in biomass stocks on lands under Article 3.4 resulting from the production of biofuels would need to be included in the accounts. For non-CO_2 greenhouse gases, the situation differs because "non-CO_2 greenhouse gas emissions from biomass as fuels are included in the Energy sector" (IPCC, 1997, Vol. 2, p. 1.15); hence, underreporting is not expected to occur.

6.3.3. Forest Harvesting and Wood Products

In the Guidelines, all CO_2 emissions and removals associated with forest harvesting and the oxidation of wood products are accounted for in the year of harvesting by the country in which the wood was grown. This approach may be inaccurate because of the underlying assumption that there is no change in the size of the wood products pool.

To resolve this issue, the IPCC (Brown *et al.*, 1999; Lim *et al.*, 1999a) identified three approaches for estimating emissions and removals of CO_2 from forest harvesting and wood products. All of these approaches are more data intensive than the current Guidelines, and concerns over their practicality have been raised (Maclaren, 1999; see also Chapter 2). Furthermore, if the Guidelines are applied, CO_2 emissions from forest harvesting are already counted; therefore, these emissions should not be counted again. If wood products are brought into the inventory under the Protocol, however, more accurate methods—such as those described below—may be elaborated. Further work, however, cannot be initiated until there is guidance from the Parties on which, if any, of the approaches below is adopted.

Stock-change approach. This approach estimates net changes in carbon stocks in the forest and wood products pool. Changes in carbon stock in forests are accounted for in the country in which the wood is grown (the producing country). Changes in the products pool are accounted for in the country where the products are used (the consuming country). These stock changes are counted within national boundaries, where and when they occur.

Production approach. This approach also estimates net changes in carbon stocks in the forest and wood products pool, but it attributes both to the producing country. This approach inventories domestically produced stocks only and does not provide a complete inventory of national stocks. Stock changes are counted when but not where they occur if wood products are traded.

Atmospheric-flow approach. This approach accounts for net emissions or removals of carbon to and from the atmosphere within national boundaries, where and when the emissions and removals occur. Removals of carbon from the atmosphere resulting from forest growth is accounted for in the producing country; emissions of carbon to the atmosphere from oxidation of harvested wood products are accounted for in the consuming country.

6.4. Adequacy of the IPCC Guidelines for Reporting Projects under the Kyoto Protocol

This section assesses the adequacy of the Guidelines for estimating changes in stocks associated with LULUCF projects. It briefly summarizes the kinds of changes to the Guidelines that may be needed, recognizing that there are aspects of reporting that are of a policy nature (e.g., certification) that are not addressed here.

Under the Kyoto Protocol, reporting of projects will be required. The Guidelines do not address projects, although many of the accounting methods can be applied to estimate emissions and removals of GHGs, changes in carbon stock, and associated activities. At this time, it is unclear which LULUCF activities will be included under Articles 6 and 12 and potentially under Articles 3.3 and 3.4. Clearly, however, many of the data and reporting requirements are similar to those of Articles 3.3 and 3.4 (see Section 6.3) if comprehensive accounting were applied. In addition, the following issues arise (see Section 6.2 and Chapter 5; Vine *et al.*, 1999).

The degree of improvements to the Guidelines depends on their intended use. Within each of the options below, Parties could choose their own methods, or internationally agreed methods could be stipulated. There are three main options of increasing complexity:

Option 1. If the Guidelines are intended to show impacts to a Party's assigned amount, a summary of changes in net GHG emissions and removals from all projects may be all that is required. The first step would involve the transfer of GHG emissions and removals from the project to the national inventory; in the second step, the GHGs from projects would be subtracted from the national inventory.

- *Option 1a*—This option would allow Parties to choose a different reporting format for projects approved by the Parties. An example would be a modification of the URF used in the Activities Implemented Jointly Pilot Project (Annex III, FCCC, 1997b; Vine *et al.*, 1999, Section 5.4.6). This option is the simplest to operationalize, but it lacks transparency in methods and data.
- *Option 1b*—This option could stipulate the use of internationally agreed methods and perhaps involve the use of guidance for the URF.

Option 2. If the Guidelines are intended to enable Parties to generate a complete and transparent record of GHG emissions and removals from each project—including a breakdown of all activities, baselines, leakage, risks, and so forth (see methods in Chapter 5)—considerable changes will be required.

Option 3. Finally, there may be a need to report all changes in net GHG emissions and removals, as well as environmental and socioeconomic impacts, especially for CDM projects (Article 12).

In option 1, a reporting format for projects could be developed with the Guidelines with minimal effort. This option would require a decision to report data at an appropriate, aggregated scale. For options 2 and 3, considerable effort may be required to qualify for carbon credits. The period involved in developing these options would depend on the decisions of the Parties.

6.4.1. Methodological Issues

Several characteristics of projects have implications for reporting. Many of these characteristics are unique to projects but are omitted from the Guidelines. Chapter 5 describes these characteristics in more detail, but we briefly discuss the reporting implications here:

Comprehensive carbon accounting. For comprehensive accounting, an extension beyond the Workbook methods may be required if additional carbon pools are to be credited.

Baselines. The Guidelines can be used to report carbon stock changes in a given year, not for estimating baselines. In the Guidelines, the change in stock is simply derived from estimates of stock at two points in time; this approach may not be sufficient, however, to capture the features discussed in Chapter 5. For example, if the GHG benefits of a project are to be estimated relative to a baseline, further development of the Guidelines would be required. To develop such methods, the Parties would have to decide whether to apply a baseline and, if so, how this baseline would be calculated. The use of baselines in the Guidelines—such as business-as-usual—would deviate from the basic principles of the Guidelines, but it would be technically feasible.

Leakage (off-site impacts). The Guidelines do not currently address issues related to off-site impacts or leakage of GHG benefits. Leakage in LULUCF projects can occur as a result of inappropriately defining project boundaries for estimation; inappropriately defining the lifetime of monitoring activities for the project; forestry and non-forestry sector policies implemented by national/international governments, industry associations, or other entities; market effects; and activity shifting. Such off-site impacts (leakage) may be more difficult to quantify than on-site impacts. Because estimating changes in carbon stock can sometimes be difficult, the level of precision and confidence levels associated with such impacts may also need to be reported. The Guidelines may need to incorporate off-site changes to facilitate estimates of net changes in carbon stocks. Addressing leakage is one of the most challenging issues of significance to project-based activities. Section 5.3.3 discusses several approaches (e.g., benchmarks, risk co-efficients, etc.) to account for and monitor leakage. Each of the approaches implies a differing level of complexity for accounting and reporting.

Permanence. The permanence of carbon sequestration is potentially important for reporting projects. Chapter 2 discusses options for developing a time equivalence factor for reporting permanence. If one of these options were adopted by Parties, there would be implications for reporting: the timeline of a project—how the endpoint of a project is defined (indefinite, 100 years, 50 years, or variable—left up to each Party); accounting for time and the use of discounting; the appropriate carbon accounting methods (carbon stock method, average stock method, and ton-C year); and liability and allocation of responsibilities for ensuring compliance. Many of the foregoing options extend beyond the approach in the Guidelines.

Sustainable development. The CDM may introduce a new set of project-level issues—namely, assessment of net environmental and socioeconomic impacts (Chapter 5). Such issues are not addressed in the current Guidelines. One option is for Parties to take a decision on developing new guidelines for these issues. The absence of such guidelines may result in inconsistent responses to the Protocol and cause projects to be favored in areas with less stringent regulations. Guidelines could be developed to ensure consistency for assessing and reporting for net environmental and socioeconomic impacts; developing a common set of criteria may be difficult, however.

Uncertainties. The ranges of uncertainty of GHG emissions and removals of projects and national inventories are likely to differ. Because the national inventory and projects may be combined in a common reporting framework, however, these differences in uncertainty should be addressed. There is concern that changes in carbon stock at the project level may not be detectable at the national level because the former are too small and fall within the error bounds of a national inventory, or they occur in geographic areas that are not recorded in a national GHG inventory. In the first case, project reporting should be kept in perspective because it may be unnecessary to identify impacts in detail when national and global measurements are much less precise. A concern arises only if the absolute uncertainty around GHG estimates of projects is greater than that of the national inventory. To assess the importance of this issue, reporting of uncertainties of GHG estimates is crucial.

Double-counting/crediting. There are key questions about double-counting/crediting the impacts of projects, as with other sectors. Double-counting/crediting can be separated into three distinct types and can occur between:

- Two sectors—Reduced fossil fuel emissions from a bioenergy project should not be credited within both the LUCF and the Energy sectors of the Guidelines.
- Two Parties—Emissions reduction may be claimed by both host and investor in the reporting of credits. This double-counting could be avoided easily if each project had an identification number, enabling an auditor to locate a corresponding credit/debit on another Party's inventory.
- National GHG inventories and projects—For Article 6 projects, credits and associated emissions reductions could be double-counted in the national inventories of Annex 1 countries.

The Guidelines were not designed for reporting of these features or for features such as project locations and boundaries, additionality, and biodiversity. The Guidelines do provide a basis, however, for reporting changes in carbon stocks for projects in a single year. To avoid double-counting, projects and national inventories should be reported a consistent way in space and time. Ideally, national reporting for the Protocol and the UNFCCC could be integrated.

6.4.2. Data Requirements

This section discusses the data requirements for projects, which are required for the accounting. Like all projects, LUCF projects have several stages: (1) project design and registration at a national, regional, or international center; (2) project implementation, involving monitoring and evaluation; and (3) project verification, leading to certification. This section focuses on the second stage (implementation), although selected data may need to be collected and reported at each stage.

Examples of data to be reported are given in Table 6-4. Further details appear elsewhere in this report (Table 6-3; Chapter 5) and are not discussed here.

The most credible project results are derived from project-specific measurements. There is some concern that an arduous project-by-project review of the project data might impose prohibitive costs. Some researchers have proposed an alternative approach based on a combination of performance benchmarks and procedural guidelines that are tied to appropriate measures of output. In all cases, measurement and verification of the actual performance of the project are required. The performance benchmarks for new projects could be chosen to represent the high-performance end of the spectrum of current commercial practice (e.g., representing roughly the top 25th percentile of best performance). In this case, the benchmark serves as a goal to be achieved. In contrast, others might want to use benchmarks as a reference or default baseline—as an extension of existing technology, and not representing the best technology or process.

6.4.3. Frequency and Period of Reporting

Currently, Annex I Parties report their national inventories annually. Much of the discussion below is also relevant to the accounting for national inventories. The frequency and period of reporting for projects has not been defined under the Protocol. For projects, the frequency of reporting could be linked to:

- Recognition of carbon credits, which could be annual
- The duration of the commitment period (currently 5 years for quantified emission limitation or reduction commitment, but could be different for projects)
- Other factors, such as duration of the project, and the adjustment of credits based on the period of reporting.

For practical purposes, Chapter 5 assumes that the reporting period equals the project lifetime. For example, if it takes 20 years to complete a reforestation project, net carbon sequestered from the project could continue to be monitored for 20 years for crediting emissions reductions. Furthermore, if annual inventories are required for assessing compliance for a commitment period, there may be a problem if project information is not available until some time after the end of a commitment period. This problem applies equally to emissions reductions for projects in other sectors. Hence, two reporting

Table 6-4: *Examples of LUCF projects and their corresponding without-project cases.*

Project-Level Features	Data Requirements
1. Baseline carbon stock	Amount of carbon stored by carbon pool
1.1. Free riders	Amount of carbon stored by carbon pool
2. Change in carbon stock due to project	Amount of carbon stored by carbon pool
2.1. Leakage	Amount of carbon lost due to leakage
2.2. Positive spillover	Amount of carbon gained due to spillover
2.3. Market transformation	Amount of carbon gained due to market transformation
3. Additionality	Additional (net) amount of carbon stored by carbon pool
4. Data collection and analysis methods	Methods used in measuring carbon stored by project; methods used in estimating leakage, spillover, and transformation
5. Verifiability	Criteria used to evaluate quality of data collection methods and analysis
6. Uncertainty	Quantitative and qualitative indicators of precision; discussion of sources of uncertainty
7. Permanence	Assessment of permanence of project
8. Environmental impacts	List of environmental impacts affected by project; relationship to environmental impact statements and legislation
9. Socioeconomic Impacts	List of socioeconomic impacts affected by project

issues arise: How is the endpoint of a project defined? What is an appropriate monitoring frequency? These reporting issues are also related to the question of permanence.

6.5. Options for Improving the Revised 1996 IPCC Guidelines for National Greenhouse Gases under the Kyoto Protocol

Depending on the decisions made by Parties, improvements to the Guidelines likely will be needed to estimate and report emissions and removals of GHGs, stock changes, and associated activities under Articles 3.3, 3.4, 3.7, 6, and 12. This section proposes options to improve the Guidelines in relation to the verifiability and transparency of reported data.

For reporting of changes in carbon stocks and GHG fluxes under the Protocol, the comprehensive approach of the Reference Manual covers most of the relevant issues, including definitions. The accounting methods and default data in the Workbook, however, are limited to two carbon pools; the Workbook and the Reporting Instructions are likely to need further work if more explicit methods and definitions are required under the Protocol. The generic areas for possible improvement include the following:

- Reporting of defined activities and changes in carbon stocks associated with them
- Methods of accounting for carbon stocks not currently provided in the Guidelines
- Methods of data collection and evaluation of data quality (uncertainties)
- Methods to integrate changes in aboveground biomass, soil carbon, and other pools and reporting of these changes spatially (geo-referencing) and by activity
- Reporting for projects that takes into account baselines, system boundaries, leakage, and additionality, as well as socioeconomic and non-GHG environmental impacts
- Reporting tables for inventories and projects that differentiate between the base year, 1990, and the commitment period. Projects that qualify under Article 6, and potentially those under Article 12, would have to be distinguished, reported, and treated differently in the national GHG inventories of Annex I and non-Annex I Parties.

6.5.1. Elements of Reporting

Article 3.3. If all carbon pools are to be included when ARD activities are reported, it would be helpful to provide accounting methods in the Guidelines for pools other than aboveground biomass and soil carbon. If methods other than those in the Guidelines are used, a description of the methods of measurement and analysis would improve transparency. If ARD activities have to be reported separately, reporting guidelines will have to be improved. For pools that the Guidelines already deal with, refinements may also be required to better reflect nonlinear decay processes in the soil, for example. Priority could be given to improving methods for carbon pools that undergo significant changes in carbon stocks, as well as pools that are to be used in assessing compliance with commitments. For improved transparency and verifiability, reporting for ARD may need to be geo-referenced.

Article 3.4. Reporting of additional activities could be similar to reporting of ARD. As before, the key decisions are generic—for example, the issue of comprehensive accounting for GHG. Unlike Article 3.3, however, potential activities under Article 3.4 are not yet defined by the FCCC. This circumstance limits the extent of the discussion here. Most broadly defined, however, additional activities (FCCC, 1998a) can be accounted by the Guidelines. The definitions of some narrowly defined activities might need to be clarified.

Articles 6 and 12. The Guidelines were not designed for reporting of projects, and the degree of their improvement would depend on their use. At a minimum, reporting of a summary of changes in net GHG emissions and removals from all projects may be all that Parties decide is required. If so, this option would be relatively easy to operationalize. If a complete record of carbon impacts of each individual project is required, however, the level of effort to develop such guidelines could be considerable. Similarly, developing guidelines to estimate environmental and socioeconomic impacts may be equally involved. Whichever options are chosen, the factors to be considered include baselines, carbon stocks conserved or changed, GHG emissions and removals, system boundaries, leakage, positive spillover, reporting for monitoring of different activities, identification of carbon stock changes for different years, and frequency of measurements and reporting. The level of accuracy required for projects is generally higher than that for national inventories, so the reporting requirements for verification and monitoring are likely to differ from those for national inventories. For each of the options identified above, a key decision is whether Parties could choose their own accounting methods or whether internationally agreed methods would be stipulated for reporting—and if so, for which of the foregoing issues.

Table 6-5: *Options (unranked) for additions/modifications to the IPCC Guidelines.*

Issue	Options
Definitions	Clarify or expand definitions of forests, ARD, and additional activities
Carbon stocks	Modify methods, worksheets, and tables so that Parties can report data by explicit geographical locations for aboveground biomass, below-ground biomass, and soil carbon, consistent with the concept of full carbon accounting, including baselines; develop additional methods, worksheets, and reporting tables to account for storage of off-site carbon, such as wood products
Non-CO_2 emissions	Identify additional or existing methods to report additional activities that have large associated sources of emissions of non-CO_2 gases; develop methods to account for non-CO_2 gases from forest fires—not from forest clearing
Reporting at different scales	Modify and develop reporting tables for project and national reporting such that they are scientifically consistent in space and time
Transparency	Modify tables to allow Parties to specify the type of method and data used, sampling strategy, etc.
Data quality	Develop methods for collecting, monitoring, and reporting of data, along with statistical assessments of accuracy (good practice) and indicative costs
Uncertainties	Develop procedures to input data variability and calculate overall uncertainties for national and/or project activities
Verification	Develop methods for verifying data, along with indicative costs and accuracy, enabling Parties to make appropriate choices
Sustainable development	Develop guidance to assess the impacts of the Protocol on sustainable development and on the environment
Assistance	Provide technical and financial assistance to data-poor countries to generate and verify data required

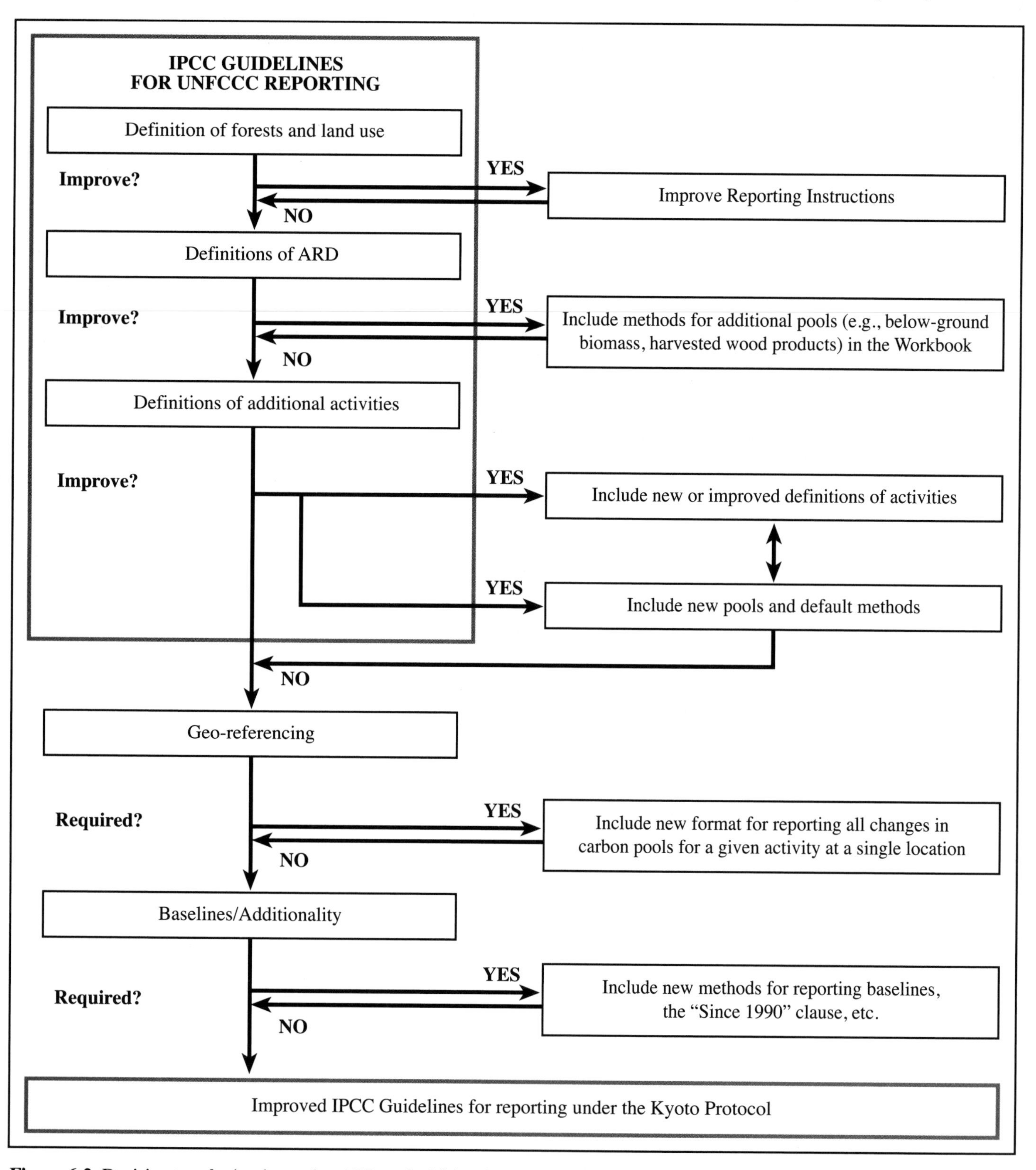

Figure 6-2: Decision tree for implementing ARD and additional activities at the national level: implications for the IPCC Guidelines.

Although the reporting needs of the Protocol differ significantly at the project and national levels, several generic issues apply. A key decision that could apply to several Articles in the Protocol is the treatment of wood products. Parties would need to indicate priorities before work could begin to improve the Guidelines, if at all. Some unranked options are included in Table 6-5.

6.5.2. *Approaches for Improving the IPCC Guidelines*

Several articles in the Protocol (e.g., Articles 3, 5, 6, 7, and 12; UNFCCC, 1997) imply the need for more detailed reporting. If Parties decide that reporting is to be different and more detailed for the Protocol, Article 5.2 of the Protocol provides a mechanism for Parties to initiate a process to improve the methods in the Guidelines.

Of importance in this context is a decision by the SBSTA at its eighth session encouraging the IPCC Inventories Program to "give high priority to completing its work on uncertainty, as well as to prepare a report on good practice in inventory management...." In addition, the "SBSTA invited the IPCC to develop a work plan addressing methodological issues raised in the IPCC special report in the context of the IPCC 1996 Revised Guidelines for National Greenhouse Gas Inventories in the areas of agriculture and land-use change and forestry, to commence as soon as practicable following completion of the special report" (FCCC, 1999). This work plan may include GPG in LUCF to be initiated in 2000.

A decision is required on what needs to be reported, and how, for each of the Articles. This decision would facilitate further assessment of the possible needs for further development of the Guidelines. Figure 6-2 illustrates a process for considering improvements to the Guidelines. It would be helpful if improved guidelines were consistent with the existing Guidelines to ensure compatibility with national reporting required under the UNFCCC. Potential constraints on reporting are the availability and quality of data, as well as the cost of generating and verifying this information. Furthermore, the demands on technical capacity and infrastructure for data collection have to be kept in mind, particularly for certain regions of the world.

Examples of several types of improvements are listed in Table 6-6. The time required for making such improvements will depend on the decisions of the Parties. A crucial decision is how to deal with project reporting. Projects are not currently covered by the Guidelines, and there are potentially more issues at the project level to be considered than at the national level.

***Table 6-6**: Examples of possible modifications to the IPCC Guidelines for reporting under the Protocol.*

Activity	Changes Required to the Guidelines	Examples of Reporting
Afforestation, reforestation, deforestation	1. Definitions in the Reporting Instructions	1. Changes in definitions of forest
	2. Reporting table	2. Changes in C stocks for areas under ARD activities as a subset of national inventory
	3. Worksheets and reporting table	3. Methods to estimate data for 1990 at national level for ARD activities
	4. Accounting approach	4. Augment carbon pools to include below-ground biomass, forest litter, soil, and wood products
	5. Accounting approach	5. Link changes in all carbon pools with geographically explicit information
	6. Accounting approach	6. Adjust nonlinear rates of change in soil carbon appropriate to commitment period
	7. Reporting table	7. Transparent reporting of sampling method used, sample size, year or periodicity of measurement, standard deviation, method of extrapolation
Additional activities	1. Worksheets and reporting table	1. Report explicitly additional activities such as sustainable forest management, crop management, agroforestry, etc.
	2. Accounting approach	2. Baselines
	3. Accounting approach	3. Augment carbon pools to include below-ground biomass, forest litter, and wood products
	4. Reporting table	4. Track changes in areas of land use to avoid double-counting
	5. Reporting table	5. Transparent reporting of sampling method used, sample size, year or periodicity of measurement, standard deviation, method of extrapolation
Project activities	1. Accounting method	1. Baseline
	2. Accounting method	2. Define and report boundary for the project and for reporting impacts on off-site C stocks
	3. Accounting method	3. Changes in C stocks due to leakage, positive spillover, and market transformation (particularly off-site)
	4. Worksheet	4. Changes in C stocks over commitment period and annually for assessing permanence
	5. Reporting table	5. Report method used, sample size, year or periodicity of measurement, standard deviation, method of extrapolation
	6. Accounting approach	6. Sustainable development

References

Brown, S., B. Lim, and B. Schlamadinger, 1999: *Evaluating Approaches for Estimating Net Emissions of Carbon Dioxide from Forest Harvesting and Wood Products.* IPCC Meeting Report, May 1998, Dakar, Senegal, International Panel on Climate Change/Organization for Economic and Commerical Development/International Energy Agency Programme on National Greenhouse Gas Inventories, Organization for Economic Cooperation and Development, Paris, France, 51 pp.

Cairns, M., S. Brown, E. Helmer, and G. Baumgardner, 1997: Root biomass allocation in the world's upland forests. *Oecologia,* **111,** 1–11.

FCCC, 1997a: *Methodological Issues. Synthesis of Information from National Communications of Annex I Parties on Sources and Sinks in the Land-use Change and Forestry Sector.* Technical Paper FCCC/TP/1997/5, 20 November 1997, 19 pp.

FCCC, 1997b: *Report of the Subsidiary Body for Scientific and Technological Advice on the Work of Its Fifth Session, Bonn, Germany, 25–28 February 1997. Annex III. Uniform Reporting Format: Activities Implemented Jointly Under the Pilot Phase.* Fifth Session of the Subsidiary Body for Scientific and Technical Advice, Bonn, Germany, 25–28 February 1997. FCCC/SBSTA/1997/4, 27 pp.

FCCC, 1998a: *Methodological Issues. Issues Related to Land-Use Change and Forestry.* Eighth Session of the Subsidiary Body for Scientific and Technical Advice, Bonn, Germany, 2–12 June 1998. FCCC/SBSTA/1998/INF.1, 21 pp.

FCCC, 1998b: *Report of the Conference of the Parties on its Fourth Session, Buenos Aires, Argentina, 2–14 November 1998: Addendum.* Fourth Session of the Conference of the Parties, Buenos Aires, Argentina, 2–14 November 1998. FCCC/CP/1998/16/ADD.1, 71 pp.

FCCC, 1999: *Report of the Subsidiary Body for Scientific and Technological Advice on its Tenth Session, Bonn, Germany, 31 May – 11 June 1999.* Tenth Session of the Subsidiary Body for Scientific and Technical Advice, Bonn, Germany, 31 May –11 June 1999. FCCC/SBSTA/1999/6, 37 pp.

FCCC, 2000: *UNFCCC Guidelines on Reporting and Review.* UNFCCC/CP/1999/7 and FCCC/SBTSA/1999/L.5, 122 pp.

Houghton, R., and K. Ramakrishna, 1999: A review of national emissions inventories from select Non-Annex I Countries: Implications for counting sources and sinks of carbon. *Annual Review of Energy and Environment,* **24,** 571–605.

IPCC, 1995: *IPCC Guidelines for National Greenhouse Gas Inventories* [Houghton, J.T., L.G. Meira Filho, B. Lim, and K. Tréanton (eds.)]. Intergovernmental Panel on Climate Change, Meteorological Office, Bracknell, United Kingdom.
- *Volume 1: Greenhouse Gas Inventory Reporting Instructions.*
- *Volume 2: Greenhouse Gas Inventory Workbook.*
- *Volume 3: Greenhouse Gas Inventory Reference Manual.*

IPCC, 1997: *Revised 1996 IPCC Guidelines for National Greenhouse Gas Inventories* [J.T. Houghton, L.G. Meira Filho, B. Lim, K. Tréanton, I. Mamaty, Y. Bonduki, D.J. Griggs, and B.A. Callander (eds.)]. Intergovernmental Panel on Climate Change, Meteorological Office, Bracknell, United Kingdom.
- *Volume 1: Greenhouse Gas Inventory Reporting Instructions.* 130 pp.
- *Volume 2: Greenhouse Gas Inventory Workbook.* 346 pp.
- *Volume 3: Greenhouse Gas Inventory Reference Manual.* 482 pp.

IPCC, 1999: *Good Practice Guidance and Uncertainty Management in National Greenhouse Gas Inventories: Draft for Government/Expert Review, December 1999.* National Greenhouse Gas Inventories Programme Technical Support Unit, Institute for Global Environmental Strategies, Kanagawa, Japan, 460 pp.

Kurz, W., S. Beukema, and M. Apps, 1996: Estimation of root biomass and dynamics for the carbon budget model of the canadian forest sector. *Canadian Journal of Forest Research,* **26(11),** 1973–79.

Lim, B., P. Frost, M. Apps, S. Brown, Y. Sokona, and Y. Bonduki, 1997: *Expert Group Meeting on Biomass Burning and Land-Use Change and Forestry.* IPCC Meeting Report, September 1997, Rockhampton, Australia, International Panel on Climate Change/Organization for Economic and Commerical Development/International Energy Agency Programme on National Greenhouse Gas Inventories, Organization for Economic Cooperation and Development, Paris, France, 21 pp.

Lim, B., S. Brown, and B. Schlamadinger, 1999a: Carbon accounting for forest harvesting and wood products: Review and evaluation of different approaches. *Environmental Science and Policy,* **2(2),** 207–216.

Lim, B., P. Boileau, Y. Bonduki, A.R. van Amstel, L.H.J.M. Janssen, J.G.J. Olivier, and C. Kroeze, 1999b: Improving the quality of national greenhouse gas inventories. *Environmental Science and Policy,* **2(3),** 335–346.

Maclaren, P., 1999: Carbon accounting and forestry: A review of the subsequent papers. *Environmental Science and Policy,* **2(2),** 207–216.

Ravindranath, N.H. and P. Sudha, 1999: *Emission Factors and Activity Data for the Land-Use Change and Forestry Sector of Greenhouse Gas Inventories in Developing Countries.* United Nations Framework Convention on Climate Change. Draft Technical Paper. 51 pp.

UNFCCC, 1994: *United Nations Framework Convention on Climate Change.*

UNFCCC, 1997: *Kyoto Protocol to the United Nations Framework Convention on Climate Change* (FCCC/CP/1997/L.7/Add. 1).

Vine, E., J. Sathaye, and W. Makundi, 1999: *Guidelines for the Monitoring, Evaluation, Reporting, Verification, and Certification of Forestry Projects for Climate Change Mitigation.* Research Report LBNL-41877, Lawrence Berkeley National Laboratory, Berkeley, CA. USA, 125 pp.

Winjum, J.K., S. Brown, and B. Schlamadinger, 1998: Forest harvests and wood products: Sources and sinks of atmospheric carbon dioxide. *Forest Science,* **44,** 272–284.

Land Use, Land-Use Change, and Forestry: Annexes

Prepared by the Intergovernmental Panel on Climate Change

A

Authors and Expert Reviewers

Argentina

Jorge Frangi	Universidad Nacional de la Plata
Carlos Scopa	Centro de Recursos Naturales CRN-INTA

Australia

Michele Barson	Bureau of Rural Sciences, Agriculture, Fisheries and Forestry
Josep. Canadell	GCTE International Project Office
Ian Carruthers	Department of the Environment, Sport and Territories
Graham Farquhar	Australian National University
Beverley Henry	Department of Natural Resources
Mark Howden	Bureau of Rural Sciences, Agriculture, Fisheries and Forestry
Mark Jackson	The Carbon Store Pty Ltd.
Rod Keenan	Bureau of Rural Sciences, Agriculture, Fisheries and Forestry
Miko Kirschbaum	CSIRO
Warren Lang	National Association of Forest Industries
Ian Noble	Australian National University
Mike Read	University of Melbourne
Penny Reyenga	Bureau of Rural Sciences, Agriculture, Fisheries and Forestry
Robert Vincin	Emission Traders Int'l Pty Ltd.

Austria

Klaus Radunsky	Federal Environment Agency
Ewald Rametsteiner	Liaison Unit of the Ministerial Conference on the Protection of Forests in Europe
Bernhard Schlamadinger	Joanneum Research
Anatoly Shvidenko	International Institute for Applied Systems Analysis
Traude Wollansky	Federal Ministry for the Environment, Youth and Family Affairs

Belgium

Danny Croon	European Confederation of Woodworking Industries
Cecile Dargnies-Peirce	European Commission
Stephan Singer	World Wildlife Federation
Oswald van Cleemput	University of Ghent
Dieter Schoene	European Commission
Guy Van Steertegem	European Confederation of Woodworking Industries
Marianne Wenning	European Commission

Benin

Epiphane Dotou Ahlonsou	Service Météorologique National

Brazil

Carlos Clemente Cerri	Universidade de Sao Paolo
Philip M. Fearnside	Instituto Nacional de Pesquisas da Amazonia
Luiz Gylvan Meira Filho	Ministry of Science and Technology of Brazil
Jose D.G. Miguez	Ministerio da Ciencia e Tecnologia, Esplanada dos Ministerios
Niro Higuchi	INTA
Thelma Krug	National Institute for Space Research
Warwick Manfrinato	University of Sao Paulo
Carlos A. Nobre	CPTEC-INPE
Newton Paciornik	Ministry of Science and Technology of Brazil
Laura Tetti	Uniao da Agroindustria Canavierira do Estado de Sao Paulo

Canada

Michael Apps	Northern Forest Centre
M. Banjaree	Agriculture and Agri-Food Canada
James P. Bruce	Canadian Climate Program Board
Julian Dumanski	Agriculture Canada
Henry Janzen	Agriculture and Agri-Food Canada
Werner Kurz	ESSA Technologies Ltd.
R. Lemke	Agriculture and Agri-Food Canada
C.W. Lindwall	Agriculture and Agri-Food Canada
Gordon McBean	Meteorological Service of Canada
B.M. McConkey	Agriculture and Agri-Food Canada
Fernando Selles	Agriculture and Agri-Food Canada
John Stewart	University of Saskatchewan (retired)
John M.R. Stone	Policy, Program and International Affairs Directorate
Martin von Mirbach	Centre for Forest and Environmental Studies

Chile

Aqiles Neuenschwander	Agrarian Innovation Foundation (FIA), Ministry of Agriculture

China

Lin Erda	Agrometeorology Institute
Wen Kegang	China Meteorological Administration
Deying Xu	Chinese Academy of Forestry

Denmark

Frans Richard Bach	National Forest and Nature Agency
Jesper Gundermann	Danish Energy Agency

Finland

Erkki J. Jatila	Finnish Meteorological Institute
Timo Karjalainen	European Forest Institute
Pirkko Kortelainen	Finnish Environment Institute
Gert Jan Nabuurs	Institute Forestry and Natural Research

France

Jean-Christophe Calvet	Meteo-France/CNRM
Philippe Ciais	LSCE/DSM
Marc Gillet	Mission Interministerielle de l'Effet de Serre
Paul-Antoine Lacour	AFOCEL
Arthur Riedacker	Mission Interministerielle de l'Effet de Serre
Nicolas Viovy	LSCE

Gambia

Bubu Pateh Jallow	Department of Water Resources

Germany

Klaus Boswald	PrimaKlima
Peter Burschel	Lehrstuhl fur Waldbau und Forsteinrichtung
Wolfgang P. Cramer	Potsdam Institute for Climate Impact Research
Harald Kohl	Federal Ministry of the Environment, Nature Conservation and Nuclear Safety
Rainer Matyssek	Technical University of Munich
Martina Mund	Max Planck Institute for Biogeochemistry
Dieter R. Sauerbeck	Counsellor to the German Federal Ministry of the Environment
Ernst-Detlef Schulze	University of Bayreuth
Eveline Trines	UNFCCC Secretariat
Jelle G. van Minnen	University of Kassel
Michael Weber	Lehrstuhl für Waldbau und Forsteinrichtung der Ludwig-Maximilians Universität München
Christian Wirth	Max-Plank-Institut fur Biogeochemie

Iceland

Halldor Thorgeirsson	Ministry for the Environment

India

M.R. Hubert	Shabnam Resources
Kilaparti Ramakrishna	Woods Hole Research Center
N.H. Ravindranath	Indian Institute of Sciences
Tejpal Singh	Tata Energy Research Institute
P. Sudha	Centre for Ecological Sciences
Raman Sukumar	Indian Institute of Science

Indonesia

A. Ngaloken Gintings	Forest Product and SocioeconomicResearch and Development Centre
Nur Masripatin	Forest Planning Agency/Ministry of Forestry and Estate Crops
Daniel Murdiyarso	BIOTROP-GCTE
Reider Persson	Center for International Forestry Research

Italy

Lorenzo Ciccarese	ANPA, National Environmental Protection Agency
Louise Fresco	Food and Agriculture Organization
Domenico Gaudioso	ANPA, National Environmental Protection Agency
Antonio Lumicisi	Ministry of the Environment
Riccardo Valentini	University of Tuscia - DISAFRI

Japan

Masahiro Amano	Forestry and Forest Products Research Institute
Mariko Handa	Parks and Recreation Foundation
Gen Inoue	National Institute for Environment Study
Kyoko Kawasaka	Citizens Alliance for Saving the Atmosphere and the Earth
Hirofumi Kazuno	Ministry of Foreign Affairs
Katsuyuki Minami	National Institute of Agro-Environmental Sciences
Osamu Mizuno	Environment Agency of Japan
Yasuhisa Tanaka	Ministry of Agriculture, Forestry and Fisheries
Toshihiko Yagi	Citizens Alliance for Saving the Atmosphere and the Earth
Yoshiki Yamagata	National Institute for Environmental Studies
Hironobu Yokota	Environment Agency of Japan

Kenya

Pedro Sanchez	ICRAF

Malaysia

Faizal Parish	Global Environment Centre

Mexico

Francisco Giner de los Rios	National Institute of Ecology
Julia Martinez	National Institute of Ecology
Omar Masera	Universidad Nacional Autonoma de Mexico
Maria Villers -Ruiz	Instituto de Geografia

New Zealand

Stuart Calman	Ministry for the Environment
Justin Ford-Robertson	Forest Research Institute
Piers Maclaren	Forest Research Institute
Martin Manning	National Institute of Water and Atmospheric Research
Helen Plume	Ministry for the Environment
Peter L. Read	Massey University
Neal Scott	Landcare Research
Ralph Sims	Massey University
Peter Sligh	Forest Industries Council

Norway

Nils Bohn	Norwegian Forest Owners Federation
Georg Borsting	Ministry of the Environment

Tore Braend — Norwegian Society of the Conservation of Nature
Oyvind Christophersen — Ministry of Environment
Birger Solberg — Institute of Forest Sciences, Agricultural University

Peru
Eduardo Calvo — Comision Nacional de Cambio Climatico
Paul Remy

Philipines
Rodel D. Lasco — University of Philippines
Wojciech Galinski — Silvatica-Research Consultants
Lech Ryszkowski — Polish Academy of Sciences

Russia
Yuri Izrael — Institute of Global Climate and Ecology
Alexey Kokorin — Institute of Global Climate and Ecology
Serguei Semenov — Institute of Global Climate & Ecology

Samoa
H.E.T. Neroni Slade — Permanent Mission of Samoa to the UN

Senegal
Youba Sokona — ENDA -TM

South Africa
R.O. Barnard — Agricultural Research Council: Institute for Soil, Climate and Water
Gerrie Coetzee — Department of Environmental Affairs and Tourism
P.J. Maritz — Directorate Agricultural Water Use Management
D.J. Pretorius — National Department of Agriculture
Robert J. Scholes — CSIR
A.T. van Coller — Directorate Agricultural Water Use Management
D.W. Van der Zel — Department of Water Affairs and Forestry (Retired)

Spain
A. Labajo — Instituto Nacional de Meteorologia
Gerardo Sanchez Pena — DGCN
Josep Penuelas — Center for Ecological Research and Forestry Applications

Sweden
Bert Bolin — University of Stockholm (retired)
Bengt Bostrom — Swedish National Energy Administration
Sune Linder — University of Agricultural Sciences
I. Colin Prentice
Will Steffen — IGBP Secretariat
Anders Turesson — Ministry of the Environment

Switzerland
Andreas Fischlin — Terrestrial Systems Ecology ETHZ
Fortunat Joos — University of Bern
Jose Romero — Office Federal de l'Environnement

Tanzania
Willy R. Makundi — Lawrence Berkeley National Laboratory
Stephen Mwakifwamba — Center for Energy, Environment, Science and Technology

Thailand
Kansri Boonpragob — Ramkhamhaeng University
Sitanon Jesdapipat — Thailand Environment Institute
Nipon Tangtham — Kasetsart University
Vute Wangwacharakul — Kasetsart University

The Netherlands
R. Aerts — Vrije Universiteit
W.L. Hare — Greenpeace International
Peter Kuikman — DLO-Research Institute for Agrobiology & Soil Fertility (AB-DLO)
Richard Sikkema — FORM Ecology
Wim Sombroek — ISRIC
Andre van Amstel — Wageningen University
J. Verbeek — Ministry of Transport, Public Works and Water Management

Trinidad
Floyd Homer — Ministry of Natural Resources

United Kingdom
Jonathan M. Anderson — University of Exeter
Mark Broadmeadow — GB Forestry Commission Research Agency
Melvin G. R. Cannell — Institute of Terrestrial Ecology
Sam Evans — GB Forestry Commission Research Agency
Paul Jarvis — Edinburgh University
Gerald Adrian Leach — Stockholm Environment Institute
Bo Lim — United Nations Development Programme
Larry Lohmann — The Corner House
Robert Matthews — GB Forestry Commission Research Agency
M. McKenzie Hedger — University of Oxford
Ronald Milne — Institute of Terrestrial Ecology
Pedro Moura-Costa — Eco Securities
J.M. Penman — Department of the Environment, Transport and the Regions
Gareth Phillips — SGS Forestry

Michelle A. Pinard — University of Aberdeen
Keith A. Smith — SAC Edinburgh
Peter Smith — IACR-Rothamsted
Richard Tipper — IERM - University of Edinburgh
David Warrilow — Department of the Environment, Transport and the Regions
Andrew Whitmore — Silsoe Research Institute

United States

Ralph Alig — U.S. Department of Agriculture
Margot Anderson — U.S. Department of Energy
Mike Anderson — U.S. Department of Agriculture
Kenneth Andrasko — U.S. Environmental Protection Agency
Dan Balzer — U.S. Department of State
Wiley Barbour — U.S. Environmental Protection Agency
Jeri Berc — U.S. Department of Agriculture
Richard Birdsey — U.S. Forest Service
Evan Bloom — U.S. Department of State
Janine Bloomfield — Environmental Defense Fund
Robert Bonnie — Environmental Defense Fund
Barbara Braatz — ICF Consulting
William Breed — U.S. Department of Energy
Jean Brennan — USAID
Sandra Brown — Winrock International
Paige Brown — Redefining Progress
Michael Buck — Hawaii Division of Forestry and Wildlife
Marilyn Buford — U.S. Department of Agriculture
Jay Chamberlin — The Pacific Forest Trust
Roger Dahlman — U.S. Department of Energy
David Darr — U.S. Department of Agriculture
Ben DeAngelo — U.S. Environmental Protection Agency
Lisa Dilling — NOAA Office of Global Programs
Craig Ditzler — U.S. Department of Agriculture
P. Doskey — Argonne National Laboratory
Jim DuBay — UtiliTree Carbon Company
Mitchell Dubensky — American Forests and Paper Association
Jerry Elwood — U.S. Department of Energy
Lauren Flejzor — U.S. Department of State
Alan J. Franzluebbers — U.S. Department of Agriculture
Peter Frumhoff — Union of Concerned Scientists
Charles T. Garten, Jr. — Oak Ridge National Laboratory
Noel Gerson — Department of the Interior
Donald Goldberg — American University
Anne Grambsch — U.S. Environmental Protection Agency
Kevin Green — U.S. Department of Transportation
Daniel Hall — American Lands
Steven Hamburg — Brown University
Richard R. Harwood — Michigan State University
Linda Heath — U.S. Department of Agriculture
John Hickman — Deere & Company Technical Center
Peter Hill — Monsanto Company
Bill Hohemstein — U.S. Environmental Protection Agency
Tom Houghtaling — UtiliTree Carbon Company
Richard Houghton — Woods Hole Research Center
Robert House — U.S. Department of Agriculture
Jim Hrubovcak — U.S. Department of Agriculture
Tony Janetos — World Resources Institute
Julie D. Jastrow — Argonne National Laboratory
Michael Jawsin — U.S. Department of Agriculture
Carol Jones — U.S. Department of Agriculture
Gary Kaster — UtiliTree Carbon Company
Holly Kaufman — U.S. Department of State
Haroon Kheshgi — Exxon Research and Engineering Company
John Kimble — U.S. Department of Agriculture
Anthony W. King — Oak Ridge National Laboratory
John Kinsman — UtiliTree Carbon Company
Anne Kinzig — Arizona State University
Rattan Lal — Ohio State University
Daniel Lashof — Natural Resources Defense Council
Eugene Lee — U.S. Environmental Protection Agency
Catherine Leining — Center for Clean Air Policy
Jan Lewandrowski — U.S. Department of Agriculture
H. Gyde Lund — Forest Information Services
Kenneth MacDicken — Center for International Forestry Research
William Mankin — Global Forest Policy Project
Gregg Marland — Oak Ridge National Laboratory
M. Mausbach — U.S. Department of Agriculture
John McCarty — U.S. Environmental Protection Agency
Stephanie Mercier — Senate Agriculture Committee
Mike Miller — Argonne National Laboratory
N.L. Miller — Department of Energy-Lawrence Berkeley National Laboratory
Jeff Miotke — U.S. Department of State
Robert A. Monserud — U.S. Department of Agriculture
Jennifer Morgan — World Wildlife Fund
Adele Morris — U.S. Department of the Treasury
Brian C. Murray — Research Triangle Institute
Tia Nelson — The Nature Conservancy
John O. Niles — Stanford University
Richard Norby — Oak Ridge National Laboratory
Marc Orlic — U.S. Environmental Protection Agency
Keith Paustian — Colorado State University
Dave Peters — U.S. Department of Defense
Annie Petsonk — Environmental Defense Fund
W.M. Post — Oak Ridge National Laboratory
Ray Prince — U.S. Department of Energy
Stephen P. Prisley — Virginia Polytechnic University
Ed Rall — U.S. Department of Agriculture
Donald C. Reicosky — U.S. Department of Agriculture
Gregory Ruark — U.S. Department of Agriculture

Marc Safley	U.S. Department of Agriculture
Neil Sampson	American Forests
Jayant Sathaye	Lawrence Berkeley National Laboratory
David L. Schertz	U.S. Department of Agriculture
William H. Schlesinger	Duke University
Jonathan Scurlock	Oak Ridge National Laboratory
Stephan Schwartzman	Environmental Defense Fund
Karn Deo Singh	Harvard University
Kenneth Skog	U.S. Department of Agriculture
David Skole	Michigan State University
Gordon Smith	Environmental Defense Fund
Jeffrey L. Smith	U.S. Department of Agriculture
Allen M. Solomon	Office of Science and Technology Policy
Susan Subak	Natural Resources Defense Council
Jeremy Symons	U.S. Environmental Protection Agency
Dennis Thompson	U.S. Department of Agriculture
Mark Trexler	Trexler and Associates
Susan Trumbore	University of California, Irvine
Edward Vine	Lawrence Berkeley National Laboratory
Rik Wanninkhof	NOAA
Laurie Wayburn	The Pacific Forest Trust
Larry Weber	Office of Science and Technology Policy
Darrell C. West	Oak Ridge National Laboratory
Tristram West	Oak Ridge National Laboratory
Diane Wickland	National Aeronautics and Space Administration
Tom Wirth	U.S. Environmental Protection Agency
Anny Wong	Global Forest Policy Project
Stan Wullschleger	Oak Ridge National Laboratory

Yugoslavia

Stanimir C. Kostadinov	Belgrade University

B

Acronyms, Abbreviations, and Units

ACRONYMS AND ABBREVIATIONS

ADB	Asian Development Bank
AEZ	Agro-Ecological Zone
AGWP	Absolute Global Warming Potential
AIJ	Activities Implemented Jointly
ARD	Afforestation, Reforestation, and Deforestation
AVHRR	Advanced Very High Resolution Radiometer
BAU	Business-as-Usual
CBD	U.N. Convention on Biological Diversity
CBL	Convective Boundary Layer
CD	U.N. Convention to Combat Desertification
CDM	Clean Development Mechanism
C-eq	Carbon-Equivalent
CER	Certified Emissions Reduction
CIFOR	Center for International Forestry Research
COP	Conference of Parties
CPC	Canopy (or crown) Projected Cover
CRP	Conservation Reserve Program
DHI	Direct Human-Induced
DOC	Dissolved Organic Carbon
DOM	Dissolved Organic Matter
EIA	Environmental Impact Assessment
ERU	Emissions Reduction Unit
ESA	European Space Agency
EU	European Union
FACE	Free Air Carbon Dioxide Enrichment; Forests Absorbing CO_2 Emissions (The Netherlands)
FAO	U.N. Food and Agriculture Organization
FCCC	Framework Convention on Climate Change
FFES	Fossil-Free Energy Scenario
FPC	Foliage Projected Cover
FSC	Forest Stewardship Council
GCM	General Circulation Model
GDP	Gross Domestic Product
GEF	Global Environment Facility
GHGs	Greenhouse Gases
GIS	Geographic Information Systems
GLASOD	Global Assessment of Soil Degradation
GPG	Good Practice Guidelines
GPP	Gross Primary Productivity
GPS	Global Positioning System
GWP	Global Warming Potential
IEA	International Energy Agency
IFF	Intergovernmental Forum on Forests
IGBP-DIS	International Geosphere-Biosphere Programme Data and Information System
INPE	Instituto Nacional de Pesquisas Espaciais (Brazil)
IPCC	Intergovernmental Panel on Climate Change
ISO	International Standards Organization
IUCN	International Union for the Conservation of Nature and Natural Resources
JI	Joint Implementation
LAI	Leaf Area Index
LUCF	Land-Use Change and Forestry
LULUCF	Land Use, Land-Use Change, and Forestry
MEA	Multilateral Environmental Agreements
MRG	Modalities, Rules, and Guidelines
MRT	Mean Residence Time
NASA	National Aeronautics and Space Administration
NBP	Net Biome Productivity
NDVI	Normalized Difference Vegetation Index
NEE	Net Ecosystem Exchange
NEP	Net Ecosystem Productivity
NGO	Nongovernmental Organization
NKCAP	Noel Kempff Climate Action Project
NMR	Nuclear Magnetic Resonance
NOAA	National Oceanographic and Atmospheric Administration
NPP	Net Primary Productivity
ODA	Official Development Assistance
OECD	Organisation for Economic Cooperation and Development
PAP	Protected Areas Project (Costa Rica)
PCSD	Philippine Council for Sustainable Development
PFP	Private Forests Project (Costa Rica)
RA	Autotrophic Respiration
RH	Heterotrophic Respiration
RIL	Reduced-Impact Logging
SAR	Second Assessment Report; Synthetic Aperture Radar
SARD	Sustainable Agriculture and Rural Development
SBSTA	Subsidiary Body for Scientific and Technological Advice
SIC	Soil Inorganic Carbon
SOC	Soil Organic Carbon
SOM	Soil Organic Matter
SR-LULUCF	Special Report on Land Use, Land-Use Change, and Forestry
TBFRA 2000	Temperate and Boreal Forest Resource Assessment 2000
TM	Thematic Mapper
UK	United Kingdom
UN	United Nations
UNCED	United Nations Conference on Environment and Development
UNCSD	United Nations Commission on Sustainable Development
UNEP	United Nations Environment Programme
UNESCO	United Nations Educational, Scientific and Cultural Organization
UNFCCC	United Nations Framework Convention on Climate Change

URF Uniform Reporting Format
USIJI U.S. Initiative for Joint Implementation
VCL Vegetation Canopy Lidar
VOC Volatile Organic Compound
WCFSD World Commission on Forests and Sustainable Development
WOCAT World Overview of Soil Conservation Approaches and Technologies

UNITS

SI (Systéme Internationale) Units

Physical Quantity	Name of Unit	Symbol
length	meter	m
mass	kilogram	kg
time	second	s
thermodynamic temperature	kelvin	K
amount of substance	mole	mol

Special Names and Symbols for Certain SI-Derived Units

Physical Quantity	Name of SI Unit	Symbol for SI Unit	Definition of Unit
force	newton	N	$kg\ m\ s^{-2}$
pressure	pascal	Pa	$kg\ m^{-1}\ s^{-2}$ (= Nm^{-2})
energy	joule	J	$kg\ m^2\ s^{-2}$
power	watt	W	$kg\ m^2\ s^{-3}$ (= Js^{-1})
frequency	hertz	Hz	s^{-1} (cycle per second)

Decimal Fractions and Multiples of SI Units Having Special Names

Physical Quantity	Name of Unit	Symbol for Unit	Definition of Unit
length	ångstrom	Å	10^{-10} m = 10^{-8}cm
length	micrometer	μm	10^{-6}m = μm
area	hectare	ha	10^4 m^2
force	dyne	dyn	10^{-5} N
pressure	bar	bar	10^5 N m^{-2}
pressure	millibar	mb	1hPa
weight	ton	t	10^3 kg

C

List of Major IPCC Reports

Climate Change—The IPCC Scientific Assessment
The 1990 Report of the IPCC Scientific Assessment Working Group (also in Chinese, French, Russian, and Spanish)

Climate Change—The IPCC Impacts Assessment
The 1990 Report of the IPCC Impacts Assessment Working Group (also in Chinese, French, Russian, and Spanish)

Climate Change—The IPCC Response Strategies
The 1990 Report of the IPCC Response Strategies Working Group (also in Chinese, French, Russian, and Spanish)

Emissions Scenarios
Prepared for the IPCC Response Strategies Working Group, 1990

Assessment of the Vulnerability of Coastal Areas to Sea Level Rise–A Common Methodology
1991 (also in Arabic and French)

Climate Change 1992—The Supplementary Report to the IPCC Scientific Assessment
The 1992 Report of the IPCC Scientific Assessment Working Group

Climate Change 1992—The Supplementary Report to the IPCC Impacts Assessment
The 1992 Report of the IPCC Impacts Assessment Working Group

Climate Change: The IPCC 1990 and 1992 Assessments
IPCC First Assessment Report Overview and Policymaker Summaries, and 1992 IPCC Supplement

Global Climate Change and the Rising Challenge of the Sea
Coastal Zone Management Subgroup of the IPCC Response Strategies Working Group, 1992

Report of the IPCC Country Studies Workshop
1992

Preliminary Guidelines for Assessing Impacts of Climate Change
1992

IPCC Guidelines for National Greenhouse Gas Inventories
Three volumes, 1994 (also in French, Russian, and Spanish)

IPCC Technical Guidelines for Assessing Climate Change Impacts and Adaptations
1995 (also in Arabic, Chinese, French, Russian, and Spanish)

Climate Change 1994—Radiative Forcing of Climate Change and an Evaluation of the IPCC IS92 Emission Scenarios
1995

Climate Change 1995—The Science of Climate Change – Contribution of Working Group I to the Second Assessment Report
1996

Climate Change 1995—Impacts, Adaptations, and Mitigation of Climate Change: Scientific-Technical Analyses – Contribution of Working Group II to the Second Assessment Report
1996

Climate Change 1995—Economic and Social Dimensions of Climate Change – Contribution of Working Group III to the Second Assessment Report
1996

Climate Change 1995—IPCC Second Assessment Synthesis of Scientific-Technical Information Relevant to Interpreting Article 2 of the UN Framework Convention on Climate Change
1996 (also in Arabic, Chinese, French, Russian, and Spanish)

Technologies, Policies, and Measures for Mitigating Climate Change – IPCC Technical Paper I
1996 (also in French and Spanish)

An Introduction to Simple Climate Models used in the IPCC Second Assessment Report – IPCC Technical Paper II
1997 (also in French and Spanish)

Stabilization of Atmospheric Greenhouse Gases: Physical, Biological and Socio-economic Implications – IPCC Technical Paper III
1997 (also in French and Spanish)

Implications of Proposed CO_2 Emissions Limitations – IPCC Technical Paper IV
1997 (also in French and Spanish)

The Regional Impacts of Climate Change: An Assessment of Vulnerability – IPCC Special Report
1998

Aviation and the Global Atmosphere – IPCC Special Report
1999

Emission Scenarios – IPCC Special Report
2000

Methodological and Technological Issues in Technology Transfer – IPCC Special Report
2000

ENQUIRIES: IPCC Secretariat, c/o World Meteorological Organization, 7 bis, Avenue de la Paix, Case Postale 2300, 1211 Geneva 2, Switzerland